**Life Sciences**

*continued on back endsheets*

# Need help with your Calculus course?

A wide range of valuable resources is available to help you master calculus.

■ *Step-by-step solutions guide your review and preparation.*

**Student Solutions Guide** (0-395-93344-7)

• This helpful guide provides detailed, step-by-step solutions to all odd-numbered section exercises and review exercises in the text.

■ *"Practice makes perfect" has never been more true.*

**Study Guide** (0-395-93345-5)

• Examples, exercises, and sample tests with answers provide additional practice in the concepts and techniques required in the applied calculus course.

■ *Graphing is easy when you have the right tools.*

**Graphing Technology Guide** (0-395-88773-9)

• Helpful keystroking instructions are provided for *Texas Instruments, Sharp, Casio,* and *Hewlett-Packard* graphing calculators to help you master concepts.
• Examples with step-by-step solutions, extensive graphics screen output, and technology tips

■ *Brush up on precalculus to succeed in calculus.*

**The Algebra of Calculus** (0-669-21885-5)

• This review of the algebra, trigonometry, and analytic geometry required for calculus gives over 200 examples with solutions, as well as pretests and exercises with answers.

*For more information, visit the Houghton Mifflin home page at **http://www.hmco.com,** or look for these resources in your bookstore. If you don't find them, check with your bookstore manager or call Houghton Mifflin toll free at **1-800-225-1464** to place an order.*

# Brief Calculus

## AN *Applied* APPROACH

### FIFTH EDITION

**RON LARSON**

The Pennsylvania State University

The Behrend College

**BRUCE H. EDWARDS**

University of Florida

with the assistance of
**David C. Falvo**
The Pennsylvania State University
The Behrend College

HOUGHTON MIFFLIN COMPANY
Boston   New York

Sponsoring Editor: Jack Shira
Managing Editor: Cathy Cantin
Senior Associate Editor: Maureen Brooks
Associate Editor: Michael J. Richards
Assistant Editor: Carolyn Johnson
Supervising Editor: Karen Carter
Project Editor: Patty Bergin
Editorial Assistant: Christine E. Lee
Art Supervisor: Gary Crespo
Marketing Manager: Ros Kane
Composition and Art: Meridian Creative Group

We have included examples and exercises that use real-life data as well as technology output from a variety of software. This would not have been possible without the help of many people and organizations. Our wholehearted thanks goes to all for their time and effort.

Trademark Acknowledgements: TI is a registered trademark of Texas Instruments, Inc. Mathcad is a registered trademark of MathSoft, Inc. Windows, Microsoft, and MS-DOS are registered trademarks of Microsoft, Inc. Mathematica is a registered trademark of Wolfram Research, Inc. DERIVE is a registered trademark of Soft Warehouse, Inc. IBM is a registered trademark of International Business Machines Corporation. Maple is a registered trademark of the University of Waterloo.

Copyright © 1999 by Houghton Mifflin Company. All rights reserved.

No part of this work may be reproduced or transmitted in any form or by any means, electronic or mechanical, including photocopying and recording, or by any information storage or retrieval system without the prior written permission of Houghton Mifflin Company unless such copying is expressly permitted by federal copyright law. Address inquiries to College Permissions, Houghton Mifflin Company, 222 Berkeley Street, Boston, MA 02116-3764.

Printed in the U.S.A.

Library of Congress Catalog Card Number: 98-71518

ISBN: 0-395-91685-2

456789–DC–02 01 00 99

# Contents

**CHAPTER 0**

## A Precalculus Review     *0-1*

**CHAPTER 1**

## Functions, Graphs, and Limits     *1*

**CHAPTER 2**

## Differentiation     *81*

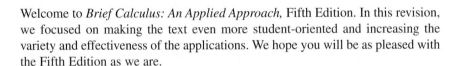

## *A word from the authors . . .*

Welcome to *Brief Calculus: An Applied Approach,* Fifth Edition. In this revision, we focused on making the text even more student-oriented and increasing the variety and effectiveness of the applications. We hope you will be as pleased with the Fifth Edition as we are.

### A student-oriented approach . . .

To encourage mastery *and* understanding, we have outlined a straightforward program of study with continual reinforcement and applicability to the real world.

This effective approach begins with Strategies for Success, a new feature found at the beginning of each chapter. In addition to outlining the key objectives of the chapter, this checklist provides page references for the various study tools in the chapter and notes additional resources that are available. The Chapter Summary and Study Strategies at the end of each chapter reinforce the Strategies for Success with a comprehensive list of the skills covered in the chapter, section references, and a correlation to the Review Exercises for guided practice.

In addition, we have added many new opportunities for algebra support throughout the text. It is crucial that a student understands an algebraic concept before attempting a related calculus concept. To help students in this area, we have added new Algebra Tips at point of use throughout the text. Many of these Algebra Tips correspond to the new two-page Algebra Review at the end of each chapter, which emphasizes the key algebraic concepts used in the chapter. Algebra Review Tutor software is readily available for additional support.

Throughout the text, Study Tips address special cases, expand upon concepts, and help students avoid common errors. Group Discussions provide opportunities for students to collaborate in mathematical discussion. The use of side comments to explain the steps of a solution has been bolstered, and many art pieces have been replaced by new images created with a state-of-the-art realism that helps students in visualization, especially with functions of several variables. Topics in Calculus Videotapes provide comprehensive coverage and explanation of key concepts.

Throughout each chapter there are many opportunities for students to assess their progress: at the end of each section (Warm Ups and section exercises), and at the end of each chapter (Chapter Summary, Study Strategies, and Review Exercises). The test items and text exercises are carefully crafted to build understanding as the student progresses through the course.

As always, we paid careful attention to the presentation—using precise mathematical language, innovative full-color design for emphasis and clarity, and a level of exposition that appeals to students—to create an effective teaching and learning tool.

## ...that applies to the changing world around us

Students taking calculus need to see how the subject matter relates to the real world. In this edition, we focused on increasing the variety of applications, especially in the life sciences, economics, and finance. All real-data applications were revised to use the most current information available, and new ones were added to increase the diversity of the exercises. We added new exercises that contain material from textbooks in other disciplines to show the relevance of calculus in other areas. In addition, exercises involving the use of spreadsheets have been incorporated throughout. An Additional Problems supplement is also available to professors who would like a wider choice of exercises.

Advances in technology are helping to change the world around us. We have updated and increased the coverage of technology to be even more readily available at point of use. Students are encouraged to use a graphing utility as a tool for exploration, discovery, and problem solving.

Reference tables in the appendix have been revised to have more relevance to the real world, and graphing utility programs have been updated for use with the most common models on the market.

We hope you will enjoy the Fifth Edition. It is a readable text with a straightforward approach, one with effective study tools and direct application to the lives and futures of calculus students.

Ron Larson

Bruce H. Edwards

## *Features*

### CHAPTER OPENERS

Each chapter opens with *Strategies for Success*, a new checklist that outlines the key objectives, lists the study tools (with page references to help students achieve the key objectives), and gives additional resources that are available. Each chapter opener also contains a list of the section topics and a photo referring students to an interesting application in the section exercises.

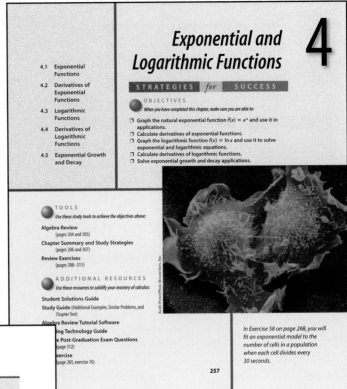

**Exponential and Logarithmic Functions**  4

4.1 Exponential Functions
4.2 Derivatives of Exponential Functions
4.3 Logarithmic Functions
4.4 Derivatives of Logarithmic Functions
4.5 Exponential Growth and Decay

**STRATEGIES** *for* **SUCCESS**

OBJECTIVES
*When you have completed this chapter, make sure you are able to:*

☐ Graph the natural exponential function $f(x) = e^x$ and use it in applications.
☐ Calculate derivatives of exponential functions.
☐ Graph the logarithmic function $f(x) = \ln x$ and use it to solve exponential and logarithmic equations.
☐ Calculate derivatives of logarithmic functions.
☐ Solve exponential growth and decay applications.

TOOLS
*Use these study tools to achieve the objectives above:*

**Algebra Review** (pages 304 and 305)
**Chapter Summary and Study Strategies** (pages 306 and 307)
**Review Exercises** (pages 308–311)

ADDITIONAL RESOURCES
*Use these resources to solidify your mastery of calculus:*

**Student Solutions Guide**
**Study Guide** (Additional Examples, Similar Problems, and Chapter Test)
**Algebra Review Tutorial Software**
**Graphing Technology Guide**
**Sample Post-Graduation Exam Questions** (page 312)
**Exercise** (page 285, exercise 76)

*In Exercise 58 on page 268, you will fit an exponential model to the number of cells in a population when each cell divides every 30 seconds.*

**257**

**FEATURES**

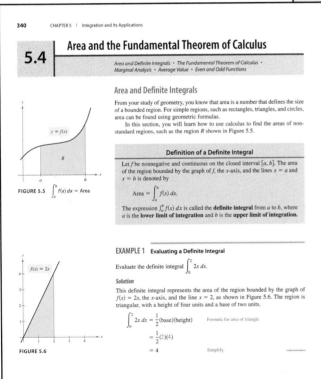

340   CHAPTER 5   |   Integration and Its Applications

**5.4**   **Area and the Fundamental Theorem of Calculus**

*Area and Definite Integrals • The Fundamental Theorem of Calculus • Marginal Analysis • Average Value • Even and Odd Functions*

**Area and Definite Integrals**

From your study of geometry, you know that area is a number that defines the size of a bounded region. For simple regions, such as rectangles, triangles, and circles, area can be found using geometric formulas.

In this section, you will learn how to use calculus to find the areas of non-standard regions, such as the region $R$ shown in Figure 5.5.

$y = f(x)$

$R$

**FIGURE 5.5** $\int_a^b f(x)\,dx = $ Area

**Definition of a Definite Integral**

Let $f$ be nonnegative and continuous on the closed interval $[a, b]$. The area of the region bounded by the graph of $f$, the $x$-axis, and the lines $x = a$ and $x = b$ is denoted by

$$\text{Area} = \int_a^b f(x)\,dx.$$

The expression $\int_a^b f(x)\,dx$ is called the **definite integral** from $a$ to $b$, where $a$ is the **lower limit of integration** and $b$ is the **upper limit of integration.**

**EXAMPLE 1**   **Evaluating a Definite Integral**

Evaluate the definite integral $\int_0^2 2x\,dx.$

**Solution**

This definite integral represents the area of the region bounded by the graph of $f(x) = 2x$, the $x$-axis, and the line $x = 2$, as shown in Figure 5.6. The region is triangular, with a height of four units and a base of two units.

$$\int_0^2 2x\,dx = \frac{1}{2}(\text{base})(\text{height}) \qquad \text{Formula for area of triangle}$$
$$= \frac{1}{2}(2)(4)$$
$$= 4 \qquad \text{Simplify.}$$

$f(x) = 2x$

**FIGURE 5.6**

### SECTION TOPICS

Each section begins with a list of topics covered in that section. This outline helps instructors with class planning and students with studying the material in the section.

### DEFINITIONS AND THEOREMS

All definitions and theorems are highlighted for emphasis and easy reference.

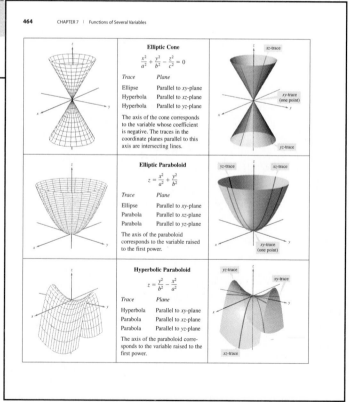

## TECHNOLOGY

Students are encouraged to use a graphing utility, computer algebra system, or spreadsheet as a tool for exploration, discovery, and problem solving. The text offers many opportunities to execute computations and programs, to visualize theoretical concepts, to discover alternative approaches, and to verify the results of other solution methods using technology. However, students are not required to have access to a graphing utility to use this text effectively. In addition to describing the benefits of using technology, the text also pays special attention to its possible misuse or misinterpretation.

## DISCOVERY

Before students are exposed to selected topics, discovery projects allow them to explore concepts on their own, making them more likely to remember the results. These optional boxed features can be omitted, if the instructor desires, with no loss of continuity in the coverage of material.

## GRAPHICS

The Fifth Edition has over 1900 figures. Computer-generated for accuracy, clarity, and realism, this art program helps students visualize mathematical concepts more easily.

## EXAMPLES

To increase the usefulness of the text as a study tool, the Fifth Edition contains a wide variety of examples, each titled for easy reference. Many of these detailed examples display solutions that are presented graphically, analytically, and /or numerically to provide further insight into mathematical concepts. Side comments clarify the steps of the solution as necessary. Examples using real-life data are identified with a 🌐 and accompanied by the type of illustration that students are used to seeing in newspapers and magazines.

---

### Applications

Inequalities are frequently used to describe conditions that occur in business and science. For instance, the inequality

$$144 \leq W \leq 180$$

describes the recommended weight $W$ for a man whose height is 5 feet 10 inches. Example 3 shows how an inequality can be used to describe the production level of a manufacturing plant.

**EXAMPLE 3**   **Production Levels**   🌐

In addition to fixed overhead costs of $500 per day, the cost of producing $x$ units of a certain item is $2.50 per unit. During the month of August, the total cost of production varied from a high of $1325 to a low of $1200 per day. Find the high and low *production levels* during the month.

**Solution**

Because it costs $2.50 to produce one unit, it costs $2.5x$ to produce $x$ units. Furthermore, because the fixed cost per day is $500, the total daily cost of producing $x$ units is

$$C = 2.5x + 500.$$

Now, because the cost ranged from $1200 to $1325, you can write the following.

| | | | |
|---|---|---|---|
| $1200 \leq$ | $2.5x + 500$ | $\leq 1325$ | Original inequality |
| $1200 - 500 \leq$ | $2.5x + 500 - 500$ | $\leq 1325 - 500$ | Subtract 500 from each side. |
| $700 \leq$ | $2.5x$ | $\leq 825$ | Simplify. |
| $\dfrac{700}{2.5} \leq$ | $\dfrac{2.5x}{2.5}$ | $\leq \dfrac{825}{2.5}$ | Divide each side by 2.5. |
| $280 \leq$ | $x$ | $\leq 330$ | Simplify. |

Thus, the daily production levels during the month varied between a low of 280 units and a high of 330 units, as pictured in Figure 0.8.

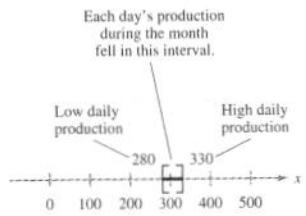

Each day's production during the month fell in this interval.

Low daily production     High daily production

280   330

0   100   200   300   400   500    $x$

**FIGURE 0.8**

---

## ALGEBRA TIPS

*Algebra Tips* offer students algebraic support at point of use. Many *Algebra Tips* are related to the *Algebra Review* at the end of the chapter, which provides additional details of examples with solutions and explanations.

## STUDY TIPS

Throughout the text, *Study Tips* help students avoid common errors, address special cases, and expand upon theoretical concepts.

---

With the five basic integration rules, you can integrate *any* polynomial function, as demonstrated in the next example.

**EXAMPLE 5**   **Integrating Polynomial Functions**

Find the following indefinite integrals.

(a) $\displaystyle\int (x + 2)\, dx$    (b) $\displaystyle\int (3x^4 - 5x^2 + x)\, dx$

**Solution**

(a) Use the Sum Rule to integrate each part separately.

$$\int (x + 2)\, dx = \int x\, dx + \int 2\, dx$$
$$= \frac{x^2}{2} + 2x + C$$

(b) Try to identify each basic integration rule used to evaluate this integral.

$$\int (3x^4 - 5x^2 + x)\, dx = 3\left(\frac{x^5}{5}\right) - 5\left(\frac{x^3}{3}\right) + \frac{x^2}{2} + C$$
$$= \frac{3}{5}x^5 - \frac{5}{3}x^3 + \frac{1}{2}x^2 + C$$

**TIP**

🌐 For help on the algebra in Example 6, see Example 1a in the *Chapter 5 Algebra Review*, on page 374.

**EXAMPLE 6**   **Rewriting Before Integrating**

Find the indefinite integral.

$$\int \frac{x + 1}{\sqrt{x}}\, dx.$$

**Solution**

Begin by rewriting the quotient in the integrand as a sum. Then rewrite each term using rational exponents.

$$\int \frac{x + 1}{\sqrt{x}}\, dx = \int \left(\frac{x}{\sqrt{x}} + \frac{1}{\sqrt{x}}\right) dx \qquad \text{Rewrite as a sum.}$$
$$= \int (x^{1/2} + x^{-1/2})\, dx \qquad \text{Use rational exponents.}$$
$$= \frac{x^{3/2}}{3/2} + \frac{x^{1/2}}{1/2} + C \qquad \text{Apply Power Rule.}$$
$$= \frac{2}{3}x^{3/2} + 2x^{1/2} + C \qquad \text{Simplify.}$$

**STUDY TIP**   When integrating quotients, don't make the mistake of integrating the numerator and denominator separately. For instance, in Example 6, be sure you see that

$$\int \frac{x + 1}{\sqrt{x}}\, dx \neq \frac{\int (x + 1)\, dx}{\int \sqrt{x}\, dx}.$$

If you have access to a symbolic integration program, try using it for this example. When we tried it, the program listed the antiderivative as $2\sqrt{x}(x + 3)/3$, which is equivalent to the result listed above.

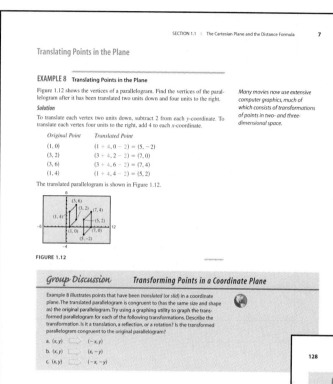

## GROUP DISCUSSIONS

Starting with Chapter 1, each section in the text closes with a *Group Discussion* problem designed to be completed as a group project in class or as a homework assignment. These problems, which encourage students to think, reason, and write about calculus, emphasize the synthesis or the further exploration of the concepts presented in the section.

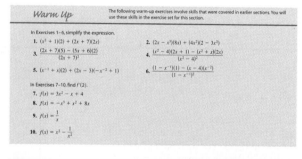

## WARM-UP EXERCISES

Starting with Chapter 1, each text section has a set of ten warm-up exercises. The *Warm Ups* enable students to review and practice the previously learned skills necessary to master the new skills presented in the section. Answers to *Warm Ups* appear in the back of the text.

## EXERCISES

The text now contains almost 6000 exercises. Each exercise set is graded, progressing from skill-development problems to more challenging problems, to build confidence, skill, and understanding. The wide variety of exercises includes many technology-oriented, real, and engaging problems. Answers to all odd-numbered and selected even exercises are included in the back of the text. To help instructors make homework assignments, many of the exercises in the text are labeled to indicate the area of application.

---

**478**  CHAPTER 7 | Functions of Several Variables

**44.** *Forestry*  The **Doyle Log Rule** is one of several methods used to determine the lumber yield of a log in board feet in terms of its diameter $d$ in inches and its length $L$ in feet. The number of board feet is given by

$$N(d, L) = \left(\frac{d-4}{4}\right)^2 L.$$

(a) Find the number of board feet of lumber in a log with a diameter of 22 inches and a length of 12 feet.

(b) Find $N(30, 12)$.

**45.** *Profit*  A corporation manufactures a product at two locations. The costs of producing $x_1$ units at location 1 and $x_2$ units at location 2 are

$$C_1(x_1) = 0.02x_1^2 + 4x_1 + 500$$

and

$$C_2(x_2) = 0.05x_2^2 + 4x_2 + 275,$$

respectively. If the product sells for $15 per unit, then the profit function for the product is

$$P(x_1, x_2) = 15(x_1 + x_2) - C_1(x_1) - C_2(x_2).$$

Find (a) $P(250, 150)$ and (b) $P(300, 200)$.

**46.** *Queuing Model*  The average amount of time that a customer waits in line for service is given by

$$W(x, y) = \frac{1}{x-y}, \qquad y < x,$$

where $y$ is the average arrival rate and $x$ is the average service rate ($x$ and $y$ are measured in the number of customers per hour). Evaluate $W$ at the following points.

(a) (15, 10)  (b) (12, 9)  (c) (12, 6)  (d) (4, 2)

**47.** *Investment*  In 1994, an investment of $1000 was made in a bond earning 10% compounded annually. The investor pays tax at rate $R$, and the annual rate of inflation is $I$. In the year 2004, the value $V$ of the bond in constant 1994 dollars is

$$V(I, R) = 1000\left[\frac{1 + 0.10(1 - R)}{1 + I}\right]^{10}.$$

Use this function of two variables to complete the table.

| | Inflation Rate, $I$ | | |
|---|---|---|---|
| Tax Rate, $R$ | 0.00 | 0.03 | 0.05 |
| 0.00 | | | |
| 0.28 | | | |
| 0.35 | | | |

**48.** *Geology: A Contour Map*  The contour map below represents color-coded seismic amplitudes of a fault horizon and a projected contour map, which is used in earthquake studies. (*Source: Adapted from Shipman/Wilson/Todd, An Introduction to Physical Science, Eighth Edition*)

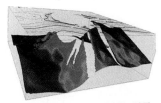

Shipman, *An Introduction to Physical Science* 8/e © 1997, Houghton Mifflin Company

(a) Discuss the use of color to represent the level curves.

(b) Do the level curves correspond to equally spaced amplitudes? Explain.

**49.** *Earnings*  The earnings per share for McDonald's Corporation from 1987 through 1996 can be modeled by

$$z = 0.186x + 0.071y - 0.396,$$

where $x$ is total revenue (in billions of dollars) and $y$ is the total net worth (in billions of dollars). (*Source: McDonald's Corporation*)

(a) Find the earnings per share when $x = 8$ and $y = 6$.

(b) Which of the two variables in this model has the greater influence on the earnings per share? Explain.

---

## VARIETY OF PROBLEMS

The text offers a rich variety of problems to accommodate a wide range of teaching and learning styles, including computational, conceptual, interpretational, matching (graphs to functions), true/false, essay, multipart, exploratory, and applied problems. A variety of relevant application problems from a wide range of disciplines—many of which use real data—clearly demonstrate the real-world usage of the mathematics. Problems are often labeled to indicate the subject of the application (as in *Break-Even Analysis* and *Lung Transplants*) or to identify a special type of question (as in *True or False* and *Writing*) for easy reference.

## GRAPHING UTILITIES

Many exercises in the text can be solved using technology; however, the symbol  identifies all exercises for which students are specifically instructed to use a graphing utility or a computer algebra system.

---

CHAPTER 1 | Functions, Graphs, and Limits

*Equilibrium*  The supply function for a product e number of units $x$ that producers are willing to r a given price per unit $p$. The supply and demand for a market are

$x + 4$  Supply

$$p = -\frac{16}{25}x + 30. \qquad \text{Demand}$$

(a) Use a graphing utility to graph the supply and demand functions in the same viewing rectangle.

(b) Use the trace feature of the graphing utility to find the *equilibrium point* for the market. That is, find the point of intersection of the two graphs.

(c) For what values of $x$ does the demand exceed the supply?

(d) For what values of $x$ does the supply exceed the demand?

**67.** *Profit*  A radio manufacturer charges $90 per unit for units that cost $60 to produce. To encourage large orders from distributors, the manufacturer will reduce the price by $0.01 per unit for each unit in excess of 100 units. (For example, an order of 101 units would have a price of $89.99 per unit, and an order of 102 units would have a price of $89.98 per unit.) This price reduction is discontinued when the price per unit drops to $75.

(a) Express the price per unit $p$ as a function of the order size $x$.

(b) Express the profit $P$ as a function of the order size $x$.

**68.** *Cost, Revenue, and Profit*  A company invests $98,000 for equipment to produce a new product. Each unit of the product costs $12.30 and is sold for $17.98. Let $x$ be the number of units produced and sold.

(a) Write the total cost $C$ as a function of $x$.

(b) Write the revenue $R$ as a function of $x$.

(c) Write the profit $P$ as a function of $x$.

**69.** *Charter Bus Fares*  For groups of 80 or more people, a charter bus company determines the rate $r$ (in dollars per person) according to the formula $r = 8 - 0.05(n - 80)$, $n \geq 80$, where $n$ is the number of people.

(a) Express the revenue $R$ for the bus company as a function of $n$.

(b) Complete the table.

| $n$ | 90 | 100 | 110 | 120 | 130 | 140 | 150 |
|---|---|---|---|---|---|---|---|
| $R$ | | | | | | | |

(c) Criticize the formula for the rate. Would you use this formula? Explain your reasoning.

**70.** *Medicine*  The temperature of a patient after being given a fever-reducing drug is given by

$$F(t) = 98 + \frac{3}{t+1},$$

where $F$ is the temperature in degrees Fahrenheit and $t$ is the time in hours. Use a graphing utility to graph the function. Be sure to choose an appropriate window setting. For what values of $t$ do you think this function would be valid? Explain.

In Exercises 71–76, use a graphing utility to graph the function. Then use the zoom and trace features to find the zeros of the function. Is the function one-to-one?

**71.** $f(x) = x\sqrt{9 - x^2}$  **72.** $f(x) = 2\left(3x^2 - \dfrac{6}{x}\right)$

**73.** $g(t) = \dfrac{t+3}{1-t}$  **74.** $h(x) = 6x^3 - 12x^2 + 4$

**75.** $f(x) = \dfrac{4 - x^2}{x}$  **76.** $g(x) = \left|\dfrac{1}{2}x^2 - 4\right|$

### Business Capsule

Cardsenders

*Cardsenders is a home-based greeting card service for businesses. It is owned by Phyllis Boverie who started the company in 1985. Start-up costs for this business ran from $5,000 to $10,000.*

**77.** *Research Project*  Use your school's library or some other reference source to find information about the start-up costs of beginning a business, such as in the business example above. Write a short paper about the company.

---

## TEXTBOOK EXERCISES

The Fifth Edition includes a number of exercises that contain material from textbooks in other disciplines, such as biology, chemistry, economics, finance, geology, physics, and psychology. These applications make the point to students that they will need to use calculus in future courses outside of the math curriculum. These exercises are identified by the icon 📖 and are labeled to indicate the subject area.

## BUSINESS CAPSULES

*Business Capsules* appear at the ends of numerous sections. These capsules and their accompanying exercises deal with business situations that are related to the mathematical concepts covered in the chapter.

**FEATURES**

## ALGEBRA REVIEW

At the end of each chapter, the *Algebra Review* illustrates the key algebraic concepts used in the chapter. Often, rudimentary steps are provided in detail for selected examples from the chapter. This review offers additional support to those students who have trouble following examples due to poor algebra skills.

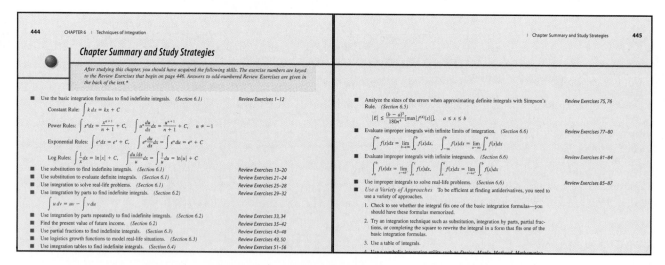

## CHAPTER SUMMARY AND STUDY STRATEGIES

The *Chapter Summary* reviews the skills covered in the chapter and correlates each skill to the *Review Exercises* that test that skill. Following each chapter summary is a short list of study strategies for addressing topics or situations specific to the chapter.

## REVIEW EXERCISES

The *Review Exercises* offer students an opportunity for additional practice as they complete each chapter. Answers to all odd-numbered *Review Exercises* appear at the end of the text.

---

446    CHAPTER 6   |   Techniques of Integration

### Chapter 6  Review Exercises

In Exercises 1–12, use a basic integration formula to find the indefinite integral.

1. $\int dt$

2. $\int (x^2 + 2x - 1)dx$

3. $\int (x + 5)^3 dx$

4. $\int \frac{2}{(x-1)^2}dx$

5. $\int e^{10x}dx$

6. $\int 3xe^{-x^2}dx$

7. $\int \frac{1}{5x}dx$

8. $\int \frac{2x^3 - x}{x^4 - x^2 + 1}dx$

9. $\int x\sqrt{x^2 + 4}\,dx$

10. $\int \frac{1}{\sqrt{2x} - 9}dx$

11. $\int \frac{2e^x}{3 + e^x}dx$

12. $\int (x^2 - 1)e^{x^3 - 3x}dx$

In Exercises 13–20, use substitution to find the indefinite integral.

13. $\int x(x-2)^3 dx$

14. $\int x(1-x)^3 dx$

15. $\int x\sqrt{x+1}\,dx$

16. $\int x^2\sqrt{x+1}\,dx$

17. $\int 2x\sqrt{x-3}\,dx$

18. $\int \frac{\sqrt{x}}{1 + \sqrt{x}}dx$

19. $\int (x+1)\sqrt{1-x}\,dx$

20. $\int \frac{x}{x-1}dx$

In Exercises 21–24, use substitution to evaluate the definite integral. Use a symbolic integration utility to check your answer.

21. $\int x\sqrt{x-2}\,dx$

22. $\int x^2\sqrt{x-2}\,dx$

23. $\int x^2(x-1)^3 dx$

24. $\int x(x+3)^4 dx$

25. *Probability*   The probability of recall in an experiment is found to be
$$\ldots \le b) = \int_a^b \frac{105}{16}x^2\sqrt{1-x}\,dx,$$
...represents the percent of recall (see figure). ...the probability that a randomly chosen individual ...ecall 80% of the material. ... is the median percent recall? That is, for what ... of $b$ is it true that $P(0 \le x \le b) = 0.5$?

Figure for 25

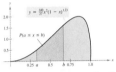

$y = \frac{105}{16}x^2\sqrt{1-x}$

$P(a \le x \le b)$

26. *Probability*   The probability of locating between $a$ and $b$ percent of oil and gas deposits in a region is
$$P(a \le x \le b) = \int_a^b \frac{140}{27}x^2(1-x)^{1/3}dx$$
(see figure).
(a) Find the probability that between 40% and 60% of the deposits will be found.
(b) Find the probability that between 0% and 50% of the deposits will be found.

$y = \frac{140}{27}x^2(1-x)^{1/3}$

$P(a \le x \le b)$

27. *Profit*   The net profit for Coca-Cola in millions of dollars per year from 1987 through 1996 can be modeled by
$$P = 449.63 + 4.15t^2 \ln t,$$
where $t$ is the time in years, with $t = 7$ corresponding to 1987.   (*Source: Coca-Cola*)
(a) Find the average net profit for the years 1987 through 1996.
(b) Find the total net profit for the years 1987 through 1996.

---

170      CHAPTER 2   |   Differentiation

### Sample Post-Graduation Exam Questions

*The following questions represent the types of questions that appear on certified public accountant (CPA) exams, graduate management admission tests (GMAT), graduate records exams (GRE), actuarial exams, and college-level academic skills tests (CLAST). The answers to the questions are given in the back of the book.*

CPA    GMAT
GRE    CLAST
Actuarial

Figure for 1

$A(-1, 4)$

$B(1, 2)$

1. What is the length of the line segment that connects $A$ to $B$ (see graph)?
   (a) 2      (b) 4      (c) $2\sqrt{2}$      (d) 6      (e) $\sqrt{3}$

For Questions 2–4, refer to the following table.

Participation in National Elections (millions of persons)

| Characteristic | 1988 Persons of voting age | 1988 Percent voted | 1992 Persons of voting age | 1992 Percent voted | 1996 Persons of voting age | 1996 Percent voted |
|---|---|---|---|---|---|---|
| **Total** | 178.1 | 57.4 | 185.7 | 61.3 | 193.7 | 54.2 |
| Male | 84.5 | 56.4 | 88.6 | 60.2 | 92.6 | 52.8 |
| Female | 93.6 | 58.3 | 97.1 | 62.3 | 101.0 | 55.5 |
| Age 18 to 20 | 10.7 | 33.2 | 9.7 | 38.5 | 10.8 | 31.2 |
| 21 to 24 | 14.8 | 38.3 | 14.6 | 45.7 | 13.9 | 33.4 |
| 25 to 34 | 42.7 | 48.0 | 41.6 | 53.2 | 40.1 | 43.1 |
| 35 to 44 | 35.2 | 61.3 | 39.7 | 63.6 | 43.3 | 54.9 |
| 45 to 64 | 45.9 | 67.9 | 49.1 | 70.0 | 53.7 | NA* |
| 65 years and over | 28.8 | 68.8 | 30.8 | 70.1 | 31.9 | NA* |

\* NA Not available

2. Which of the following groups had the highest percent of voters in 1992?
   (a) male      (b) age 35 to 44      (c) age 25 to 34
   (d) age 18 to 20      (e) female

3. In 1996, what percent (to the nearest percent) of persons of voting age were female?
   (a) 47      (b) 53      (c) 55      (d) 57      (e) 52

4. In 1988, how many males of voting age voted?
   (a) 47,658,000      (b) 48,503,000      (c) 102,229,400
   (d) 47,377,000      (e) 54,568,800

For Questions 5 and 6, refer to the following example.
Suppose the position $s$ at time $t$ of an accelerating object is given by $s(t) = t(t^2 - 2)^2$.

5. The acceleration function is
   (a) $5t^4 - 12t^2 + 4$      (b) $5t^4 - 8t$      (c) $20t^3 - 24t$
   (d) $20t^3 - 8$      (e) $t^5 - 4t^2 + 4$

6. The velocity of the object at time $t = 3$ is
   (a) 301      (b) 381      (c) 147      (d) 532      (e) 468

## SAMPLE POST-GRADUATION EXAM QUESTIONS

To emphasize the relevance of calculus, every chapter concludes with sample questions representative of the types of questions on certified public accountant (CPA) exams, graduate management admissions tests (GMAT), graduate records exams (GRE), actuarial exams, and college-level academic skills tests (CLAST). The answers to all *Sample Post-Graduation Exam Questions* are given in the back of the text.

**FEATURES**

## Supplements

*Calculus: An Applied Approach,* Fifth Edition, by Larson and Edwards, is accompanied by a comprehensive supplements package with ancillaries for students, for instructors, and for use as classroom resources. Most items are keyed directly to the textbook.

### PRINTED RESOURCES FOR THE INSTRUCTOR

*Instructor's Resource Guide* by Bruce H. Edwards, University of Florida
- Notes to the New Teacher
- List of text exercises requiring technology
- Chapter summaries
- Detailed solutions to all even-numbered Section and Review Exercises
- Answers to all Warm-Up Exercises and Sample Post-Graduation Exam Questions
- Suggested solutions and discussion guidelines for all Discovery, Technology, and Group Discussion features

*Test Item File*
- Printed testbank
- Over 2500 multiple-choice and open-ended test items coded by level of difficulty
- Mid-Chapter Quizzes, Chapter Tests, Cumulative Tests, and Final Exams
- Also available as test-generating software

### PRINTED RESOURCES FOR THE STUDENT

*Student Solutions Guide* by Bruce H. Edwards, University of Florida
- Step-by-step solutions to all odd-numbered text exercises
- Practice chapter tests with solutions

*Graphing Technology Guide* by Benjamin N. Levy and Laurel Technical Services
- Keystroke instructions for a wide variety of Texas Instruments, Casio, Sharp, and Hewlett-Packard graphing calculators, including the most current models
- Examples with step-by-step solutions
- Extensive graphics screen output
- Technology tips

*The Algebra of Calculus* by Eric J. Braude
  - Review of the algebra, trigonometry, and analytical geometry required for calculus
  - Over 200 examples with solutions
  - Pretests and exercises with answers

*Study Guide* by Ronnie Khuri, University of Florida
  - Numerous additional examples and exercises with answers
  - Practice tests with answers

## MEDIA RESOURCES FOR THE INSTRUCTOR

Applied Calculus Website
  - Additional applications: life sciences, case studies, spreadsheets, and other challenging problems
  - Selected research exercises from the text extended with web links and additional information and questions
  - Algebra review self-tests
  - Instructor resources

Computerized Testing
  - Test-generating software
  - Over 2500 test items
  - Also available as a printed testbank

Transparency Package
  - Color transparencies

## MEDIA RESOURCES FOR THE STUDENT

*Topics in Calculus Videotapes* by Dana Mosely
  - Comprehensive coverage of key topics
  - Explanation of concepts, sample problems, and applications

*Graphing Calculator Instructional Videotapes* by Dana Mosely
  - Introduces students to a wide variety of Texas Instruments, Casio, Sharp, and Hewlett-Packard graphing calculators, including the most current models
  - Covers basic capabilities shared by many graphing utilities

*Algebra Review Tutorial Software*
  - Interactive tutorial software
  - Diagnostic feedback
  - Self-tests
  - Guided exercises with step-by-step solutions
  - Glossary

*Note: Visit the Houghton Mifflin home page at http://www.hmco.com*

SUPPLEMENTS

# Acknowledgements

We would like to thank the many people who have helped us at various stages of this project during the past 18 years. Their encouragement, criticisms, and suggestions have been invaluable to us.

## FIFTH EDITION REVIEWERS

David Bregenzer
Utah State University

Joseph Chance
University of Texas—Pan American

Karabi Datta
Northern Illinois University

Betty Givan
Eastern Kentucky University

Mark Greenhalgh
Fullerton College

Raymond Heitmann
University of Texas at Austin

Duane Kouba
University of California—Davis

Yuhlong Lio
University of South Dakota

Richard Semmler
Northern Virginia Community College—
    Annandale

Bernard Shapiro
University of Massachusetts, Lowell

Jane Y. Smith
University of Florida

Charles W. Zimmerman
Robert Morris College

## PREVIOUS EDITION REVIEWERS

Carol Achs, Mesa Community College; Mary Chabot, Mt. San Antonio College; John Chuchel, University of California; Miriam E. Connellan, Marquette University; William Conway, University of Arizona; Bruce H. Edwards, University of Florida; Roger A. Engle, Clarion University of Pennsylvania; Karen Hay, Mesa Community College; William C. Huffman, Loyola University of Chicago; Arlene Jesky, Rose State College; Ronnie Khuri, University of Florida; James A. Kurre, The Pennsylvania State University; Melvin Lax, California State University—Long Beach; Norbert Lerner, State University of New York at Cortland; Peter J. Livorsi, Oakton Community College; Samuel A. Lynch, Southwest Missouri State University; Kevin McDonald, Mt. San Antonio College; Earl H. McKinney, Ball State University; Philip R. Montgomery, University of Kansas; Mike Nasab, Long Beach City College; Karla Neal, Louisiana State University; James Osterburg, University of Cincinnati; Rita Richards, Scottsdale Community College; Stephen B. Rodi, Austin Community College; Yvonne Sandoval-Brown, Pima Community College; Jane Y. Smith, University of Florida; DeWitt L. Sumners, Florida State University; Jonathan Wilkin, Northern Virginia Community College; Carol G. Williams, Pepperdine University; Melvin R. Woodard, Indiana University of Pennsylvania; Carlton Woods, Auburn University at Montgomery; Jan E. Wynn, Brigham Young University; Robert A. Yawin, Springfield Technical Community College

Our thanks to Robert Hostetler, The Behrend College, The Pennsylvania State University, for his significant contributions to previous editions of this text.

We would also like to thank the staff at Larson Texts, Inc., who assisted with proofreading the manuscript, preparing and proofreading the art package, and checking the supplements.

A special note of thanks goes to all the students who have used the previous editions of the text.

On a personal level, we are grateful to our spouses, Deanna Gilbert Larson and Consuelo Edwards, for their love, patience, and support. Also, a special thanks goes to R. Scott O'Neil.

If you have suggestions for improving this text, please feel free to write to us. Over the past two decades we have received many useful comments from both instructors and students, and we value these very much.

Ron Larson
Bruce H. Edwards

# A Plan for You as a Student

Studying mathematics is a building process. You use what you know to learn the next topic. Plan on putting a good deal of effort into the course now, and provide yourself with constant checks to make sure you are keeping up.

## How to Prepare for Class

Be ready for class by reading the syllabus and always staying on top of what is next. If you are prepared for class, you have a much better chance of succeeding. Before class,

✓ Review your notes from the previous class.

✓ Look over the portion of the text that is to be covered in class.

✓ Pay special attention to the definitions, theorems, and rules highlighted in the text.

✓ Read the solved examples, trying to follow along.

✓ Make a note of anything you don't understand, and pay special attention to that topic when your instructor explains it.

✓ Try the Technology and Discovery exercises to get a better grasp of the material before the instructor presents it.

## Keeping Up

Math is not a spectator sport. There are no shortcuts—the key to success is keeping up with the course.

✓ Go to class, track down any lecture notes you do miss, and review them carefully so you don't fall behind.

✓ While you are in class, take good notes.

✓ Do the homework as soon as possible after class, when the concepts are still fresh in your mind. As a good refresher, work the Warm-Up exercises at the beginning of each exercise set and use the answers in the back of the text to check your work.

✓ As you work the exercises, refer to your notes and reread the section. Look for an Example that is similar to the exercises you are trying to do. For example, Exercises 1–4 on page 275 are similar to Example 1 on page 270.

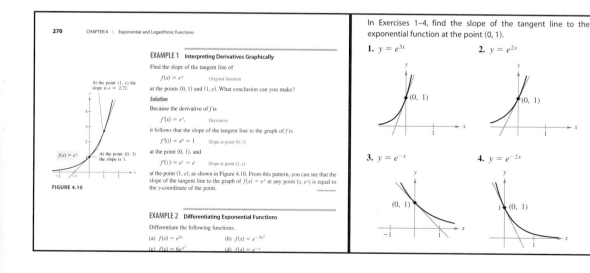

## If You Get Stuck

There are plenty of resources available to guide you as you study calculus. For example,

✓ If you get stuck on an exercise, look back at the Study Tips and Algebra Tips in that section.

✓ Find a study partner and work through problems together or consult with each other when you're stuck.

✓ Work with a tutor for regular assistance.

✓ Consult one of the many supplements available for this text: *Student Solutions Guide, Study Guide, Algebra Review Tutorial Software, The Algebra of Calculus,* and *Graphing Technology Guide.* (Descriptions of each of these can be found in the *Supplements* section on pages xvi–xvii.)

## How to Prepare for an Exam

The best way to prepare for an exam is to keep up with the course—cramming seldom works. Numerous features in this text provide an opportunity for you to prepare for an exam.

✓ Review the Chapter Summary and Study Strategies.

✓ Work the Review Exercises, which are correlated to each skill listed in the Chapter Summary and Study Strategies, as shown on pages 74 and 76 below.

✓ Take practice tests offered in the Study Guide.

✓ Rework homework problems as necessary.

✓ Read the Algebra Review if you would like some assistance in that area.

✓ When you take the exam, go in with a clear mind and a positive attitude.

✓ When you get the exam back, make sure you understand any errors that you've made, so that you can build on the knowledge you've gained to go on to the next topic.

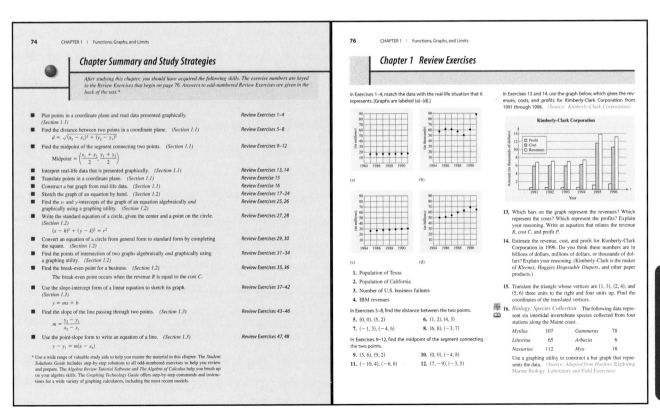

**STUDENT PLAN**

# 0

# A Precalculus Review

## STRATEGIES *for* SUCCESS

This initial chapter, **A Precalculus Review,** is just that—a review chapter. Make sure that you have a solid understanding of all the material in this chapter before beginning Chapter 1. You can also use it as a reference as you progress through the text, coming back to brush up on an algebra skill that you may have forgotten over the course of the semester. As with all math courses, calculus is a building process—that is, you use what you know to go on to the next topic.

At the beginning of each chapter in this book, we present *Strategies for Success,* which will help you learn the material in that chapter. We provide a list of the primary OBJECTIVES you need to achieve by the end of the chapter, a list of TOOLS that will help you achieve those objectives, and a list of ADDITIONAL RESOURCES that are helpful as well. Use these lists to succeed in calculus.

L.S. Stepanowicz/Visuals Unlimited

*In Exercise 29 on page 0-7, you will graph on a real number line the pH values of familiar solutions such as coffee and oven cleaner.*

# 0.1 The Real Line and Order

*The Real Line • Order and Intervals on the Real Line • Solving Inequalities • Applications*

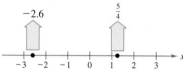

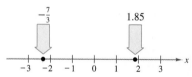

Every point on the real line corresponds to a real number.

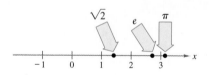

Every real number corresponds to a point on the real line.

**FIGURE 0.2**

## The Real Line

Real numbers can be represented with a coordinate system called the **real line** (or *x*-axis), as shown in Figure 0.1. The **positive direction** (to the right) is denoted by an arrowhead and indicates the direction of increasing values of *x*. The real number corresponding to a particular point on the real line is called the **coordinate** of the point. As shown in Figure 0.1, it is customary to label those points whose coordinates are integers.

The point on the real line corresponding to zero is called the **origin.** Numbers to the right of the origin are **positive,** and numbers to the left of the origin are **negative.** The term **nonnegative** describes a number that is either positive or zero.

The importance of the real line is that it provides you with a conceptually perfect picture of the real numbers. That is, each point on the real line corresponds to one and only one real number, and each real number corresponds to one and only one point on the real line. This type of relationship is called a **one-to-one correspondence** and is illustrated in Figure 0.2.

Each of the four points in Figure 0.2 corresponds to a real number that can be expressed as the ratio of two integers. $\left(\text{Note that } 1.85 = \frac{37}{20} \text{ and } -2.6 = -\frac{13}{5}.\right)$ Such numbers are called **rational.** Rational numbers have either terminating or infinitely repeating decimal representations.

| *Terminating Decimals* | *Infinitely Repeating Decimals* |
|---|---|
| $\dfrac{2}{5} = 0.4$ | $\dfrac{1}{3} = 0.333\ldots = 0.3\overline{3}\,^{*}$ |
| $\dfrac{7}{8} = 0.875$ | $\dfrac{12}{7} = 1.714285714285\ldots = 1.\overline{714285}$ |

Real numbers that are not rational are called **irrational,** and they cannot be represented as the ratio of two integers (or as terminating or infinitely repeating decimals). To represent an irrational number, people usually resort to a decimal approximation. Some irrational numbers occur so frequently in applications that mathematicians have invented special symbols to represent them. For example, the symbols $\sqrt{2}$, $\pi$, and $e$ represent irrational numbers whose decimal approximations are as follows. (See Figure 0.3.)

$$\sqrt{2} \approx 1.4142135623$$
$$\pi \approx 3.1415926535$$
$$e \approx 2.7182818284$$

*The bar indicates which digit or digits repeat infinitely.

## Order and Intervals on the Real Line

One important property of the real numbers is that they are **ordered:** 0 is less than 1, $-3$ is less than $-2.5$, $\pi$ is less than $\frac{22}{7}$, and so on. You can visualize this property on the real line by observing that $a$ is less than $b$ if and only if $a$ lies to the left of $b$. Symbolically, "$a$ is less than $b$" is denoted by the inequality

$\qquad a < b.$

For example, the inequality $\frac{3}{4} < 1$ follows from the fact that $\frac{3}{4}$ lies to the left of 1 on the real line, as shown in Figure 0.4.

When three real numbers $a$, $x$, and $b$ are ordered such that $a < x$ and $x < b$, we say that $x$ is **between** $a$ and $b$ and write

$\qquad a < x < b.$ $\qquad$ *x is between a and b.*

The set of *all* real numbers between $a$ and $b$ is called the **open interval** between $a$ and $b$ and is denoted by $(a, b)$. An interval of the form $(a, b)$ does not contain the "endpoints" $a$ and $b$. Intervals that include their endpoints are called **closed** and are denoted by $[a, b]$. Intervals of the form $[a, b)$ and $(a, b]$ are neither open nor closed. Figure 0.5 shows the nine types of intervals on the real line.

$\frac{3}{4}$ lies to the left of 1, so $\frac{3}{4} < 1$

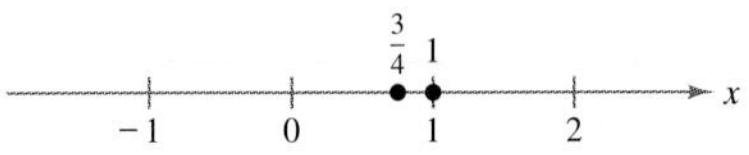

**FIGURE 0.4**

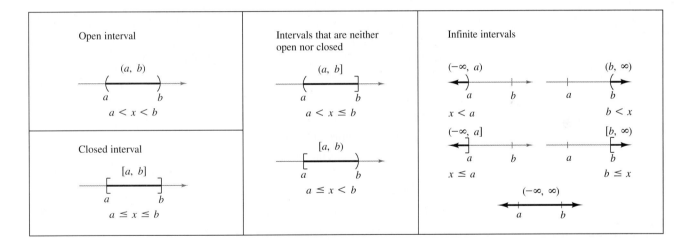

**FIGURE 0.5** Intervals on the Real Line

> **STUDY TIP** Note that a square bracket is used to denote "less than or equal to" ($\leq$). Furthermore, the symbols $\infty$ and $-\infty$ denote positive and negative infinity. These symbols do not denote real numbers; they merely let you describe unbounded conditions more concisely. For instance, the interval $[b, \infty)$ is unbounded to the right because it includes *all* real numbers that are greater than or equal to $b$.

## Solving Inequalities

In calculus, you are frequently required to "solve inequalities" involving variable expressions such as $3x - 4 < 5$. The number $a$ is a **solution** of an inequality if the inequality is true when $a$ is substituted for $x$. The set of all values of $x$ that satisfy an equality is called the **solution set** of the inequality. The following properties are useful for solving inequalities. (Similar properties are obtained if $<$ is replaced by $\leq$ and $>$ is replaced by $\geq$.)

ALGEBRA

**TIP**

Once you have solved an inequality, it is a good idea to check some $x$-values in your solution set to see whether they satisfy the original inequality. You might also check some values outside your solution set to verify that they do *not* satisfy the inequality. For example, Figure 0.6 shows that when $x = 0$ or $x = 2$ the inequality is satisfied, but when $x = 4$ the inequality is not satisfied.

### Properties of Inequalities

1. Transitive property: $a < b$ and $b < c$ ⟹ $a < c$
2. Adding inequalities: $a < b$ and $c < d$ ⟹ $a + c < b + d$
3. Multiplying by a (positive) constant: $a < b$ ⟹ $ac < bc, \quad c > 0$
4. Multiplying by a (negative) constant: $a < b$ ⟹ $ac > bc, \quad c < 0$
5. Adding a constant: $a < b$ ⟹ $a + c < b + c$
6. Subtracting a constant: $a < b$ ⟹ $a - c < b - c$

Note that you *reverse the inequality* when you multiply by a negative number. For example, if $x < 3$, then $-4x > -12$. This principle also applies to division by a negative number. Thus, if $-2x > 4$, then $x < -2$.

### EXAMPLE 1   Solving an Inequality

Find the solution set of the inequality $3x - 4 < 5$.

**Solution**

$$3x - 4 < 5 \qquad \text{Original inequality}$$
$$3x - 4 + 4 < 5 + 4 \qquad \text{Add 4 to each side.}$$
$$3x < 9 \qquad \text{Simplify.}$$
$$\frac{1}{3}(3x) < \frac{1}{3}(9) \qquad \text{Multiply each side by } \tfrac{1}{3}.$$
$$x < 3 \qquad \text{Simplify.}$$

Thus, the solution set is the interval $(-\infty, 3)$, as shown in Figure 0.6.

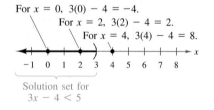

For $x = 0$, $3(0) - 4 = -4$.
For $x = 2$, $3(2) - 4 = 2$.
For $x = 4$, $3(4) - 4 = 8$.

Solution set for
$3x - 4 < 5$

**FIGURE 0.6**

In Example 1, all five inequalities listed as steps in the solution have the same solution set, and they are called **equivalent inequalities.**

The inequality in Example 1 involves a first-degree polynomial. To solve inequalities involving polynomials of higher degree, you can use the fact that a polynomial can change signs *only* at its real zeros (the real numbers that make the polynomial zero). Between two consecutive real zeros, a polynomial must be entirely positive or entirely negative. This means that when the real zeros of a polynomial are put in order, they divide the real line into **test intervals** in which the polynomial has no sign changes. That is, if a polynomial has the factored form

$$(x - r_1)(x - r_2), \ldots, (x - r_n), \qquad r_1 < r_2 < r_3 < \cdots < r_n,$$

then the test intervals are

$$(-\infty, r_1), \quad (r_1, r_2), \quad \ldots, \quad (r_{n-1}, r_n), \quad \text{and} \quad (r_n, \infty).$$

For example, the polynomial

$$x^2 - x - 6 = (x - 3)(x + 2)$$

can change signs only at $x = -2$ and $x = 3$. To determine the sign of the polynomial in the intervals $(-\infty, -2)$, $(-2, 3)$, and $(3, \infty)$, you need to test only *one value* from each interval.

## EXAMPLE 2    Solving a Polynomial Inequality

Find the solution set of the inequality $x^2 < x + 6$.

**Solution**

$$\begin{aligned} x^2 &< x + 6 & \text{Original inequality} \\ x^2 - x - 6 &< 0 & \text{Polynomial form} \\ (x - 3)(x + 2) &< 0 & \text{Factor.} \end{aligned}$$

Thus, the polynomial $x^2 - x - 6$ has $x = -2$ and $x = 3$ as its zeros. You can solve the inequality by testing the sign of the polynomial in each of the following intervals.

$$x < -2, \quad -2 < x < 3, \quad x > 3$$

To test an interval, choose a representative number in the interval and compute the sign of each factor. For example, for any $x < -2$, both of the factors $(x - 3)$ and $(x + 2)$ are negative. Consequently, the product (of two negative numbers) is positive, and the inequality is *not* satisfied in the interval

$$x < -2.$$

We suggest the testing format shown in Figure 0.7. Because the inequality is satisfied only by the center test interval, you conclude that the solution set is given by the interval

$$-2 < x < 3. \qquad \text{Solution set}$$

Sign of $(x - 3)(x + 2)$

| $x$ | Sign | < 0? |
|-----|------|------|
| $-3$ | $(-)(-)$ | no |
| $-2$ | $(-)(0)$ | no |
| $-1$ | $(-)(+)$ | yes |
| $0$ | $(-)(+)$ | yes |
| $1$ | $(-)(+)$ | yes |
| $2$ | $(-)(+)$ | yes |
| $3$ | $(0)(+)$ | no |
| $4$ | $(+)(+)$ | no |

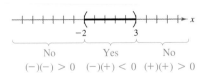

**FIGURE 0.7**    Is $(x - 3)(x + 2) < 0$?

## Applications

Inequalities are frequently used to describe conditions that occur in business and science. For instance, the inequality

$$144 \leq W \leq 180$$

describes the recommended weight $W$ for a man whose height is 5 feet 10 inches. Example 3 shows how an inequality can be used to describe the production level of a manufacturing plant.

### EXAMPLE 3   Production Levels

In addition to fixed overhead costs of $500 per day, the cost of producing $x$ units of a certain item is $2.50 per unit. During the month of August, the total cost of production varied from a high of $1325 to a low of $1200 per day. Find the high and low *production levels* during the month.

### Solution

Because it costs $2.50 to produce one unit, it costs $2.5x$ to produce $x$ units. Furthermore, because the fixed cost per day is $500, the total daily cost of producing $x$ units is

$$C = 2.5x + 500.$$

Now, because the cost ranged from $1200 to $1325, you can write the following.

| | | | |
|---|---|---|---|
| $1200 \leq$ | $2.5x + 500$ | $\leq 1325$ | Original inequality |
| $1200 - 500 \leq$ | $2.5x + 500 - 500 \leq$ | $1325 - 500$ | Subtract 500 from each side. |
| $700 \leq$ | $2.5x$ | $\leq 825$ | Simplify. |
| $\dfrac{700}{2.5} \leq$ | $\dfrac{2.5x}{2.5}$ | $\leq \dfrac{825}{2.5}$ | Divide each side by 2.5. |
| $280 \leq$ | $x$ | $\leq 330$ | Simplify. |

Thus, the daily production levels during the month varied between a low of 280 units and a high of 330 units, as pictured in Figure 0.8.

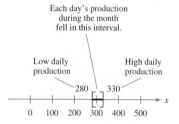

**FIGURE 0.8**

## EXERCISES 0.1

In Exercises 1–10, determine whether the real number is rational or irrational.

* **1.** $0.7$

**2.** $-3678$

**3.** $\dfrac{3\pi}{2}$

**4.** $3\sqrt{2} - 1$

**5.** $4.3\overline{451}$

**6.** $\dfrac{22}{7}$

**7.** $\sqrt[3]{64}$

**8.** $0.\overline{8177}$

**9.** $\sqrt[3]{60}$

**10.** $2e$

In Exercises 11–14, determine whether the given value of $x$ satisfies the inequality.

**11.** $5x - 12 > 0$

    (a) $x = 3$   (b) $x = -3$   (c) $x = \frac{5}{2}$   (d) $x = \frac{3}{2}$

**12.** $x + 1 < \dfrac{2x}{3}$

    (a) $x = 0$   (b) $x = 4$   (c) $x = -4$   (d) $x = -3$

**13.** $0 < \dfrac{x - 2}{4} < 2$

    (a) $x = 4$   (b) $x = 10$   (c) $x = 0$   (d) $x = \frac{7}{2}$

**14.** $-1 < \dfrac{3 - x}{2} \le 1$

    (a) $x = 0$   (b) $x = \sqrt{5}$   (c) $x = 1$   (d) $x = 5$

In Exercises 15–28, solve the inequality and sketch the graph of the solution on the real number line.

**15.** $x - 5 \ge 7$

**16.** $2x > 3$

**17.** $4x + 1 < 2x$

**18.** $2x + 7 < 3$

**19.** $4 - 2x < 3x - 1$

**20.** $x - 4 \le 2x + 1$

**21.** $-4 < 2x - 3 < 4$

**22.** $0 \le x + 3 < 5$

**23.** $\dfrac{3}{4} > x + 1 > \dfrac{1}{4}$

**24.** $-1 < -\dfrac{x}{3} < 1$

**25.** $\dfrac{x}{2} + \dfrac{x}{3} > 5$

**26.** $\dfrac{x}{2} - \dfrac{x}{3} > 5$

**27.** $x^2 - x \le 0$

**28.** $2x^2 + 1 < 9x - 3$

* The *answers* to the odd-numbered and selected even exercises are given in the back of the text. *Worked-out solutions* to the odd-numbered exercises are given in the *Student Solutions Guide*.

**29.** *Biology: pH values*   The pH scale measures the concentration of hydrogen ions in a solution. Strong acids produce low pH values, while strong bases produce high pH values. Represent the following approximate pH values on a real number line: hydrochloric acid, 0.0; lemon juice, 2.0; oven cleaner, 13.0; baking soda, 9.0; pure water, 7.0; black coffee, 5.0. *(Source: Adapted from Levine/Miller, Biology: Discovering Life, Second Edition)*

**30.** *Sales*   A doughnut shop at a shopping mall sells a dozen doughnuts for $2.95. Beyond the fixed costs (for rent, utilities, and insurance) of $150 per day, it costs $1.45 for enough materials (flour, sugar, etc.) and labor to produce each dozen doughnuts. If the daily profit *varies between* $50 and $200, between what levels (in dozens) do the daily sales vary?

**31.** *Profit*   The revenue for selling $x$ units of a product is

$$R = 115.95x,$$

and the cost of producing $x$ units is

$$C = 95x + 750.$$

To obtain a profit, the revenue must be *greater than* the cost. For what values of $x$ will this product return a profit?

**32.** *Physiology*   The maximum heart rate of a person in normal health is related to the person's age by the equation

$$r = 220 - A,$$

where $r$ is the maximum heart rate in beats per minute and $A$ is the person's age in years. Some physiologists recommend that during physical activity a person should strive to increase his or her heart rate to at least 60% of the maximum heart rate for sedentary people and at most 90% of the maximum heart rate for highly fit people. Express as an interval the range of the target heart rate for a 20-year-old.

**33.** *Area*   A square region is to have an area of *at least* 500 square meters. What must the length of the sides of the region be?

In Exercises 34 and 35, determine whether each statement is true or false, given $a < b$.

**34.** (a) $-2a < -2b$        (b) $a + 2 < b + 2$

    (c) $6a < 6b$          (d) $\dfrac{1}{a} < \dfrac{1}{b}$

**35.** (a) $a - 4 < b - 4$    (b) $4 - a < 4 - b$

    (c) $-3b < -3a$       (d) $\dfrac{a}{4} < \dfrac{b}{4}$

## 0.2 | Absolute Value and Distance on the Real Line

*Absolute Value of a Real Number • Distance on the Real Line •*
*Intervals Defined by Absolute Value • Applications*

### Absolute Value of a Real Number

## Technology

The absolute value key on most calculators is denoted by **ABS.** For instance, to find the absolute value of $-4$, you can use the following keystroke sequence.

| Keystrokes | Display |
|---|---|
| ABS (−) 4 ENTER | 4 |

#### Definition of Absolute Value

The **absolute value** of a real number $a$ is

$$|a| = \begin{cases} a, & \text{if } a \geq 0 \\ -a, & \text{if } a < 0. \end{cases}$$

At first glance it may appear from this definition that the absolute value of a real number can be negative, but this is not possible. For example, let $a = -3$. Then, because $-3 < 0$, you have

$$|a| = |-3| = -(-3) = 3.$$

The following properties are useful for working with absolute values.

#### Properties of Absolute Value

1. Multiplication:    $|ab| = |a||b|$

2. Division:    $\left|\dfrac{a}{b}\right| = \dfrac{|a|}{|b|}, \quad b \neq 0$

3. Power:    $|a^n| = |a|^n$

4. Square root:    $\sqrt{a^2} = |a|$

Be sure you understand the fourth property in this list. A common error in algebra is to imagine that by squaring a number and then taking the square root, you come back to the original number. But this is true only if the original number is nonnegative. For instance, if $a = 2$, then

$$\sqrt{2^2} = \sqrt{4} = 2,$$

but if $a = -2$, then

$$\sqrt{(-2)^2} = \sqrt{4} = 2.$$

The reason for this is that (by definition) the square root symbol $\sqrt{\phantom{x}}$ denotes only the nonnegative root.

## Distance on the Real Line

Consider two distinct points on the real line, as shown in Figure 0.9.

1. The **directed distance from $a$ to $b$** is $b - a$.
2. The **directed distance from $b$ to $a$** is $a - b$.
3. The **distance between $a$ and $b$** is $|a - b|$ or $|b - a|$.

In Figure 0.9, note that because $b$ is to the right of $a$, the directed distance from $a$ to $b$ (moving to the right) is positive. Moreover, because $a$ is to the left of $b$, the directed distance from $b$ to $a$ (moving to the left) is negative. The distance *between* two points on the real line can never be negative.

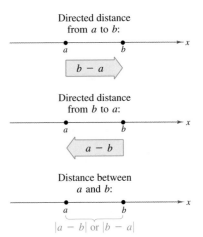

**FIGURE 0.9**

### Distance Between Two Points on the Real Line

The distance $d$ between points $x_1$ and $x_2$ on the real line is given by

$$d = |x_2 - x_1| = \sqrt{(x_2 - x_1)^2}.$$

Note that the order of subtraction with $x_1$ and $x_2$ does not matter because

$$|x_2 - x_1| = |x_1 - x_2| \quad \text{and} \quad (x_2 - x_1)^2 = (x_1 - x_2)^2.$$

## EXAMPLE 1   Finding Distance on the Real Line

Determine the distance between $-3$ and 4 on the real line. What is the directed distance from $-3$ to 4? From 4 to $-3$?

### Solution

The distance between $-3$ and 4 is given by

$$|-3 - 4| = |-7| = 7 \quad \text{or} \quad |4 - (-3)| = |7| = 7. \qquad |a - b| \text{ or } |b - a|$$

The directed distance from $-3$ to 4 is

$$4 - (-3) = 7. \qquad b - a$$

The directed distance from 4 to $-3$ is

$$-3 - 4 = -7. \qquad a - b$$

(See Figure 0.10.)

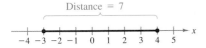

**FIGURE 0.10**

## Intervals Defined by Absolute Value

### EXAMPLE 2   Defining an Interval on the Real Line

Find the interval on the real line that contains all numbers that lie no more than two units from 3.

**Solution**

Let $x$ be any point in this interval. You need to find all $x$ such that the distance between $x$ and 3 is less than or equal to 2. This implies that

$$|x - 3| \le 2.$$

Requiring the absolute value of $x - 3$ to be less than or equal to 2 means that $x - 3$ must lie between $-2$ and 2. Hence, you can write

$$-2 \le x - 3 \le 2.$$

Solving this pair of inequalities, you have

$$-2 + 3 \le x - 3 + 3 \le 2 + 3$$
$$1 \le \quad x \quad \le 5. \qquad \text{Solution set}$$

Therefore, the interval is $[1, 5]$, as shown in Figure 0.11.

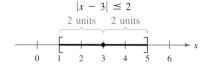

$|x - 3| \le 2$
2 units    2 units

**FIGURE 0.11**

---

### Two Basic Types of Inequalities Involving Absolute Value

Let $a$ and $d$ be real numbers, where $d > 0$.

$|x - a| \le d$ if and only if $a - d \le x \le a + d$.

$|x - a| \ge d$ if and only if $x \le a - d$ or $a + d \le x$.

| Inequality | Interpretation | Graph |
|---|---|---|
| $|x - a| \le d$ | All numbers $x$ whose distance from $a$ is less than or equal to $d$. | 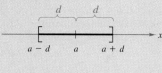 |
| $|x - a| \ge d$ | All numbers $x$ whose distance from $a$ is greater than or equal to $d$. | 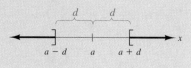 |

**TIP**

**ALGEBRA**

Be sure you see that inequalities of the form $|x - a| \ge d$ have solution sets consisting of two intervals. To describe the two intervals without using absolute values, you must use *two* separate inequalities, connected by an "or" to indicate union.

# Applications

## EXAMPLE 3  Quality Control

A large manufacturer hired a quality control firm to determine the reliability of a certain product. Using statistical methods, the firm determined that the manufacturer could expect $0.35\% \pm 0.17\%$ of the units to be defective. If the manufacturer offers a money-back guarantee on this product, how much should be budgeted to cover the refunds on 100,000 units? (Assume that the retail price is $8.95.)

### Solution

Let $r$ represent the percent of defective units (written in decimal form). You know that $r$ will differ from 0.0035 by at most 0.0017.

$$0.0035 - 0.0017 \leq r \leq 0.0035 + 0.0017$$
$$0.0018 \leq r \leq 0.0052 \qquad \text{Figure 0.12(a)}$$

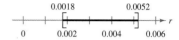

(a) Percent of defective units

Now, letting $x$ be the number of defective units out of 100,000, it follows that $x = 100{,}000r$ and you have

$$0.0018(100{,}000) \leq 100{,}000r \leq 0.0052(100{,}000)$$
$$180 \leq \quad x \quad \leq 520. \qquad \text{Figure 0.12(b)}$$

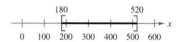

(b) Number of defective units

Finally, letting $C$ be the cost of refunds, you have $C = 8.95x$. Thus, the total cost of refunds for 100,000 units should fall within the interval given by

$$180(8.95) \leq 8.95x \leq 520(8.95)$$
$$\$1611 \leq \quad C \quad \leq \$4654. \qquad \text{Figure 0.12(c)}$$

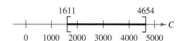

(c) Cost of refunds

**FIGURE 0.12**

In Example 3, the manufacturer should expect to spend between $1611 and $4654 for refunds. Of course, the safer budget figure for refunds would be the higher of these estimates. However, from a statistical point of view, the most representative estimate would be the average of these two extremes. Graphically, the average of two numbers is the **midpoint** of the interval with the two numbers as endpoints, as shown in Figure 0.13.

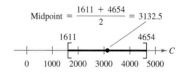

**FIGURE 0.13**

## Midpoint of an Interval

The **midpoint** of the interval with endpoints $a$ and $b$ is found by taking the average of the endpoints.

$$\text{Midpoint} = \frac{a + b}{2}$$

## EXERCISES 0.2

In Exercises 1–4, find (a) the directed distance from $a$ to $b$, (b) the directed distance from $b$ to $a$, and (c) the distance between $a$ and $b$.

**1.** $a = 126, b = 75$      **2.** $a = -126, b = -75$

**3.** $a = 9.34, b = -5.65$      **4.** $a = \frac{16}{5}, b = \frac{112}{75}$

In Exercises 5–8, find the midpoint of the given interval.

**5.** $[7, 21]$      **6.** $[8.6, 11.4]$

**7.** $[-6.85, 9.35]$      **8.** $[-4.6, -1.3]$

In Exercises 9–22, solve the inequality and sketch the graph of the solution on the real line.

**9.** $|x| < 5$      **10.** $|2x| < 6$

**11.** $\left|\dfrac{x}{2}\right| > 3$      **12.** $|5x| > 10$

**13.** $|x + 2| < 5$      **14.** $|3x + 1| \geq 4$

**15.** $\left|\dfrac{x - 3}{2}\right| \geq 5$      **16.** $|2x + 1| < 5$

**17.** $|10 - x| > 4$      **18.** $|25 - x| \geq 20$

**19.** $|9 - 2x| < 1$      **20.** $\left|1 - \dfrac{2x}{3}\right| < 1$

**21.** $|x - a| \leq b$      **22.** $|2x - a| \geq b$

In Exercises 23–34, use absolute values to describe the given interval (or pair of intervals) on the real line.

**23.** $[-2, 2]$      **24.** $(-3, 3)$

**25.** $(-\infty, -2) \cup (2, \infty)$      **26.** $(-\infty, -3] \cup [3, \infty)$

**27.** $[2, 6]$      **28.** $(-7, -1)$

**29.** $(-\infty, 0) \cup (4, \infty)$      **30.** $(-\infty, 20) \cup (24, \infty)$

**31.** All numbers *less than* two units from 4.

**32.** All numbers *more than* six units from 3.

**33.** $y$ is *at most* two units from $a$.

**34.** $y$ is *less than* $h$ units from $c$.

**35.** *Statistics*   The heights $h$ of two-thirds of the members of a certain population satisfy the inequality

$$\left|\frac{h - 68.5}{2.7}\right| \leq 1,$$

where $h$ is measured in inches. Determine the interval on the real line in which these heights lie.

**36.** *Show Collies*   The American Kennel Club has developed guidelines for judging the features of various breeds of dogs. For collies, the guidelines specify that the weights for males satisfy the inequality

$$\left|\frac{w - 57.5}{7.5}\right| \leq 1,$$

where $w$ is measured in pounds. Determine the interval on the real line in which these weights lie.

**37.** *Production*   The estimated daily production $x$ at a refinery is given by

$$|x - 200,000| \leq 25,000,$$

where $x$ is measured in barrels of oil. Determine the high and low production levels.

**38.** *Stock Price*   A stock market analyst predicts that over the next year the price $p$ of a certain stock will not change from its current price of $\$33\frac{1}{8}$ by more than $\$2$. Use absolute values to write this prediction as an inequality.

*Budget Variance*   In Exercises 39–42, (a) use absolute value notation to represent the two intervals in which expenses must lie if they are to be within $\$500$ and within 5% of the specified budget amount, and (b) using the more stringent constraint, determine whether the given expense is at variance with the budget restriction.

| | Item | Budget | Expense |
|---|---|---|---|
| **39.** | Utilities | $4750.00 | $5116.37 |
| **40.** | Insurance | $15,000.00 | $14,695.00 |
| **41.** | Maintenance | $20,000.00 | $22,718.35 |
| **42.** | Taxes | $7500.00 | $8691.00 |

# 0.3 | Exponents and Radicals

*Expressions Involving Exponents or Radicals • Operations with Exponents •*
*Domain of an Algebraic Expression*

## Expressions Involving Exponents or Radicals

### Properties of Exponents

1. Whole-number exponents:     $x^n = \underbrace{x \cdot x \cdot x \cdots x}_{n \text{ factors}}$

2. Zero exponent:     $x^0 = 1, \quad x \neq 0$

3. Negative exponents:     $x^{-n} = \dfrac{1}{x^n}, \quad x \neq 0$

4. Radicals (principal $n$th root):     $\sqrt[n]{x} = a \implies x = a^n$
5. Rational exponents $(1/n)$:     $x^{1/n} = \sqrt[n]{x}$
6. Rational exponents $(m/n)$:     $x^{m/n} = (x^{1/n})^m = \left(\sqrt[n]{x}\right)^m$
   $x^{m/n} = (x^m)^{1/n} = \sqrt[n]{x^m}$

7. Special convention (square root):     $\sqrt[2]{x} = \sqrt{x}$

> **TIP**
>
> If $n$ is even, then the principal $n$th root is positive. For example, $\sqrt{4} = +2$ and $\sqrt[4]{81} = +3$.
>
> *ALGEBRA*

---

### EXAMPLE 1   Evaluating Expressions

| *Expression* | *x-Value* | *Substitution* |
|---|---|---|
| (a) $y = -2x^2$ | $x = 4$ | $y = -2(4^2) = -2(16) = -32$ |
| (b) $y = 3x^{-3}$ | $x = -1$ | $y = 3(-1)^{-3} = \dfrac{3}{(-1)^3} = \dfrac{3}{-1} = -3$ |
| (c) $y = (-x)^2$ | $x = \dfrac{1}{2}$ | $y = \left(-\dfrac{1}{2}\right)^2 = \dfrac{1}{4}$ |
| (d) $y = \dfrac{2}{x^{-2}}$ | $x = 3$ | $y = \dfrac{2}{3^{-2}} = 2(3^2) = 18$ |

---

### EXAMPLE 2   Evaluating Expressions

| *Expression* | *x-Value* | *Substitution* |
|---|---|---|
| (a) $y = 2x^{1/2}$ | $x = 4$ | $y = 2\sqrt{4} = 2(2) = 4$ |
| (b) $y = \sqrt[3]{x^2}$ | $x = 8$ | $y = 8^{2/3} = (8^{1/3})^2 = 2^2 = 4$ |

## Operations with Exponents

**Technology**

Graphing utilities perform the established order of operations when evaluating an expression. To see this, try entering the expression

$$1200\left(1 + \frac{0.09}{12}\right)^{12 \cdot 6}$$

into your graphing utility, using the keystrokes below.

1200 ⊠ ⟮ 1 ⊞ .09 ⊟ 12 ⟯ ⌃ ⟮ 12 ⊠ 6 ⟯

What do you get?

### Operations with Exponents

1. Multiplying like bases: $x^n x^m = x^{n+m}$     Add exponents.

2. Dividing like bases: $\dfrac{x^n}{x^m} = x^{n-m}$     Subtract exponents.

3. Removing parentheses: $(xy)^n = x^n y^n$

$$\left(\frac{x}{y}\right)^n = \frac{x^n}{y^n}$$

$$(x^n)^m = x^{nm}$$

4. Special conventions: $-x^n = -(x^n), \quad -x^n \neq (-x)^n$

$$cx^n = c(x^n), \quad cx^n \neq (cx)^n$$

$$x^{n^m} = x^{(n^m)}, \quad x^{n^m} \neq (x^n)^m .$$

### EXAMPLE 3   Simplifying Expressions with Exponents

Simplify the following expressions.

(a) $2x^2(x^3)$     (b) $(3x)^2 \sqrt[3]{x}$     (c) $\dfrac{3x^2}{(x^{1/2})^3}$

(d) $\dfrac{5x^4}{(x^2)^3}$     (e) $x^{-1}(2x^2)$     (f) $\dfrac{-\sqrt{x}}{5x^{-1}}$

**Solution**

(a) $2x^2(x^3) = 2x^{2+3} = 2x^5$     $x^n x^m = x^{n+m}$

(b) $(3x)^2 \sqrt[3]{x} = 9x^2 x^{1/3} = 9x^{2+(1/3)} = 9x^{7/3}$     $x^n x^m = x^{n+m}$

(c) $\dfrac{3x^2}{(x^{1/2})^3} = 3\left(\dfrac{x^2}{x^{3/2}}\right) = 3x^{2-(3/2)} = 3x^{1/2}$     $(x^n)^m = x^{nm}, \quad \dfrac{x^n}{x^m} = x^{n-m}$

(d) $\dfrac{5x^4}{(x^2)^3} = \dfrac{5x^4}{x^6} = 5x^{4-6} = 5x^{-2} = \dfrac{5}{x^2}$     $(x^n)^m = x^{nm}, \quad \dfrac{x^n}{x^m} = x^{n-m}$

(e) $x^{-1}(2x^2) = 2x^{-1}x^2 = 2x^{2-1} = 2x$     $x^n x^m = x^{n+m}$

(f) $\dfrac{-\sqrt{x}}{5x^{-1}} = -\dfrac{1}{5}\left(\dfrac{x^{1/2}}{x^{-1}}\right) = -\dfrac{1}{5}x^{(1/2)+1} = -\dfrac{1}{5}x^{3/2}$     $\dfrac{x^n}{x^m} = x^{n-m}$

Note in Example 3 that one characteristic of simplified expressions is the absence of negative exponents. Another characteristic of simplified expressions is that people usually prefer to write sums and differences in *factored form*. To do this, you can use the **distributive property.**

$$abx^n + acx^{n+m} = ax^n(b + cx^m)$$

Study the next example carefully to be sure that you understand the concepts involved in the factoring process.

---

## EXAMPLE 4   Simplifying by Factoring

(a) $2x^2 - x^3 = x^2(2 - x)$

(b) $2x^3 + x^2 = x^2(2x + 1)$

(c) $2x^{1/2} + 4x^{5/2} = 2x^{1/2}(1 + 2x^2)$

(d) $2x^{-1/2} + 3x^{5/2} = x^{-1/2}(2 + 3x^3) = \dfrac{2 + 3x^3}{\sqrt{x}}$

Many algebraic expressions obtained in calculus occur in unsimplified form. For instance, the two expressions shown in the following example are the result of an operation in calculus called **differentiation.** [The first is the derivative of $2(x + 1)^{3/2}(2x - 3)^{5/2}$, and the second is the derivative of $2(x + 1)^{1/2}(2x - 3)^{5/2}$.]

---

## EXAMPLE 5   Simplifying by Factoring

Simplify the following expressions by factoring.

(a) $3(x + 1)^{1/2}(2x - 3)^{5/2} + 10(x + 1)^{3/2}(2x - 3)^{3/2}$

(b) $(x + 1)^{-1/2}(2x - 3)^{5/2} + 10(x + 1)^{1/2}(2x - 3)^{3/2}$

**Solution**

(a) $3(x + 1)^{1/2}(2x - 3)^{5/2} + 10(x + 1)^{3/2}(2x - 3)^{3/2}$

$$= (x + 1)^{1/2}(2x - 3)^{3/2}[3(2x - 3) + 10(x + 1)]$$
$$= (x + 1)^{1/2}(2x - 3)^{3/2}(6x - 9 + 10x + 10)$$
$$= (x + 1)^{1/2}(2x - 3)^{3/2}(16x + 1)$$

(b) $(x + 1)^{-1/2}(2x - 3)^{5/2} + 10(x + 1)^{1/2}(2x - 3)^{3/2}$

$$= (x + 1)^{-1/2}(2x - 3)^{3/2}[(2x - 3) + 10(x + 1)]$$
$$= (x + 1)^{-1/2}(2x - 3)^{3/2}(2x - 3 + 10x + 10)$$
$$= (x + 1)^{-1/2}(2x - 3)^{3/2}(12x + 7)$$
$$= \dfrac{(2x - 3)^{3/2}(12x + 7)}{(x + 1)^{1/2}}$$

Example 6 shows some additional types of expressions that can occur in calculus. [The expression in part c is an antiderivative of $(x + 1)^{2/3}(2x + 3)$, and the expression in part d is the derivative of $(x + 2)^3/(x - 1)^3$.]

## EXAMPLE 6   Factors Involving Quotients

Simplify the following expressions by factoring.

(a) $\dfrac{3x^2 + x^4}{2x}$

(b) $\dfrac{\sqrt{x} + x^{3/2}}{x}$

(c) $\dfrac{3}{5}(x + 1)^{5/3} + \dfrac{3}{4}(x + 1)^{8/3}$

(d) $\dfrac{3(x + 2)^2(x - 1)^3 - 3(x + 2)^3(x - 1)^2}{[(x - 1)^3]^2}$

*Solution*

(a) $\dfrac{3x^2 + x^4}{2x} = \dfrac{x^2(3 + x^2)}{2x} = \dfrac{x^{2-1}(3 + x^2)}{2} = \dfrac{x(3 + x^2)}{2}$

(b) $\dfrac{\sqrt{x} + x^{3/2}}{x} = \dfrac{x^{1/2}(1 + x)}{x} = \dfrac{1 + x}{x^{1-(1/2)}} = \dfrac{1 + x}{\sqrt{x}}$

(c) $\dfrac{3}{5}(x + 1)^{5/3} + \dfrac{3}{4}(x + 1)^{8/3} = \dfrac{12}{20}(x + 1)^{5/3} + \dfrac{15}{20}(x + 1)^{8/3}$

$\qquad\qquad = \dfrac{3}{20}(x + 1)^{5/3}[4 + 5(x + 1)]$

$\qquad\qquad = \dfrac{3}{20}(x + 1)^{5/3}(4 + 5x + 5)$

$\qquad\qquad = \dfrac{3}{20}(x + 1)^{5/3}(5x + 9)$

(d) $\dfrac{3(x + 2)^2(x - 1)^3 - 3(x + 2)^3(x - 1)^2}{[(x - 1)^3]^2}$

$\qquad = \dfrac{3(x + 2)^2(x - 1)^2[(x - 1) - (x + 2)]}{(x - 1)^6}$

$\qquad = \dfrac{3(x + 2)^2(x - 1 - x - 2)}{(x - 1)^{6-2}}$

$\qquad = \dfrac{-9(x + 2)^2}{(x - 1)^4}$

## Domain of an Algebraic Expression

When working with algebraic expressions involving $x$, you face the potential difficulty of substituting a value of $x$ for which the expression is not defined (does not produce a real number). For example, the expression $\sqrt{2x + 3}$ is *not defined* when $x = -2$ because $\sqrt{2(-2) + 3}$ is not a real number.

The set of all values for which an expression is defined is called its **domain.** Thus, the domain of $\sqrt{2x + 3}$ is the set of all values of $x$ such that $\sqrt{2x + 3}$ is a real number. In order for $\sqrt{2x + 3}$ to represent a real number, it is necessary that $2x + 3 \geq 0$. In other words, $\sqrt{2x + 3}$ is defined only for those values of $x$ that lie in the interval $\left[-\frac{3}{2}, \infty\right)$, as shown in Figure 0.14.

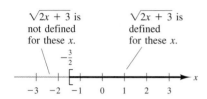

$\sqrt{2x + 3}$ is        $\sqrt{2x + 3}$ is
not defined              defined
for these $x$.           for these $x$.

**FIGURE 0.14**

---

## EXAMPLE 7    Finding the Domain of an Expression

Find the domain of each of the following expressions.

(a) $\sqrt{3x - 2}$

(b) $\dfrac{1}{\sqrt{3x - 2}}$

(c) $\sqrt[3]{9x + 1}$

*Solution*

(a) The domain of $\sqrt{3x - 2}$ consists of all $x$ such that

$$3x - 2 \geq 0, \qquad \text{Expression must be nonnegative.}$$

which implies that $x \geq \frac{2}{3}$. Therefore, the domain is $\left[\frac{2}{3}, \infty\right)$.

(b) The domain of $1/\sqrt{3x - 2}$ is the same as the domain of $\sqrt{3x - 2}$, except that $1/\sqrt{3x - 2}$ is not defined when $3x - 2 = 0$. Because this occurs when $x = \frac{2}{3}$, we conclude that the domain is $\left(\frac{2}{3}, \infty\right)$.

(c) Because $\sqrt[3]{9x + 1}$ is defined for *all* real numbers, we conclude that its domain is $(-\infty, \infty)$.

## EXERCISES 0.3

In Exercises 1–20, evaluate the expression for the indicated value of $x$.

| Expression | $x$-Value | Expression | $x$-Value |
|---|---|---|---|
| **1.** $-3x^3$ | $x = 2$ | **2.** $\dfrac{x^2}{2}$ | $x = 6$ |
| **3.** $4x^{-3}$ | $x = 2$ | **4.** $7x^{-2}$ | $x = 4$ |
| **5.** $\dfrac{1 + x^{-1}}{x^{-1}}$ | $x = 2$ | **6.** $x - 4x^{-2}$ | $x = 3$ |
| **7.** $3x^2 - 4x^3$ | $x = -2$ | **8.** $5(-x)^3$ | $x = 3$ |
| **9.** $6x^0 - (6x)^0$ | $x = 10$ | **10.** $\dfrac{1}{(-x)^{-3}}$ | $x = 4$ |
| **11.** $\sqrt[3]{x^2}$ | $x = 27$ | **12.** $\sqrt{x^3}$ | $x = \frac{1}{9}$ |
| **13.** $x^{-1/2}$ | $x = 4$ | **14.** $x^{-3/4}$ | $x = 16$ |
| **15.** $x^{-2/5}$ | $x = -32$ | **16.** $(x^{2/3})^3$ | $x = 10$ |
| **17.** $500x^{60}$ | $x = 1.01$ | **18.** $\dfrac{10{,}000}{x^{120}}$ | $x = 1.075$ |
| **19.** $\sqrt[3]{x}$ | $x = -154$ | **20.** $\sqrt[6]{x}$ | $x = 325$ |

In Exercises 21–34, simplify the expression.

**21.** $5x^4(x^2)$

**22.** $(8x^4)(2x^3)$

**23.** $6y^2(2y^4)^2$

**24.** $z^{-3}(3z^4)$

**25.** $10(x^2)^2$

**26.** $(4x^3)^2$

**27.** $\dfrac{7x^2}{x^{-3}}$

**28.** $\dfrac{r^4}{r^6}$

**29.** $\dfrac{12(x + y)^3}{9(x + y)}$

**30.** $\left(\dfrac{12s^2}{9s}\right)^3$

**31.** $\dfrac{3x\sqrt{x}}{x^{1/2}}$

**32.** $\left(\sqrt[3]{x^2}\right)^3$

**33.** $\left(\dfrac{\sqrt{2}\sqrt{x^3}}{\sqrt{x}}\right)^4$

**34.** $(2x^2yz^5)^0$

In Exercises 35–40, simplify by removing all possible factors from the radical.

**35.** (a) $\sqrt{8}$     (b) $\sqrt{18}$

**36.** (a) $\sqrt[3]{\frac{16}{27}}$     (b) $\sqrt[3]{\frac{24}{125}}$

The symbol ⚙ indicates an exercise in which you are instructed to use graphing technology or a symbolic computer algebra system. The solutions of other exercises may also be facilitated by use of appropriate technology.

**37.** (a) $\sqrt[3]{16x^5}$     (b) $\sqrt[4]{32x^4z^5}$

**38.** (a) $\sqrt[4]{(3x^2y^3)^4}$     (b) $\sqrt[3]{54x^7}$

**39.** (a) $\sqrt{75x^2y^{-4}}$     (b) $\sqrt{5(x - y)^3}$

**40.** (a) $\sqrt[5]{96b^6c^3}$     (b) $\sqrt{72(x + 1)^4}$

In Exercises 41–50, find the domain of the given expression.

**41.** $\sqrt{x - 1}$

**42.** $\sqrt{5 - 2x}$

**43.** $\sqrt{x^2 + 3}$

**44.** $\sqrt[5]{1 - x}$

**45.** $\dfrac{1}{\sqrt[3]{x - 1}}$

**46.** $\dfrac{1}{\sqrt{x + 4}}$

**47.** $\dfrac{1}{\sqrt[4]{2x - 6}}$

**48.** $\dfrac{\sqrt{x - 1}}{x + 1}$

**49.** $\sqrt{x - 1} + \sqrt{5 - x}$

**50.** $\dfrac{1}{\sqrt{2x + 3}} + \sqrt{6 - 4x}$

⚙ **Compound Interest** In Exercises 51 and 52, a certificate of deposit has a principal of $P$ and an annual percentage rate of $r$ (expressed as a decimal) compounded $n$ times per year. Enter the compound interest formula

$$A = P\left(1 + \frac{r}{n}\right)^N$$

into a graphing utility and use it to find the balance after $N$ compoundings.

**51.** $P = \$10{,}000$,   $r = 6.5\%$,   $n = 12$,   $N = 120$

**52.** $P = \$7000$,   $r = 5\%$,   $n = 365$,   $N = 1000$

**53.** *The Period of a Pendulum*   The period of a pendulum is

$$T = 2\pi\sqrt{\frac{L}{32}},$$

where $T$ is the period in seconds and $L$ is the length of the pendulum in feet. Find the period of a pendulum whose length is 4 feet.

**54.** *Annuity*   A balance $A$, after $n$ annual payments of $P$ dollars have been made into an annuity earning an annual percentage rate of $r$ compounded annually, is given by

$$A = P(1 + r) + P(1 + r)^2 + \cdots + P(1 + r)^n.$$

Rewrite this formula by completing the following factorization: $A = P(1 + r)(\quad)$.

# Factoring Polynomials

## 0.4

## Factorization Techniques

The Fundamental Theorem of Algebra states that every $n$th-degree polynomial

$$a_n x^n + a_{n-1} x^{n-1} + \cdots + a_1 x + a_0$$

has precisely $n$ **zeros.** (The zeros may be repeated or imaginary.) The problem of finding the zeros of a polynomial is equivalent to the problem of factoring the polynomial into linear factors.

---

### Special Products and Factorization Techniques

*Quadratic Formula*

$$ax^2 + bx + c = 0 \implies x = \frac{-b \pm \sqrt{b^2 - 4ac}}{2a}$$

*Example*

$$x^2 + 3x - 1 = 0 \implies x = \frac{-3 \pm \sqrt{13}}{2}$$

*Special Products*

$$x^2 - a^2 = (x - a)(x + a)$$
$$x^3 - a^3 = (x - a)(x^2 + ax + a^2)$$
$$x^3 + a^3 = (x + a)(x^2 - ax + a^2)$$
$$x^4 - a^4 = (x - a)(x + a)(x^2 + a^2)$$

*Examples*

$$x^2 - 9 = (x - 3)(x + 3)$$
$$x^3 - 8 = (x - 2)(x^2 + 2x + 4)$$
$$x^3 + 64 = (x + 4)(x^2 - 4x + 16)$$
$$x^4 - 16 = (x - 2)(x + 2)(x^2 + 4)$$

*Binomial Theorem**

$$(x + a)^2 = x^2 + 2ax + a^2$$
$$(x - a)^2 = x^2 - 2ax + a^2$$
$$(x + a)^3 = x^3 + 3ax^2 + 3a^2x + a^3$$
$$(x - a)^3 = x^3 - 3ax^2 + 3a^2x - a^3$$
$$(x + a)^4 = x^4 + 4ax^3 + 6a^2x^2 + 4a^3x + a^4$$
$$(x - a)^4 = x^4 - 4ax^3 + 6a^2x^2 - 4a^3x + a^4$$

*Examples*

$$(x + 3)^2 = x^2 + 6x + 9$$
$$(x^2 - 5)^2 = x^4 - 10x^2 + 25$$
$$(x + 2)^3 = x^3 + 6x^2 + 12x + 8$$
$$(x - 1)^3 = x^3 - 3x^2 + 3x - 1$$
$$(x + 2)^4 = x^4 + 8x^3 + 24x^2 + 32x + 16$$
$$(x - 4)^4 = x^4 - 16x^3 + 96x^2 - 256x + 256$$

$$(x + a)^n = x^n + nax^{n-1} + \frac{n(n-1)}{2!} a^2 x^{n-2} + \frac{n(n-1)(n-2)}{3!} a^3 x^{n-3} + \cdots + na^{n-1}x + a^n$$

$$(x - a)^n = x^n - nax^{n-1} + \frac{n(n-1)}{2!} a^2 x^{n-2} - \frac{n(n-1)(n-2)}{3!} a^3 x^{n-3} + \cdots \pm na^{n-1}x \mp a^n$$

*Factoring by Grouping*

$$acx^3 + adx^2 + bcx + bd = ax^2(cx + d) + b(cx + d)$$
$$= (ax^2 + b)(cx + d)$$

*Example*

$$3x^3 - 2x^2 - 6x + 4 = x^2(3x - 2) - 2(3x - 2)$$
$$= (x^2 - 2)(3x - 2)$$

\* The *factorial* symbol ! is defined as follows: $0! = 1$, $1! = 1$, $2! = 2 \cdot 1 = 2$,
$3! = 3 \cdot 2 \cdot 1 = 6$, $4! = 4 \cdot 3 \cdot 2 \cdot 1 = 24$, and so on.

## EXAMPLE 1  Applying the Quadratic Formula

Use the Quadratic Formula to find all real zeros of the following polynomials.

(a) $4x^2 + 6x + 1$       (b) $x^2 + 6x + 9$       (c) $2x^2 - 6x + 5$

*Solution*

(a) Using $a = 4$, $b = 6$, and $c = 1$, you can write

$$x = \frac{-b \pm \sqrt{b^2 - 4ac}}{2a}$$

$$= \frac{-6 \pm \sqrt{36 - 16}}{8}$$

$$= \frac{-6 \pm \sqrt{20}}{8}$$

$$= \frac{-6 \pm 2\sqrt{5}}{8}$$

$$= \frac{2\left(-3 \pm \sqrt{5}\right)}{2(4)}$$

$$= \frac{-3 \pm \sqrt{5}}{4}.$$

Thus, there are two real zeros:

$$x = \frac{-3 - \sqrt{5}}{4} \approx -1.309 \quad \text{and} \quad x = \frac{-3 + \sqrt{5}}{4} \approx -0.191.$$

(b) In this case, $a = 1$, $b = 6$, and $c = 9$, and the Quadratic Formula yields

$$x = \frac{-b \pm \sqrt{b^2 - 4ac}}{2a} = \frac{-6 \pm \sqrt{36 - 36}}{2} = -\frac{6}{2} = -3.$$

Thus, there is one (repeated) real zero: $x = -3$.

(c) For this quadratic equation, $a = 2$, $b = -6$, and $c = 5$. Thus,

$$x = \frac{-b \pm \sqrt{b^2 - 4ac}}{2a} = \frac{6 \pm \sqrt{36 - 40}}{4} = \frac{6 \pm \sqrt{-4}}{4}.$$

Because $\sqrt{-4}$ is imaginary, there are no real zeros.

In Example 1, the zeros in part a are irrational, and the zeros in part c are imaginary. In both of these cases the quadratic is said to be **irreducible** because it cannot be factored into linear factors with rational coefficients. The next example shows how to find the zeros associated with *reducible* quadratics. In this example, factoring is used to find the zeros of each quadratic. Try using the Quadratic Formula to obtain the same zeros.

## EXAMPLE 2 Factoring Quadratics

Find the zeros of the following quadratic polynomials.

(a) $x^2 - 5x + 6$     (b) $x^2 - 6x + 9$     (c) $2x^2 + 5x - 3$

*Solution*

(a) Because

$$x^2 - 5x + 6 = (x - 2)(x - 3),$$

the zeros are $x = 2$ and $x = 3$.

(b) Because

$$x^2 - 6x + 9 = (x - 3)^2,$$

the only zero is $x = 3$.

(c) Because

$$2x^2 + 5x - 3 = (2x - 1)(x + 3),$$

the zeros are $x = \frac{1}{2}$ and $x = -3$.

> **TIP**
>
> The zeros of a polynomial in $x$ are the values of $x$ that make the polynomial zero. To find the zeros, factor the polynomial into linear factors and set each factor equal to zero. For instance, the zeros of $(x - 2)(x - 3)$ occur when $x - 2 = 0$ and $x - 3 = 0$.

**ALGEBRA**

## EXAMPLE 3 Finding the Domain of a Radical Expression

Find the domain of $\sqrt{x^2 - 3x + 2}$.

*Solution*

Because

$$x^2 - 3x + 2 = (x - 1)(x - 2),$$

you know that the zeros of the quadratic are $x = 1$ and $x = 2$. Thus, you need to test the sign of the quadratic in the three intervals $(-\infty, 1)$, $(1, 2)$, and $(2, \infty)$, as shown in Figure 0.15. After testing each of these intervals, you can see that the quadratic is negative in the center interval and positive in the outer two intervals. Moreover, because the quadratic is zero when $x = 1$ and $x = 2$, you can conclude that the domain of $\sqrt{x^2 - 3x + 2}$ is

$$(-\infty, 1] \cup [2, \infty). \qquad \text{Domain}$$

Values of $\sqrt{x^2 - 3x + 2}$

| $x$ | $\sqrt{x^2 - 3x + 2}$ |
|---|---|
| 0 | $\sqrt{2}$ |
| 1 | 0 |
| 1.5 | Undefined |
| 2 | 0 |
| 3 | $\sqrt{2}$ |

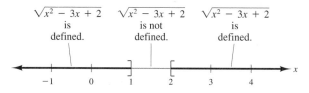

$\sqrt{x^2 - 3x + 2}$ is defined.　$\sqrt{x^2 - 3x + 2}$ is not defined.　$\sqrt{x^2 - 3x + 2}$ is defined.

**FIGURE 0.15**

## Factoring Polynomials of Degree Three or More

It can be difficult to find the zeros of polynomials of degree three or more. However, if one of the zeros of a polynomial is known, then you can use that zero to reduce the degree of the polynomial. For example, if you know that $x = 2$ is a zero of $x^3 - 4x^2 + 5x - 2$, then you know that $(x - 2)$ is a factor, and you can use long division to factor the polynomial as follows.

$$x^3 - 4x^2 + 5x - 2 = (x - 2)(x^2 - 2x + 1)$$
$$= (x - 2)(x - 1)(x - 1)$$

As an alternative to long division, many people prefer to use **synthetic division** to reduce the degree of a polynomial.

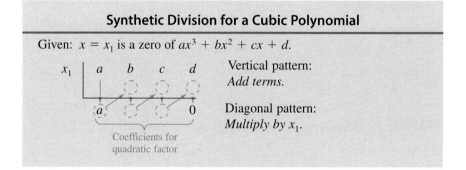

**Synthetic Division for a Cubic Polynomial**

Given: $x = x_1$ is a zero of $ax^3 + bx^2 + cx + d$.

Vertical pattern:
*Add terms.*

Diagonal pattern:
*Multiply by $x_1$.*

Coefficients for
quadratic factor

Performing synthetic division on the polynomial $x^3 - 4x^2 + 5x - 2$ using the given zero, $x = 2$, produces the following.

$$
\begin{array}{r|rrrr}
2 & 1 & -4 & 5 & -2 \\
  &   & 2 & -4 & 2 \\
\hline
  & 1 & -2 & 1 & 0
\end{array}
$$

$(x - 2)(x^2 - 2x + 1) = x^3 - 4x^2 + 5x - 2$

When you use synthetic division, remember to take *all* coefficients into account—*even if some of them are zero.* For instance, if you know that $x = -2$ is a zero of $x^3 + 3x + 14$, you can apply synthetic division as follows.

$$
\begin{array}{r|rrrr}
-2 & 1 & 0 & 3 & 14 \\
   &   & -2 & 4 & -14 \\
\hline
   & 1 & -2 & 7 & 0
\end{array}
$$

$(x + 2)(x^2 - 2x + 7) = x^3 + 3x + 14$

# The Rational Zero Theorem

There is a systematic way to find the *rational* zeros of a polynomial. You can use the **Rational Zero Theorem** (also called the Rational Root Theorem).

---

**Rational Zero Theorem**

---

If a polynomial

$$a_n x^n + a_{n-1} x^{n-1} + \cdots + a_1 x + a_0$$

has integer coefficients, then every *rational* zero is of the form $x = p/q$, where $p$ is a factor of $a_0$, and $q$ is a factor of $a_n$.

---

## EXAMPLE 4 Using the Rational Zero Theorem

Find all real zeros of $2x^3 + 3x^2 - 8x + 3$.

**Solution**

$$\underset{\text{Factors of constant term: } \pm 1, \pm 3}{\underset{\text{Factors of leading coefficient: } \pm 1, \pm 2}{2x^3 + 3x^2 - 8x + 3}}$$

The possible rational zeros are the factors of the constant term divided by the factors of the leading coefficient.

$$1, -1, 3, -3, \frac{1}{2}, -\frac{1}{2}, \frac{3}{2}, -\frac{3}{2}$$

By testing these possible zeros, you can see that $x = 1$ works.

$$2(1)^3 + 3(1)^2 - 8(1) + 3 = 2 + 3 - 8 + 3 = 0$$

Now, by synthetic division you have the following.

$$
\begin{array}{r|rrrr}
1 & 2 & 3 & -8 & 3 \\
  &   & 2 & 5 & -3 \\
\hline
  & 2 & 5 & -3 & 0
\end{array}
$$

$$(x - 1)(2x^2 + 5x - 3) = 2x^3 + 3x^2 - 8x + 3$$

Finally, by factoring the quadratic, $2x^2 + 5x - 3 = (2x - 1)(x + 3)$, you have

$$2x^3 + 3x^2 - 8x + 3 = (x - 1)(2x - 1)(x + 3)$$

and you can conclude that the zeros are $x = 1$, $x = \frac{1}{2}$, and $x = -3$.

**STUDY TIP** In Example 4, you can check that the zeros are correct by substituting into the original polynomial.

*Check that $x = 1$ is a zero.*

$$2(1)^3 + 3(1)^2 - 8(1) + 3$$
$$= 2 + 3 - 8 + 3$$
$$= 0$$

*Check that $x = \frac{1}{2}$ is a zero.*

$$2\left(\frac{1}{2}\right)^3 + 3\left(\frac{1}{2}\right)^2 - 8\left(\frac{1}{2}\right) + 3$$
$$= \frac{1}{4} + \frac{3}{4} - 4 + 3$$
$$= 0$$

*Check that $x = -3$ is a zero.*

$$2(-3)^3 + 3(-3)^2 - 8(-3) + 3$$
$$= -54 + 27 + 24 + 3$$
$$= 0$$

# EXERCISES 0.4

In Exercises 1–6, use the Quadratic Formula to find all real zeros of the second-degree polynomial.

**1.** $6x^2 - x - 1$

**2.** $8x^2 - 2x - 1$

**3.** $4x^2 - 12x + 9$

**4.** $9x^2 + 12x + 4$

**5.** $y^2 + 4y + 1$

**6.** $x^2 + 6x - 1$

In Exercises 7–16, write the second-degree polynomial as the product of two linear factors.

**7.** $x^2 - 4x + 4$

**8.** $x^2 + 10x + 25$

**9.** $4x^2 + 4x + 1$

**10.** $9x^2 - 12x + 4$

**11.** $x^2 + x - 2$

**12.** $2x^2 - x - 1$

**13.** $3x^2 - 5x + 2$

**14.** $x^2 - xy - 2y^2$

**15.** $x^2 - 4xy + 4y^2$

**16.** $a^2b^2 - 2abc + c^2$

In Exercises 17–30, completely factor the polynomial.

**17.** $81 - y^4$

**18.** $x^4 - 16$

**19.** $x^3 - 8$

**20.** $y^3 - 64$

**21.** $y^3 + 64$

**22.** $z^3 + 125$

**23.** $x^3 - 27$

**24.** $(x - a)^3 + b^3$

**25.** $x^3 - 4x^2 - x + 4$

**26.** $x^3 - x^2 - x + 1$

**27.** $2x^3 - 3x^2 + 4x - 6$

**28.** $x^3 - 5x^2 - 5x + 25$

**29.** $2x^3 - 4x^2 - x + 2$

**30.** $x^3 - 7x^2 - 4x + 28$

In Exercises 31–48, find all real solutions of the equation.

**31.** $x^2 - 5x = 0$

**32.** $2x^2 - 3x = 0$

**33.** $x^2 - 9 = 0$

**34.** $x^2 - 25 = 0$

**35.** $x^2 - 3 = 0$

**36.** $x^2 - 8 = 0$

**37.** $(x - 3)^2 - 9 = 0$

**38.** $(x + 1)^2 - 8 = 0$

**39.** $x^2 + x - 2 = 0$

**40.** $x^2 + 5x + 6 = 0$

**41.** $x^2 - 5x + 6 = 0$

**42.** $x^2 + x - 20 = 0$

**43.** $x^3 + 64 = 0$

**44.** $x^3 - 216 = 0$

**45.** $x^4 - 16 = 0$

**46.** $x^4 - 625 = 0$

**47.** $x^3 - x^2 - 4x + 4 = 0$

**48.** $2x^3 + x^2 + 6x + 3 = 0$

In Exercises 49–52, find the interval (or intervals) on which the given expression is defined.

**49.** $\sqrt{x^2 - 7x + 12}$

**50.** $\sqrt{x^2 - 4}$

**51.** $\sqrt{4 - x^2}$

**52.** $\sqrt{144 - 9x^2}$

In Exercises 53–56, use synthetic division to complete the indicated factorization.

**53.** $x^3 + 8 = (x + 2)(\quad)$

**54.** $x^3 - 2x^2 - x + 2 = (x + 1)(\quad)$

**55.** $2x^3 - x^2 - 2x + 1 = (x - 1)(\quad)$

**56.** $x^4 - 16x^3 + 96x^2 - 256x + 256 = (x - 4)(\quad)$

In Exercises 57–64, use the Rational Zero Theorem as an aid in finding all real solutions of the given equation.

**57.** $x^3 - x^2 - x + 1 = 0$

**58.** $x^3 - x^2 - 4x + 4 = 0$

**59.** $x^3 - 6x^2 + 11x - 6 = 0$

**60.** $x^3 + 2x^2 - 5x - 6 = 0$

**61.** $4x^3 - 4x^2 - x + 1 = 0$

**62.** $18x^3 - 9x^2 - 8x + 4 = 0$

**63.** $x^3 - 3x^2 - 3x - 4 = 0$

**64.** $4x^3 - 6x^2 + 2x + 3 = 0$

**65.** *Average Cost*   The minimum average cost of producing $x$ units of a certain product occurs when the production level is set at the (positive) solution of

$$0.0003x^2 - 1200 = 0.$$

Determine this production level.

**66.** *Profit*   The profit $P$ from sales is given by

$$P = -200x^2 + 2000x - 3800,$$

where $x$ is the number of units sold per day (in hundreds). Determine the interval for $x$ such that the profit will be greater than 1000.

**67.** *Chemistry: Finding Concentrations*   Use the Quadratic Formula to solve the expression

$$1.8 \times 10^{-5} = \frac{x^2}{1.0 \times 10^{-4} - x},$$

which is needed to determine the quantity of hydrogen ions ($[H^+]$) in a solution of $1.0 \times 10^{-4}$M acetic acid. Because $x$ represents a concentration of $[H^+]$, only positive values of $x$ are possible solutions. *(Source: Adapted from Zumdahl, Chemistry, Fourth Edition)*

# 0.5 | Fractions and Rationalization

## Operations with Fractions

In the final section of this chapter you will review operations involving fractional expressions such as

$$\frac{2}{x}, \quad \frac{x^2 + 2x - 4}{x + 6}, \quad \text{and} \quad \frac{1}{\sqrt{x^2 + 1}}.$$

The first two expressions have polynomials as both numerator and denominator and are called **rational expressions.** A rational expression is **proper** if the degree of the numerator is less than the degree of the denominator. For example, $x/(x^2 + 1)$ is proper. If the degree of the numerator is greater than or equal to the degree of the denominator, then the rational expression is **improper.** For example, $x^2/(x^2 + 1)$ is improper.

---

### Operations with Fractions

1. Add fractions (find a common denominator):

$$\frac{a}{b} + \frac{c}{d} = \frac{a}{b}\left(\frac{d}{d}\right) + \frac{c}{d}\left(\frac{b}{b}\right) = \frac{ad}{bd} + \frac{bc}{bd} = \frac{ad + bc}{bd}$$

2. Subtract fractions (find a common denominator):

$$\frac{a}{b} - \frac{c}{d} = \frac{a}{b}\left(\frac{d}{d}\right) - \frac{c}{d}\left(\frac{b}{b}\right) = \frac{ad}{bd} - \frac{bc}{bd} = \frac{ad - bc}{bd}$$

3. Multiply fractions:

$$\left(\frac{a}{b}\right)\left(\frac{c}{d}\right) = \frac{ac}{bd}$$

4. Divide fractions (invert and multiply):

$$\frac{a/b}{c/d} = \left(\frac{a}{b}\right)\left(\frac{d}{c}\right) = \frac{ad}{bc}, \quad \frac{a/b}{c} = \frac{a/b}{c/1} = \left(\frac{a}{b}\right)\left(\frac{1}{c}\right) = \frac{a}{bc}$$

5. Cancel:

$$\frac{\cancel{a}b}{\cancel{a}c} = \frac{b}{c}, \quad \frac{ab + ac}{ad} = \frac{\cancel{a}(b + c)}{\cancel{a}d} = \frac{b + c}{d}$$

---

## EXAMPLE 1   Adding and Subtracting Rational Expressions

Perform the indicated operations and simplify.

(a) $x + \dfrac{1}{x}$    (b) $\dfrac{1}{x + 1} - \dfrac{2}{2x - 1}$

*Solution*

(a) $x + \dfrac{1}{x} = \dfrac{x^2}{x} + \dfrac{1}{x}$        Write with common denominator.

$\qquad = \dfrac{x^2 + 1}{x}$        Add fractions.

(b) $\dfrac{1}{x + 1} - \dfrac{2}{2x - 1} = \dfrac{(2x - 1)}{(x + 1)(2x - 1)} - \dfrac{2(x + 1)}{(x + 1)(2x - 1)}$

$\qquad = \dfrac{2x - 1 - 2x - 2}{2x^2 + x - 1}$

$\qquad = \dfrac{-3}{2x^2 + x - 1}$

In adding (or subtracting) fractions whose denominators have no common factors, it is convenient to use the following pattern.

$$\frac{a}{b} + \frac{c}{d} = \frac{a \quad c}{b \quad d} = \frac{ad + bc}{bd}$$

For instance, in part b of Example 1, you could have used this pattern as follows.

$$\frac{1}{x + 1} - \frac{2}{2x - 1} = \frac{(2x - 1) - 2(x + 1)}{(x + 1)(2x - 1)}$$

$$= \frac{2x - 1 - 2x - 2}{(x + 1)(2x - 1)}$$

$$= \frac{-3}{2x^2 + x - 1}$$

In Example 1, the denominators of the rational expressions have no common factors. When the denominators do have common factors, it is best to find the least common denominator before adding or subtracting. For instance, when adding $1/x$ and $2/x^2$, you can recognize that the least common denominator is $x^2$ and write

$$\frac{1}{x} + \frac{2}{x^2} = \frac{x}{x^2} + \frac{2}{x^2}$$        Write with common denominator.

$$= \frac{x + 2}{x^2}.$$        Add fractions.

This is further demonstrated in Example 2.

## EXAMPLE 2   Adding and Subtracting Rational Expressions

Perform the indicated operations and simplify.

(a) $\dfrac{x}{x^2 - 1} + \dfrac{3}{x + 1}$   (b) $\dfrac{1}{2(x^2 + 2x)} - \dfrac{1}{4x}$

### Solution

(a) Because $x^2 - 1 = (x + 1)(x - 1)$, the least common denominator is $x^2 - 1$.

$$\dfrac{x}{x^2 - 1} + \dfrac{3}{x + 1} = \dfrac{x}{(x - 1)(x + 1)} + \dfrac{3}{x + 1} \qquad \text{Factor.}$$

$$= \dfrac{x}{(x - 1)(x + 1)} + \dfrac{3(x - 1)}{(x - 1)(x + 1)} \qquad \begin{array}{l}\text{Write with} \\ \text{common} \\ \text{denominator.}\end{array}$$

$$= \dfrac{x + 3x - 3}{(x - 1)(x + 1)} \qquad \text{Add fractions.}$$

$$= \dfrac{4x - 3}{x^2 - 1} \qquad \text{Simplify.}$$

(b) In this case, the least common denominator is $4x(x + 2)$.

$$\dfrac{1}{2(x^2 + 2x)} - \dfrac{1}{4x} = \dfrac{1}{2x(x + 2)} - \dfrac{1}{2(2x)} \qquad \text{Factor.}$$

$$= \dfrac{2}{2(2x)(x + 2)} - \dfrac{x + 2}{2(2x)(x + 2)} \qquad \begin{array}{l}\text{Write with} \\ \text{common} \\ \text{denominator.}\end{array}$$

$$= \dfrac{2 - x - 2}{4x(x + 2)} \qquad \begin{array}{l}\text{Subtract} \\ \text{fractions}\end{array}$$

$$= \dfrac{-x}{4x(x + 2)} \qquad \begin{array}{l}\text{Cancel} \\ \text{common factor.}\end{array}$$

$$= \dfrac{-1}{4(x + 2)} \qquad \text{Simplify.}$$

**ALGEBRA TIP**   To add more than two fractions you must find a denominator that is common to all the fractions. For instance, to add $\frac{1}{2}$, $\frac{1}{3}$, and $\frac{1}{5}$, use a (least) common denominator of 30 and write

$$\dfrac{1}{2} + \dfrac{1}{3} + \dfrac{1}{5} = \dfrac{15}{30} + \dfrac{10}{30} + \dfrac{6}{30} \qquad \text{Write with common denominator.}$$

$$= \dfrac{31}{30}. \qquad \text{Add fractions.}$$

To add more than two rational expressions, use a similar procedure, as demonstrated in Example 3. (Expressions such as those shown in this example are used in calculus to perform an integration technique called integration by partial fractions.)

## EXAMPLE 3    Adding More than Two Rational Expressions

Perform the indicated addition of rational expressions.

(a) $\dfrac{A}{x + 2} + \dfrac{B}{x - 3} + \dfrac{C}{x + 4}$

(b) $\dfrac{A}{x + 2} + \dfrac{B}{(x + 2)^2} + \dfrac{C}{x - 1}$

**Solution**

(a) The least common denominator is $(x + 2)(x - 3)(x + 4)$.

$$\frac{A}{x + 2} + \frac{B}{x - 3} + \frac{C}{x + 4}$$

$$= \frac{A(x - 3)(x + 4) + B(x + 2)(x + 4) + C(x + 2)(x - 3)}{(x + 2)(x - 3)(x + 4)}$$

$$= \frac{A(x^2 + x - 12) + B(x^2 + 6x + 8) + C(x^2 - x - 6)}{(x + 2)(x - 3)(x + 4)}$$

$$= \frac{Ax^2 + Bx^2 + Cx^2 + Ax + 6Bx - Cx - 12A + 8B - 6C}{(x + 2)(x - 3)(x + 4)}$$

$$= \frac{(A + B + C)x^2 + (A + 6B - C)x + (-12A + 8B - 6C)}{(x + 2)(x - 3)(x + 4)}$$

(b) Here the least common denominator is $(x + 2)^2(x - 1)$.

$$\frac{A}{x + 2} + \frac{B}{(x + 2)^2} + \frac{C}{x - 1}$$

$$= \frac{A(x + 2)(x - 1) + B(x - 1) + C(x + 2)^2}{(x + 2)^2(x - 1)}$$

$$= \frac{A(x^2 + x - 2) + B(x - 1) + C(x^2 + 4x + 4)}{(x + 2)^2(x - 1)}$$

$$= \frac{Ax^2 + Cx^2 + Ax + Bx + 4Cx - 2A - B + 4C}{(x + 2)^2(x - 1)}$$

$$= \frac{(A + C)x^2 + (A + B + 4C)x + (-2A - B + 4C)}{(x + 2)^2(x - 1)}$$

## Expressions Involving Radicals

In calculus, the operation of differentiation tends to produce "messy" expressions when applied to fractional expressions. This is especially true when the fractional expression involves radicals. When differentiation is used, it is important to be able to simplify these expressions so that you can obtain more manageable forms. All of the expressions in Examples 4 and 5 are the results of differentiation. In each case, note how much *simpler* the simplified form is than the original form.

---

### EXAMPLE 4   Simplifying an Expression with Radicals

Simplify the following expressions.

(a) $\dfrac{\sqrt{x+1} - \dfrac{x}{2\sqrt{x+1}}}{x+1}$

(b) $\left(\dfrac{1}{x + \sqrt{x^2+1}}\right)\left(1 + \dfrac{2x}{2\sqrt{x^2+1}}\right)$

**Solution**

(a) $\dfrac{\sqrt{x+1} - \dfrac{x}{2\sqrt{x+1}}}{x+1} = \dfrac{\dfrac{2(x+1)}{2\sqrt{x+1}} - \dfrac{x}{2\sqrt{x+1}}}{x+1}$      Write with common denominator.

$= \dfrac{\dfrac{2x + 2 - x}{2\sqrt{x+1}}}{\dfrac{x+1}{1}}$      Subtract fractions.

$= \dfrac{x+2}{2\sqrt{x+1}}\left(\dfrac{1}{x+1}\right)$      To divide, invert and multiply.

$= \dfrac{x+2}{2(x+1)^{3/2}}$      Multiply.

(b) $\left(\dfrac{1}{x + \sqrt{x^2+1}}\right)\left(1 + \dfrac{2x}{2\sqrt{x^2+1}}\right)$

$= \left(\dfrac{1}{x + \sqrt{x^2+1}}\right)\left(1 + \dfrac{x}{\sqrt{x^2+1}}\right)$

$= \left(\dfrac{1}{x + \sqrt{x^2+1}}\right)\left(\dfrac{\sqrt{x^2+1}}{\sqrt{x^2+1}} + \dfrac{x}{\sqrt{x^2+1}}\right)$

$= \left(\dfrac{1}{x + \sqrt{x^2+1}}\right)\left(\dfrac{x + \sqrt{x^2+1}}{\sqrt{x^2+1}}\right)$

$= \dfrac{1}{\sqrt{x^2+1}}$

---

## EXAMPLE 5   Simplifying an Expression with Radicals

Simplify the following expression.

$$\frac{-x\left(\dfrac{2x}{2\sqrt{x^2+1}}\right)+\sqrt{x^2+1}}{x^2}+\left(\frac{1}{x+\sqrt{x^2+1}}\right)\left(1+\frac{2x}{2\sqrt{x^2+1}}\right)$$

### Solution

From part b of Example 4, you already know that the second part of this sum simplifies to $1/\sqrt{x^2+1}$. The first part simplifies as follows.

$$\frac{-x\left(\dfrac{2x}{2\sqrt{x^2+1}}\right)+\sqrt{x^2+1}}{x^2}=\frac{-x^2}{x^2\sqrt{x^2+1}}+\frac{\sqrt{x^2+1}}{x^2}$$

$$=\frac{-x^2}{x^2\sqrt{x^2+1}}+\frac{x^2+1}{x^2\sqrt{x^2+1}}$$

$$=\frac{-x^2+x^2+1}{x^2\sqrt{x^2+1}}$$

$$=\frac{1}{x^2\sqrt{x^2+1}}$$

Therefore, the sum is

$$\frac{-x\left(\dfrac{2x}{2\sqrt{x^2+1}}\right)+\sqrt{x^2+1}}{x^2}+\left(\frac{1}{x+\sqrt{x^2+1}}\right)\left(1+\frac{2x}{2\sqrt{x^2+1}}\right)$$

$$=\frac{1}{x^2\sqrt{x^2+1}}+\frac{1}{\sqrt{x^2+1}}$$

$$=\frac{1}{x^2\sqrt{x^2+1}}+\frac{x^2}{x^2\sqrt{x^2+1}}$$

$$=\frac{x^2+1}{x^2\sqrt{x^2+1}}$$

$$=\frac{\sqrt{x^2+1}}{x^2}.$$

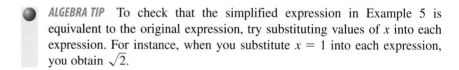 **ALGEBRA TIP**   To check that the simplified expression in Example 5 is equivalent to the original expression, try substituting values of $x$ into each expression. For instance, when you substitute $x = 1$ into each expression, you obtain $\sqrt{2}$.

## Rationalization Techniques

In working with quotients involving radicals, it is often convenient to move the radical expression from the denominator to the numerator, or vice versa. For example, you can move $\sqrt{2}$ from the denominator to the numerator in the following quotient by multiplying by $\sqrt{2}/\sqrt{2}$.

*Radical in Denominator*    *Rationalize*    *Radical in Numerator*

$$\frac{1}{\sqrt{2}} \implies \frac{1}{\sqrt{2}}\left(\frac{\sqrt{2}}{\sqrt{2}}\right) \implies \frac{\sqrt{2}}{2}$$

This process is called **rationalizing the denominator.** A similar process is used to **rationalize the numerator.**

### Rationalizing Techniques

1. If the denominator is $\sqrt{a}$, multiply by $\dfrac{\sqrt{a}}{\sqrt{a}}$.

2. If the denominator is $\sqrt{a} - \sqrt{b}$, multiply by $\dfrac{\sqrt{a} + \sqrt{b}}{\sqrt{a} + \sqrt{b}}$.

3. If the denominator is $\sqrt{a} + \sqrt{b}$, multiply by $\dfrac{\sqrt{a} - \sqrt{b}}{\sqrt{a} - \sqrt{b}}$.

The same guidelines apply to rationalizing numerators.

**TIP**

The success of the second and third rationalizing techniques stems from the following.

$$\left(\sqrt{a} - \sqrt{b}\right)\left(\sqrt{a} + \sqrt{b}\right)$$
$$= a - b$$

*ALGEBRA*

---

## EXAMPLE 6    Rationalizing Denominators and Numerators

(a) $\dfrac{3}{\sqrt{12}} = \dfrac{3}{2\sqrt{3}} = \dfrac{3}{2\sqrt{3}}\left(\dfrac{\sqrt{3}}{\sqrt{3}}\right) = \dfrac{3\sqrt{3}}{2(3)} = \dfrac{\sqrt{3}}{2}$

(b) $\dfrac{\sqrt{x+1}}{2} = \dfrac{\sqrt{x+1}}{2}\left(\dfrac{\sqrt{x+1}}{\sqrt{x+1}}\right) = \dfrac{x+1}{2\sqrt{x+1}}$

(c) $\dfrac{1}{\sqrt{5}+\sqrt{2}} = \dfrac{1}{\sqrt{5}+\sqrt{2}}\left(\dfrac{\sqrt{5}-\sqrt{2}}{\sqrt{5}-\sqrt{2}}\right) = \dfrac{\sqrt{5}-\sqrt{2}}{5-2} = \dfrac{\sqrt{5}-\sqrt{2}}{3}$

(d) $\dfrac{1}{\sqrt{x}-\sqrt{x+1}} = \dfrac{1}{\sqrt{x}-\sqrt{x+1}}\left(\dfrac{\sqrt{x}+\sqrt{x+1}}{\sqrt{x}+\sqrt{x+1}}\right)$

$$= \dfrac{\sqrt{x}+\sqrt{x+1}}{x-(x+1)} = -\sqrt{x}-\sqrt{x+1}$$

## EXERCISES 0.5

In Exercises 1–28, perform the indicated operations and simplify your answer.

**1.** $\dfrac{5}{x-1} + \dfrac{x}{x-1}$

**2.** $\dfrac{2x-1}{x+3} + \dfrac{1-x}{x+3}$

**3.** $\dfrac{2x}{x^2+2} - \dfrac{1-3x}{x^2+2}$

**4.** $\dfrac{5x+10}{2x-1} - \dfrac{2x+10}{2x-1}$

**5.** $\dfrac{4}{x} - \dfrac{3}{x^2}$

**6.** $\dfrac{5}{x-1} + \dfrac{3}{x}$

**7.** $\dfrac{2}{x+2} - \dfrac{1}{x-2}$

**8.** $\dfrac{x}{x^2+x-2} - \dfrac{1}{x+2}$

**9.** $\dfrac{5}{x-3} + \dfrac{3}{3-x}$

**10.** $\dfrac{x}{2-x} + \dfrac{2}{x-2}$

**11.** $\dfrac{A}{x-6} + \dfrac{B}{x+3}$

**12.** $\dfrac{Ax+B}{x^2+2} + \dfrac{C}{x-4}$

**13.** $-\dfrac{1}{x} + \dfrac{2}{x^2+1}$

**14.** $\dfrac{2}{x+1} + \dfrac{1-x}{x^2-2x+3}$

**15.** $\dfrac{-x}{(x+1)^{3/2}} + \dfrac{2}{(x+1)^{1/2}}$

**16.** $2\sqrt{x}(x-2) + \dfrac{(x-2)^2}{2\sqrt{x}}$

**17.** $\dfrac{2-t}{2\sqrt{1+t}} - \sqrt{1+t}$

**18.** $-\dfrac{\sqrt{x^2+1}}{x^2} + \dfrac{1}{\sqrt{x^2+1}}$

**19.** $\dfrac{1}{x^2-x-2} - \dfrac{x}{x^2-5x+6}$

**20.** $\dfrac{x-1}{x^2+5x+4} + \dfrac{2}{x^2-x-2} + \dfrac{10}{x^2+2x-8}$

**21.** $\dfrac{A}{x+1} + \dfrac{B}{(x+1)^2} + \dfrac{C}{x-2}$

**22.** $\dfrac{A}{x-5} + \dfrac{B}{x+5} + \dfrac{C}{(x+5)^2}$

**23.** $\left(2x\sqrt{x^2+1} - \dfrac{x^3}{\sqrt{x^2+1}}\right) \div (x^2+1)$

**24.** $\left(\sqrt{x^3+1} - \dfrac{3x^3}{2\sqrt{x^3+1}}\right) \div (x^3+1)$

**25.** $\dfrac{(x^2+2)^{1/2} - x^2(x^2+2)^{-1/2}}{x^2}$

**26.** $\dfrac{x(x+1)^{-1/2} - (x+1)^{1/2}}{x^2}$

**27.** $\dfrac{\dfrac{\sqrt{x+1}}{\sqrt{x}} - \dfrac{\sqrt{x}}{\sqrt{x+1}}}{2(x+1)}$

**28.** $\dfrac{\dfrac{2x^2}{3(x^2-1)^{2/3}} - (x^2-1)^{1/3}}{x^2}$

In Exercises 29–48, rationalize the numerator or denominator and simplify.

**29.** $\dfrac{3}{\sqrt{27}}$

**30.** $\dfrac{5}{\sqrt{10}}$

**31.** $\dfrac{\sqrt{2}}{3}$

**32.** $\dfrac{\sqrt{26}}{2}$

**33.** $\dfrac{x}{\sqrt{x}-4}$

**34.** $\dfrac{4y}{\sqrt{y}+8}$

**35.** $\dfrac{\sqrt{y^3}}{6y}$

**36.** $\dfrac{x\sqrt{x^2+4}}{3}$

**37.** $\dfrac{49(x-3)}{\sqrt{x^2-9}}$

**38.** $\dfrac{10(x+2)}{\sqrt{x^2-x-6}}$

**39.** $\dfrac{5}{\sqrt{14}-2}$

**40.** $\dfrac{13}{6+\sqrt{10}}$

**41.** $\dfrac{2x}{5-\sqrt{3}}$

**42.** $\dfrac{x}{\sqrt{2}+\sqrt{3}}$

**43.** $\dfrac{1}{\sqrt{6}+\sqrt{5}}$

**44.** $\dfrac{\sqrt{15}+3}{12}$

**45.** $\dfrac{\sqrt{3}-\sqrt{2}}{x}$

**46.** $\dfrac{x-1}{\sqrt{x}+x}$

**47.** $\dfrac{2x-\sqrt{4x-1}}{2x-1}$

**48.** $\dfrac{10}{\sqrt{x}+\sqrt{x+5}}$

**49. *Installment Loan*** The monthly payment $M$ for an installment loan is given by the formula

$$M = P\left[\dfrac{r/12}{1 - \left(\dfrac{1}{(r/12)+1}\right)^N}\right],$$

where $P$ is the amount of the loan, $r$ is the annual percentage rate, and $N$ is the number of monthly payments. Enter the formula into a graphing utility, and use it to find the monthly payment for a loan of \$10,000 at an annual percentage rate of $14\%\,(r = 0.14)$ for 5 years $(N = 60$ monthly payments).

**50. *Inventory*** A retailer has determined that the cost $C$ of ordering and storing $x$ units of a product is

$$C = 6x + \dfrac{900{,}000}{x}.$$

(a) Write the expression for cost as a single fraction.

(b) Determine the cost for ordering and storing $x = 240$ units of this product.

# Functions, Graphs, and Limits

**1**

## STRATEGIES *for* SUCCESS

 **OBJECTIVES**

*When you have completed this chapter, make sure you are able to:*

❏ Plot points in the Cartesian plane and find the distance between two points.
❏ Sketch the graph of an equation, and find the *x*- and *y*-intercepts.
❏ Write equations of lines in slope-intercept form and sketch the lines.
❏ Evaluate functions and simplify.
❏ Find the limit of a function graphically, analytically, and numerically.
❏ Discuss the continuity of a variety of functions.

 **TOOLS**

*Use these study tools to achieve the objectives above:*

**Algebra Review**
(pages 72 and 73)

**Chapter Summary and Study Strategies**
(pages 74 and 75)

**Review Exercises**
(pages 76–79)

 **ADDITIONAL RESOURCES**

*Use these resources to solidify your mastery of calculus:*

**Student Solutions Guide**

**Study Guide** (Additional Examples, Similar Problems, and Chapter Test)

**Algebra Review Tutorial Software**

**Graphing Technology Guide**

**Sample Post-Graduation Exam Questions**
(page 80)

**Web Exercises**
(page 23, exercise 70; page 34, exercise 65)

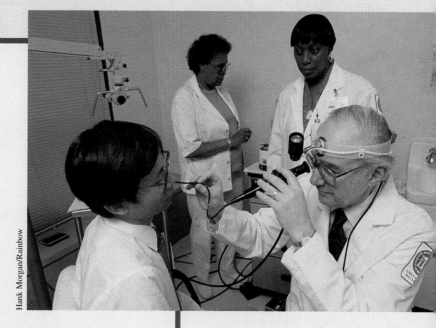

Hank Morgan/Rainbow

*In Exercise 41 on page 10, you will graph and compare the relationship between the number of doctors at an HMO and the number of ear infections treated.*

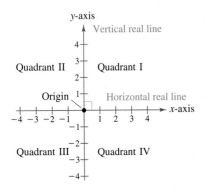

FIGURE 1.1   The Cartesian Plane

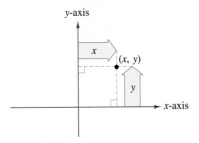

FIGURE 1.2

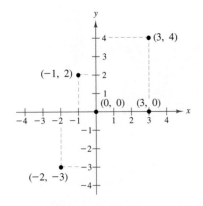

FIGURE 1.3

# The Cartesian Plane and the Distance Formula

*The Cartesian Plane • The Distance Formula • The Midpoint Formula • Translating Points in the Plane*

## The Cartesian Plane

Just as you can represent real numbers by points on a real number line, you can represent ordered pairs of real numbers by points in a plane called the **rectangular coordinate system,** or the **Cartesian plane,** after the French mathematician René Descartes (1596–1650).

The Cartesian plane is formed by using two real lines intersecting at right angles, as shown in Figure 1.1. The horizontal real line is usually called the **x-axis,** and the vertical real line is usually called the **y-axis.** The point of intersection of these two axes is the **origin,** and the two axes divide the plane into four parts called **quadrants.**

Each point in the plane corresponds to an **ordered pair** $(x, y)$ of real numbers $x$ and $y$, called **coordinates** of the point. The **x-coordinate** represents the directed distance from the $y$-axis to the point, and the **y-coordinate** represents the directed distance from the $x$-axis to the point, as shown in Figure 1.2.

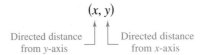

**STUDY TIP**   The notation $(x, y)$ denotes both a point in the plane and an open interval on the real line. The context will tell you which meaning is intended.

---

## EXAMPLE 1   Plotting Points in the Cartesian Plane

Plot the points $(-1, 2)$, $(3, 4)$, $(0, 0)$, $(3, 0)$, and $(-2, -3)$.

### Solution

To plot the point

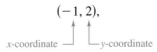

imagine a vertical line through $-1$ on the $x$-axis and a horizontal line through 2 on the $y$-axis. The intersection of these two lines is the point $(-1, 2)$. The other four points can be plotted in a similar way, and are shown in Figure 1.3.

The beauty of a rectangular coordinate system is that it allows you to visualize relationships between two variables. It would be difficult to overestimate the importance of Descartes's introduction of coordinates to the plane. Today, his ideas are in common use in virtually every scientific and business-related field. In Example 2, notice how much your intuition is enhanced by the use of a graphical presentation.

**STUDY TIP**   In Example 2, you could have let $t = 1$ represent the year 1988. In that case, the horizontal axis would not have been broken, and the tick marks would have been labeled 1 through 10 (instead of 1988 through 1997).

## EXAMPLE 2   Sketching a Scatter Plot

The amount A (in millions of dollars) spent on water skis in the United States from 1988 through 1997 is given in the table, where $t$ represents the year. Sketch a scatter plot of the data.   *(Source: National Sporting Goods Association)*

| $t$ | 1988 | 1989 | 1990 | 1991 | 1992 | 1993 | 1994 | 1995 | 1996 | 1997 |
|---|---|---|---|---|---|---|---|---|---|---|
| $A$ | 160 | 96 | 88 | 63 | 55 | 51 | 51 | 54 | 54 | 54 |

### Solution

To sketch a *scatter plot* of the data given in the table, you simply represent each pair of values by an ordered pair $(t, A)$, and plot the resulting points, as shown in Figure 1.4. For instance, the first pair of values is represented by the ordered pair (1988, 160). Note that the break in the $t$-axis indicates that the numbers between 0 and 1988 have been omitted.

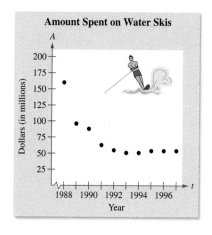

**FIGURE 1.4**

## Technology

The scatter plot in Example 2 is only one way to represent the given data graphically. Two other techniques are shown at right. The first is a *bar graph* and the second is a *line graph*. All three graphical representations were created with a computer. If you have access to computer graphing software, try using it to represent graphically the data given in Example 2.

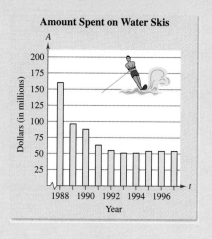

Bar graph

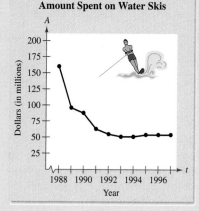

Line graph

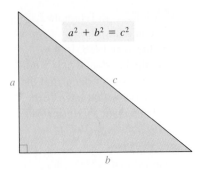

**FIGURE 1.5** Pythagorean Theorem

## The Distance Formula

Recall from the Pythagorean Theorem that, for a right triangle with hypotenuse of length $c$ and sides of lengths $a$ and $b$, you have

$$a^2 + b^2 = c^2, \qquad \text{Pythagorean Theorem}$$

as shown in Figure 1.5. (The converse is also true. That is, if $a^2 + b^2 = c^2$, then the triangle is a right triangle.)

Suppose you want to determine the distance $d$ between two points $(x_1, y_1)$ and $(x_2, y_2)$ in the plane. With these two points, a right triangle can be formed, as shown in Figure 1.6. The length of the vertical side of the triangle is $|y_2 - y_1|$, and the length of the horizontal side is $|x_2 - x_1|$. By the Pythagorean Theorem, you can write

$$d^2 = |x_2 - x_1|^2 + |y_2 - y_1|^2$$
$$d = \sqrt{|x_2 - x_1|^2 + |y_2 - y_1|^2}$$
$$d = \sqrt{(x_2 - x_1)^2 + (y_2 - y_1)^2}.$$

This result is the **Distance Formula.**

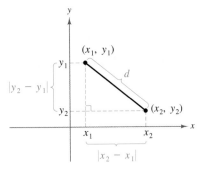

**FIGURE 1.6** Distance Between Two Points

### The Distance Formula

The distance $d$ between the points $(x_1, y_1)$ and $(x_2, y_2)$ in the plane is

$$d = \sqrt{(x_2 - x_1)^2 + (y_2 - y_1)^2}.$$

## EXAMPLE 3   Finding a Distance

Find the distance between the points $(-2, 1)$ and $(3, 4)$.

**Solution**

Let $(x_1, y_1) = (-2, 1)$ and $(x_2, y_2) = (3, 4)$. Then apply the Distance Formula as follows.

$$
\begin{aligned}
d &= \sqrt{(x_2 - x_1)^2 + (y_2 - y_1)^2} && \text{Distance Formula} \\
&= \sqrt{[3 - (-2)]^2 + (4 - 1)^2} && \text{Substitute for } x_1, y_1, x_2, \text{ and } y_2. \\
&= \sqrt{(5)^2 + (3)^2} && \text{Simplify.} \\
&= \sqrt{34} && \\
&\approx 5.83 && \text{Use a calculator.}
\end{aligned}
$$

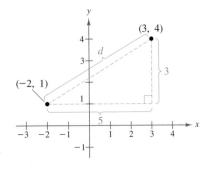

**FIGURE 1.7**

Note in Figure 1.7 that a distance of 5.83 looks about right.

## EXAMPLE 4   Verifying a Right Triangle

Use the Distance Formula to show that the points $(2, 1)$, $(4, 0)$, and $(5, 7)$ are vertices of a right triangle.

### Solution

The three points are plotted in Figure 1.8. Using the Distance Formula, you can find the lengths of the three sides as follows.

$$d_1 = \sqrt{(5 - 2)^2 + (7 - 1)^2} = \sqrt{9 + 36} = \sqrt{45}$$
$$d_2 = \sqrt{(4 - 2)^2 + (0 - 1)^2} = \sqrt{4 + 1} = \sqrt{5}$$
$$d_3 = \sqrt{(5 - 4)^2 + (7 - 0)^2} = \sqrt{1 + 49} = \sqrt{50}$$

Because

$$d_1^2 + d_2^2 = 45 + 5 = 50 = d_3^2,$$

you can apply the converse of the Pythagorean Theorem to conclude that the triangle must be a right triangle.

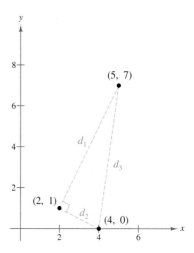

**FIGURE 1.8**

The figures provided with Examples 3 and 4 were not really essential to the solution. *Nevertheless,* we strongly recommend that you develop the habit of including sketches with your solutions—even if they are not required.

## EXAMPLE 5   Finding the Length of a Pass

In a football game, a quarterback throws a pass from the 5-yard line, 20 yards from the sideline. The pass is caught by a wide receiver on the 45-yard line, 50 yards from the same sideline, as shown in Figure 1.9. How long was the pass?

### Solution

You can find the length of the pass by finding the distance between the points $(20, 5)$ and $(50, 45)$.

$$d = \sqrt{(50 - 20)^2 + (45 - 5)^2} \qquad \text{Distance Formula}$$
$$= \sqrt{900 + 1600}$$
$$= 50 \qquad \text{Simplify.}$$

Thus, the pass was 50 yards long.

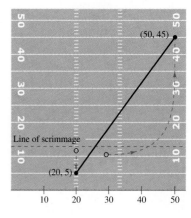

**FIGURE 1.9**

*STUDY TIP*   In Example 5, the scale along the goal line does not normally appear on a football field. However, when you use coordinate geometry to solve real-life problems, you are free to place the coordinate system in any way that is convenient to the solution of the problem.

## The Midpoint Formula

To find the **midpoint** of the line segment that joins two points in a coordinate plane, you can simply find the average values of the respective coordinates of the two endpoints.

### The Midpoint Formula

The midpoint of the segment joining the points $(x_1, y_1)$ and $(x_2, y_2)$ is

$$\text{Midpoint} = \left( \frac{x_1 + x_2}{2}, \frac{y_1 + y_2}{2} \right).$$

---

### EXAMPLE 6    Finding a Segment's Midpoint

Find the midpoint of the line segment joining the points $(-5, -3)$ and $(9, 3)$, as shown in Figure 1.10.

**Solution**

Let $(x_1, y_1) = (-5, -3)$ and $(x_2, y_2) = (9, 3)$.

$$\text{Midpoint} = \left( \frac{x_1 + x_2}{2}, \frac{y_1 + y_2}{2} \right) \qquad \text{Midpoint Formula}$$

$$= \left( \frac{-5 + 9}{2}, \frac{-3 + 3}{2} \right) \qquad \text{Substitute for } x_1, y_1, x_2, \text{ and } y_2.$$

$$= (2, 0) \qquad \text{Simplify.}$$

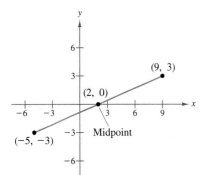

**FIGURE 1.10**

---

### EXAMPLE 7    Estimating Annual Sales

Eastman Kodak had annual sales of $13.6 billion in 1994 and $16.0 billion in 1996. Without knowing any additional information, what would you estimate the 1995 sales to have been?    *(Source: Eastman Kodak Company)*

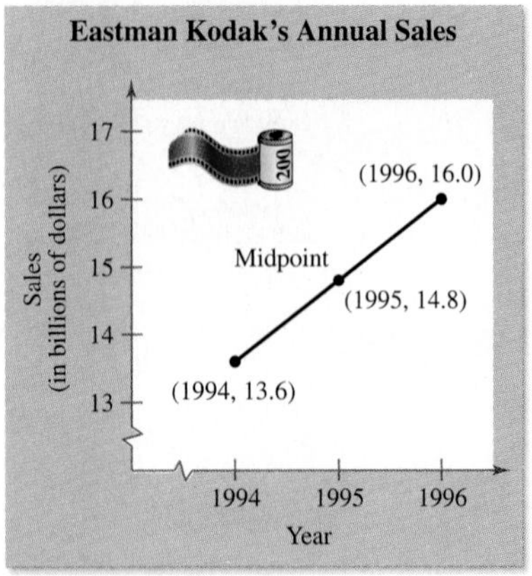

**Solution**

One solution to the problem is to assume that sales followed a linear pattern. With this assumption, you can estimate the 1995 sales by finding the midpoint of the segment connecting the points (1994, 13.6) and (1996, 16.0).

$$\text{Midpoint} = \left( \frac{1994 + 1996}{2}, \frac{13.6 + 16.0}{2} \right) = (1995, 14.8)$$

Hence, you would estimate the 1995 sales to have been about $14.8 billion, as shown in Figure 1.11. (The actual 1995 sales were $15.0 billion.)

**FIGURE 1.11**

# Translating Points in the Plane

---

## EXAMPLE 8 Translating Points in the Plane

Figure 1.12 shows the vertices of a parallelogram. Find the vertices of the parallelogram after it has been translated two units down and four units to the right.

### Solution

To translate each vertex two units down, subtract 2 from each $y$-coordinate. To translate each vertex four units to the right, add 4 to each $x$-coordinate.

*Many movies now use extensive computer graphics, much of which consists of transformations of points in two- and three-dimensional space.*

| Original Point | Translated Point |
|---|---|
| $(1, 0)$ | $(1 + 4, 0 - 2) = (5, -2)$ |
| $(3, 2)$ | $(3 + 4, 2 - 2) = (7, 0)$ |
| $(3, 6)$ | $(3 + 4, 6 - 2) = (7, 4)$ |
| $(1, 4)$ | $(1 + 4, 4 - 2) = (5, 2)$ |

The translated parallelogram is shown in Figure 1.12.

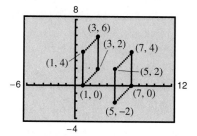

**FIGURE 1.12**

---

## *Group Discussion*　　*Transforming Points in a Coordinate Plane*

Example 8 illustrates points that have been *translated* (or *slid*) in a coordinate plane. The translated parallelogram is congruent to (has the same size and shape as) the original parallelogram. Try using a graphing utility to graph the transformed parallelogram for each of the following transformations. Describe the transformation. Is it a translation, a reflection, or a rotation? Is the transformed parallelogram congruent to the original parallelogram?

a. $(x, y) \implies (-x, y)$

b. $(x, y) \implies (x, -y)$

c. $(x, y) \implies (-x, -y)$

## Warm Up

The following warm-up exercises involve skills that were covered in earlier sections. You will use these skills in the exercise set for this section.

In Exercises 1–6, simplify the expression.

**1.** $\sqrt{(3-6)^2 + [1-(-5)]^2}$

**2.** $\sqrt{(-2-0)^2 + [-7-(-3)]^2}$

**3.** $\dfrac{5+(-4)}{2}$

**4.** $\dfrac{-3+(-1)}{2}$

**5.** $\sqrt{27} + \sqrt{12}$

**6.** $\sqrt{8} - \sqrt{18}$

In Exercises 7–10, solve for x or y.

**7.** $\sqrt{(3-x)^2 + (7-4)^2} = \sqrt{45}$

**8.** $\sqrt{(6-2)^2 + (-2-y)^2} = \sqrt{52}$

**9.** $\dfrac{x+(-5)}{2} = 7$

**10.** $\dfrac{-7+y}{2} = -3$

## EXERCISES 1.1

In Exercises 1–4, (a) find the length of each side of the right triangle, and (b) show that these lengths satisfy the Pythagorean Theorem.

***1.**

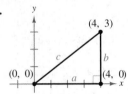

**2.**

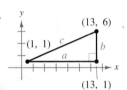

**3.**

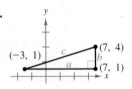

**4.**

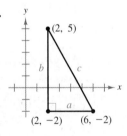

In Exercises 5–12, (a) plot the points, (b) find the distance between the points, and (c) find the midpoint of the line segment joining the points.

**5.** $(3, 1), (5, 5)$

**6.** $(-3, 2), (3, -2)$

**7.** $\left(\frac{1}{2}, 1\right), \left(-\frac{3}{2}, -5\right)$

**8.** $\left(\frac{2}{3}, -\frac{1}{3}\right), \left(\frac{5}{6}, 1\right)$

**9.** $(2, 2), (4, 14)$

**10.** $(-3, 7), (1, -1)$

**11.** $\left(1, \sqrt{3}\right), (-1, 1)$

**12.** $(-2, 0), \left(0, \sqrt{2}\right)$

In Exercises 13–16, show that the points form the vertices of the indicated figure. (A rhombus is a quadrilateral whose sides have the same length.)

| Vertices | Figure |
|---|---|
| **13.** $(0, 1), (3, 7), (4, -1)$ | Right triangle |
| **14.** $(1, -3), (3, 2), (-2, 4)$ | Isosceles triangle |
| **15.** $(0, 0), (1, 2), (2, 1), (3, 3)$ | Rhombus |
| **16.** $(0, 1), (3, 7), (4, 4), (1, -2)$ | Parallelogram |

In Exercises 17–20, use the Distance Formula to determine whether the points are collinear.

**17.** $(0, -4), (2, 0), (3, 2)$

**18.** $(0, 4), (7, -6), (-5, 11)$

**19.** $(-2, -6), (1, -3), (5, 2)$

**20.** $(-1, 1), (3, 3), (5, 5)$

In Exercises 21 and 22, find x such that the distance between the points is 5.

**21.** $(1, 0), (x, -4)$

**22.** $(2, -1), (x, 2)$

* The answers to the odd-numbered and selected even exercises are given in the back of the text. Worked-out solutions to the odd-numbered exercises are given in the *Student Solutions Guide*. The symbol ⚗ indicates an exercise in which you are instructed to use graphing technology or a symbolic computer algebra system. The solutions of other exercises may also be facilitated by use of appropriate technology.

In Exercises 23 and 24, find $y$ such that the distance between the points is 8.

**23.** $(0, 0), (3, y)$      **24.** $(5, 1), (5, y)$

**25.** Use the Midpoint Formula repeatedly to find the three points that divide the segment joining $(x_1, y_1)$ and $(x_2, y_2)$ into four equal parts.

**26.** Show that $\left(\frac{1}{3}[2x_1 + x_2], \frac{1}{3}[2y_1 + y_2]\right)$ is one of the points of trisection of the line segment joining $(x_1, y_1)$ and $(x_2, y_2)$. Then, find the second point of trisection by finding the midpoint of the segment joining

$$\left(\frac{1}{3}[2x_1 + x_2], \frac{1}{3}[2y_1 + y_2]\right) \text{ and } (x_2, y_2).$$

**27.** Use Exercise 25 to find the points that divide the line segment joining the given points into four equal parts.

  (a) $(1, -2), (4, -1)$      (b) $(-2, -3), (0, 0)$

**28.** Use Exercise 26 to find the points of trisection of the line segment joining the given points.

  (a) $(1, -2), (4, 1)$      (b) $(-2, -3), (0, 0)$

**29.** *Building Dimensions*    The base and height of the trusses for the roof of a house are 32 feet and 5 feet, respectively (see figure).

  (a) Find the distance from the eaves to the peak of the roof.
  (b) The length of the house is 40 feet. Use the result of part (a) to find the number of square feet of roofing.

Figure for 29          Figure for 30

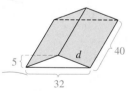

**30.** *Wire Length*    A guy wire is stretched from a broadcasting tower at a point 200 feet above the ground to an anchor 125 feet from the base (see figure). How long is the wire?

In Exercises 31 and 32, use a graphing utility to graph a scatter plot, a bar graph, or a line graph to represent the data. Describe any trends that appear.

**31.** *Cable TV*    The numbers (in millions) of cable television subscribers in the United States for 1985–1994 are given in the table.    *(Source: Nielsen Media Research)*

Table for 31

| Year | 1985 | 1986 | 1987 | 1988 | 1989 |
|------|------|------|------|------|------|
| Subscribers | 39.9 | 42.2 | 44.9 | 48.6 | 52.6 |

| Year | 1990 | 1991 | 1992 | 1993 | 1994 |
|------|------|------|------|------|------|
| Subscribers | 54.9 | 55.8 | 56.4 | 57.2 | 60.5 |

**32.** *Cellular Telephones*    The numbers (in millions) of cellular telephone subscribers in the United States for 1988–1995 are given in the table.    *(Source: Cellular Telecommunications Industry Association)*

| Year | 1988 | 1989 | 1990 | 1991 |
|------|------|------|------|------|
| Subscribers | 2.07 | 3.51 | 5.28 | 7.56 |

| Year | 1992 | 1993 | 1994 | 1995 |
|------|------|------|------|------|
| Subscribers | 11.03 | 16.01 | 24.13 | 33.79 |

*Dow Jones Average*    In Exercises 33 and 34, use the figure below showing the Dow Jones Industrial Average for common stocks.    *(Source: Dow Jones, Inc.)*

**33.** Estimate the Dow Jones Average for the following dates.

  (a) August 1994        (b) December 1994
  (c) November 1995      (d) June 1996

**34.** Estimate the percent increase or decrease in the Dow Jones Industrial Average (a) from August 1994 to November 1994 and (b) from December 1995 to June 1996.

Figure for 33 and 34

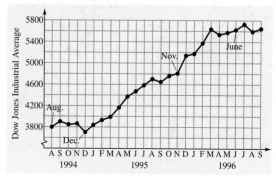

*Construction Contracts*    In Exercises 35 and 36, use the figure, which shows the value of residential building construction contracts (in billions of dollars) in the United States from 1980 to 1995.  *(Source: Dodge Construction Potentials)*

**35.** Estimate the value of construction contracts for the following years.

(a) 1982                      (b) 1987

(c) 1992                      (d) 1995

**36.** Estimate the percent increase or decrease in the value of construction contracts (a) from 1985 to 1986 and (b) from 1990 to 1991.

Figure for 35 and 36

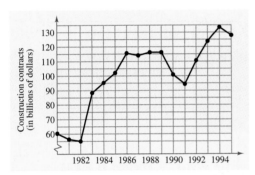

*Research Project*    In Exercises 37 and 38, (a) use the Midpoint Formula to estimate the revenue and profit of the company in 1994. (b) Then use your school's library or some other reference source to find the actual revenue and profit for 1994. (c) Did the revenue and profit increase in a linear pattern from 1992 to 1996? Explain your reasoning. (d) What were the company's expenses during each of the given years? (e) How would you rate the company's growth from 1992 to 1996?  *(Source: Coca-Cola and Pepsico)*

**37.** Coca-Cola

| Year | 1992 | 1994 | 1996 |
|---|---|---|---|
| Revenue (millions of $) | 13,074 | | 18,546 |
| Profit (millions of $) | 1883.8 | | 3492.0 |

**38.** Pepsico

| Year | 1992 | 1994 | 1996 |
|---|---|---|---|
| Revenue (millions of $) | 21,970 | | 31,645 |
| Profit (millions of $) | 1301.7 | | 1865.0 |

*Computer Graphics*    In Exercises 39 and 40, the red figure is translated to a new position in the plane to form the blue figure. (a) Find the vertices of the transformed figure. (b) Then use a graphing utility to draw both figures.

**39.**                      **40.**

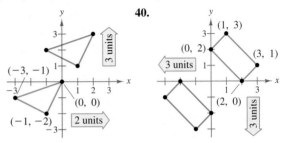

**41.** *Economics: HMOs*    The table shows the numbers of ear infections treated by doctors at an HMO with clinics of three different sizes: small, medium, and large. The larger the clinic, the more patients the doctors can handle.

| Cases per Small Clinic | Cases per Medium Clinic | Cases per Large Clinic | Number of Doctors |
|---|---|---|---|
| 0 | 0 | 0 | 0 |
| 20 | 30 | 35 | 1 |
| 28 | 42 | 49 | 2 |
| 35 | 53 | 62 | 3 |
| 40 | 60 | 70 | 4 |

(a) Show the relationship between doctors and ear infections treated with *three* curves where the number of doctors is on the horizontal axis and the number of ear infections treated is on the vertical axis.

(b) Compare the three relationships.

*(Source: Adapted from Taylor,* Economics, *First Edition)*

# Graphs of Equations

## 1.2

*The Graph of an Equation • Intercepts of a Graph • Circles • Points of Intersection • Mathematical Models*

## The Graph of an Equation

In Section 1.1, you used a coordinate system to represent graphically the relationship between two quantities. There, the graphical picture consisted of a collection of points in a coordinate plane (see Example 2 in Section 1.1).

Frequently, a relationship between two quantities is expressed as an equation. For instance, degrees on the Fahrenheit scale are related to degrees on the Celsius scale by the equation $F = \frac{9}{5}C + 32$. In this section, you will study some basic procedures for sketching the graph of such an equation. The **graph** of an equation is the set of all points that are solutions of the equation.

---

**EXAMPLE 1** **Sketching the Graph of an Equation**

Sketch the graph of $y = 7 - 3x$.

### Solution

The simplest way to sketch the graph of an equation is the *point-plotting method.* With this method, you construct a table of values that consists of several solution points of the equation, as shown in the table below. For instance, when $x = 0$,

$$y = 7 - 3(0) = 7,$$

which implies that $(0, 7)$ is a solution point of the graph.

| $x$ | 0 | 1 | 2 | 3 | 4 |
|---|---|---|---|---|---|
| $y = 7 - 3x$ | 7 | 4 | 1 | $-2$ | $-5$ |

From the table, it follows that $(0, 7)$, $(1, 4)$, $(2, 1)$, $(3, -2)$, and $(4, -5)$ are solution points of the equation. After plotting these points, you can see that they appear to lie on a line, as shown in Figure 1.13. The graph of the equation is the line that passes through the five plotted points.

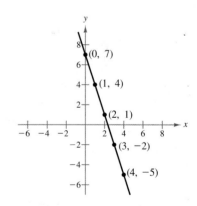

**FIGURE 1.13**  Solution Points for $y = 7 - 3x$

*STUDY TIP*  Even though we refer to the sketch shown in Figure 1.13 as the graph of $y = 7 - 3x$, it actually represents only a *portion* of the graph. The entire graph is a line that would extend off the page.

*STUDY TIP* The graph shown in Example 2 is a **parabola.** The graph of any second-degree equation of the form

$$y = ax^2 + bx + c, \; a \neq 0$$

has a similar shape. If $a > 0$, the parabola opens upward, as in Figure 1.14(b), and if $a < 0$, the parabola opens downward.

## EXAMPLE 2 Sketching the Graph of an Equation

Sketch the graph of $y = x^2 - 2$.

### Solution

Begin by constructing a table of values, as shown below.

| $x$ | $-2$ | $-1$ | 0 | 1 | 2 | 3 |
|---|---|---|---|---|---|---|
| $y = x^2 - 2$ | 2 | $-1$ | $-2$ | $-1$ | 2 | 7 |

Next, plot the points given in the table, as shown in Figure 1.14(a). Finally, connect the points with a smooth curve, as shown in Figure 1.14(b).

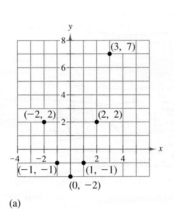

(a)

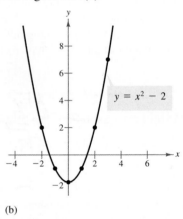

(b)

**FIGURE 1.14**

The point-plotting technique demonstrated in Examples 1 and 2 is easy to use, but it does have some shortcomings. With too few solution points, you can badly misrepresent the graph of a given equation. For instance, how would you connect the four points in Figure 1.15? Without further information, any one of the three graphs in Figure 1.16 would be reasonable.

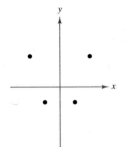

**FIGURE 1.15**

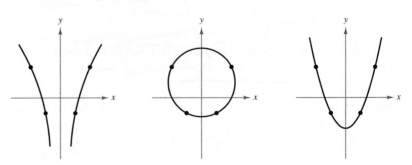

**FIGURE 1.16**

## Technology

### Creating a Viewing Rectangle

A **viewing rectangle** for a graph is a rectangular portion of the coordinate plane. A viewing rectangle is determined by six values: the minimum $x$-value, the maximum $x$-value, the $x$-scale, the minimum $y$-value, the maximum $y$-value, and the $y$-scale. When you enter these six values into a graphing utility, you are setting the **range** or **window.** Some graphing utilities have a standard viewing rectangle, as shown at the right.

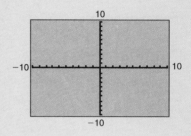

By choosing different viewing rectangles for a graph, it is possible to obtain very different impressions of the graph's shape. For instance, below are four different viewing rectangles for the graph of

$$y = 0.1x^4 - x^3 + 2x^2.$$

Of these, the view shown in part (a) is the most complete.

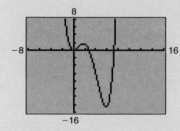

(a)

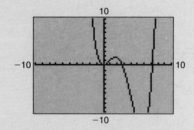

(b)

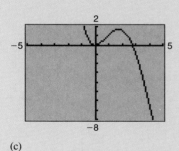

(c)

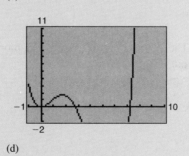

(d)

On most graphing utilities, the display screen is two-thirds as high as it is wide. On such screens, you can obtain a graph with a true geometric perspective by using a **square setting**—one in which

$$\frac{Y_{max} - Y_{min}}{X_{max} - X_{min}} = \frac{2}{3}.$$

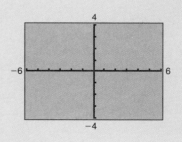

One such setting is shown at the right. Notice that the $x$- and $y$-tick marks are equally spaced on a square setting, but not on a standard setting.

## Intercepts of a Graph

It is often easy to determine the solution points that have zero as either the $x$-coordinate or the $y$-coordinate. These points are called **intercepts** because they are the points at which the graph intersects the $x$- or $y$-axis.

A graph may have no intercepts or several intercepts, as shown in Figure 1.17.

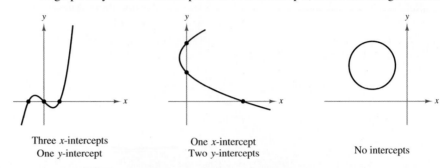

No $x$-intercept
One $y$-intercept

Three $x$-intercepts
One $y$-intercept

One $x$-intercept
Two $y$-intercepts

No intercepts

**FIGURE 1.17**

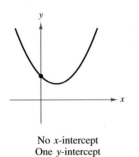

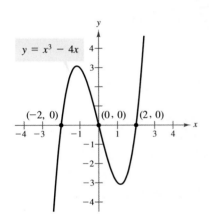

$y = x^3 - 4x$

$(-2, 0)$  $(0, 0)$  $(2, 0)$

**FIGURE 1.18**

---

### Finding Intercepts

1. To find $x$-intercepts, let $y$ be zero and solve the equation for $x$.
2. To find $y$-intercepts, let $x$ be zero and solve the equation for $y$.

---

*ALGEBRA TIP*  Finding intercepts involves solving equations. For a review of some techniques for solving equations, see page 73.

---

## EXAMPLE 3  Finding $x$- and $y$-Intercepts

Find the $x$- and $y$-intercepts for the graphs of the following equations.

(a) $y = x^3 - 4x$        (b) $y^2 - 3 = x$

*Solution*

(a) Let $y = 0$. Then $0 = x(x^2 - 4)$ has solutions $x = 0$ and $x = \pm 2$.

$x$-intercepts:  $(0, 0), (2, 0), (-2, 0)$

Let $x = 0$. Then $y = 0$.

$y$-intercept:  $(0, 0)$        (See Figure 1.18.)

(b) Let $y = 0$. Then $-3 = x$.

$x$-intercept:  $(-3, 0)$

Let $x = 0$. Then $y^2 - 3 = 0$ has solutions $y = \pm\sqrt{3}$.

$y$-intercepts:  $\left(0, \sqrt{3}\right), \left(0, -\sqrt{3}\right)$        (See Figure 1.19.)

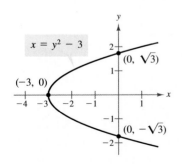

$x = y^2 - 3$

$(-3, 0)$

$(0, \sqrt{3})$

$(0, -\sqrt{3})$

**FIGURE 1.19**

## Technology

### Zooming In to Find Intercepts

You can use the **zoom** feature of a graphing utility to approximate the $x$-intercepts of a graph. Suppose you want to approximate the $x$-intercept(s) of the graph of $y = 2x^3 - 3x + 2$. Begin by graphing the equation, as shown below in part (a). From the viewing rectangle shown, the graph appears to have only one $x$-intercept. This intercept lies between $-2$ and $-1$. By zooming in on the intercept, you can improve the approximation, as shown in part (b). To three decimal places, the solution is $x \approx -1.476$.

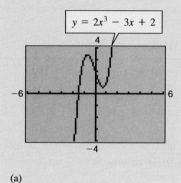

(a)

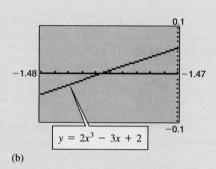

(b)

*STUDY TIP* Some graphing utilities have a built-in program that can find the $x$-intercepts of a graph. If your graphing utility has this feature, try using it to find the $x$-intercept of the graph shown at the left. (Your calculator may call this the *root* or *zero* function.)

Here are some suggestions for using the zoom feature.

1. With each successive zoom-in, adjust the $x$-scale so that the viewing rectangle shows at least one tick mark on each side of the $x$-intercept.

2. The error in your approximation will be less than the distance between two scale marks.

3. The **trace** feature can usually be used to add one more decimal place of accuracy without changing the viewing rectangle.

Part (a) below shows the graph of $y = x^2 - 5x + 3$. Parts (b) and (c) show "zoom-in views" of the two intercepts. From these views, you can approximate the $x$-intercepts to be $x \approx 0.697$ and $x \approx 4.303$.

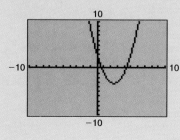

(a)

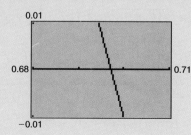

(b)

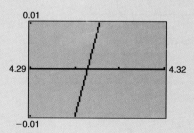

(c)

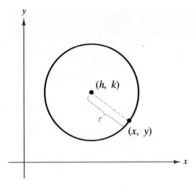

**FIGURE 1.20**

## Circles

Throughout this course, you will learn to recognize several types of graphs from their equations. For instance, you should recognize that the graph of a second-degree equation of the form $y = ax^2 + bx + c$ is a parabola (see Example 2). Another easily recognized graph is that of a **circle.**

Consider the circle shown in Figure 1.20. A point $(x, y)$ is on the circle if and only if its distance from the center $(h, k)$ is $r$. By the Distance Formula,

$$\sqrt{(x - h)^2 + (y - k)^2} = r.$$

By squaring both sides of this equation, you obtain the **standard form of the equation of a circle.**

---

### Standard Form of the Equation of a Circle

The point $(x, y)$ lies on the circle of radius $r$ and center $(h, k)$ if and only if

$$(x - h)^2 + (y - k)^2 = r^2.$$

---

From this result, you can see that the standard form of the equation of a circle with its center at the origin, $(h, k) = (0, 0)$, is simply

$$x^2 + y^2 = r^2. \qquad \text{Circle with center at origin}$$

---

## EXAMPLE 4   Finding the Equation of a Circle

The point $(3, 4)$ lies on a circle whose center is at $(-1, 2)$, as shown in Figure 1.21. Find the standard form of the equation of this circle.

### Solution

The radius of the circle is the distance between $(-1, 2)$ and $(3, 4)$.

$$
\begin{aligned}
r &= \sqrt{[3 - (-1)]^2 + (4 - 2)^2} && \text{Distance Formula} \\
&= \sqrt{16 + 4} && \text{Simplify.} \\
&= \sqrt{20} && \text{Radius}
\end{aligned}
$$

Using $(h, k) = (-1, 2)$, the standard form of the equation of the circle is

$$
\begin{aligned}
(x - h)^2 + (y - k)^2 &= r^2 \\
[x - (-1)]^2 + (y - 2)^2 &= \left(\sqrt{20}\right)^2 && \text{Substitute for } h, k, \text{ and } r. \\
(x + 1)^2 + (y - 2)^2 &= 20. && \text{Standard form}
\end{aligned}
$$

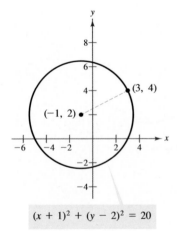

$(x + 1)^2 + (y - 2)^2 = 20$

**FIGURE 1.21**

The general form of the equation of a circle is

$$Ax^2 + Ay^2 + Dx + Ey + F = 0, \qquad A \neq 0 \qquad \text{General form}$$

To change from general form to standard form, you can use a process called **completing the square,** as demonstrated in Example 5.

---

## EXAMPLE 5 Completing the Square

Sketch the graph of the circle whose general equation is

$$4x^2 + 4y^2 + 20x - 16y + 37 = 0.$$

### Solution

First divide by 4 so that the coefficients of $x^2$ and $y^2$ are both 1.

$$4x^2 + 4y^2 + 20x - 16y + 37 = 0 \qquad \text{General form}$$

$$x^2 + y^2 + 5x - 4y + \tfrac{37}{4} = 0 \qquad \text{Divide by 4.}$$

$$(x^2 + 5x + \quad) + (y^2 - 4y + \quad) = -\tfrac{37}{4} \qquad \text{Group terms.}$$

$$\left(x^2 + 5x + \tfrac{25}{4}\right) + (y^2 - 4y + 4) = -\tfrac{37}{4} + \tfrac{25}{4} + 4 \qquad \begin{array}{l}\text{Complete the}\\\text{square.}\end{array}$$

$$\underbrace{\qquad}_{(\text{Half})^2} \qquad \underbrace{\qquad}_{(\text{Half})^2}$$

$$\left(x + \tfrac{5}{2}\right)^2 + (y - 2)^2 = 1 \qquad \text{Standard form}$$

From the standard form, you can see that the circle is centered at $\left(-\tfrac{5}{2}, 2\right)$ and has a radius of 1, as shown in Figure 1.22.

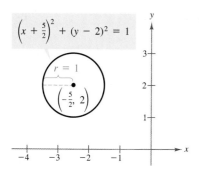

**FIGURE 1.22**

---

The general equation $Ax^2 + Ay^2 + Dx + Ey + F = 0$ may not always represent a circle. In fact, such an equation will have no solution points if the procedure of completing the square yields the impossible result

$$(x - h)^2 + (y - k)^2 = \text{negative number.} \qquad \text{No solution points}$$

---

## Technology

To graph a circle on a graphing utility, you can solve its equation for $y$ and sketch the top and bottom halves of the circle separately. For instance, you can graph the circle $(x + 1)^2 + (y - 2)^2 = 20$ by graphing the following equations.

$$y = 2 + \sqrt{20 - (x + 1)^2}$$

$$y = 2 - \sqrt{20 - (x + 1)^2}$$

If you want the result to appear circular, you need to use a square setting, as shown below.

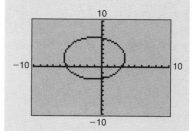

Standard setting

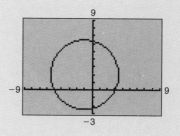

Square setting

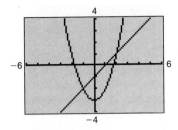

**FIGURE 1.23**

*STUDY TIP*   The *Technology* note on page 15 described how to use a graphing utility to find the *x*-intercepts of a graph. A similar procedure can be used to find the points of intersection of two graphs. (Your calculator may call this the *intersect* function.)

## Points of Intersection

A **point of intersection** of two graphs is an ordered pair that is a solution point of both graphs. For instance, Figure 1.23 shows that the graphs of

$$y = x^2 - 3 \quad \text{and} \quad y = x - 1$$

have two points of intersection: $(2, 1)$ and $(-1, -2)$. To find the points analytically, set the two *y*-values equal to each other and solve the equation

$$x^2 - 3 = x - 1$$

for *x*.

A common business application that involves points of intersection is **break-even analysis.** The marketing of a new product typically requires an initial investment. When sufficient units have been sold so that the total revenue has offset the total cost, the sale of the product has reached the **break-even point.** The **total cost** of producing *x* units of a product is denoted by *C*, and the **total revenue** from the sale of *x* units of the product is denoted by *R*. Thus, you can find the break-even point by setting the cost *C* equal to the revenue *R*, and solving for *x*.

## EXAMPLE 6   Finding a Break-Even Point

A business manufactures a product at a cost of $0.65 per unit and sells the product for $1.20 per unit. The company's initial investment to produce the product was $10,000. How many units must the company sell to break even?

### Solution

The total cost of producing *x* units of the product is given by

$$C = 0.65x + 10,000. \qquad \text{Cost equation}$$

The total revenue from the sale of *x* units is given by

$$R = 1.2x. \qquad \text{Revenue equation}$$

To find the break-even point, set the cost equal to the revenue and solve for *x*.

| | |
|---|---|
| $R = C$ | Set revenue equal to cost. |
| $1.2x = 0.65x + 10,000$ | Substitute for *R* and *C*. |
| $0.55x = 10,000$ | Subtract $0.65x$ from both sides. |
| $x = \dfrac{10,000}{0.55}$ | Divide both sides by 0.55. |
| $x \approx 18,182$ | Use a calculator. |

Thus, the company must sell 18,182 units before it breaks even. This result is shown graphically in Figure 1.24.

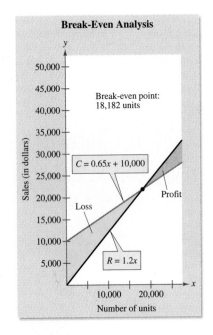

**FIGURE 1.24**

## Mathematical Models

In this text, you will see many examples of the use of equations as **mathematical models** of real-life phenomena. In developing a mathematical model to represent actual data, you should strive for two (often conflicting) goals—accuracy and simplicity.

### EXAMPLE 7    Using Mathematical Models

The table shows the annual sales (in millions of dollars) for Dell Computers and Sun Microsystems for 1992 through 1996. In the summer of 1997, the publication *Value Line* listed projected 1998 sales for the companies as $15,000 million and $10,350 million. How do you think these projections were obtained? *(Source: Dell Computers and Sun Microsystems)*

| Year | 1992 | 1993 | 1994 | 1995 | 1996 |
|------|------|------|------|------|------|
| $t$  | 2    | 3    | 4    | 5    | 6    |
| Dell | 2014 | 2873 | 3475 | 5296 | 7759 |
| Sun  | 3589 | 4309 | 4690 | 5902 | 7095 |

> **TIP**
> For help in evaluating the expressions in Example 7, see the review of Order of Operations on page 72.

*ALGEBRA*

### Solution

The projections were obtained by using past sales to predict future sales. The past revenues were modeled by equations that were found by a statistical procedure called least squares regression analysis.

$$R = 316t^2 - 1138t + 3145, \quad 2 \le t \le 6 \qquad \text{Dell}$$
$$R = 127t^2 - 155t + 3452, \quad 2 \le t \le 6 \qquad \text{Sun}$$

Using $t = 8$ to represent 1998, you can predict the 1998 sales to be

$$R = 316(8)^2 - 1138(8) + 3145 = 14{,}265 \qquad \text{Dell}$$
$$R = 127(8)^2 - 155(8) + 3452 = 10{,}340. \qquad \text{Sun}$$

These two projections are close to those projected by *Value Line*. The graphs of the two models are shown in Figure 1.25.

To test the accuracy of a model, you can compare the actual data with the values given by the model. For instance, the table below compares the actual Dell sales with those given by the model.

| Year   | 1992 | 1993 | 1994 | 1995 | 1996 |
|--------|------|------|------|------|------|
| Actual | 2014 | 2873 | 3475 | 5296 | 7759 |
| Model  | 2133 | 2575 | 3649 | 5355 | 7693 |

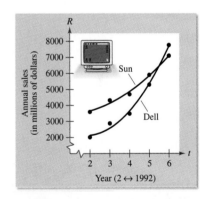

**FIGURE 1.25**

Much of your study of calculus will center around the behavior of the graphs of mathematical models. Figure 1.26 shows the graphs of six basic algebraic equations. Familiarity with these graphs will help you in the creation and use of mathematical models.

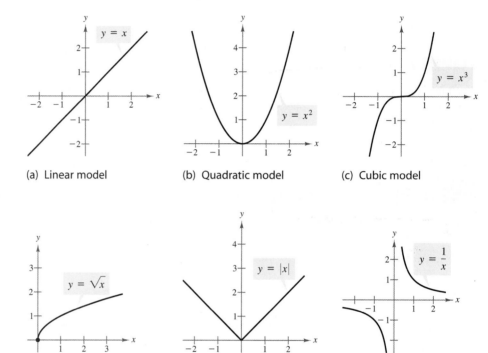

(a) Linear model    (b) Quadratic model    (c) Cubic model

(d) Square root model    (e) Absolute value model    (f) Rational model

**FIGURE 1.26**

---

*Group Discussion*    **Graphical, Numerical, and Analytical Solutions**

Most problems in calculus can be solved in a variety of ways. Often, you can solve a problem graphically, numerically (using a table), and analytically. For instance, Example 6 compares a graphical and analytical approach to finding points of intersection.

In Example 7, suppose you were asked to find the point in time at which Dell's sales exceeded Sun's sales. Explain how to use *each* of the three approaches to answer the question. For this question, which approach do you think is best? Explain.

Suppose you answered the question and obtained $t = 5.48$. What date does this represent—June 1995 or June 1996? Explain.

## Warm Up

The following warm-up exercises involve skills that were covered in earlier sections. You will use these skills in the exercise set for this section.

In Exercises 1–6, solve for y.

**1.** $5y - 12 = x$

**2.** $-y = 15 - x$

**3.** $x^3y + 2y = 1$

**4.** $x^2 + x - y^2 - 6 = 0$

**5.** $(x - 2)^2 + (y + 1)^2 = 9$

**6.** $(x + 6)^2 + (y - 5)^2 = 81$

In Exercises 7–10, factor the expression.

**7.** $x^2 - 3x + 2$

**8.** $x^2 + 5x + 6$

**9.** $y^2 - 3y + \frac{9}{4}$

**10.** $y^2 - 7y + \frac{49}{4}$

## EXERCISES 1.2

In Exercises 1–4, determine whether the points are solution points of the given equation.

**1.** $2x - y - 3 = 0$

  (a) $(1, 2)$    (b) $(1, -1)$    (c) $(4, 5)$

**2.** $x^2 + y^2 = 4$

  (a) $\left(1, -\sqrt{3}\right)$    (b) $\left(\frac{1}{2}, -1\right)$    (c) $\left(\frac{3}{2}, \frac{7}{2}\right)$

**3.** $x^2y + x^2 - 5y = 0$

  (a) $\left(0, \frac{1}{5}\right)$    (b) $(2, 4)$    (c) $(-2, -4)$

**4.** $x^2 - xy + 4y = 3$

  (a) $(0, 2)$    (b) $\left(-2, -\frac{1}{6}\right)$    (c) $(3, -6)$

In Exercises 5–10, match the equation with its graph. Use a graphing utility, set for a square setting, to confirm your result. [The graphs are labeled (a)–(f).]

**5.** $y = x - 2$

**6.** $y = -\frac{1}{2}x + 2$

**7.** $y = x^2 + 2x$

**8.** $y = \sqrt{9 - x^2}$

**9.** $y = |x| - 2$

**10.** $y = x^3 - x$

(a)
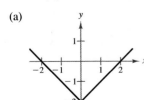

(b)
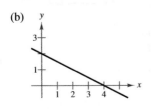

(c)
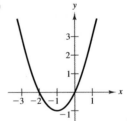

(d)
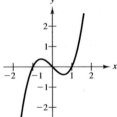

(e)
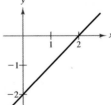

(f)
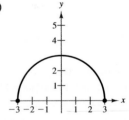

In Exercises 11–20, find the intercepts of the graph of the given equation.

**11.** $2x - y - 3 = 0$

**12.** $y = (x - 1)(x - 3)$

**13.** $y = x^2 + x - 2$

**14.** $y^2 = x^3 - 4x$

**15.** $y = x^2\sqrt{9 - x^2}$

**16.** $xy = 4$

**17.** $y = \dfrac{x^2 - 4}{x - 2}$

**18.** $y = \dfrac{x^2 + 3x}{(3x + 1)^2}$

**19.** $x^2y - x^2 + 4y = 0$

**20.** $y = x^2 + 1$

In Exercises 21–34, sketch the graph of the equation and plot the intercepts. Use a graphing utility to verify your results.

**21.** $y = x + 3$

**22.** $y = -3x + 2$

**23.** $y = 2x - 3$

**24.** $y = x^2 + 3$

**25.** $y = x^3 + 2$

**26.** $y = 1 - x^3$

**27.** $y = (x + 2)^2$

**28.** $y = \sqrt{x + 1}$

**29.** $y = -\sqrt[3]{x}$

**30.** $y = -|x - 2|$

**31.** $y = \dfrac{1}{x - 3}$

**32.** $y = \dfrac{1}{x^2 + 1}$

**33.** $x = y^2 - 4$

**34.** $x = 4 - y^2$

In Exercises 35–42, write the general form of the equation of the circle.

**35.** Center: $(0, 0)$; radius: 3

**36.** Center: $(0, 0)$; radius: 5

**37.** Center: $(2, -1)$; radius: 4

**38.** Center: $(-4, 3)$; radius: $\frac{5}{8}$

**39.** Center: $(-1, 2)$; solution point: $(0, 0)$

**40.** Center: $(3, -2)$; solution point: $(-1, 1)$

**41.** Endpoints of a diameter: $(3, 3), (-3, 3)$

**42.** Endpoints of a diameter: $(-4, -1), (4, 1)$

In Exercises 43–50, complete the square to write the equation of the circle in standard form. Then use a graphing utility to graph the circle.

**43.** $x^2 + y^2 - 2x + 6y + 6 = 0$

**44.** $x^2 + y^2 - 2x + 6y - 15 = 0$

**45.** $x^2 + y^2 + 4x + 6y - 3 = 0$

**46.** $3x^2 + 3y^2 - 6y - 1 = 0$

**47.** $2x^2 + 2y^2 - 2x - 2y - 3 = 0$

**48.** $4x^2 + 4y^2 - 4x + 2y - 1 = 0$

**49.** $16x^2 + 16y^2 + 16x + 40y - 7 = 0$

**50.** $x^2 + y^2 - 4x + 2y + 3 = 0$

In Exercises 51–58, find the points of intersection (if any) of the graphs of the equations. Use a graphing utility to check your results.

**51.** $x + y = 2, \ 2x - y = 1$

**52.** $x + y = 7, \ 3x - 2y = 11$

**53.** $x^2 + y^2 = 25, \ 2x + y = 10$

**54.** $x^2 + y = 4, \ 2x - y = 1$

**55.** $y = x^3, \ y = 2x$

**56.** $y = \sqrt{x}, \ y = x$

**57.** $y = x^4 - 2x^2 + 1, \ y = 1 - x^2$

**58.** $y = x^3 - 2x^2 + x - 1, \ y = -x^2 + 3x - 1$

**59.** *Break-Even Analysis*   You are setting up a part-time business with an initial investment of $5000. The unit cost of the product is $11.80, and the selling price is $19.30.

(a) Find equations for the total cost $C$ and total revenue $R$ for $x$ units.

(b) Find the break-even point by finding the point of intersection of the cost and revenue equations.

(c) How many units would yield a profit of $100?

**60.** *Break-Even Analysis*   A certain car model costs $14,500 with a gasoline engine and $15,450 with a diesel engine. The numbers of miles per gallon of fuel for cars with these two engines are 22 and 31, respectively. Assume that the price of each type of fuel is $1.389 per gallon.

(a) Show that the cost $C_g$ of driving the gasoline-powered car $x$ miles is

$$C_g = 14{,}500 + \frac{1.389x}{22}$$

and the cost $C_d$ of driving the diesel model $x$ miles is

$$C_d = 15{,}450 + \frac{1.389x}{31}.$$

(b) Find the break-even point. That is, find the mileage at which the diesel-powered car becomes more economical than the gasoline-powered car.

*Break-Even Analysis*   In Exercises 61–64, find the sales necessary to break even for the given cost and revenue equations. (Round your answer up to the nearest whole unit.) Use a graphing utility to graph the equations and then find the break-even point.

**61.** $C = 0.85x + 35{,}000, \ R = 1.55x$

**62.** $C = 6x + 500{,}000, \ R = 35x$

**63.** $C = 8650x + 250{,}000, \ R = 9950x$

**64.** $C = 5.5\sqrt{x} + 10{,}000, \ R = 3.29x$

**65.** Use a graphing utility to sketch the graphs of the equation $y = cx + 1$ for $c = 1, 2, 3, 4,$ and 5. Then make a conjecture about the $x$-coefficient and the graph of the equation.

**66.** *College Text Expenses*   The amount of money $y$ (in millions of dollars) spent on college textbooks in the United States is given in the table. *(Source: Book Industry Study Group, Inc.)*

| Year | 1985 | 1988 | 1990 | 1993 | 1994 |
|------|------|------|------|------|------|
| Expense | 1575 | 1998 | 2319 | 2498 | 2536 |

A mathematical model for the data is

$$y = 0.537t^4 - 22.3t^3 + 324t^2 - 1820t + 5030,$$

where $t$ represents the year, with $t = 5$ corresponding to 1985.

(a) Compare the actual expenses with those given by the model. How good is the model? Explain your reasoning.

(b) Use the model to predict the expenses in 2000.

**67.** *Farm Work Force*   The farm work force in the United States, as a percent of the total work force, is given in the table. *(Source: Department of Commerce)*

| Year | 1900 | 1910 | 1920 | 1930 | 1940 |
|------|------|------|------|------|------|
| Percent | 37.5 | 31.0 | 27.0 | 21.4 | 17.4 |

| Year | 1950 | 1960 | 1970 | 1980 | 1990 |
|------|------|------|------|------|------|
| Percent | 11.6 | 6.0 | 3.1 | 2.2 | 1.6 |

A mathematical model for the data is

$$y = \frac{37.18 - 0.43t}{1 + 0.01t},$$

where $y$ represents the percent and $t$ represents the year, with $t = 0$ corresponding to 1900.

(a) Compare the actual percent with that given by the model. How good is the model?

(b) Use the model to predict the farm work force as a percent of the total work force in 2000.

(c) Discuss the validity of your prediction in part (b).

**68.** *Weekly Salary*   A mathematical model for the average weekly salary $y$ of a person in finance, insurance, or real estate is $y = 1.729t^2 + 40.24t + 803.3 + 2176/t$, where $t$ represents the year, with $t = 0$ corresponding to 1980. *(Source: U.S. Bureau of Labor Statistics)*

Table for 68

| Year | 1985 | 1990 | 1992 | 1994 | 1995 | 1996 |
|------|------|------|------|------|------|------|
| Salary | | | | | | |

(a) Use the model to complete the table.

(b) This model was created using actual data from 1985 through 1995. How accurate do you think the model is in predicting the 1996 average salary? Explain your reasoning.

(c) What does this model predict the average salary to be in 2000? Do you think this prediction is valid?

**69.** Define the break-even point for a business marketing a new product. Give examples of a linear cost equation and a linear revenue equation for which the break-even point is 10,000 units.

**70.** *Lung Transplants*   A mathematical model for the number of lung transplants performed in the United States is $y = -10.1t^2 + 177.0t + 217.5$, where $y$ is the number of transplants and $t$ is the time in years, with $t = 0$ corresponding to 1990. *(Source: U.S. Department of Health and Human Services)*

| Year | 1990 | 1991 | 1992 | 1993 | 1994 | 1995 |
|------|------|------|------|------|------|------|
| Transplants | | | | | | |

(a) Enter the model into a graphing utility and use it to complete the table.

(b) Use your school's library or some other reference source to find the actual number of lung transplants for the years 1990 to 1995. Compare the actual numbers with those given by the model. How good is the model? Explain your reasoning.

(c) Using this model, what is the prediction for the number of transplants in the year 2000? How valid do you think the prediction is? What factors could affect this model's accuracy?

 In Exercises 71–74, use a graphing utility to graph the equation. Use the graphing utility to approximate the intercepts of the graph.

**71.** $y = 0.24x^2 + 1.32x + 5.36$

**72.** $y = -0.56x^2 - 5.34x + 6.25$

**73.** $y = \sqrt{0.3x^2 - 4.3x + 5.7}$

**74.** $y = (0.2x^2 + 1)/(0.1x + 2.4)$

# Lines in the Plane and Slope

## 1.3

*Using Slope • Finding the Slope of a Line • Writing Linear Equations • Parallel and Perpendicular Lines • Extended Application: Linear Depreciation*

## Using Slope

The simplest mathematical model for relating two variables is the **linear equation** $y = mx + b$. The equation is called *linear* because its graph is a line. (In this text, we use the term *line* to mean *straight line*.) By letting $x = 0$, you can see that the line crosses the $y$-axis at $y = b$, as shown in Figure 1.27. In other words, the $y$-intercept is $(0, b)$. The steepness or slope of the line is $m$.

$$y = mx + b$$

Slope ⌐      ⌐ $y$-intercept

The **slope** of a line is the number of units the line rises (or falls) vertically for each unit of horizontal change from left to right, as shown in Figure 1.27.

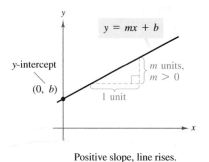

Positive slope, line rises.

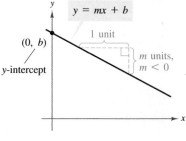

Negative slope, line falls.

**FIGURE 1.27**

A linear equation that is written in the form $y = mx + b$ is said to be written in **slope-intercept form.**

---

### The Slope-Intercept Form of the Equation of a Line

The graph of the equation

$$y = mx + b$$

is a line whose slope is $m$ and whose $y$-intercept is $(0, b)$.

---

*Discovery*

Use a graphing utility to compare the slopes of the lines $y = mx$, where $m = 0.5, 1, 2,$ and $4$. Which line rises most quickly? Now, let $m = -0.5, -1, -2,$ and $-4$. Which line falls most quickly? Let $m = 0.01, 0.001,$ and $0.0001$. What is the slope of a horizontal line? Use a square setting to obtain a true geometric perspective.

Once you have determined the slope and the $y$-intercept of a line, it is a relatively simple matter to sketch its graph.

In the following example, note that none of the lines is vertical. A vertical line has an equation of the form

$x = a$.          Vertical line

Because such an equation cannot be written in the form $y = mx + b$, it follows that the slope of a vertical line is undefined, as indicated in Figure 1.28.

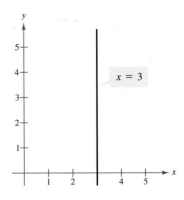

**FIGURE 1.28**   When the line is vertical, the slope is undefined.

## EXAMPLE 1   Graphing a Linear Equation

Sketch the graphs of the following linear equations.

(a) $y = 2x + 1$      (b) $y = 2$      (c) $x + y = 2$

### Solution

(a) Because $b = 1$, the $y$-intercept is $(0, 1)$. Moreover, because the slope is $m = 2$, the line *rises* two units for each unit the line moves to the right, as shown in Figure 1.29(a).

(b) By writing this equation in the form $y = (0)x + 2$, you can see that the $y$-intercept is $(0, 2)$ and the slope is zero. A zero slope implies that the line is horizontal—that is, it doesn't rise *or* fall, as shown in Figure 1.29(b).

(c) By writing this equation in slope-intercept form

$x + y = 2$                    Original equation

$y = -x + 2$                  Subtract $x$ from both sides.

$y = (-1)x + 2$            Slope-intercept form

you can see that the $y$-intercept is $(0, 2)$. Moreover, because the slope is $m = -1$, this line *falls* one unit for each unit the line moves to the right, as shown in Figure 1.29(c).

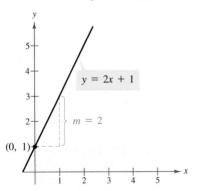

(a) When $m$ is positive, the line rises.

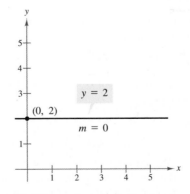

(b) When $m$ is zero, the line is horizontal.

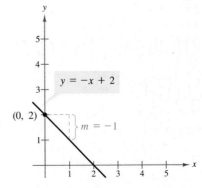

(c) When $m$ is negative, the line falls.

**FIGURE 1.29**

In real-life problems, the slope of a line can be interpreted as either a *ratio* or a *rate*. If the *x*-axis and *y*-axis have the same unit of measure, then the slope has no units and is a **ratio.** If the *x*-axis and *y*-axis have different units of measure, then the slope is a **rate** or **rate of change.**

### EXAMPLE 2  Using Slope as a Ratio

The maximum recommended slope of a wheelchair ramp is $\frac{1}{12} \approx 0.083$. A business is installing a wheelchair ramp that rises 22 inches over a horizontal length of 24 feet, as shown in Figure 1.30. Is the ramp steeper than recommended? *(Source:* American Disabilities Act Handbook*)*

**Solution**

The horizontal length of the ramp is 12(24) or 288 inches. Thus, the slope of the ramp is

$$\text{Slope} = \frac{\text{vertical change}}{\text{horizontal change}} = \frac{22 \text{ in.}}{288 \text{ in.}} \approx 0.076.$$

Thus, the slope is not steeper than recommended.

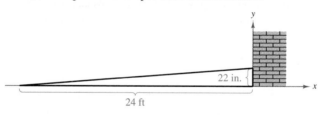

22 in.

24 ft

**FIGURE 1.30**

### EXAMPLE 3  Using Slope as a Rate of Change

A manufacturing company determines that the total cost in dollars of producing *x* units of a product is

$$C = 25x + 3500. \qquad \text{Cost equation}$$

Describe the practical significance of the *y*-intercept and slope of the line given by this equation.

**Solution**

The *y*-intercept (0, 3500) tells you that the cost of producing zero units is $3500. This is the **fixed cost** of production—it includes costs that must be paid regardless of the number of units produced. The slope of $m = 25$ tells you that the cost of producing each unit is $25, as shown in Figure 1.31. Economists call the cost per unit the **marginal cost.** If the production increases by one unit, then the "margin" or extra amount of cost is $25.

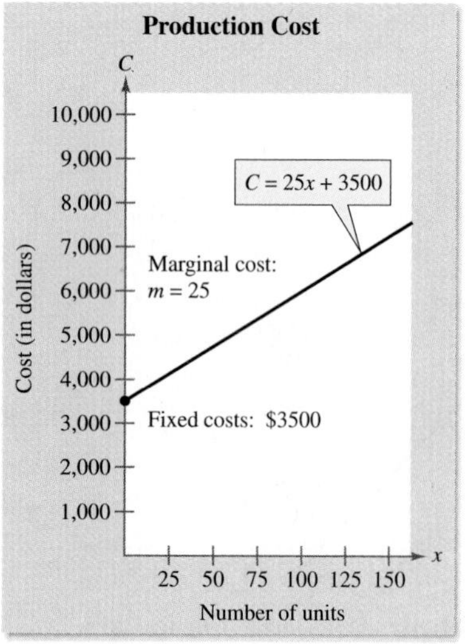

**Production Cost**

*C*

10,000
9,000
8,000 — $C = 25x + 3500$
7,000
Marginal cost: $m = 25$
6,000
5,000
Cost (in dollars)
4,000
3,000 — Fixed costs: $3500
2,000
1,000

25  50  75  100 125 150    *x*

Number of units

**FIGURE 1.31**

## Finding the Slope of a Line

Given an equation of a line, you can find its slope by writing the equation in slope-intercept form. If you are not given an equation, you can still find the slope of a line. For instance, suppose you want to find the slope of the line passing through the points $(x_1, y_1)$ and $(x_2, y_2)$, as shown in Figure 1.32. As you move from left to right along this line, a change of $(y_2 - y_1)$ units in the vertical direction corresponds to a change of $(x_2 - x_1)$ units in the horizontal direction. These two changes are denoted by the symbols

$$\Delta y = y_2 - y_1 = \text{the change in } y$$

and

$$\Delta x = x_2 - x_1 = \text{the change in } x.$$

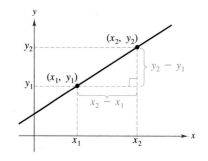

**FIGURE 1.32**

($\Delta$ is the Greek capital letter delta, and the symbols $\Delta y$ and $\Delta x$ are read as "delta $y$" and "delta $x$.") The ratio of $\Delta y$ to $\Delta x$ represents the slope of the line that passes through the points $(x_1, y_1)$ and $(x_2, y_2)$.

$$\text{Slope} = \frac{\Delta y}{\Delta x} = \frac{y_2 - y_1}{x_2 - x_1}$$

Be sure you see that $\Delta x$ represents a single number, not the product of two numbers ($\Delta$ and $x$). The same is true for $\Delta y$.

---

### The Slope of a Line Passing Through Two Points

The **slope** $m$ of the line passing through $(x_1, y_1)$ and $(x_2, y_2)$ is

$$m = \frac{\Delta y}{\Delta x} = \frac{y_2 - y_1}{x_2 - x_1}$$

where $x_1 \neq x_2$.

---

When this formula is used for slope, the *order of subtraction* is important. Given two points on a line, you are free to label either one of them as $(x_1, y_1)$ and the other as $(x_2, y_2)$. However, once you have done this, you must form the numerator and denominator using the same order of subtraction.

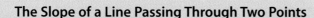

$$m = \frac{y_2 - y_1}{x_2 - x_1} \qquad m = \frac{y_1 - y_2}{x_1 - x_2} \qquad m = \frac{y_2 - y_1}{x_1 - x_2}$$

Correct  Correct  Incorrect

For instance, the slope of the line passing through the points $(3, 4)$ and $(5, 7)$ can be calculated as

$$m = \frac{7 - 4}{5 - 3} = \frac{3}{2} \qquad \text{or} \qquad m = \frac{4 - 7}{3 - 5} = \frac{-3}{-2} = \frac{3}{2}.$$

## EXAMPLE 4  Finding the Slope of a Line

Find the slope of the line passing through the pairs of points.

(a) $(-2, 0)$ and $(3, 1)$        (b) $(-1, 2)$ and $(2, 2)$

(c) $(0, 4)$ and $(1, -1)$        (d) $(3, 4)$ and $(3, 1)$

*Solution*

(a) Letting $(x_1, y_1) = (-2, 0)$ and $(x_2, y_2) = (3, 1)$, you obtain a slope of

$$m = \frac{y_2 - y_1}{x_2 - x_1} = \frac{1 - 0}{3 - (-2)} = \frac{1}{5}. \quad \xleftarrow{\text{Difference in }y\text{-values}} $$
$$\xleftarrow{\text{Difference in }x\text{-values}}$$

(b) The slope of the line passing through $(-1, 2)$ and $(2, 2)$ is

$$m = \frac{2 - 2}{2 - (-1)} = \frac{0}{3} = 0.$$

(c) The slope of the line passing through $(0, 4)$ and $(1, -1)$ is

$$m = \frac{-1 - 4}{1 - 0} = \frac{-5}{1} = -5.$$

(d) The slope of the vertical line passing through $(3, 4)$ and $(3, 1)$ is not defined because division by zero is undefined. (See Figure 1.33.)

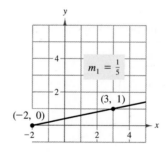

(a) Positive slope; line rises.

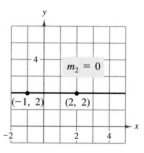

(b) Zero slope; line is horizontal.

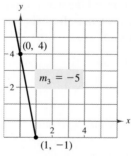

(c) Negative slope; line falls.

(d) Vertical line; undefined slope.

**FIGURE 1.33**

## Writing Linear Equations

If $(x_1, y_1)$ is a point lying on a line of slope $m$ and $(x, y)$ is *any other* point on the line, then

$$\frac{y - y_1}{x - x_1} = m.$$

This equation, involving the variables $x$ and $y$, can be rewritten in the form $y - y_1 = m(x - x_1)$, which is the **point-slope form** of the equation of a line.

---

### Point-Slope Form of the Equation of a Line

The equation of the line with slope $m$ passing through the point $(x_1, y_1)$ is

$$y - y_1 = m(x - x_1).$$

---

The point-slope form is most useful for *finding* the equation of a line. You should remember this formula—it is used repeatedly throughout the text.

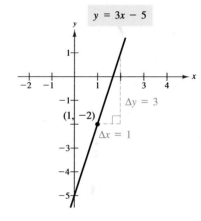

$y = 3x - 5$

### EXAMPLE 5   Using the Point-Slope Form

Find the equation of the line that has a slope of 3 and passes through the point $(1, -2)$, as shown in Figure 1.34.

#### Solution

Use the point-slope form with $m = 3$ and $(x_1, y_1) = (1, -2)$.

| | |
|---|---|
| $y - y_1 = m(x - x_1)$ | Point-slope form |
| $y - (-2) = 3(x - 1)$ | Substitute for $m$, $x_1$, and $y_1$. |
| $y + 2 = 3x - 3$ | Simplify. |
| $y = 3x - 5$ | Slope-intercept form |

**FIGURE 1.34**

The point-slope form can be used to find an equation of the line passing through points $(x_1, y_1)$ and $(x_2, y_2)$. To do this, first find the slope of the line

$$m = \frac{y_2 - y_1}{x_2 - x_1}, \qquad x_1 \neq x_2,$$

and then use the point-slope form to obtain the equation

$$y - y_1 = \frac{y_2 - y_1}{x_2 - x_1}(x - x_1). \qquad \text{Two-point form}$$

This is sometimes called the **two-point form** of the equation of a line.

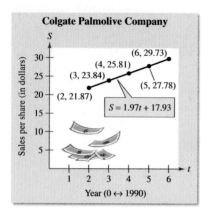

**FIGURE 1.35**

## EXAMPLE 6  Predicting Cash Flow Per Share

The sales per share for Colgate Palmolive Company was $21.87 in 1992 and $29.73 in 1996. Using only this information, write a linear equation that gives the cash flow per share in terms of the year.   *(Source: Colgate Palmolive Company)*

### Solution

Let $t = 0$ represent 1990. Then the two given values are represented by the ordered pairs $(2, 21.87)$ and $(6, 29.73)$. The slope of the line passing through these points is

$$m = \frac{29.73 - 21.87}{6 - 2} \approx 1.97. \qquad m = \frac{y_2 - y_1}{t_2 - t_1}$$

Using the point-slope form, you can find the equation that relates the sales per share $S$ and the year $t$ to be

$$S = 1.97t + 17.93, \qquad 2 \le t \le 6.$$

You can use this model to estimate the sales per share in other years. For instance, the model estimates the sales per share in 1993, 1994, and 1995 to be $23.84, $25.81, and $27.78, respectively, as shown in Figure 1.35. (In this case, the predictions are reasonably good—the actual sales per share in these years were $23.92, $26.27, and $28.65.)

The estimation method illustrated in Example 6 is called **linear interpolation.** Note in Figure 1.36(a) that an interpolated point lies between the two given points. When the estimated point lies beyond the given points, as shown in Figure 1.36(b), the procedure is called **linear extrapolation.**

Because the slope of a vertical line is not defined, its equation cannot be written in slope-intercept form. However, every line has an equation that can be written in the **general form**

$$Ax + By + C = 0, \qquad \text{General form}$$

where $A$ and $B$ are not both zero. For instance, the vertical line given by $x = a$ can be represented by the general form $x - a = 0$. The five most common forms of equations of lines are summarized below.

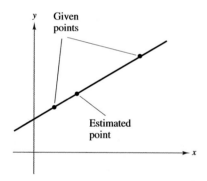

(a)  Linear Interpolation

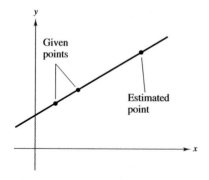

(b)  Linear Extrapolation

**FIGURE 1.36**

| Equations of Lines | |
|---|---|
| 1.  General form: | $Ax + By + C = 0$ |
| 2.  Vertical line: | $x = a$ |
| 3.  Horizontal line: | $y = b$ |
| 4.  Slope-intercept form: | $y = mx + b$ |
| 5.  Point-slope form: | $y - y_1 = m(x - x_1)$ |

## Parallel and Perpendicular Lines

The slope of a line is a convenient tool for determining whether two lines are parallel or perpendicular.

 **Technology**

On a graphing utility, lines will not appear to have the correct slope unless you use a viewing rectangle that has a "square setting." For instance, try graphing the lines in Example 7 using the standard setting $-10 \leq x \leq 10$ and $-10 \leq y \leq 10$. Then reset the viewing rectangle with the square setting $-9 \leq x \leq 9$ and $-6 \leq y \leq 6$. On which setting do the lines $y = \frac{2}{3}x - \frac{5}{3}$ and $y = -\frac{3}{2}x + 2$ appear perpendicular?

---

### Parallel and Perpendicular Lines

1. Two distinct nonvertical lines are **parallel** if and only if their slopes are equal. That is, $m_1 = m_2$.

2. Two nonvertical lines are **perpendicular** if and only if their slopes are negative reciprocals of each other. That is, $m_1 = -1/m_2$.

---

**EXAMPLE 7** **Finding Parallel and Perpendicular Lines**

Find an equation of the line that passes through the point $(2, -1)$ and is

(a) parallel to the line $2x - 3y = 5$

(b) perpendicular to the line $2x - 3y = 5$.

*Solution*

By writing the given equation in slope-intercept form

$$2x - 3y = 5 \qquad \text{Given equation}$$
$$-3y = -2x + 5 \qquad \text{Subtract } 2x \text{ from both sides.}$$
$$y = \tfrac{2}{3}x - \tfrac{5}{3} \qquad \text{Slope-intercept form}$$

you can see that it has a slope of $m = \frac{2}{3}$, as shown in Figure 1.37.

(a) Any line parallel to the given line must also have a slope of $\frac{2}{3}$. Thus, the line through $(2, -1)$ that is parallel to the given line has the following equation.

$$y - (-1) = \tfrac{2}{3}(x - 2) \qquad \text{Point-slope form}$$
$$3(y + 1) = 2(x - 2) \qquad \text{Multiply both sides by 3.}$$
$$3y + 3 = 2x - 4 \qquad \text{Distribute.}$$
$$3y = 2x - 7 \qquad \text{Simplify.}$$
$$y = \tfrac{2}{3}x - \tfrac{7}{3} \qquad \text{Slope-intercept form}$$

(b) Any line perpendicular to the given line must have a slope of $-1/(2/3)$ or $-\frac{3}{2}$. Therefore, the line through $(2, -1)$ that is perpendicular to the given line has the following equation.

$$y - (-1) = -\tfrac{3}{2}(x - 2) \qquad \text{Point-slope form}$$
$$2(y + 1) = -3(x - 2) \qquad \text{Multiply both sides by 2.}$$
$$3x + 2y = 4 \qquad \text{General form}$$
$$y = -\tfrac{3}{2}x + 2 \qquad \text{Slope-intercept form}$$

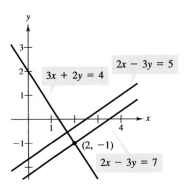

**FIGURE 1.37**

## Extended Application:　Linear Depreciation

Most business expenses can be deducted the same year they occur. One exception to this is the cost of property that has a useful life of more than 1 year, such as buildings, cars, or equipment. Such costs must be **depreciated** over the useful life of the property. If the *same amount* is depreciated each year, the procedure is called **linear depreciation.**

### EXAMPLE 8　Depreciating Equipment

Your company has purchased a $12,000 machine that has a useful life of 8 years. The salvage value at the end of 8 years is $2000. Write a linear equation that describes the *nondepreciated* value of the machine each year.

### Solution

Let $V$ represent the value of the machine at the end of year $t$. You can represent the initial value of the machine by the ordered pair (0, 12,000) and the salvage value of the machine by the ordered pair (8, 2000). The slope of the line is

$$m = \frac{2000 - 12{,}000}{8 - 0} = -\$1250, \qquad m = \frac{y_2 - y_1}{t_2 - t_1}$$

which represents the annual depreciation in *dollars per year*. Using the point-slope form, you can write the equation of the line as follows.

$$V - 12{,}000 = -1250(t - 0) \qquad \text{Point-slope form}$$
$$V = -1250t + 12{,}000 \qquad \text{Slope-intercept form}$$

The table shows the nondepreciated value of the machine at the end of each year.

| $t$ | 0 | 1 | 2 | 3 | 4 | 5 | 6 | 7 | 8 |
|---|---|---|---|---|---|---|---|---|---|
| $V$ | 12,000 | 10,750 | 9500 | 8250 | 7000 | 5750 | 4500 | 3250 | 2000 |

The graph of this equation is shown in Figure 1.38.

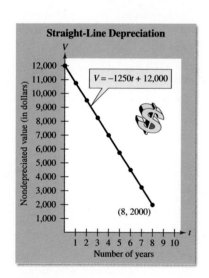

**Straight-Line Depreciation**

$V = -1250t + 12{,}000$

(8, 2000)

Nondepreciated value (in dollars)

Number of years

**FIGURE 1.38**

---

*Group Discussion*　　*Comparing Different Types of Depreciation*

The Internal Revenue Service allows businesses to choose different types of depreciation. Another type is

$$\text{Uniform Declining Balances: } V = 12{,}000\left(\frac{n - 1.605}{n}\right)^t, \quad n = 8.$$

Construct a table that compares this type of depreciation with linear depreciation. What are the advantages of each type?

## Warm Up

The following warm-up exercises involve skills that were covered in earlier sections. You will use these skills in the exercise set for this section.

In Exercises 1 and 2, simplify the expression.

**1.** $\dfrac{5 - (-2)}{-3 - 4}$

**2.** $\dfrac{-7 - (-0)}{4 - 1}$

**3.** Evaluate $-\dfrac{1}{m}$ when $m = -3$.

**4.** Evaluate $-\dfrac{1}{m}$ when $m = \dfrac{6}{7}$.

In Exercises 5–10, solve for $y$ in terms of $x$.

**5.** $-4x + y = 7$

**6.** $3x - y = 7$

**7.** $y - 2 = 3(x - 4)$

**8.** $y - (-5) = -1[x - (-2)]$

**9.** $y - (-3) = \dfrac{4 - (-3)}{2 - 1}(x - 2)$

**10.** $y - 1 = \dfrac{-3 - 1}{-7 - (-1)}[x - (-1)]$

## EXERCISES 1.3

In Exercises 1–4, estimate the slope of the line.

**1.**
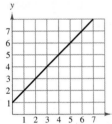

**2.**
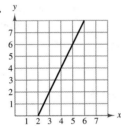

**3.**
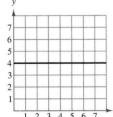

**4.**
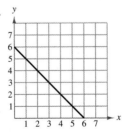

In Exercises 5–12, plot the points and find the slope of the line passing through the points.

**5.** $(3, -4), (5, 2)$

**6.** $(-2, 1), (4, -3)$

**7.** $\left(\tfrac{1}{2}, 2\right), (6, 2)$

**8.** $\left(-\tfrac{3}{2}, -5\right), \left(\tfrac{5}{6}, 4\right)$

**9.** $(-8, -3), (-8, -5)$

**10.** $(2, 1), (2, 5)$

**11.** $(1, 2), (-2, 2)$

**12.** $\left(\tfrac{7}{8}, \tfrac{3}{4}\right), \left(\tfrac{5}{4}, -\tfrac{1}{4}\right)$

In Exercises 13–20, use the point on the line and slope of the line to find three additional points on the line. (There are many correct solutions.)

| Point | Slope | Point | Slope |
|---|---|---|---|
| **13.** $(2, 1)$ | $m = 0$ | **14.** $(-3, 4)$ | $m$ undefined |
| **15.** $(6, -4)$ | $m = -2$ | **16.** $(-2, -2)$ | $m = 2$ |
| **17.** $(1, 7)$ | $m = -3$ | **18.** $(10, -6)$ | $m = -1$ |
| **19.** $(-8, 1)$ | $m$ undefined | **20.** $(-3, -1)$ | $m = 0$ |

In Exercises 21–26, find the slope and $y$-intercept (if possible) of the line.

**21.** $x + 5y = 20$

**22.** $2x + y = 40$

**23.** $7x - 5y = 15$

**24.** $6x - 5y = 15$

**25.** $x = 4$

**26.** $y = -1$

In Exercises 27–34, write an equation of the line that passes through the points. Then use the equation to sketch the line.

**27.** $(4, 3), (0, -5)$

**28.** $(-3, -4), (1, 4)$

**29.** $(0, 0), (-1, 3)$

**30.** $(-3, 6), (1, 2)$

**31.** $(2, 3), (2, -2)$

**32.** $(6, 1), (10, 1)$

**33.** $(1, -2), (3, -2)$

**34.** $\left(\tfrac{7}{8}, \tfrac{3}{4}\right), \left(\tfrac{5}{4}, -\tfrac{1}{4}\right)$

In Exercises 35–44, write an equation of the indicated line. Then use a graphing utility to graph the equation.

| Point | Slope | Point | Slope |
|-------|-------|-------|-------|
| **35.** $(0, 3)$ | $m = \frac{3}{4}$ | **36.** $(-1, 2)$ | $m$ undefined |
| **37.** $(0, 0)$ | $m = \frac{2}{3}$ | **38.** $(-1, -4)$ | $m = \frac{1}{4}$ |
| **39.** $(-2, 7)$ | $m = -3$ | **40.** $(-2, 4)$ | $m = -\frac{3}{5}$ |
| **41.** $(0, 2)$ | $m = 4$ | **42.** $\left(0, -\frac{2}{3}\right)$ | $m = \frac{1}{6}$ |
| **43.** $\left(0, \frac{2}{3}\right)$ | $m = \frac{3}{4}$ | **44.** $(0, 4)$ | $m = 0$ |

In Exercises 45 and 46, explain how to use the concept of slope to determine whether the three points are collinear. Then explain how to use the Distance Formula to determine whether the points are collinear.

**45.** $(-2, 1), (-1, 0), (2, -2)$

**46.** $(0, 4), (7, -6), (-5, 11)$

**47.** Write an equation of the vertical line with $x$-intercept at 3.

**48.** Write an equation of the horizontal line through $(0, -5)$.

 In Exercises 49–56, write an equation of the line through the given point (a) parallel to the given line and (b) perpendicular to the given line. Then use a graphing utility to graph all three equations on the same set of coordinate axes.

| Point | Line |
|-------|------|
| **49.** $(-3, 2)$ | $x + y = 7$ |
| **50.** $(2, 1)$ | $4x - 2y = 3$ |
| **51.** $(-6, 4)$ | $3x + 4y = 7$ |
| **52.** $\left(\frac{7}{8}, \frac{3}{4}\right)$ | $5x + 3y = 0$ |
| **53.** $(-1, 0)$ | $y = -3$ |
| **54.** $(2, 5)$ | $x = 4$ |
| **55.** $(1, 1)$ | $-2x + 3y = -3$ |
| **56.** $(12, -3)$ | $x + 4y = 4$ |

 In Exercises 57–64, sketch the graph of the equation. Use a graphing utility to verify your sketch.

**57.** $y = -2$

**58.** $x = 4$

**59.** $2x - y - 3 = 0$

**60.** $x + 2y + 6 = 0$

**61.** $y = -2x + 1$

**62.** $y - 1 = 3(x + 4)$

**63.** $y + 2 = -4(x + 1)$

**64.** $4x + 5y = 20$

**65.** *Population*   The resident populations of Oklahoma (in thousands) were 3146 in 1990 and 3278 in 1995. Assume that the relationship between the population $y$ and the year $t$ is linear. Let $t = 0$ represent 1990.   *(Source: U.S. Bureau of Census)*

(a) Write a linear model for the data. What is the slope and what does it tell you about the population?

(b) Estimate the population in 1992.

(c) Use your model to estimate the population in 1996.

(d) Use your school's library or some other reference source to find the actual populations in 1992 and 1996. How close were your estimates?

(e) Do you think your model could be used to predict the population in 2000? Explain.

**66.** *Temperature Conversion*   Write a linear equation that expresses the relationship between the temperature in degrees Celsius $C$ and degrees Fahrenheit $F$. Use the fact that water freezes at 0°C (32°F) and boils at 100°C (212°F).

**67.** *Chemistry: Temperature Conversion*   Use the result of Exercise 66 to answer the following:

(a) A person has a temperature of 102.5°F. What is this temperature on the Celsius scale?

(b) Gallium has a melting point of 29.8°C. Is gallium a liquid or a solid at 68°F?

*(Source: Adapted from Zumdahl,* Chemistry, *Fourth Edition)*

**68.** *Reimbursed Expenses*   A company reimburses its sales representatives $150 per day for lodging and meals, plus $0.31 per mile driven. Write a linear equation giving the daily cost $C$ in terms of $x$, the number of miles driven.

**69.** *Union Negotiation*   You are on a negotiating panel in a union hearing for a large corporation. The union is asking for a base pay of $8.75 per hour *plus* an additional piecework rate of $0.80 per unit produced. The corporation is offering a base pay of $6.35 per hour *plus* a piecework rate of $1.15.

(a) Write a linear equation for the hourly wages $W$ in terms of $x$, the number of units produced per hour, for each pay schedule.

(b) Use a graphing utility to graph each linear equation and find the point of intersection.

(c) Interpret the meaning of the point of intersection of the graphs. How would you use this information to advise the corporation and the union?

**70.** *Chemistry: Temperature Conversion*   Ethylene glycol is the main component in automobile antifreeze. To monitor the temperature of an auto cooling system, you intend to use a meter that reads from 0 to 100. You devise a new temperature scale (°A) based on the approximate melting and boiling points of a typical antifreeze solution ($-45°C$ and $115°C$). You wish these points to correspond to $0°A$ and $100°A$, respectively.

(a) Derive an expression for converting between °A and °C.
(b) Derive an expression for converting between °F and °A.
(c) At what temperature would your thermometer and a Celsius thermometer give the same numerical reading?
(d) Your thermometer reads $86°A$. What is the temperature in °C and in °F?
(e) What is a temperature of $45°C$ in °A?

*(Source: Zumdahl, Chemistry, Fourth Edition)*

**71.** *Linear Depreciation*   A company constructs a warehouse for $825,000. The warehouse has an estimated useful life of 25 years, after which its value is expected to be $75,000. Write a linear equation giving the value $y$ of the warehouse during its 25 years of useful life. (Let $t$ represent the time in years.)

**72.** *Linear Depreciation*   A small business purchases a piece of equipment for $1025. After 5 years the equipment will be outdated, having no value.

(a) Write a linear equation giving the value of the equipment in terms of the time $t$, $0 \le t \le 5$.
(b) Use a graphing utility to graph the equation.
(c) Move the cursor along the graph and estimate (to two-decimal-place accuracy) the value of the equipment when $t = 3$.
(d) Move the cursor along the graph and estimate (to two-decimal-place accuracy) the time when the value of the equipment will be $600.

**73.** *Apartment Rental*   A real estate office handles an apartment complex with 50 units. When the rent is $380 per month, all 50 units are occupied. When the rent is $425, however, the average number of occupied units drops to 47. Assume that the relationship between the monthly rent $p$ and the demand $x$ is linear. (The term *demand* refers to the number of occupied units.)

(a) Write a linear equation expressing $x$ in terms of $p$.
(b) *Linear Extrapolation*   Predict the number of occupied units when the rent is set at $455.
(c) *Linear Interpolation*   Predict the number of occupied units when the rent is set at $395.

**74.** *Profit*   You are a contractor and have purchased a piece of equipment for $26,500. The equipment costs an average of $5.25 per hour for fuel and maintenance, and the operator is paid $9.50 per hour.

(a) Write a linear equation giving the total cost $C$ of operating the equipment for $t$ hours.
(b) You charge your customers $25 per hour of machine use. Write an equation for the revenue $R$ derived from $t$ hours of use.
(c) Use the formula for profit, $P = R - C$, to write an equation for the profit derived from $t$ hours of use.
(d) Find the number of hours you must operate the equipment before you break even.

**75.** *Research Project*   Personal income (in billions of dollars) in the United States was 4792 in 1990 and 6102 in 1995. Assume that the relationship between the personal income $Y$ and the time $t$ (in years) is linear. Let $t = 0$ correspond to 1990.   *(Source: U.S. Bureau of Economic Analysis)*

(a) Write a linear model for the data.
(b) *Linear Interpolation*   Estimate the personal income in 1992.
(c) *Linear Extrapolation*   Estimate the personal income in 1996.
(d) Use your school's library or some other reference source to find the actual personal income in 1992 and 1996. How close were your estimates?

**76.** *Sales Commission*   As a salesperson, you receive a monthly salary of $2000, plus a commission of 7% of sales. You are offered a new job at $2300 per month, plus a commission of 5% of sales.

(a) Write a linear equation for your current monthly wage $W$ in terms of your monthly sales $S$.
(b) Write a linear equation for the monthly wage $W$ of your job offer in terms of the monthly sales $S$.
(c) Use a graphing utility to graph both equations on the same viewing rectangle. Find the point of intersection. What does it signify?
(d) You think you can sell $20,000 per month. Should you change jobs? Explain.

In Exercises 77–82, use a graphing utility to graph the cost function. Determine the maximum production level $x$, given that the cost $C$ cannot exceed $100,000.

**77.** $C = 23,500 + 3100x$

**78.** $C = 30,000 + 575x$

**79.** $C = 18,375 + 1150x$

**80.** $C = 24,900 + 1785x$

**81.** $C = 75,500 + 89x$

**82.** $C = 83,620 + 67x$

# 1.4

## Functions

*Functions • The Graph of a Function • Function Notation •*
*Combinations of Functions • Inverse Functions*

## Functions

In many common relationships between two variables, the value of one of the variables depends on the value of the other variable. For example, the sales tax on an item depends on its selling price, the distance an object moves in a given amount of time depends on its speed, the price of mailing a package with an overnight delivery service depends on the package's weight, and the area of a circle depends on its radius.

Consider the relationship between the area of a circle and its radius. This relationship can be expressed by the equation

$$A = \pi r^2.$$

In this equation, the value of $A$ depends on the choice of $r$. Because of this, $A$ is the **dependent variable** and $r$ is the **independent variable.**

Most of the relationships that you will study in this course have the property that for a given value of the independent variable, there corresponds exactly one value of the dependent variable. Such a relationship is a **function.**

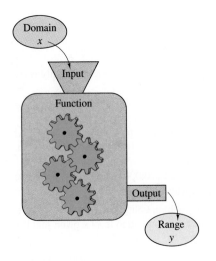

**FIGURE 1.39**

### Definition of a Function

A **function** is a relationship between two variables such that to each value of the independent variable there corresponds exactly one value of the dependent variable.

The **domain** of the function is the set of all values of the independent variable for which the function is defined. The **range** of the function is the set of all values taken on by the dependent variable.

In Figure 1.39, notice that you can think of a function as a machine that inputs values of the independent variable and outputs values of the dependent variable.

Although functions can be described by various means such as tables, graphs, and diagrams, they are most often specified by formulas or equations. For instance, the equation $y = 4x^2 + 3$ describes $y$ as a function of $x$. For this function, $x$ is the independent variable and $y$ is the dependent variable.

# EXAMPLE 1   Deciding Whether Relations Are Functions

Which of the following equations define $y$ as a function of $x$?

(a) $x + y = 1$        (b) $x^2 + y^2 = 1$

(c) $x^2 + y = 1$        (d) $x + y^2 = 1$

### Solution

To decide whether an equation defines a function, it is helpful to isolate the dependent variable on the left side. For instance, to decide whether the equation $x + y = 1$ defines $y$ as a function of $x$, write the equation in the form

$$y = 1 - x.$$

From this form, you can see that for any value of $x$, there is exactly one value of $y$. Thus, $y$ is a function of $x$.

| Original Equation | Rewritten Equation | Test: Is y a function of x? |
|---|---|---|
| (a) $x + y = 1$ | $y = 1 - x$ | Yes, each value of $x$ determines exactly one value of $y$. |
| (b) $x^2 + y^2 = 1$ | $y = \pm\sqrt{1 - x^2}$ | No, some values of $x$ determine two values of $y$. |
| (c) $x^2 + y = 1$ | $y = 1 - x^2$ | Yes, each value of $x$ determines exactly one value of $y$. |
| (d) $x + y^2 = 1$ | $y = \pm\sqrt{1 - x}$ | No, some values of $x$ determine two values of $y$. |

Note that the equations that assign two values ($\pm$) to the dependent variable for a given value of the independent variable do not define functions of $x$. For instance, in part b, when $x = 0$, the equation

$$y = \pm\sqrt{1 - x^2}$$

indicates that $y = +1$ or $y = -1$. Figure 1.40 shows the graphs of the four equations.

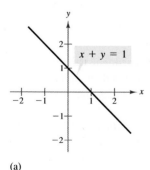

(a)

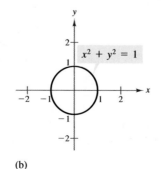

(b)

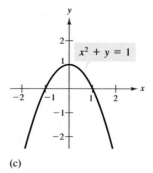

(c)

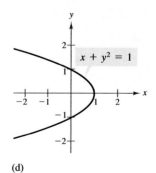

(d)

**FIGURE 1.40**

## Technology

The procedure used in Example 1, isolating the dependent variable on the left side, is also useful for graphing equations with a graphing utility. In fact, the standard graphing program on most graphing utilities is a "function grapher." To graph an equation in which $y$ is not a function of $x$, such as a circle, you usually have to enter two or more equations into the graphing utility.

## The Graph of a Function

When the graph of a function is sketched, the standard convention is to let the horizontal axis represent the independent variable. When this convention is used, the test described in Example 1 has a nice graphical interpretation called the **vertical line test.** This test states that if every vertical line intersects the graph of an equation at most once, then the equation defines $y$ as a function of $x$. For instance, in Figure 1.40, the graphs in parts (a) and (c) pass the vertical line test, but those in parts (b) and (d) do not.

The domain of a function may be described explicitly, or it may be *implied* by an equation used to define the function. For example, the function given by $y = 1/(x^2 - 4)$ has an implied domain that consists of all real $x$ except $x = \pm 2$. These two values are excluded from the domain because division by zero is undefined.

Another type of implied domain is that used to avoid even roots of negative numbers, as indicated in Example 2.

### EXAMPLE 2   Finding the Domain and Range of a Function

Find the domain and range of each of the following functions.

(a) $y = \sqrt{x - 1}$

(b) $y = \begin{cases} 1 - x, & x < 1 \\ \sqrt{x - 1}, & x \geq 1 \end{cases}$

**Solution**

(a) Because $\sqrt{x - 1}$ is not defined for $x - 1 < 0$ (that is, for $x < 1$), it follows that the domain of the function is the interval $x \geq 1$ or $[1, \infty)$. To find the range, observe that $\sqrt{x - 1}$ is never negative. Moreover, as $x$ takes on the various values in the domain, $y$ takes on all nonnegative values. Thus, the range is the interval $y \geq 0$ or $[0, \infty)$. The graph of the function, shown in Figure 1.41(a), confirms these results.

(b) Because this **compound** function is defined for $x < 1$ *and* for $x \geq 1$, the domain is the entire set of real numbers. When $x \geq 1$, the function behaves as in part a. For $x < 1$, the value of $1 - x$ is positive, and therefore the range of the function is $y \geq 0$ or $[0, \infty)$, as shown in Figure 1.41(b).

A function is **one-to-one** if to each value of the dependent variable in the range there corresponds exactly one value of the independent variable. For instance, the function in Example 2a is one-to-one, whereas the function in Example 2b is not one-to-one.

Geometrically, a function is one-to-one if every horizontal line intersects the graph of the function at most once. This geometrical interpretation is the **horizontal line test** for one-to-one functions. Thus, a graph that represents a one-to-one function must satisfy *both* the vertical line test and the horizontal line test.

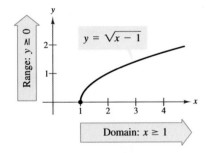

(a)

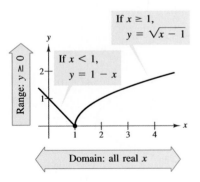

(b)

**FIGURE 1.41**

## Function Notation

When using an equation to define a function, you generally isolate the dependent variable on the left. For instance, writing the equation $x + 2y = 1$ as

$$y = \frac{1 - x}{2}$$

indicates that $y$ is the dependent variable. In **function notation,** this equation has the form

$$f(x) = \frac{1 - x}{2}. \qquad \text{Function notation}$$

The independent variable is $x$, and the name of the function is "$f$." The symbol $f(x)$ is read as "$f$ of $x$," and it denotes the value of the dependent variable. For instance, the value of $f$ when $x = 3$ is

$$f(3) = \frac{1 - (3)}{2} = \frac{-2}{2} = -1.$$

The value $f(3)$ is called a **function value,** and it lies in the range of $f$. This means that the point $(3, f(3))$ lies on the graph of $f$. One of the advantages of function notation is that it allows you to be less wordy. For instance, instead of asking "What is the value of $y$ when $x = 3$?" you can ask "What is $f(3)$?"

---

### EXAMPLE 3   Evaluating a Function

Find the value of the function

$$f(x) = 2x^2 - 4x + 1$$

when $x$ is $-1$, $0$, and $2$. Is $f$ one-to-one?

**Solution**

When $x = -1$, the value of $f$ is

$$f(-1) = 2(-1)^2 - 4(-1) + 1 = 2 + 4 + 1 = 7.$$

When $x = 0$, the value of $f$ is

$$f(0) = 2(0)^2 - 4(0) + 1 = 0 - 0 + 1 = 1.$$

When $x = 2$, the value of $f$ is

$$f(2) = 2(2)^2 - 4(2) + 1 = 8 - 8 + 1 = 1.$$

Because two different values of $x$ yield the same value of $f(x)$, the function is *not* one-to-one, as shown in Figure 1.42.

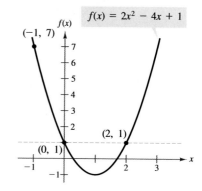

**FIGURE 1.42**

> **STUDY TIP**   You can use the horizontal line test to determine whether the function in Example 3 is one-to-one. Because the line $y = 1$ intersects the graph of the function twice, the function is *not* one-to-one.

## Technology

Most graphing utilities can be programmed to evaluate functions. The program depends on the calculator used. Here is a sample program for the *TI-82* or *TI-83*. See the appendix for other calculator models.

PROGRAM: EVALUATE
:Lbl 1
:Input "ENTER X", X
:Disp Y₁
:Goto 1

To use this program, enter a function in Y₁. Then run the program—it will allow you to evaluate the function at several values of *x*.

Example 3 suggests that the role of the variable $x$ in the equation

$$f(x) = 2x^2 - 4x + 1$$

is simply that of a placeholder. Informally, $f$ could be defined by the equation

$$f(\;\;\;) = 2(\;\;\;)^2 - 4(\;\;\;) + 1.$$

To evaluate $f(-2)$, simply place $-2$ in each set of parentheses.

$$f(-2) = 2(-2)^2 - 4(-2) + 1 = 8 + 8 + 1 = 17$$

The ratio in part b of the next example is called a *difference quotient*. In Section 2.1, you will see that it has special significance in calculus.

## EXAMPLE 4   Evaluating a Function

For the function $f(x) = x^2 - 4x + 7$, evaluate the following.

(a) $f(x + \Delta x)$     (b) $\dfrac{f(x + \Delta x) - f(x)}{\Delta x}$

### Solution

(a) To evaluate $f$ at $x + \Delta x$, substitute $x + \Delta x$ into each set of parentheses, as follows.

$$f(x + \Delta x) = (x + \Delta x)^2 - 4(x + \Delta x) + 7$$
$$= x^2 + 2x\Delta x + (\Delta x)^2 - 4x - 4\Delta x + 7$$

(b) Using the result of part a, you can write the following.

$$\frac{f(x + \Delta x) - f(x)}{\Delta x}$$
$$= \frac{[(x + \Delta x)^2 - 4(x + \Delta x) + 7] - [x^2 - 4x + 7]}{\Delta x}$$
$$= \frac{x^2 + 2x\Delta x + (\Delta x)^2 - 4x - 4\Delta x + 7 - x^2 + 4x - 7}{\Delta x}$$
$$= \frac{2x\Delta x + (\Delta x)^2 - 4\Delta x}{\Delta x}$$
$$= 2x + \Delta x - 4, \quad \Delta x \neq 0$$

Although $f$ is often used as a convenient function name and $x$ as the independent variable, you can use other symbols. For instance, the following equations all define the same function.

$$f(x) = x^2 - 4x + 7$$
$$f(t) = t^2 - 4t + 7$$
$$g(s) = s^2 - 4s + 7$$

## Combinations of Functions

Two functions can be combined in various ways to create new functions. For instance, if $f(x) = 2x - 3$ and $g(x) = x^2 + 1$, you can form the following functions.

$$f(x) + g(x) = (2x - 3) + (x^2 + 1) = x^2 + 2x - 2 \qquad \text{Sum}$$

$$f(x) - g(x) = (2x - 3) - (x^2 + 1) = -x^2 + 2x - 4 \qquad \text{Difference}$$

$$f(x)g(x) = (2x - 3)(x^2 + 1) = 2x^3 - 3x^2 + 2x - 3 \qquad \text{Product}$$

$$\frac{f(x)}{g(x)} = \frac{2x - 3}{x^2 + 1} \qquad \text{Quotient}$$

You can combine two functions in yet another way called the **composition.** The resulting function is a **composite function.**

---

### Definition of Composite Function

The function given by $(f \circ g)(x) = f(g(x))$ is the **composite** of $f$ with $g$. The **domain** of $(f \circ g)$ is the set of all $x$ in the domain of $g$ such that $g(x)$ is in the domain of $f$, as indicated in Figure 1.43.

---

The composite of $f$ with $g$ may not be equal to the composite of $g$ with $f$, as illustrated in the following example.

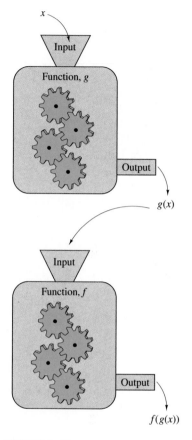

**FIGURE 1.43**

---

## EXAMPLE 5 Forming Composite Functions

Given $f(x) = 2x - 3$ and $g(x) = x^2 + 1$, find the following.

(a) $f(g(x))$  (b) $g(f(x))$

### Solution

(a) The composite of $f$ with $g$ is given by

$$f(g(x)) = 2(g(x)) - 3 \qquad \text{Evaluate } f \text{ at } g(x).$$

$$= 2(x^2 + 1) - 3 \qquad \text{Substitute } x^2 + 1 \text{ for } g(x).$$

$$= 2x^2 - 1. \qquad \text{Simplify.}$$

(b) The composite of $g$ with $f$ is given by

$$g(f(x)) = (f(x))^2 + 1 \qquad \text{Evaluate } g \text{ at } f(x).$$

$$= (2x - 3)^2 + 1 \qquad \text{Substitute } 2x - 3 \text{ for } f(x).$$

$$= 4x^2 - 12x + 10. \qquad \text{Simplify.}$$

## Inverse Functions

Informally, the inverse of a function $f$ is another function $g$ that "undoes" what $f$ has done.

$$x \quad\quad f(x) \quad\quad g(f(x)) = x$$

### Definition of Inverse Function

Two functions, $f$ and $g$, are **inverses** of each other if

$$f(g(x)) = x$$

for each $x$ in the domain of $g$ and

$$g(f(x)) = x$$

for each $x$ in the domain of $f$. The function $g$ is denoted by $f^{-1}$, which is read as "$f$ inverse." For $f$ and $g$ to be inverses of each other, the range of $g$ must be equal to the domain of $f$, and vice versa.

*ALGEBRA TIP* Don't be confused by the "exponential notation" for inverse functions. Whenever we write $f^{-1}(x)$, we will *always* be referring to the inverse of the function $f$, not to the reciprocal of $f$.

### EXAMPLE 6   Finding Inverse Functions

Several functions and their inverses are shown below. In each case, note that the inverse function "undoes" the original function. For instance, to undo multiplication by 2, you should divide by 2.

(a) $f(x) = 2x$      $f^{-1}(x) = \frac{1}{2}x$

(b) $f(x) = \frac{1}{3}x$      $f^{-1}(x) = 3x$

(c) $f(x) = x + 4$      $f^{-1}(x) = x - 4$

(d) $f(x) = 2x - 5$      $f^{-1}(x) = \frac{1}{2}(x + 5)$

(e) $f(x) = x^3$      $f^{-1}(x) = \sqrt[3]{x}$

(f) $f(x) = \dfrac{1}{x}$      $f^{-1}(x) = \dfrac{1}{x}$

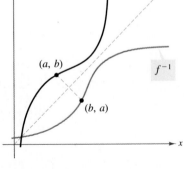

**FIGURE 1.44**   The graph of $f^{-1}$ is a reflection of the graph of $f$ in the line $y = x$.

The graphs of $f$ and $f^{-1}$ are mirror images of each other (with respect to the line $y = x$), as shown in Figure 1.44. Try using a graphing utility to confirm this for each of the functions given in Example 6.

The functions in Example 6 are simple enough so that their inverses can be found by inspection. The next example demonstrates a strategy for finding the inverses of more complicated functions.

## EXAMPLE 7   Finding the Inverse of a Function

Find the inverse of the function $f(x) = \sqrt{2x - 3}$.

**Solution**

Begin by substituting $y$ for $f(x)$. Then, interchange $x$ and $y$ and solve for $y$.

$$f(x) = \sqrt{2x - 3} \qquad \text{Original function}$$

$$y = \sqrt{2x - 3} \qquad \text{Substitute } y \text{ for } f(x).$$

$$x = \sqrt{2y - 3} \qquad \text{Interchange } x \text{ and } y.$$

$$x^2 = 2y - 3 \qquad \text{Square both sides.}$$

$$x^2 + 3 = 2y \qquad \text{Add 3 to both sides.}$$

$$\frac{x^2 + 3}{2} = y \qquad \text{Divide both sides by 2.}$$

Thus, the inverse function has the form

$$f^{-1}(\boxed{\phantom{x}}) = \frac{(\boxed{\phantom{x}})^2 + 3}{2}.$$

Using $x$ as the independent variable, you can write

$$f^{-1}(x) = \frac{x^2 + 3}{2}, \qquad x \geq 0.$$

In Figure 1.45, note that the domain of $f^{-1}$ coincides with the range of $f$.

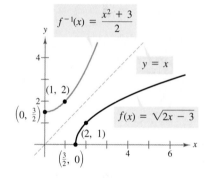

$$f^{-1}(x) = \frac{x^2 + 3}{2}$$

$$y = x$$

$$(1, 2)$$

$$\left(0, \tfrac{3}{2}\right)$$

$$f(x) = \sqrt{2x - 3}$$

$$(2, 1)$$

$$\left(\tfrac{3}{2}, 0\right)$$

**FIGURE 1.45**

After you have found the inverse of a function, you should check your results. You can check your results *graphically* by observing that the graphs of $f$ and $f^{-1}$ are reflections of each other in the line $y = x$. You can check your results *algebraically* by evaluating $f(f^{-1}(x))$ and $f^{-1}(f(x))$—both should be equal to $x$.

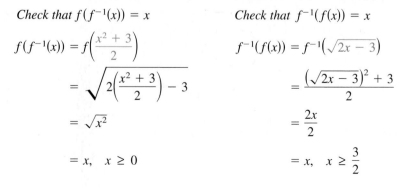

*Check that* $f(f^{-1}(x)) = x$

$$f(f^{-1}(x)) = f\!\left(\frac{x^2 + 3}{2}\right)$$

$$= \sqrt{2\!\left(\frac{x^2 + 3}{2}\right) - 3}$$

$$= \sqrt{x^2}$$

$$= x, \quad x \geq 0$$

*Check that* $f^{-1}(f(x)) = x$

$$f^{-1}(f(x)) = f^{-1}\!\left(\sqrt{2x - 3}\right)$$

$$= \frac{\left(\sqrt{2x - 3}\right)^2 + 3}{2}$$

$$= \frac{2x}{2}$$

$$= x, \quad x \geq \frac{3}{2}$$

### Technology

A graphing utility can help you check that the graphs of $f$ and $f^{-1}$ are reflections of each other in the line $y = x$. To do this, graph $y = f(x)$, $y = f^{-1}(x)$, and $y = x$ on the same viewing rectangle, using a *square setting*.

Not every function has an inverse. In fact, for a function to have an inverse, it must be one-to-one.

## EXAMPLE 8    A Function That Has No Inverse

Show that the function

$$f(x) = x^2 - 1$$

has no inverse. (Assume that the domain of $f$ is the set of all real numbers.)

**Solution**

Begin by sketching the graph of $f$, as shown in Figure 1.46. Note that

$$f(2) = (2)^2 - 1 = 3$$

and

$$f(-2) = (-2)^2 - 1 = 3.$$

Thus $f$ does not pass the horizontal line test, which implies that it is not one-to-one, and therefore has no inverse. The same conclusion can by obtained by trying to find the inverse of $f$.

| | |
|---|---|
| $f(x) = x^2 - 1$ | Original function |
| $y = x^2 - 1$ | Substitute $y$ for $f(x)$. |
| $x = y^2 - 1$ | Interchange $x$ and $y$. |
| $x + 1 = y^2$ | Add 1 to both sides. |
| $\pm\sqrt{x+1} = y$ | Take square root of both sides. |

The last equation does not define $y$ as a function of $x$, and thus $f$ has no inverse.

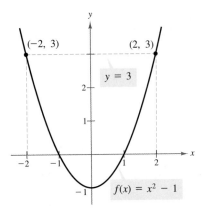

**FIGURE 1.46**    $f$ is not one-to-one and has no inverse.

*Group Discussion*    *Comparing Composition Orders*

You are buying an automobile whose price is $18,500. Which of the following options would you choose? Explain.

a.  You are given a factory rebate of $2000, followed by a dealer discount of 10%.

b.  You are given a dealer discount of 10%, followed by a factory rebate of $2000.

Let $f(x) = x - 2000$ and let $g(x) = 0.9x$. Which option is represented by the composite $f(g(x))$? Which is represented by the composite $g(f(x))$?

## Warm Up

The following warm-up exercises involve skills that were covered in earlier sections. You will use these skills in the exercise set for this section.

In Exercises 1–6, simplify the expression.

**1.** $5(-1)^2 - 6(-1) + 9$

**2.** $(-2)^3 + 7(-2)^2 - 10$

**3.** $(x - 2)^2 + 5x - 10$

**4.** $(3 - x) + (x + 3)^3$

**5.** $\dfrac{1}{1 - (1 - x)}$

**6.** $1 + \dfrac{x - 1}{x}$

In Exercises 7–10, solve for $y$ in terms of $x$.

**7.** $2x + y - 6 = 11$

**8.** $5y - 6x^2 - 1 = 0$

**9.** $(y - 3)^2 = 5 + (x + 1)^2$

**10.** $y^2 - 4x^2 = 2$

## EXERCISES 1.4

In Exercises 1–4, evaluate the function at the specified values of the independent variable. Simplify the result.

**1.** $f(x) = 2x - 3$

(a) $f(0)$  (b) $f(-3)$
(c) $f(x - 1)$  (d) $f(1 + \Delta x)$

**2.** $f(x) = x^2 - 2x + 2$

(a) $f\left(\frac{1}{2}\right)$  (b) $f(-1)$
(c) $f(c)$  (d) $f(x + \Delta x)$

**3.** $g(x) = \dfrac{1}{x}$

(a) $g(2)$  (b) $g\left(\frac{1}{4}\right)$
(c) $g(x + \Delta x)$  (d) $g(x + \Delta x) - g(x)$

**4.** $f(x) = |x| + 4$

(a) $f(2)$  (b) $f(-2)$
(c) $f(x^2)$  (d) $f(x + \Delta x) - f(x)$

In Exercises 5–8, evaluate the difference quotient and simplify the result.

**5.** $f(x) = 3x - 1$

$\dfrac{f(x + \Delta x) - f(x)}{\Delta x}$

**6.** $h(x) = x^2 - x + 1$

$\dfrac{h(2 + \Delta x) - h(2)}{\Delta x}$

**7.** $f(x) = x^3 - x$

$\dfrac{f(x + \Delta x) - f(x)}{\Delta x}$

**8.** $f(x) = \dfrac{1}{\sqrt{x - 1}}$

$\dfrac{f(x) - f(2)}{x - 2}$

In Exercises 9–16, decide whether the equation defines $y$ as a function of $x$.

**9.** $x^2 + y^2 = 4$

**10.** $3x - 2y + 5 = 0$

**11.** $\frac{1}{2}x - 6y = -3$

**12.** $x + y^2 = 4$

**13.** $x^2 + y = 4$

**14.** $x^2 + y^2 - 2x - 4y + 1 = 0$

**15.** $y^2 = x^2 - 1$

**16.** $x^2y - x^2 + 4y = 0$

 In Exercises 17–22, use a graphing utility to graph the function. Then determine the domain and range of the function.

**17.** $f(x) = \sqrt{9 - x^2}$

**18.** $f(x) = 5x^3 + 6x^2 - 1$

**19.** $f(x) = \dfrac{|x|}{x}$

**20.** $f(x) = \dfrac{1}{\sqrt{x}}$

**21.** $f(x) = \dfrac{x - 2}{x + 4}$

**22.** $f(x) = \dfrac{x^2}{1 - x}$

In Exercises 23–26, find the domain and range of the function. Use interval notation to write your result.

**23.** $f(x) = x^3$

**24.** $f(x) = \sqrt{2x - 3}$

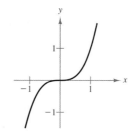

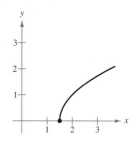

**25.** $f(x) = 4 - x^2$

**26.** $f(x) = |x - 2|$

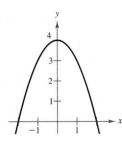

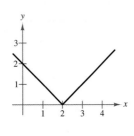

In Exercises 27–30, use the vertical line test to determine whether $y$ is a function of $x$.

**27.** $x^2 + y^2 = 9$

**28.** $x - xy + y + 1 = 0$

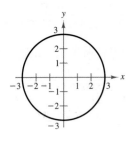

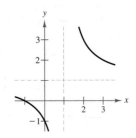

**29.** $x^2 = xy - 1$

**30.** $x = |y|$

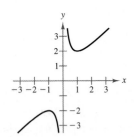

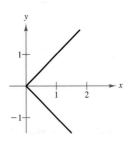

In Exercises 31–34, find (a) $f(x) + g(x)$, (b) $f(x) \cdot g(x)$, and (c) $f(x)/g(x)$. Also, find (d) $f(g(x))$ and (e) $g(f(x))$ if defined.

**31.** $f(x) = x^2 + 1$       $g(x) = x - 1$

**32.** $f(x) = 2x - 5$       $g(x) = 5$

**33.** $f(x) = x^2 + 5$       $g(x) = \sqrt{1 - x}$

**34.** $f(x) = \dfrac{x}{x + 1}$       $g(x) = x^3$

In Exercises 35–38, show that $f$ and $g$ are inverse functions by showing that $f(g(x)) = x$ and $g(f(x)) = x$. Then sketch the graphs of $f$ and $g$ on the same coordinate axes.

**35.** $f(x) = 5x + 1$;      $g(x) = \dfrac{x - 1}{5}$

**36.** $f(x) = \dfrac{1}{x}$;      $g(x) = \dfrac{1}{x}$

**37.** $f(x) = 9 - x^2$,   $x \geq 0$;     $g(x) = \sqrt{9 - x}$,   $x \leq 9$

**38.** $f(x) = 1 - x^3$;      $g(x) = \sqrt[3]{1 - x}$

In Exercises 39–46, find the inverse of $f$. Then sketch the graphs of $f$ and $f^{-1}$ on the same coordinate axes.

**39.** $f(x) = 2x - 3$        **40.** $f(x) = 6 - 3x$

**41.** $f(x) = x^5$           **42.** $f(x) = x^3 + 1$

**43.** $f(x) = \sqrt{9 - x^2}$,   $0 \leq x \leq 3$

**44.** $f(x) = \sqrt{x^2 - 4}$,   $x \geq 2$

**45.** $f(x) = x^{2/3}$,   $x \geq 0$

**46.** $f(x) = x^{3/5}$

In Exercises 47–52, use a graphing utility to graph the function. Then use the horizontal line test to decide whether the function is one-to-one. If it is, find its inverse.

**47.** $f(x) = 3 - 7x$       **48.** $f(x) = \sqrt{x - 2}$

**49.** $f(x) = x^2$           **50.** $f(x) = x^4$

**51.** $f(x) = |x - 2|$       **52.** $f(x) = 3$

**53.** Given $f(x) = \sqrt{x}$ and $g(x) = x^2 - 1$, find the composite functions.

   (a) $f(g(1))$            (b) $g(f(1))$
   (c) $g(f(0))$            (d) $f(g(-4))$
   (e) $f(g(x))$            (f) $g(f(x))$

**54.** Given $f(x) = 1/x$ and $g(x) = x^2 - 1$, find the composite functions.

   (a) $f(g(2))$            (b) $g(f(2))$
   (c) $f\left(g\left(1/\sqrt{2}\right)\right)$      (d) $g\left(f\left(1/\sqrt{2}\right)\right)$
   (e) $f(g(x))$            (f) $g(f(x))$

In Exercises 55–58, select a function from (a) $f(x) = cx$, (b) $g(x) = cx^2$, (c) $h(x) = c\sqrt{|x|}$, and (d) $r(x) = c/x$ and determine the value of the constant $c$ such that the function fits the data in the table.

**55.**

| $x$ | $-4$ | $-1$ | $0$ | $1$ | $4$ |
|---|---|---|---|---|---|
| $y$ | $-32$ | $-2$ | $0$ | $-2$ | $-32$ |

**56.**

| x | −4 | −1 | 0 | 1 | 4 |
|---|---|---|---|---|---|
| y | −1 | −$\frac{1}{4}$ | 0 | $\frac{1}{4}$ | 1 |

**57.**

| x | −4 | −1 | 0 | 1 | 4 |
|---|---|---|---|---|---|
| y | −8 | −32 | undef. | 32 | 8 |

**58.**

| x | −4 | −1 | 0 | 1 | 4 |
|---|---|---|---|---|---|
| y | 6 | 3 | 0 | 3 | 6 |

**59.** Use the graph of $f(x) = \sqrt{x}$ below to sketch the graph of each function.

(a) $y = \sqrt{x} + 2$
(b) $y = -\sqrt{x}$
(c) $y = \sqrt{x - 2}$
(d) $y = \sqrt{x + 3}$
(e) $y = \sqrt{x - 4}$
(f) $y = 2\sqrt{x}$

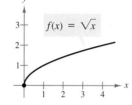

**60.** Use the graph of $f(x) = |x|$ below to sketch the graph of each function.

(a) $y = |x| + 3$
(b) $y = -\frac{1}{2}|x|$
(c) $y = |x - 2|$
(d) $y = |x + 1| - 1$
(e) $y = 2|x|$

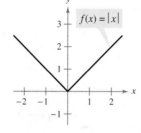

**61.** Use the graph of $f(x) = x^2$ to find a formula for each of the functions whose graphs are shown.

(a)

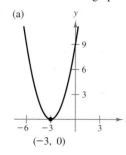

(−3, 0)

(b)

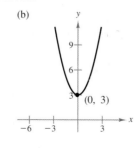

(0, 3)

(c)

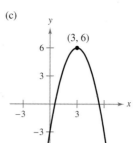

(3, 6)

(d)

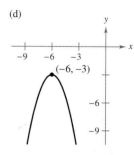

(−6, −3)

**62.** *Real Estate* Express the value $V$ of a real estate firm in terms of $x$, the number of acres of property owned. Each acre is valued at $2500 and other company assets total $750,000.

**63.** *Cost* The inventor of a new game believes that the variable cost for producing the game is $0.95 per unit. The fixed cost is $6000.

(a) Express the total cost $C$ as a function of $x$, the number of games sold.
(b) Find a formula for the average cost per unit $\overline{C} = C/x$.
(c) The selling price for each game is $1.69. How many units must be sold before the average cost per unit falls below the selling price?

**64.** *Demand* The demand function for a commodity is

$$p = \frac{14.75}{1 + 0.01x}, \qquad x \geq 0,$$

where $p$ is the price per unit and $x$ is the number of units sold.

(a) Find $x$ as a function of $p$.
(b) Find the number of units sold when the price is $10.

**65.** *Cost* A power station is on one side of a river that is $\frac{1}{2}$ mile wide. A factory is 3 miles downstream on the other side of the river (see figure). It costs $10/ft to run the power lines on land and $15/ft to run them underwater. Express the cost $C$ of running the lines from the power station to the factory as a function of $x$.

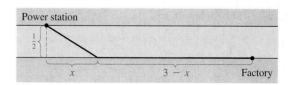

**66.** *Market Equilibrium*   The supply function for a product relates the number of units $x$ that producers are willing to supply for a given price per unit $p$. The supply and demand functions for a market are

$$p = \frac{2}{5}x + 4 \qquad \text{Supply}$$

$$p = -\frac{16}{25}x + 30. \qquad \text{Demand}$$

(a) Use a graphing utility to graph the supply and demand functions in the same viewing rectangle.

(b) Use the trace feature of the graphing utility to find the *equilibrium point* for the market. That is, find the point of intersection of the two graphs.

(c) For what values of $x$ does the demand exceed the supply?

(d) For what values of $x$ does the supply exceed the demand?

**67.** *Profit*   A radio manufacturer charges $90 per unit for units that cost $60 to produce. To encourage large orders from distributors, the manufacturer will reduce the price by $0.01 per unit for each unit in excess of 100 units. (For example, an order of 101 units would have a price of $89.99 per unit, and an order of 102 units would have a price of $89.98 per unit.) This price reduction is discontinued when the price per unit drops to $75.

(a) Express the price per unit $p$ as a function of the order size $x$.

(b) Express the profit $P$ as a function of the order size $x$.

**68.** *Cost, Revenue, and Profit*   A company invests $98,000 for equipment to produce a new product. Each unit of the product costs $12.30 and is sold for $17.98. Let $x$ be the number of units produced and sold.

(a) Write the total cost $C$ as a function of $x$.

(b) Write the revenue $R$ as a function of $x$.

(c) Write the profit $P$ as a function of $x$.

**69.** *Charter Bus Fares*   For groups of 80 or more people, a charter bus company determines the rate $r$ (in dollars per person) according to the formula $r = 8 - 0.05(n - 80)$, $n \geq 80$, where $n$ is the number of people.

(a) Express the revenue $R$ for the bus company as a function of $n$.

(b) Complete the table.

| $n$ | 90 | 100 | 110 | 120 | 130 | 140 | 150 |
|-----|----|-----|-----|-----|-----|-----|-----|
| $R$ |    |     |     |     |     |     |     |

(c) Criticize the formula for the rate. Would you use this formula? Explain your reasoning.

**70.** *Medicine*   The temperature of a patient after being given a fever-reducing drug is given by

$$F(t) = 98 + \frac{3}{t + 1},$$

where $F$ is the temperature in degrees Fahrenheit and $t$ is the time in hours. Use a graphing utility to graph the function. Be sure to choose an appropriate window setting. For what values of $t$ do you think this function would be valid? Explain.

In Exercises 71–76, use a graphing utility to graph the function. Then use the zoom and trace features to find the zeros of the function. Is the function one-to-one?

**71.** $f(x) = x\sqrt{9 - x^2}$

**72.** $f(x) = 2\left(3x^2 - \frac{6}{x}\right)$

**73.** $g(t) = \frac{t + 3}{1 - t}$

**74.** $h(x) = 6x^3 - 12x^2 + 4$

**75.** $f(x) = \frac{4 - x^2}{x}$

**76.** $g(x) = \left|\frac{1}{2}x^2 - 4\right|$

## Business Capsule

Cardsenders

Cardsenders *is a home-based greeting card service for businesses. It is owned by Phyllis Boverie who started the company in 1985. Start-up costs for this business ran from $5,000 to $10,000.*

**77.** *Research Project*   Use your school's library or some other reference source to find information about the start-up costs of beginning a business, such as in the business example above. Write a short paper about the company.

## 1.5 | Limits

*The Limit of a Function • Properties of Limits • Techniques for Evaluating Limits • One-Sided Limits • Unbounded Behavior*

## The Limit of a Function

In everyday language, people refer to a speed limit, a wrestler's weight limit, the limit of one's endurance, or stretching a spring to its limit. These phrases all suggest that a limit is a bound, which on some occasions may not be reached but on other occasions may be reached or exceeded.

Consider a spring that will break only if a weight of 10 pounds or more is attached. To determine how far the spring will stretch without breaking, you could attach increasingly heavier weights and measure the spring length $s$ for each weight $w$, as shown in Figure 1.47. If the spring length approaches a value of $L$, then we say that "the limit of $s$ as $w$ approaches 10 is $L$." A mathematical limit is much like the limit of a spring. The notation for a limit is

$$\lim_{x \to c} f(x) = L,$$

which is read as "the limit of $f(x)$ as $x$ approaches $c$ is $L$."

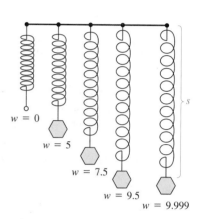

**FIGURE 1.47**  What is the limit of $s$ as $w$ approaches 10 lb?

---

## EXAMPLE 1  Finding a Limit

Find the limit: $\lim_{x \to 1} (x^2 + 1)$.

### Solution

Let $f(x) = x^2 + 1$. From the graph of $f$ in Figure 1.48, it appears that $f(x)$ approaches 2 as $x$ approaches 1 from either side, and you can write

$$\lim_{x \to 1} (x^2 + 1) = 2.$$

The table yields the same conclusion. Notice that as $x$ gets closer and closer to 1, $f(x)$ gets closer and closer to 2.

|       | *x* approaches 1 | | | | *x* approaches 1 | | |
|-------|-------|-------|-------|-------|-------|-------|-------|
| $x$    | 0.900 | 0.990 | 0.999 | 1.000 | 1.001 | 1.010 | 1.100 |
| $f(x)$ | 1.810 | 1.980 | 1.998 | 2.000 | 2.002 | 2.020 | 2.200 |

*f(x)* approaches 2         *f(x)* approaches 2

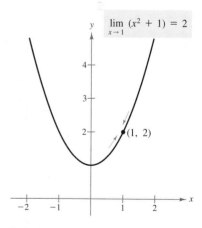

$$\lim_{x \to 1} (x^2 + 1) = 2$$

**FIGURE 1.48**

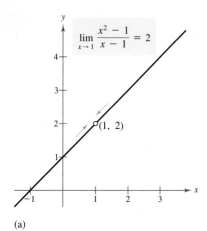

(a)

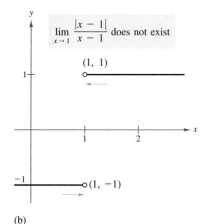

(b)

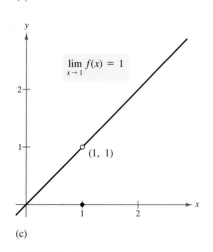

(c)

**FIGURE 1.49**

## EXAMPLE 2 Finding Limits Graphically and Numerically

Find the limit: $\lim_{x \to 1} f(x)$.

(a) $f(x) = \dfrac{x^2 - 1}{x - 1}$    (b) $f(x) = \dfrac{|x - 1|}{x - 1}$    (c) $f(x) = \begin{cases} x, & x \neq 1 \\ 0, & x = 1 \end{cases}$

**Solution**

(a) From the graph of $f$, in Figure 1.49(a), it appears that $f(x)$ approaches 2 as $x$ approaches 1 from either side. A missing point is denoted by the open dot on the graph. This conclusion is reinforced by the table. Be sure you see that *it does not matter that $f(x)$ is undefined when $x = 1$. The limit depends only on values of $f(x)$ near 1, not at 1.*

| | x approaches 1 | | | | x approaches 1 | | |
|---|---|---|---|---|---|---|---|
| $x$ | 0.900 | 0.990 | 0.999 | 1.000 | 1.001 | 1.010 | 1.100 |
| $f(x)$ | 1.900 | 1.990 | 1.999 | ? | 2.001 | 2.010 | 2.100 |

$f(x)$ approaches 2        $f(x)$ approaches 2

(b) From the graph of $f$, in Figure 1.49(b), you can see that $f(x) = -1$ for all values to the left of $x = 1$ and $f(x) = 1$ for all values to the right of $x = 1$. Thus, $f(x)$ is approaching a different value from the left of $x = 1$ than it is from the right of $x = 1$. In such situations, we say that the *limit does not exist*. This conclusion is reinforced by the table.

| | x approaches 1 | | | | x approaches 1 | | |
|---|---|---|---|---|---|---|---|
| $x$ | 0.900 | 0.990 | 0.999 | 1.000 | 1.001 | 1.010 | 1.100 |
| $f(x)$ | $-1.000$ | $-1.000$ | $-1.000$ | ? | 1.000 | 1.000 | 1.000 |

$f(x)$ approaches $-1$        $f(x)$ approaches 1

(c) From the graph of $f$, in Figure 1.49(c), it appears that $f(x)$ approaches 1 as $x$ approaches 1 from either side. This conclusion is reinforced by the table. It does not matter that $f(1) = 0$. The limit depends only on values of $f(x)$ near 1, not at 1.

| | x approaches 1 | | | | x approaches 1 | | |
|---|---|---|---|---|---|---|---|
| $x$ | 0.900 | 0.990 | 0.999 | 1.000 | 1.001 | 1.010 | 1.100 |
| $f(x)$ | 0.900 | 0.990 | 0.999 | 0.000 | 1.001 | 1.010 | 1.100 |

$f(x)$ approaches 1        $f(x)$ approaches 1

There are three important ideas to learn from Examples 1 and 2.

1. Saying that the limit of $f(x)$ approaches $L$ as $x$ approaches $c$ means that the value of $f(x)$ may be made *arbitrarily close* to the number $L$ by choosing $x$ closer and closer to $c$.

2. For a limit to exist, you must allow $x$ to approach $c$ from *either side* of $c$. If $f(x)$ approaches a different number as $x$ approaches $c$ from the left than it does as $x$ approaches $c$ from the right, then the limit *does not exist*. (See Example 2b.)

3. The value of $f(x)$ when $x = c$ has no bearing on the existence or nonexistence of the limit of $f(x)$ as $x$ approaches $c$. For instance, in Example 2a, the limit of $f(x)$ exists as $x$ approaches 1 even though the function $f$ is not defined at $x = 1$.

### Technology

Try using a graphing utility to determine the following limit.

$$\lim_{x \to 1} \frac{x^3 + 4x - 5}{x - 1}$$

You can do this by graphing

$$f(x) = \frac{x^3 + 4x - 5}{x - 1}$$

and zooming in near $x = 1$. From the graph, what does the limit appear to be?

---

### Definition of the Limit of a Function

If $f(x)$ becomes arbitrarily close to a single number $L$ as $x$ approaches $c$ from either side, then

$$\lim_{x \to c} f(x) = L,$$

which is read as "the **limit** of $f(x)$ as $x$ approaches $c$ is $L$."

## Properties of Limits

Many times the limit of $f(x)$ as $x$ approaches $c$ is simply $f(c)$, as illustrated in Example 1. Whenever the limit of $f(x)$ as $x$ approaches $c$ is

$$\lim_{x \to c} f(x) = f(c), \qquad \text{Direct substitution}$$

we say that the limit can be evaluated by **direct substitution.** (In the next section, you will learn that a function that has this property is *continuous at c*.) It is important that you learn to recognize the types of functions that have this property. Some basic ones are given in the following list.

---

### Properties of Limits

Let $b$ and $c$ be real numbers, and let $n$ be a positive integer.

1. $\lim\limits_{x \to c} b = b$ 2. $\lim\limits_{x \to c} x = c$

3. $\lim\limits_{x \to c} x^n = c^n$ 4. $\lim\limits_{x \to c} \sqrt[n]{x} = \sqrt[n]{c}$

In Property 4, if $n$ is even, then $c$ must be positive.

## Technology

Symbolic algebra utilities, such as *Derive, Mathematica, Mathcad, Maple,* and the *TI-92,* are capable of evaluating limits. For instance, *Derive* for Windows can evaluate the limit in Example 3 as follows.

#1: $x^2 + 2x - 3$       Given

#2: $\lim_{x \to 2} (x^2 + 2x - 3)$    Limit

#3: 5                     Simplify.

## *Discovery*

Use a graphing utility to graph $y_1 = 1/x^2$. Does $y_1$ approach a limit as $x$ approaches 0? Evaluate $y_1 = 1/x^2$ at several positive and negative values of $x$ near 0 to confirm your answer. Does $\lim_{x \to 1} 1/x^2$ exist?

By combining the properties of limits with the following rules for operating with limits, you can find limits for a wide variety of algebraic functions.

### Operations with Limits

Let $b$ and $c$ be real numbers and let $n$ be a positive integer. If the limits of $f(x)$ and $g(x)$ exist as $x$ approaches $c$, then the following operations are valid.

1. Constant multiple: $\lim_{x \to c} [bf(x)] = b\left[\lim_{x \to c} f(x)\right]$

2. Addition: $\lim_{x \to c} [f(x) \pm g(x)] = \lim_{x \to c} f(x) \pm \lim_{x \to c} g(x)$

3. Multiplication: $\lim_{x \to c} [f(x) \cdot g(x)] = \left[\lim_{x \to c} f(x)\right]\left[\lim_{x \to c} g(x)\right]$

4. Division: $\lim_{x \to c} \dfrac{f(x)}{g(x)} = \dfrac{\lim_{x \to c} f(x)}{\lim_{x \to c} g(x)}, \qquad \lim_{x \to c} g(x) \neq 0$

5. Power: $\lim_{x \to c} [f(x)]^n = \left[\lim_{x \to c} f(x)\right]^n$

6. Radical: $\lim_{x \to c} \sqrt[n]{f(x)} = \sqrt[n]{\lim_{x \to c} f(x)}$

In Property 6, if $n$ is even and $\lim_{x \to c} f(x) = L$, then $L$ must be positive.

---

### EXAMPLE 3    Evaluating the Limit of a Polynomial

$$\lim_{x \to 2} (x^2 + 2x - 3) = \lim_{x \to 2} x^2 + \lim_{x \to 2} 2x - \lim_{x \to 2} 3 \qquad \text{Addition Property}$$

$$= 2^2 + 2(2) - 3 \qquad \text{Direct substitution}$$

$$= 4 + 4 - 3 \qquad \text{Simplify.}$$

$$= 5$$

---

Example 3 is an illustration of the following important result, which states that the limit of a polynomial can be evaluated by direct substitution.

### The Limit of a Polynomial Function

If $p$ is a polynomial function and $c$ is any real number, then

$$\lim_{x \to c} p(x) = p(c).$$

## Techniques for Evaluating Limits

Many techniques for evaluating limits are based on the following important theorem. Basically, the theorem states that if two functions agree at all but a single point $c$, then they have identical limit behavior at $x = c$.

---

### The Replacement Theorem

Let $c$ be a real number and $f(x) = g(x)$ for all $x \neq c$. If the limit of $g(x)$ exists as $x \to c$, then the limit of $f(x)$ also exists and

$$\lim_{x \to c} f(x) = \lim_{x \to c} g(x).$$

---

To apply the Replacement Theorem, you can use a result from algebra which states that for a polynomial function $p$, $p(c) = 0$ if and only if $(x - c)$ is a factor of $p(x)$. This concept is demonstrated in Example 4.

---

### EXAMPLE 4   Evaluating a Limit

Evaluate the limit: $\lim\limits_{x \to 1} \dfrac{x^3 - 1}{x - 1}$.

#### Solution

Note that the numerator and denominator are zero when $x = 1$. This implies that $(x - 1)$ is a factor of both, and you can cancel this common factor.

$$\frac{x^3 - 1}{x - 1} = \frac{(x - 1)(x^2 + x + 1)}{x - 1} \qquad \text{Factor numerator.}$$

$$= \frac{(x - 1)(x^2 + x + 1)}{x - 1} \qquad \text{Cancel common factor.}$$

$$= x^2 + x + 1, \quad x \neq 1 \qquad \text{Simplify.}$$

Thus, the rational function $(x^3 - 1)/(x - 1)$ and the polynomial function $x^2 + x + 1$ agree for all values of $x$ other than $x = 1$, and you can apply the Replacement Theorem.

$$\lim_{x \to 1} \frac{x^3 - 1}{x - 1} = \lim_{x \to 1} (x^2 + x + 1) = 1^2 + 1 + 1 = 3$$

Figure 1.50 illustrates this result graphically. Note that the two graphs are identical except that the graph of $g$ contains the point $(1, 3)$, whereas this point is missing on the graph of $f$. (In Figure 1.50, the missing point is denoted by an open dot.)

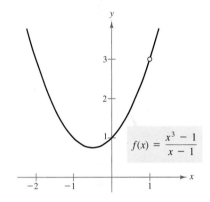

$$f(x) = \frac{x^3 - 1}{x - 1}$$

$$g(x) = x^2 + x + 1$$

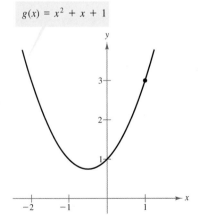

**FIGURE 1.50**

The technique used to evaluate the limit in Example 4 is called the **cancellation technique.** This technique is further demonstrated in the next example.

*Discovery*

Use a graphing utility to graph

$$y = \frac{x^2 + x - 6}{x + 3}.$$

Is the graph a line? Why or why not?

### EXAMPLE 5  Using the Cancellation Technique

Evaluate the limit:  $\lim\limits_{x \to -3} \dfrac{x^2 + x - 6}{x + 3}$.

**Solution**

Direct substitution fails because both the numerator and the denominator are zero when $x = -3$.

$$\lim_{x \to -3} \frac{x^2 + x - 6}{x + 3} \quad \begin{array}{l} \longleftarrow\ \lim\limits_{x \to -3} (x^2 + x - 6) = 0 \\ \longleftarrow\ \lim\limits_{x \to -3} (x + 3) = 0 \end{array}$$

However, because the limits of both the numerator and denominator are zero, you know that they have a *common factor* of $(x + 3)$. Thus, for all $x \neq -3$, you can cancel this factor to obtain the following.

$$\lim_{x \to -3} \frac{x^2 + x - 6}{x + 3} = \lim_{x \to -3} \frac{(x - 2)(x + 3)}{x + 3} \qquad \text{Factor numerator.}$$

$$= \lim_{x \to -3} \frac{(x - 2)(x + 3)}{x + 3} \qquad \text{Cancel common factor.}$$

$$= \lim_{x \to -3} (x - 2) \qquad \text{Simplify.}$$

$$= -5 \qquad \text{Direct substitution}$$

This result is shown graphically in Figure 1.51. Note that the graph of $f$ coincides with the graph of $g(x) = x - 2$, except that the graph of $f$ has a hole at $(-3, -5)$.

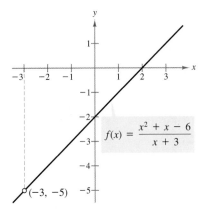

$$f(x) = \frac{x^2 + x - 6}{x + 3}$$

$(-3, -5)$

**FIGURE 1.51**  *f* is undefined when $x = -3$.

### EXAMPLE 6  Evaluating a Limit

Evaluate the limit:  $\lim\limits_{x \to -2} \dfrac{x^2 + x - 2}{x - 2}$.

**Solution**

Because the limit of the denominator is not zero, you can use the limit of a quotient (Property 4 of Operations with Limits) to obtain

$$\lim_{x \to -2} \frac{x^2 + x - 2}{x - 2} = \lim_{x \to -2} \frac{(-2)^2 - 2 - 2}{-2 - 2} \qquad \text{Direct substitution}$$

$$= \frac{0}{-4} \qquad \text{Simplify.}$$

$$= 0. \qquad \text{The limit is 0.}$$

## One-Sided Limits

In Example 2b, you saw that one way in which a limit can fail to exist is when a function approaches a different value from the left of $c$ than it approaches from the right of $c$. This type of behavior can be described more concisely with the concept of a **one-sided limit.**

$$\lim_{x \to c^-} f(x) = L \qquad \text{Limit from the left}$$

$$\lim_{x \to c^+} f(x) = L \qquad \text{Limit from the right}$$

The first of these two limits is read as "the limit of $f(x)$ as $x$ approaches $c$ from the left is $L$." The second is read as "the limit of $f(x)$ as $x$ approaches $c$ from the right is $L$."

### EXAMPLE 7   Evaluating One-Sided Limits

Find the limit as $x \to 0$ from the left and the limit as $x \to 0$ from the right for the function

$$f(x) = \frac{|2x|}{x}.$$

*Solution*

From the graph of $f$, shown in Figure 1.52, you can see that $f(x) = -2$ for all $x < 0$. Therefore, the limit from the left is

$$\lim_{x \to 0^-} \frac{|2x|}{x} = -2. \qquad \text{Limit from the left}$$

Because $f(x) = 2$ for all $x > 0$, the limit from the right is

$$\lim_{x \to 0^+} \frac{|2x|}{x} = 2. \qquad \text{Limit from the right}$$

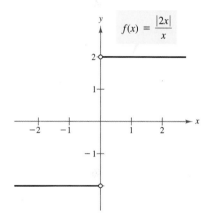

**FIGURE 1.52**

In Example 7, note that the function approaches different limits from the left and from the right. In such cases, the limit of $f(x)$ as $x \to c$ does not exist. For the limit of a function to exist as $x \to c$, *both* one-sided limits must exist and must be equal.

---

### Existence of a Limit

---

If $f$ is a function and $c$ and $L$ are real numbers, then

$$\lim_{x \to c} f(x) = L$$

if and only if both the left and right limits are equal to $L$.

 **Technology**

On most graphing utilities, the absolute value function is denoted by *abs*. You can verify the result in Example 7 by graphing

$$y = \frac{abs(2x)}{x}$$

in the viewing rectangle $-3 \le x \le 3$ and $-3 \le y \le 3$.

## EXAMPLE 8   Evaluating One-Sided Limits

Find the limit of $f(x)$ as $x$ approaches 1.

$$f(x) = \begin{cases} 4 - x, & x < 1 \\ 4x - x^2, & x > 1 \end{cases}$$

**Solution**

Remember that you are concerned about the value of $f$ near $x = 1$ rather than at $x = 1$. Thus, for $x < 1$, $f(x)$ is given by $4 - x$, and you can use direct substitution to obtain

$$\lim_{x \to 1^-} f(x) = \lim_{x \to 1^-} (4 - x) = 4 - 1 = 3.$$

For $x > 1$, $f(x)$ is given by $4x - x^2$, and you can use direct substitution to obtain

$$\lim_{x \to 1^+} f(x) = \lim_{x \to 1^+} (4x - x^2) = 4(1) - 1^2 = 4 - 1 = 3.$$

Because both one-sided limits exist and are equal to 3, it follows that

$$\lim_{x \to 1} f(x) = 3.$$

The graph in Figure 1.53 confirms this conclusion.

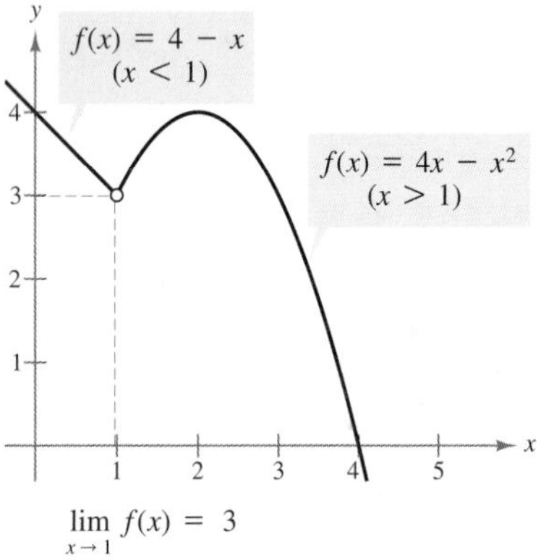

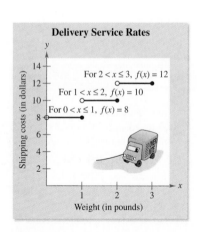

$f(x) = 4 - x$
$(x < 1)$

$f(x) = 4x - x^2$
$(x > 1)$

$$\lim_{x \to 1} f(x) = 3$$

**FIGURE 1.53**

## EXAMPLE 9   Comparing One-Sided Limits

An overnight delivery service charges \$8 for the first pound and \$2 for each additional pound. Let $x$ represent the weight of a parcel and let $f(x)$ represent the shipping cost.

$$f(x) = \begin{cases} 8, & 0 < x \le 1 \\ 10, & 1 < x \le 2 \\ 12, & 2 < x \le 3 \end{cases}$$

Show that the limit of $f(x)$ as $x \to 2$ does not exist.

**Solution**

The graph of $f$ is shown in Figure 1.54. The limit of $f(x)$ as $x$ approaches 2 from the left is

$$\lim_{x \to 2^-} f(x) = 10,$$

whereas the limit of $f(x)$ as $x$ approaches 2 from the right is

$$\lim_{x \to 2^+} f(x) = 12.$$

Because these one-sided limits are not equal, the limit of $f(x)$ as $x \to 2$ does not exist.

**Delivery Service Rates**

Shipping costs (in dollars)

For $2 < x \le 3$, $f(x) = 12$
For $1 < x \le 2$, $f(x) = 10$
For $0 < x \le 1$, $f(x) = 8$

Weight (in pounds)

**FIGURE 1.54**

# Unbounded Behavior

Example 9 illustrates a limit that fails to exist because the limits from the left and right differ. Another important way in which a limit can fail to exist is when $f(x)$ increases or decreases without bound as $x$ approaches $c$.

## EXAMPLE 10 An Unbounded Function

Evaluate the limit (if possible): $\lim\limits_{x \to 2} \dfrac{3}{x-2}$.

*Solution*

From Figure 1.55, you can see that $f(x)$ decreases without bound as $x$ approaches 2 from the left and $f(x)$ increases without bound as $x$ approaches 2 from the right. Symbolically, you can write this as

$$\lim_{x \to 2^-} \frac{3}{x-2} = -\infty$$

and

$$\lim_{x \to 2^+} \frac{3}{x-2} = \infty.$$

Because $f$ is unbounded as $x$ approaches 2, the limit does not exist.

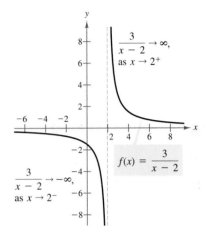

**FIGURE 1.55**

> **STUDY TIP**  The equal sign in the statement $\lim\limits_{x \to c^+} f(x) = \infty$ does not mean that the limit exists. On the contrary, it tells you how the limit *fails to exist* by denoting the unbounded behavior of $f(x)$ as $x$ approaches $c$.

---

## *Group Discussion*        ***Evaluating a Limit***

Consider the following limit.

$$\lim_{x \to 0} \frac{\sqrt{x+1} - 1}{x}$$

a. Approximate the limit graphically, using a graphing utility.

b. Approximate the limit numerically, by constructing a table.

c. Evaluate the limit analytically, using the following equation.

$$\frac{\sqrt{x+1} - 1}{x} = \left(\frac{\sqrt{x+1} - 1}{x}\right)\left(\frac{\sqrt{x+1} + 1}{\sqrt{x+1} + 1}\right) = \frac{1}{\sqrt{x+1} + 1}$$

Which method do you prefer? Explain.

*Warm Up*

The following warm-up exercises involve skills that were covered in earlier sections. You will use these skills in the exercise set for this section.

In Exercises 1–4, evaluate the expression and simplify.

**1.** $f(x) = x^2 - 3x + 3$      (a) $f(-1)$     (b) $f(c)$     (c) $f(x + h)$

**2.** $f(x) = \begin{cases} 2x - 2, & x < 1 \\ 3x + 1, & x \geq 1 \end{cases}$      (a) $f(-1)$     (b) $f(3)$     (c) $f(t^2 + 1)$

**3.** $f(x) = x^2 - 2x + 2$      $\dfrac{f(1 + h) - f(1)}{h}$

**4.** $f(x) = 4x$      $\dfrac{f(2 + h) - f(2)}{h}$

In Exercises 5–8, find the domain and range of the function and sketch its graph.

**5.** $h(x) = -\dfrac{5}{x}$             **6.** $g(x) = \sqrt{25 - x^2}$

**7.** $f(x) = |x - 3|$          **8.** $f(x) = \dfrac{|x|}{x}$

In Exercises 9 and 10, determine whether $y$ is a function of $x$.

**9.** $9x^2 + 4y^2 = 49$           **10.** $2x^2y + 8x = 7y$

## EXERCISES 1.5

In Exercises 1–8, complete the table and estimate the limit.

**1.** $\lim\limits_{x \to 2} (5x + 4)$

| $x$ | 1.9 | 1.99 | 1.999 | 2 | 2.001 | 2.01 | 2.1 |
|---|---|---|---|---|---|---|---|
| $f(x)$ | | | | | | | |

**2.** $\lim\limits_{x \to 2} \dfrac{x - 2}{x^2 - x - 2}$

| $x$ | 1.9 | 1.99 | 1.999 | 2 | 2.001 | 2.01 | 2.1 |
|---|---|---|---|---|---|---|---|
| $f(x)$ | | | | | | | |

**3.** $\lim\limits_{x \to 2} \dfrac{x - 2}{x^2 - 4}$

| $x$ | 1.9 | 1.99 | 1.999 | 2 | 2.001 | 2.01 | 2.1 |
|---|---|---|---|---|---|---|---|
| $f(x)$ | | | | | | | |

**4.** $\lim\limits_{x \to 2} \dfrac{x^5 - 32}{x - 2}$

| $x$ | 1.9 | 1.99 | 1.999 | 2 | 2.001 | 2.01 | 2.1 |
|---|---|---|---|---|---|---|---|
| $f(x)$ | | | | | | | |

**5.** $\lim\limits_{x \to 0} \dfrac{\sqrt{x + 3} - \sqrt{3}}{x}$

| $x$ | $-0.1$ | $-0.01$ | $-0.001$ | 0 | 0.001 | 0.01 | 0.1 |
|---|---|---|---|---|---|---|---|
| $f(x)$ | | | | | | | |

**6.** $\lim\limits_{x \to 0} \dfrac{\sqrt{x + 2} - \sqrt{2}}{x}$

| $x$ | $-0.1$ | $-0.01$ | $-0.001$ | 0 | 0.001 | 0.01 | 0.1 |
|---|---|---|---|---|---|---|---|
| $f(x)$ | | | | | | | |

**7.** $\lim\limits_{x\to2^-} \dfrac{2 - x}{\sqrt{4 - x^2}}$

| $x$ | 1.5 | 1.9 | 1.99 | 1.999 | 2 |
|---|---|---|---|---|---|
| $f(x)$ | | | | | |

**8.** $\lim\limits_{x\to0^+} \dfrac{[1/(2 + x)] - (1/2)}{2x}$

| $x$ | 0.5 | 0.1 | 0.01 | 0.001 | 0 |
|---|---|---|---|---|---|
| $f(x)$ | | | | | |

In Exercises 9–12, use the graph to determine the limit visually.

**9.**

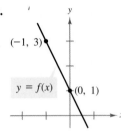

**10.**

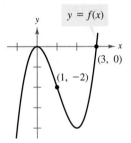

(a) $\lim\limits_{x\to0} f(x)$

(b) $\lim\limits_{x\to-1} f(x)$

(a) $\lim\limits_{x\to1} f(x)$

(b) $\lim\limits_{x\to3} f(x)$

**11.**

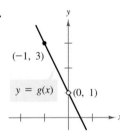

**12.**

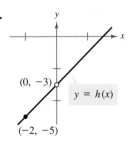

(a) $\lim\limits_{x\to0} g(x)$

(b) $\lim\limits_{x\to-1} g(x)$

(a) $\lim\limits_{x\to-2} h(x)$

(b) $\lim\limits_{x\to0} h(x)$

In Exercises 13 and 14, find the limit of (a) $f(x) + g(x)$, (b) $f(x)g(x)$, and (c) $f(x)/g(x)$ as $x$ approaches $c$.

**13.** $\lim\limits_{x\to c} f(x) = 3$

$\lim\limits_{x\to c} g(x) = 9$

**14.** $\lim\limits_{x\to c} f(x) = \frac{3}{2}$

$\lim\limits_{x\to c} g(x) = \frac{1}{2}$

In Exercises 15–20, use the graph to determine the limit visually (if it exists).

(a) $\lim\limits_{x\to c^+} f(x)$

(b) $\lim\limits_{x\to c^-} f(x)$

(c) $\lim\limits_{x\to c} f(x)$

**15.**

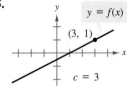

**16.**

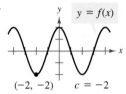

**17.**

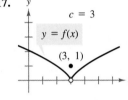

**18.**

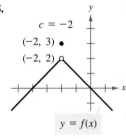

**19.**

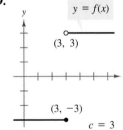

**20.**

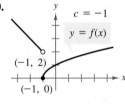

In Exercises 21–30, find the limit.

**21.** $\lim\limits_{x\to-3} (3x + 2)$

**22.** $\lim\limits_{x\to0} (2x - 3)$

**23.** $\lim\limits_{x\to1} (1 - x^2)$

**24.** $\lim\limits_{x\to2} (-x^2 + x - 2)$

**25.** $\lim\limits_{x\to3} \sqrt{x + 1}$

**26.** $\lim\limits_{x\to4} \sqrt[3]{x + 4}$

**27.** $\lim\limits_{x\to-3} \dfrac{2}{x + 2}$

**28.** $\lim\limits_{x\to-2} \dfrac{3x + 1}{2 - x}$

**29.** $\lim\limits_{x\to-2} \dfrac{x^2 - 1}{2x}$

**30.** $\lim\limits_{x\to-1} \dfrac{4x - 5}{3 - x}$

In Exercises 31–48, find the limit (if it exists).

**31.** $\lim\limits_{x \to -1} \dfrac{x^2 - 1}{x + 1}$

**32.** $\lim\limits_{x \to -1} \dfrac{2x^2 - x - 3}{x + 1}$

**33.** $\lim\limits_{x \to 2} \dfrac{x - 2}{x^2 - 4x + 4}$

**34.** $\lim\limits_{x \to 2} \dfrac{2 - x}{x^2 - 4}$

**35.** $\lim\limits_{t \to 5} \dfrac{t - 5}{t^2 - 25}$

**36.** $\lim\limits_{t \to 1} \dfrac{t^2 + t - 2}{t^2 - 1}$

**37.** $\lim\limits_{x \to -2} \dfrac{x^3 + 8}{x + 2}$

**38.** $\lim\limits_{x \to 1} \dfrac{x^3 - 1}{x - 1}$

**39.** $\lim\limits_{x \to 0} \dfrac{|x|}{x}$

**40.** $\lim\limits_{x \to 2} \dfrac{|x - 2|}{x - 2}$

**41.** $\lim\limits_{x \to 3} f(x)$, where $f(x) = \begin{cases} \frac{1}{3}x - 2, & x \le 3 \\ -2x + 5, & x > 3 \end{cases}$

**42.** $\lim\limits_{s \to 1} f(s)$, where $f(s) = \begin{cases} s, & s \le 1 \\ 1 - s, & s > 1 \end{cases}$

**43.** $\lim\limits_{\Delta x \to 0} \dfrac{2(x + \Delta x) - 2x}{\Delta x}$

**44.** $\lim\limits_{\Delta x \to 0} \dfrac{(1 + \Delta x)^3 - 1}{\Delta x}$

**45.** $\lim\limits_{\Delta x \to 0} \dfrac{(x + \Delta x)^3 - x^3}{\Delta x}$

**46.** $\lim\limits_{\Delta x \to 0} \dfrac{\sqrt{x + \Delta x} - \sqrt{x}}{\Delta x}$

**47.** $\lim\limits_{\Delta t \to 0} \dfrac{(t + \Delta t)^2 - 5(t + \Delta t) - (t^2 - 5t)}{\Delta t}$

**48.** $\lim\limits_{x \to 4} \dfrac{\sqrt{x} - 2}{x - 4}$

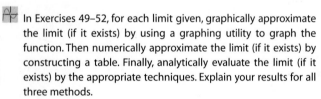 In Exercises 49–52, for each limit given, graphically approximate the limit (if it exists) by using a graphing utility to graph the function. Then numerically approximate the limit (if it exists) by constructing a table. Finally, analytically evaluate the limit (if it exists) by the appropriate techniques. Explain your results for all three methods.

**49.** $\lim\limits_{x \to 1^-} \dfrac{2}{x^2 - 1}$

**50.** $\lim\limits_{x \to 1^+} \dfrac{5}{1 - x}$

**51.** $\lim\limits_{x \to -2^-} \dfrac{1}{x + 2}$

**52.** $\lim\limits_{x \to 0^-} \dfrac{x + 1}{x}$

**53.** *Water Pollution*   The cost (in dollars) of removing $p\%$ of the pollutants from the water in a small lake is given by

$$C = \dfrac{25{,}000p}{100 - p} \quad \text{for} \quad 0 \le p < 100,$$

where $C$ is the cost and $p$ is the percent of pollutants.

(a) Find the cost of removing 50% of the pollutants.

(b) What percent of the pollutants can be removed for $100,000?

(c) Evaluate $\lim\limits_{p \to 100^-} C$. Explain your results.

**54.** Find $\lim\limits_{x \to 0} f(x)$, given

$$4 - x^2 \le f(x) \le 4 + x^2, \text{ for all } x.$$

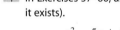 **55.** The limit of $f(x) = (1 + x)^{1/x}$ is a natural base for many business applications.

$$\lim\limits_{x \to 0} (1 + x)^{1/x} = e \approx 2.718$$

(a) Show the reasonableness of this limit by completing the table.

| $x$ | $-0.01$ | $-0.001$ | $-0.0001$ | 0 | 0.0001 | 0.001 | 0.01 |
|---|---|---|---|---|---|---|---|
| $f(x)$ | | | | | | | |

(b) Use a graphing utility to graph $f$ and to confirm the answer in part (a).

(c) Find the domain and range of the function.

**56.** *Compound Interest*   You deposit $1000 in an account that is compounded quarterly at an annual rate of $r$ (in decimal form). The balance $A$ after 10 years is

$$A = 1000\left(1 + \dfrac{r}{4}\right)^{40}.$$

Does the limit of $A$ exist as the interest rate approaches 6%? If so, what is the limit?

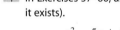 In Exercises 57–60, use a graphing utility to estimate the limit (if it exists).

**57.** $\lim\limits_{x \to 2} \dfrac{x^2 - 5x + 6}{x^2 - 4x + 4}$

**58.** $\lim\limits_{x \to 1} \dfrac{x^2 + 6x - 7}{x^3 - x^2 + 2x - 2}$

**59.** $\lim\limits_{x \to -4} \dfrac{x^3 + 4x^2 + x + 4}{2x^2 + 7x - 4}$

**60.** $\lim\limits_{x \to -2} \dfrac{4x^3 + 7x^2 + x + 6}{3x^2 - x - 14}$

**61.** *Compound Interest*   Consider a certificate of deposit that pays 10% (annual percentage rate) on an initial deposit of $500. The balance $A$ after 10 years is

$$A = 500(1 + 0.1x)^{10/x},$$

where $x$ is the length of the compounding period (in years).

(a) Use a graphing utility to graph $A$, where $0 \le x \le 1$.

(b) Use the zoom and trace features to estimate the balance for quarterly compounding and daily compounding.

(c) Use the zoom and trace features to estimate

$$\lim\limits_{x \to 0^+} A.$$

What do you think this limit represents? Explain your reasoning.

# Continuity

## 1.6

Continuity · Continuity on a Closed Interval · The Greatest Integer Function ·
Extended Application: Compound Interest

## Continuity

In mathematics, the term "continuous" has much the same meaning as it does in everyday use. To say that a function is continuous at $x = c$ means that there is no interruption in the graph of $f$ at $c$. The graph of $f$ is unbroken at $c$, and there are no holes, jumps, or gaps. As simple as this concept may seem, its precise definition eluded mathematicians for many years. In fact, it was not until the early 1800s that a precise definition was finally developed.

Before looking at this definition, consider the function whose graph is shown in Figure 1.56. This figure identifies three values of $x$ at which the function $f$ is not continuous.

1. At $x = c_1, f(c_1)$ is not defined.

2. At $x = c_2$, $\lim\limits_{x \to c_2} f(x)$ does not exist.

3. At $x = c_3, f(c_3) \neq \lim\limits_{x \to c_3} f(x)$.

At all other points in the interval $(a, b)$, the graph of $f$ is uninterrupted, which implies that the function $f$ is continuous at all other points in the interval $(a, b)$.

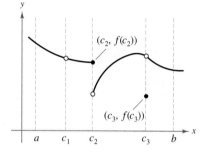

**FIGURE 1.56**   $f$ is not continuous when $x = c_1, c_2, c_3$.

### Definition of Continuity

Let $c$ be a number in the interval $(a, b)$, and let $f$ be a function whose domain contains the interval $(a, b)$. The function $f$ is **continuous at the point** $c$ if the following conditions are true.

1. $f(c)$ is defined.

2. $\lim\limits_{x \to c} f(x)$ exists.

3. $\lim\limits_{x \to c} f(x) = f(c)$.

If $f$ is continuous at every point in the interval $(a, b)$, then it is **continuous on the interval** $(a, b)$.

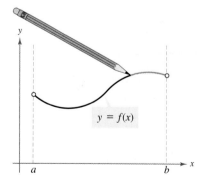

**FIGURE 1.57**   On the interval $(a, b)$, the graph of $f$ can be traced with a pencil.

Roughly, you can say that a function is continuous on an interval if its graph on the interval can be traced using a pencil and paper without lifting the pencil from the paper, as shown in Figure 1.57.

## Technology

Most graphing utilities can draw graphs in two different modes: connected mode and dot mode. The connected mode works well as long as the function is continuous on the entire interval represented by the viewing rectangle. If, however, the function is not continuous at one or more $x$-values in the viewing rectangle, then the connected mode may try to "connect" parts of the graphs that should not be connected. For instance, try graphing the function $y_1 = (x + 3)/(x - 2)$ on the viewing rectangle $-8 \leq x \leq 8$ and $-6 \leq y \leq 6$. Do you notice any problems?

In Section 1.5, you studied several types of functions that meet the three conditions for continuity. Specifically, if *direct substitution* can be used to evaluate the limit of a function at $c$, then the function is continuous at $c$. Two types of functions that have this property are polynomial functions and rational functions.

### Continuity of Polynomial and Rational Functions

1. A polynomial function is continuous at every real number.

2. A rational function is continuous at every number in its domain.

### EXAMPLE 1 Determining Continuity of a Function

Discuss the continuity of the following functions.

(a) $f(x) = x^2 - 2x + 3$    (b) $f(x) = x^3 - x$

*Solution*

Each of these functions is a *polynomial function*. Therefore, each is continuous on the entire real line, as indicated in Figure 1.58.

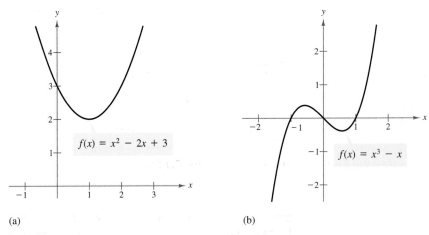

(a)                                    (b)

**FIGURE 1.58**   Both functions are continuous on $(-\infty, \infty)$.

Polynomial functions are one of the most important types of functions used in calculus. Be sure you see from Example 1 that the graph of a polynomial function is continuous on the entire real line, and therefore has no holes, jumps, or gaps. Rational functions, on the other hand, need not be continuous on the entire real line, as shown in Example 2.

## EXAMPLE 2  Determining Continuity of a Function

Discuss the continuity of the following functions.

(a) $f(x) = \dfrac{1}{x}$       (b) $f(x) = \dfrac{x^2 - 1}{x - 1}$       (c) $f(x) = \dfrac{1}{x^2 + 1}$

**Solution**

Each of these functions is a rational function and is therefore continuous at every number in its domain.

(a) The domain of $f(x) = 1/x$ consists of all real numbers other than $x = 0$. Therefore, this function is continuous on the intervals $(-\infty, 0)$ and $(0, \infty)$. (See Figure 1.59a.)

(b) The domain of $f(x) = (x^2 - 1)/(x - 1)$ consists of all real numbers other than $x = 1$. Therefore, this function is continuous on the intervals $(-\infty, 1)$ and $(1, \infty)$. (See Figure 1.59b.)

(c) The domain of $f(x) = 1/(x^2 + 1)$ consists of all real numbers. Therefore, this function is continuous on the entire real line. (See Figure 1.59c.)

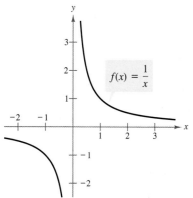
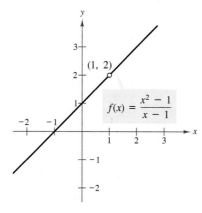
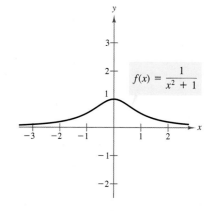
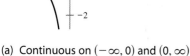
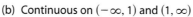

(a) Continuous on $(-\infty, 0)$ and $(0, \infty)$    (b) Continuous on $(-\infty, 1)$ and $(1, \infty)$    (c) Continuous on $(-\infty, \infty)$

**FIGURE 1.59**

If a function is continuous at every number in an open interval except $c$, then $c$ is a **discontinuity** of the function. Discontinuities fall into two categories: removable and nonremovable. The number $c$ is a **removable** discontinuity of $f$ if the graph of $f$ has a hole at the point $(c, L)$ where $L$ is the limit of $f(x)$ as $x$ approaches $c$. For instance, the function in Example 2b has a removable discontinuity at $(1, 2)$. To remove the discontinuity, all you need to do is redefine the function so that $f(1) = 2$.

A discontinuity at $x = c$ is **nonremovable** if the function cannot be made continuous at $x = c$ by defining or redefining the function at $x = c$. For instance, the function in Example 2a has a nonremovable discontinuity at $x = 0$.

## Continuity on a Closed Interval

The intervals discussed in Examples 1 and 2 are open. To discuss continuity on a closed interval, you can use the concept of one-sided limits, as defined in Section 1.5.

---

### Definition of Continuity on a Closed Interval

Let $f$ be defined on a closed interval $[a, b]$. If $f$ is continuous on the open interval $(a, b)$ and

$$\lim_{x \to a^+} f(x) = f(a) \qquad \text{and} \qquad \lim_{x \to b^-} f(x) = f(b),$$

then $f$ is **continuous on the closed interval** $[a, b]$. Moreover, $f$ is **continuous from the right** at $a$ and **continuous from the left** at $b$.

---

Similar definitions can be made to cover continuity on intervals of the form $(a, b]$ and $[a, b)$, or on infinite intervals. For example, the function

$$f(x) = \sqrt{x}$$

is continuous on the infinite interval $[0, \infty)$.

---

### EXAMPLE 3   Examining Continuity at an Endpoint

Discuss the continuity of

$$f(x) = \sqrt{3 - x}.$$

*Solution*

Notice that the domain of $f$ is the set $(-\infty, 3]$. Moreover, $f$ is continuous from the left at $x = 3$ because

$$\lim_{x \to 3^-} f(x) = \lim_{x \to 3^-} \sqrt{3 - x}$$
$$= 0$$
$$= f(3).$$

For all $x < 3$, the function $f$ satisfies the three conditions for continuity. Thus, you can conclude that $f$ is continuous on the interval $(-\infty, 3]$, as shown in Figure 1.60.

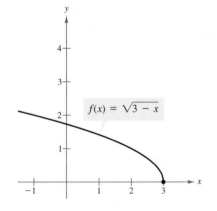

$f(x) = \sqrt{3 - x}$

**FIGURE 1.60**

*STUDY TIP*   When working with radical functions of the form $f(x) = \sqrt{g(x)}$, remember that the domain of $f$ coincides with the solution of $g(x) \geq 0$.

## EXAMPLE 4  Examining Continuity on a Closed Interval

Discuss the continuity of

$$g(x) = \begin{cases} 5 - x, & -1 \leq x \leq 2 \\ x^2 - 1, & 2 < x \leq 3. \end{cases}$$

### Solution

The polynomial functions $5 - x$ and $x^2 - 1$ are continuous on the intervals $[-1, 2)$ and $(2, 3]$, respectively. Thus, to conclude that $g$ is continuous on the entire interval $[-1, 3]$, you need only check the behavior of $g$ when $x = 2$. You can do this by taking the one-sided limits when $x = 2$.

$$\lim_{x \to 2^-} g(x) = \lim_{x \to 2^-} (5 - x) = 3 \qquad \text{Limit from the left}$$

and

$$\lim_{x \to 2^+} g(x) = \lim_{x \to 2^+} (x^2 - 1) = 3 \qquad \text{Limit from the right}$$

Because these two limits are equal,

$$\lim_{x \to 2} g(x) = g(2) = 3.$$

Thus, $g$ is continuous at $x = 2$ and, consequently, it is continuous on the entire interval $[-1, 3]$. The graph of $g$ is shown in Figure 1.61.

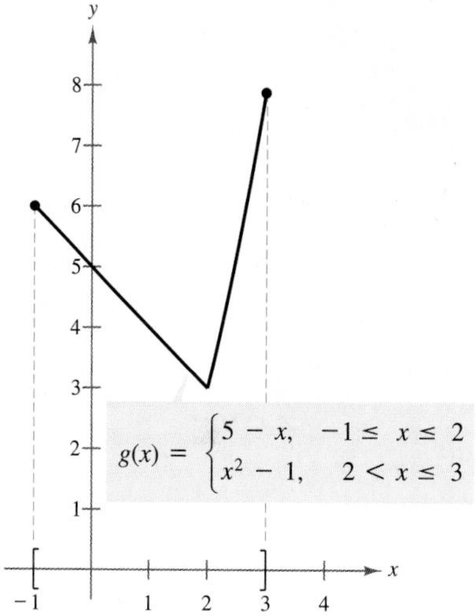

**FIGURE 1.61**

## The Greatest Integer Function

Many functions that are used in business applications are **step functions.** For instance, the function in Example 9 in Section 1.5 is a step function. The **greatest integer function** is another example of a step function. This function is denoted by

$$[\![x]\!] = \text{greatest integer less than or equal to } x.$$

For example,

$$[\![-2.1]\!] = \text{greatest integer less than or equal to } -2.1 = -3$$
$$[\![-2]\!] = \text{greatest integer less than or equal to } -2 = -2$$
$$[\![1.5]\!] = \text{greatest integer less than or equal to } 1.5 = 1.$$

Note that the graph of the greatest integer function (Figure 1.62) jumps up one unit at each integer. This implies that the function is not continuous at each integer.

In real-life applications, the domain of the greatest integer function is often restricted to nonnegative values of $x$. In such cases this function serves the purpose of **truncating** the decimal portion of $x$. For example, 1.345 is truncated to 1 and 3.57 is truncated to 3. That is,

$$[\![1.345]\!] = 1 \quad \text{and} \quad [\![3.57]\!] = 3.$$

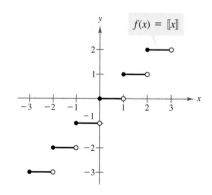

**FIGURE 1.62**  Greatest Integer Function

R. R. Donnelley & Sons

*R. R. Donnelley & Sons Company is the world's largest commercial printer. It prints and binds a major share of the national publications in the United States, including* Time, Newsweek, *and* TV Guide. *In 1996, printing and binding of books accounted for 11% of Donnelley's business. The other part of its business came from printing and binding of catalogs, magazines, directories, and other types of publications. In the photo, an employee is aligning a roll of paper on a printing press in one of Donnelley's plants.*

## EXAMPLE 5   Modeling a Cost Function

A bookbinding company produces 10,000 books in an 8-hour shift. The fixed costs *per shift* amount to $5000, and the unit cost *per book* is $3. Using the greatest integer function, you can write the cost of producing $x$ books as

$$C = 5000\left(1 + \left[\!\left[\frac{x - 1}{10,000}\right]\!\right]\right) + 3x.$$

Sketch the graph of this cost function.

### Solution

Note that during the first 8-hour shift

$$\left[\!\left[\frac{x - 1}{10,000}\right]\!\right] = 0, \qquad 1 \le x \le 10,000,$$

which implies

$$C = 5000\left(1 + \left[\!\left[\frac{x - 1}{10,000}\right]\!\right]\right) + 3x = 5000 + 3x.$$

During the second 8-hour shift

$$\left[\!\left[\frac{x - 1}{10,000}\right]\!\right] = 1, \qquad 10,001 \le x \le 20,000,$$

which implies

$$C = 5000\left(1 + \left[\!\left[\frac{x - 1}{10,000}\right]\!\right]\right) + 3x = 10,000 + 3x.$$

The graph of $C$ is shown in Figure 1.63. Note the graph's discontinuities.

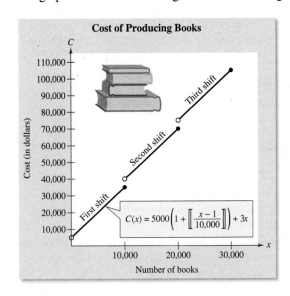

**FIGURE 1.63**

# Technology

## Step Functions and Compound Functions

To graph a step function or compound function with a graphing utility, you must be familiar with the utility's programming language. For instance, different *Texas Instruments (TI)* graphing calculators have different "integer truncation" functions. One is IPart($x$), and it yields the truncated integer part of $x$. For example, IPart($-1.2$) $= -1$ and IPart($3.4$) $= 3$. The other function is Int($x$), which is the greatest integer function. The graphs of these two functions are shown below. When graphing a step function, you should set your graphing utility to dot mode.

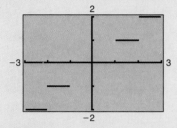

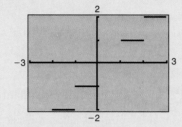

Graph of $f(x) = $ IPart($x$)       Graph of $f(x) = $ Int($x$)

On *TI* graphing calculators, you can graph a compound function such as

$$f(x) = \begin{cases} x^2 - 4, & x \le 2 \\ -x + 2, & 2 < x \end{cases}$$

by entering

$$Y_1 = (X^2 - 4)(X \le 2) + (-X + 2)(2 < X).$$

The graph of this function is shown below.

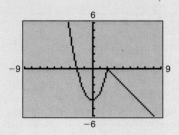

**Technology**

You can use a spreadsheet or the table feature of a graphing utility to create a table. Try doing this for the data shown at the right.

# Extended Application: Compound Interest

Banks and other financial institutions differ on how interest is paid to an account. If the interest is added to the account so that future interest is paid on previously earned interest, then the interest is said to be **compounded.** Suppose, for example, that you deposited $10,000 in an account that pays 6% interest, compounded quarterly. Because the 6% is the annual interest rate, the quarterly rate is $\frac{1}{4}(0.06) = 0.015$ or 1.5%. The balances during the first five quarters are shown below.

| Quarter | Balance |
|---------|---------|
| 1st | $10,000.00 |
| 2nd | 10,000.00 + (0.015)(10,000.00) = $10,150.00 |
| 3rd | 10,150.00 + (0.015)(10,150.00) = $10,302.25 |
| 4th | 10,302.25 + (0.015)(10,302.25) = $10,456.78 |
| 5th | 10,456.78 + (0.015)(10,456.78) = $10,613.63 |

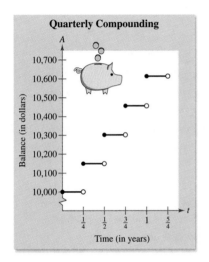

**Quarterly Compounding**

Balance (in dollars)

Time (in years)

**FIGURE 1.64**

## EXAMPLE 6 Graphing Compound Interest

Sketch the graph of the balance in the account described above.

**Solution**

Let $A$ represent the balance in the account and let $t$ represent the time, in years. You can use the greatest integer function to represent the balance, as follows.

$$A = 10,000(1 + 0.015)^{[\![4t]\!]}$$

From the graph shown in Figure 1.64, notice that the function has a discontinuity at each quarter.

---

 *Group Discussion*

## Compound Interest

If $P$ dollars is deposited in an account, compounded $n$ times per year, with an annual rate of $r$ (in decimal form), then the balance $A$ after $t$ years is given by

$$A = P\left(1 + \frac{r}{n}\right)^{nt}.$$

Sketch the graph of each of the following. Which function is continuous? Describe the differences in policy between a bank that uses the first formula and a bank that uses the second formula.

a. $A = P\left(1 + \dfrac{r}{n}\right)^{nt}$    b. $A = P\left(1 + \dfrac{r}{n}\right)^{[\![nt]\!]}$

*Warm Up*

The following warm-up exercises involve skills that were covered in earlier sections. You will use these skills in the exercise set for this section.

In Exercises 1–4, simplify the expression.

**1.** $\dfrac{x^2 + 6x + 8}{x^2 - 6x - 16}$

**2.** $\dfrac{x^2 - 5x - 6}{x^2 - 9x + 18}$

**3.** $\dfrac{2x^2 - 2x - 12}{4x^2 - 24x + 36}$

**4.** $\dfrac{x^3 - 16x}{x^3 + 2x^2 - 8x}$

In Exercises 5–8, solve for x.

**5.** $x^2 + 7x = 0$

**6.** $x^2 + 4x - 5 = 0$

**7.** $3x^2 + 8x + 4 = 0$

**8.** $x^3 + 5x^2 - 24x = 0$

In Exercises 9 and 10, find the limit.

**9.** $\lim\limits_{x\to 3} (2x^2 - 3x + 4)$

**10.** $\lim\limits_{x\to -2} (3x^3 - 8x + 7)$

## EXERCISES 1.6

In Exercises 1–4, decide whether the function is continuous on the entire real line. Explain your reasoning.

**1.** $f(x) = 5x^3 - x^2 + 2$

**2.** $f(x) = (x^2 - 1)^3$

**3.** $f(x) = \dfrac{1}{x^2 - 4}$

**4.** $f(x) = \dfrac{1}{4 + x^2}$

In Exercises 5–28, find the intervals on which the function is continuous.

**5.** $f(x) = -\dfrac{x^3}{2}$

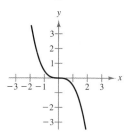

**6.** $f(x) = \dfrac{x^2 - 1}{x}$

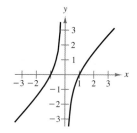

**7.** $f(x) = \dfrac{x^2 - 1}{x + 1}$

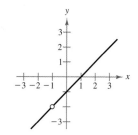

**8.** $f(x) = \dfrac{1}{x^2 - 4}$

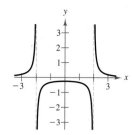

**9.** $f(x) = x^2 - 2x + 1$

**10.** $f(x) = \dfrac{1}{x^2 + 1}$

**11.** $f(x) = \dfrac{1}{x - 1}$

**12.** $f(x) = \dfrac{x}{x^2 - 1}$

**13.** $f(x) = \dfrac{x}{x^2 + 1}$

**14.** $f(x) = \dfrac{x - 3}{x^2 - 9}$

**15.** $f(x) = \dfrac{x - 5}{x^2 - 9x + 20}$

**16.** $f(x) = \dfrac{x - 1}{x^2 + x - 2}$

**17.** $f(x) = \begin{cases} -x, & x < 1 \\ 1, & x = 1 \\ x, & x > 1 \end{cases}$

**18.** $f(x) = \dfrac{[\![x]\!]}{2} + x$

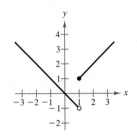

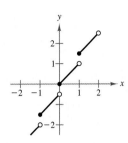

**19.** $f(x) = \begin{cases} -2x + 3, & x < 1 \\ x^2, & x \geq 1 \end{cases}$

**20.** $f(x) = \begin{cases} \frac{1}{2}x + 1, & x \leq 2 \\ 3 - x, & x > 2 \end{cases}$

**21.** $f(x) = \begin{cases} 3 + x, & x \leq 2 \\ x^2 + 1, & x > 2 \end{cases}$

**22.** $f(x) = \begin{cases} |x - 2| + 3, & x < 0 \\ x + 5, & x \geq 0 \end{cases}$

**23.** $f(x) = \dfrac{|x + 1|}{x + 1}$

**24.** $f(x) = \dfrac{|4 - x|}{4 - x}$

**25.** $f(x) = [\![x - 1]\!]$

**26.** $f(x) = x - [\![x]\!]$

**27.** $h(x) = f(g(x)), \quad f(x) = \dfrac{1}{\sqrt{x}}, \quad g(x) = x - 1, x > 1$

**28.** $h(x) = f(g(x)), \quad f(x) = \dfrac{1}{x - 1}, \quad g(x) = x^2 + 5$

In Exercises 29–32, discuss the continuity of the function on the closed interval. If there are any discontinuities, determine whether they are removable.

| Function | Interval |
|---|---|
| **29.** $f(x) = x^2 - 4x - 5$ | $[-1, 5]$ |
| **30.** $f(x) = \dfrac{5}{x^2 + 1}$ | $[-2, 2]$ |
| **31.** $f(x) = \dfrac{1}{x - 2}$ | $[1, 4]$ |
| **32.** $f(x) = \dfrac{x}{x^2 - 4x + 3}$ | $[0, 4]$ |

In Exercises 33–38, sketch the graph of the function and describe the intervals on which the function is continuous.

**33.** $f(x) = \dfrac{x^2 - 16}{x - 4}$

**34.** $f(x) = \dfrac{2x^2 + x}{x}$

**35.** $f(x) = \dfrac{x^3 + x}{x}$

**36.** $f(x) = \dfrac{x^4 - 1}{x^2 - 1}$

**37.** $f(x) = \begin{cases} x^2 + 1, & x < 0 \\ x - 1, & x \geq 0 \end{cases}$

**38.** $f(x) = x - 2[\![x]\!]$

**39.** Determine the constant $a$ such that the function is continuous on the entire real line.

$$f(x) = \begin{cases} x^3, & x \leq 2 \\ ax^2, & x > 2 \end{cases}$$

**40.** Determine the constants $a$ and $b$ such that the function is continuous on the entire real line.

$$f(x) = \begin{cases} 2, & x \leq -1 \\ ax + b, & -1 < x < 3 \\ -2, & x \geq 3 \end{cases}$$

In Exercises 41 and 42, use a graphing utility to graph the function. Then use the graph to determine any $x$-values at which the function is not continuous.

**41.** $h(x) = \dfrac{1}{x^2 - x - 2}$

**42.** $f(x) = \begin{cases} 2x - 4, & x \leq 3 \\ x^2 - 2x, & x > 3 \end{cases}$

In Exercises 43–46, find the interval(s) on which the function is continuous.

**43.** $f(x) = \dfrac{x}{x^2 + 1}$

**44.** $f(x) = x\sqrt{x + 3}$

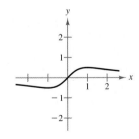

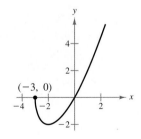

**45.** $f(x) = \frac{1}{2}[\![2x]\!]$

**46.** $f(x) = \dfrac{x + 1}{\sqrt{x}}$

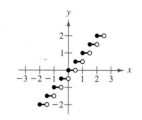

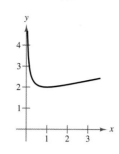

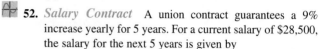

*Writing* In Exercises 47 and 48, use a graphing utility to graph the function on the interval $[-4, 4]$. Does the graph of the function appear continuous on this interval? *Is* the function continuous on $[-4, 4]$? Write a short paragraph about the importance of examining a function analytically as well as graphically.

**47.** $f(x) = \dfrac{x^2 + x}{x}$

**48.** $f(x) = \dfrac{x^3 - 8}{x - 2}$

**49.** *Compound Interest* A deposit of $7500 is made in an account that pays 6% compounded quarterly. The amount $A$ in the account after $t$ years is

$$A = 7500(1.015)^{[\![4t]\!]}, \qquad t \geq 0.$$

(a) Sketch the graph of $A$. Is the graph continuous? Explain.

(b) What is the balance after 7 years?

**50.** *Environmental Cost* The cost $C$ (in millions of dollars) of removing $x$ percent of the pollutants emitted from the smokestack of a factory can be modeled by

$$C = \dfrac{2x}{100 - x}.$$

(a) What is the implied domain of $C$? Explain.

(b) Use a graphing utility to graph the cost function. Is the function continuous on its domain? Explain.

(c) Find the cost of removing 75% of the pollutants from the smokestack.

**51.** *Telephone Rates* A dial-direct long distance call between two cities costs $1.04 for the first two minutes and $0.36 for each additional minute or fraction thereof.

(a) Use the greatest integer function to write the cost $C$ of a call in terms of the time $t$ (in minutes). Graph the cost function and discuss its continuity.

(b) Find the cost of a 9-minute call.

**52.** *Salary Contract* A union contract guarantees a 9% increase yearly for 5 years. For a current salary of $28,500, the salary for the next 5 years is given by

$$S = 28,500(1.09)^{[\![t]\!]},$$

where $t = 0$ represents the present year.

(a) Use the greatest integer function of a graphing utility to graph the salary function and discuss its continuity.

(b) Find the salary during the fifth year (when $t = 5$).

**53.** *Inventory Management* The number of units in inventory in a small company is

$$N = 25\left(2\left[\!\!\left[\dfrac{t + 2}{2}\right]\!\!\right] - t\right), \qquad 0 \leq t \leq 12,$$

where the real number $t$ is the time in months.

(a) Use the greatest integer function of a graphing utility to graph this function and discuss its continuity.

(b) How often must the company replenish its inventory?

**54.** *Owning a Franchise* You have purchased a franchise. You have determined a linear model for your revenue as a function of time. Is the model a continuous function? Would your actual revenue be a continuous function of time? Explain.

**55.** *Animal Reproduction* The gestation period of rabbits is only 26 to 30 days. Therefore, the population of a form (rabbit's home) can increase dramatically in a short period of time. The table gives the population of a form, where $t$ is the time in months and $N$ is the rabbit population.

| $t$ | 0 | 1 | 2 | 3 | 4 | 5 | 6 |
|---|---|---|---|---|---|---|---|
| $N$ | 2 | 8 | 10 | 14 | 10 | 15 | 12 |

Graph the population as a function of time. Find any points of discontinuity in the function. Explain.

# Chapter 1  *Algebra Review*

## Order of Operations

**Technology**

Most scientific and graphing calculators use the same order of operations listed at the right. Try entering the expressions at the right into your calculator. Do you get the same results?

Much of the algebra in Chapter 1 involves evaluation of algebraic expressions. When you evaluate an algebraic expression, you need to know the priorities assigned to different operations. These priorities are called the *order of operations.*

1. Perform operations inside *symbols of grouping* or *absolute value symbols*, starting with the innermost symbol.

2. Evaluate all *exponential* expressions.

3. Perform all *multiplications* and *divisions* from left to right.

4. Perform all *additions* and *subtractions* from left to right.

### EXAMPLE 1    Using Order of Operations

(a) $7 - [(5 \cdot 3) + 2^3] = 7 - [15 + 2^3]$        Multiply inside parentheses.

$\qquad\qquad\qquad\quad = 7 - [15 + 8]$        Evaluate exponential expression.

$\qquad\qquad\qquad\quad = 7 - 23$        Add inside brackets.

$\qquad\qquad\qquad\quad = -16$        Subtract.

(b) $[36 \div (3^2 \cdot 2)] + 6 = [36 \div (9 \cdot 2)] + 6$        Evaluate exponential expression.

$\qquad\qquad\qquad\qquad\; = [36 \div 18] + 6$        Multiply inside parentheses.

$\qquad\qquad\qquad\qquad\; = 2 + 6$        Divide inside brackets.

$\qquad\qquad\qquad\qquad\; = 8$        Add.

(c) $36 - [3^2 \cdot (2 \div 6)] = 36 - \left[3^2 \cdot \frac{1}{3}\right]$        Divide inside parentheses.

$\qquad\qquad\qquad\qquad\; = 36 - \left[9 \cdot \frac{1}{3}\right]$        Evaluate exponential expression.

$\qquad\qquad\qquad\qquad\; = 36 - 3$        Multiply inside brackets.

$\qquad\qquad\qquad\qquad\; = 33$        Subtract.

(d) $10 - 2(8 + |5 - 7|) = 10 - 2(8 + |-2|)$        Subtract inside absolute value symbols.

$\qquad\qquad\qquad\qquad\;\; = 10 - 2(8 + 2)$        Evaluate absolute value.

$\qquad\qquad\qquad\qquad\;\; = 10 - 2(10)$        Add inside parentheses.

$\qquad\qquad\qquad\qquad\;\; = 10 - 20$        Multiply.

$\qquad\qquad\qquad\qquad\;\; = -10$        Subtract.

## Solving Equations

A second algebraic skill in Chapter 1 is solving an equation in one variable.

1. To solve a *linear equation*, you can add or subtract the same quantity from both sides of the equation. You can also multiply or divide both sides of the equation by the same *nonzero* quantity.

2. To solve a *quadratic equation*, you can take the square root of both sides, use factoring, or use the Quadratic Formula.

3. To solve a radical equation, isolate the radical on one side of the equation and square both sides of the equation.

4. To solve an absolute value equation, use the definition of absolute value to rewrite the equation as two equations.

**Technology**

The equations at the left are solved algebraically. Most graphing calculators have a "solve" key that allows you to solve the equation graphically. If you have a graphing calculator, try using it to solve graphically the equations shown at the left.

### EXAMPLE 2   Solving Equations

(a)  $3x - 3 = 5x - 7$ — Original (linear) equation

$-3 = 2x - 7$ — Subtract $3x$ from both sides.

$4 = 2x$ — Add 7 to both sides.

$2 = x$ — Divide both sides by 2.

(b)  $2x^2 = 10$ — Original (quadratic) equation

$x^2 = 5$ — Divide both sides by 2.

$x = \pm\sqrt{5}$ — Take the square root of both sides.

(c)  $2x^2 + 5x - 6 = 6$ — Original (quadratic) equation

$2x^2 + 5x - 12 = 0$ — Write in standard form.

$(2x - 3)(x + 4) = 0$ — Factor.

$2x - 3 = 0$ ⟹ $x = \frac{3}{2}$ — Set first factor equal to zero.

$x + 4 = 0$ ⟹ $x = -4$ — Set second factor equal to zero.

(d)  $\sqrt{2x - 7} = 5$ — Original (radical) equation

$2x - 7 = 25$ — Square both sides.

$2x = 32$ — Add 7 to both sides.

$x = 16$ — Divide both sides by 2.

**TIP**

You should be aware that solving radical equations can sometimes lead to *extraneous solutions* (those that do not satisfy the original equation). For example, squaring both sides of the following equation yields two possible solutions, one of which is extraneous.

$\sqrt{x} = x - 2$

$x = x^2 - 4x + 4$

$0 = x^2 - 5x + 4$

$= (x - 4)(x - 1)$

$x - 4 = 0$ ⟹ $x = 4$  (solution)

$x - 1 = 0$ ⟹ $x = 1$  (extraneous)

ALGEBRA

# Chapter Summary and Study Strategies

*After studying this chapter, you should have acquired the following skills. The exercise numbers are keyed to the Review Exercises that begin on page 76. Answers to odd-numbered Review Exercises are given in the back of the text.* \*

■ Plot points in a coordinate plane and read data presented graphically.     *Review Exercises 1–4*
    *(Section 1.1)*

■ Find the distance between two points in a coordinate plane.   *(Section 1.1)*     *Review Exercises 5–8*
$$d = \sqrt{(x_2 - x_1)^2 + (y_2 - y_1)^2}$$

■ Find the midpoint of the segment connecting two points.   *(Section 1.1)*     *Review Exercises 9–12*
$$\text{Midpoint} = \left(\frac{x_1 + x_2}{2}, \frac{y_1 + y_2}{2}\right)$$

■ Interpret real-life data that is presented graphically.   *(Section 1.1)*     *Review Exercises 13, 14*

■ Translate points in a coordinate plane.   *(Section 1.1)*     *Review Exercise 15*

■ Construct a bar graph from real-life data.   *(Section 1.1)*     *Review Exercise 16*

■ Sketch the graph of an equation by hand.   *(Section 1.2)*     *Review Exercises 17–24*

■ Find the $x$- and $y$-intercepts of the graph of an equation algebraically *and*     *Review Exercises 25, 26*
    graphically using a graphing utility.   *(Section 1.2)*

■ Write the standard equation of a circle, given the center and a point on the circle.     *Review Exercises 27, 28*
    *(Section 1.2)*
$$(x - h)^2 + (y - k)^2 = r^2$$

■ Convert an equation of a circle from general form to standard form by completing     *Review Exercises 29, 30*
    the square.   *(Section 1.2)*

■ Find the points of intersection of two graphs algebraically *and* graphically using     *Review Exercises 31–34*
    a graphing utility.   *(Section 1.2)*

■ Find the break-even point for a business.   *(Section 1.2)*     *Review Exercises 35, 36*

    The break-even point occurs when the revenue $R$ is equal to the cost $C$.

■ Use the slope-intercept form of a linear equation to sketch its graph.     *Review Exercises 37–42*
    *(Section 1.3)*
$$y = mx + b$$

■ Find the slope of the line passing through two points.   *(Section 1.3)*     *Review Exercises 43–46*
$$m = \frac{y_2 - y_1}{x_2 - x_1}$$

■ Use the point-slope form to write an equation of a line.   *(Section 1.3)*     *Review Exercises 47, 48*
$$y - y_1 = m(x - x_1)$$

\* Use a wide range of valuable study aids to help you master the material in this chapter. The *Student Solutions Guide* includes step-by-step solutions to all odd-numbered exercises to help you review and prepare. The *Algebra Review Tutorial Software* and *The Algebra of Calculus* help you brush up on your algebra skills. The *Graphing Technology Guide* offers step-by-step commands and instructions for a wide variety of graphing calculators, including the most recent models.

■ Find equations of parallel and perpendicular lines. *(Section 1.3)*     *Review Exercises 49, 50*

Parallel lines: $m_1 = m_2$     Perpendicular lines: $m_1 = -\dfrac{1}{m_2}$

■ Use linear equations to solve real-life problems such as predicting future sales or creating a linear depreciation schedule. *(Section 1.3)*     *Review Exercises 51, 52*

■ Use the vertical line test to decide whether an equation defines a function. *(Section 1.4)*     *Review Exercises 53–56*

■ Use function notation to evaluate functions. *(Section 1.4)*     *Review Exercises 57, 58*

■ Find the domain and range of a function. *(Section 1.4)*     *Review Exercises 59–64*

■ Combine functions to form other functions. *(Section 1.4)*     *Review Exercises 65, 66*

■ Use the horizontal line test to determine whether a function has an inverse. If it does, find the inverse. *(Section 1.4)*     *Review Exercises 67–70*

■ Determine whether a limit exists. If it does, approximate the limit graphically and numerically using a table. *(Section 1.5)*     *Review Exercises 71–88*

■ Evaluate limits analytically and evaluate one-sided limits. *(Section 1.5)*     *Review Exercises 89–94*

■ Determine whether a function is continuous at a point, in an open interval, and in a closed interval. *(Section 1.6)*     *Review Exercises 95–104*

■ Use analytical and graphical models of real-life data to solve real-life problems. *(Section 1.6)*     *Review Exercises 105–107*

*On pages xx–xxi of the preface, we included a feature called* A Plan for You as a Student. *If you have not already read this, we encourage you to do so now. Here are some other strategies that can help you succeed in this course.*

■ *Use a Graphing Utility*    A graphing calculator or graphing software for a computer can help you in this course in two important ways. As an *exploratory device*, a graphing utility allows you to learn concepts by allowing you to compare graphs of equations. For instance, sketching the graphs of $y = x^2$, $y = x^2 + 1$, and $y = x^2 - 1$ helps confirm that adding (or subtracting) a constant from a function shifts the graph of the function vertically. As a *problem-solving tool*, a graphing utility frees you of some of the drudgery of sketching complicated graphs by hand. The time that you save can be spent using mathematics to solve real-life problems.

■ *Use the Warm-Up Exercises*    Each exercise set in this text begins with a set of ten warm-up exercises. We urge you to begin each homework session by quickly working all ten warm-up exercises (all are answered in the back of the text). The "old" skills covered in the warm-up exercises are needed to master the "new" skills in the section-exercise set. The warm-up exercises remind you that mathematics is cumulative—to be successful in this course, you must retain "old" skills.

■ *Use the Additional Study Aids*    The additional study aids were prepared specifically to help you master the concepts discussed in the text. They are the *Student Solutions Guide*, the *Algebra Review Tutorial Software*, *The Algebra of Calculus*, the *Graphing Technology Guide*, and the *Study Guide*.

# Chapter 1  Review Exercises

In Exercises 1–4, match the data with the real-life situation that it represents. [Graphs are labeled (a)–(d).]

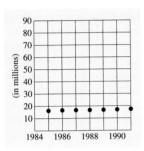

(a)

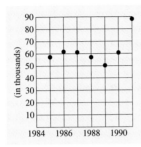

(b)

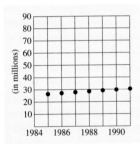

(c)

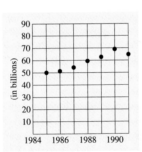

(d)

**1.** Population of Texas

**2.** Population of California

**3.** Number of U.S. business failures

**4.** IBM revenues

In Exercises 5–8, find the distance between the two points.

**5.** $(0, 0), (5, 2)$

**6.** $(1, 2), (4, 3)$

**7.** $(-1, 3), (-4, 6)$

**8.** $(6, 8), (-3, 7)$

In Exercises 9–12, find the midpoint of the segment connecting the two points.

**9.** $(5, 6), (9, 2)$

**10.** $(0, 0), (-4, 8)$

**11.** $(-10, 4), (-6, 8)$

**12.** $(7, -9), (-3, 5)$

In Exercises 13 and 14, use the graph below, which gives the revenues, costs, and profits for Kimberly-Clark Corporation from 1991 through 1996.  *(Source: Kimberly-Clark Corporation)*

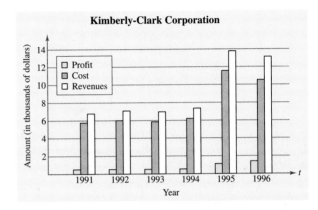

**13.** Which bars on the graph represent the revenues? Which represent the costs? Which represent the profits? Explain your reasoning. Write an equation that relates the revenue $R$, cost $C$, and profit $P$.

**14.** Estimate the revenue, cost, and profit for Kimberly-Clark Corporation in 1996. Do you think these numbers are in billions of dollars, millions of dollars, or thousands of dollars? Explain your reasoning. (Kimberly-Clark is the maker of *Kleenex, Huggies Disposable Diapers*, and other paper products.)

**15.** Translate the triangle whose vertices are $(1, 3), (2, 4)$, and $(5, 6)$ three units to the right and four units up. Find the coordinates of the translated vertices.

**16.** *Biology: Species Collection*    The following data represent six intertidal invertebrate species collected from four stations along the Maine coast.

| | | | |
|---|---|---|---|
| *Mytilus* | 107 | *Gammarus* | 78 |
| *Littorina* | 65 | *Arbacia* | 6 |
| *Nassarius* | 112 | *Mya* | 18 |

Use a graphing utility to construct a bar graph that represents the data.  *(Source: Adapted from Haefner, Exploring Marine Biology: Laboratory and Field Exercises)*

In Exercises 17–24, sketch the graph of the equation.

**17.** $y = 4x - 12$

**18.** $y = 4 - 3x$

**19.** $y = x^2 + 5$

**20.** $y = 1 - x^2$

**21.** $y = x^2 + 5x + 6$

**22.** $y = |2x - 3|$

**23.** $y = x^3 + 4$

**24.** $y = \sqrt{2x}$

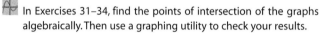 In Exercises 25 and 26, find the intercepts of the graph of the equation algebraically. Check your results with a graphing utility.

**25.** $4x + y + 3 = 0$

**26.** $y = (x - 1)^3 + 2(x - 1)^2$

In Exercises 27 and 28, write the standard form of the circle.

**27.** Center: $(0, 0)$;
Solution point: $(2, \sqrt{5})$

**28.** Center: $(2, -1)$;
Solution point: $(-1, 7)$

In Exercises 29 and 30, complete the square to write the equation of the circle in standard form. Determine the radius and center of the circle. Then sketch the circle.

**29.** $x^2 + y^2 - 6x + 8y = 0$

**30.** $x^2 + y^2 + 10x + 4y - 7 = 0$

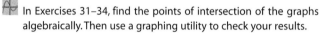 In Exercises 31–34, find the points of intersection of the graphs algebraically. Then use a graphing utility to check your results.

**31.** $x + y = 2$, $2x - y = 1$

**32.** $x^2 + y^2 = 5$, $x - y = 1$

**33.** $y = x^3$, $y = x$

**34.** $y = \sqrt{x}$, $y = x$

**35.** *Break-Even Analysis*  The student government association wants to raise money by having a T-shirt sale. Each shirt costs $8. The silk screening costs $200 for the design, plus $2 per shirt. Each shirt will sell for $14.

(a) Find equations for the total cost $C$ and the total revenue $R$ for selling $x$ shirts.

(b) Find the break-even point.

**36.** *Break-Even Analysis*  You are starting a part-time business. You make an initial investment of $6000. The unit cost of the product is $6.50, and the selling price is $13.90.

(a) Find equations for the total cost $C$ and the total revenue $R$ for selling $x$ units of the product.

(b) Find the break-even point.

In Exercises 37–42, find the slope and $y$-intercept (if possible) of the linear equation. Then sketch the graph of the equation.

**37.** $3x + y = -2$

**38.** $-\frac{1}{3}x + \frac{5}{6}y = 1$

**39.** $y = -\frac{5}{3}$

**40.** $x = -3$

**41.** $-2x - 5y - 5 = 0$

**42.** $3.2x - 0.8y + 5.6 = 0$

In Exercises 43–46, find the slope of the line passing through the two points.

**43.** $(0, 0)$, $(7, 6)$

**44.** $(-1, 5)$, $(-5, 7)$

**45.** $(10, 17)$, $(-11, -3)$

**46.** $(-11, -3)$, $(-1, -3)$

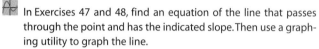 In Exercises 47 and 48, find an equation of the line that passes through the point and has the indicated slope. Then use a graphing utility to graph the line.

**47.** Point: $(3, -1)$;  Slope: $m = -2$

**48.** Point: $(-3, -3)$;  Slope: $m = \frac{1}{2}$

In Exercises 49 and 50, find the general form of the equation of the line passing through the point and satisfying the given condition.

**49.** Point: $(-3, 6)$

(a) Slope is $\frac{7}{8}$.

(b) Parallel to the line $4x + 2y = 7$.

(c) Passes through the origin.

(d) Perpendicular to the line $3x - 2y = 2$.

**50.** Point: $(1, -3)$

(a) Parallel to the $x$-axis.

(b) Perpendicular to the $x$-axis.

(c) Parallel to the line $-4x + 5y = -3$.

(d) Perpendicular to the line $5x - 2y = 3$.

**51.** *Demand*  When a wholesaler sold a product at $32 per unit, sales were 750 units per week. After a price increase of $5 per unit, however, the sales dropped to 700 units per week.

(a) Write the quantity demanded $x$ as a linear function of the price $p$.

(b) *Linear Interpolation*  Predict the number of units sold at a price of $34.50 per unit.

(c) *Linear Extrapolation*  Predict the number of units sold at a price of $42.00 per unit.

**52.** *Linear Depreciation* A small business purchases a typesetting system for $117,000. After 9 years, the system will be obsolete and have no value.

(a) Write a linear equation giving the value $v$ of the system in terms of the time $t$.

(b) Use a graphing utility to graph the function.

(c) Use a graphing utility to estimate the value of the system after 4 years.

(d) Use a graphing utility to estimate the time when the system's value will be $84,000.

In Exercises 53–56, use the vertical line test to determine whether $y$ is a function of $x$.

**53.** $y = -x^2 + 2$

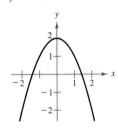

**54.** $x^2 + y^2 = 4$

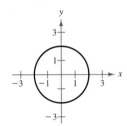

**55.** $y^2 - \frac{1}{4}x^2 = 4$

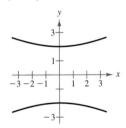

**56.** $y = |x + 4|$

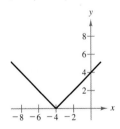

**57.** Given $f(x) = 3x + 4$, find the following.

(a) $f(1)$    (b) $f(x + 1)$    (c) $f(2 + \Delta x)$

**58.** Given $f(x) = x^2 + 4x + 3$, find the following.

(a) $f(0)$    (b) $f(x - 1)$    (c) $f(x + \Delta x) - f(x)$

In Exercises 59–64, use a graphing utility to graph the function. Then find the domain and range of the function.

**59.** $f(x) = x^2 + 3x + 2$

**60.** $f(x) = 2$

**61.** $f(x) = \sqrt{x + 1}$

**62.** $f(x) = \dfrac{x - 3}{x^2 + x - 12}$

**63.** $f(x) = -|x| + 3$

**64.** $f(x) = -\frac{12}{13}x - \frac{7}{8}$

In Exercises 65 and 66, use $f$ and $g$ to find the following.

(a) $f(x) + g(x)$    (b) $f(x) - g(x)$    (c) $f(x)g(x)$

(d) $\dfrac{f(x)}{g(x)}$    (e) $f(g(x))$    (f) $g(f(x))$

**65.** $f(x) = 1 + x^2$,   $g(x) = 2x - 1$

**66.** $f(x) = 2x - 3$,   $g(x) = \sqrt{x + 1}$

In Exercises 67–70, find the inverse of $f$ (if it exists).

**67.** $f(x) = \frac{3}{2}x$

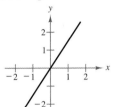

**68.** $f(x) = |x + 1|$

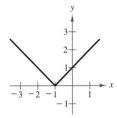

**69.** $f(x) = -x^2 + \frac{1}{2}$

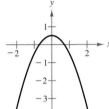

**70.** $f(x) = x^3 - 1$

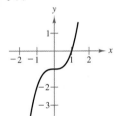

In Exercises 71–86, find the limit (if it exists).

**71.** $\lim\limits_{x \to 2} (5x - 3)$

**72.** $\lim\limits_{x \to 2} (2x + 9)$

**73.** $\lim\limits_{x \to 2} (5x - 3)(2x + 3)$

**74.** $\lim\limits_{x \to 2} \dfrac{5x - 3}{2x + 9}$

**75.** $\lim\limits_{t \to 3} \dfrac{t^2 + 1}{t}$

**76.** $\lim\limits_{t \to 1} \dfrac{t + 1}{t - 2}$

**77.** $\lim\limits_{t \to 0} \dfrac{t^2 + 1}{t}$

**78.** $\lim\limits_{t \to 2} \dfrac{t + 1}{t - 2}$

**79.** $\lim\limits_{x \to -2} \dfrac{x + 2}{x^2 - 4}$

**80.** $\lim\limits_{x \to 3^-} \dfrac{x^2 - 9}{x - 3}$

**81.** $\lim\limits_{x \to 0^+} \left(x - \dfrac{1}{x}\right)$

**82.** $\lim\limits_{x \to 1/2} \dfrac{2x - 1}{6x - 3}$

**83.** $\lim\limits_{x \to 0} \dfrac{[1/(x - 2)] - 1}{x}$

**84.** $\lim\limits_{s \to 0} \dfrac{\left(1/\sqrt{1 + s}\right) - 1}{s}$

**85.** $\lim\limits_{\Delta x \to 0} \dfrac{(x + \Delta x)^3 - (x + \Delta x) - (x^3 - x)}{\Delta x}$

**86.** $\lim\limits_{\Delta x \to 0} \dfrac{1 - (x + \Delta x)^2 - (1 - x^2)}{\Delta x}$

In Exercises 87 and 88, use a table to estimate the limit.

**87.** $\lim\limits_{x \to 1^+} \dfrac{\sqrt{2x + 1} - \sqrt{3}}{x - 1}$    **88.** $\lim\limits_{x \to 1^+} \dfrac{1 - \sqrt[3]{x}}{x - 1}$

In Exercises 89–94, determine whether the statement is true or false.

**89.** $\lim\limits_{x \to 0} \dfrac{|x|}{x} = 1$    **90.** $\lim\limits_{x \to 0} x^3 = 0$

**91.** $\lim\limits_{x \to 0} \sqrt{x} = 0$    **92.** $\lim\limits_{x \to 0} \sqrt[3]{x} = 0$

**93.** $\lim\limits_{x \to 2} f(x) = 3, \quad f(x) = \begin{cases} 3, & x \le 2 \\ 0, & x > 2 \end{cases}$

**94.** $\lim\limits_{x \to 3} f(x) = 1, \quad f(x) = \begin{cases} x - 2, & x \le 3 \\ -x^2 + 8x - 14, & x > 3 \end{cases}$

In Exercises 95–102, find the intervals on which the function is continuous.

**95.** $f(x) = \dfrac{1}{(x + 4)^2}$    **96.** $f(x) = \dfrac{x + 2}{x}$

**97.** $f(x) = \dfrac{3}{x + 1}$    **98.** $f(x) = \dfrac{x + 1}{2x + 2}$

**99.** $f(x) = [\![x + 3]\!]$    **100.** $f(x) = \dfrac{3x^2 - x - 2}{x - 1}$

**101.** $f(x) = \begin{cases} x, & x \le 0 \\ x + 1, & x > 0 \end{cases}$

**102.** $f(x) = \begin{cases} x, & x \le 0 \\ x^2, & x > 0 \end{cases}$

In Exercises 103 and 104, find $a$ such that $f$ is continuous on $(-\infty, \infty)$.

**103.** $f(x) = \begin{cases} -x + 1, & x \le 3 \\ ax - 8, & x > 3 \end{cases}$

**104.** $f(x) = \begin{cases} x + 1, & x < 1 \\ 2x + a, & x \ge 1 \end{cases}$

**105.** *National Debt*  The table lists the national debt $D$ (in billions of dollars) for selected years. A mathematical model for the national debt is

$$D = 0.2t^3 + 3.05t^2 + 5.88t + 387.94,$$

where $t = 0$ represents 1970. *(Source: U.S. Office of Management and Budget)*

Table for 105

| $t$ | 0 | 5 | 10 | 15 |
|---|---|---|---|---|
| $D$ | 381 | 542 | 914 | 1817 |

| $t$ | 20 | 21 | 22 | 23 |
|---|---|---|---|---|
| $D$ | 3266 | 3599 | 4083 | 4436 |

(a) Use a graphing utility to graph the model.
(b) Create a table that compares the values given by the model with the actual data.
(c) Use the model to estimate the national debt in 1999.

**106.** *Recycling*  A recycling center pays \$0.25 for each pound of aluminum cans. Twenty-four aluminum cans weigh 1 pound. A mathematical model for the amount paid $A$ by the recycling center is

$$A = \frac{1}{4}\left[\!\!\left[\frac{x}{24}\right]\!\!\right],$$

where $x$ is the number of cans.

(a) Use a graphing utility to graph the function and then discuss its continuity.
(b) How much does a recycling center pay out for 1500 cans?

**107.** *Biology: Cross Fertilization*  A researcher experimenting with strains of corn produced the results in the figure below. From the figure, visually estimate the $x$- and $y$-intercepts, and use these points to write an equation for each line.

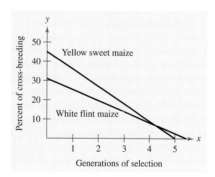

*(Source: Adapted from Levine/Miller, Biology: Discovering Life, Second Edition)*

# Sample Post-Graduation Exam Questions

CPA    GMAT
GRE    CLAST
          Actuarial

*The following questions represent the types of questions that appear on certified public accountant (CPA) exams, graduate management admission tests (GMAT), graduate records exams (GRE), actuarial exams, and college-level academic skills tests (CLAST). The answers to the questions are given in the back of the book.*

Figure for 1–5

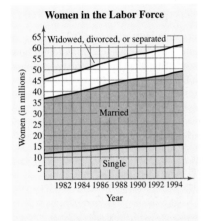

**Women in the Labor Force**

**Percent of Total Labor Force**

In Questions 1–5, use the data given in the graph.   *(Source: U.S. Bureau of the Census)*

**1.** The total labor force in 1990 was about $y$ million with $y$ equal to

(a) 100      (b) 108      (c) 115      (d) 124      (e) 132

**2.** In 1985, the percent of women in the labor force who were married was about

(a) 19      (b) 32      (c) 45      (d) 60      (e) 82

**3.** What was the first year when more than 50 million women were in the labor force?

(a) 1980      (b) 1984      (c) 1985      (d) 1990      (e) 1992

**4.** Between 1980 and 1990, the number of women in the labor force

(a) increased by about 25%      (b) increased by about 50%
(c) increased by about 100%      (d) increased by about 125%
(e) increased by about 150%

**5.** Which of the following statements about the labor force can be inferred from the graphs?

   I. Between 1980 and 1995, there were no years when more than 15 million widowed, divorced, or separated women were in the labor force.
  II. In every year between 1980 and 1995, the number of married women in the labor force increased.
 III. In every year between 1980 and 1995, women made up at least $\frac{2}{5}$ of the total labor force.

(a) I only                          (b) II only                          (c) I and II only
(d) II and III only          (e) I, II, and III

**6.** What is the length of the line segment connecting $(1, 3)$ to $(-1, 5)$?

(a) $\sqrt{3}$      (b) 2      (c) $2\sqrt{2}$      (d) 4      (e) 8

**7.** The interest charged on a loan is $p$ dollars per \$1000 for the first month and $q$ dollars per \$1000 for each month after the first month. How much interest will be charged during the first 3 months on a loan of \$10,000?

(a) $30p$                          (b) $30q$                          (c) $p + 2q$
(d) $20p + 10q$          (e) $10p + 20q$

**8.** If $x + y > 5$ and $x - y > 3$, which of the following describes the $x$ solutions?

(a) $x > 3$      (b) $x > 4$      (c) $x > 5$      (d) $x < 5$      (e) $x < 3$

# Differentiation 2

STRATEGIES *for* SUCCESS

OBJECTIVES

*When you have completed this chapter, make sure you are able to:*

❏ Find the slope of a graph and calculate derivatives using the limit definition.
❏ Use the Constant Rule, Power Rule, Constant Multiple Rule, and Sum and Difference Rules.
❏ Find rates of change: velocity, marginal profit, marginal revenue, and marginal cost.
❏ Use the Product, Quotient, Chain, and General Power Rules.
❏ Calculate higher-order derivatives and derivatives using implicit differentiation.
❏ Solve related-rate problems and applications.

TOOLS

*Use these study tools to achieve the objectives above:*

**Algebra Review**
(pages 162 and 163)

**Chapter Summary and Study Strategies**
(pages 164 and 165)

**Review Exercises**
(pages 166–169)

ADDITIONAL RESOURCES

*Use these resources to solidify your mastery of calculus:*

**Student Solutions Guide**

**Study Guide** (Additional Examples, Similar Problems, and Chapter Test)

**Algebra Review Tutorial Software**

**Graphing Technology Guide**

**Sample Post-Graduation Exam Questions**
(page 170)

**Web Exercise**
(page 130, exercise 62)

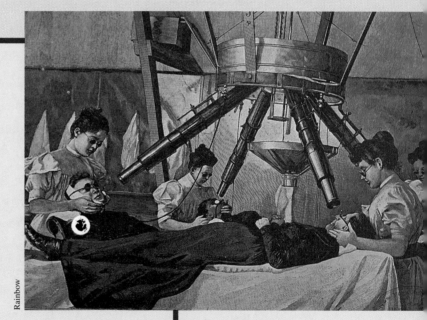

Rainbow

*In Exercise 58 on page 104, you will make observations about the prevalence of migraines. This engraving shows the use of magnified sunlight in a method once used in treating migraines.*

# 2.1  The Derivative and the Slope of a Graph

*Tangent Line to a Graph  •  Slope of a Graph  •  Slope and the Limit Process  •
The Derivative of a Function  •  Differentiability and Continuity*

## Tangent Line to a Graph

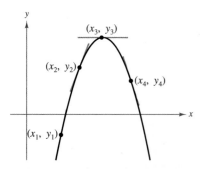

**FIGURE 2.1**  The slope of a graph changes from one point to another.

Calculus is a branch of mathematics that studies rates of change of functions. In this course, you will learn that rates of change have many applications in real life. In Section 1.3, you learned how the slope of a line indicates the rate at which a line rises or falls. For a line, this rate (or slope) is the same at every point on the line. For graphs other than lines, the rate at which the graph rises or falls changes from point to point. For instance, in Figure 2.1, the parabola is rising more quickly at the point $(x_1, y_1)$ than it is at the point $(x_2, y_2)$. At the vertex $(x_3, y_3)$, the graph levels off, and at the point $(x_4, y_4)$, the graph is falling.

To determine the rate at which a graph rises or falls at a *single point,* you can find the slope of the **tangent line** at the point. In simple terms, the tangent line to the graph of a function $f$ at a point $P(x_1, y_1)$ is the line that best approximates the graph at that point, as shown in Figure 2.1. Figure 2.2 shows other examples of tangent lines.

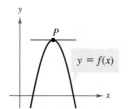 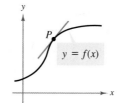 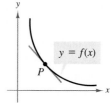

**FIGURE 2.2**  Tangent Line to a Graph at a Point

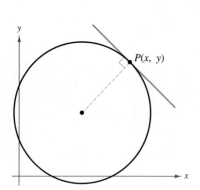

**FIGURE 2.3**  Tangent Line to a Circle

When Isaac Newton (1642–1727) was working on the "tangent line problem," he realized that it is difficult to define precisely what is meant by a tangent to a general curve. From geometry, you know that a line is tangent to a circle if the line intersects the circle at only one point, as shown in Figure 2.3. Tangent lines to a noncircular graph, however, can intersect the graph at more than one point. For instance, in the second graph in Figure 2.2, if the tangent line were extended, it would intersect the graph at a point other than the point of tangency. In this section, you will see how the notion of a limit can be used to define a general tangent line.

## Slope of a Graph

Because a tangent line approximates the graph at a point, the problem of finding the slope of a graph at a point becomes one of finding the slope of the tangent line at the point.

### EXAMPLE 1   Approximating the Slope of a Graph

Use the graph in Figure 2.4 to approximate the slope of the graph of $f(x) = x^2$ at the point $(1, 1)$.

*Solution*

From the graph of $f(x) = x^2$, you can see that the tangent line at $(1, 1)$ rises approximately two units for each unit change in $x$. Thus, the slope of the tangent line at $(1, 1)$ is given by

$$\text{Slope} = \frac{\text{change in } y}{\text{change in } x} \approx \frac{2}{1} = 2.$$

Because the tangent line at the point $(1, 1)$ has a slope of about 2, you can conclude that the graph has a slope of about 2 at the point $(1, 1)$.

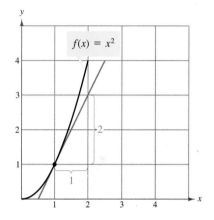

**FIGURE 2.4**

> *STUDY TIP*   When visually approximating the slope of a graph, note that the scales on the horizontal and vertical axes may differ. When this happens (as it frequently does in applications), the slope of the tangent line is distorted, and you must be careful to account for the difference in scales.

### EXAMPLE 2   Interpreting Slope

Figure 2.5 graphically depicts the average daily temperature (in degrees Fahrenheit) in Duluth, Minnesota. Estimate the slope of this graph at the indicated point and give a physical interpretation of the result.   *(Source: National Oceanic and Atmospheric Administration)*

*Solution*

From the graph, you can see that the tangent line at the given point falls approximately 27 units for each two-unit change in $x$. Thus, you can estimate the slope at the given point to be

$$\text{Slope} = \frac{\text{change in } y}{\text{change in } x} \approx \frac{-27}{2} = -13.5 \text{ degrees per month.}$$

This means that you can expect the average daily temperatures in November to be about 13.5 degrees *lower* than the corresponding temperatures in October.

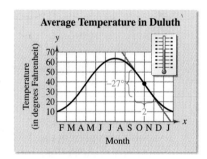

**FIGURE 2.5**

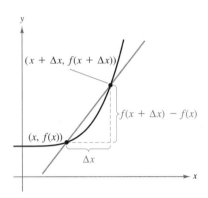

**FIGURE 2.6**  The Secant Line Through the Two Points $(x, f(x))$ and $(x + \Delta x, f(x + \Delta x))$

## Slope and the Limit Process

In Examples 1 and 2, you approximated the slope of a graph at a point by making a careful graph and then "eyeballing" the tangent line at the point of tangency. A more precise method of approximating tangent lines makes use of a **secant line** through the point of tangency and a second point on the graph, as shown in Figure 2.6. If $(x, f(x))$ is the point of tangency and $(x + \Delta x, f(x + \Delta x))$ is a second point on the graph of $f$, then the slope of the secant line through the two points is

$$m_{\text{sec}} = \frac{f(x + \Delta x) - f(x)}{\Delta x}.$$    Slope of secant line

The right side of this equation is called the **difference quotient.** The denominator $\Delta x$ is the **change in $x$,** and the numerator is the **change in $y$.** The beauty of this procedure is that you obtain better and better approximations of the slope of the tangent line by choosing the second point closer and closer to the point of tangency, as shown in Figure 2.7.

Using the limit process, you can find the *exact* slope of the tangent line at $(x, f(x))$.

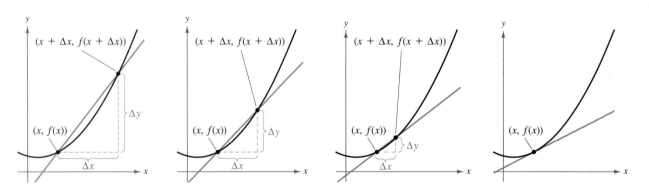

**FIGURE 2.7**    As $\Delta x$ approaches 0, the secant lines approach the tangent line.

**STUDY TIP**    $\Delta x$ is used as a variable to represent the change in $x$ in the definition of the slope of a graph. Other variables may also be used. For instance, this definition is sometimes written as

$$m = \lim_{h \to 0} \frac{f(x + h) - f(x)}{h}.$$

### Definition of the Slope of a Graph

The **slope** $m$ of the graph of $f$ at the point $(x, f(x))$ is equal to the slope of its tangent line at $(x, f(x))$, and is given by

$$m = \lim_{\Delta x \to 0} m_{\text{sec}} = \lim_{\Delta x \to 0} \frac{f(x + \Delta x) - f(x)}{\Delta x}$$

provided this limit exists.

## EXAMPLE 3   Finding Slope by the Limit Process

Find the slope of the graph of $f(x) = x^2$ at the point $(-2, 4)$.

### Solution

Begin by finding an expression that represents the slope of a secant line at the point $(-2, 4)$.

$$\begin{aligned}
m_{sec} &= \frac{f(-2 + \Delta x) - f(-2)}{\Delta x} & \text{Set up difference quotient.}\\[2mm]
&= \frac{(-2 + \Delta x)^2 - (-2)^2}{\Delta x} & \text{Use } f(x) = x^2.\\[2mm]
&= \frac{4 - 4\Delta x + (\Delta x)^2 - 4}{\Delta x} & \text{Expand terms.}\\[2mm]
&= \frac{-4\Delta x + (\Delta x)^2}{\Delta x} & \text{Simplify.}\\[2mm]
&= \frac{\Delta x(-4 + \Delta x)}{\Delta x} & \text{Factor and cancel.}\\[2mm]
&= -4 + \Delta x, \ \Delta x \neq 0 & \text{Simplify.}
\end{aligned}$$

Next, take the limit of $m_{sec}$ as $\Delta x \to 0$.

$$m = \lim_{\Delta x \to 0} m_{sec} = \lim_{\Delta x \to 0} (-4 + \Delta x) = -4$$

Thus, the graph of $f$ has a slope of $-4$ at the point $(-2, 4)$, as shown in Figure 2.8.

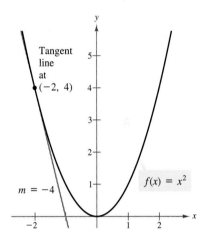

**FIGURE 2.8**

## EXAMPLE 4   Finding the Slope of a Graph

Find the slope of $f(x) = -2x + 4$.

### Solution

You know from your study of linear functions that the line given by $f(x) = -2x + 4$ has a slope of $-2$, as shown in Figure 2.9. This conclusion is consistent with the limit definition of slope.

$$\begin{aligned}
m &= \lim_{\Delta x \to 0} \frac{f(x + \Delta x) - f(x)}{\Delta x}\\[2mm]
&= \lim_{\Delta x \to 0} \frac{[-2(x + \Delta x) + 4] - [-2x + 4]}{\Delta x}\\[2mm]
&= \lim_{\Delta x \to 0} \frac{-2x - 2\Delta x + 4 + 2x - 4}{\Delta x}\\[2mm]
&= \lim_{\Delta x \to 0} \frac{-2\Delta x}{\Delta x}\\[2mm]
&= -2
\end{aligned}$$

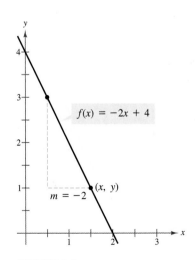

**FIGURE 2.9**

*Discovery*

Use a graphing utility to graph the function $y_1 = x^2 + 1$ and the three lines $y_2 = 3x - 1$, $y_3 = 4x - 3$, and $y_4 = 5x - 5$. Which of these lines appears to be tangent to $y_1$ at the point $(2, 5)$? Confirm your answer by showing that the graphs of $y_1$ and its tangent line have only one point of intersection, whereas the graphs of $y_1$ and the other lines each have two points of intersection.

It is important that you see the distinction between the ways the difference quotients were set up in Examples 3 and 4. In Example 3, you were finding the slope of a graph at a specific point $(c, f(c))$. To find the slope, you can use the following form of a difference quotient.

$$m = \lim_{\Delta x \to 0} \frac{f(c + \Delta x) - f(c)}{\Delta x} \qquad \text{Slope at specific point}$$

In Example 4, however, you were finding a formula for the slope at *any* point on the graph. In such cases, you should use $x$, rather than $c$, in the difference quotient.

$$m = \lim_{\Delta x \to 0} \frac{f(x + \Delta x) - f(x)}{\Delta x} \qquad \text{Formula for slope}$$

Except for linear functions, this form will always produce a function of $x$, which can then be evaluated to find the slope at any desired point.

## EXAMPLE 5   Finding a Formula for the Slope of a Graph

Find a formula for the slope of the graph of $f(x) = x^2 + 1$. What is the slope at the points $(-1, 2)$ and $(2, 5)$?

*Solution*

$$
\begin{aligned}
m_{\text{sec}} &= \frac{f(x + \Delta x) - f(x)}{\Delta x} && \text{Set up difference quotient.} \\[2mm]
&= \frac{[(x + \Delta x)^2 + 1] - (x^2 + 1)}{\Delta x} && \text{Use } f(x) = x^2 + 1. \\[2mm]
&= \frac{x^2 + 2x\Delta x + (\Delta x)^2 + 1 - x^2 - 1}{\Delta x} && \text{Expand terms.} \\[2mm]
&= \frac{2x\Delta x + (\Delta x)^2}{\Delta x} && \text{Simplify.} \\[2mm]
&= \frac{\Delta x(2x + \Delta x)}{\Delta x} && \text{Factor and cancel.} \\[2mm]
&= 2x + \Delta x, \quad \Delta x \neq 0 && \text{Simplify.}
\end{aligned}
$$

Next, take the limit of $m_{\text{sec}}$ as $\Delta x \to 0$.

$$
\begin{aligned}
m &= \lim_{\Delta x \to 0} m_{\text{sec}} \\[2mm]
&= \lim_{\Delta x \to 0} (2x + \Delta x) \\[2mm]
&= 2x
\end{aligned}
$$

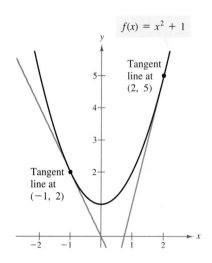

$f(x) = x^2 + 1$

Tangent line at $(2, 5)$

Tangent line at $(-1, 2)$

**FIGURE 2.10**

Using the formula $m = 2x$, you can find the slope at the specified points. At $(-1, 2)$, the slope is $m = 2(-1) = -2$, and at $(2, 5)$, the slope is $m = 2(2) = 4$. The graph of $f$ is shown in Figure 2.10.

## The Derivative of a Function

In Example 5, you started with the function $f(x) = x^2 + 1$, and used the limit process to derive another function, $m = 2x$, that represents the slope of the graph of $f$ at the point $(x, f(x))$. This derived function is called the **derivative** of $f$ at $x$. It is denoted by $f'(x)$, which is read as "$f$ prime of $x$."

---

### Definition of the Derivative

The **derivative of $f$ at $x$** is given by

$$f'(x) = \lim_{\Delta x \to 0} \frac{f(x + \Delta x) - f(x)}{\Delta x}$$

provided this limit exists. A function is **differentiable** at $x$ if its derivative exists at $x$. The process of finding derivatives is called **differentiation.**

---

In addition to $f'(x)$, other notations can be used to denote the derivative of $y = f(x)$. The most common are

$$\frac{dy}{dx}, \quad y', \quad \frac{d}{dx}[f(x)], \quad \text{and} \quad D_x[y].$$

**STUDY TIP**  The notation $dy/dx$ is read "the derivative of $y$ with respect to $x$," and using limit notation, you can write

$$\frac{dy}{dx} = \lim_{\Delta x \to 0} \frac{\Delta y}{\Delta x}$$

$$= \lim_{\Delta x \to 0} \frac{f(x + \Delta x) - f(x)}{\Delta x}$$

$$= f'(x).$$

---

## EXAMPLE 6   Finding a Derivative

Find the derivative of $f(x) = 3x^2 - 2x$.

**Solution**

$$f'(x) = \lim_{\Delta x \to 0} \frac{f(x + \Delta x) - f(x)}{\Delta x}$$

$$= \lim_{\Delta x \to 0} \frac{[3(x + \Delta x)^2 - 2(x + \Delta x)] - (3x^2 - 2x)}{\Delta x}$$

$$= \lim_{\Delta x \to 0} \frac{3x^2 + 6x\Delta x + 3(\Delta x)^2 - 2x - 2\Delta x - 3x^2 + 2x}{\Delta x}$$

$$= \lim_{\Delta x \to 0} \frac{6x\Delta x + 3(\Delta x)^2 - 2\Delta x}{\Delta x}$$

$$= \lim_{\Delta x \to 0} \frac{\Delta x(6x + 3\Delta x - 2)}{\Delta x}$$

$$= \lim_{\Delta x \to 0} (6x + 3\Delta x - 2)$$

$$= 6x - 2$$

Thus, the derivative of $f(x) = 3x^2 - 2x$ is $f'(x) = 6x - 2$.

### Technology

You can use a graphing utility to confirm the result given in Example 7. One way to do this is to choose a point on the graph of $y = 2/t$, such as $(1, 2)$, and find the equation of the tangent line at that point. Using the derivative found in the example, you know that the slope of the tangent line when $t = 1$ is $m = -2$. This means that the tangent line at the point $(1, 2)$ is

$$y - 2 = -2(t - 1) \text{ or}$$
$$y = -2t + 4.$$

By graphing $y = 2/t$ and $y = -2t + 4$ on the same screen, as shown below, you can confirm that the line is tangent to the graph at the point $(1, 2)$.

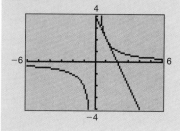

In many applications, it is convenient to use a variable other than $x$ as the independent variable. Example 7 shows a function that uses $t$ as the independent variable.

## EXAMPLE 7   Finding a Derivative

Find the derivative of $y$ with respect to $t$ for the function

$$y = \frac{2}{t}.$$

**Solution**

Consider $y = f(t)$, and use the limit process as follows.

$$\frac{dy}{dt} = \lim_{\Delta t \to 0} \frac{f(t + \Delta t) - f(t)}{\Delta t} \qquad \text{Set up difference quotient.}$$

$$= \lim_{\Delta t \to 0} \frac{\left( \dfrac{2}{t + \Delta t} - \dfrac{2}{t} \right)}{\Delta t} \qquad \text{Use } f(t) = 2/t.$$

$$= \lim_{\Delta t \to 0} \frac{\left[ \dfrac{2t - 2t - 2\Delta t}{t(t + \Delta t)} \right]}{\Delta t} \qquad \text{Expand terms.}$$

$$= \lim_{\Delta t \to 0} \frac{-2\Delta t}{t(\Delta t)(t + \Delta t)} \qquad \text{Factor and cancel.}$$

$$= \lim_{\Delta t \to 0} \frac{-2}{t(t + \Delta t)} \qquad \text{Simplify.}$$

$$= -\frac{2}{t^2} \qquad \text{Evaluate the limit.}$$

Thus, the derivative of $y$ with respect to $t$ is

$$\frac{dy}{dt} = -\frac{2}{t^2}.$$

Remember that the derivative of a function gives you a formula for finding the slope of the tangent at any point on the function's graph. For example, the slope of the tangent line to the graph of $f$ at the point $(1, 2)$ is given by

$$f'(1) = -\frac{2}{1^2} = -2.$$

To find the slope of the graph at other points, substitute the $t$-coordinate of the point into the derivative, as shown below.

| Point | $t$-Coordinate | Slope |
|---|---|---|
| $(2, 1)$ | $t = 2$ | $m = f'(2) = -\dfrac{2}{2^2} = -\dfrac{1}{2}$ |
| $(-2, -1)$ | $t = -2$ | $m = f'(-2) = -\dfrac{2}{(-2)^2} = -\dfrac{1}{2}$ |

## Differentiability and Continuity

Not every function is differentiable. Figure 2.11 shows some common situations in which a function will not be differentiable at a point—vertical tangent lines, discontinuities, and sharp turns in the graph. Each of the functions shown in Figure 2.11 is differentiable at every value of *x except x* = 0.

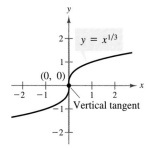
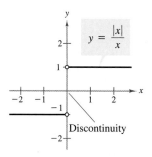
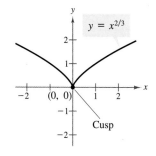
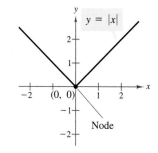

**FIGURE 2.11**    Functions That Are Not Differentiable at (0, 0)

From Figure 2.11, you can see that continuity is not a strong enough condition to guarantee differentiability. [All but one of the functions are continuous at (0, 0), but none is differentiable there.] On the other hand, if a function is differentiable at a point, then it must be continuous. This important result is stated in the following theorem.

---

### Differentiability Implies Continuity

If a function is differentiable at $x = c$, then it is continuous at $x = c$.

---

### *Group Discussion*    *A Graphing Utility Experiment*

Use a graphing utility to sketch the graph of $f(x) = x^3 - 3x$. On the same screen, graph the following lines.

a. $y = 2$

b. $y = x - 3$

c. $y = -3x$

d. $y = -x + 1$

Which of these lines appear to be tangent to the graph of $f$? If the line appears to be tangent, what is the point of tangency? Explain how you can verify your conclusion analytically.

*Warm Up*

The following warm-up exercises involve skills that were covered in earlier sections. You will use these skills in the exercise set for this section.

In Exercises 1 and 2, find an equation of the line containing $P$ and $Q$.

**1.** $P(2, 1)$, $Q(2, 4)$

**2.** $P(2, 2)$, $Q(-5, 2)$

In Exercises 3–6, evaluate the limit.

**3.** $\lim\limits_{\Delta x \to 0} \dfrac{2x\Delta x + (\Delta x)^2}{\Delta x}$

**4.** $\lim\limits_{\Delta x \to 0} \dfrac{3x^2\Delta x + 3x(\Delta x)^2 + (\Delta x)^3}{\Delta x}$

**5.** $\lim\limits_{\Delta x \to 0} \dfrac{1}{x(x + \Delta x)}$

**6.** $\lim\limits_{\Delta x \to 0} \dfrac{(x + \Delta x)^2 - x^2}{\Delta x}$

In Exercises 7–10, find the domain of the function.

**7.** $f(x) = \dfrac{1}{x - 1}$

**8.** $f(x) = \dfrac{1}{5}x^3 - 2x^2 + \dfrac{1}{3}x - 1$

**9.** $f(x) = \dfrac{6x}{x^3 + x}$

**10.** $f(x) = \dfrac{x^2 - 2x - 24}{x^2 + x - 12}$

## EXERCISES 2.1

In Exercises 1–4, trace the graph and sketch the tangent lines at $(x_1, y_1)$ and $(x_2, y_2)$.

In Exercises 5–10, estimate the slope of the graph at the point $(x, y)$. (Each square on the grid is 1 unit by 1 unit.)

**1.**
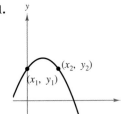

**2.**
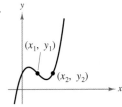

**5.**
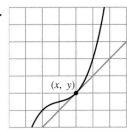

**6.**
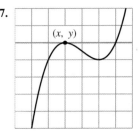

**3.**
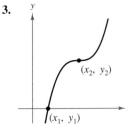

**4.**
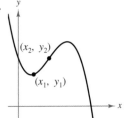

**7.**

**8.**
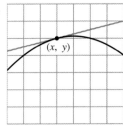

**9.**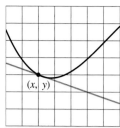

**10.**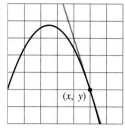

**11.** *Revenue*   The graph represents the revenue $R$ (in millions of dollars per year) for Nike from 1990 through 1996. Estimate the slope of the graph in 1991 and in 1995. *(Source: Nike, Inc.)*

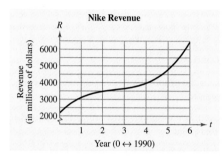

**12.** *Revenue*   The graph represents the revenue $R$ (in millions of dollars per year) for Kmart Corporation from 1990 through 1996. Estimate the slope of the graph in 1992 and in 1995. *(Source: Kmart Corporation)*

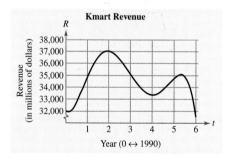

In Exercises 13–26, use the limit definition to find the derivative of the function.

**13.** $f(x) = 3$

**14.** $f(x) = -4$

**15.** $f(x) = -5x + 3$

**16.** $f(x) = \frac{1}{2}x$

**17.** $f(x) = x^2$

**18.** $f(x) = 1 - x^2$

**19.** $f(x) = 3x^2 - 5x - 2$

**20.** $f(x) = x - x^2$

**21.** $h(t) = \sqrt{t - 1}$

**22.** $f(x) = \sqrt{2x}$

**23.** $f(t) = t^3 - 12t$

**24.** $f(t) = t^3 + t^2$

**25.** $f(x) = \dfrac{1}{x^2}$

**26.** $g(s) = \dfrac{1}{s - 1}$

In Exercises 27–36, find the slope of the tangent line to the graph of $f$ at the indicated point.

| *Function* | *Point of Tangency* |
|---|---|
| **27.** $f(x) = 6 - 2x$ | $(2, 2)$ |
| **28.** $f(x) = 2x + 4$ | $(1, 6)$ |
| **29.** $f(x) = -x$ | $(0, 0)$ |
| **30.** $f(x) = 6$ | $(-2, 6)$ |
| **31.** $f(x) = x^2 - 2$ | $(2, 2)$ |
| **32.** $f(x) = x^2 + 2x + 1$ | $(-3, 4)$ |
| **33.** $f(x) = x^3$ | $(2, 8)$ |
| **34.** $f(x) = x^3 + 3$ | $(-2, -5)$ |
| **35.** $f(x) = x^3 + 2x$ | $(1, 3)$ |
| **36.** $f(x) = \sqrt{2x - 2}$ | $(9, 4)$ |

In Exercises 37–42, find an equation of the tangent line to the graph of $f$ at the indicated point. Then verify your result by sketching the graph of $f$ and the tangent line.

| *Function* | *Point of Tangency* |
|---|---|
| **37.** $f(x) = \frac{1}{2}x^2$ | $(2, 2)$ |
| **38.** $f(x) = -x^2$ | $(-1, -1)$ |
| **39.** $f(x) = (x - 1)^2$ | $(-2, 9)$ |
| **40.** $f(x) = 2x^2 - 1$ | $(0, -1)$ |
| **41.** $f(x) = \sqrt{x} + 1$ | $(4, 3)$ |
| **42.** $f(x) = \dfrac{1}{x}$ | $(1, 1)$ |

In Exercises 43–46, find an equation of the line that is tangent to the graph of $f$ and parallel to the given line.

| *Function* | *Line* |
|---|---|
| **43.** $f(x) = -\frac{1}{4}x^2$ | $x + y = 0$ |
| **44.** $f(x) = x^2 + 1$ | $2x + y = 0$ |
| **45.** $f(x) = -\frac{1}{2}x^3$ | $6x + y + 4 = 0$ |
| **46.** $f(x) = \dfrac{1}{\sqrt{x}}$ | $x + 2y - 6 = 0$ |

In Exercises 47–54, describe the x-values at which the function is differentiable. Explain your reasoning.

**47.** $y = |x + 3|$

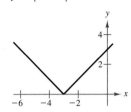

**48.** $y = |x^2 - 9|$

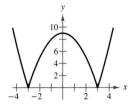

**49.** $y = (x - 3)^{2/3}$

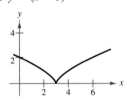

**50.** $y = x^{2/5}$

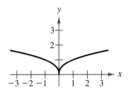

**51.** $y = \sqrt{x - 1}$

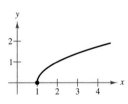

**52.** $y = \dfrac{x^2}{x^2 - 4}$

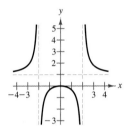

**53.** $y = \begin{cases} x^3 + 3, & x < 0 \\ x^3 - 3, & x \geq 0 \end{cases}$

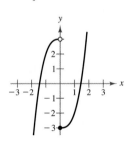

**54.** $y = \begin{cases} x^2, & x \leq 1 \\ -x^2, & x > 1 \end{cases}$

In Exercises 55–58, use a graphing utility to graph f over the interval $[-2, 2]$. Then complete the table by *graphically* estimating the slope of the graph at the indicated points. Finally, evaluate the slope *analytically* and compare your results with those obtained graphically.

| $x$ | $-2$ | $-\frac{3}{2}$ | $-1$ | $-\frac{1}{2}$ | $0$ | $\frac{1}{2}$ | $1$ | $\frac{3}{2}$ | $2$ |
|---|---|---|---|---|---|---|---|---|---|
| $f(x)$ | | | | | | | | | |
| $f'(x)$ | | | | | | | | | |

**55.** $f(x) = \frac{1}{4}x^3$

**56.** $f(x) = \frac{1}{2}x^2$

**57.** $f(x) = -\frac{1}{2}x^3$

**58.** $f(x) = -\frac{3}{2}x^2$

*True or False* In Exercises 59–62, determine whether the statement is true or false. If it is false, explain why or give an example that shows it is false.

**59.** The slope of the graph of $y = x^2$ is different at every point on the graph of $f$.

**60.** If a function is continuous at a point, then it is differentiable at that point.

**61.** If a function is differentiable at a point, then it is continuous at that point.

**62.** A tangent line to a graph can intersect the graph at more than one point.

In Exercises 63 and 64, find the derivative of the given function f. Then use a graphing utility to graph f and its derivative on the same set of coordinate axes. What does the x-intercept of the derivative indicate about the graph of f?

**63.** $f(x) = x^2 - 4x$

**64.** $f(x) = 2 + 6x - x^2$

**65.** *Writing* Use a graphing utility to graph the two functions $f(x) = x^2 + 1$ and $g(x) = |x| + 1$ in the same viewing rectangle. Use the zoom and trace features to analyze the graphs near the point $(0, 1)$. What do you observe? Which function is differentiable at this point? Write a short paragraph describing the geometric significance of differentiability at a point.

# 2.2 Some Rules for Differentiation

*The Constant Rule • The Power Rule • The Constant Multiple Rule • The Sum and Difference Rules • Extended Application: Increasing Sales*

## The Constant Rule

In Section 2.1, you found derivatives by the limit process. This process is tedious, even for simple functions, but fortunately there are rules that greatly simplify differentiation. These rules allow you to calculate derivatives without the *direct* use of limits.

---

### The Constant Rule

The derivative of a constant function is zero. That is,

$$\frac{d}{dx}[c] = 0, \quad c \text{ is a constant.}$$

---

***Proof*** Let $f(x) = c$. Then, by the limit definition of the derivative, you can write

$$f'(x) = \lim_{\Delta x \to 0} \frac{f(x + \Delta x) - f(x)}{\Delta x} = \lim_{\Delta x \to 0} \frac{c - c}{\Delta x} = \lim_{\Delta x \to 0} 0 = 0.$$

Therefore,

$$\frac{d}{dx}[c] = 0.$$

———

***STUDY TIP*** Note in Figure 2.12 that the Constant Rule is equivalent to saying that the slope of a horizontal line is zero.

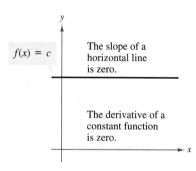

**FIGURE 2.12**

---

## EXAMPLE 1 Finding Derivatives of Constant Functions

(a) $\dfrac{d}{dx}[7] = 0$

(b) If $f(x) = 0$, then $f'(x) = 0$.

(c) If $y = 2$, then $\dfrac{dy}{dx} = 0$.

(d) If $g(t) = -\dfrac{3}{2}$, then $g'(t) = 0$.

———

## The Power Rule

The binomial expansion process is used to prove the Power Rule.

$$(x + \Delta x)^2 = x^2 + 2x\Delta x + (\Delta x)^2$$
$$(x + \Delta x)^3 = x^3 + 3x^2\Delta x + 3x(\Delta x)^2 + (\Delta x)^3$$
$$(x + \Delta x)^n = x^n + nx^{n-1}\Delta x + \underbrace{\frac{n(n-1)x^{n-2}}{2}(\Delta x)^2 + \cdots + (\Delta x)^n}$$

$(\Delta x)^2$ is a factor of these terms.

### The (Simple) Power Rule

$$\frac{d}{dx}[x^n] = nx^{n-1}, \qquad n \text{ is any real number.}$$

**Proof**   We prove only the case in which $n$ is a positive integer. Let $f(x) = x^n$. Using the binomial expansion, you can write

$$f'(x) = \lim_{\Delta x \to 0} \frac{f(x + \Delta x) - f(x)}{\Delta x}$$

$$= \lim_{\Delta x \to 0} \frac{(x + \Delta x)^n - x^n}{\Delta x}$$

$$= \lim_{\Delta x \to 0} \frac{x^n + nx^{n-1}\Delta x + \dfrac{n(n-1)x^{n-2}}{2}(\Delta x)^2 + \cdots + (\Delta x)^n - x^n}{\Delta x}$$

$$= \lim_{\Delta x \to 0} \left[ nx^{n-1} + \frac{n(n-1)x^{n-2}}{2}(\Delta x) + \cdots + (\Delta x)^{n-1} \right]$$

$$= nx^{n-1} + 0 + \cdots + 0$$

$$= nx^{n-1}.$$

For the Power Rule, the case in which $n = 1$ is worth remembering as a separate differentiation rule. That is,

$$\frac{d}{dx}[x] = 1. \qquad \text{The derivative of } x \text{ is } 1.$$

This rule is consistent with the fact that the slope of the line given by $y = x$ is 1, as shown in Figure 2.13.

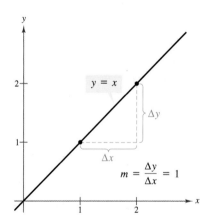

**FIGURE 2.13**   The slope of the line $y = x$ is 1.

## EXAMPLE 2   Applying the Power Rule

| Function | Derivative |
|---|---|
| (a) $f(x) = x^3$ | $f'(x) = 3x^2$ |
| (b) $y = \dfrac{1}{x^2} = x^{-2}$ | $\dfrac{dy}{dx} = (-2)x^{-3} = -\dfrac{2}{x^3}$ |
| (c) $g(t) = t$ | $g'(t) = 1$ |
| (d) $R = x^4$ | $\dfrac{dR}{dx} = 4x^3$ |

In Example 2b, note that *before* differentiating, you should rewrite $1/x^2$ as $x^{-2}$. Rewriting is the first step in *many* differentiation problems.

| Function: | Rewrite: | Differentiate: | Simplify: |
|---|---|---|---|
| $y = \dfrac{1}{x^2}$ | $y = x^{-2}$ | $\dfrac{dy}{dx} = (-2)x^{-3}$ | $\dfrac{dy}{dx} = -\dfrac{2}{x^3}$ |

Remember that the derivative of a function $f$ is another function that gives the slope of the graph of $f$ at any point at which $f$ is differentiable. Thus, you can use the derivative to find slopes, as shown in Example 3.

## EXAMPLE 3   Finding the Slope of a Graph

Find the slope of the graph of

$$f(x) = x^2 \qquad \text{Original function}$$

when $x = -2, -1, 0, 1,$ and 2.

### Solution

Begin by using the Power Rule to find the derivative of $f$.

$$f'(x) = 2x \qquad \text{Derivative}$$

You can use the derivative to find the slope of the graph of $f$, as follows.

| x-Value | Slope of Graph of f |
|---|---|
| $x = -2$ | $m = f'(-2) = 2(-2) = -4$ |
| $x = -1$ | $m = f'(-1) = 2(-1) = -2$ |
| $x = 0$ | $m = f'(0) = 2(0) = 0$ |
| $x = 1$ | $m = f'(1) = 2(1) = 2$ |
| $x = 2$ | $m = f'(2) = 2(2) = 4$ |

The graph of $f$ is shown in Figure 2.14.

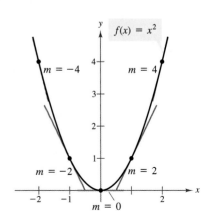

**FIGURE 2.14**

## The Constant Multiple Rule

To prove the Constant Multiple Rule, the following property of limits is used.

$$\lim_{x \to a} cg(x) = c\left[\lim_{x \to a} g(x)\right]$$

### The Constant Multiple Rule

If $f$ is a differentiable function of $x$, and $c$ is a real number, then

$$\frac{d}{dx}[cf(x)] = cf'(x), \qquad c \text{ is a constant.}$$

**Proof**   Apply the definition of the derivative to produce

$$\frac{d}{dx}[cf(x)] = \lim_{\Delta x \to 0} \frac{cf(x + \Delta x) - cf(x)}{\Delta x}$$

$$= \lim_{\Delta x \to 0} c\left[\frac{f(x + \Delta x) - f(x)}{\Delta x}\right]$$

$$= c\left[\lim_{\Delta x \to 0} \frac{f(x + \Delta x) - f(x)}{\Delta x}\right]$$

$$= cf'(x).$$

Informally, the Constant Multiple Rule states that constants can be factored out of the differentiation process.

$$\frac{d}{dx}[cf(x)] = c\frac{d}{dx}[\bigcirc f(x)] = cf'(x)$$

The usefulness of this rule is often overlooked, especially when the constant appears in the denominator, as follows.

$$\frac{d}{dx}\left[\frac{f(x)}{c}\right] = \frac{d}{dx}\left[\frac{1}{c}f(x)\right] = \frac{1}{c}\left(\frac{d}{dx}[\bigcirc f(x)]\right) = \frac{1}{c}f'(x)$$

To use the Constant Multiple Rule efficiently, look for constants that can be factored out *before* differentiating. For example,

$$\frac{d}{dx}[5x^2] = 5\frac{d}{dx}[x^2] = 5(2x) = 10x$$

and

$$\frac{d}{dx}\left[\frac{x^2}{5}\right] = \frac{1}{5}\left(\frac{d}{dx}[x^2]\right) = \frac{1}{5}(2x) = \frac{2}{5}x.$$

# EXAMPLE 4  Using the Power and Constant Multiple Rules

Differentiate the following functions.

(a) $y = 2x^{1/2}$     (b) $f(t) = \dfrac{4t^2}{5}$

**Solution**

(a) Using the Constant Multiple Rule and the Power Rule, you can write

$$\frac{dy}{dx} = \frac{d}{dx}[2x^{1/2}] = \underbrace{2\frac{d}{dx}[x^{1/2}]}_{\text{Constant Multiple Rule}} = \underbrace{2\left(\frac{1}{2}x^{-1/2}\right)}_{\text{Power Rule}} = x^{-1/2} = \frac{1}{\sqrt{x}}.$$

(b) Begin by rewriting $f(t)$ as

$$f(t) = \frac{4t^2}{5} = \frac{4}{5}t^2.$$

Then, use the Constant Multiple Rule and the Power Rule to obtain

$$f'(t) = \frac{d}{dt}\left[\frac{4}{5}t^2\right] = \frac{4}{5}\left[\frac{d}{dt}(t^2)\right] = \frac{4}{5}(2t) = \frac{8}{5}t.$$

You may find it helpful to combine the Constant Multiple Rule and the Power Rule into one combined rule.

$$\frac{d}{dx}[cx^n] = cnx^{n-1}, \qquad n \text{ is a real number, } c \text{ is a constant.}$$

For instance, in Example 4b, you can apply this combined rule to obtain

$$\frac{d}{dt}\left[\frac{4}{5}t^2\right] = \left(\frac{4}{5}\right)(2)(t) = \frac{8}{5}t.$$

The three functions in the next example are simple, yet errors are frequently made in differentiating functions involving a constant multiple of the first power of $x$. Keep in mind that

$$\frac{d}{dx}[cx] = c, \qquad c \text{ is a constant.}$$

# EXAMPLE 5  Applying the Constant Multiple Rule

|   | *Original Function* | *Derivative* |
|---|---|---|
| (a) | $y = -\dfrac{3x}{2}$ | $y' = -\dfrac{3}{2}$ |
| (b) | $y = 3\pi x$ | $y' = 3\pi$ |
| (c) | $y = -\dfrac{x}{2}$ | $y' = -\dfrac{1}{2}$ |

**Technology**

If you have access to a symbolic differentiation utility, such as *Derive, Mathematica, Mathcad, Maple,* or the *TI-92,* try using it to confirm the derivatives shown in this section. For instance, if you use *Derive* for Windows to differentiate the function in Example 4b, you will obtain the following.

Given, Expression (4t^2)/5

#1: $\dfrac{4t^2}{5}$

Calculus, Differentiate

#2: $\dfrac{d}{dt}\dfrac{4t^2}{5}$

Simplify, Basic

#3: $\dfrac{8t}{5}$

## Technology

Most graphing utilities have a built-in program to approximate the derivative of a function at a specific point. Here is a sample program for the *TI-83* that lets you conveniently evaluate the derivative of a function at several points $x$.

```
PROGRAM : DERIV
: Lbl 1
: Input "ENTER X", X
: Deriv(Y₁, X, X, .001)→D
: Disp "DERIVATIVE", D
: Goto 1
: End
```

To use this program, enter the function in $Y_1$. Then run the program—it will allow you to approximate the derivative of the function at several values of $x$.

**ALGEBRA**

### TIP

When differentiating functions involving radicals, you should rewrite the function with rational exponents. For instance, you should rewrite $y = \sqrt[3]{x}$ as $y = x^{1/3}$, and you should rewrite

$$y = \frac{1}{\sqrt[3]{x^4}} \text{ as } y = x^{-4/3}.$$

Parentheses can play an important role in the use of the Constant Multiple Rule and the Power Rule. In Example 6, be sure you understand the mathematical conventions involving the use of parentheses.

## EXAMPLE 6    Using Parentheses When Differentiating

Find the derivative of each function.

(a) $y = \dfrac{5}{2x^3}$    (b) $y = \dfrac{5}{(2x)^3}$    (c) $y = \dfrac{7}{3x^{-2}}$    (d) $y = \dfrac{7}{(3x)^{-2}}$

*Solution*

| Function | Rewrite | Differentiate | Simplify |
|---|---|---|---|
| (a) $y = \dfrac{5}{2x^3}$ | $y = \dfrac{5}{2}(x^{-3})$ | $y' = \dfrac{5}{2}(-3x^{-4})$ | $y' = -\dfrac{15}{2x^4}$ |
| (b) $y = \dfrac{5}{(2x)^3}$ | $y = \dfrac{5}{8}(x^{-3})$ | $y' = \dfrac{5}{8}(-3x^{-4})$ | $y' = -\dfrac{15}{8x^4}$ |
| (c) $y = \dfrac{7}{3x^{-2}}$ | $y = \dfrac{7}{3}(x^2)$ | $y' = \dfrac{7}{3}(2x)$ | $y' = \dfrac{14x}{3}$ |
| (d) $y = \dfrac{7}{(3x)^{-2}}$ | $y = 63(x^2)$ | $y' = 63(2x)$ | $y' = 126x$ |

## EXAMPLE 7    Differentiating Radical Functions

Find the derivative of each function.

(a) $y = \sqrt{x}$    (b) $y = \dfrac{1}{2\sqrt[3]{x^2}}$    (c) $y = \sqrt{2x}$

*Solution*

| Function | Rewrite | Differentiate | Simplify |
|---|---|---|---|
| (a) $y = \sqrt{x}$ | $y = x^{1/2}$ | $y' = \left(\dfrac{1}{2}\right)x^{-1/2}$ | $y' = \dfrac{1}{2\sqrt{x}}$ |
| (b) $y = \dfrac{1}{2\sqrt[3]{x^2}}$ | $y = \dfrac{1}{2}x^{-2/3}$ | $y' = \dfrac{1}{2}\left(-\dfrac{2}{3}\right)x^{-5/3}$ | $y' = -\dfrac{1}{3x^{5/3}}$ |
| (c) $y = \sqrt{2x}$ | $y = \sqrt{2}(x^{1/2})$ | $y' = \sqrt{2}\left(\dfrac{1}{2}\right)x^{-1/2}$ | $y' = \dfrac{1}{\sqrt{2x}}$ |

## The Sum and Difference Rules

The next two rules are ones that you might expect to be true, and you may have used them without thinking about it. For instance, if you were asked to differentiate $y = 3x + 2x^3$, you would probably write

$$y' = 3 + 6x^2$$

without questioning your answer. The validity of differentiating a sum term by term is given by the Sum and Difference Rules.

---

### The Sum and Difference Rules

The derivative of the sum (or difference) of two differentiable functions is the sum (or difference) of their derivatives.

$$\frac{d}{dx}[f(x) + g(x)] = f'(x) + g'(x) \qquad \text{Sum Rule}$$

$$\frac{d}{dx}[f(x) - g(x)] = f'(x) - g'(x) \qquad \text{Difference Rule}$$

---

**Proof**   Let $h(x) = f(x) + g(x)$. Then, you can prove the Sum Rule as follows.

$$h'(x) = \lim_{\Delta x \to 0} \frac{h(x + \Delta x) - h(x)}{\Delta x}$$

$$= \lim_{\Delta x \to 0} \frac{f(x + \Delta x) + g(x + \Delta x) - f(x) - g(x)}{\Delta x}$$

$$= \lim_{\Delta x \to 0} \frac{f(x + \Delta x) - f(x) + g(x + \Delta x) - g(x)}{\Delta x}$$

$$= \lim_{\Delta x \to 0} \left[ \frac{f(x + \Delta x) - f(x)}{\Delta x} + \frac{g(x + \Delta x) - g(x)}{\Delta x} \right]$$

$$= \lim_{\Delta x \to 0} \frac{f(x + \Delta x) - f(x)}{\Delta x} + \lim_{\Delta x \to 0} \frac{g(x + \Delta x) - g(x)}{\Delta x}$$

$$= f'(x) + g'(x)$$

Thus,

$$\frac{d}{dx}[f(x) + g(x)] = f'(x) + g'(x).$$

The Difference Rule can be proved in a similar manner.

*STUDY TIP*   Look back at Example 6 on page 87. Notice that the example asks for the derivative of the difference of two functions. Verify this result by using the Difference Rule.

The Sum and Difference Rules can be extended to the sum or difference of any finite number of functions. For instance, if $y = f(x) + g(x) + h(x)$, then $y' = f'(x) + g'(x) + h'(x)$.

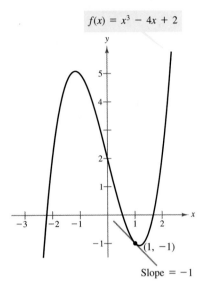

$f(x) = x^3 - 4x + 2$

Slope = $-1$

**FIGURE 2.15**

With the four differentiation rules listed in this section, you can differentiate *any* polynomial function.

### EXAMPLE 8   Using the Sum and Difference Rules

Find the slope of the graph of $f(x) = x^3 - 4x + 2$ at the point $(1, -1)$.

**Solution**

The derivative of $f(x)$ is

$$f'(x) = 3x^2 - 4.$$

Thus, the slope of the graph of $f$ at $(1, -1)$ is

$$\text{Slope} = f'(1) = 3(1)^2 - 4 = -1,$$

as shown in Figure 2.15.

Example 8 illustrates the use of the derivative for determining the shape of a graph. A rough sketch of the graph of $f(x) = x^3 - 4x + 2$ might lead you to think that the point $(1, -1)$ is a minimum point of the graph. After finding the slope at this point to be $-1$, however, you can conclude that the minimum point (where the slope is 0) is farther to the right. (You will study techniques for finding minimum and maximum points in Section 3.2.)

### EXAMPLE 9   Using the Sum and Difference Rules

Find an equation of the tangent line to the graph of

$$g(x) = -\frac{1}{2}x^4 + 3x^3 - 2x$$

at the point $\left(-1, -\frac{3}{2}\right)$.

**Solution**

The derivative of $g(x)$ is

$$g'(x) = -2x^3 + 9x^2 - 2,$$

which implies that the slope of the graph at the point $\left(-1, -\frac{3}{2}\right)$ is

$$\text{Slope} = g'(-1) = -2(-1)^3 + 9(-1)^2 - 2$$
$$= 2 + 9 - 2$$
$$= 9,$$

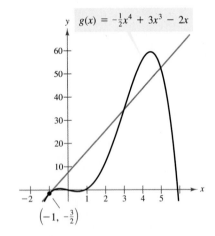

$g(x) = -\frac{1}{2}x^4 + 3x^3 - 2x$

$\left(-1, -\frac{3}{2}\right)$

**FIGURE 2.16**

as shown in Figure 2.16. Using the point-slope form, you can write the equation of the tangent line at $\left(-1, -\frac{3}{2}\right)$ as follows.

$$y - \left(-\frac{3}{2}\right) = 9[x - (-1)] \qquad \text{Point-slope form}$$

$$y = 9x + \frac{15}{2} \qquad \text{Equation of tangent line}$$

# Extended Application: Increasing Sales

## EXAMPLE 10 Modeling Sales

From 1991 through 1996, the revenue $R$ (in millions of dollars) for Microsoft Corporation can be modeled by

$$R = 76.9t^3 - 624.2t^2 + 2427.8t - 81.5, \qquad 1 \le t \le 6,$$

where $t = 1$ represents 1991. At what rate was Microsoft's revenue changing in 1992? *(Source: Microsoft Corporation)*

### Solution

One way to answer this question is to find the derivative of the revenue model with respect to time.

$$\frac{dR}{dt} = 230.7t^2 - 1248.4t + 2427.8, \qquad 1 \le t \le 6$$

In 1992 (when $t = 2$), the rate of change of the revenue with respect to time is given by

$$\frac{dR}{dt} = 230.7(2)^2 - 1248.4(2) + 2427.8$$

$$= 853.8$$

Because $R$ is measured in millions of dollars and $t$ is measured in years, it follows that the derivative $dR/dt$ is measured in millions of dollars per year. Thus, at the end of 1992, Microsoft's revenues were increasing at a rate of about $850 million per year, as shown in Figure 2.17.

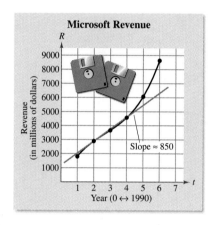

**FIGURE 2.17**

---

*Group Discussion*     *Units for Rates of Change*

In Example 10, the units for $R$ are millions of dollars and the units for $dR/dt$ are millions of dollars per year. State the units for the derivatives of each of the following models.

a. A population model that gives the population $P$ (in millions of people) of the United States in terms of the year, $t$. What are the units for $dP/dt$?

b. A position model that gives the height $s$ (in feet) of an object in terms of the time $t$ (in seconds). What are the units for $ds/dt$?

c. A demand model that gives the price per unit $p$ (in dollars) of a product in terms of the number of units $x$ sold. What are the units for $dp/dx$?

*Warm Up*

The following warm-up exercises involve skills that were covered in earlier sections. You will use these skills in the exercise set for this section.

In Exercises 1 and 2, evaluate each expression when $x = 2$.

**1.** (a) $2x^2$ (b) $(2x)^2$ (c) $2x^{-2}$

**2.** (a) $\dfrac{1}{(3x)^2}$ (b) $\dfrac{1}{4x^3}$ (c) $\dfrac{(2x)^{-3}}{4x^{-2}}$

In Exercises 3–6, simplify the expression.

**3.** $4(3)x^3 + 2(2)x$

**4.** $\frac{1}{2}(3)x^2 - \frac{3}{2}x^{1/2}$

**5.** $\left(\frac{1}{4}\right)x^{-3/4}$

**6.** $\frac{1}{3}(3)x^2 - 2\left(\frac{1}{2}\right)x^{-1/2} + \frac{1}{3}x^{-2/3}$

In Exercises 7–10, solve the equation.

**7.** $3x^2 + 2x = 0$

**8.** $x^3 - x = 0$

**9.** $x^2 + 8x - 20 = 0$

**10.** $x^2 - 10x - 24 = 0$

## EXERCISES 2.2

In Exercises 1–4, find the slope of the tangent line to $y = x^n$ at the point $(1, 1)$.

**1.** (a) $y = x^2$        (b) $y = x^{1/2}$

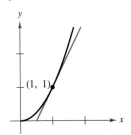
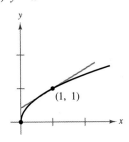

**2.** (a) $y = x^{3/2}$        (b) $y = x^3$

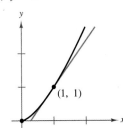
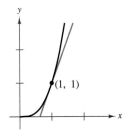

**3.** (a) $y = x^{-1}$        (b) $y = x^{-1/3}$

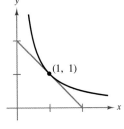
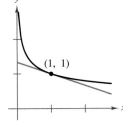

**4.** (a) $y = x^{-1/2}$        (b) $y = x^{-2}$

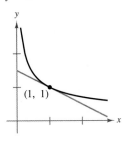
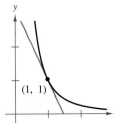

In Exercises 5–20, use the differentiation rules in this section to find the derivative of the function.

**5.** $y = 3$

**6.** $f(x) = -2$

**7.** $f(x) = x + 1$

**8.** $g(x) = 3x - 1$

**9.** $g(x) = x^2 + 4$

**10.** $y = t^2 + 2t - 3$

**11.** $f(t) = -3t^2 + 2t - 4$

**12.** $y = x^3 - 9$

**13.** $s(t) = t^3 - 2t + 4$

**14.** $y = 2x^3 - x^2 + 3x - 1$

**15.** $y = 4t^{4/3}$

**16.** $h(x) = x^{5/2}$

**17.** $f(x) = 4\sqrt{x}$

**18.** $g(x) = 4\sqrt[3]{x} + 2$

**19.** $y = 4x^{-2} + 2x^2$

**20.** $s(t) = 4t^{-1} + 1$

In Exercises 21–26, use Example 6 as a model to find the derivative.

| Function | Rewrite | Differentiate | Simplify |
|---|---|---|---|
| **21.** $y = \dfrac{1}{4x^3}$ | | | |
| **22.** $y = \dfrac{2}{3x^2}$ | | | |
| **23.** $y = \dfrac{1}{(4x)^3}$ | | | |
| **24.** $y = \dfrac{\pi}{(3x)^2}$ | | | |
| **25.** $y = \dfrac{\sqrt{x}}{x}$ | | | |
| **26.** $y = \dfrac{4}{x^{-3}}$ | | | |

In Exercises 27–32, find the value of the derivative of the function at the indicated point.

| Function | Point |
|---|---|
| **27.** $f(x) = \dfrac{1}{x}$ | $(1, 1)$ |
| **28.** $f(x) = -\frac{1}{2} + \frac{7}{5}x^3$ | $\left(0, -\frac{1}{2}\right)$ |
| **29.** $f(t) = 4 - \dfrac{4}{3t}$ | $\left(\frac{1}{2}, \frac{4}{3}\right)$ |
| **30.** $y = 3x\left(x^2 - \dfrac{2}{x}\right)$ | $(2, 18)$ |
| **31.** $y = (2x + 1)^2$ | $(0, 1)$ |
| **32.** $f(x) = 3(5 - x)^2$ | $(5, 0)$ |

In Exercises 33–42, find $f'(x)$.

**33.** $f(x) = x^2 - \dfrac{4}{x}$

**34.** $f(x) = x^2 - 3x - 3x^{-2} + 5x^{-3}$

**35.** $f(x) = x^2 - 2x - \dfrac{2}{x^4}$

**36.** $f(x) = x^2 + 4x + \dfrac{1}{x}$

**37.** $f(x) = \dfrac{2x^3 - 4x^2 + 3}{x^2}$

**38.** $f(x) = \dfrac{2x^2 - 3x + 1}{x}$

**39.** $f(x) = x(x^2 + 1)$

**40.** $f(x) = (x^2 + 2x)(x + 1)$

**41.** $f(x) = x^{4/5}$

**42.** $f(x) = x^{1/3} - 1$

In Exercises 43–46, find an equation of the tangent line to the graph of the function at the indicated point.

| Function | Point |
|---|---|
| **43.** $y = -2x^4 + 5x^2 - 3$ | $(1, 0)$ |
| **44.** $y = x^3 + x$ | $(-1, -2)$ |
| **45.** $f(x) = \sqrt[3]{x} + \sqrt[5]{x}$ | $(1, 2)$ |
| **46.** $f(x) = \dfrac{1}{\sqrt[3]{x^2}}$ | $(-1, 1)$ |

In Exercises 47–50, determine the point(s), if any, at which the graph of the function has a horizontal tangent line.

**47.** $y = -x^4 + 3x^2 - 1$

**48.** $y = \frac{1}{2}x^2 + 5x$

**49.** $y = x^3 + x$

**50.** $y = x^2 + 1$

In Exercises 51 and 52, perform the following.

(a) Sketch the graphs of $f, g,$ and $h$ on the same set of coordinate axes.

(b) Find $f'(1), g'(1),$ and $h'(1)$.

(c) Sketch the graph of the tangent lines to each graph when $x = 1$.

**51.** $f(x) = x^3$
   $g(x) = x^3 + 3$
   $h(x) = x^3 - 2$

**52.** $f(x) = \sqrt{x}$
   $g(x) = \sqrt{x} + 4$
   $h(x) = \sqrt{x} - 2$

**53.** Use the Constant Rule, the Constant Multiple Rule, and the Sum Rule to find $h'(1)$ given that $f'(1) = 3$.

(a) $h(x) = f(x) - 2$

(b) $h(x) = 2f(x)$

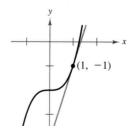

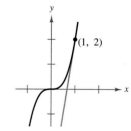

(c) $h(x) = -f(x)$         (d) $h(x) = -1 + 2f(x)$

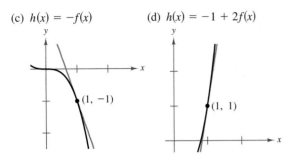

(1, −1)                    (1, 1)

**54. *Revenue***   The revenue $R$ (in millions of dollars per year) for Nike from 1990 through 1996 can be modeled by

$$R = 58.16t^3 - 449.37t^2 + 1311.9t + 2184.49,$$

where $t = 0$ represents 1990.   *(Source: Nike, Inc.)*

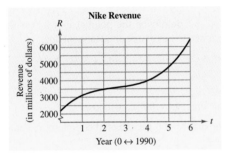

**Nike Revenue**

(a) Find the slope of the graph in 1991 and in 1995.

(b) Compare your results with those obtained in Exercise 11 in Section 2.1.

(c) What are the units for the slope of the graph? Interpret the slope of the graph in the context of the problem.

**55. *Revenue***   The revenue $R$ (in millions of dollars per year) for Kmart Corporation from 1990 through 1996 can be modeled by

$$R = -76.5t^5 + 1088.7t^4 - 5195.3t^3 + 8780.8t^2 - 1763.4t + 32{,}023.7,$$

where $t = 0$ represents 1990. *(Source: Kmart Corporation)*

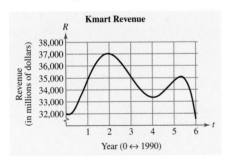

**Kmart Revenue**

(a) Find the slope of the graph in 1992 and in 1995.

(b) Compare your results with those obtained in Exercise 12 in Section 2.1.

(c) What are the units for the slope of the graph? Interpret the slope of the graph in the context of the problem.

**56. *Cost***   The variable cost for manufacturing an electrical component is $7.75 per unit, and the fixed cost is $500. Write the cost $C$ as a function of $x$, the number of units produced. Show that the derivative of this cost function is constant and equal to the variable cost.

**57. *Cost***   A college club raises funds by selling candy bars for $1.00 each. The club pays $0.60 for each candy bar and has annual fixed costs of $250. Write the profit $P$ as a function of $x$, the number of candy bars sold. Show that the derivative of the profit function is a constant and that it is equal to the profit on each candy bar sold.

**58. *Psychology: Migraine Prevalence***   The graph illustrates the prevalence of migraine headaches in men and women of selected income groups.  *(Source: Adapted from Sue/Sue/ Sue, Understanding Abnormal Behavior, Fifth Edition)*

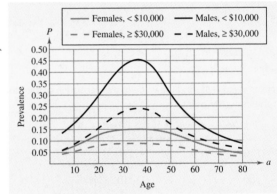

(a) Write a short paragraph describing your general observations about the prevalence of migraines in women and men with respect to age group and income bracket.

(b) Describe the graph of the derivative of each curve, and explain the significance of each derivative. Include an explanation of the units of the derivatives, and indicate the time intervals in which the derivatives would be positive and negative.

In Exercises 59 and 60, use a graphing utility to graph $f$ and $f'$ over the given interval. Determine any points at which the graph of $f$ has horizontal tangents.

| Function | Interval |
|---|---|
| **59.** $f(x) = 4.1x^3 - 12x^2 + 2.5x$ | $[0, 3]$ |
| **60.** $f(x) = x^3 - 1.4x^2 - 0.96x + 1.44$ | $[-2, 2]$ |

# 2.3 Rates of Change: Velocity and Marginals

*Average Rate of Change • Instantaneous Rate of Change and Velocity • Rates of Change in Economics: Marginals*

## Average Rate of Change

In Sections 2.1 and 2.2, you studied the two primary applications of derivatives.

1. **Slope** The derivative of $f$ is a function that gives the slope of the graph of $f$ at a point $(x, f(x))$.

2. **Rate of Change** The derivative of $f$ is a function that gives the rate of change of $f(x)$ with respect to $x$ at the point $(x, f(x))$.

In this section, you will see that there are many real-life applications of rates of change. A few are velocity, acceleration, population growth rates, unemployment rates, production rates, and water flow rates. Although rates of change often involve change with respect to time, you can investigate the rate of change of one variable with respect to any other related variable.

When determining the rate of change of one variable with respect to another, you must be careful to distinguish between *average* and *instantaneous* rates of change. The distinction between these two rates of change is comparable to the distinction between the slope of the secant line through two points on a graph and the slope of the tangent line at one point on the graph.

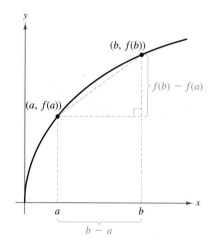

**FIGURE 2.18**

### Definition of Average Rate of Change

If $y = f(x)$, then the **average rate of change** of $y$ with respect to $x$ on the interval $[a, b]$ is

$$\text{Average rate of change} = \frac{f(b) - f(a)}{b - a} = \frac{\Delta y}{\Delta x}.$$

Note that $f(a)$ is the value of the function at the *left* endpoint of the interval, $f(b)$ is the value of the function at the *right* endpoint of the interval, and $b - a$ is the width of the interval, as shown in Figure 2.18.

*STUDY TIP* In real-life problems, it is important to list the units of measure for a rate of change. The units for $\Delta y/\Delta x$ are "$y$-units" per "$x$-units." For example, if $y$ is measured in miles and $x$ is measured in hours, then $\Delta y/\Delta x$ is measured in *miles per hour.*

### EXAMPLE 1  Drug Concentration

The concentration $C$ (in milligrams per milliliter) of a drug in a patient's blood-stream is monitored over 10-minute intervals for 2 hours, where $t$ is measured in minutes, as shown in the table. Find the average rates of change over the following intervals.

(a) $[0, 10]$    (b) $[0, 20]$    (c) $[100, 110]$

| $t$ | 0 | 10 | 20 | 30 | 40 | 50 | 60 | 70 | 80 | 90 | 100 | 110 | 120 |
|---|---|---|---|---|---|---|---|---|---|---|---|---|---|
| $C$ | 0 | 2 | 17 | 37 | 55 | 73 | 89 | 103 | 111 | 113 | 113 | 103 | 68 |

**STUDY TIP**  In Example 1, the average rate of change is positive when the concentration increases and negative when the concentration decreases, as shown in Figure 2.19.

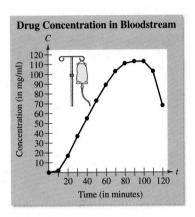

**FIGURE 2.19**

**Solution**

(a) For the interval $[0, 10]$, the average rate of change is

Value of $C$ at right endpoint
Value of $C$ at left endpoint

$$\frac{\Delta C}{\Delta t} = \frac{2 - 0}{10 - 0} = \frac{2}{10} = 0.2 \text{ mg per ml/min.}$$

Width of interval

(b) For the interval $[0, 20]$, the average rate of change is

$$\frac{\Delta C}{\Delta t} = \frac{17 - 0}{20 - 0} = \frac{17}{20} = 0.85 \text{ mg per ml/min.}$$

(c) For the interval $[100, 110]$, the average rate of change is

$$\frac{\Delta C}{\Delta t} = \frac{103 - 113}{110 - 100} = \frac{-10}{10} = -1 \text{ mg per ml/min.}$$

The rates of change in Example 1 are in milligrams per milliliter per minute because the concentration is measured in milligrams per milliliter and the time is measured in minutes.

Concentration is measured in milligrams per milliliter.

Rate of change is measured in milligrams per milliliter per minute.

$$\frac{\Delta C}{\Delta t} = \frac{2 - 0}{10 - 0} = \frac{2}{10} = 0.2 \text{ mg per ml/min}$$

Time is measured in minutes.

A common application of an average rate of change is to find the **average velocity** of an object that is moving in a straight line. That is,

$$\text{Average velocity} = \frac{\text{change in distance}}{\text{change in time}}.$$

This formula is demonstrated in Example 2.

## EXAMPLE 2    Finding an Average Velocity

If a free-falling object is dropped from a height of 100 feet, and *air resistance is neglected,* the height $h$ (in feet) of the object at time $t$ (in seconds) is given by

$$h = -16t^2 + 100,$$

as indicated in Figure 2.20. Find the average velocity of the object over the following intervals.

(a) $[1, 2]$        (b) $[1, 1.5]$        (c) $[1, 1.1]$

### Solution

You can use the position equation $h = -16t^2 + 100$ to determine the heights at $t = 1$, $t = 1.1$, $t = 1.5$, and $t = 2$, as shown in the table.

| $t$ (in seconds) | 0 | 1 | 1.1 | 1.5 | 2 |
|---|---|---|---|---|---|
| $h$ (in feet) | 100 | 84 | 80.64 | 64 | 36 |

(a) For the interval $[1, 2]$, the object falls from a height of 84 feet to a height of 36 feet. Thus, the average velocity is

$$\frac{\Delta h}{\Delta t} = \frac{36 - 84}{2 - 1} = \frac{-48}{1} = -48 \text{ ft/sec.}$$

(b) For the interval $[1, 1.5]$, the average velocity is

$$\frac{\Delta h}{\Delta t} = \frac{64 - 84}{1.5 - 1} = \frac{-20}{0.5} = -40 \text{ ft/sec.}$$

(c) For the interval $[1, 1.1]$, the average velocity is

$$\frac{\Delta h}{\Delta t} = \frac{80.64 - 84}{1.1 - 1} = \frac{-3.36}{0.1} = -33.6 \text{ ft/sec.}$$

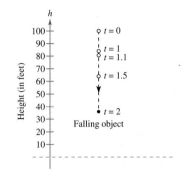

**FIGURE 2.20    Some falling objects have considerable air resistance. Other falling objects have negligible air resistance. When modeling a falling-body problem, you must decide whether to account for air resistance or neglect it.**

***STUDY TIP***    In Example 2, the average velocities are negative because the object is moving downward.

## Instantaneous Rate of Change and Velocity

Suppose in Example 2 you wanted to find the rate of change of $h$ at the instant $t = 1$ second. Such a rate is called an **instantaneous rate of change.** You can approximate the instantaneous rate of change at $t = 1$ by calculating the average rate of change over smaller and smaller intervals of the form $[1, 1 + \Delta t]$, as shown in the table. From the table, it seems reasonable to conclude that the instantaneous rate of change of the height when $t = 1$ is $-32$ feet per second.

$\Delta t$ approaches 0

| $\Delta t$ | 1 | 0.5 | 0.1 | 0.01 | 0.001 | 0.0001 | 0 |
|---|---|---|---|---|---|---|---|
| $\dfrac{\Delta h}{\Delta t}$ | $-48$ | $-40$ | $-33.6$ | $-32.16$ | $-32.016$ | $-32.0016$ | $-32$ |

$\dfrac{\Delta h}{\Delta t}$ approaches $-32$

*STUDY TIP* The limit in this definition is the same as the limit in the definition of the derivative of $f$ at $x$. This is the second major interpretation of the derivative—as an *instantaneous rate of change in one variable with respect to another.* Recall that the first interpretation of the derivative is as the slope of the graph of $f$ at $x$.

### Definition of Instantaneous Rate of Change

The **instantaneous rate of change** (or simply **rate of change**) of $y = f(x)$ at $x$ is the limit of the average rate of change on the interval $[x, x + \Delta x]$, as $\Delta x$ approaches 0.

$$\lim_{\Delta x \to 0} \frac{\Delta y}{\Delta x} = \lim_{\Delta x \to 0} \frac{f(x + \Delta x) - f(x)}{\Delta x}$$

If $y$ is a length and $x$ is time, then the rate of change is a **velocity.**

## EXAMPLE 3  Finding a Rate of Change

Find the velocity of the object in Example 2 when $t = 1$.

*Solution*

From Example 2, you know that the height of the falling object is given by

$h = -16t^2 + 100.$     Position function

By taking the derivative of this position function, you obtain the velocity function.

$h'(t) = -32t$     Velocity function

The velocity function gives the velocity at *any* time. Thus, when $t = 1$, the velocity is $h'(1) = -32(1) = -32$ feet per second.

The general **position function** for a free-falling object, neglecting air resistance, is

$$h = -16t^2 + v_0 t + h_0,$$ Position function

where $h$ is the height (in feet), $t$ is the time (in seconds), $v_0$ is the initial velocity (in feet per second), and $h_0$ is the initial height (in feet). Remember that the model assumes that positive velocities indicate upward motion and negative velocities indicate downward motion. The derivative $h' = -32t + v_0$ is the **velocity function.** The absolute value of the velocity is the **speed** of the object.

## EXAMPLE 4   Finding the Velocity of a Diver

At time $t = 0$, a diver jumps from a diving board that is 32 feet high, as shown in Figure 2.21. Because the diver's initial velocity is 16 feet per second, his position function is

$$h = -16t^2 + 16t + 32.$$ Position function

(a)  When does the diver hit the water?

(b)  What is the diver's velocity at impact?

*Solution*

(a)  To find the time at which the diver hits the water, let $h = 0$ and solve for $t$.

$$-16t^2 + 16t + 32 = 0 \qquad \text{Set } h \text{ equal to 0.}$$
$$-16(t^2 - t - 2) = 0 \qquad \text{Factor out common factor.}$$
$$-16(t + 1)(t - 2) = 0 \qquad \text{Factor trinomial.}$$
$$t = -1 \text{ or } t = 2 \qquad \text{Solve for } t.$$

The solution $t = -1$ doesn't make sense in the problem, so you can conclude that the diver hits the water when $t = 2$ seconds.

(b)  The velocity at time $t$ is given by the derivative

$$h' = -32t + 16.$$ Velocity function

The velocity at time $t = 2$ is $h' = -32(2) + 16 = -48$ feet per second.

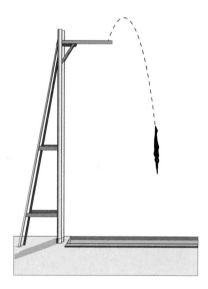

**FIGURE 2.21**

In Example 4, note that the diver's initial velocity is $v_0 = 16$ feet per second (upward) and his initial height is $h_0 = 32$ feet.

Initial velocity is 16 feet per second.
Initial height is 32 feet.

$$h = -16t^2 + 16t + 32$$

*Discovery*

Graph the polynomial function $h = -16t^2 + 16t + 32$ from Example 4 on the domain $0 \le t \le 2$. What is the maximum value of $h$? What is the derivative of $h$ at this maximum point? In general, discuss how the derivative can be used to find the maximum or minimum values of a function.

## Rates of Change in Economics:  Marginals

Another important use of rates of change is in the field of economics. Economists refer to *marginal profit, marginal revenue,* and *marginal cost* as the rates of change of the profit, revenue, and cost with respect to the number $x$ of units produced or sold. An equation that relates these three quantities is

$$P = R - C,$$

where $P$, $R$, and $C$ represent the following quantities.

$$P = \text{total profit}, \quad R = \text{total revenue}, \quad \text{and} \quad C = \text{total cost}$$

The derivatives of these quantities are called the **marginal profit, marginal revenue,** and **marginal cost,** respectively.

$$\frac{dP}{dx} = \text{marginal profit}$$

$$\frac{dR}{dx} = \text{marginal revenue}$$

$$\frac{dC}{dx} = \text{marginal cost}$$

In many business and economics problems, the number of units produced or sold is restricted to positive integer values, as indicated in Figure 2.22(a). (Of course, it could happen that a sale involves half or quarter units, but it is hard to conceive of a sale involving $\sqrt{2}$ units.) The variable that denotes such units is called a **discrete variable.** To analyze a function of a discrete variable $x$, you can temporarily assume that $x$ is a **continuous variable** and is able to take on any real value in a given interval, as indicated in Figure 2.22(b). Then, you can use the methods of calculus to find the $x$-value that corresponds to the marginal revenue, maximum profit, minimum cost, or whatever is called for. Finally, you should round the solution to the nearest sensible $x$-value—cents, dollars, units, or days, depending on the context of the problem.

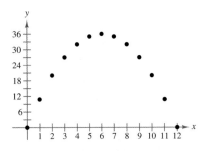

(a)  Function of a Discrete Variable

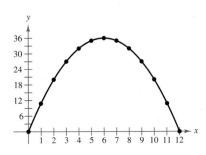

(b)  Function of a Continuous Variable

**FIGURE 2.22**

## EXAMPLE 5   Finding the Marginal Profit

The profit derived from selling $x$ units of an item is given by

$$P = 0.0002x^3 + 10x.$$

(a) Find the marginal profit for a production level of 50 units.

(b) Compare this with the actual gain in profit obtained by increasing the production level from 50 to 51 units.

### Solution

(a) Because the profit is $P = 0.0002x^3 + 10x$, the marginal profit is given by the derivative

$$\frac{dP}{dx} = 0.0006x^2 + 10.$$

When $x = 50$, the marginal profit is

$$\frac{dP}{dx} = 0.0006(50)^2 + 10 \qquad \text{Substitute 50 for } x.$$

$$= 1.5 + 10$$

$$= \$11.50 \text{ per unit.} \qquad \text{Marginal profit for } x = 50$$

(b) For $x = 50$, the actual profit is

$$P = (0.0002)(50)^3 + 10(50) \qquad \text{Substitute 50 for } x.$$

$$= 25 + 500$$

$$= \$525.00, \qquad \text{Actual profit for } x = 50$$

and for $x = 51$, the actual profit is

$$P = (0.0002)(51)^3 + 10(51) \qquad \text{Substitute 51 for } x.$$

$$\approx 26.53 + 510$$

$$= \$536.53. \qquad \text{Actual profit for } x = 51$$

Thus, the additional profit obtained by increasing the production level from 50 to 51 units is

$$536.53 - 525.00 = \$11.53. \qquad \text{Extra profit for one unit}$$

Note that the actual profit increase of \$11.53 (when $x$ increases from 50 to 51 units) can be approximated by the marginal profit of \$11.50 per unit (when $x = 50$), as shown in Figure 2.23.

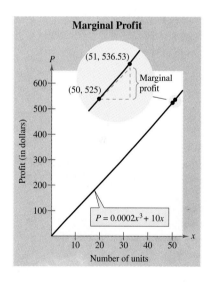

**Marginal Profit**

(51, 536.53)

Marginal profit

(50, 525)

Profit (in dollars)

$P = 0.0002x^3 + 10x$

Number of units

**FIGURE 2.23**

> *STUDY TIP*   The reason the marginal profit gives a good approximation of the actual change in profit is that the graph of $P$ is nearly straight over the interval $50 \le x \le 51$. You will study more about the use of marginals to approximate actual changes in Section 3.8.

The profit function in Example 5 is unusual in that the profit continues to increase as long as the number of units sold increases. In practice, it is more common to encounter situations in which sales can be increased only by lowering the price per item. Such reductions in price will ultimately cause the profit to decline.

The number of units $x$ that consumers are willing to purchase at a given price per unit $p$ is given by the **demand function**

$$p = f(x).$$      Demand function

The total revenue $R$ is then related to the price per unit and the quantity demanded (or sold) by the equation

$$R = xp.$$      Revenue function

## EXAMPLE 6   Finding a Demand Function

A business sells 2000 items per month at a price of $10 each. It is estimated that monthly sales will increase 250 units for each $0.25 reduction in price. Use this information to find the demand function and total revenue function.

### Solution

From the given estimate, $x$ increases 250 units each time $p$ drops $0.25 from the original cost of $10. This is described by the equation

$$x = 2000 + 250\left(\frac{10 - p}{0.25}\right)$$

$$= 2000 + 10{,}000 - 1000p$$

$$= 12{,}000 - 1000p.$$

Solving for $p$ in terms of $x$ produces

$$p = 12 - \frac{x}{1000}.$$      Demand function

This, in turn, implies that the revenue function is

$$R = xp$$      Formula for revenue

$$= x\left(12 - \frac{x}{1000}\right)$$

$$= 12x - \frac{x^2}{1000}.$$      Revenue function

The graph of the demand function is shown in Figure 2.24. Notice that the price decreases as the quantity demanded increases.

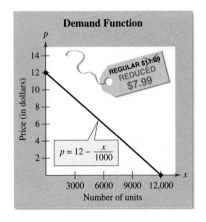

**FIGURE 2.24**

# Technology

## Modeling a Demand Function

To model a demand function, you need data that indicates how many units of a product will sell at a given price. As you might imagine, this type of data is not easy to obtain for a new product. After a product has been on the market a while, however, its sales history can provide the necessary data.

As an example, consider the two bar graphs shown below. From these graphs, you can see that from 1981 through 1996, the price of prerecorded videocassettes dropped from an average price of about $57 to an average price of about $14. During that time, the number of units sold increased from about 1 million to about 392 million. *(Source: Paul Kagan Associates, Inc.)*

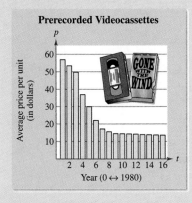

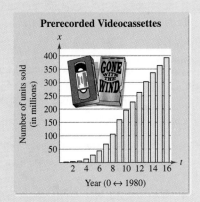

The information in the two bar graphs is combined in the table, where $p$ represents the price (in dollars) and $x$ represents the units sold (in millions).

| $t$ | 1 | 2 | 3 | 4 | 5 | 6 | 7 | 8 |
|---|---|---|---|---|---|---|---|---|
| $x$ | 1.1 | 2.0 | 4.6 | 13.3 | 28.0 | 43.7 | 69.0 | 105.9 |
| $p$ | 56.97 | 53.38 | 49.65 | 36.78 | 29.85 | 22.05 | 17.15 | 15.60 |

| $t$ | 9 | 10 | 11 | 12 | 13 | 14 | 15 | 16 |
|---|---|---|---|---|---|---|---|---|
| $x$ | 160.0 | 196.0 | 226.4 | 263.2 | 303.1 | 336.5 | 363.4 | 392.5 |
| $p$ | 14.45 | 14.40 | 14.26 | 14.21 | 13.92 | 13.80 | 13.66 | 13.59 |

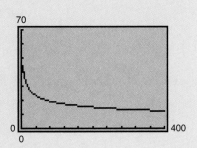

By entering the ordered pairs $(x, p)$ into a calculator or computer that has a power regression program, you can find the following power model for the demand for prerecorded videocassettes: $p = 65.7x^{-0.278}$, $1.1 \leq x \leq 392.5$. A graph of this demand function is shown at the right.

**Demand Function**

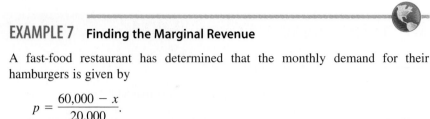

$p = \dfrac{60{,}000 - x}{20{,}000}$

**FIGURE 2.25**   As the price decreases, more hamburgers are sold.

# EXAMPLE 7   Finding the Marginal Revenue

A fast-food restaurant has determined that the monthly demand for their hamburgers is given by

$$p = \frac{60{,}000 - x}{20{,}000}.$$

Figure 2.25 shows that as the price decreases, the quantity demanded increases. The table shows the demand for hamburgers at various prices.

| $x$ | 60,000 | 50,000 | 40,000 | 30,000 | 20,000 | 10,000 | 0 |
|---|---|---|---|---|---|---|---|
| $p$ | $0.00 | $0.50 | $1.00 | $1.50 | $2.00 | $2.50 | $3.00 |

Find the increase in revenue per hamburger for monthly sales of 20,000 hamburgers. In other words, find the marginal revenue when $x = 20{,}000$.

## Solution

Because the demand is given by

$$p = \frac{60{,}000 - x}{20{,}000},$$

the revenue is given by $R = xp$, and you have

$$R = xp = x\left(\frac{60{,}000 - x}{20{,}000}\right)$$

$$= \frac{1}{20{,}000}(60{,}000x - x^2).$$

By differentiating, you can find the marginal revenue to be

$$\frac{dR}{dx} = \frac{1}{20{,}000}(60{,}000 - 2x).$$

Thus, when $x = 20{,}000$, the marginal revenue is

$$\frac{dR}{dx} = \frac{1}{20{,}000}[60{,}000 - 2(20{,}000)]$$

$$= \frac{20{,}000}{20{,}000}$$

$$= \$1 \text{ per unit.}$$

*STUDY TIP*   Writing a demand function in the form $p = f(x)$ is a convention used in economics. From a consumer's point of view, it might seem more reasonable to think that the quantity demanded is a function of the price. Mathematically, however, the two points of view are equivalent because a typical demand function is one-to-one and therefore possesses an inverse. For instance, in Example 7, you could write the demand function as $x = 60{,}000 - 20{,}000p$.

## EXAMPLE 8   Finding the Marginal Profit

Suppose in Example 7 that the cost of producing $x$ hamburgers is

$$C = 5000 + 0.56x, \quad 0 \le x \le 50{,}000.$$

Find the profit and the marginal profit for the following production levels.

(a) $x = 20{,}000$     (b) $x = 24{,}400$     (c) $x = 30{,}000$

*Solution*

From Example 7, you know that the total revenue from selling $x$ units is

$$R = \frac{1}{20{,}000}(60{,}000x - x^2).$$

Because the total profit is given by $P = R - C$, you have

$$P = \frac{1}{20{,}000}(60{,}000x - x^2) - (5000 + 0.56x)$$

$$= 3x - \frac{x^2}{20{,}000} - 5000 - 0.56x$$

$$= 2.44x - \frac{x^2}{20{,}000} - 5000.$$

(See Figure 2.26.) Thus, the marginal profit is

$$\frac{dP}{dx} = 2.44 - \frac{x}{10{,}000}.$$

Using these formulas, you can compute the profit and marginal profit.

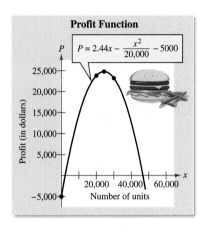

**Profit Function**

$$P = 2.44x - \frac{x^2}{20{,}000} - 5000$$

**FIGURE 2.26**

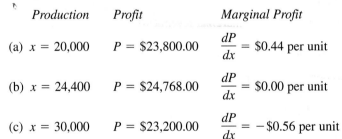

|  Production  |  Profit  |  Marginal Profit  |
|---|---|---|
| (a) $x = 20{,}000$ | $P = \$23{,}800.00$ | $\dfrac{dP}{dx} = \$0.44$ per unit |
| (b) $x = 24{,}400$ | $P = \$24{,}768.00$ | $\dfrac{dP}{dx} = \$0.00$ per unit |
| (c) $x = 30{,}000$ | $P = \$23{,}200.00$ | $\dfrac{dP}{dx} = -\$0.56$ per unit |

## *Group Discussion*     *Interpreting Marginal Profit*

In Example 8 and Figure 2.26, notice that when 20,000 hamburgers are sold, the marginal profit is positive, when 24,400 are sold, the marginal profit is zero, and when 30,000 are sold, the marginal profit is negative. What does this mean? If you owned this business, how much would you charge for hamburgers? Explain your reasoning.

---

*Warm Up*

The following warm-up exercises involve skills that were covered in earlier sections. You will use these skills in the exercise set for this section.

In Exercises 1 and 2, evaluate the expression.

**1.** $\dfrac{-63 - (-105)}{21 - 7}$

**2.** $\dfrac{-37 - 54}{16 - 3}$

In Exercises 3–10, find the derivative of the function.

**3.** $y = 4x^2 - 2x + 7$

**4.** $y = -3t^3 + 2t^2 - 8$

**5.** $s = -16t^2 + 24t + 30$

**6.** $y = -16x^2 + 54x + 70$

**7.** $A = \frac{1}{10}(-2r^3 + 3r^2 + 5r)$

**8.** $y = \frac{1}{9}(6x^3 - 18x^2 + 63x - 15)$

**9.** $y = 12x - \dfrac{x^2}{5000}$

**10.** $y = 138 + 74x - \dfrac{x^3}{10,000}$

---

# EXERCISES 2.3

**1.** *Research and Development*  The graph shows the amount $A$ (in billions of constant 1987 dollars per year) spent on R&D in the United States from 1975 through 1995. Approximate the average rate of change of $A$ during the following periods. *(Source: National Science Foundation)*

(a) 1975–1980    (b) 1980–1985
(c) 1985–1990    (d) 1975–1995

Research and Development

**2.** *Trade Deficit*  The graph shows the value $I$ (in billions of dollars per year) of goods imported to the United States and the value $E$ (in billions of dollars per year) of goods exported from the United States from 1970 through 1995. Approximate the indicated average rate of change.
*(Source: U.S. Bureau of Census)*

(a) Imports: 1970–1985    (b) Exports: 1970–1985
(c) Imports: 1985–1995    (d) Exports: 1985–1995

Figure for 2

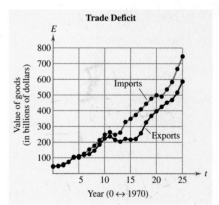

Trade Deficit

In Exercises 3–8, sketch the graph of the function and find its average rate of change on the interval. Compare this rate with the instantaneous rates of change at the endpoints of the interval.

| Function | Interval |
|---|---|
| **3.** $f(t) = 2t + 7$ | $[1, 2]$ |
| **4.** $h(x) = 1 - x^2$ | $[0, 1]$ |
| **5.** $h(x) = x^2 - 4$ | $[-2, 2]$ |
| **6.** $f(x) = x^2 - 6x - 1$ | $[-1, 3]$ |
| **7.** $f(x) = \dfrac{1}{x}$ | $[1, 4]$ |
| **8.** $f(x) = x^{-1/2}$ | $[1, 4]$ |

In Exercises 9 and 10, use a graphing utility to graph the function and find its average rate of change on the interval. Compare this rate with the instantaneous rates of change at the endpoints of the interval.

| Function | Interval |
|---|---|
| **9.** $g(x) = x^4 - x^2 + 2$ | $[1, 3]$ |
| **10.** $g(x) = x^3 - 1$ | $[-1, 1]$ |

**11.** *Drug Effectiveness* The effectiveness $E$ (on a scale from 0 to 1) of a pain-killing drug $t$ hours after entering the bloodstream is given by

$$E = \frac{1}{27}(9t + 3t^2 - t^3), \quad 0 \le t \le 4.5.$$

Find the average rate of change of $E$ on the indicated intervals and compare this rate with the instantaneous rates of change at the endpoints of the intervals.

(a) $[0, 1]$    (b) $[1, 2]$    (c) $[2, 3]$    (d) $[3, 4]$

**12.** *Wind Chill* At $0°$ Celsius, the heat loss $H$ (in kilocalories per square meter per hour) from a person's body can be modeled by

$$H = 33(10\sqrt{v} - v + 10.45),$$

where $v$ is the wind speed (in meters per second).

(a) Find $\dfrac{dH}{dv}$ and interpret its meaning for this situation.

(b) Find the rate of change of $H$ when $v = 2$ and when $v = 5$.

**13.** *Velocity* The height $s$ (in feet) at time $t$ (in seconds) of a silver dollar dropped from the top of the Washington Monument is

$$s = -16t^2 + 555.$$

(a) Find the average velocity on the interval $[2, 3]$.

(b) Find the instantaneous velocity when $t = 2$ and when $t = 3$.

(c) How long will it take the dollar to hit the ground?

(d) Find the velocity of the dollar when it hits the ground.

**14.** *Velocity* The height $s$ (in feet) of an object fired straight up from ground level with an initial velocity of 200 feet per second is given by

$$s = -16t^2 + 200t,$$

where $t$ is the time (in seconds).

(a) How fast is the object moving after 1 second?

(b) During which interval of time is the speed decreasing?

(c) During which interval of time is the speed increasing?

**15.** *Velocity* The position $s$ (in feet) of an accelerating car is

$$s = 10t^{3/2}, \quad 0 \le t \le 10,$$

where $t$ is the time (in seconds). Find the velocity of the car at the following times.

(a) $t = 0$    (b) $t = 1$    (c) $t = 4$    (d) $t = 9$

**16.** *Physics: Velocity* A race car travels northward on a straight, level track at a constant speed, traveling 0.750 kilometers in 20.0 seconds. The return trip over the same track is made in 25.0 seconds.

(a) What is the average velocity of the car in meters per second for the first leg of the run?

(b) What is the average velocity for the total trip?

*(Source: Shipman/Wilson/Todd,* An Introduction to Physical Science, *Eighth Edition)*

*Marginal Cost* In Exercises 17–20, find the marginal cost for producing $x$ units. (The cost is measured in dollars.)

**17.** $C = 4500 + 1.47x$

**18.** $C = 104,000 + 7200x$

**19.** $C = 55,000 + 470x - 0.25x^2, \quad 0 \le x \le 940$

**20.** $C = 100(9 + 3\sqrt{x})$

*Marginal Revenue* In Exercises 21–24, find the marginal revenue for producing $x$ units. (The revenue is measured in dollars.)

**21.** $R = 50x - 0.5x^2$

**22.** $R = 30x - x^2$

**23.** $R = -6x^3 + 8x^2 + 200x$

**24.** $R = 50(20x - x^{3/2})$

*Marginal Profit* In Exercises 25–28, find the marginal profit for producing $x$ units. (The profit is measured in dollars.)

**25.** $P = -2x^2 + 72x - 145$

**26.** $P = -0.25x^2 + 2000x - 1,250,000$

**27.** $P = -0.00025x^2 + 12.2x - 25,000$

**28.** $P = -0.5x^3 + 30x^2 - 164.25x - 1000$

**29.** *Marginal Revenue* The revenue (in dollars) from renting $x$ apartments can be modeled by

$$R = 2x(900 + 32x - x^2).$$

(a) Find the additional revenue when the number of rentals is increased from 14 to 15.

(b) Find the marginal revenue when $x = 14$.

(c) Compare the results of parts (a) and (b).

**30.** *Marginal Cost*   The cost (in dollars) of producing $x$ units of a product is given by

$$C = 3.6\sqrt{x} + 500.$$

(a) Find the additional cost when the production increases from nine to ten units.

(b) Find the marginal cost when $x = 9$.

(c) Compare the results of parts (a) and (b).

**31.** *Marginal Profit*   The profit (in dollars) from selling $x$ units of a product is

$$P = -0.1x^2 + 2x - 100.$$

(a) Find the additional profit when the sales increase from seven to eight units.

(b) Find the marginal profit when $x = 7$.

(c) Compare the results of parts (a) and (b).

**32.** *Population Growth*   The population of a developing rural area has been growing according to the model

$$P = 22t^2 + 52t + 10{,}000,$$

where $t$ is time in years, with $t = 0$ representing 1970.

(a) Evaluate $P$ for $t = 0$, 10, 20, and 25. Explain these values.

(b) Determine the population growth rate, $dP/dt$.

(c) Evaluate $dP/dt$ for the same values as in part (a). Explain your results.

**33.** *Body Temperature*   The temperature of a person during an illness is given by

$$T = -0.01t^2 - 0.2t + 102.5,$$

where $t$ is time in hours.

(a) Use a graphing utility to graph the function. Be sure to choose an appropriate window.

(b) Do the slopes of the tangent lines appear to be positive or negative? What does this tell you about this situation?

(c) Evaluate the function for $t = 0$, 4, 8, and 12.

(d) Find $dT/dt$ and explain its meaning for this situation.

(e) Evaluate $dT/dt$ for $t = 0$, 4, 8, and 12.

**34.** *Marginal Profit*   The profit (in dollars) from selling $x$ units of a product is given by

$$P = 36{,}000 + 2048\sqrt{x} - \frac{1}{8x^2}, \quad 150 \le x \le 275.$$

Find the marginal profit for the following sales.

(a) $x = 150$   (b) $x = 175$   (c) $x = 200$
(d) $x = 225$   (e) $x = 250$   (f) $x = 275$

**35.** *Profit*   The monthly demand function and cost function for $x$ newspapers at a newsstand are

$$p = 5 - 0.001x \quad \text{and} \quad C = 35 + 1.5x.$$

(a) Find the monthly revenue as a function of $x$.

(b) Find the monthly profit as a function of $x$.

(c) Complete the table.

| $x$ | 600 | 1200 | 1800 | 2400 | 3000 |
|---|---|---|---|---|---|
| $dR/dx$ | | | | | |
| $dP/dx$ | | | | | |
| $P$ | | | | | |

**36.** *Economics: Marginal Revenue*   Use the table to answer the questions below.

| Quantity Produced and Sold $(Q)$ | Price $(P)$ | Total Revenue $(TR)$ | Marginal Revenue $(MR)$ |
|---|---|---|---|
| 0 | 160 | 0 | — |
| 2 | 140 | 280 | 130 |
| 4 | 120 | 480 | 90 |
| 6 | 100 | 600 | 50 |
| 8 | 80 | 640 | 10 |
| 10 | 60 | 600 | −30 |

(a) Use the regression capabilities of a graphing utility to find a quadratic model that relates the total revenue $(TR)$ to the quantity produced and sold $(Q)$.

(b) Using derivatives, find a model for marginal revenue from the model you found in part (a).

(c) Calculate the marginal revenue for all values of $Q$ using the model you found in part (b), and compare these values with the actual values given in the table. How good was your model?

*(Source: Adapted from Taylor, Economics, First Edition)*

**37.** *Marginal Profit*   When the admission price to a baseball game was $6 per ticket, 36,000 tickets were sold. When the price was raised to $7, only 33,000 tickets were sold. Assume that the demand function is linear and that the variable and fixed costs for the ballpark owners are $0.20 and $85,000, respectively.

(a) Find the profit $P$ as a function of $x$, the number of tickets sold.

(b) Use a graphing utility to graph $P$, and comment about the slope of $P$ when $x = 18,000$.

(c) Find the marginal profit when 18,000 tickets are sold.

**38. *Marginal Profit*** In Exercise 37, suppose ticket sales decreased to 30,000 when the price increased to $7. How does this change the answers to Exercise 37?

**39. *Marginal Cost*** The cost $C$ of producing $x$ units is modeled by $C = v(x) + k$, where $v$ represents the variable cost and $k$ represents the fixed cost. Show that the marginal cost is independent of the fixed cost.

**40. *Profit*** The demand function for a product is

$$p = \frac{50}{\sqrt{x}}, \quad 1 \le x \le 8000,$$

and the cost function is

$$C = 0.5x + 500, \quad 0 \le x \le 8000.$$

Find the marginal profit for the following sales.

(a) $x = 900$       (b) $x = 1600$

(c) $x = 2500$     (d) $x = 3600$

If you were in charge of setting the price for this product, what price would you set? Explain your reasoning.

**41. *Inventory Management*** The annual inventory cost for a manufacturer is

$$C = \frac{1,008,000}{Q} + 6.3Q,$$

where $Q$ is the order size when the inventory is replenished. Find the change in annual cost when $Q$ is increased from 350 to 351, and compare this with the instantaneous rate of change when $Q = 350$.

**42. *Center of Population*** Since 1790, the center of population of the United States has been gradually moving westward. Use the figure to estimate the rate (in miles per year) at which the center of population was moving *westward* during the given period. (*Source: U.S. Bureau of Census*)

(a) From 1790 to 1900      (b) From 1900 to 1990

(c) From 1790 to 1990

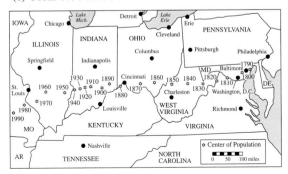

**43. *Fuel Cost*** A car is driven 15,000 miles a year and gets $x$ miles per gallon. The average fuel cost is $1.30 per gallon. Find the annual cost $C$ of fuel as a function of $x$ and use this function to complete the table.

| $x$ | 10 | 15 | 20 | 25 | 30 | 35 | 40 |
|---|---|---|---|---|---|---|---|
| $C$ | | | | | | | |
| $dC/dx$ | | | | | | | |

Who would benefit more from a 1 mile per gallon increase in fuel efficiency—the driver that gets 15 miles per gallon or the driver that gets 35 miles per gallon? Explain.

**44.** Use a graphing utility to sketch the graph of $f$ and $f'$ for

$$f(x) = \frac{4}{x}, \quad 0 < x \le 5.$$

(a) Complete the table.

| $x$ | $\frac{1}{8}$ | $\frac{1}{4}$ | $\frac{1}{2}$ | 1 | 2 | 3 | 4 | 5 |
|---|---|---|---|---|---|---|---|---|
| $f(x)$ | | | | | | | | |
| $f'(x)$ | | | | | | | | |

(b) Find the average rate of change of $f$ over the intervals determined by consecutive $x$-values in the table.

In Exercises 45 and 46, use a graphing utility to sketch the graphs of $f$ and $f'$. Then determine the points (if any) at which $f$ has a horizontal tangent.

**45.** $f(x) = \frac{1}{4}x^3, \quad -2 \le x \le 2$

**46.** $f(x) = x^4 - 12x^3 + 52x^2 - 96x + 64, \quad 1 \le x \le 5$

**47. *Biology: Population Growth*** Many populations in nature experience logistic growth, which consists of four phases, as shown in the figure. Describe the rate of growth of the population in each phase, and give possible reasons as to why the rates might be changing from phase to phase. (*Source: Adapted from Levine/Miller,* Biology: Discovering Life, *Second Edition*)

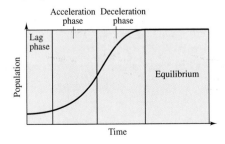

## 2.4 The Product and Quotient Rules

*The Product Rule • The Quotient Rule • Simplifying Derivatives •
Extended Application: Blood Pressure*

### The Product Rule

In Section 2.2 you saw that the derivative of a sum or difference of two functions is simply the sum or difference of their derivatives. The rules for the derivative of a product or quotient of two functions are not as simple.

---

#### The Product Rule

The derivative of the product of two differentiable functions is equal to the first function times the derivative of the second plus the second function times the derivative of the first.

$$\frac{d}{dx}[f(x)g(x)] = f(x)g'(x) + g(x)f'(x)$$

---

*STUDY TIP*   Rather than trying to remember the formula for the Product Rule, we suggest that you remember its verbal statement: *the first function times the derivative of the second plus the second function times the derivative of the first.*

**Proof**

Some mathematical proofs, such as the proof of the Sum Rule, are straightforward. Others involve clever steps that may appear unmotivated. The following proof involves such a step—adding and subtracting the same quantity. (This step is shown in color.) Let $F(x) = f(x)g(x)$.

$$F'(x) = \lim_{\Delta x \to 0} \frac{F(x + \Delta x) - F(x)}{\Delta x}$$

$$= \lim_{\Delta x \to 0} \frac{f(x + \Delta x)g(x + \Delta x) - f(x)g(x)}{\Delta x}$$

$$= \lim_{\Delta x \to 0} \frac{f(x + \Delta x)g(x + \Delta x) - f(x + \Delta x)g(x) + f(x + \Delta x)g(x) - f(x)g(x)}{\Delta x}$$

$$= \lim_{\Delta x \to 0} \left[ f(x + \Delta x)\frac{g(x + \Delta x) - g(x)}{\Delta x} + g(x)\frac{f(x + \Delta x) - f(x)}{\Delta x} \right]$$

$$= \lim_{\Delta x \to 0} f(x + \Delta x)\frac{g(x + \Delta x) - g(x)}{\Delta x} + \lim_{\Delta x \to 0} g(x)\frac{f(x + \Delta x) - f(x)}{\Delta x}$$

$$= \left[ \lim_{\Delta x \to 0} f(x + \Delta x) \right]\left[ \lim_{\Delta x \to 0} \frac{g(x + \Delta x) - g(x)}{\Delta x} \right]$$

$$\qquad + \left[ \lim_{\Delta x \to 0} g(x) \right]\left[ \lim_{\Delta x \to 0} \frac{f(x + \Delta x) - f(x)}{\Delta x} \right]$$

$$= f(x)g'(x) + g(x)f'(x)$$

## EXAMPLE 1   Finding the Derivative of a Product

Find the derivative of $y = (3x - 2x^2)(5 + 4x)$.

### Solution

Using the Product Rule, you can write

$$\frac{dy}{dx} = \underbrace{(3x - 2x^2)}_{(\text{First})} \underbrace{\frac{d}{dx}[5 + 4x]}_{\left(\substack{\text{Derivative} \\ \text{of second}}\right)} + \underbrace{(5 + 4x)}_{(\text{Second})} \underbrace{\frac{d}{dx}[3x - 2x^2]}_{\left(\substack{\text{Derivative} \\ \text{of first}}\right)}$$

$$= (3x - 2x^2)(4) + (5 + 4x)(3 - 4x)$$
$$= (12x - 8x^2) + (15 - 8x - 16x^2)$$
$$= 15 + 4x - 24x^2.$$

> **STUDY TIP**   In general, the derivative of the product of two functions is not equal to the product of the derivatives of the two functions. To see this, compare the product of the derivatives of $f(x) = 3x - 2x^2$ and $g(x) = 5 + 4x$ with the derivative found in Example 1.

In the next example, notice that the first step in differentiating is *rewriting the original function*.

## EXAMPLE 2   Finding the Derivative of a Product

Find the derivative of

$$f(x) = \left(\frac{1}{x} + 1\right)(x - 1). \qquad \text{Original function}$$

### Solution

Rewrite the function. Then use the Product Rule to find the derivative.

$$f(x) = (x^{-1} + 1)(x - 1) \qquad \text{Rewrite function.}$$

$$f'(x) = (x^{-1} + 1)\frac{d}{dx}[x - 1] + (x - 1)\frac{d}{dx}[x^{-1} + 1] \qquad \text{Product Rule}$$

$$= (x^{-1} + 1)(1) + (x - 1)(-x^{-2})$$

$$= \frac{1}{x} + 1 - \frac{x - 1}{x^2}$$

$$= \frac{x + x^2 - x + 1}{x^2} \qquad \text{Write with common denominator.}$$

$$= \frac{x^2 + 1}{x^2} \qquad \text{Simplify.}$$

## Technology

If you have access to a symbolic differentiation utility, such as *Derive, Maple, Mathcad, Mathematica,* or the *TI-92*, try using it to confirm several of the derivatives in this section. The form of the derivative can depend on how you use software. For instance, if you use *Derive* for Windows to differentiate the function in Example 2, you will obtain the same form listed in the example. But if you use *Derive* for Windows to first expand the function, and then differentiate, you will obtain

$$f'(x) = \frac{1}{x^2} + 1.$$

You now have two differentiation rules that deal with products—the Constant Multiple Rule and the Product Rule. The difference between these two rules is that the Constant Multiple Rule deals with the product of a constant and a variable quantity,

$$F(x) = c\,f(x),$$    Use Constant Multiple Rule.

whereas the Product Rule deals with the product of two variable quantities,

$$F(x) = f(x)\,g(x).$$    Use Product Rule.

The next example compares these two rules.

## EXAMPLE 3    Comparing Differentiation Rules

Find the derivatives of the functions.

(a)  $y = 2x(x^2 + 3x)$

(b)  $y = 2(x^2 + 3x)$

**Solution**

(a)  By the Product Rule,

$$\frac{dy}{dx} = (2x)\frac{d}{dx}\left[x^2 + 3x\right] + (x^2 + 3x)\frac{d}{dx}\left[2x\right]$$    Product Rule

$$= (2x)(2x + 3) + (x^2 + 3x)(2)$$

$$= 4x^2 + 6x + 2x^2 + 6x$$

$$= 6x^2 + 12x.$$

(b)  By the Constant Multiple Rule,

$$\frac{dy}{dx} = 2\frac{d}{dx}\left[x^2 + 3x\right]$$    Constant Multiple Rule

$$= 2(2x + 3)$$

$$= 4x + 6.$$

The Product Rule can be extended to products that have more than two factors. For example, if $f$, $g$, and $h$ are differentiable functions of $x$, then

$$\frac{d}{dx}\left[f(x)g(x)h(x)\right] = f'(x)g(x)h(x) + f(x)g'(x)h(x) + f(x)g(x)h'(x).$$

*STUDY TIP*    You could calculate the derivatives in Example 3 without the Product Rule. For part (a),

$$y = 2x(x^2 + 3x) = 2x^3 + 6x^2$$

and

$$\frac{dy}{dx} = 6x^2 + 12x.$$

## The Quotient Rule

In Section 2.2, you saw that by using the Constant Rule, Power Rule, Constant Multiple Rule, and Sum and Difference Rules, you were able to differentiate any polynomial function. By combining these rules with the Quotient Rule, you can now differentiate any *rational* function.

---

### The Quotient Rule

The derivative of the quotient of two differentiable functions is equal to the denominator times the derivative of the numerator minus the numerator times the derivative of the denominator, all divided by the square of the denominator.

$$\frac{d}{dx}\left[\frac{f(x)}{g(x)}\right] = \frac{g(x)f'(x) - f(x)g'(x)}{[g(x)]^2}, \qquad g(x) \neq 0$$

---

*STUDY TIP* From this differentiation rule, you can see that the derivative of a quotient is not, in general, the quotient of the derivatives. That is,

$$\frac{d}{dx}\left[\frac{f(x)}{g(x)}\right] \neq \frac{f'(x)}{g'(x)}.$$

*Proof*

Let $F(x) = f(x)/g(x)$. As in the proof of the Product Rule, a key step in this proof is adding and subtracting the same quantity.

$$F'(x) = \lim_{\Delta x \to 0} \frac{F(x + \Delta x) - F(x)}{\Delta x}$$

$$= \lim_{\Delta x \to 0} \frac{\dfrac{f(x + \Delta x)}{g(x + \Delta x)} - \dfrac{f(x)}{g(x)}}{\Delta x}$$

$$= \lim_{\Delta x \to 0} \frac{g(x)f(x + \Delta x) - f(x)g(x + \Delta x)}{\Delta x g(x)g(x + \Delta x)}$$

$$= \lim_{\Delta x \to 0} \frac{g(x)f(x + \Delta x) - f(x)g(x) + f(x)g(x) - f(x)g(x + \Delta x)}{\Delta x g(x)g(x + \Delta x)}$$

$$= \frac{\displaystyle\lim_{\Delta x \to 0} \frac{g(x)[f(x + \Delta x) - f(x)]}{\Delta x} - \lim_{\Delta x \to 0} \frac{f(x)[g(x + \Delta x) - g(x)]}{\Delta x}}{\displaystyle\lim_{\Delta x \to 0} [g(x)g(x + \Delta x)]}$$

$$= \frac{g(x)\left[\displaystyle\lim_{\Delta x \to 0} \dfrac{f(x + \Delta x) - f(x)}{\Delta x}\right] - f(x)\left[\displaystyle\lim_{\Delta x \to 0} \dfrac{g(x + \Delta x) - g(x)}{\Delta x}\right]}{\displaystyle\lim_{\Delta x \to 0} [g(x)g(x + \Delta x)]}$$

$$= \frac{g(x)f'(x) - f(x)g'(x)}{[g(x)]^2}$$

*STUDY TIP* As suggested for the Product Rule, we recommend that you remember the verbal statement of the Quotient Rule rather than trying to remember the formula for the rule.

## EXAMPLE 4   Finding the Derivative of a Quotient

Find the derivative of $y = \dfrac{x - 1}{2x + 3}$.

**Solution**

Apply the Quotient Rule, as follows.

$$\frac{dy}{dx} = \frac{(2x + 3)\dfrac{d}{dx}[x - 1] - (x - 1)\dfrac{d}{dx}[2x + 3]}{(2x + 3)^2}$$

$$= \frac{(2x + 3)(1) - (x - 1)(2)}{(2x + 3)^2}$$

$$= \frac{2x + 3 - 2x + 2}{(2x + 3)^2}$$

$$= \frac{5}{(2x + 3)^2}$$

**ALGEBRA**

### TIP

When applying the Quotient Rule, we suggest that you enclose all factors and derivatives in parentheses. Also, pay special attention to the subtraction required in the numerator.

## EXAMPLE 5   Finding an Equation of a Tangent Line

Find the equation of the tangent line to the graph of

$$y = \frac{2x^2 - 4x + 3}{2 - 3x}$$

when $x = 1$.

**Solution**

Apply the Quotient Rule, as follows.

$$\frac{dy}{dx} = \frac{(2 - 3x)\dfrac{d}{dx}[2x^2 - 4x + 3] - (2x^2 - 4x + 3)\dfrac{d}{dx}[2 - 3x]}{(2 - 3x)^2}$$

$$= \frac{(2 - 3x)(4x - 4) - (2x^2 - 4x + 3)(-3)}{(2 - 3x)^2}$$

$$= \frac{-12x^2 + 20x - 8 - (-6x^2 + 12x - 9)}{(2 - 3x)^2}$$

$$= \frac{-12x^2 + 20x - 8 + 6x^2 - 12x + 9}{(2 - 3x)^2}$$

$$= \frac{-6x^2 + 8x + 1}{(2 - 3x)^2}$$

When $x = 1$, the value of the function is $y = -1$ and the slope is $m = 3$. Using the point-slope form of a line, you can find the equation of the tangent line to be $y = 3x - 4$. The graph of the function and the tangent line is shown in Figure 2.27.

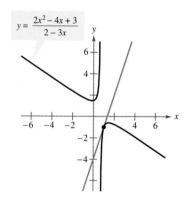

$y = \dfrac{2x^2 - 4x + 3}{2 - 3x}$

**FIGURE 2.27**

## EXAMPLE 6  Finding the Derivative of a Quotient

Find the derivative of

$$y = \frac{3 - (1/x)}{x + 5}.$$

### Solution

Begin by rewriting the original function. Then apply the Quotient Rule and simplify the result.

$$y = \frac{3 - (1/x)}{x + 5} \qquad \text{Original function}$$

$$= \frac{3x - 1}{x(x + 5)} \qquad \text{Multiply numerator and denominator by } x.$$

$$= \frac{3x - 1}{x^2 + 5x} \qquad \text{Rewritten function}$$

$$\frac{dy}{dx} = \frac{(x^2 + 5x)(3) - (3x - 1)(2x + 5)}{(x^2 + 5x)^2} \qquad \text{Apply Quotient Rule.}$$

$$= \frac{(3x^2 + 15x) - (6x^2 + 13x - 5)}{(x^2 + 5x)^2}$$

$$= \frac{-3x^2 + 2x + 5}{(x^2 + 5x)^2} \qquad \text{Simplify.}$$

Not every quotient needs to be differentiated by the Quotient Rule. For instance, each of the quotients in the next example can be considered as the product of a constant and a function of $x$. In such cases, the Constant Multiple Rule is more efficient than the Quotient Rule.

## EXAMPLE 7  Rewriting Before Differentiating

| Original Function | Rewrite | Differentiate | Simplify |
|---|---|---|---|
| (a) $y = \dfrac{x^2 + 3x}{6}$ | $y = \dfrac{1}{6}(x^2 + 3x)$ | $y' = \dfrac{1}{6}(2x + 3)$ | $y' = \dfrac{1}{3}x + \dfrac{1}{2}$ |
| (b) $y = \dfrac{5x^4}{8}$ | $y = \dfrac{5}{8}x^4$ | $y' = \dfrac{5}{8}(4x^3)$ | $y' = \dfrac{5}{2}x^3$ |
| (c) $y = \dfrac{-3(3x - 2x^2)}{7x}$ | $y = -\dfrac{3}{7}(3 - 2x)$ | $y' = -\dfrac{3}{7}(-2)$ | $y' = \dfrac{6}{7}$ |
| (d) $y = \dfrac{9}{5x^2}$ | $y = \dfrac{9}{5}(x^{-2})$ | $y' = \dfrac{9}{5}(-2x^{-3})$ | $y' = -\dfrac{18}{5x^3}$ |

*STUDY TIP*  To see the efficiency of using the Constant Multiple Rule in Example 7, try using the Quotient Rule to find the derivatives of the four functions.

## Simplifying Derivatives

### EXAMPLE 8   Combining the Product and Quotient Rules

Find the derivative of

$$y = \frac{(1 - 2x)(3x + 2)}{5x - 4}.$$

*Solution*

This function contains a product within a quotient. You could first multiply the factors in the numerator and then apply the Quotient Rule. However, to gain practice in using the Product Rule within the Quotient Rule, try differentiating as follows.

$$
\begin{aligned}
y' &= \frac{(5x - 4)\dfrac{d}{dx}[(1 - 2x)(3x + 2)] - (1 - 2x)(3x + 2)\dfrac{d}{dx}[5x - 4]}{(5x - 4)^2} \\[2mm]
&= \frac{(5x - 4)[(1 - 2x)(3) + (3x + 2)(-2)] - (1 - 2x)(3x + 2)(5)}{(5x - 4)^2} \\[2mm]
&= \frac{(5x - 4)(-12x - 1) - (1 - 2x)(15x + 10)}{(5x - 4)^2} \\[2mm]
&= \frac{(-60x^2 + 43x + 4) - (-30x^2 - 5x + 10)}{(5x - 4)^2} \\[2mm]
&= \frac{-30x^2 + 48x - 6}{(5x - 4)^2}
\end{aligned}
$$

In the examples in this section, much of the work in obtaining the final form of the derivative occurs *after* the differentiation. As summarized in the following list, direct application of differentiation rules often yields results that are not in simplified form. Note that two characteristics of simplified form are the absence of negative exponents and the combining of like terms.

| | *f'(x) After Differentiating* | *f'(x) After Simplifying* |
|---|---|---|
| *Example 1* | $(3x - 2x^2)(4) + (5 + 4x)(3 - 4x)$ | $15 + 4x - 24x^2$ |
| *Example 2* | $(x^{-1} + 1)(1) + (x - 1)(-x^{-2})$ | $\dfrac{x^2 + 1}{x^2}$ |
| *Example 5* | $\dfrac{(2 - 3x)(4x - 4) - (2x^2 - 4x + 3)(-3)}{(2 - 3x)^2}$ | $\dfrac{-6x^2 + 8x + 1}{(2 - 3x)^2}$ |
| *Example 8* | $\dfrac{(5x - 4)[(1 - 2x)(3) + (3x + 2)(-2)] - (1 - 2x)(3x + 2)(5)}{(5x - 4)^2}$ | $\dfrac{-30x^2 + 48x - 6}{(5x - 4)^2}$ |

## Extended Application:  Blood Pressure

### EXAMPLE 9  Rate of Change of Blood Pressure

As blood moves from the heart through the major arteries out to the capillaries and back through the veins, the systolic pressure continuously drops. Consider a person whose blood pressure $P$ (in millimeters of mercury) is given by

$$P = \frac{25t^2 + 125}{t^2 + 1}, \qquad 0 \le t \le 10,$$

where $t$ is measured in seconds. At what rate is the blood pressure changing 5 seconds after blood leaves the heart?

#### Solution

Begin by applying the Quotient Rule.

$$\frac{dP}{dt} = \frac{(t^2 + 1)(50t) - (25t^2 + 125)(2t)}{(t^2 + 1)^2} \qquad \text{Quotient Rule}$$

$$= \frac{50t^3 + 50t - 50t^3 - 250t}{(t^2 + 1)^2}$$

$$= -\frac{200t}{(t^2 + 1)^2} \qquad \text{Simplify.}$$

When $t = 5$, the rate of change is

$$\frac{dP}{dt} = -\frac{200(5)}{26^2}$$

$$\approx -1.48 \text{ mm/sec.}$$

Therefore, the pressure is *dropping* at a rate of 1.48 millimeters per second when $t = 5$ seconds.

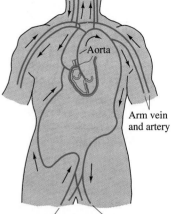

Aorta

Arm vein and artery

Leg vein and artery

---

### *Group Discussion*    **Testing Differentiation Rules**

What is wrong with the following differentiations?

$$f(x) = (x^3)(x^4)$$
$$f'(x) = (3x^2)(4x^3)$$

$$g(x) = \frac{x^3}{x^4}$$
$$g'(x) = \frac{3x^2}{4x^3}$$

If you were teaching a calculus class, how could you convince your class that these derivatives are incorrect?

## Warm Up

The following warm-up exercises involve skills that were covered in earlier sections. You will use these skills in the exercise set for this section.

In Exercises 1–6, simplify the expression.

**1.** $(x^2 + 1)(2) + (2x + 7)(2x)$

**2.** $(2x - x^3)(8x) + (4x^2)(2 - 3x^2)$

**3.** $\dfrac{(2x + 7)(5) - (5x + 6)(2)}{(2x + 7)^2}$

**4.** $\dfrac{(x^2 - 4)(2x + 1) - (x^2 + x)(2x)}{(x^2 - 4)^2}$

**5.** $(x^{-1} + x)(2) + (2x - 3)(-x^{-2} + 1)$

**6.** $\dfrac{(1 - x^{-1})(1) - (x - 4)(x^{-2})}{(1 - x^{-1})^2}$

In Exercises 7–10, find $f'(2)$.

**7.** $f(x) = 3x^2 - x + 4$

**8.** $f(x) = -x^3 + x^2 + 8x$

**9.** $f(x) = \dfrac{1}{x}$

**10.** $f(x) = x^2 - \dfrac{1}{x^2}$

## EXERCISES 2.4

In Exercises 1–10, find the value of the derivative of the function at the indicated point.

| Function | Point |
|---|---|
| **1.** $f(x) = x^2(3x^3 - 1)$ | $(1, 2)$ |
| **2.** $f(x) = (x^2 + 1)(2x + 5)$ | $(-1, 6)$ |
| **3.** $f(x) = \frac{1}{3}(2x^3 - 4)$ | $\left(0, -\frac{4}{3}\right)$ |
| **4.** $f(x) = \frac{1}{7}(5 - 6x^2)$ | $\left(1, -\frac{1}{7}\right)$ |
| **5.** $g(x) = (x^2 - 4x + 3)(x - 2)$ | $(4, 6)$ |
| **6.** $g(x) = (x^2 - 2x + 1)(x^3 - 1)$ | $(1, 0)$ |
| **7.** $h(x) = \dfrac{x}{x - 5}$ | $(6, 6)$ |
| **8.** $h(x) = \dfrac{x^2}{x + 3}$ | $\left(-1, \frac{1}{2}\right)$ |
| **9.** $f(t) = \dfrac{2t^2 - 3}{3t}$ | $\left(2, \frac{5}{6}\right)$ |
| **10.** $f(x) = \dfrac{x + 1}{x - 1}$ | $(2, 3)$ |

In Exercises 11–18, find the derivative of the function. Use Example 7 as a model.

| Function | Rewrite | Differentiate | Simplify |
|---|---|---|---|
| **11.** $y = \dfrac{x^2 + 2x}{x}$ | | | |
| **12.** $y = \dfrac{4x^{3/2}}{x}$ | | | |
| **13.** $y = \dfrac{7}{3x^3}$ | | | |
| **14.** $y = \dfrac{4}{5x^2}$ | | | |
| **15.** $y = \dfrac{4x^2 - 3x}{8\sqrt{x}}$ | | | |
| **16.** $y = \dfrac{x^2 - 4}{x + 2}$ | | | |
| **17.** $y = \dfrac{x^2 - 4x + 3}{x - 1}$ | | | |
| **18.** $y = \dfrac{3x^2 - 4x}{6x}$ | | | |

In Exercises 19–34, differentiate the function.

**19.** $f(x) = (x^3 - 3x)(2x^2 + 3x + 5)$

**20.** $h(t) = (t^5 - 1)(4t^2 - 7t - 3)$

**21.** $g(t) = (2t^2 - 3)(4 - t^2 - t^4)$

**22.** $h(p) = (p^3 - 2)^2$

**23.** $f(x) = \sqrt[3]{x}(\sqrt{x} + 3)$

**24.** $f(x) = \sqrt[3]{x}(x + 1)$

**25.** $f(x) = \dfrac{3x - 2}{2x - 3}$

**26.** $f(x) = \dfrac{x^3 + 3x + 2}{x^2 - 1}$

**27.** $f(x) = \dfrac{3 - 2x - x^2}{x^2 - 1}$

**28.** $f(x) = (x^5 - 3x)\left(\dfrac{1}{x^2}\right)$

**29.** $f(x) = x\left(1 - \dfrac{2}{x + 1}\right)$

**30.** $h(t) = \dfrac{t + 2}{t^2 + 5t + 6}$

**31.** $g(s) = \dfrac{5}{s^2 + 2s + 2}$

**32.** $f(x) = \dfrac{x + 1}{\sqrt{x}}$

**33.** $g(x) = \left(\dfrac{x - 3}{x + 4}\right)(x^2 + 2x + 1)$

**34.** $f(x) = (3x^3 + 4x)(x - 5)(x + 1)$

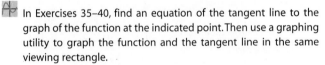

In Exercises 35–40, find an equation of the tangent line to the graph of the function at the indicated point. Then use a graphing utility to graph the function and the tangent line in the same viewing rectangle.

| *Function* | *Point* |
|---|---|
| **35.** $f(x) = (x - 1)(x^2 - 3x + 2)$ | $(0, -2)$ |
| **36.** $h(x) = (x^2 - 1)^2$ | $(-2, 9)$ |
| **37.** $f(x) = x^3\left(\dfrac{x - 2}{x + 1}\right)$ | $\left(1, -\frac{1}{2}\right)$ |
| **38.** $f(x) = \dfrac{\sqrt{x}}{x + 1}$ | $\left(1, \frac{1}{2}\right)$ |
| **39.** $f(x) = \left(\dfrac{x^2 - x - 3}{x^2 + 1}\right)(x^2 + x + 1)$ | $(0, -3)$ |
| **40.** $f(x) = (x^2 - x)(x^2 + 1)(x^2 + x + 1)$ | $(1, 0)$ |

In Exercises 41–44, find the point(s), if any, at which the graph of $f$ has a horizontal tangent.

**41.** $f(x) = \dfrac{x^2}{x - 1}$

**42.** $f(x) = \dfrac{x^2}{x^2 + 1}$

**43.** $f(x) = \dfrac{x^4}{x^3 + 1}$

**44.** $f(x) = \dfrac{x^4 + 3}{x^2 + 1}$

In Exercises 45–48, use a graphing utility to graph $f$ and $f'$ on the interval $[-2, 2]$.

**45.** $f(x) = x(x + 1)$

**46.** $f(x) = x^2(x + 1)$

**47.** $f(x) = x(x + 1)(x - 1)$

**48.** $f(x) = x^2(x + 1)(x - 1)$

*Demand* In Exercises 49 and 50, use the demand function to find the rate of change in the demand $x$ for the given price $p$.

**49.** $x = 275\left(1 - \dfrac{3p}{5p + 1}\right), p = \$4$

**50.** $x = 300 - p - \dfrac{2p}{p + 1}, p = \$3$

**51.** *Oxygen Level* The model
$$f(t) = \dfrac{t^2 - t + 1}{t^2 + 1}$$
measures the percent of the normal level of oxygen in a pond, where $t$ is the time (in weeks) after organic waste is dumped into the pond. Find the rate of change of $f$ with respect to $t$ when (a) $t = 0.5$, (b) $t = 2$, and (c) $t = 8$.

**52.** *Refrigeration* The temperature $T$ of food placed in a refrigerator is modeled by
$$T = 10\left(\dfrac{4t^2 + 16t + 75}{t^2 + 4t + 10}\right),$$
where $t$ is the time (in hours). What is the initial temperature of the food? Find the rate of change of $T$ with respect to $t$ when (a) $t = 1$, (b) $t = 3$, (c) $t = 5$, and (d) $t = 10$.

**53.** *Population Growth* A population of bacteria is introduced into a culture. The number of bacteria $P$ can be modeled by
$$P = 500\left(1 + \dfrac{4t}{50 + t^2}\right),$$
where $t$ is the time (in hours). Find the rate of change of the population when $t = 2$.

**54.** *Quality Control* The percent $P$ of defective parts produced by a new employee $t$ days after the employee starts work can be modeled by
$$P = \dfrac{t + 1750}{50(t + 2)}.$$
Find the rate of change of $P$ when (a) $t = 1$ and (b) $t = 10$.

**55.** *Profit* The demand $x$ for a product is inversely proportional to the square of the price for $x \geq 5$.

(a) The price is \$1000 and the demand is 16 units. Find the demand function.

(b) The product costs \$250 per unit and the fixed cost is \$10,000. Find the cost function.

(c) Find the profit function and use a graphing utility to graph it. From the graph, what price would you set for this product? Explain your reasoning.

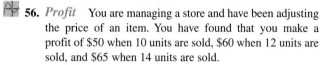

**56.** *Profit*   You are managing a store and have been adjusting the price of an item. You have found that you make a profit of $50 when 10 units are sold, $60 when 12 units are sold, and $65 when 14 units are sold.

(a) Fit this data to the model $P = ax^2 + bx + c$.

(b) Use a graphing utility to graph $P$.

(c) Find the point on the graph at which the marginal profit is zero. Interpret this point in the context of the problem.

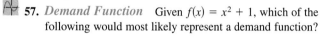

**57.** *Demand Function*   Given $f(x) = x^2 + 1$, which of the following would most likely represent a demand function?

(a) $p = f(x)$      (b) $p = xf(x)$      (c) $p = 1/f(x)$

Explain your reasoning. Use a graphing utility to graph each function, and use each graph as part of your explanation.

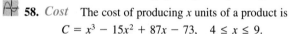

**58.** *Cost*   The cost of producing $x$ units of a product is

$$C = x^3 - 15x^2 + 87x - 73, \quad 4 \le x \le 9.$$

(a) Use a graphing utility to graph the marginal cost function and the average cost function, $C/x$, on the same set of axes.

(b) Find the point of intersection of the graphs of $dC/dx$ and $C/x$. Does this point have any significance?

**59.** *Inventory Replenishment*   The ordering and transportation cost $C$ (in thousands of dollars) of the components used in manufacturing a product is

$$C = 100\left(\frac{200}{x^2} + \frac{x}{x + 30}\right), \quad 1 \le x,$$

where $x$ is the order size (in hundreds). Find the rate of change of $C$ with respect to $x$ for the following order sizes.

(a) $x = 10$      (b) $x = 15$      (c) $x = 20$

What do these rates of change imply about increasing the size of an order?

**60.** *Sales Analysis*   The monthly sale of memberships $M$ at a newly built fitness center is modeled by

$$M(t) = \frac{300t}{t^2 + 1} + 8,$$

where $t$ is the number of months that it has been open.

(a) Find $M'(t)$.

(b) Find $M(3)$ and $M'(3)$ and interpret the results.

(c) Find $M(24)$ and $M'(24)$ and interpret the results.

**61.** *Car Costs*   The cost per mile in cents to own a car from 1980 to 1995 can be modeled by the equation

$$C = \frac{27.9 - 5.3t + 0.3t^2}{1 - 0.18t + 0.01t^2},$$

where $t$ is the year and $t = 0$ corresponds to 1980. Find $dC/dt$ and evaluate it for $t = 0, 5, 10,$ and $15$. Interpret the meaning of these values.   *(Source: American Automobile Manufacturers Association)*

## Business Capsule

Hispanic Market Connections, Inc.

*Isabel Valdés is the founder of Hispanic Market Connections, Inc. The database compiled by her company was considered so useful that the company won an award from* American Demographics *magazine as one of America's "Best 100 Sources of Marketing Information." Isabel Valdés, who has two Master's degrees from Stanford University, co-authored a 450-page book entitled* Handbook on Marketing to Hispanics. *Of her book, Valdés says that "it offers a lot more than the traditional demographic tables and statistics."*

**62.** *Research Project*   Use your school's library or some other reference source to find information about a company that performs market research. Write a short paper about the company. (One such company is described above.)

# 2.5 | The Chain Rule

*The Chain Rule • The General Power Rule • Simplification Techniques • Extended Application: Revenue Per Share • Summary of Differentiation Rules*

## The Chain Rule

In this section, you will study one of the most powerful rules of differential calculus—the **Chain Rule.** This differentiation rule deals with composite functions and adds versatility to the rules presented in Sections 2.2 and 2.4. For example, compare the following functions. Those on the left can be differentiated without the Chain Rule, whereas those on the right are best done with the Chain Rule.

| *Without the Chain Rule* | *With the Chain Rule* |
|---|---|
| $y = x^2 + 1$ | $y = \sqrt{x^2 + 1}$ |
| $y = x + 1$ | $y = (x + 1)^{-1/2}$ |
| $y = 3x + 2$ | $y = (3x + 2)^5$ |

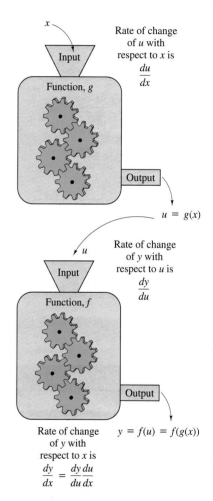

Rate of change of $u$ with respect to $x$ is

$$\frac{du}{dx}$$

$u = g(x)$

Rate of change of $y$ with respect to $u$ is

$$\frac{dy}{du}$$

$y = f(u) = f(g(x))$

Rate of change of $y$ with respect to $x$ is

$$\frac{dy}{dx} = \frac{dy}{du}\frac{du}{dx}$$

**FIGURE 2.28**

---

**The Chain Rule**

If $y = f(u)$ is a differentiable function of $u$, and $u = g(x)$ is a differentiable function of $x$, then $y = f(g(x))$ is a differentiable function of $x$, and

$$\frac{dy}{dx} = \frac{dy}{du}\frac{du}{dx}$$

or, equivalently,

$$\frac{d}{dx}[f(g(x))] = f'(g(x))g'(x).$$

---

Basically, the Chain Rule states that if $y$ changes $dy/du$ times as fast as $u$, and $u$ changes $du/dx$ times as fast as $x$, then $y$ changes

$$\frac{dy}{du}\frac{du}{dx}$$

times as fast as $x$, as illustrated in Figure 2.28. One advantage of the $dy/dx$ notation for derivatives is that it helps you remember differentiation rules, such as the Chain Rule. For instance, in the formula $dy/dx = (dy/du)(du/dx)$, you can imagine that the $du$'s cancel.

## EXAMPLE 4  Rewriting Before Differentiating

Find the tangent line to the graph of

$$y = \sqrt[3]{(x^2 + 4)^2}$$   Original function

when $x = 2$.

### Solution

Begin by rewriting the function in rational exponent form.

$$y = (x^2 + 4)^{2/3}$$   Rewrite original function.

Then, using the inside function, $u = x^2 + 4$, apply the General Power Rule.

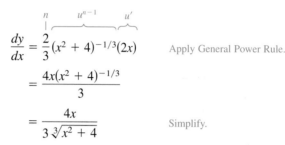

$$\frac{dy}{dx} = \frac{2}{3}(x^2 + 4)^{-1/3}(2x)$$   Apply General Power Rule.

$$= \frac{4x(x^2 + 4)^{-1/3}}{3}$$

$$= \frac{4x}{3\sqrt[3]{x^2 + 4}}$$   Simplify.

When $x = 2$, $y = 4$ and $dy/dx = \frac{4}{3}$. Using the point-slope form, you can find the equation of the tangent line to be $y = \frac{4}{3}x + \frac{4}{3}$. The graph of the function and the tangent line is shown in Figure 2.29.

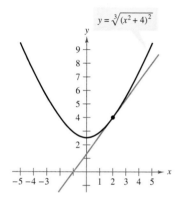

$y = \sqrt[3]{(x^2 + 4)^2}$

**FIGURE 2.29**

**STUDY TIP**   The derivative of a quotient can sometimes be found more easily with the General Power Rule than with the Quotient Rule. This is especially true when the numerator is a constant, as illustrated in Example 5.

## EXAMPLE 5  Finding the Derivative of a Quotient

Find the derivative of each function.

(a) $y = \dfrac{3}{x^2 + 1}$      (b) $y = \dfrac{3}{(x + 1)^2}$

### Solution

(a) Begin by rewriting the function as

$$y = 3(x^2 + 1)^{-1}.$$   Rewrite original function.

Then apply the General Power Rule to obtain

$$\frac{dy}{dx} = -3(x^2 + 1)^{-2}(2x) = -\frac{6x}{(x^2 + 1)^2}.$$   Apply General Power Rule.

(b) Begin by rewriting the function as

$$y = 3(x + 1)^{-2}.$$   Rewrite original function.

Then apply the General Power Rule to obtain

$$\frac{dy}{dx} = -6(x + 1)^{-3}(1) = -\frac{6}{(x + 1)^3}.$$   Apply General Power Rule.

## Simplification Techniques

Throughout this chapter we have emphasized writing derivatives in simplified form. The reason for this is that most applications of derivatives require a simplified form. The next two examples illustrate some useful simplification techniques.

---

### EXAMPLE 6   Simplifying by Factoring Out Least Powers

$y = x^2\sqrt{1 - x^2}$                                    Original function

$\quad = x^2(1 - x^2)^{1/2}$                              Rewrite function.

$y' = x^2\dfrac{d}{dx}[(1 - x^2)^{1/2}] + (1 - x^2)^{1/2}\dfrac{d}{dx}[x^2]$    Product Rule

$\quad = x^2\left[\dfrac{1}{2}(1 - x^2)^{-1/2}(-2x)\right] + (1 - x^2)^{1/2}(2x)$    Power Rule

$\quad = -x^3(1 - x^2)^{-1/2} + 2x(1 - x^2)^{1/2}$

$\quad = x(1 - x^2)^{-1/2}[-x^2(1) + 2(1 - x^2)]$          Factor.

$\quad = x(1 - x^2)^{-1/2}(2 - 3x^2)$

$\quad = \dfrac{x(2 - 3x^2)}{\sqrt{1 - x^2}}$             Simplify.

**TIP**

In Example 6, note that you subtract exponents when factoring. That is, when $(1 - x^2)^{-1/2}$ is factored out of $(1 - x^2)^{1/2}$, the *remaining* factor has an exponent of $\frac{1}{2} - \left(-\frac{1}{2}\right)$ or 1. Thus,

$$(1 - x^2)^{1/2} = (1 - x^2)^{-1/2}(1 - x^2)^1.$$

*ALGEBRA*

---

### EXAMPLE 7   Differentiating a Quotient Raised to a Power

Find the derivative of

$$f(x) = \left(\dfrac{3x - 1}{x^2 + 3}\right)^2.$$

#### Solution

$$f'(x) = \overset{n}{2}\left(\overset{u^{n-1}}{\dfrac{3x - 1}{x^2 + 3}}\right)\overset{u'}{\dfrac{d}{dx}\left[\dfrac{3x - 1}{x^2 + 3}\right]}$$

$$= \left[\dfrac{2(3x - 1)}{x^2 + 3}\right]\left[\dfrac{(x^2 + 3)(3) - (3x - 1)(2x)}{(x^2 + 3)^2}\right]$$

$$= \dfrac{2(3x - 1)(3x^2 + 9 - 6x^2 + 2x)}{(x^2 + 3)^3}$$

$$= \dfrac{2(3x - 1)(-3x^2 + 2x + 9)}{(x^2 + 3)^3}$$

*STUDY TIP*    In Example 7, try to find $f'(x)$ by applying the Quotient Rule to

$$f(x) = \dfrac{(3x - 1)^2}{(x^2 + 3)^2}.$$

Which method do you prefer?

PhotoDisc

*After AT&T was required to split up its national telephone system, several new telecommunications companies started and prospered. The most successful of these new companies is MCI. In 1997, it had 3556 million circuit-miles of microwave and fiber optics in its network. The photo shows an example of fiber optics.*

## Extended Application: Revenue Per Share

### EXAMPLE 8   Finding Rates of Change

From 1987 through 1996, the revenue $R$ (in billions of dollars) for MCI Communications Corporation can be modeled by

$$R = (0.247t + 0.255)^2, \qquad 7 \le t \le 16,$$

where $t = 7$ represents 1987. Use the model to approximate the rate of change in the revenue in 1990, 1992, and 1994. If you had been an MCI stockholder from 1987 through 1996, would you have been satisfied with the performance of this stock?   *(Source: MCI Communications Corporation)*

**Solution**

The rate of change in $R$ is given by the derivative $dR/dt$. You can use the General Power Rule to find the derivative.

$$\frac{dR}{dt} = 2(0.247t + 0.255)^1(0.247) = 0.494(0.247t + 0.255)$$

In 1990, the revenue was changing at a rate of

$$\frac{dR}{dt} = 0.494[0.247(10) + 0.255] \approx \$1.35 \text{ billion per year.} \qquad \text{1990}$$

In 1992, the revenue was changing at a rate of

$$\frac{dR}{dt} = 0.494[0.247(12) + 0.255] \approx \$1.59 \text{ billion per year.} \qquad \text{1992}$$

In 1994, the revenue was changing at a rate of

$$\frac{dR}{dt} = 0.494[0.247(14) + 0.255] \approx \$1.83 \text{ billion per year.} \qquad \text{1994}$$

The graph of the revenue function is shown in Figure 2.30. For most investors, the performance of MCI stock would be considered to be very good.

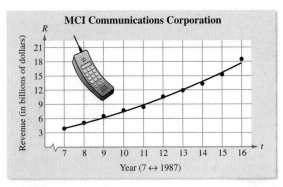

**FIGURE 2.30**

# Summary of Differentiation Rules

You now have all the rules you need to differentiate *any* algebraic function. For your convenience, they are summarized below.

## Summary of Differentiation Rules

1.  Constant Rule
$$\frac{d}{dx}[c] = 0, \qquad c \text{ is a constant.}$$

2.  Constant Multiple Rule
$$\frac{d}{dx}[cu] = c\frac{du}{dx}, \qquad c \text{ is a constant.}$$

3.  Sum and Difference Rules
$$\frac{d}{dx}[u \pm v] = \frac{du}{dx} \pm \frac{dv}{dx}$$

4.  Product Rule
$$\frac{d}{dx}[uv] = u\frac{dv}{dx} + v\frac{du}{dx}$$

5.  Quotient Rule
$$\frac{d}{dx}\left[\frac{u}{v}\right] = \frac{v\dfrac{du}{dx} - u\dfrac{dv}{dx}}{v^2}$$

6.  Power Rules
$$\frac{d}{dx}[x^n] = nx^{n-1}$$
$$\frac{d}{dx}[u^n] = nu^{n-1}\frac{du}{dx}$$

7.  Chain Rule
$$\frac{dy}{dx} = \frac{dy}{du}\frac{du}{dx}$$

## *Group Discussion*      **Comparing Power Rules**

You now have two power rules for differentiation.

$$\frac{d}{dx}[x^n] = nx^{n-1} \qquad \text{Simple Power Rule}$$

$$\frac{d}{dx}[u^n] = nu^{n-1}\frac{du}{dx} \qquad \text{General Power Rule}$$

Explain how you can tell which rule to use. Then state whether you would use the simple or general rule for each of the following.

a. $y = x^4$     b. $y = (x^2 + 1)$     c. $y = (x^2 + 1)^2$

d. $y = \dfrac{1}{x^4}$     e. $y = \dfrac{1}{x - 1}$     f. $y = \dfrac{1}{\sqrt{2x - 1}}$

*Warm Up*

The following warm-up exercises involve skills that were covered in earlier sections. You will use these skills in the exercise set for this section.

In Exercises 1–6, rewrite the expression with rational exponents.

**1.** $\sqrt[5]{(1 - 5x)^2}$

**2.** $\sqrt[4]{(2x - 1)^3}$

**3.** $\dfrac{1}{\sqrt{4x^2 + 1}}$

**4.** $\dfrac{1}{\sqrt[3]{x - 6}}$

**5.** $\dfrac{\sqrt{x}}{\sqrt[3]{1 - 2x}}$

**6.** $\dfrac{\sqrt{(3 - 7x)^3}}{2x}$

In Exercises 7–10, factor the expression.

**7.** $3x^3 - 6x^2 + 5x - 10$

**8.** $5x\sqrt{x} - x - 5\sqrt{x} + 1$

**9.** $4(x^2 + 1)^2 - x(x^2 + 1)^3$

**10.** $-x^5 + 3x^3 + x^2 - 3$

# EXERCISES 2.5

In Exercises 1–8, identify the inside function, $u = g(x)$, and the outside function, $y = f(u)$.

| $y = f(g(x))$ | $u = g(x)$ | $y = f(u)$ |
|---|---|---|

**1.** $y = (6x - 5)^4$

**2.** $y = (x^2 - 2x + 3)^3$

**3.** $y = (4 - x^2)^{-1}$

**4.** $y = (x^2 + 1)^{4/3}$

**5.** $y = \sqrt{5x - 2}$

**6.** $y = \sqrt{9 - x^2}$

**7.** $y = (3x + 1)^{-1}$

**8.** $y = (x + 1)^{-1/2}$

In Exercises 9–12, match the function with the rule that you would use to *most efficiently* find the derivative.

(a) Simple Power Rule  
(b) Constant Rule  
(c) General Power Rule  
(d) Quotient Rule

**9.** $f(x) = \dfrac{2}{1 - x^3}$

**10.** $f(x) = \dfrac{2x}{1 - x^3}$

**11.** $f(x) = \sqrt[3]{8^2}$

**12.** $f(x) = \sqrt[3]{x^2}$

In Exercises 13–30, use the General Power Rule to find the derivative of the function.

**13.** $y = (2x - 7)^3$

**14.** $y = (3x^2 + 1)^4$

**15.** $g(x) = (4 - 2x)^3$

**16.** $h(t) = (1 - t^2)^4$

**17.** $h(x) = (6x - x^3)^2$

**18.** $f(x) = (4x - x^2)^3$

**19.** $f(x) = (x^2 - 9)^{2/3}$

**20.** $f(t) = (9t + 2)^{2/3}$

**21.** $f(t) = \sqrt{t + 1}$

**22.** $g(x) = \sqrt{2x + 3}$

**23.** $s(t) = \sqrt{2t^2 + 5t + 2}$

**24.** $y = \sqrt[3]{3x^3 + 4x}$

**25.** $y = \sqrt[3]{9x^2 + 4}$

**26.** $y = 2\sqrt{4 - x^2}$

**27.** $f(x) = -3\sqrt[4]{2 - 9x}$

**28.** $f(x) = (25 + x^2)^{-1/2}$

**29.** $h(x) = (4 - x^3)^{-4/3}$

**30.** $f(x) = (4 - 3x)^{-5/2}$

In Exercises 31–36, find an equation of the tangent line to the graph of $f$ at the point $(2, f(2))$. Use a graphing utility to check your result by graphing the original function and the tangent line in the same viewing rectangle.

**31.** $f(x) = 2(x^2 - 1)^3$

**32.** $g(x) = 3(9x - 4)^4$

**33.** $f(x) = \sqrt{4x^2 - 7}$

**34.** $f(x) = x\sqrt{x^2 + 5}$

**35.** $g(x) = \sqrt{x^2 - 2x + 1}$

**36.** $g(x) = (4 - 3x^2)^{-2/3}$

In Exercises 37–40, use a symbolic differentiation utility to find the derivative of the function. Graph the function and its derivative in the same viewing rectangle. Describe the behavior of the function when the derivative is zero.

**37.** $f(x) = \dfrac{\sqrt{x} + 1}{x^2 + 1}$

**38.** $f(x) = \sqrt{\dfrac{2x}{x + 1}}$

**39.** $f(x) = \sqrt{\dfrac{x + 1}{x}}$

**40.** $f(x) = \sqrt{x}(2 - x^2)$

In Exercises 41–60, find the derivative of the function.

**41.** $y = \dfrac{1}{x - 2}$

**42.** $s(t) = \dfrac{1}{t^2 + 3t - 1}$

**43.** $y = -\dfrac{4}{(t + 2)^2}$

**44.** $f(x) = \dfrac{3}{x^3 - 4}$

**45.** $f(x) = \dfrac{1}{(x^2 - 3x)^2}$

**46.** $y = \dfrac{1}{\sqrt{x + 2}}$

**47.** $g(t) = \dfrac{1}{t^2 - 2}$

**48.** $g(x) = \dfrac{3}{\sqrt[3]{x^3 - 1}}$

**49.** $y = x\sqrt{2x + 3}$

**50.** $y = t\sqrt{t + 1}$

**51.** $y = t^2\sqrt{t - 2}$

**52.** $f(x) = x^3(x - 4)^2$

**53.** $f(x) = x(3x - 9)^3$

**54.** $y = \sqrt{x}(x - 2)^2$

**55.** $f(x) = \sqrt{\dfrac{3 - 2x}{4x}}$

**56.** $g(t) = \dfrac{3t^2}{\sqrt{t^2 + 2t - 1}}$

**57.** $f(x) = \dfrac{\sqrt[3]{x^3 + 1}}{x}$

**58.** $y = \sqrt{x - 1} + \sqrt{x + 1}$

**59.** $y = \left(\dfrac{6 - 5x}{x^2 - 1}\right)^2$

**60.** $y = \left(\dfrac{4x^2}{3 - x}\right)^3$

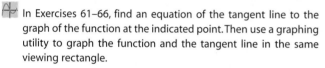

In Exercises 61–66, find an equation of the tangent line to the graph of the function at the indicated point. Then use a graphing utility to graph the function and the tangent line in the same viewing rectangle.

| Function | Point |
|---|---|
| **61.** $f(t) = \left(\dfrac{6}{3 - t}\right)^2$ | $(0, 4)$ |
| **62.** $s(x) = \dfrac{1}{\sqrt{x^2 - 3x + 4}}$ | $\left(3, \tfrac{1}{2}\right)$ |
| **63.** $f(t) = (t^2 - 9)\sqrt{t + 2}$ | $(-1, -8)$ |
| **64.** $y = \sqrt{\dfrac{2x}{x + 1}}$ | $(1, 1)$ |
| **65.** $f(x) = \dfrac{x + 1}{2x - 3}$ | $(2, 3)$ |
| **66.** $y = \dfrac{x}{\sqrt{25 + x^2}}$ | $(0, 0)$ |

**67.** *Compound Interest*   You deposit $1000 in an account with an annual interest rate of $r$ (in decimal form) compounded monthly. At the end of 5 years, the balance is

$$A = 1000\left(1 + \dfrac{r}{12}\right)^{60}.$$

Find the rate of change of $A$ with respect to $r$ when (a) $r = 0.08$, (b) $r = 0.10$, and (c) $r = 0.12$.

**68.** *Pollution Levels*   An environmental study indicates that the average daily level $P$ of a certain pollutant in the air in parts per million can be modeled by the equation

$$P = 0.25\sqrt{0.5n^2 + 5n + 25},$$

where $n$ is the number of residents of the community in thousands. Find the rate at which the level of pollutant is increasing when the population of the community is 12,000.

**69.** *Population Growth*   The number $N$ of bacteria in a culture after $t$ days is modeled by

$$N = 400\left[1 - \dfrac{3}{(t^2 + 2)^2}\right].$$

Complete the following table. What can you conclude?

| $t$ | 0 | 1 | 2 | 3 | 4 |
|---|---|---|---|---|---|
| $dN/dt$ |  |  |  |  |  |

**70.** *Depreciation*   The value $V$ of a machine $t$ years after it is purchased is inversely proportional to the square root of $t + 1$. The initial value of the machine is $10,000.

(a) Write $V$ as a function of $t$.
(b) Find the rate of depreciation when $t = 1$.
(c) Find the rate of depreciation when $t = 3$.

**71.** *Depreciation*   Repeat Exercise 70 given that the value of the machine $t$ years after it is purchased is inversely proportional to the cube root of $t + 1$.

**72.** *Credit-Card Rate*   The average annual rate $r$ (in percent form) for commercial bank credit cards from 1990 through 1995 can be modeled by

$$r = \sqrt{1.01t^4 - 6.72t^3 + 4.85t^2 + 2.4t + 330.26},$$

where $t = 0$ represents 1990. (*Source: Federal Reserve Bulletin*)

(a) Find the derivative of this model. Which differentiation rules did you use?
(b) Use a graphing utility to graph the derivative. Use the interval $0 \le t \le 5$.
(c) Use the trace feature to find the years during which the finance rate was changing the most.
(d) Use the trace feature to find the years during which the finance rate was changing the least.

# 2.6  Higher-Order Derivatives

*Second, Third, and Higher-Order Derivatives •
Acceleration*

## Second, Third, and Higher-Order Derivatives

The derivative of $f'$ is the **second derivative** of $f$ and is denoted by $f''$.

$$\frac{d}{dx}[f'(x)] = f''(x) \qquad \text{Second derivative}$$

The derivative of $f''$ is the **third derivative** of $f$ and is denoted by $f'''$.

$$\frac{d}{dx}[f''(x)] = f'''(x) \qquad \text{Third derivative}$$

By continuing this process, you obtain **higher-order derivatives** of $f$.

**EXAMPLE 1   Finding Higher-Order Derivatives**

| | |
|---|---|
| $f(x) = 2x^4 - 3x^2$ | Original function |
| $f'(x) = 8x^3 - 6x$ | First derivative |
| $f''(x) = 24x^2 - 6$ | Second derivative |
| $f'''(x) = 48x$ | Third derivative |
| $f^{(4)}(x) = 48$ | Fourth derivative |
| $f^{(5)}(x) = 0$ | Fifth derivative |

*STUDY TIP*   In the context
of higher-order derivatives,
the "standard" derivative $f'$
is often called the **first
derivative** of $f$.

### Notation for Higher-Order Derivatives

| | | | | | |
|---|---|---|---|---|---|
| 1. 1st derivative: | $y'$, | $f'(x)$, | $\dfrac{dy}{dx}$, | $\dfrac{d}{dx}[f(x)]$, | $D_x[y]$ |
| 2. 2nd derivative: | $y''$, | $f''(x)$, | $\dfrac{d^2y}{dx^2}$, | $\dfrac{d^2}{dx^2}[f(x)]$, | $D_x^2[y]$ |
| 3. 3rd derivative: | $y'''$, | $f'''(x)$, | $\dfrac{d^3y}{dx^3}$, | $\dfrac{d^3}{dx^3}[f(x)]$, | $D_x^3[y]$ |
| 4. 4th derivative: | $y^{(4)}$, | $f^{(4)}(x)$, | $\dfrac{d^4y}{dx^4}$, | $\dfrac{d^4}{dx^4}[f(x)]$, | $D_x^4[y]$ |
| 5. $n$th derivative: | $y^{(n)}$, | $f^{(n)}(x)$, | $\dfrac{d^ny}{dx^n}$, | $\dfrac{d^n}{dx^n}[f(x)]$, | $D_x^n[y]$ |

## EXAMPLE 2    Finding Higher-Order Derivatives

Find the value of $g'''(2)$ for the function

$$g(t) = -t^4 + 2t^3 + t + 4. \qquad \text{Original function}$$

### Solution

Begin by differentiating three times.

$$g'(t) = -4t^3 + 6t^2 + 1 \qquad \text{First derivative}$$

$$g''(t) = -12t^2 + 12t \qquad \text{Second derivative}$$

$$g'''(t) = -24t + 12 \qquad \text{Third derivative}$$

Then, evaluate the third derivative of $g$ at $t = 2$.

$$g'''(2) = -24(2) + 12 = -36 \qquad \text{Value of third derivative}$$

Examples 1 and 2 show how to find higher-order derivatives of *polynomial* functions. Note that with each successive differentiation, the degree of the polynomial drops by one. Eventually, higher-order derivatives of polynomial functions degenerate to a constant function. Specifically, the $n$th-order derivative of an $n$th-degree polynomial function

$$f(x) = a_n x^n + a_{n-1} x^{n-1} + \cdots + a_1 x + a_0$$

is the constant function

$$f^{(n)}(x) = n! a_n,$$

where $n! = 1 \cdot 2 \cdot 3 \cdots n$. Each derivative of order higher than $n$ is the zero function. Polynomial functions are the *only* functions with this characteristic. For other functions, successive differentiation never produces a constant function.

## EXAMPLE 3   Finding Higher-Order Derivatives

$$y = x^{-1} = \frac{1}{x} \qquad \text{Original derivative}$$

$$y' = (-1)x^{-2} = -\frac{1}{x^2} \qquad \text{First derivative}$$

$$y'' = (-1)(-2)x^{-3} = \frac{2}{x^3} \qquad \text{Second derivative}$$

$$y''' = (-1)(-2)(-3)x^{-4} = -\frac{6}{x^4} \qquad \text{Third derivative}$$

$$y^{(4)} = (-1)(-2)(-3)(-4)x^{-5} = \frac{24}{x^5} \qquad \text{Fourth derivative}$$

---

### Technology

Higher-order derivatives of non-polynomial functions can be difficult to find by hand. If you have access to a symbolic differentiation utility, such as *Derive*, *Maple*, *Mathcad*, *Mathematica*, or the *TI-92*, try using it to find higher-order derivatives. For instance, if you use *Derive* for Windows to find the third derivative of $y = 1/(x^2 + 1)$, you will obtain the following.

$$\#1: \left[\frac{d}{dx}\right]^3 \frac{1}{x^2 + 1}$$

$$\#2: \frac{24x(1 - x^2)}{(x^2 + 1)^4}$$

## Acceleration

*STUDY TIP*    Acceleration
is measured in units of length
per unit of time squared.
For instance, if the velocity is
measured in feet per second,
then the acceleration is
measured in "feet per second
squared," or more formally in
"feet per second per second."

In Section 2.3, you saw that the velocity of an object moving in a straight path is
given by the derivative of its position function. In other words, the rate of change
of the position with respect to time is defined to be the velocity. In a similar way,
the rate of change of the velocity with respect to time is defined to be the **acceleration** of the object.

$$s = f(t)$$    Position function

$$\frac{ds}{dt} = f'(t)$$    Velocity function

$$\frac{d^2s}{dt^2} = f''(t)$$    Acceleration function

To find the position, velocity, or acceleration at a particular time $t$, substitute the
given value of $t$ into the appropriate function, as illustrated in Example 4.

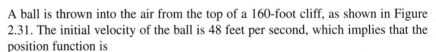

### EXAMPLE 4    Finding Acceleration

A ball is thrown into the air from the top of a 160-foot cliff, as shown in Figure
2.31. The initial velocity of the ball is 48 feet per second, which implies that the
position function is

$$s = -16t^2 + 48t + 160,$$

where the time $t$ is measured in seconds. Find the height, the velocity, and the
acceleration of the ball when $t = 3$.

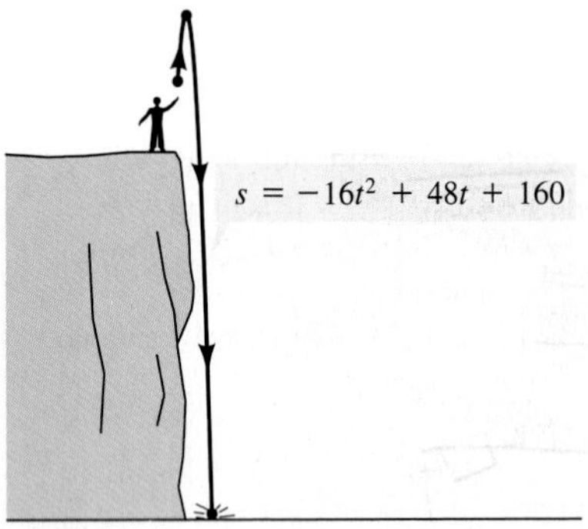

*Solution*

Begin by differentiating to find the velocity and acceleration functions.

$$s = -16t^2 + 48t + 160$$    Position function

$$\frac{ds}{dt} = -32t + 48$$    Velocity function

$$\frac{d^2s}{dt^2} = -32$$    Acceleration function

To find the height, velocity, and acceleration when $t = 3$, substitute $t = 3$ into
each of the above functions.

$$s = -16(3)^2 + 48(3) + 160 = 160 \text{ ft}$$    Position when $t = 3$

$$\frac{ds}{dt} = -32(3) + 48 = -48 \text{ ft/sec}$$    Velocity when $t = 3$

$$\frac{d^2s}{dt^2} = -32 \text{ ft/sec}^2$$    Acceleration when $t = 3$

$$s = -16t^2 + 48t + 160$$

**FIGURE 2.31**

In Example 4, notice that the acceleration of the ball is $-32$ ft/sec$^2$ at any time $t$. This constant acceleration is due to the gravitational force of the earth and is called the **acceleration due to gravity.** Note that the negative value indicates that the ball is being pulled *down*—toward the earth.

Although the acceleration exerted on a falling object is relatively constant near the earth's surface, it varies greatly throughout our solar system. Large planets exert a much greater gravitational pull than do small planets or moons. The next example describes the motion of a free-falling object on the moon.

NASA

### EXAMPLE 5 Finding Acceleration on the Moon

An astronaut standing on the surface of the moon throws a rock into the air. The height $s$ (in feet) of the rock is given by

$$s = -\frac{27}{10}t^2 + 27t + 6,$$

where $t$ is measured in seconds. How does the acceleration due to gravity on the moon compare with that on the earth?

*Solution*

$$s = -\frac{27}{10}t^2 + 27t + 6 \qquad \text{Position function}$$

$$\frac{ds}{dt} = -\frac{27}{5}t + 27 \qquad \text{Velocity function}$$

$$\frac{d^2s}{dt^2} = -\frac{27}{5} \qquad \text{Acceleration function}$$

Thus, the acceleration at any time is $-\frac{27}{5} = -5.4$ ft/sec$^2$—about one-sixth of the acceleration due to gravity on the earth.

*The acceleration due to gravity on the surface of the moon is only about one-sixth that exerted by the earth. Thus, if you were on the moon and threw an object into the air, it would rise to a greater height than it would on the earth's surface.*

The position function described in Example 5 neglects air resistance, which is appropriate because the moon has no atmosphere—and *no air resistance.* This means that the position function for any free-falling object on the moon is given by

$$s = -\frac{27}{10}t^2 + v_0t + h_0,$$

where $s$ is the height (in feet), $t$ is the time (in seconds), $v_0$ is the initial velocity, and $h_0$ is the initial height. For instance, the rock in Example 5 was thrown upward with an initial velocity of 27 feet per second and had an initial height of 6 feet. This position function is valid for all objects—heavy ones such as hammers and light ones such as feathers.

In 1971, astronaut David R. Scott demonstrated the lack of atmosphere on the moon by dropping a hammer and a feather from the same height. Both took exactly the same time to fall to the ground. If they were dropped from a height of 6 feet, how long did each take to hit the ground?

*This drawing depicts the first moon landing on July 20, 1969. The event was witnessed live by millions of observers on the earth via satellite television. Can you see what is wrong with the drawing?*

## EXAMPLE 6  Finding Velocity and Acceleration

The velocity $v$ (in feet per second) of a certain automobile starting from rest is

$$v = \frac{80t}{t + 5},$$   Velocity function

where $t$ is the time (in seconds). The position of the automobile is shown in Figure 2.32. Find the velocity and acceleration of the automobile at 10-second intervals from $t = 0$ to $t = 60$.

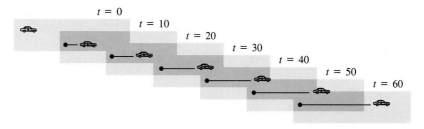

**FIGURE 2.32**

*Solution*

To find the acceleration function, differentiate the velocity function.

$$\frac{dv}{dt} = \frac{(t + 5)(80) - (80t)(1)}{(t + 5)^2} = \frac{400}{(t + 5)^2}$$   Acceleration function

| $t$ (seconds) | 0 | 10 | 20 | 30 | 40 | 50 | 60 |
|---|---|---|---|---|---|---|---|
| $v$ (ft/sec) | 0 | 53.3 | 64.0 | 68.6 | 71.1 | 72.7 | 73.8 |
| $\frac{dv}{dt}$ (ft/sec²) | 16 | 1.78 | 0.64 | 0.33 | 0.20 | 0.13 | 0.09 |

In the table, note that the acceleration approaches zero as the velocity levels off. This observation should agree with your experience—when riding in an accelerating automobile, you do not feel the velocity, but you do feel the acceleration. In other words, you feel changes in velocity.

## *Group Discussion*  **Acceleration Due to Gravity**

Newton's law of universal gravitation states that the gravitational attraction of two objects is directly proportional to their masses and inversely proportional to the square of the distance between their centers. The earth has a mass of $5.979 \times 10^{24}$ kilograms and a radius of 6371 kilometers. The moon has a mass of $7.354 \times 10^{22}$ kilograms and a radius of 1738 kilometers. Discuss how you could find the ratio of the earth's gravity to the moon's gravity. What is this ratio?

*Warm Up*

The following warm-up exercises involve skills that were covered in earlier sections. You will use these skills in the exercise set for this section.

In Exercises 1–4, solve the equation.

**1.** $-16t^2 + 24t = 0$

**2.** $-16t^2 + 80t + 224 = 0$

**3.** $-16t^2 + 128t + 320 = 0$

**4.** $-16t^2 + 9t + 1440 = 0$

In Exercises 5–8, find $dy/dx$.

**5.** $y = x^2(2x + 7)$

**6.** $y = (x^2 + 3x)(2x^2 - 5)$

**7.** $y = \dfrac{x^2}{2x + 7}$

**8.** $y = \dfrac{x^2 + 3x}{2x^2 - 5}$

In Exercises 9 and 10, find the domain and range of $f$.

**9.** $f(x) = x^2 - 4$

**10.** $f(x) = \sqrt{x - 7}$

# EXERCISES 2.6

In Exercises 1–14, find the second derivative of the function.

**1.** $f(x) = 5 - 4x$

**2.** $f(x) = 3x - 1$

**3.** $f(x) = x^2 + 7x - 4$

**4.** $f(x) = 3x^2 + 4x$

**5.** $g(t) = \frac{1}{3}t^3 - 4t^2 + 2t$

**6.** $g(t) = t^{-1/3}$

**7.** $f(t) = \dfrac{3}{4t^2}$

**8.** $f(x) = 4(x^2 - 1)^2$

**9.** $f(x) = 3(2 - x^2)^3$

**10.** $f(x) = x\sqrt[3]{x}$

**11.** $f(x) = \dfrac{x + 1}{x - 1}$

**12.** $g(t) = -\dfrac{4}{(t + 2)^2}$

**13.** $y = x^2(x^2 + 4x + 8)$

**14.** $h(s) = s^3(s^2 - 2s + 1)$

In Exercises 15–20, find the third derivative of the function.

**15.** $f(x) = x^5 - 3x^4$

**16.** $f(x) = x^4 - 2x^3$

**17.** $f(x) = 5x(x + 4)^3$

**18.** $f(x) = (x - 1)^2$

**19.** $f(x) = \dfrac{3}{(4x)^2}$

**20.** $f(x) = \dfrac{1}{x}$

In Exercises 21–24, find the indicated value.

| Function | Value |
| --- | --- |
| **21.** $g(t) = 5t^4 + 10t^2 + 3$ | $g''(2)$ |
| **22.** $f(x) = \sqrt{9 - x^2}$ | $f''(-\sqrt{5})$ |
| **23.** $f(x) = \sqrt{4 - x}$ | $f'''(-5)$ |
| **24.** $f(t) = \sqrt{2t + 3}$ | $f'''\left(\frac{1}{2}\right)$ |

In Exercises 25–30, find the indicated derivative.

| Given | Derivative |
| --- | --- |
| **25.** $f'(x) = 2x^2$ | $f''(x)$ |
| **26.** $f''(x) = 20x^3 - 36x^2$ | $f'''(x)$ |
| **27.** $f''(x) = \dfrac{2x - 2}{x}$ | $f'''(x)$ |
| **28.** $f'''(x) = 2\sqrt{x - 1}$ | $f^{(4)}(x)$ |
| **29.** $f^{(4)}(x) = (x + 1)^2$ | $f^{(6)}(x)$ |
| **30.** $f(x) = x^3 - 2x$ | $f''(x)$ |

In Exercises 31–38, find the second derivative and solve the equation $f''(x) = 0$.

**31.** $f(x) = x^3 - 9x^2 + 27x - 27$

**32.** $f(x) = 3x^3 - 9x + 1$

**33.** $f(x) = (x + 3)(x - 4)(x + 5)$

**34.** $f(x) = x^4 - 8x^3 + 18x^2 - 16x + 2$

**35.** $f(x) = 3x^4 - 18x^2$

**36.** $f(x) = x\sqrt{4 - x^2}$

**37.** $f(x) = \dfrac{x}{x^2 + 3}$

**38.** $f(x) = \dfrac{x}{x^2 + 1}$

**39.** *Velocity and Acceleration* A ball is propelled straight up from ground level with an initial velocity of 144 feet per second.

(a) Write the position function of the ball.
(b) Write the velocity and acceleration functions.
(c) When is the ball at its highest point? How high is this point?
(d) How fast is the ball traveling when it hits the ground? How is this speed related to the initial velocity?

**40.** *Velocity and Acceleration* A brick becomes dislodged from the top of the Empire State Building (at a height of 1250 feet) and falls to the sidewalk below.

(a) Write the position function of the brick.
(b) Write the velocity and acceleration functions.
(c) How long does it take the brick to hit the sidewalk?
(d) How fast is the brick traveling when it hits the sidewalk?

**41.** *Velocity and Acceleration* The velocity (in feet per second) of an automobile starting from rest is modeled by

$$\frac{ds}{dt} = \frac{90t}{t + 10}.$$

Complete the table showing the velocity and acceleration at 10-second intervals during the first minute of travel. What can you conclude?

| $t$ | 0 | 10 | 20 | 30 | 40 | 50 | 60 |
|---|---|---|---|---|---|---|---|
| $\dfrac{ds}{dt}$ | | | | | | | |
| $\dfrac{d^2s}{dt^2}$ | | | | | | | |

**42.** *Stopping Distance* A car is traveling at a rate of 66 feet per second (45 miles per hour) when the brakes are applied. The position function for the car is

$$s = -8.25t^2 + 66t,$$

where $s$ is measured in feet and $t$ is measured in seconds. Complete the table showing the position, velocity, and acceleration for the given values of $t$. What can you conclude?

| $t$ | 0 | 1 | 2 | 3 | 4 | 5 |
|---|---|---|---|---|---|---|
| $s(t)$ | | | | | | |
| $v(t)$ | | | | | | |
| $a(t)$ | | | | | | |

 In Exercises 43 and 44, use a graphing utility to graph $f$, $f'$, and $f''$ in the same viewing rectangle. What is the relationship between the degree of $f$ and the degrees of its successive derivatives? In general, what is the relationship between the degree of a polynomial function and the degrees of its successive derivatives?

**43.** $f(x) = x^2 - 6x + 6$

**44.** $f(x) = 3x^3 - 9x$

In Exercises 45 and 46, the graphs of $f$, $f'$, and $f''$ are shown together. Which is which? Explain your reasoning.

**45.** 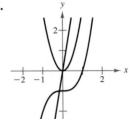  **46.**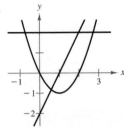

*True or False* In Exercises 47–53, determine whether the statement is true or false. Explain your reasoning.

**47.** If $y = f(x)g(x)$, then $y' = f'(x)g'(x)$.

**48.** If $y = (x + 1)(x + 2)(x + 3)(x + 4)$, then $y^{(5)} = 0$.

**49.** If $f'(c)$ and $g'(c)$ are zero and $h(x) = f(x)g(x)$, then $h'(c) = 0$.

**50.** If $f(x)$ is an $n$th-degree polynomial, then $f^{(n+1)}(x) = 0$.

**51.** The second derivative represents the rate of change of the first derivative.

**52.** If the velocity of an object is constant, then its acceleration is zero.

**53.** If $f'(x) = g'(x)$, then $f(x) = g(x)$.

**54.** *Projectile Motion* An object is thrown upward from the top of a 64-foot building with an initial velocity of 48 feet per second.

(a) Write the position function of the object.
(b) Write the velocity and acceleration functions.
(c) When will the object hit the ground?
(d) When is the velocity of the object zero?
(e) How high does the object go?
(f) Use a graphing utility to graph the position, velocity, and acceleration functions in the same viewing rectangle. Write a short paragraph that describes the relationship among these functions.

## Implicit Differentiation

**2.7**

*Implicit and Explicit Functions • Implicit Differentiation • Extended Application: Demand Function*

## Implicit and Explicit Functions

So far in the text, functions involving two variables have generally been expressed in the **explicit form** $y = f(x)$. That is, one of the two variables has been explicitly given in terms of the other. For example,

$$y = 3x - 5, \quad s = -16t^2 + 20t, \quad \text{and} \quad u = 3w - w^2$$

are each written in explicit form, and we say that $y$, $s$, and $u$ are functions of $x$, $t$, and $w$, explicitly. Many functions, however, are not given explicitly and are only implied by a given equation, as illustrated in Example 1.

---

### EXAMPLE 1 Finding a Derivative Explicitly

Find $dy/dx$ for the equation $xy = 1$.

**Solution**

In this equation, $y$ is **implicitly** defined as a function of $x$. One way to find $dy/dx$ is to first solve the equation for $y$, then differentiate as usual.

| | |
|---|---|
| $xy = 1$ | Implicit form |
| $y = \dfrac{1}{x}$ | Solve for $y$. |
| $= x^{-1}$ | Rewrite. |
| $\dfrac{dy}{dx} = -x^{-2}$ | Differentiate with respect to $x$. |
| $= -\dfrac{1}{x^2}$ | Simplify. |

The procedure shown in Example 1 works well whenever you can easily write the given function explicitly. You cannot, however, use this procedure when you are unable to solve for $y$ as a function of $x$. For instance, how would you find $dy/dx$ in the equation

$$x^2 - 2y^3 + 4y = 2,$$

where it is very difficult to express $y$ as a function of $x$ explicitly? To do this, you can use a procedure called **implicit differentiation.**

## Implicit Differentiation

To understand how to find $dy/dx$ implicitly, you must realize that the differentiation is taking place *with respect to x*. This means that when you differentiate terms involving $x$ alone, you can differentiate as usual. *But* when you differentiate terms involving $y$, you must apply the Chain Rule because you are assuming that $y$ is defined implicitly as a function of $x$. Study the next example carefully. Note in particular how the Chain Rule is used to introduce the $dy/dx$ factors in parts b and d.

---

### EXAMPLE 2   Applying the Chain Rule

Differentiate the following with respect to $x$.

(a) $3x^2$     (b) $2y^3$     (c) $x + 3y$     (d) $xy^2$

*Solution*

(a) The only variable in this expression is $x$. Thus, to differentiate with respect to $x$, you can use the Simple Power Rule and the Constant Multiple Rule to obtain

$$\frac{d}{dx}[3x^2] = 6x.$$

(b) This case is different. The variable in the expression is $y$, and yet you are asked to differentiate with respect to $x$. To do this, assume that $y$ is a differentiable function of $x$ and use the Chain Rule.

$$\frac{d}{dx}[2y^3] = \overset{c}{2} \; \overset{n}{(3)} \; \overset{u^{n-1}}{y^2} \; \overset{u'}{\frac{dy}{dx}} \qquad \text{Chain Rule}$$

$$= 6y^2\frac{dy}{dx}$$

(c) This expression involves both $x$ and $y$. By the Sum Rule and the Constant Multiple Rule, you can write

$$\frac{d}{dx}[x + 3y] = 1 + 3\frac{dy}{dx}.$$

(d) By the Product Rule and the Chain Rule, you can write

$$\frac{d}{dx}[xy^2] = x\frac{d}{dy}[y^2] + y^2\frac{d}{dx}[x] \qquad \text{Product Rule}$$

$$= x\left(2y\frac{dy}{dx}\right) + y^2(1) \qquad \text{Chain Rule}$$

$$= 2xy\frac{dy}{dx} + y^2.$$

## Implicit Differentiation

Consider an equation involving $x$ and $y$ in which $y$ is a differentiable function of $x$. You can use the following steps to find $dy/dx$.

1. Differentiate both sides of the equation *with respect to x.*
2. Write the result so that all terms involving $dy/dx$ are on the left side of the equation and all other terms are on the right side of the equation.
3. Factor $dy/dx$ out of the terms on the left side of the equation.
4. Solve for $dy/dx$ by dividing both sides of the equation by the left-hand factor that does not contain $dy/dx$.

In Example 3, note that implicit differentiation can produce an expression for $dy/dx$ that contains both $x$ and $y$.

## EXAMPLE 3    Finding the Slope of a Graph Implicitly

Find the slope of the tangent line to the ellipse given by $x^2 + 4y^2 = 4$ at the point $\left(\sqrt{2}, -1/\sqrt{2}\right)$, as shown in Figure 2.33.

### Solution

$$x^2 + 4y^2 = 4 \qquad\qquad \text{Original equation}$$

$$\frac{d}{dx}[x^2 + 4y^2] = \frac{d}{dx}[4] \qquad\qquad \text{Differentiate with respect to } x.$$

$$2x + 8y\left(\frac{dy}{dx}\right) = 0 \qquad\qquad \text{Implicit differentiation}$$

$$8y\left(\frac{dy}{dx}\right) = -2x \qquad\qquad \text{Subtract } 2x \text{ from both sides.}$$

$$\frac{dy}{dx} = \frac{-2x}{8y} \qquad\qquad \text{Divide both sides by } 8y.$$

$$\frac{dy}{dx} = -\frac{x}{4y} \qquad\qquad \text{Simplify.}$$

To find the slope at the given point, substitute $x = \sqrt{2}$ and $y = -1/\sqrt{2}$ into the derivative.

$$\frac{dy}{dx} = -\frac{x}{4y} \qquad\qquad \text{Derivative}$$

$$= -\frac{\sqrt{2}}{4\left(-1/\sqrt{2}\right)} \qquad\qquad \text{Substitute for } x \text{ and } y.$$

$$= \frac{1}{2} \qquad\qquad \text{Simplify.}$$

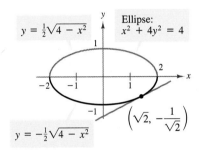

**FIGURE 2.33**   Slope of tangent line is $\frac{1}{2}$.

**STUDY TIP**   To see the benefit of implicit differentiation, try reworking Example 3 using the explicit function

$$y = -\frac{1}{2}\sqrt{4 - x^2}.$$

The graph of this function is the lower half of the ellipse.

## EXAMPLE 4   Using Implicit Differentiation

Find $dy/dx$ for the equation $y^3 + y^2 - 5y - x^2 = -4$.

**Solution**

$$y^3 + y^2 - 5y - x^2 = -4 \qquad \text{Original equation}$$

$$\frac{d}{dx}[y^3 + y^2 - 5y - x^2] = \frac{d}{dx}[-4] \qquad \text{Differentiate both sides.}$$

$$3y^2\frac{dy}{dx} + 2y\frac{dy}{dx} - 5\frac{dy}{dx} - 2x = 0 \qquad \text{Implicit differentiation}$$

$$3y^2\frac{dy}{dx} + 2y\frac{dy}{dx} - 5\frac{dy}{dx} = 2x \qquad \text{Collect } dy/dx \text{ terms.}$$

$$\frac{dy}{dx}(3y^2 + 2y - 5) = 2x \qquad \text{Factor.}$$

$$\frac{dy}{dx} = \frac{2x}{3y^2 + 2y - 5}$$

The graph of the original equation is shown in Figure 2.34. What is the slope of the graph at the points $(1, -3)$, $(2, 0)$, and $(1, 1)$?

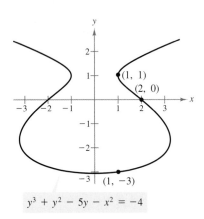

$$y^3 + y^2 - 5y - x^2 = -4$$

**FIGURE 2.34**

## EXAMPLE 5   Finding the Slope of a Graph Implicitly

Find the slope of the graph of $2x^2 - y^2 = 1$ at the point $(1, 1)$.

**Solution**

Begin by finding $dy/dx$ implicitly.

$$2x^2 - y^2 = 1 \qquad \text{Original equation}$$

$$4x - 2y\left(\frac{dy}{dx}\right) = 0 \qquad \text{Differentiate both sides.}$$

$$-2y\left(\frac{dy}{dx}\right) = -4x \qquad \text{Subtract } 4x \text{ from each side.}$$

$$\frac{dy}{dx} = \frac{2x}{y} \qquad \text{Divide each side by } -2y.$$

At the point $(1, 1)$, the slope of the graph is

$$\frac{dy}{dx} = \frac{2x}{y}$$

$$= \frac{2(1)}{1}$$

$$= 2,$$

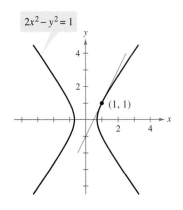

$$2x^2 - y^2 = 1$$

**FIGURE 2.35**   Hyperbola

as shown in Figure 2.35. The graph is called a **hyperbola.**

## Extended Application: Demand Function

### EXAMPLE 6  Using a Demand Function

The demand function for a product is modeled by

$$p = \frac{3}{0.000001x^3 + 0.01x + 1},$$

where $p$ is measured in dollars and $x$ is measured in thousands of units, as shown in Figure 2.36. Find the rate of change of the demand $x$ with respect to the price $p$ when $x = 100$.

**Demand Function**

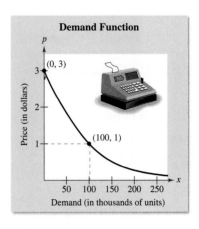

**FIGURE 2.36**

### Solution

To simplify the differentiation, begin by rewriting the function. Then, differentiate *with respect to p.*

$$p = \frac{3}{0.000001x^3 + 0.01x + 1}$$

$$0.000001x^3 + 0.01x + 1 = \frac{3}{p}$$

$$0.000003x^2 \frac{dx}{dp} + 0.01\frac{dx}{dp} = -\frac{3}{p^2}$$

$$(0.000003x^2 + 0.01)\frac{dx}{dp} = -\frac{3}{p^2}$$

$$\frac{dx}{dp} = -\frac{3}{p^2(0.000003x^2 + 0.01)}$$

When $x = 100$, the price is

$$p = \frac{3}{0.000001(100)^3 + 0.01(100) + 1} = \$1.$$

Thus, when $x = 100$ and $p = 1$, the rate of change of the demand with respect to the price is

$$\frac{dx}{dp} = -\frac{3}{(1)^2[0.000003(100)^2 + 0.01]} = -75.$$

This means that when $x = 100$, the demand is dropping at the rate of 75 thousand units for each dollar increase in price.

## *Group Discussion*     **Comparing Derivatives**

In Example 6, the derivative $dx/dp$ does not represent the slope of the graph of the demand function. Because the demand function is given by $p = f(x)$, the slope of the graph is given by $dp/dx$. Find $dp/dx$. Show that the two derivatives are related by $dx/dp = 1/(dp/dx)$. What does $dp/dx$ represent?

## Warm Up

The following warm-up exercises involve skills that were covered in earlier sections. You will use these skills in the exercise set for this section.

In Exercises 1–6, solve the equation for y.

**1.** $x - \dfrac{y}{x} = 2$

**2.** $\dfrac{4}{x-3} = \dfrac{1}{y}$

**3.** $xy - x + 6y = 6$

**4.** $12 + 3y = 4x^2 + x^2y$

**5.** $x^2 + y^2 = 5$

**6.** $x = \pm\sqrt{6 - y^2}$

In Exercises 7–10, evaluate the expression at the indicated point.

**7.** $\dfrac{3x^2 - 4}{3y^2}, \quad (2, 1)$

**8.** $\dfrac{x^2 - 2}{1 - y}, \quad (0, -3)$

**9.** $\dfrac{5x}{3y^2 - 12y + 5}, \quad (-1, 2)$

**10.** $\dfrac{1}{y^2 - 2xy + x^2}, \quad (4, 3)$

## EXERCISES 2.7

In Exercises 1–6, find $dy/dx$.

**1.** $5xy = 1$

**2.** $\frac{1}{2}x^2 - y = 6x$

**3.** $y^2 = 1 - x^2, \; 0 \le x \le 1$

**4.** $4x^2y - \dfrac{3}{y} = 0$

**5.** $\dfrac{2 - x}{y - 3} = 5$

**6.** $\dfrac{xy - y^2}{y - x} = 1$

In Exercises 7–18, find $dy/dx$ by implicit differentiation and evaluate the derivative at the indicated point.

|  | Equation | Point |
|---|---|---|
| **7.** | $x^2 + y^2 = 49$ | $(0, 7)$ |
| **8.** | $x^2 - y^2 = 16$ | $(4, 0)$ |
| **9.** | $y + xy = 4$ | $(-5, -1)$ |
| **10.** | $x^2 - y^3 = 3$ | $(2, 1)$ |
| **11.** | $x^3 - xy + y^2 = 4$ | $(0, -2)$ |

|  | Equation | Point |
|---|---|---|
| **12.** | $x^2y + y^2x = -2$ | $(2, -1)$ |
| **13.** | $x^3y^3 - y = x$ | $(0, 0)$ |
| **14.** | $x^3 + y^3 = 2xy$ | $(1, 1)$ |
| **15.** | $x^{1/2} + y^{1/2} = 9$ | $(16, 25)$ |
| **16.** | $\sqrt{xy} = x - 2y$ | $(4, 1)$ |
| **17.** | $x^{2/3} + y^{2/3} = 5$ | $(8, 1)$ |
| **18.** | $(x + y)^3 = x^3 + y^3$ | $(-1, 1)$ |

In Exercises 19–22, find the slope of the graph at the indicated point.

**19.** $3x^2 - 2y + 5 = 0$

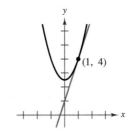

**20.** $x^2 + y^2 = 4$

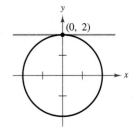

**21.** $4x^2 + 9y^2 = 36$

**22.** $x^2 - y^3 = 0$

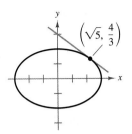

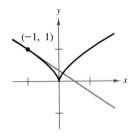

In Exercises 23–26, find $dy/dx$ implicitly and explicitly (the explicit functions are shown on the graph) and show that the results are equivalent. Use the graph to estimate the slope of the tangent line at the labeled point. Then verify your result analytically by evaluating $dy/dx$ at the point.

**23.** $x^2 + y^2 = 25$

$y = \sqrt{25 - x^2}$

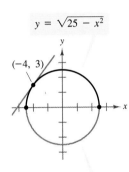

$y = -\sqrt{25 - x^2}$

**24.** $x - y^2 - 1 = 0$

$y = \sqrt{x - 1}$

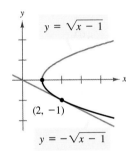

$y = -\sqrt{x - 1}$

**25.** $9x^2 + 16y^2 = 144$

$y = \dfrac{\sqrt{144 - 9x^2}}{4}$

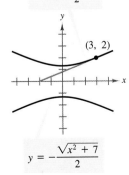

$y = -\dfrac{\sqrt{144 - 9x^2}}{4}$

**26.** $4y^2 - x^2 = 7$

$y = \dfrac{\sqrt{x^2 + 7}}{2}$

$y = -\dfrac{\sqrt{x^2 + 7}}{2}$

In Exercises 27–32, find equations of the tangent lines to the graph at the given points. Graph the equation and the tangent lines in the same viewing rectangle of your graphing utility.

| Equation | Points |
|---|---|
| **27.** $x^2 + y^2 = 169$ | $(5, 12)$ and $(-12, 5)$ |
| **28.** $x^2 + y^2 = 9$ | $(0, 3)$ and $\left(2, \sqrt{5}\right)$ |
| **29.** $y^2 = 5x^3$ | $\left(1, \sqrt{5}\right)$ and $\left(1, -\sqrt{5}\right)$ |
| **30.** $4xy + x^2 = 5$ | $(1, 1)$ and $(5, -1)$ |
| **31.** $x^3 + y^3 = 8$ | $(0, 2)$ and $(2, 0)$ |
| **32.** $y^2 = \dfrac{x^3}{4 - x}$ | $(2, 2)$ and $(2, -2)$ |

*Demand* In Exercises 33–36, find the rate of change of $x$ with respect to $p$.

**33.** $p = 0.006x^4 + 0.02x^2 + 10, \quad x \geq 0$

**34.** $p = \dfrac{725}{0.02x}$

**35.** $p = \dfrac{200 - x}{2x}, \quad 0 < x \leq 200$

**36.** $p = \sqrt{\dfrac{500 - x}{2x}}, \quad 0 < x \leq 500$

**37.** *Production*  Let $x$ represent the units of labor and $y$ the capital invested in a manufacturing process. When 135,540 units are produced, the relationship between labor and capital can be modeled by $100x^{0.75}y^{0.25} = 135{,}540$.

(a) Find the rate of change of $y$ with respect to $x$ when $x = 1500$ and $y = 1000$.

(b) The model used in the problem is called the *Cobb-Douglas production function*. Graph the model on a graphing utility and describe the relationship between labor and capital.

**38.** *U.S. AIDS Epidemic*  The number of cases $y$ of AIDS reported from 1985 to 1995 can be modeled by

$$\sqrt{79.23y + 1} = -0.088t^4 + 27.03t^2 + 291.14,$$
$$5 \leq t \leq 15,$$

where $t = 5$ represents 1985.

(a) Graph the model on a graphing utility and describe the results.

(b) Use the graph to determine the year during which the number of reported cases increased most rapidly.

(c) Complete the table to confirm your estimate.

| $t$ | 5 | 6 | 7 | 8 | 9 | 10 | 11 | 12 | 13 | 14 | 15 |
|---|---|---|---|---|---|---|---|---|---|---|---|
| $y$ | | | | | | | | | | | |
| $y'$ | | | | | | | | | | | |

## Warm Up

The following warm-up exercises involve skills that were covered in earlier sections. You will use these skills in the exercise set for this section.

In Exercises 1–6, write a formula for the given quantity.

**1.** Area of a circle

**2.** Volume of a sphere

**3.** Surface area of a cube

**4.** Volume of a cube

**5.** Volume of a cone

**6.** Area of a triangle

In Exercises 7–10, find $dy/dx$ by implicit differentiation.

**7.** $x^2 + y^2 = 9$

**8.** $3xy - x^2 = 6$

**9.** $x^2 + 2y + xy = 12$

**10.** $x + xy^2 - y^2 = xy$

## EXERCISES 2.8

In Exercises 1–4, find the indicated values of $dy/dt$ and $dx/dt$.

| Equation | Find | Given |
|---|---|---|
| **1.** $y = x^2 - \sqrt{x}$ | (a) $\dfrac{dy}{dt}$ | $x = 4, \dfrac{dx}{dt} = 8$ |
| | (b) $\dfrac{dx}{dt}$ | $x = 16, \dfrac{dy}{dt} = 12$ |
| **2.** $y = x^2 - 3x$ | (a) $\dfrac{dy}{dt}$ | $x = 3, \dfrac{dx}{dt} = 2$ |
| | (b) $\dfrac{dx}{dt}$ | $x = 1, \dfrac{dy}{dt} = 5$ |
| **3.** $xy = 4$ | (a) $\dfrac{dy}{dt}$ | $x = 8, \dfrac{dx}{dt} = 10$ |
| | (b) $\dfrac{dx}{dt}$ | $x = 1, \dfrac{dy}{dt} = -6$ |
| **4.** $x^2 + y^2 = 25$ | (a) $\dfrac{dy}{dt}$ | $x = 3, y = 4, \dfrac{dx}{dt} = 8$ |
| | (b) $\dfrac{dx}{dt}$ | $x = 4, y = 3, \dfrac{dy}{dt} = -2$ |

**5.** *Area*   The radius $r$ of a circle is increasing at a rate of 2 inches per minute. Find the rate of change of the area when (a) $r = 6$ inches and (b) $r = 24$ inches.

**6.** *Volume*   The radius $r$ of a sphere is increasing at a rate of 2 inches per minute. Find the rate of change of the volume when (a) $r = 6$ inches and (b) $r = 24$ inches.

**7.** *Area*   Let $A$ be the area of a circle of radius $r$ that is changing with respect to time. If $dr/dt$ is constant, is $dA/dt$ constant? Explain your reasoning.

**8.** *Volume*   Let $V$ be the volume of a sphere of radius $r$ that is changing with respect to time. If $dr/dt$ is constant, is $dV/dt$ constant? Explain your reasoning.

**9.** *Inflating Balloon*   A spherical balloon is inflated with gas at the rate of 20 cubic feet per minute. How fast is the radius of the balloon changing at the instant the radius is (a) 1 foot and (b) 2 feet?

**10.** *Volume of Cone*   The radius $r$ of a right circular cone is increasing at a rate of 2 inches per minute. The height $h$ of the cone is related to the radius by $h = 3r$. Find the rate of change of the volume when (a) $r = 6$ inches and (b) $r = 24$ inches.

**11.** *Advertising Costs*   A retail sporting goods store estimates that weekly sales $S$ and weekly advertising costs $x$ are related by the equation $S = 2250 + 50x + 0.35x^2$. The current weekly advertising costs are \$1500, and these costs are increasing at a rate of \$125 per week. Find the current rate of change of weekly sales.

**12.** *Pet Toys*   A company that manufactures pet toys calculates that its costs and revenue can be modeled by the equations

$$C = 75,000 + 1.05x \quad \text{and} \quad R = 500x - \frac{x^2}{25},$$

where $x$ is the number of toys produced in one week. If production in one particular week is 5000 toys and is increasing at a rate of 250 toys per week, find the following:

(a) the rate at which the cost is changing.

(b) the rate at which revenue is changing.

(c) the rate at which profit is changing.

**13.** *Expanding Cube* All edges of a cube are expanding at a rate of 3 centimeters per second. How fast is the volume changing when each edge is (a) 1 centimeter and (b) 10 centimeters?

**14.** *Expanding Cube* All edges of a cube are expanding at a rate of 3 centimeters per second. How fast is the surface area changing when each edge is (a) 1 centimeter and (b) 10 centimeters?

**15.** *Moving Point* A point is moving along the graph of $y = x^2$ so that $dx/dt$ is 2 centimeters per minute. Find $dy/dt$ for the following.

(a) $x = -3$  (b) $x = 0$  (c) $x = 1$  (d) $x = 3$

**16.** *Moving Point* A point is moving along the graph of $y = 1/(1 + x^2)$ so that $dx/dt$ is 2 centimeters per minute. Find $dy/dt$ for the following.

(a) $x = -2$  (b) $x = 2$  (c) $x = 0$  (d) $x = 10$

**17.** *Sliding Ladder* A 25-foot ladder is leaning against a house (see figure). The base of the ladder is pulled away from the house at a rate of 2 feet per second. How fast is the top of the ladder moving down the wall when the base is (a) 7 feet, (b) 15 feet, and (c) 24 feet from the house?

Figure for 17          Figure for 18

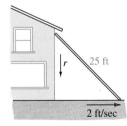

25 ft

r

2 ft/sec

4 ft/sec

12 ft  13 ft

**18.** *Boat Speed* A boat is pulled by a winch on a dock, and the winch is 12 feet above the deck of the boat (see figure). The winch pulls the rope at a rate of 4 feet per second. Find the speed of the boat when 13 feet of rope is out. What happens to the speed of the boat as it gets closer and closer to the dock?

**19.** *Air Traffic Control* An air traffic controller spots two airplanes at the same altitude converging to a point as they fly at right angles to each other. One airplane is 150 miles from the point and has a speed of 450 miles per hour. The other is 200 miles from the point and has a speed of 600 miles per hour.

(a) At what rate is the distance between the planes changing?

(b) How much time does the controller have to get one of the airplanes on a different flight path?

**20.** *Airplane Speed* An airplane flying at an altitude of 6 miles passes directly over a radar antenna (see figure). When the airplane is 10 miles away ($s = 10$), the radar detects that the distance $s$ is changing at a rate of 240 miles per hour. What is the speed of the airplane?

Figure for 20                    Figure for 21

y

6 mi

x

s

x

2nd

3rd      x      1st

s

90

Home

**21.** *Baseball* A (square) baseball diamond has sides that are 90 feet long (see figure). A player 26 feet from third base is running at a speed of 30 feet per second. At what rate is the player's distance from home plate changing?

**22.** *Environment* An accident at an oil drilling platform is causing a circular oil slick. The slick is 0.08 foot thick, and when the radius is 750 feet, the slick is increasing at the rate of 0.5 foot per minute. At what rate (in cubic feet per minute) is oil flowing from the site of the accident?

**23.** *Profit* A company is increasing the production of a product at the rate of 25 units per week. The demand and cost functions for the product are

$$p = 50 - 0.01x \quad \text{and} \quad C = 4000 + 40x - 0.02x^2.$$

Find the rate of change of the profit with respect to time when the weekly sales are $x = 800$ units. Use a graphing utility to graph the profit function, and use the zoom and trace features of the graphing utility to verify your result.

**24.** *Sales* The profit for a product is increasing at a rate of $6384 per week. The demand and cost functions for the product are

$$p = 6000 - 0.4x^2 \quad \text{and} \quad C = 2400x + 5200.$$

Find the rate of change of sales with respect to time when the weekly sales are $x = 44$ units.

**25.** *Drug Costs* The annual cost (in millions of dollars) for a government agency to seize $p\%$ of an illegal drug is

$$C = \frac{528p}{100 - p}, \quad 0 \le p < 100.$$

The agency's goal is to increase $p$ by 5% per year. Find the rate of change of the cost when (a) $p = 30\%$ and (b) $p = 60\%$. Use a graphing utility to graph $C$. What happens to the graph of $C$ as $p$ approaches 100?

# Chapter Summary and Study Strategies

*After studying this chapter, you should have acquired the following skills. The exercise numbers are keyed to the Review Exercises that begin on page 166. Answers to odd-numbered Review Exercises are given in the back of the text.\**

■ Approximate the slope of the tangent line to a graph at a point.   *(Section 2.1)*     *Review Exercises 1–4*

■ Interpret the slope of a graph in a real-life setting.   *(Section 2.1)*     *Review Exercises 5, 6*

■ Use the limit definition to find the slope of a graph at a point.   *(Section 2.1)*     *Review Exercises 7–10*

■ Use the limit definition to find the derivative of a function.   *(Section 2.1)*     *Review Exercises 11–14*

$$f'(x) = \lim_{\Delta x \to 0} \frac{f(x + \Delta x) - f(x)}{\Delta x}$$

■ Use the derivative to find the slope of a graph at a point.   *(Section 2.1)*     *Review Exercises 15–18*

■ Use the derivative to find an equation of a tangent line to a graph at a point.   *(Section 2.1)*     *Review Exercises 19, 20*

$$y - y_1 = m(x - x_1)$$

■ Use the graph of a function to recognize points at which the function is not differentiable.   *(Section 2.1)*     *Review Exercises 21–24*

■ Use the Constant Rule for differentiation.   *(Section 2.2)*     *Review Exercises 25, 26*

$$\frac{d}{dx}[c] = 0$$

■ Use the Simple Power Rule for differentiation.   *(Section 2.2)*     *Review Exercises 27–30*

$$\frac{d}{dx}[x^n] = nx^{n-1}$$

■ Use the Constant Multiple Rule for differentiation.   *(Section 2.2)*     *Review Exercises 31–34*

$$\frac{d}{dx}[cf(x)] = cf'(x)$$

■ Use the Sum and Difference Rules for differentiation.   *(Section 2.2)*     *Review Exercises 35–38*

$$\frac{d}{dx}[f(x) \pm g(x)] = f'(x) \pm g'(x)$$

■ Find the average rate of change of a function over an interval and the instantaneous rate of change at a point.   *(Section 2.3)*     *Review Exercises 39, 40*

$$\text{Average rate of change} = \frac{f(b) - f(a)}{b - a}$$

$$\text{Instantaneous rate of change} = \lim_{\Delta x \to 0} \frac{f(x + \Delta x) - f(x)}{\Delta x}$$

■ Find the average and instantaneous rates of change of a quantity in a real-life problem.   *(Section 2.3)*     *Review Exercises 41–44*

\* Use a wide range of valuable study aids to help you master the material in this chapter. The *Student Solutions Guide* includes step-by-step solutions to all odd-numbered exercises to help you review and prepare. The *Algebra Review Tutorial Software* and *The Algebra of Calculus* help you brush up on your algebra skills. The *Graphing Technology Guide* offers step-by-step commands and instructions for a wide variety of graphing calculators, including the most recent models.

■ Find the velocity of an object that is moving in a straight line.  *(Section 2.3)*        *Review Exercises 45, 46*

■ Create mathematical models for the revenue, cost, and profit for a product.  *(Section 2.3)*    *Review Exercises 47, 48*

$$P = R - C, \qquad R = xp$$

■ Find the marginal revenue, marginal cost, and marginal profit for a product.  *(Section 2.3)*    *Review Exercises 49–54*

■ Use the Product Rule for differentiation.  *(Section 2.4)*        *Review Exercises 55–58*

$$\frac{d}{dx}[f(x)g(x)] = f(x)g'(x) + g(x)f'(x)$$

■ Use the Quotient Rule for differentiation.  *(Section 2.4)*        *Review Exercises 59, 60*

$$\frac{d}{dx}\left[\frac{f(x)}{g(x)}\right] = \frac{g(x)f'(x) - f(x)g'(x)}{[g(x)]^2}$$

■ Use the General Power Rule for differentiation.  *(Section 2.5)*        *Review Exercises 61–64*

$$\frac{d}{dx}[u^n] = nu^{n-1}u'$$

■ Use differentiation rules efficiently to find the derivative of any algebraic function.    *Review Exercises 65–74*
Then simplify the result.  *(Section 2.5)*

■ Use derivatives to answer questions about real-life situations.  *(Sections 2.1–2.5)*    *Review Exercises 75, 76*

■ Find higher-order derivatives.  *(Section 2.6)*        *Review Exercises 77–84*

■ Find and use the position function to determine the velocity and acceleration of a    *Review Exercises 85, 86*
moving object.  *(Section 2.6)*

■ Find derivatives implicitly.  *(Section 2.7)*        *Review Exercises 87–94*

■ Solve related-rate problems.  *(Section 2.8)*        *Review Exercises 95, 96*

■ *Simplify Your Derivatives*    Often our students ask if they have to simplify their derivatives. Our answer is "Yes, if you expect to use them." In the next chapter, you will see that almost all applications of derivatives require that the derivatives be written in simplified form. It is not difficult to see the advantage of a derivative in simplified form. Consider, for instance, the derivative of $f(x) = x/\sqrt{x^2 + 1}$. The "raw form" produced by the Quotient and Chain Rules

$$f'(x) = \frac{(x^2 + 1)^{1/2}(1) - (x)(\frac{1}{2})(x^2 + 1)^{-1/2}(2x)}{\left(\sqrt{x^2 + 1}\right)^2}$$

is obviously much more difficult to use than the simplified form

$$f'(x) = \frac{1}{(x^2 + 1)^{3/2}}.$$

■ *List Units of Measure in Applied Problems*    When using derivatives in real-life applications, be sure to list the units of measure for each variable. For instance, if $R$ is measured in dollars and $t$ is measured in years, then the derivative $dR/dt$ is measured in dollars per year.

# Chapter 2  Review Exercises

In Exercises 1–4, approximate the slope of the tangent line to the graph at $(x, y)$.

**1.**
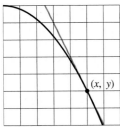

**2.**
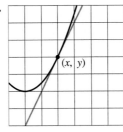

**3.**
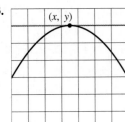

**4.**
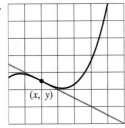

**5.** *Revenue*   The graph approximates the annual revenue (in millions of dollars per year) of Reebok, International for the years 1990–1996, with $t = 0$ corresponding to 1990. Estimate the slope of the graph when $t = 2$ and $t = 4$. Interpret each slope in the context of the problem. *(Source: Reebok, International)*

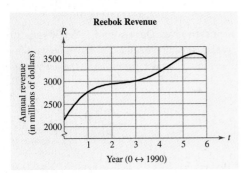

**6.** *Cellular Telephones*   The graph approximates the number of subscribers (in thousands per year) of cellular telephones for 1988–1995, with $t = 0$ corresponding to 1988. Estimate the slope of the graph when $t = 2$ and $t = 6$. Interpret each slope in the context of the problem. *(Source: Cellular Telecommunications Industry Association)*

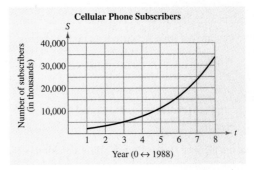

In Exercises 7–10, use the limit definition to find the slope of the tangent line to the graph of $f$ at the indicated point.

**7.** $f(x) = -3x - 5;\ (-2, 1)$     **8.** $f(x) = x^2 + 10;\ (2, 14)$

**9.** $f(x) = \sqrt{x + 9};\ (-5, 2)$     **10.** $f(x) = \dfrac{x + 1}{x};\ (1, 2)$

In Exercises 11–14, use the limit definition to find the derivative of the function.

**11.** $f(x) = 7x + 3$     **12.** $f(x) = x^2 - 7x - 8$

**13.** $f(x) = \dfrac{1}{x - 5}$     **14.** $f(x) = \sqrt{x - 5}$

In Exercises 15–18, find the slope of the graph of $f$ at the indicated point.

**15.** $f(x) = 8 - 5x;\ (3, -7)$

**16.** $f(x) = -\frac{1}{2}x^2 + 2x;\ (2, 2)$

**17.** $f(x) = \sqrt{x} + 2;\ (9, 5)$     **18.** $f(x) = \dfrac{5}{x};\ (1, 5)$

In Exercises 19 and 20, use the derivative to find an equation of the tangent line to the graph of $f$ at the indicated point.

**19.** $f(x) = \dfrac{x^2 + 3}{x};\ (1, 4)$

**20.** $f(x) = -x^2 - 4x - 4;\ (-4, -4)$

In Exercises 21–24, determine the $x$-value at which the function is not differentiable.

**21.** $y = \dfrac{x + 1}{x - 1}$

**22.** $y = -|x| + 3$

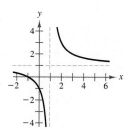

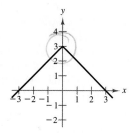

**23.** $y = \begin{cases} -x - 2, & x \leq 0 \\ x^3 + 2, & x > 0 \end{cases}$

**24.** $y = (x + 1)^{2/3}$

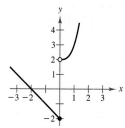

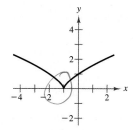

In Exercises 25–32, find the derivative of the function.

**25.** $f(x) = \sqrt{5}$

**26.** $f(x) = -10^2$

**27.** $y = x^5$

**28.** $g(x) = -\dfrac{1}{x^3}$

**29.** $y = \sqrt{x}$

**30.** $h(t) = \dfrac{1}{\sqrt{t}}$

**31.** $f(x) = 3x^4$

**32.** $f(x) = \dfrac{4}{x^4}$

In Exercises 33–38, find the equation of the tangent line at the indicated point. Then use a graphing utility to graph the function and the equation of the tangent line in the same viewing rectangle.

| Function | Point |
|----------|-------|
| **33.** $g(t) = \dfrac{2}{3t^2}$ | $\left(1, \dfrac{2}{3}\right)$ |
| **34.** $h(x) = \dfrac{2}{(3x)^2}$ | $\left(2, \dfrac{1}{18}\right)$ |

| Function | Point |
|----------|-------|
| **35.** $y = 11x^4 - 5x^2 + 1$ | $(-1, 7)$ |
| **36.** $y = x^3 - 5 + \dfrac{3}{x^3}$ | $(-1, -9)$ |
| **37.** $f(x) = \sqrt{x} - \dfrac{1}{\sqrt{x}}$ | $(1, 0)$ |
| **38.** $f(x) = 2x^{-3} + 4 - \sqrt{x}$ | $(1, 5)$ |

In Exercises 39 and 40, find the average rate of change of the function over the indicated interval. Then compare the average rate of change with the instantaneous rates of change at the endpoints of the interval.

**39.** $f(x) = x^2 + 3x - 4; \quad [0, 1]$

**40.** $f(x) = x^3 + x; \quad [-2, 2]$

**41.** *Revenue* The annual revenue $R$ (in millions of dollars per year) of Reebok, International for the years 1990–1996 can be modeled by

$$R = -1.58t^5 + 12.62t^4 + 18.12t^3 - 338.46t^2 + 922.55t + 2153.78,$$

where $t = 0$ corresponds to 1990. *(Source: Reebok, International)*

(a) Find the average rate of change for the interval from 1992 to 1996.

(b) Find the instantaneous rate of change of the model in 1992 and 1996.

(c) Interpret the results of parts (a) and (b) in the context of the problem.

**42.** *Cellular Telephones* The number of subscribers $S$ (in thousands per year) of cellular telephones for the years 1988–1995 can be modeled by

$$S = \dfrac{2163.9 + 666.4t}{1 - 0.18t + 0.01t^2},$$

where $t = 0$ corresponds to 1988. *(Source: Cellular Telecommunications Industry Association)*

(a) Find the average rate of change for the interval from 1991 to 1995.

(b) Find the instantaneous rate of change of the model in 1991 and 1995.

(c) Interpret the results of parts (a) and (b) in the context of the problem.

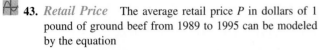

**43.** *Retail Price*  The average retail price $P$ in dollars of 1 pound of ground beef from 1989 to 1995 can be modeled by the equation

$$P = 0.0025t^5 - 0.027t^4 + 0.09t^3 - 0.088t^2 - 0.067t + 2.02,$$

where $t$ is the year, with $t = 0$ representing 1990. (*Source: U.S. Bureau of Labor Statistics*)

(a) Find the rate of change of the price with respect to the year.

(b) At what rate is the price of beef changing in 1992? in 1995?

(c) Use a graphing utility to graph the function for $-1 \le t \le 5$. During which years is the price increasing? decreasing?

(d) For what years do the slopes of the tangent lines appear to be positive? negative?

(e) Compare your answers for parts (c) and (d).

**44.** *Recycling*  The amount $T$ of recycled paper products in millions of tons from 1970 to 1995 can be modeled by the equation

$$T = \sqrt{54.89 + 36.11t - 4.63t^2 + 0.19t^3},$$

where $t$ is the year, with $t = 0$ corresponding to 1970. (*Source: Franklin Associates, Ltd*)

(a) Use a graphing utility to graph the equation. Be sure to choose an appropriate window.

(b) Determine $dT/dt$. Evaluate $dT/dt$ for 1970, 1980, and 1990.

(c) Is $dT/dt$ positive for $t \ge 0$? Does this agree with the graph of the function? What does this tell you about this situation? Explain.

**45.** *Velocity*  A rock is dropped from a tower on the Brooklyn Bridge, 276 feet above the East River. Let $t$ represent the time in seconds.

(a) Write a model for the position function (assume that air resistance is negligible).

(b) Find the average velocity during the first 2 seconds.

(c) Find the instantaneous velocity when $t = 2$ and $t = 3$.

(d) How long will it take for the rock to hit the water?

(e) When it hits the water, what is the rock's speed?

**46.** *Velocity*  The straight-line distance $s$ (in feet) traveled by an accelerating bicyclist can be modeled by

$$s = 2t^{3/2}, \qquad 0 \le t \le 8,$$

where $t$ is the time (in seconds). Complete the table showing the velocity of the bicyclist at 2-second intervals.

Table for 46

| Time, $t$ | 0 | 2 | 4 | 6 | 8 |
|-----------|---|---|---|---|---|
| Velocity  |   |   |   |   |   |

**47.** *Revenue, Cost, and Profit*  The fixed cost of operating a small flower shop is $2500 per month. The average cost of a floral arrangement is $15 and the average price is $27.50. Write the monthly revenue, cost, and profit functions for the floral shop in terms of $x$, the number of arrangements sold.

**48.** *Profit*  The weekly demand and cost functions for a product are

$$p = 1.89 - 0.0083x \qquad \text{and} \qquad C = 21 + 0.65x.$$

Write the profit function for this product.

*Marginal Cost*  In Exercises 49 and 50, find the marginal cost function.

**49.** $C = 2500 + 320x$  **50.** $C = 475 + 5.25x^{2/3}$

*Marginal Revenue*  In Exercises 51 and 52, find the marginal revenue function.

**51.** $R = \dfrac{35x}{\sqrt{x-2}}, \quad x \ge 6$  **52.** $R = x\left(5 + \dfrac{10}{\sqrt{x}}\right)$

*Marginal Profit*  In Exercises 53 and 54, find the marginal profit function.

**53.** $P = -0.0002x^3 + 6x^2 - x - 2000$

**54.** $P = -\frac{1}{15}x^3 + 4000x^2 - 120x - 144{,}000$

In Exercises 55–74, find the derivative of the function. Simplify your result.

**55.** $f(x) = x^3(5 - 3x^2)$  **56.** $y = (3x^2 + 7)(x^2 - 2x)$

**57.** $y = (4x - 3)(x^3 - 2x^2)$  **58.** $s = \left(4 - \dfrac{1}{t^2}\right)(t^2 - 3t)$

**59.** $f(x) = \dfrac{6x - 5}{x^2 + 1}$  **60.** $f(x) = \dfrac{x^2 + x - 1}{x^2 - 1}$

**61.** $f(x) = (5x^2 + 2)^3$  **62.** $f(x) = \sqrt[3]{x^2 - 1}$

**63.** $h(x) = \dfrac{2}{\sqrt{x+1}}$  **64.** $g(x) = \sqrt{x^6 - 12x^3 + 9}$

**65.** $g(x) = x\sqrt{x^2 + 1}$  **66.** $g(t) = \dfrac{t}{(1-t)^3}$

**67.** $f(x) = -2(1 - 4x^2)^2$  **68.** $f(x) = \left(x^2 + \dfrac{1}{x}\right)^5$

**69.** $h(x) = [x^2(2x + 3)]^3$

**70.** $f(x) = [(x - 2)(x + 4)]^2$

**71.** $f(x) = x^2(x - 1)^5$

**72.** $f(s) = s^3(s^2 - 1)^{5/2}$

**73.** $h(t) = \dfrac{\sqrt{3t + 1}}{(1 - 3t)^2}$

**74.** $g(x) = \dfrac{(3x + 1)^2}{(x^2 + 1)^2}$

 **75.** *Refrigeration*  The temperature $T$ (in degrees Fahrenheit) of food placed in a freezer can be modeled by

$$T = \frac{1300}{t^2 + 2t + 25},$$

where $t$ is the time (in hours).

(a) Find the rate of change of $T$ when $t = 1$, $t = 3$, $t = 5$, and $t = 10$.

(b) Graph the model on a graphing utility and describe the rate at which the temperature is changing.

**76.** *Forestry*  According to the *Doyle Log Rule*, the volume $V$ (in board feet) of a log of length $L$ (feet) and diameter $D$ (inches) at the small end is

$$V = \left(\frac{D - 4}{4}\right)^2 L.$$

Find the rate at which the volume is changing for a 12-foot-long log whose smallest diameter is (a) 8 inches, (b) 16 inches, (c) 24 inches, and (d) 36 inches.

In Exercises 77–84, find the indicated derivative.

**77.** Given $f(x) = 3x^2 + 7x + 1$, find $f''(x)$.

**78.** Given $f'(x) = 5x^4 - 6x^2 + 2x$, find $f'''(x)$.

**79.** Given $f'''(x) = -\dfrac{6}{x^4}$, find $f^{(5)}(x)$.

**80.** Given $f(x) = \sqrt{x}$, find $f^{(4)}(x)$.

**81.** Given $f'(x) = 7x^{5/2}$, find $f''(x)$.

**82.** Given $f(x) = x^2 + \dfrac{3}{x}$, find $f''(x)$.

**83.** Given $f''(x) = 6\sqrt[3]{x}$, find $f'''(x)$.

**84.** Given $f'''(x) = 20x^4 - \dfrac{2}{x^3}$, find $f^{(5)}(x)$.

**85.** *Diver*  A person dives from a 30-foot platform with an initial velocity of 5 feet per second (upward).

(a) Find the position function of the diver.

(b) How long will it take for the diver to hit the water?

(c) What is the diver's velocity at impact?

(d) What is the diver's acceleration at impact?

**86.** *Velocity and Acceleration*  The position function of a particle is given by

$$s = \frac{1}{t^2 + 2t + 1},$$

where $s$ is the height (in feet) and $t$ is the time (in seconds). Find the velocity and acceleration functions.

In Exercises 87–90, use implicit differentiation to find $dy/dx$.

**87.** $x^2 + 3xy + y^3 = 10$

**88.** $x^2 + 9y^2 - 4x + 3y - 7 = 0$

**89.** $y^2 - x^2 = 49$

**90.** $y^2 + x^2 - 6y - 2x - 5 = 0$

In Exercises 91–94, use implicit differentiation to find an equation of the tangent line at the indicated point.

| *Equation* | *Point* |
| --- | --- |
| **91.** $y^2 = x - y$ | $(2, 1)$ |
| **92.** $y^2 - 3 = 2\sqrt{xy}$ | $\left(\frac{1}{8}, 2\right)$ |
| **93.** $2\sqrt[3]{x} + 3\sqrt{y} = 10$ | $(8, 4)$ |
| **94.** $y^3 - 2x^2y + 3xy^2 = -1$ | $(0, -1)$ |

**95.** *Water Level*  A swimming pool is 40 feet long, 20 feet wide, 4 feet deep at the shallow end, and 9 feet deep at the deep end (see figure). Water is being pumped into the pool at the rate of 10 cubic feet per minute. How fast is the water level rising when there is 4 feet of water in the deep end?

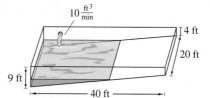

 **96.** *Profit*  The demand and cost functions for a product can be modeled by $p = 211 - 0.002x$ and $C = 30x + 1{,}500{,}000$, where $x$ is the number of units produced.

(a) Write the profit function for this product.

(b) Find the marginal profit when 80,000 units are produced.

(c) Graph the profit function on a graphing utility and use the graph to determine the price you would charge for the product. Explain your reasoning.

# Sample Post-Graduation Exam Questions

CPA   GMAT
GRE   CLAST
      Actuarial

*The following questions represent the types of questions that appear on certified public accountant (CPA) exams, graduate management admission tests (GMAT), graduate records exams (GRE), actuarial exams, and college-level academic skills tests (CLAST). The answers to the questions are given in the back of the book.*

Figure for 1

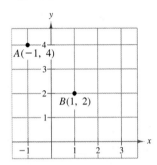

$A(-1, 4)$

$B(1, 2)$

**1.** What is the length of the line segment that connects $A$ to $B$ (see graph)?

   (a) 2     (b) 4     (c) $2\sqrt{2}$     (d) 6     (e) $\sqrt{3}$

For Questions 2–4, refer to the following table.

Participation in National Elections (millions of persons)

|  | 1988 | | 1992 | | 1996 | |
|---|---|---|---|---|---|---|
| Characteristic | Persons of voting age | Percent voted | Persons of voting age | Percent voted | Persons of voting age | Percent voted |
| **Total** | 178.1 | 57.4 | 185.7 | 61.3 | 193.7 | 54.2 |
| Male | 84.5 | 56.4 | 88.6 | 60.2 | 92.6 | 52.8 |
| Female | 93.6 | 58.3 | 97.1 | 62.3 | 101.0 | 55.5 |
| Age 18 to 20 | 10.7 | 33.2 | 9.7 | 38.5 | 10.8 | 31.2 |
| 21 to 24 | 14.8 | 38.3 | 14.6 | 45.7 | 13.9 | 33.4 |
| 25 to 34 | 42.7 | 48.0 | 41.6 | 53.2 | 40.1 | 43.1 |
| 35 to 44 | 35.2 | 61.3 | 39.7 | 63.6 | 43.3 | 54.9 |
| 45 to 64 | 45.9 | 67.9 | 49.1 | 70.0 | 53.7 | NA* |
| 65 years and over | 28.8 | 68.8 | 30.8 | 70.1 | 31.9 | NA* |

\* NA Not available

**2.** Which of the following groups had the highest percent of voters in 1992?

   (a) male           (b) age 35 to 44     (c) age 25 to 34
   (d) age 18 to 20    (e) female

**3.** In 1996, what percent (to the nearest percent) of persons of voting age were female?

   (a) 47     (b) 53     (c) 55     (d) 57     (e) 52

**4.** In 1988, how many males of voting age voted?

   (a) 47,658,000     (b) 48,503,000     (c) 102,229,400
   (d) 47,377,000     (e) 54,568,800

For Questions 5 and 6, refer to the following example.
Suppose the position $s$ at time $t$ of an accelerating object is given by $s(t) = t(t^2 - 2)^2$.

**5.** The acceleration function is

   (a) $5t^4 - 12t^2 + 4$     (b) $5t^4 - 8t$     (c) $20t^3 - 24t$
   (d) $20t^3 - 8$     (e) $t^5 - 4t^2 + 4$

**6.** The velocity of the object at time $t = 3$ is

   (a) 301     (b) 381     (c) 147     (d) 532     (e) 468

# *Applications of the Derivative* 3

## STRATEGIES *for* SUCCESS

### OBJECTIVES
*When you have completed this chapter, make sure you are able to:*

❏ Find the open intervals on which a function is increasing or decreasing.
❏ Determine relative and absolute extrema of a function.
❏ Determine the concavity and points of inflection of a graph.
❏ Solve real-life optimization problems.
❏ Determine the vertical and horizontal asymptotes of a graph.
❏ Use calculus to analyze the shape of the graph of a function.
❏ Use differentials in marginal analysis applications.

### TOOLS
*Use these study tools to achieve the objectives above:*

**Algebra Review**
(pages 248 and 249)

**Chapter Summary and Study Strategies**
(pages 250 and 251)

**Review Exercises**
(pages 252–255)

### ADDITIONAL RESOURCES
*Use these resources to solidify your mastery of calculus:*

**Student Solutions Guide**

**Study Guide** (Additional Examples, Similar Problems, and Chapter Test)

**Algebra Review Tutorial Software**

**Graphing Technology Guide**

**Sample Post-Graduation Exam Questions**
(page 256)

**Web Exercise**
(page 200, exercise 63)

CoCo McCoy/Rainbow

*In Exercise 93 on page 255, you will find the total revenue and marginal revenue of a monopoly as the quantity of output varies.*

# Increasing and Decreasing Functions

## 3.1

*Increasing and Decreasing Functions · Critical Numbers and Their Use ·*
*Extended Application:  Profit, Revenue, and Cost*

## Increasing and Decreasing Functions

A function is **increasing** if its graph moves up as $x$ moves to the right and **decreasing** if its graph moves down as $x$ moves to the right. The following definition states this more formally.

---

### Definition of Increasing and Decreasing Functions

A function $f$ is **increasing** on an interval if for any $x_1$ and $x_2$ in the interval

$$x_2 > x_1 \quad \text{implies} \quad f(x_2) > f(x_1).$$

A function $f$ is **decreasing** on an interval if for any $x_1$ and $x_2$ in the interval

$$x_2 > x_1 \quad \text{implies} \quad f(x_2) < f(x_1).$$

---

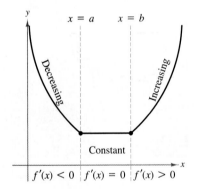

**FIGURE 3.1**

The function in Figure 3.1 is decreasing on the interval $(-\infty, a)$, constant on the interval $(a, b)$, and increasing on the interval $(b, \infty)$. Actually, from the definition of increasing and decreasing functions, the function shown in Figure 3.1 is decreasing on the interval $(-\infty, a]$ and increasing on the interval $[b, \infty)$. In this text, however, we restrict the discussion to finding *open* intervals on which a function is increasing or decreasing.

The derivative of a function can be used to determine whether the function is increasing or decreasing on an interval.

---

### Test for Increasing and Decreasing Functions

Let $f$ be differentiable on the interval $(a, b)$.

1. If $f'(x) > 0$ for all $x$ in $(a, b)$, then $f$ is increasing on $(a, b)$.

2. If $f'(x) < 0$ for all $x$ in $(a, b)$, then $f$ is decreasing on $(a, b)$.

3. If $f'(x) = 0$ for all $x$ in $(a, b)$, then $f$ is constant on $(a, b)$.

---

## EXAMPLE 1  Testing for Increasing and Decreasing Functions

Show that the function

$$f(x) = x^2$$

is decreasing on the open interval $(-\infty, 0)$ and increasing on the open interval $(0, \infty)$.

### Solution

The derivative of $f$ is

$$f'(x) = 2x.$$

On the open interval $(-\infty, 0)$, the fact that $x$ is negative implies that $f'(x) = 2x$ is also negative. Hence, by the test for a decreasing function, you can conclude that $f$ is *decreasing* on this interval. Similarly, on the open interval $(0, \infty)$, the fact that $x$ is positive implies that $f'(x) = 2x$ is also positive. Hence, it follows that $f$ is *increasing* on this interval, as shown in Figure 3.2.

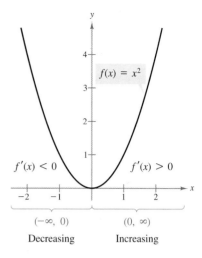

**FIGURE 3.2**

## EXAMPLE 2  Modeling Poultry Consumption

From 1985 through 1994, the consumption $C$ of chicken (in boneless pounds per person per year) can be modeled by

$$C = 0.076t^2 + 0.0776t + 34.13, \qquad 5 \le t \le 14,$$

where $t = 5$ represents 1985 (see Figure 3.3). Show that the consumption of chicken was increasing from 1985 through 1994. *(Source: U.S. Department of Agriculture)*

### Solution

The derivative of this model is $dC/dt = 0.152t + 0.0776$. As long as $t$ is positive, the derivative is also positive. Therefore, the function is increasing, which implies that the consumption of chicken was increasing from 1985 through 1994.

*Discovery*

Use a graphing utility to graph $f(x) = 2 - x^2$ and $f'(x) = -2x$ in the same coordinate plane. On what interval is $f$ increasing? On what interval is $f'$ positive? Describe how the first derivative can be used to determine where a function is increasing and decreasing. Repeat this analysis for $g(x) = x^3 - x$ and $g'(x) = 3x^2 - 1$.

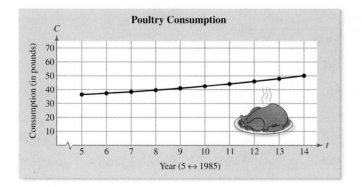

**FIGURE 3.3**

## Critical Numbers and Their Use

In Example 1, you were given two intervals—one on which the function was decreasing and one on which it was increasing. Suppose you had been asked to determine these intervals. To do this, you could have used the fact that for a continuous function, $f'(x)$ can change signs only at $x$-values where $f'(x) = 0$ or at $x$-values where $f'(x)$ is undefined, as shown in Figure 3.4. These two types of numbers are called the **critical numbers** of $f$.

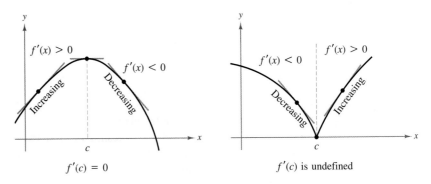

**FIGURE 3.4**

*STUDY TIP* This definition requires that a critical number be in the domain of the function. For example, $x = 0$ is not a critical number of the function $f(x) = 1/x$.

### Definition of a Critical Number

If $f$ is defined at $c$, then $c$ is a critical number of $f$ if $f'(c) = 0$ or if $f'$ is undefined at $c$.

To determine the intervals on which a continuous function is increasing or decreasing, you can use the following guidelines.

### Guidelines for Applying Increasing/Decreasing Test

1. Find the derivative of $f$.

2. Locate the critical numbers of $f$ and use these numbers to determine test intervals. That is, find all $x$ for which $f'(x) = 0$ or $f'(x)$ is undefined.

3. Test the sign of $f'(x)$ at an arbitrary number in each of the test intervals.

4. Use the test for increasing and decreasing functions to decide whether $f$ is increasing or decreasing on each interval.

## EXAMPLE 3   Finding Increasing and Decreasing Intervals

Find the open intervals on which the function

$$f(x) = x^3 - \frac{3}{2}x^2$$

is increasing or decreasing.

### Solution

Begin by finding the derivative of $f$. Then set the derivative equal to zero and solve for the critical numbers.

| | |
|---|---|
| $f'(x) = 3x^2 - 3x$ | Differentiate original function. |
| $3x^2 - 3x = 0$ | Set derivative equal to 0. |
| $3(x)(x - 1) = 0$ | Factor. |
| $x = 0, x = 1$ | Critical numbers |

Because there are no $x$-values for which $f'$ is undefined, it follows that $x = 0$ and $x = 1$ are the *only* critical numbers. Thus, the intervals that need to be tested are $(-\infty, 0)$, $(0, 1)$, and $(1, \infty)$. The table summarizes the testing of these three intervals.

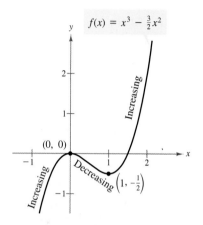

| Interval | $-\infty < x < 0$ | $0 < x < 1$ | $1 < x < \infty$ |
|---|---|---|---|
| Test value | $x = -1$ | $x = \frac{1}{2}$ | $x = 2$ |
| Sign of $f'(x)$ | $f'(-1) = 6 > 0$ | $f'\left(\frac{1}{2}\right) = -\frac{3}{4} < 0$ | $f'(2) = 6 > 0$ |
| Conclusion | Increasing | Decreasing | Increasing |

The graph of $f$ is shown in Figure 3.5. Note that the test values in the intervals were chosen for convenience—other $x$-values could have been used.

**FIGURE 3.5**

## Technology

You can use the trace feature of a graphing utility to confirm the result of Example 3. Begin by graphing the function, as shown at the right. Then activate the trace feature and move the cursor from left to right. In intervals in which the function is increasing, note that the $y$-values increase as the $x$-values increase, whereas in intervals in which the function is decreasing, the $y$-values decrease as the $x$-values increase.

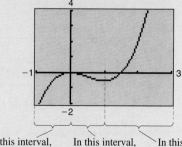

In this interval, the $y$-values increase as the $x$-values increase.   In this interval, the $y$-values decrease as the $x$-values increase.   In this interval, the $y$-values increase as the $x$-values increase.

Not only is the function in Example 3 continuous on the entire real line, it is also differentiable there. For such functions, the only critical numbers are those for which $f'(x) = 0$. The next example considers a continuous function that has *both* types of critical numbers—those for which $f'(x) = 0$ and those for which $f'$ is undefined.

---

## EXAMPLE 4  Finding Increasing and Decreasing Intervals

**TIP**

For help on the algebra in Example 4, see Example 2d in the *Chapter 3 Algebra Review*, on page 249.

Find the open intervals on which the function

$$f(x) = (x^2 - 4)^{2/3}$$

is increasing or decreasing.

### Solution

Begin by finding the derivative of the function.

$$f'(x) = \frac{2}{3}(x^2 - 4)^{-1/3}(2x) \qquad \text{Differentiate.}$$

$$= \frac{4x}{3(x^2 - 4)^{1/3}} \qquad \text{Simplify.}$$

From this, you can see that the derivative is zero when $x = 0$ and the derivative is undefined when $x = \pm 2$. Thus, the critical numbers are

$$x = -2, \quad x = 0, \quad \text{and} \quad x = 2. \qquad \text{Critical numbers}$$

This implies that the test intervals are

$$(-\infty, -2), \quad (-2, 0), \quad (0, 2), \quad \text{and} \quad (2, \infty). \qquad \text{Test intervals}$$

The table summarizes the testing of these four intervals, and the graph of the function is shown in Figure 3.6.

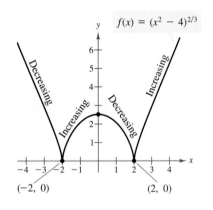

$f(x) = (x^2 - 4)^{2/3}$

$(-2, 0)$     $(2, 0)$

**FIGURE 3.6**

| Interval | $-\infty < x < -2$ | $-2 < x < 0$ | $0 < x < 2$ | $2 < x < \infty$ |
|---|---|---|---|---|
| Test value | $x = -3$ | $x = -1$ | $x = 1$ | $x = 3$ |
| Sign of $f'(x)$ | $f'(-3) < 0$ | $f'(-1) > 0$ | $f'(1) < 0$ | $f'(3) > 0$ |
| Conclusion | Decreasing | Increasing | Decreasing | Increasing |

**ALGEBRA TIP**  In the table, it is not necessary to *evaluate* $f'(x)$ at the test values—you need only to determine its sign. For example, you can determine the sign of $f'(-3)$ as follows.

$$f'(-3) = \frac{4(-3)}{3(9 - 4)^{1/3}} = \frac{\text{negative}}{\text{positive}} = \text{negative}$$

The functions in Examples 1–4 are continuous on the entire real line. If there are isolated $x$-values at which a function is not continuous, then these $x$-values should be used along with the critical numbers to determine the test intervals. For example, the function

$$f(x) = \frac{x^4 + 1}{x^2}$$

is not continuous when $x = 0$. Because the derivative of $f$,

$$f'(x) = \frac{2(x^4 - 1)}{x^3},$$

is zero when $x = \pm 1$, you should use the following numbers to determine the test intervals.

$x = -1, x = 1$     Critical numbers

$x = 0$     Discontinuity

After testing $f'(x)$, you can determine that the function is decreasing on the intervals $(-\infty, -1)$ and $(0, 1)$, and increasing on the intervals $(-1, 0)$ and $(1, \infty)$, as shown in Figure 3.7.

The converse of the test for increasing and decreasing functions is *not* true. For instance, it is possible for a function to be increasing on an interval even though its derivative is not positive at every point in the interval.

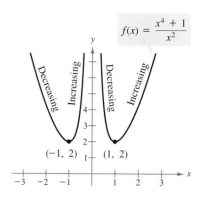

FIGURE 3.7

---

## EXAMPLE 5   Testing an Increasing Function

Show that $f(x) = x^3 - 3x^2 + 3x$ is increasing on the entire real line.

### Solution

From the derivative of $f$,

$$f'(x) = 3x^2 - 6x + 3 = 3(x - 1)^2,$$

you can see that the only critical number is $x = 1$. Thus, the test intervals are $(-\infty, 1)$ and $(1, \infty)$. The table summarizes the testing of these two intervals. From Figure 3.8, you can see that $f$ is increasing on the entire real line—even though $f'(1) = 0$. To convince yourself of this, look back at the definition of an increasing function.

| Interval | $-\infty < x < 1$ | $1 < x < \infty$ |
|---|---|---|
| Test value | $x = 0$ | $x = 2$ |
| Sign of $f'(x)$ | $f'(0) = 3(-1)^2 > 0$ | $f'(2) = 3(1)^2 > 0$ |
| Conclusion | Increasing | Increasing |

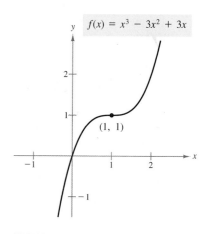

FIGURE 3.8

## Extended Application: Profit, Revenue, and Cost

### EXAMPLE 6  Profit Analysis

A national toy distributor determines the following cost and revenue models for one of its games.

$$C = 2.4x - 0.0002x^2, \qquad 0 \le x \le 6000$$
$$R = 7.2x - 0.001x^2, \qquad 0 \le x \le 6000$$

Determine the interval on which the profit function is increasing.

### Solution

The profit for producing $x$ units is

$$
\begin{aligned}
P &= R - C \\
&= (7.2x - 0.001x^2) - (2.4x - 0.0002x^2) \\
&= 4.8x - 0.0008x^2.
\end{aligned}
$$

To find the intervals on which the profit is increasing, set the marginal profit $P'$ equal to zero and solve for $x$.

$$P' = 4.8 - 0.0016x \qquad \text{Differentiate profit function.}$$
$$4.8 - 0.0016x = 0 \qquad \text{Set } P' \text{ equal to 0.}$$
$$-0.0016x = -4.8 \qquad \text{Subtract 4.8 from both sides.}$$
$$x = \frac{-4.8}{-0.0016} \qquad \text{Divide both sides by } -0.0016.$$
$$x = 3000 \text{ units} \qquad \text{Simplify.}$$

In the interval $(0, 3000)$, $P'$ is positive and the profit is *increasing*. In the interval $(3000, 6000)$, $P'$ is negative and the profit is *decreasing*. The graphs of the cost, revenue, and profit functions are shown in Figure 3.9.

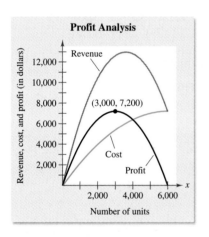

**FIGURE 3.9**

## Group Discussion

## Comparing Cost, Revenue, and Profit

Use the models given in Example 6 to answer the following questions.

1. What is the demand function for the product described in the example?

2. What price would you set to obtain a maximum profit?

3. What price would you set to obtain a maximum revenue?

4. Why doesn't the maximum revenue occur at the same $x$-value as the maximum profit?

## Warm Up

The following warm-up exercises involve skills that were covered in earlier sections. You will use these skills in the exercise set for this section.

In Exercises 1–4, solve the equation.

**1.** $x^2 = 8x$

**2.** $15x = \dfrac{5}{8}x^2$

**3.** $\dfrac{x^2 - 25}{x^3} = 0$

**4.** $\dfrac{2x}{\sqrt{1 - x^2}} = 0$

In Exercises 5–8, find the domain of the expression.

**5.** $\dfrac{x + 3}{x - 3}$

**6.** $\dfrac{2}{\sqrt{1 - x}}$

**7.** $\dfrac{2x + 1}{x^2 - 3x - 10}$

**8.** $\dfrac{3x}{\sqrt{9 - 3x^2}}$

In Exercises 9 and 10, evaluate the expression when $x = -2, 0$, and 2.

**9.** $\dfrac{4}{3}x^3 - 6x^2 + \dfrac{1}{3}$

**10.** $\dfrac{2x - 3}{x^2 - 3x - 1}$

## EXERCISES 3.1

In Exercises 1–4, evaluate the derivative of the function at the indicated points on the graph.

**1.** $f(x) = \dfrac{x^2}{x^2 + 4}$

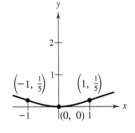

**2.** $f(x) = x + \dfrac{32}{x^2}$

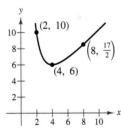

**3.** $f(x) = (x + 2)^{2/3}$

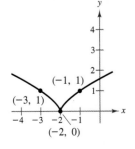

**4.** $f(x) = -3x\sqrt{x + 1}$

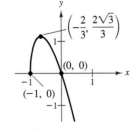

In Exercises 5–8, identify the open intervals on which the function is increasing or decreasing.

**5.** $f(x) = -(x + 1)^2$

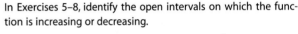
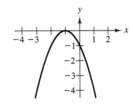

**6.** $f(x) = \dfrac{x^3}{4} - 3x$

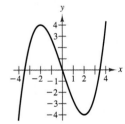

**7.** $f(x) = x^4 - 2x^2$

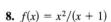
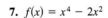

**8.** $f(x) = x^2/(x + 1)$

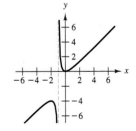
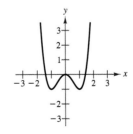

In Exercises 9–18, find the critical numbers and the open intervals on which the function is increasing or decreasing. Sketch the graph of the function.

**9.** $f(x) = 2x - 3$

**10.** $f(x) = 5 - 3x$

**11.** $g(x) = -(x - 1)^2$

**12.** $g(x) = (x + 2)^2$

**13.** $y = x^2 - 5x$

**14.** $y = -(x^2 - 2x)$

**15.** $y = x^3 - 6x^2$

**16.** $y = (x - 2)^3$

**17.** $f(x) = -(x + 1)^3$

**18.** $f(x) = \sqrt{4 - x^2}$

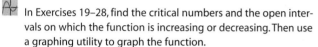 In Exercises 19–28, find the critical numbers and the open intervals on which the function is increasing or decreasing. Then use a graphing utility to graph the function.

**19.** $f(x) = -2x^2 + 4x + 3$

**20.** $f(x) = x^2 + 8x + 10$

**21.** $y = 3x^3 + 12x^2 + 15x$

**22.** $y = (x - 1)^2(x + 2)$

**23.** $h(x) = x^{2/3}$

**24.** $h(x) = \sqrt[3]{x - 1}$

**25.** $f(x) = x^4 - 2x^3$

**26.** $f(x) = x\sqrt{x + 1}$

**27.** $f(x) = \dfrac{x}{x^2 + 4}$

**28.** $f(x) = \dfrac{x^2}{x^2 + 4}$

In Exercises 29–34, find the critical numbers and the open intervals on which the function is increasing or decreasing. (*Hint:* Check for discontinuities.) Sketch the graph of the function.

**29.** $f(x) = x + \dfrac{1}{x}$

**30.** $f(x) = \dfrac{x}{x + 1}$

**31.** $f(x) = \dfrac{2x}{16 - x^2}$

**32.** $y = \begin{cases} 2x + 1, & x \le -1 \\ x^2 - 2, & x > -1 \end{cases}$

**33.** $y = \begin{cases} 4 - x^2, & x \le 0 \\ -2x, & x > 0 \end{cases}$

**34.** $y = \begin{cases} -x^3 + 1, & x \le 0 \\ -x^2 + 2x, & x > 0 \end{cases}$

 **35.** *Ordering and Transportation Cost*  The ordering and transportation cost $C$ (in hundreds of dollars) for an automobile dealership is

$$C = 10 - \left( \frac{1}{x} + 1 - \frac{3}{x + 3} \right), \qquad 1 \le x,$$

where $x$ is the number of automobiles ordered.

(a) Find the intervals on which $C$ is increasing or decreasing.

(b) Use a graphing utility to graph the cost function.

(c) Use the trace feature to determine the order sizes for which the cost is $900. Assuming that the revenue function is increasing for $x \ge 0$, which order size would you use? Explain your reasoning.

**36.** *Chemistry: Molecular Velocity*  A plot of the relative number of $N_2$ (nitrogen) molecules that have a given velocity at three temperatures is shown in the figure. Comment on the differences in the average velocity (indicated by the curve's peak) for the various temperatures, and the intervals where the velocity is increasing and decreasing for the various temperatures.  (*Source: Adapted from Zumdahl, Chemistry, Fourth Edition*)

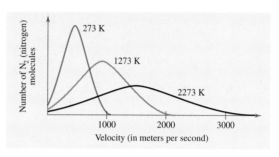

*Position Function*  In Exercises 37 and 38, the position function gives the height $s$ (in feet) of a ball, where the time $t$ is measured in seconds. Find the time interval in which the ball is rising and the interval in which it is falling.

**37.** $s = 96t - 16t^2, \qquad 0 \le t \le 6$

**38.** $s = -16t^2 + 64t, \qquad 0 \le t \le 4$

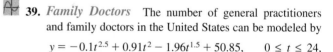 **39.** *Family Doctors*  The number of general practitioners and family doctors in the United States can be modeled by

$$y = -0.1t^{2.5} + 0.91t^2 - 1.96t^{1.5} + 50.85, \qquad 0 \le t \le 24,$$

where $y$ is the number of general practitioners in thousands, and $t$ is the time in years, with $t = 0$ corresponding to 1970.  (*Source: American Medical Association*)

(a) Use a graphing utility to graph the model. Then graphically estimate the years during which the model is increasing and the years during which it is decreasing.

(b) Use the test for increasing and decreasing functions to verify the result of part (a).

**40.** *Profit*  The profit $P$ made by a cinema from selling $x$ bags of popcorn can be modeled by

$$P = 2.36x - \frac{x^2}{25,000} - 3500, \qquad 0 \le x \le 50,000.$$

(a) Find the intervals on which $P$ is increasing or decreasing.

(b) If you owned the cinema, what price would you charge to obtain a maximum profit for popcorn? Explain.

## 3.2 Extrema and the First-Derivative Test

*Relative Extrema • The First-Derivative Test • Absolute Extrema • Applications of Extrema*

### Relative Extrema

In the preceding section, you used the derivative to determine the intervals on which a function is increasing or decreasing. In this section, you will examine the points at which a function changes from increasing to decreasing, or vice versa. At such a point, the function has a **relative extremum.** (The plural of extremum is *extrema*.) The **relative extrema** of a function include the **relative minima** and **relative maxima** of the function. For instance, the function shown in Figure 3.10 has two relative extrema—the left point is a relative maximum and the right point is a relative minimum.

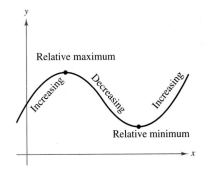

**FIGURE 3.10**

---

### Definition of Relative Extrema

Let $f$ be a function defined at $c$.

1. $f(c)$ is a **relative maximum** of $f$ if there exists an interval $(a, b)$ containing $c$ such that $f(x) \leq f(c)$ for all $x$ in $(a, b)$.

2. $f(c)$ is a **relative minimum** of $f$ if there exists an interval $(a, b)$ containing $c$ such that $f(x) \geq f(c)$ for all $x$ in $(a, b)$.

If $f(c)$ is a relative extremum of $f$, then the relative extremum is said to *occur* at $x = c$.

---

For a continuous function, the relative extrema must occur at critical numbers of the function, as shown in Figure 3.11.

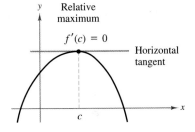

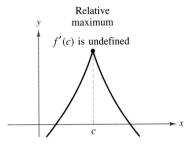

---

### Occurrence of Relative Extrema

If $f$ has a relative minimum or relative maximum when $x = c$, then $c$ is a critical number of $f$. That is, either

$$f'(c) = 0 \quad \text{or} \quad f'(c) \text{ is undefined.}$$

---

**FIGURE 3.11**

## The First-Derivative Test

The result on the previous page implies that in your search for relative extrema of a continuous function, you need only test the critical numbers of the function. Once you have determined that $c$ is a critical number of a function $f$, the **First-Derivative Test** for relative extrema enables you to classify $f(c)$ as a relative minimum, a relative maximum, or neither.

### *Discovery*

Use a graphing utility to graph the function $f(x) = x^2$ and its first derivative $f'(x) = 2x$ in the same coordinate plane. Where does $f$ have a relative minimum? What is the sign of $f'$ to the left of this relative minimum? What is the sign of $f'$ to the right? Describe how the sign of $f'$ can be used to determine the relative extrema of a function.

---

### First-Derivative Test for Relative Extrema

Let $f$ be continuous on the interval $(a, b)$ in which $c$ is the only critical number. If $f$ is differentiable on the interval (except possibly at $c$), then $f(c)$ can be classified as a relative minimum, a relative maximum, or neither, as follows.

1. In the interval $(a, b)$, if $f'(x)$ is negative to the left of $x = c$ and positive to the right of $x = c$, then $f(c)$ is a relative minimum.

2. In the interval $(a, b)$, if $f'(x)$ is positive to the left of $x = c$ and negative to the right of $x = c$, then $f(c)$ is a relative maximum.

3. In the interval $(a, b)$, if $f'(x)$ has the same sign to the left and right of $x = c$, then $f(c)$ is not a relative extremum of $f$.

---

A graphical interpretation of the First-Derivative Test is shown in Figure 3.12.

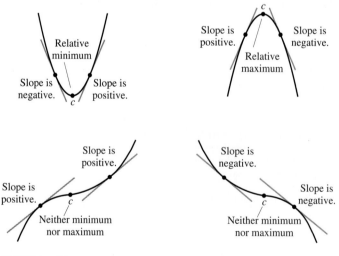

**FIGURE 3.12**

## EXAMPLE 1 Finding Relative Extrema

Find all relative extrema of the function

$$f(x) = 2x^3 - 3x^2 - 36x + 14.$$

### Solution

Begin by finding the critical numbers of $f$.

$$f'(x) = 6x^2 - 6x - 36 \qquad \text{Find derivative of } f.$$
$$6x^2 - 6x - 36 = 0 \qquad \text{Set derivative equal to 0.}$$
$$6(x^2 - x - 6) = 0 \qquad \text{Factor out common factor.}$$
$$6(x - 3)(x + 2) = 0 \qquad \text{Factor.}$$
$$x = -2, 3 \qquad \text{Critical numbers}$$

Because $f'(x)$ is defined for all $x$, the only critical numbers of $f$ are $x = -2$ and $x = 3$. Using these numbers, you can form the three test intervals $(-\infty, -2)$, $(-2, 3)$, and $(3, \infty)$. The testing of the three intervals is shown in the table.

| Interval | $-\infty < x < -2$ | $-2 < x < 3$ | $3 < x < \infty$ |
|---|---|---|---|
| Test value | $x = -3$ | $x = 0$ | $x = 4$ |
| Sign of $f'(x)$ | $f'(-3) = 36 > 0$ | $f'(0) = -36 < 0$ | $f'(4) = 36 > 0$ |
| Conclusion | Increasing | Decreasing | Increasing |

Using the First-Derivative Test, you can conclude that the critical number $-2$ yields a relative maximum [$f'(x)$ changes sign from positive to negative], and the critical number 3 yields a relative minimum [$f'(x)$ changes sign from negative to positive.]

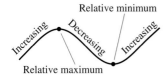

Relative minimum

Increasing  Decreasing  Increasing

Relative maximum

The graph of $f$ is shown in Figure 3.13. To find the $y$-coordinates of the relative extrema, substitute the $x$-coordinates into the function. For instance, the relative maximum is $f(-2) = 58$ and the relative minimum is $f(3) = -67$.

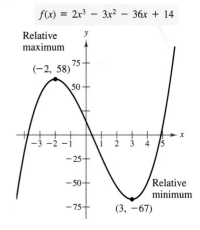

$f(x) = 2x^3 - 3x^2 - 36x + 14$

Relative maximum $(-2, 58)$

Relative minimum $(3, -67)$

**FIGURE 3.13**

*STUDY TIP* In Section 2.2, Example 8, you examined the graph of the function $f(x) = x^3 - 4x + 2$ and discovered that it does *not* have a relative minimum at the point $(1, -1)$. Try using the First-Derivative Test to find the point at which the graph *does* have a relative minimum.

In Example 1, both critical numbers yielded relative extrema. In the next example, only one of the two critical numbers yields a relative extremum.

**TIP**

For help on the algebra in Example 2, see Example 2c in the *Chapter 3 Algebra Review*, on page 249.

## EXAMPLE 2   Finding Relative Extrema

Find all relative extrema of the function

$$f(x) = x^4 - x^3.$$

**Solution**

From the derivative of the function,

$$f'(x) = 4x^3 - 3x^2 = x^2(4x - 3),$$

you can see that the function has only two critical numbers: $x = 0$ and $x = \frac{3}{4}$. These numbers produce the test intervals $(-\infty, 0)$, $\left(0, \frac{3}{4}\right)$, and $\left(\frac{3}{4}, \infty\right)$, which are tested in the table.

| Interval | $-\infty < x < 0$ | $0 < x < \frac{3}{4}$ | $\frac{3}{4} < x < \infty$ |
|---|---|---|---|
| Test value | $x = -1$ | $x = \frac{1}{2}$ | $x = 1$ |
| Sign of $f'(x)$ | $f'(-1) = -7 < 0$ | $f'\left(\frac{1}{2}\right) = -\frac{1}{4} < 0$ | $f'(1) = 1 > 0$ |
| Conclusion | Decreasing | Decreasing | Increasing |

By the First-Derivative Test, it follows that $f$ has a relative minimum when $x = \frac{3}{4}$, as shown in Figure 3.14. Note that the critical number $x = 0$ does not yield a relative extremum.

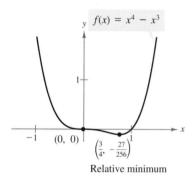

$f(x) = x^4 - x^3$

$(0, 0)$   $\left(\frac{3}{4}, -\frac{27}{256}\right)$

Relative minimum

**FIGURE 3.14**

## EXAMPLE 3   Finding Relative Extrema

Find all relative extrema of the function

$$f(x) = 2x - 3x^{2/3}.$$

**Solution**

From the derivative of the function,

$$f'(x) = 2 - \frac{2}{x^{1/3}} = \frac{2(x^{1/3} - 1)}{x^{1/3}},$$

you can see that $f'(1) = 0$ and $f'$ is undefined at $x = 0$. Therefore, the function has two critical numbers: $x = 1$ and $x = 0$. These numbers produce the test intervals

$$(-\infty, 0), \quad (0, 1), \quad \text{and} \quad (1, \infty).$$

By testing these intervals, you can conclude that $f$ has a relative maximum at $(0, 0)$ and a relative minimum at $(1, -1)$, as shown in Figure 3.15.

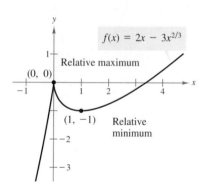

$f(x) = 2x - 3x^{2/3}$

Relative maximum

$(0, 0)$

$(1, -1)$   Relative minimum

**FIGURE 3.15**

# Technology

## Finding Relative Extrema

There are several ways to use technology to find relative extrema of a function. One way is to use a graphing utility to graph the function, and then use the zoom and trace features to find the relative minimum and relative maximum points. For instance, consider the graph of $f(x) = 3.1x^3 - 7.3x^2 + 1.2x + 2.5$, as shown at the right. From the graph, you can see that the function has one relative maximum and one relative minimum. You can approximate the coordinates of these points by zooming in and using the trace feature, as shown below.

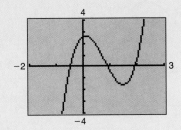

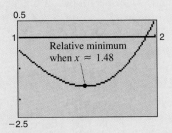

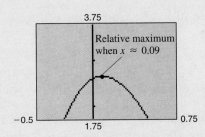

A second way to use technology to find relative extrema is to perform the First-Derivative Test with a symbolic differentiation utility. For instance, you can use *Derive* for Windows to differentiate the function, set the derivative equal to zero, and solve the resulting equation. After obtaining the critical numbers, 1.48287 and 0.0870148, you can graph the function and observe that the first yields a relative minimum and the second yields a relative maximum.

#1: $3.1 \cdot x^3 - 7.3 \cdot x^2 + 1.2 \cdot x + 2.5$     Author.

#2: $\dfrac{d}{dx}(3.1 \cdot x^3 - 7.3 \cdot x^2 + 1.2 \cdot x + 2.5)$     Differentiate.

#3: $\dfrac{93 \cdot x^2}{10} - \dfrac{73 \cdot x}{5} + \dfrac{6}{5}$     Simplify.

#4: $\left[ x = \dfrac{\sqrt{4213}}{93} + \dfrac{73}{93},\ x = \dfrac{73}{93} - \dfrac{\sqrt{4213}}{93} \right]$     Solve.

#5: $[x = 1.48287,\ x = 0.0870148]$     Approximate.

> *STUDY TIP*   Some graphing calculators have a special feature that allows you to find the minimum or maximum of a function on an interval. For instance, on the *TI-82* or *TI-83*, try using the following keys.
>
> [2nd] [CALC] [MAXIMUM]

## Absolute Extrema

The terms *relative minimum* and *relative maximum* describe the *local* behavior of a function. To describe the *global* behavior of the function on an entire interval, you can use the terms **absolute maximum** and **absolute minimum.**

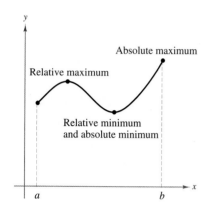

**FIGURE 3.16**

### Definition of Absolute Extrema

Let $f$ be defined on an interval $I$ containing $c$.

1. $f(c)$ is an **absolute minimum of** $f$ on $I$ if $f(c) \le f(x)$ for every $x$ in $I$.

2. $f(c)$ is an **absolute maximum of** $f$ on $I$ if $f(c) \ge f(x)$ for every $x$ in $I$.

The absolute minimum and absolute maximum values of a function on an interval are sometimes simply called the **minimum** and **maximum** of $f$ on $I$.

Be sure that you understand the distinction between relative extrema and absolute extrema. For instance, in Figure 3.16, the function has a relative minimum that also happens to be an absolute minimum on the interval $[a, b]$. The relative maximum of $f$, however, is not the absolute maximum on the interval $[a, b]$. The next theorem points out that if a continuous function has a closed interval as its domain, then it *must* have both an absolute minimum and an absolute maximum on the interval. From Figure 3.16, note that these extrema can occur at endpoints of the interval.

### Extreme Value Theorem

If $f$ is continuous on $[a, b]$, then $f$ takes on both a minimum value and a maximum value on $[a, b]$.

Although a continuous function has just one minimum and one maximum value on a closed interval, either of these values can occur for more than one $x$-value. For instance, on the interval $[-3, 3]$, the function

$$f(x) = 9 - x^2$$

has a minimum value of zero when $x = -3$ *and* when $x = 3$, as shown in Figure 3.17.

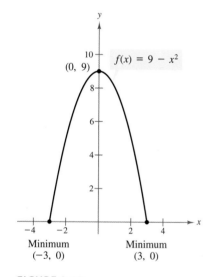

**FIGURE 3.17**

When looking for the extreme value of a function on a *closed* interval, remember that you must consider the values of the function at the endpoints as well as at the critical numbers of the function. You can use the following guidelines to find extrema on a closed interval.

**Technology**

A graphing utility can help you locate the extrema of a function on a closed interval. For instance, try using a graphing utility to confirm the results of Example 4. (Set the viewing rectangle to $-1 \le x \le 6$ and $-8 \le y \le 4$.) Use the trace feature to check that the minimum $y$-value occurs when $x = 3$ and the maximum $y$-value occurs when $x = 0$.

---

### Guidelines for Finding Extrema on a Closed Interval

To find the extrema of a continuous function $f$ on a closed interval $[a, b]$, use the following steps.

1. Evaluate $f$ at each of its critical numbers in $(a, b)$.

2. Evaluate $f$ at each endpoint, $a$ and $b$.

3. The least of these values is the minimum, and the greatest is the maximum.

---

## EXAMPLE 4 Finding Extrema on a Closed Interval

Find the minimum and maximum values of

$$f(x) = x^2 - 6x + 2$$

on the interval $[0, 5]$.

### Solution

Begin by finding the critical numbers of the function.

$$f'(x) = 2x - 6 \qquad \text{Find derivative of } f.$$
$$2x - 6 = 0 \qquad \text{Set derivative equal to 0.}$$
$$2x = 6 \qquad \text{Add 6 to both sides.}$$
$$x = 3 \qquad \text{Solve for } x.$$

From this, you can see that the only critical number of $f$ is $x = 3$. Because this number lies in the interval under question, you should test the values of $f(x)$ at this number *and* at the endpoints of the interval, as shown in the table.

| $x$-value | Endpoint: $x = 0$ | Critical number: $x = 3$ | Endpoint: $x = 5$ |
|-----------|-------------------|--------------------------|-------------------|
| $f(x)$ | $f(0) = 2$ | $f(3) = -7$ | $f(5) = -3$ |
| Conclusion | Maximum | Minimum | |

From the table, you can see that the minimum of $f$ on the interval $[0, 5]$ is $f(3) = -7$. Moreover, the maximum of $f$ on the interval $[0, 5]$ is $f(0) = 2$. This is confirmed by the graph of $f$, as shown in Figure 3.18.

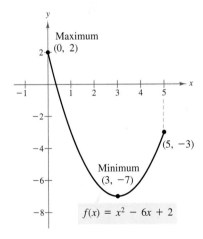

**FIGURE 3.18**

## Applications of Extrema

Finding the minimum and maximum values of a function is one of the most common applications of calculus.

### EXAMPLE 5   Finding the Maximum Profit

In Examples 7 and 8 in Section 2.3, we discussed a fast-food restaurant whose profit function for hamburgers is

$$P = 2.44x - \frac{x^2}{20,000} - 5000, \quad 0 \le x \le 50,000.$$

Find the production level that produces a maximum profit.

**Solution**

Begin by setting the marginal profit equal to zero and solving for $x$.

$$P' = 2.44 - \frac{x}{10,000} \qquad \text{Find marginal profit.}$$

$$2.44 - \frac{x}{10,000} = 0 \qquad \text{Set marginal profit equal to 0.}$$

$$-\frac{x}{10,000} = -2.44 \qquad \text{Subtract 2.44 from both sides.}$$

$$x = 24,400 \text{ units} \qquad \text{Critical number}$$

From Figure 3.19, you can see that the critical number $x = 24,400$ corresponds to the production level that produces a maximum profit. To find the maximum profit, substitute $x = 24,400$ into the profit function.

$$P = 2.44x - \frac{x^2}{20,000} - 5000$$

$$= 2.44(24,400) - \frac{(24,400)^2}{20,000} - 5000$$

$$= \$24,768$$

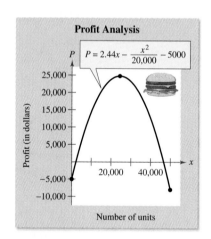

**Profit Analysis**

$P = 2.44x - \dfrac{x^2}{20,000} - 5000$

Profit (in dollars)

Number of units

**FIGURE 3.19**

---

*Group Discussion*

### Setting the Price of a Product

In Example 5, you discovered that a production level of 24,400 units corresponds to a maximum profit. Remember that this model assumes that the quantity demanded and the price per unit are related by a demand function. Thus, the only way to sell more hamburgers is to lower the price, and consequently lower the profit. What is the demand function for Example 5? What is the price per unit that produces a maximum profit?

*Warm Up*

The following warm-up exercises involve skills that were covered in earlier sections. You will use these skills in the exercise set for this section.

In Exercises 1–4, solve the equation $f'(x) = 0$.

**1.** $f(x) = 4x^4 - 2x^2 + 1$

**2.** $f(x) = \frac{1}{3}x^3 - \frac{3}{2}x^2 - 10x$

**3.** $f(x) = 5x^{4/5} - 4x$

**4.** $f(x) = \frac{1}{2}x^2 - 3x^{5/3}$

In Exercises 5–8, use $g(x) = -x^5 - 2x^4 + 4x^3 + 2x - 1$ to determine the sign of the derivative.

**5.** $g'(-4)$

**6.** $g'(0)$

**7.** $g'(1)$

**8.** $g'(3)$

In Exercises 9 and 10, decide whether the function is increasing or decreasing on the given interval.

**9.** $f(x) = 2x^2 - 11x - 6$, $(3, 6)$

**10.** $f(x) = x^3 + 2x^2 - 4x - 8$, $(-2, 0)$

## EXERCISES 3.2

In Exercises 1–4, use a table similar to that in Example 1 to find all relative extrema of the function.

**1.** $f(x) = -2x^2 + 4x + 3$

**2.** $f(x) = x^2 + 8x + 10$

**3.** $f(x) = x^2 - 6x$

**4.** $f(x) = (x - 1)^2(x + 2)$

In Exercises 5–12, find all relative extrema of the function.

**5.** $g(x) = 6x^3 - 15x^2 + 12x$

**6.** $g(x) = \frac{1}{5}(x^5 - 5x)$

**7.** $h(x) = -(x + 4)^3$

**8.** $h(x) = (x - 3)^3$

**9.** $f(x) = x^3 - 6x^2 + 15$

**10.** $f(x) = x^4 - 1$

**11.** $f(x) = x^4 - 2x^3$

**12.** $f(x) = x^4 - 32x + 4$

In Exercises 13–18, use a graphing utility to graph the function. Then find all relative extrema of the function.

**13.** $f(x) = x^{1/5} + 2$

**14.** $f(t) = (t - 1)^{1/3}$

**15.** $g(t) = t^{2/3}$

**16.** $f(x) = x + \dfrac{1}{x}$

**17.** $f(x) = \dfrac{x}{x + 1}$

**18.** $h(x) = \dfrac{4}{x^2 + 1}$

In Exercises 19–26, find the absolute extrema of the function on the closed interval.

| Function | Interval |
|---|---|
| **19.** $f(x) = 2(3 - x)$ | $[-1, 2]$ |
| **20.** $f(x) = \frac{1}{3}(2x + 5)$ | $[0, 5]$ |
| **21.** $f(x) = 5 - 2x^2$ | $[0, 3]$ |
| **22.** $f(x) = x^2 + 2x - 4$ | $[-1, 1]$ |
| **23.** $f(x) = x^3 - 3x^2$ | $[-1, 3]$ |
| **24.** $f(x) = 3x^{2/3} - 2x$ | $[-1, 2]$ |
| **25.** $h(s) = \dfrac{1}{3 - s}$ | $[0, 2]$ |
| **26.** $h(t) = \dfrac{t}{t - 2}$ | $[3, 5]$ |

In Exercises 27–30, find the absolute extrema of the function on the closed interval. Use a graphing utility to confirm your results.

| Function | Interval |
|---|---|
| **27.** $f(x) = x^3 - 12x$ | $[0, 4]$ |
| **28.** $g(t) = \dfrac{t^2}{t^2 + 3}$ | $[-1, 1]$ |
| **29.** $h(t) = (t - 1)^{2/3}$ | $[-7, 2]$ |
| **30.** $g(x) = 4\left(1 + \dfrac{1}{x} + \dfrac{1}{x^2}\right)$ | $[-4, 5]$ |

In Exercises 31–34, use a graphing utility to graphically find the absolute extrema of the function on the closed interval.

| Function | Interval |
|---|---|
| **31.** $f(x) = 0.4x^3 - 1.8x^2 + x - 3$ | $[0, 5]$ |
| **32.** $f(x) = \frac{4}{3}x\sqrt{3 - x}$ | $[0, 3]$ |
| **33.** $f(x) = 3.2x^5 + 5x^3 - 3.5x$ | $[0, 1]$ |
| **34.** $f(x) = 4\sqrt{x} - 2x + 1$ | $[0, 6]$ |

In Exercises 35–38, find the absolute extrema of the function on the interval $[0, \infty)$.

**35.** $f(x) = \dfrac{4x}{x^2 + 1}$       **36.** $f(x) = \dfrac{8}{x + 1}$

**37.** $f(x) = \dfrac{2x}{x^2 + 4}$       **38.** $f(x) = 8 - \dfrac{4x}{x^2 + 1}$

In Exercises 39 and 40, find the maximum value of $|f''(x)|$ on the closed interval. (You will use this skill in Section 6.5 to estimate the error in the Trapezoidal Rule.)

| Function | Interval |
|---|---|
| **39.** $f(x) = x^3(3x^2 - 10)$ | $[0, 1]$ |
| **40.** $f(x) = \dfrac{1}{x^2 + 1}$ | $[0, 3]$ |

 In Exercises 41 and 42, find the maximum value of $|f^{(4)}(x)|$ on the closed interval. (You will use this skill in Section 6.5 to estimate the error in Simpson's Rule.) Use a graphing utility to check your answer.

| Function | Interval |
|---|---|
| **41.** $f(x) = 15x^4 - \left(\dfrac{2x - 1}{2}\right)^6$ | $[0, 1]$ |
| **42.** $f(x) = \dfrac{1}{x^2}$ | $[1, 2]$ |

 **43.** *Cost*   A retailer has determined the cost $C$ for ordering and storing $x$ units of a product to be

$$C = 3x + \frac{20,000}{x}, \quad 0 < x \le 200.$$

The delivery truck can bring at most 200 units per order. Find the order size that will minimize the cost. Use a graphing utility to verify your result.

**44.** *Profit*   The quantity demanded $x$ for a product is inversely proportional to the cube of the price $p$ for $p > 1$. When the price is $10 per unit, the quantity demanded is eight units. The initial cost is $100 and the cost per unit is $4. What price will yield a maximum profit?

**45.** *Profit*   When soft drinks sold for $0.80 per can at football games, approximately 6000 cans were sold. When the price was raised to $1.00 per can, the quantity demanded dropped to 5600. The initial cost is $5000 and the cost per unit is $0.40. Assuming that the demand function is linear, use the table feature of a graphing utility to determine the price that will yield a maximum profit.

**46.** *Trachea Contraction*   Coughing forces the trachea (windpipe) to contract, which in turn affects the velocity of the air through the trachea. The velocity of the air during coughing can be modeled by

$$v = k(R - r)r^2, \quad 0 \le r < R,$$

where $k$ is a constant, $R$ is the normal radius of the trachea, and $r$ is the radius during coughing. What radius $r$ will produce the maximum air velocity?

**47.** *Demographics*   From 1950 to 1995, the number $r$ of males to 100 females in the U.S. can be modeled by

$$r = 0.00006t^3 + 0.0004t^2 - 0.21t + 98.7,$$

where $t = 0$ corresponds to 1950. *(Source: U.S. Bureau of the Census)*   Determine the year in which the number $r$ was a minimum. In that year, were there more females or more males in the population? Explain.

**48.** *Biology: Fertility Rates*   The graph of the U.S. fertility rate shows fluctuations influenced by economic and social trends. Use it to answer the following questions.

(a) Around what year was the fertility rate the highest, and to how many births per woman did this rate correspond?

(b) During which time periods was the fertility rate increasing most rapidly? Most slowly?

(c) During which time periods was the fertility rate decreasing most rapidly? Most slowly?

(d) Give some possible real-life reasons for fluctuations in the fertility rate. *(Source: Adapted from Levine/Miller, Biology: Discovering Life, Second Edition)*

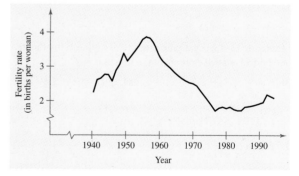

# Concavity and the Second-Derivative Test

**3.3**

## Concavity

You already know that locating the intervals over which a function $f$ increases or decreases is helpful in determining its graph. In this section, you will see that locating the intervals in which $f'$ increases or decreases can determine where the graph of $f$ is curving upward or curving downward. This property of curving upward or downward is defined formally as the **concavity** of the graph of the function.

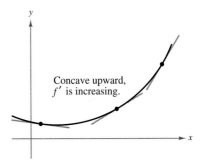

---

### Definition of Concavity

Let $f$ be differentiable on an open interval $I$. The graph of $f$ is

1. **concave upward** on $I$ if $f'$ is increasing on the interval.

2. **concave downward** on $I$ if $f'$ is decreasing on the interval.

---

From Figure 3.20, you can observe the following graphical interpretation of concavity.

1. A curve that is concave upward lies *above* its tangent line.

2. A curve that is concave downward lies *below* its tangent line.

This visual test for concavity is useful when the graph of a function is given. To determine concavity without seeing a graph, you need an analytic test. It turns out that you can use the second derivative to determine these intervals in much the same way that you used the first derivative to determine the intervals in which $f$ is increasing or decreasing.

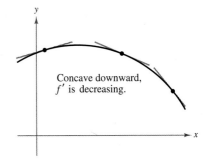

**FIGURE 3.20**

---

### Test for Concavity

Let $f$ be a function whose second derivative exists on an open interval $I$.

1. If $f''(x) > 0$ for all $x$ in $I$, then $f$ is concave upward on $I$.

2. If $f''(x) < 0$ for all $x$ in $I$, then $f$ is concave downward on $I$.

---

*Discovery*

Use a graphing utility to graph the function $f(x) = x^3 - x$ and its second derivative $f''(x) = 6x$ in the same coordinate plane. On what interval is $f$ concave upward? On what interval is $f''$ positive? Describe how the second derivative can be used to determine where a function is concave upward and concave downward. Repeat this analysis for the functions $g(x) = x^4 - 6x^2$ and $g''(x) = 12x^2 - 12$.

For a *continuous* function $f$, you can find the intervals on which the graph of $f$ is concave upward and concave downward as follows. [For a function that is not continuous, the test intervals should be formed using points of discontinuity, along with the points at which $f''(x)$ is zero or undefined.]

### Guidelines for Applying Concavity Test

1. Locate the $x$-values at which $f''(x) = 0$ or $f''(x)$ is undefined.

2. Use these $x$-values to determine the test intervals.

3. Test the sign of $f''(x)$ in each test interval.

## EXAMPLE 1    Applying the Test for Concavity

(a) The graph of the function

$$f(x) = x^2 \qquad \text{Function}$$

is concave upward on the entire real line because its second derivative

$$f''(x) = 2 \qquad \text{Second derivative}$$

is positive for all $x$. (See Figure 3.21.)

(b) The graph of the function

$$f(x) = \sqrt{x} \qquad \text{Function}$$

is concave downward for $x > 0$, because its second derivative

$$f''(x) = -\frac{1}{4}x^{-3/2} \qquad \text{Second derivative}$$

is negative for all $x > 0$. (See Figure 3.22.)

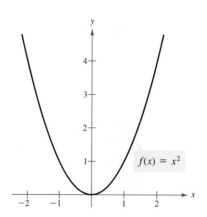

**FIGURE 3.21**    Concave Upward

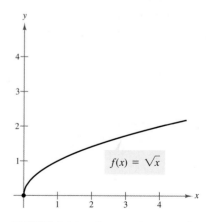

**FIGURE 3.22**    Concave Downward

## EXAMPLE 2  Determining Concavity

Determine the intervals in which the graph of the function is concave upward or concave downward.

$$f(x) = \frac{6}{x^2 + 3}$$

**TIP**

For help on the algebra in Example 2, see Example 1a in the *Chapter 3 Algebra Review*, on page 248.

ALGEBRA

### Solution

Begin by finding the second derivative of $f$.

$$f(x) = 6(x^2 + 3)^{-1} \qquad \text{Original function}$$

$$f'(x) = (-6)(2x)(x^2 + 3)^{-2} \qquad \text{Chain Rule}$$

$$= \frac{-12x}{(x^2 + 3)^2} \qquad \text{Simplify.}$$

$$f''(x) = \frac{(x^2 + 3)^2(-12) - (-12x)(2)(2x)(x^2 + 3)}{(x^2 + 3)^4} \qquad \text{Quotient Rule}$$

$$= \frac{-12(x^2 + 3) + (48x^2)}{(x^2 + 3)^3} \qquad \text{Simplify.}$$

$$= \frac{36(x^2 - 1)}{(x^2 + 3)^3} \qquad \text{Simplify.}$$

From this, you can see that $f''(x)$ is defined for all real numbers and $f''(x) = 0$ when $x = \pm 1$. Thus, you can test the concavity of $f$ by testing the intervals $(-\infty, -1)$, $(-1, 1)$, and $(1, \infty)$, as shown in the table. The graph of $f$ is shown in Figure 3.23.

| Interval | $-\infty < x < -1$ | $-1 < x < 1$ | $1 < x < \infty$ |
|---|---|---|---|
| Test value | $x = -2$ | $x = 0$ | $x = 2$ |
| Sign of $f''(x)$ | $f''(-2) > 0$ | $f''(0) < 0$ | $f''(2) > 0$ |
| Conclusion | Concave upward | Concave downward | Concave upward |

**STUDY TIP**   In Example 2, $f'$ is increasing on the interval $(1, \infty)$ even though $f$ is decreasing there. Be sure you see that the increasing or decreasing of $f'$ does not necessarily correspond to the increasing or decreasing of $f$.

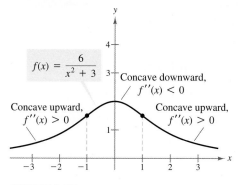

**FIGURE 3.23**

### *Discovery*

Use a graphing utility to graph $f(x) = x^3 - 6x^2 + 12x - 6$ and $f''(x) = 6x - 12$ in the same coordinate plane. At what $x$-value does $f''(x) = 0$? At what $x$-value does the point of inflection occur? Repeat this analysis for $g(x) = x^4 - 5x^2 + 7$ and $g''(x) = 12x^2 - 10$. Make a general statement about the relationship of the point of inflection of a function and the second derivative of a function.

## Points of Inflection

If the tangent line to a graph exists at a point at which the concavity changes, then the point is a **point of inflection.** Three examples of inflection points are shown in Figure 3.24. (Note that the third graph has a vertical tangent line at its point of inflection.)

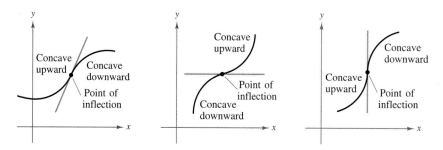

**FIGURE 3.24**    The graph *crosses* its tangent line at a point of inflection.

---

### Definition of Point of Inflection

If the graph of a continuous function possesses a tangent line at a point where its concavity changes from upward to downward (or vice versa), then the point is a **point of inflection.**

---

**STUDY TIP**    As shown in Figure 3.24, a graph crosses its tangent line at a point of inflection.

Because a point of inflection occurs where the concavity of a graph changes, it must be true that at such points the sign of $f''$ changes. Thus, to locate possible points of inflection, you need only determine the values of $x$ for which $f''(x) = 0$ or for which $f''(x)$ does not exist. This parallels the procedure for locating the relative extrema of $f$ by determining the critical numbers of $f$.

---

### Property of Points of Inflection

If $(c, f(c))$ is a point of inflection of the graph of $f$, then either

$$f''(c) = 0 \quad \text{or} \quad f''(c) \text{ does not exist.}$$

**EXAMPLE 3**   **Finding Points of Inflection**

Discuss the concavity of the graph of $f$ and find its points of inflection.

$$f(x) = x^4 + x^3 - 3x^2 + 1$$

*Solution*

Begin by finding the second derivative of $f$.

$$f'(x) = 4x^3 + 3x^2 - 6x \qquad \text{Find first derivative.}$$

$$f''(x) = 12x^2 + 6x - 6 \qquad \text{Find second derivative.}$$

$$= 6(2x - 1)(x + 1) \qquad \text{Factor.}$$

From this, you can see that the possible points of inflection occur at $x = \frac{1}{2}$ and $x = -1$. After testing the intervals

$$(-\infty, -1), \qquad \left(-1, \frac{1}{2}\right), \qquad \text{and} \qquad \left(\frac{1}{2}, \infty\right),$$

you can determine that the graph is concave upward in $(-\infty, -1)$, concave downward in $\left(-1, \frac{1}{2}\right)$, and concave upward in $\left(\frac{1}{2}, \infty\right)$. Because the concavity changes at $x = -1$ and $x = \frac{1}{2}$, you can conclude that the graph has points of inflection at these $x$-values, as shown in Figure 3.25.

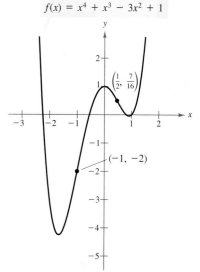

$f(x) = x^4 + x^3 - 3x^2 + 1$

**FIGURE 3.25**   Two Points of Inflection

It is possible for the second derivative to be zero at a point that is *not* a point of inflection. For example, compare the graphs of $f(x) = x^3$ and $g(x) = x^4$, as shown in Figure 3.26. Both second derivatives are zero when $x = 0$, but only the graph of $f$ has a point of inflection at $x = 0$. This shows that before concluding that a point of inflection exists at a value of $x$ for which $f''(x) = 0$, you must test to be certain that the concavity actually changes at the point.

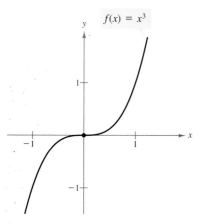

$f(x) = x^3$

$f''(0) = 0$, and $(0, 0)$ is a point of inflection.

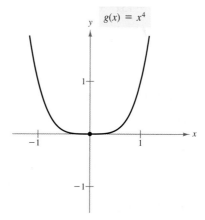

$g(x) = x^4$

$g''(0) = 0$, but $(0, 0)$ is not a point of inflection.

**FIGURE 3.26**

FIGUR

(−1, −

Relativ
minim

FIGUR

## Warm Up

The following warm-up exercises involve skills that were covered in earlier sections. You will use these skills in the exercise set for this section.

In Exercises 1–6, find the second derivative of the function.

**1.** $f(x) = 4x^4 - 9x^3 + 5x - 1$

**2.** $g(s) = (s^2 - 1)(s^2 - 3s + 2)$

**3.** $g(x) = (x^2 + 1)^4$

**4.** $f(x) = (x - 3)^{4/3}$

**5.** $h(x) = \dfrac{5}{(2x)^3}$

**6.** $f(x) = \dfrac{2x - 1}{3x + 2}$

In Exercises 7–10, find the critical numbers of the function.

**7.** $f(x) = 5x^3 - 5x + 11$

**8.** $f(x) = x^4 - 4x^3 - 10$

**9.** $g(t) = \dfrac{16 + t^2}{t}$

**10.** $h(x) = \dfrac{x^4 - 50x^2}{8}$

## EXERCISES 3.3

In Exercises 1–6, find the intervals on which the graph is concave upward and those on which it is concave downward.

**5.** $f(x) = \dfrac{24}{x^2 + 12}$

**6.** $y = x^5 + 5x^4 - 40x^2$

**1.** $y = x^2 - x - 2$

**2.** $y = -x^3 + 3x^2 - 2$

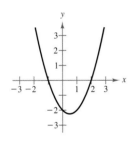

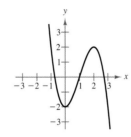

**3.** $f(x) = \dfrac{x^2 - 1}{2x + 1}$

**4.** $f(x) = \dfrac{x^2 + 4}{4 - x^2}$

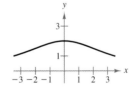

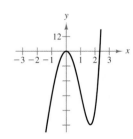

In Exercises 7–14, find all relative extrema of the function. Use the Second-Derivative Test when applicable.

**7.** $f(x) = 6x - x^2$

**8.** $f(x) = (x - 5)^2$

**9.** $f(x) = x^3 - 5x^2 + 7x$

**10.** $f(x) = x^4 - 4x^3 + 2$

**11.** $f(x) = x^{2/3} - 3$

**12.** $f(x) = \sqrt{x^2 + 1}$

**13.** $f(x) = x + \dfrac{4}{x}$

**14.** $f(x) = \dfrac{x}{x - 1}$

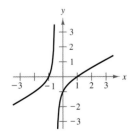

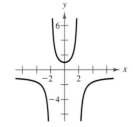

In Exercises 15–18, use a graphing utility to estimate graphically all relative extrema of the function.

**15.** $f(x) = -(x - 5)^2$

**16.** $f(x) = x^2 + 3x - 8$

**17.** $f(x) = 5 + 3x^2 - x^3$

**18.** $f(x) = 3x^3 + 5x^2 - 2$

In Exercises 19–22, state the sign of $f'(x)$ and $f''(x)$ in the interval $(0, 2)$.

**19.**

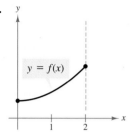

**20.**

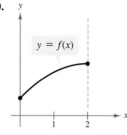

**21.**

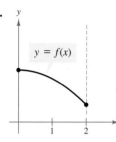

**22.**

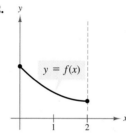

In Exercises 23–30, find the points of inflection of the graph.

**23.** $f(x) = x^3 - 9x^2 + 24x - 18$

**24.** $f(x) = x(6 - x)^2$

**25.** $f(x) = (x - 1)^3(x - 5)$

**26.** $f(x) = x^4 - 18x^2 + 5$

**27.** $g(x) = 2x^4 - 8x^3 + 12x^2 + 12x$

**28.** $f(x) = -4x^3 - 8x^2 + 32$

**29.** $h(x) = (x - 2)^3(x - 1)$

**30.** $f(t) = (1 - t)(t - 4)(t^2 - 4)$

 In Exercises 31–42, use a graphing utility to graph the function and identify all relative extrema and points of inflection.

**31.** $f(x) = x^3 - 12x$  **32.** $f(x) = x^3 + 1$

**33.** $f(x) = x^3 - 6x^2 + 12x$  **34.** $f(x) = x^3 - \frac{3}{2}x^2 - 6x$

**35.** $f(x) = \frac{1}{4}x^4 - 2x^2$  **36.** $f(x) = 2x^4 - 8x + 3$

**37.** $g(x) = (x - 2)(x + 1)^2$  **38.** $g(x) = (x - 6)(x + 2)^3$

**39.** $g(x) = x\sqrt{x + 3}$  **40.** $g(x) = \frac{x^3}{16}(x + 4)$

**41.** $f(x) = \dfrac{4}{1 + x^2}$  **42.** $f(x) = x - 4\sqrt{x + 1}$

In Exercises 43 and 44, sketch a graph of a function $f$ having the given characteristics.

|  | Function | First Derivative | Second Derivative |
|---|---|---|---|
| **43.** | $f(2) = 0$ | $f'(x) < 0, \ x < 3$ | $f''(x) > 0$ |
|  | $f(4) = 0$ | $f'(3) = 0$ |  |
|  |  | $f'(x) > 0, \ x > 3$ |  |
| **44.** | $f(2) = 0$ | $f'(x) > 0, \ x < 3$ | $f''(x) > 0, \ x \neq 3$ |
|  | $f(4) = 0$ | $f'(3)$ is undefined |  |
|  |  | $f'(x) < 0, \ x > 3$ |  |

In Exercises 45 and 46, use the graph to sketch the graph of $f'$. Find the intervals on which (a) $f'(x)$ is positive, (b) $f'(x)$ is negative, (c) $f'$ is increasing, and (d) $f'$ is decreasing. For each of these intervals, describe the corresponding behavior of $f$.

**45.**

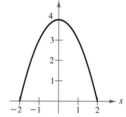

**46.**

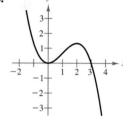

 *Point of Diminishing Returns*  In Exercises 47 and 48, identify the point of diminishing returns for the input-output function. For each function, $R$ is the revenue and $x$ is the amount spent on advertising. Use a graphing utility to check your results.

**47.** $R = \dfrac{1}{50,000}(600x^2 - x^3), \quad 0 \leq x \leq 400$

**48.** $R = -\frac{4}{9}(x^3 - 9x^2 - 27), \quad 0 \leq x \leq 5$

 *Average Cost*  In Exercises 49 and 50, you are given the total cost of producing $x$ units. Find the production level that minimizes the average cost per unit. Use a graphing utility to check your results.

**49.** $C = 0.5x^2 + 15x + 5000$

**50.** $C = 0.002x^3 + 20x + 500$

*Productivity*  In Exercises 51 and 52, consider a college student who works from 7 P.M. to 11 P.M. assembling mechanical components. The number $N$ of components assembled after $t$ hours is given by the function. At what time is the student assembling components at the greatest rate?

**51.** $N = -0.12t^3 + 0.54t^2 + 8.22t,$     $0 \le t \le 4$

**52.** $N = \dfrac{20t^2}{4 + t^2},$     $0 \le t \le 4$

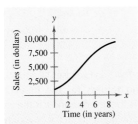 *Sales Growth*  In Exercises 53 and 54, find the time $t$ in years when the annual sales $x$ of a new product are increasing at the greatest rate. Use a graphing utility to verify your results.

**53.** $x = \dfrac{10,000t^2}{9 + t^2}$

**54.** $x = \dfrac{500,000t^2}{36 + t^2}$

**55.** *Sales*  A company introduces a new product line. The annual sales of the product increase over time, as shown in the graph. Visually locate the point of inflection of the graph. Describe its significance.

**56.** Show that the point of inflection of the graph of $f(x) = x(x - 6)^2$ lies midway between the relative extrema of $f$.

In Exercises 57–60, use a graphing utility to graph $f, f',$ and $f''$ in the same viewing rectangle. Graphically locate the relative extrema and points of inflection of the graph.

|  | *Function* | *Interval* |
|---|---|---|
| **57.** | $f(x) = \frac{1}{2}x^3 - x^2 + 3x - 5$ | $[0, 3]$ |
| **58.** | $f(x) = -\frac{1}{20}x^5 - \frac{1}{12}x^2 - \frac{1}{3}x + 1$ | $[-2, 2]$ |
| **59.** | $f(x) = \dfrac{2}{x^2 + 1}$ | $[-3, 3]$ |
| **60.** | $f(x) = \dfrac{x^2}{x^2 + 1}$ | $[-3, 3]$ |

**61.** *Tax Returns*  The number $N$ (in thousands) of tax returns for different years can be modeled by

$$N = -\frac{8816.69}{t^2} + \frac{3489.9}{t} - 1.14t^2 + 42.77t - 464.1$$

where $t$ is the time in years, with $t = 8$ corresponding to 1988. Use a graphing utility to graph the function. *(Source: Internal Revenue Service)*

**62.** *Epidemic*  The spread of a virus can be modeled by

$$N = -t^3 + 12t^2,     0 \le t \le 12,$$

where $N$ is the number of people infected in hundreds, and $t$ is the time in weeks.

(a) What is the maximum number of people projected to be infected?

(b) When will the virus be spreading most rapidly?

(c) Use a graphing utility to confirm your results.

## Business Capsule

Microsoft Corporation

*Bill Gates, president of Microsoft Corporation, was the wealthiest person in America in 1996 with a net worth of $39.8 billion. He wrote his first computer program at age 13, and scored a perfect 800 in math on his SAT test.*

**63.** *Research Project*  Use your school's library or some other reference source to research the financial history of a business person, such as Bill Gates or Ted Turner. Gather the data on the person's taxable income over a period of time, and use a graphing utility to graph a scatter plot of the data. Fit a model to the data. Does the model appear to be concave up or down? Does it appear to be increasing or decreasing? Discuss the implications of your answers.

# 3.4 | Optimization Problems

Solving Optimization Problems

## Solving Optimization Problems

One of the most common applications of calculus is the determination of optimum (minimum or maximum) values. Before learning a general method for solving optimization problems, study the following example.

### EXAMPLE 1 Finding the Maximum Volume

A manufacturer wants to design an open box that has a square base and a surface area of 108 square inches, as shown in Figure 3.31. What dimensions will produce a box with a maximum volume?

**Solution**

Because the base of the box is square, the volume is

$V = x^2h.$         Primary equation

This equation is called the **primary equation** because it gives a formula for the quantity to be optimized. The surface area of the box is

$S = $ (area of base) $+$ (area of four sides)

$108 = x^2 + 4xh.$       Secondary equation

Because $V$ is to be optimized, it helps to express $V$ as a function of just one variable. To do this, solve the secondary equation for $h$ in terms of $x$ to obtain $h = (108 - x^2)/4x$ and substitute into the primary equation.

$$V = x^2h = x^2\left(\frac{108 - x^2}{4x}\right) = 27x - \frac{1}{4}x^3$$    Function of one variable

Before finding which $x$-value yields a maximum value of $V$, you need to determine the **feasible domain** of the function. That is, what values of $x$ make sense in the problem? Because $x$ must be nonnegative and the area of the base $(A = x^2)$ is at most 108, you can conclude that the feasible domain is

$0 \leq x \leq \sqrt{108}.$       Feasible domain

Using the techniques described in the first three sections of this chapter, you can determine that (in the interval $0 \leq x \leq \sqrt{108}$) this function has an absolute maximum when $x = 6$ inches and $h = 3$ inches.

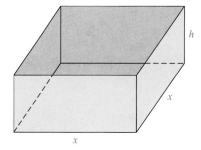

**FIGURE 3.31** Open Box with Square Base $S = x^2 + 4xh = 108$

In studying Example 1, be sure that you understand the basic question that it asks. Some students have trouble with optimization problems because they are too eager to start solving the problem by using a pat formula. For instance, in Example 1, you should realize that there are infinitely many open boxes having 108 square inches of surface area. You might begin to solve this problem by asking yourself which basic shape would seem to yield a maximum volume. Should the box be tall, cubical, or squat? You might even try calculating a few volumes, as shown in Figure 3.32, to see if you can get a good feeling for what the optimum dimensions should be.

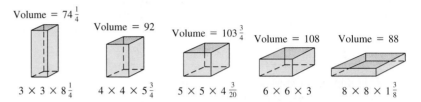

Volume $= 74\frac{1}{4}$     Volume $= 92$     Volume $= 103\frac{3}{4}$     Volume $= 108$     Volume $= 88$

$3 \times 3 \times 8\frac{1}{4}$     $4 \times 4 \times 5\frac{3}{4}$     $5 \times 5 \times 4\frac{3}{20}$     $6 \times 6 \times 3$     $8 \times 8 \times 1\frac{3}{8}$

**FIGURE 3.32**     Which size box has the maximum volume?

Remember that you are not ready to begin solving an optimization problem until you have clearly identified what the problem is. Once you are sure you understand what is being asked, you are ready to begin considering a method for solving the problem.

There are several steps in the solution of Example 1. The first step is to sketch a diagram and assign symbols to all *known* quantities and to all quantities *to be determined.* The second step is to write a primary equation for the quantity to be optimized. Then, a secondary equation is used to rewrite the primary equation as a function of one variable. Finally, calculus is used to determine the optimum value. These steps are summarized below.

*STUDY TIP*   When performing Step 5, remember that to determine the maximum or minimum value of a continuous function $f$ on a closed interval, you need to compare the values of $f$ at its critical numbers with the values of $f$ at the endpoints of the interval. The largest of these values is the desired maximum and the smallest is the desired minimum.

### Guidelines for Solving Optimization Problems

1. Assign symbols to all given quantities and to all quantities to be determined. When feasible, sketch a diagram.

2. Write a **primary equation** for the quantity to be maximized or minimized. (A summary of several common formulas is given in the appendix.)

3. Reduce the primary equation to one having a single independent variable: this may involve the use of a **secondary equation** that relates the independent variables of the primary equation.

4. Determine the **feasible domain** of the primary equation. That is, determine the values for which the stated problem makes sense.

5. Use calculus to find the desired maximum or minimum value.

## EXAMPLE 2 Finding a Maximum Sum

The product of two positive numbers is 288. Minimize the sum of the second number and twice the first number.

### Solution

1. Let $x$ be the first number, $y$ the second, and $S$ the sum to be minimized.
2. Because you want to minimize $S$, the primary equation is

$$S = 2x + y. \qquad \text{Primary equation}$$

3. Because the product of the two numbers is 288, you can write the following secondary equation.

$$xy = 288 \qquad \text{Secondary equation}$$

$$y = \frac{288}{x}$$

Using this result, you can rewrite the primary equation as a function of one variable.

$$S = 2x + \frac{288}{x} \qquad \text{Function of one variable}$$

4. Because the numbers are positive, the feasible domain is

$$0 < x. \qquad \text{Feasible domain}$$

5. To find the maximum value of $S$, begin by finding its critical numbers.

$$\frac{dS}{dx} = 2 - \frac{288}{x^2} \qquad \text{Find derivative of } S.$$

$$0 = 2 - \frac{288}{x^2} \qquad \text{Set derivative equal to 0.}$$

$$x^2 = 144 \qquad \text{Simplify.}$$

$$x = \pm 12 \qquad \text{Critical numbers}$$

Choosing the positive $x$-value, you can use the First-Derivative Test to conclude that $S$ is decreasing in the interval $(0, 12)$ and increasing in the interval $(12, \infty)$, as shown in the table. Therefore, $x = 12$ yields a minimum, and the two numbers are

$$x = 12 \quad \text{and} \quad y = \frac{288}{12} = 24.$$

| Interval | $0 < x < 12$ | $12 < x < \infty$ |
|---|---|---|
| Test value | $x = 1$ | $x = 13$ |
| Sign of $\dfrac{dS}{dx}$ | $\dfrac{dS}{dx} < 0$ | $\dfrac{dS}{dx} > 0$ |
| Conclusion | $S$ is decreasing | $S$ is increasing |

**TIP**

For help on the algebra in Example 2, see Example 1b in the *Chapter 3 Algebra Review*, on page 248.

## Technology

After you have written the primary equation as a function of a single variable, you can estimate the optimum value by graphing the function. For instance, the graph of

$$S = 2x + \frac{288}{x},$$

shown below, indicates that the minimum value of $S$ occurs when $x$ is about 12.

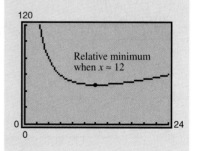

## EXAMPLE 3   Finding a Minimum Distance

Find the points on the graph of $y = 4 - x^2$ that are closest to $(0, 2)$.

*Solution*

1. Figure 3.33 indicates that there are two points at a minimum distance from the point $(0, 2)$.

2. You are asked to minimize the distance $d$. Hence, you can use the Distance Formula to obtain a primary equation.

$$d = \sqrt{(x - 0)^2 + (y - 2)^2} \qquad \text{Primary equation}$$

3. Using the secondary equation $y = 4 - x^2$, you can rewrite the primary equation as a function of a single variable.

$$d = \sqrt{x^2 + (4 - x^2 - 2)^2} \qquad \text{Substitute } 4 - x^2 \text{ for } y.$$
$$= \sqrt{x^4 - 3x^2 + 4} \qquad \text{Simplify.}$$

Because $d$ is smallest when the expression under the radical is smallest, you simplify the problem by finding the minimum value of

$$f(x) = x^4 - 3x^2 + 4.$$

4. The domain of $f$ is the entire real line.

5. To find the minimum value of $f(x)$, first find the critical numbers of $f$.

$$f'(x) = 4x^3 - 6x \qquad \text{Find derivative of } f.$$
$$0 = 4x^3 - 6x \qquad \text{Set derivative equal to 0.}$$
$$0 = 2x(2x^2 - 3) \qquad \text{Factor.}$$
$$x = 0, \sqrt{\tfrac{3}{2}}, -\sqrt{\tfrac{3}{2}} \qquad \text{Critical numbers}$$

By the First-Derivative Test, you can conclude that $x = 0$ yields a relative maximum, whereas both $\sqrt{3/2}$ and $-\sqrt{3/2}$ yield a minimum. Hence, on the graph of $y = 4 - x^2$, the points that are closest to the point $(0, 2)$ are

$$\left(\sqrt{\tfrac{3}{2}}, \tfrac{5}{2}\right) \qquad \text{and} \qquad \left(-\sqrt{\tfrac{3}{2}}, \tfrac{5}{2}\right).$$

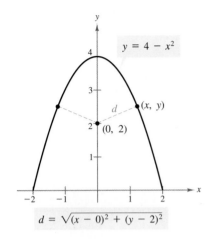

$$d = \sqrt{(x - 0)^2 + (y - 2)^2}$$

**FIGURE 3.33**

**TIP**

ALGEBRA

For help on the algebra in Example 3, see Example 1c in the *Chapter 3 Algebra Review*, on page 248.

*STUDY TIP*   To confirm the result in Example 3, try computing the distance between several points on the graph of $y = 4 - x^2$ and the point $(0, 2)$. For instance, the distance between $(1, 3)$ and $(0, 2)$ is

$$d = \sqrt{(0 - 1)^2 + (2 - 3)^2} = \sqrt{2} \approx 1.414.$$

Note that this is greater than the distance between $\left(\sqrt{3/2}, 5/2\right)$ and $(0, 2)$, which is

$$d = \sqrt{\left(0 - \sqrt{\tfrac{3}{2}}\right)^2 + \left(2 - \tfrac{5}{2}\right)^2} = \sqrt{\tfrac{7}{4}} \approx 1.323.$$

## EXAMPLE 4   Finding a Minimum Area

A rectangular page is to contain 24 square inches of print. The margins at the top and bottom of the page are $1\frac{1}{2}$ inches wide. The margins on each side are 1 inch wide. What should the dimensions of the page be so that the least amount of paper is used?

### Solution

1. A diagram of the page is shown in Figure 3.34.

2. Letting $A$ be the area to be minimized, the primary equation is

$$A = (x + 3)(y + 2).$$   Primary equation

3. The printed area inside the margins is given by

$$24 = xy.$$   Secondary equation

Solving this equation for $y$ produces

$$y = \frac{24}{x}.$$

By substituting this into the primary equation, you obtain

$$A = (x + 3)\left(\frac{24}{x} + 2\right)$$   Function of one variable

$$= (x + 3)\left(\frac{24 + 2x}{x}\right)$$   Rewrite as fraction.

$$= \frac{2x^2}{x} + \frac{30x}{x} + \frac{72}{x}$$   Multiply and separate into terms.

$$= 2x + 30 + \frac{72}{x}.$$   Simplify.

4. Because $x$ must be positive, the feasible domain is $x > 0$.

5. To find the minimum area, begin by finding the critical numbers of $A$.

$$\frac{dA}{dx} = 2 - \frac{72}{x^2}$$   Find derivative of $A$.

$$0 = 2 - \frac{72}{x^2}$$   Set derivative equal to 0.

$$-2 = -\frac{72}{x^2}$$   Subtract 2 from each side.

$$x^2 = 36$$   Simplify.

$$x = \pm 6$$   Critical numbers

Because $x = -6$ is not in the feasible domain, you need only consider the critical number $x = 6$. Using the First-Derivative Test, it follows that $A$ is a minimum when $x = 6$. Thus, the dimensions of the page should be $x + 3 = 6 + 3 = 9$ inches by $y + 2 = 24/6 + 2 = 6$ inches.

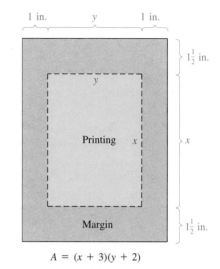

$$A = (x + 3)(y + 2)$$

**FIGURE 3.34**

As applications go, the four examples described in this section are fairly simple, and yet the resulting primary equations are quite complicated. Real-life applications often involve equations that are at least as complex as these four, and you should expect this. Remember that one of the main goals of this course is to enable you to use the power of calculus to analyze equations that at first glance seem formidable.

Also remember that once you have found the primary equation, you can use the graph of the equation as an aid to solving the optimization problem. For instance, the graphs of the primary equations in Examples 1 through 4 are shown in Figure 3.35.

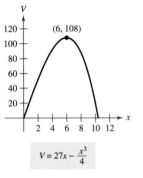

$$V = 27x - \frac{x^3}{4}$$

Example 1

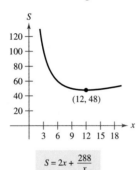

$$S = 2x + \frac{288}{x}$$

Example 2

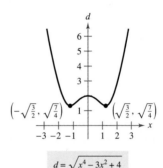

$$d = \sqrt{x^4 - 3x^2 + 4}$$

Example 3

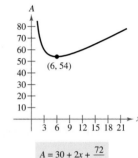

$$A = 30 + 2x + \frac{72}{x}$$

Example 4

**FIGURE 3.35**

## *Group Discussion*          *Comparing Graphical, Numerical, and Analytic Approaches*

In Examples 1, 2, 3, and 4, an analytic approach was used to find the minimum or maximum value. Divide your class into eight groups and ask the groups to resolve the examples using a numerical approach (table) or a graphical approach (using a graphing utility).

**Group 1:** Use a *numerical* approach to solve Example 1.

**Group 2:** Use a *graphical* approach to solve Example 1.

**Group 3:** Use a *numerical* approach to solve Example 2.

**Group 4:** Use a *graphical* approach to solve Example 2.

**Group 5:** Use a *numerical* approach to solve Example 3.

**Group 6:** Use a *graphical* approach to solve Example 3.

**Group 7:** Use a *numerical* approach to solve Example 4.

**Group 8:** Use a *graphical* approach to solve Example 4.

When all the groups are done, the groups working on the same example should compare all three approaches. What are the advantages of each approach? What are the disadvantages?

*Warm Up*

The following warm-up exercises involve skills that were covered in earlier sections. You will use these skills in the exercise set for this section.

In Exercises 1–4, write a formula for the written statement.

**1.** The sum of one number and half a second number is 12.

**2.** The product of one number and twice another is 24.

**3.** The area of a rectangle is 24 square units.

**4.** The distance between two points is ten units.

In Exercises 5–10, find the critical numbers of the function.

**5.** $y = x^2 + 6x - 9$

**6.** $y = 2x^3 - x^2 - 4x$

**7.** $y = 5x + 125/x$

**8.** $y = 3x + \dfrac{96}{x^2}$

**9.** $y = \dfrac{x^2 + 1}{x}$

**10.** $y = \dfrac{x}{x^2 + 9}$

## EXERCISES 3.4

In Exercises 1–6, find two positive numbers satisfying the given requirements.

**1.** The sum is 110 and the product is a maximum.

**2.** The sum is $S$ and the product is a maximum.

**3.** The sum of the first and twice the second is 36 and the product is a maximum.

**4.** The sum of the first and twice the second is 100 and the product is a maximum.

**5.** The product is 192 and the sum is a minimum.

**6.** The product is 192 and the sum of the first plus three times the second is a minimum.

**7.** What positive number $x$ minimizes the sum of $x$ and its reciprocal?

**8.** The difference of two numbers is 50. Find the two numbers so that their product is a minimum.

**9.** Verify that each rectangular solid has a surface area of 150 square inches (see figure). Find the volume of each.

**10.** Verify that each right circular cylinder has a surface area of $24\pi$ (see figure). Find the volume of each.

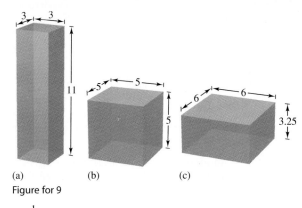

(a)     (b)     (c)

Figure for 9

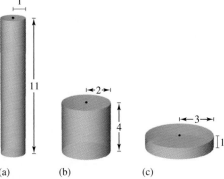

(a)     (b)     (c)

Figure for 10

In Exercises 11 and 12, find the length and width of the specified rectangle.

**11.** Maximum area; Perimeter: 100 feet

**12.** Minimum perimeter; Area: 64 square feet

**13.** *Area*   A rancher has 200 feet of fencing to enclose two adjacent rectangular corrals (see figure). What dimensions should be used so that the enclosed area will be a maximum?

Figure for 13          Figure for 15

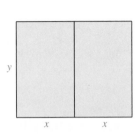          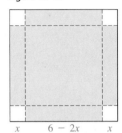

**14.** *Fence Length*   A dairy farmer plans to enclose a rectangular pasture adjacent to a river. To provide enough grass for the herd, the pasture must contain 180,000 square meters. No fencing is required along the river. What dimensions use the least amount of fencing?

**15.** *Volume*   An open box is to be made from a 6-inch by 6-inch square piece of material by cutting equal squares from the corners and turning up the sides (see figure). Find the volume of the largest box that can be made in this manner.

**16.** *Volume*   An open box is to be made from a 2-foot by 3-foot rectangular piece of material by cutting equal squares from the corners and turning up the sides. Find the volume of the largest box that can be made in this manner.

**17.** *Surface Area*   A net enclosure for golf practice is open at one end (see figure). The volume of the enclosure is $83\frac{1}{3}$ cubic meters. Find the dimensions that require the least amount of netting.

Figure for 17          Figure for 21

          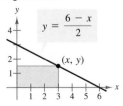

**18.** *Apple Trees*   A tree farmer estimates that if 80 apple trees are planted, the average yield will be 400 apples per tree. But because of the size of the farm, for each additional tree planted the yield will decrease by 32 apples per tree. How many trees should be planted to maximize the total yield of apples? What is the maximum yield?

**19.** *Area*   An indoor physical fitness room consists of a rectangular region with a semicircle on each end. The perimeter of the room is to be a 200-meter running track. Find the dimensions that will make the area of the rectangular region as large as possible.

**20.** *Area*   A page is to contain 30 square inches of print. The margins at the top and bottom of the page are each 2 inches wide. The margins on each side are 1 inch wide. What dimensions will minimize the amount of paper used?

**21.** *Area*   A rectangle is bounded by the $x$- and $y$-axes and the graph of $y = (6 - x)/2$ (see figure). What length and width should the rectangle have so that its area is a maximum?

**22.** *Area*   A right triangle is formed in the first quadrant by the $x$-and $y$-axes and a line through the point $(2, 3)$. Find the vertices of the triangle so that its area is a minimum.

**23.** *Area*   A rectangle is bounded by the $x$-axis and the semicircle $y = \sqrt{25 - x^2}$ (see figure). What length and width should the rectangle have so that its area is a maximum?

Figure for 23          Figure for 26

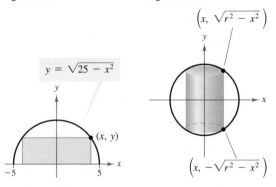

**24.** *Area*   Find the dimensions of the largest rectangle that can be inscribed in a semicircle of radius $r$. (See Exercise 23.)

**25.** *Volume*   You are designing a soft drink container that has the shape of a right circular cylinder. The container is supposed to hold 12 fluid ounces (1 fluid ounce is approximately 1.80469 cubic inches). Find the dimensions that will use a minimum amount of construction material.

**26.** *Volume* Find the volume of the largest right circular cylinder that can be inscribed in a sphere of radius $r$ (see figure).

**27.** *Closest Point* Find the point on the graph of $y = x^2 + 1$ that is closest to the point $(0, 4)$.

**28.** *Closest Point* Find the point on the graph of $y = x^2$ that is closest to the point $\left(2, \frac{1}{2}\right)$.

**29.** *Volume* A rectangular package to be sent by a postal service can have a maximum combined length and girth of 108 inches. Find the dimensions of the package with maximum volume. Assume that the package's dimensions are $x$ by $x$ by $y$ (see figure).

Figure for 29    Figure for 33

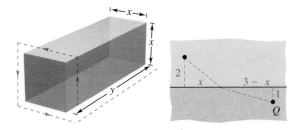

**30.** *Surface Area* A container is formed by adjoining two hemispheres to each end of a right circular cylinder. The total volume of the container is 12 cubic inches. Find the radius of the cylinder that produces the minimum surface area.

**31.** *Area* The combined perimeter of a circle and a square is 16. Find the dimensions of the circle and square that produce a minimum total area.

**32.** *Area* The combined perimeter of an equilateral triangle and a square is 10. Find the dimensions of the triangle and square that produce a minimum total area.

**33.** *Time* You are in a boat 2 miles from the nearest point on the coast. You are to go to point $Q$, 3 miles down the coast and 1 mile inland (see figure). You can row at a rate of 2 miles per hour and can walk at a rate of 4 miles per hour. Toward which point on the coast should you row in order to reach point $Q$ in the least time?

**34.** *Strawberry Farms* A strawberry farmer will receive $4 per bushel of strawberries during the first week of harvesting. Each week after that, the value drops $0.10 per bushel. The farmer estimates that there are approximately 120 bushels of strawberries in the fields, and that the crop is increasing at a rate of 4 bushels per week. When should the farmer harvest the strawberries to maximize their value? How many bushels of strawberries will yield the maximum value? What is the maximum value of the strawberries?

**35.** *Strength of a Beam* A wooden beam has a rectangular cross section of height $h$ and width $w$ (see figure). The strength $S$ of the beam is directly proportional to its width and the square of its height. What are the dimensions of the strongest beam that can be cut from a round log of diameter 24 inches? (*Hint:* $S = kh^2w$, where $k$ is the proportionality constant.)

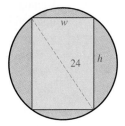

Figure for 35

**36.** *Area* Use a graphing utility to graph the primary equation and its first derivative to find the dimensions of the rectangle of maximum area that can be inscribed in a semicircle of radius 10.

**37.** *Area* Four feet of wire is to be used to form a square and a circle.

(a) Express the sum of the areas of the square and the circle as a function $A(x)$ of the side of the square $x$.
(b) What is the domain of $A(x)$?
(c) Use a graphing utility to graph $A$ on its domain.
(d) How much wire should be used for the square and how much for the circle in order to enclose the least total area?
(e) How much wire should be used for the square and how much for the circle in order to enclose the most total area?

# Business and Economics Applications

*Optimization in Business and Economics • Price Elasticity of Demand • Business Terms and Formulas*

## Optimization in Business and Economics

The problems in this section are primarily optimization problems. Hence, the five-step procedure used in Section 3.4 is an appropriate strategy to follow.

### EXAMPLE 1 Finding the Maximum Revenue

A company has determined that its total revenue (in dollars) for a product can be modeled by

$$R = -x^3 + 450x^2 + 52{,}500x,$$

where $x$ is the number of units produced (and sold). What production level will yield a maximum revenue?

*Solution*

1. A sketch of the revenue function is given in Figure 3.36.

2. The primary equation is the given revenue function.

$$R = -x^3 + 450x^2 + 52{,}500x \qquad \text{Primary equation}$$

3. Because $R$ is already given as a function of one variable, you do not need a secondary equation.

4. The feasible domain of the primary equation is

$$0 \le x \le 546. \qquad \text{Feasible domain}$$

This is determined by finding the $x$-intercepts of the revenue function, as shown in Figure 3.36.

5. To maximize the revenue, find the critical numbers.

$$\frac{dR}{dx} = -3x^2 + 900x + 52{,}500 = 0 \qquad \text{Set derivative equal to 0.}$$

$$-3(x - 350)(x + 50) = 0 \qquad \text{Factor.}$$

$$x = 350, \ -50 \qquad \text{Critical numbers}$$

The only critical number in the feasible domain is $x = 350$. From the graph of the function, you can see that the production level of 350 units corresponds to a maximum revenue.

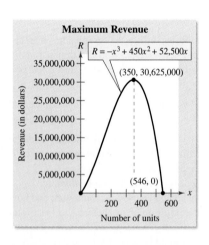

**Maximum Revenue**

$R = -x^3 + 450x^2 + 52{,}500x$

(350, 30,625,000)

(546, 0)

Revenue (in dollars)

Number of units

**FIGURE 3.36** Maximum revenue occurs when $dR/dx = 0$.

To study the effects of production levels on cost, economists use the **average cost function** $\overline{C}$, which is defined as

$$\overline{C} = \frac{C}{x},$$   Average cost function

where $C = f(x)$ is the total cost function and $x$ is the number of units produced.

## EXAMPLE 2   Finding the Minimum Average Cost

A company estimates that the cost (in dollars) of producing $x$ units of a product can be modeled by

$$C = 800 + 0.04x + 0.0002x^2.$$

Find the production level that minimizes the average cost per unit.

**Solution**

1. $C$ represents the total cost, $x$ represents the number of units produced, and $\overline{C}$ represents the average cost per unit.

2. The primary equation is

$$\overline{C} = \frac{C}{x}.$$   Primary equation

3. Substituting the given equation for $C$ produces

$$\overline{C} = \frac{800 + 0.04x + 0.0002x^2}{x}$$   Substitute for $C$.

$$= \frac{800}{x} + 0.04 + 0.0002x.$$   Function of one variable

4. The feasible domain for this function is

$$0 < x.$$   Feasible domain

5. You can find the critical numbers as follows.

$$\frac{d\overline{C}}{dx} = -\frac{800}{x^2} + 0.0002 = 0$$   Set derivative equal to 0.

$$0.0002 = \frac{800}{x^2}$$

$$x^2 = \frac{800}{0.0002}$$   Multiply by $x^2$ and divide by 0.0002.

$$x^2 = 4,000,000$$

$$x = \pm 2000$$   Critical numbers

By choosing the positive value of $x$ and sketching the graph of $\overline{C}$, as shown in Figure 3.37, you can see that a production level of $x = 2000$ minimizes the average cost per unit.

**STUDY TIP**   To see that $x = 2000$ corresponds to a minimum average cost in Example 2, try evaluating $\overline{C}$ for several values of $x$. For instance, when $x = 400$, the average cost per unit is $\overline{C} = \$2.12$, but when $x = 2000$, the average cost per unit is $\overline{C} = \$0.84$.

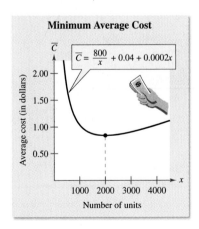

**FIGURE 3.37**   Minimum average cost occurs when $d\overline{C}/dx = 0$.

## EXAMPLE 3 Finding the Maximum Revenue

A business sells 2000 units per month at a price of $10 each. It can sell 250 more items per month for each $0.25 reduction in price. What price per unit will maximize the monthly revenue?

### Solution

1. Let $x$ represent the number of units sold in a month, let $p$ represent the price per unit, and let $R$ represent the monthly revenue.

2. Because the revenue is to be maximized, the primary equation is

$$R = xp. \qquad \text{Primary equation}$$

3. A price of $p = \$10$ corresponds to $x = 2000$ and a price of $p = \$9.75$ corresponds to $x = 2250$. Using this information, you can use the point-slope form to create the demand equation.

$$p - 10 = \frac{10 - 9.75}{2000 - 2250}(x - 2000) \qquad \text{Point-slope form}$$

$$p - 10 = -0.001(x - 2000) \qquad \text{Simplify.}$$

$$p = -0.001x + 12 \qquad \text{Secondary equation}$$

Substituting this value into the revenue equation produces

$$R = x(-0.001x + 12) \qquad \text{Substitute for } p.$$

$$= -0.001x^2 + 12x. \qquad \text{Function of one variable}$$

4. The feasible domain of the revenue function is

$$0 \le x \le 12{,}000. \qquad \text{Feasible domain}$$

5. To maximize the revenue, find the critical numbers.

$$\frac{dR}{dx} = 12 - 0.002x = 0 \qquad \text{Set derivative equal to 0.}$$

$$-0.002x = -12$$

$$x = 6000 \qquad \text{Critical number}$$

From the graph of $R$ in Figure 3.38, you can see that this production level yields a maximum revenue. The price that corresponds to this production level is

$$p = 12 - 0.001(x) \qquad \text{Demand function}$$

$$= 12 - 0.001(6000) \qquad \text{Substitute 6000 for } x.$$

$$= \$6. \qquad \text{Price per unit}$$

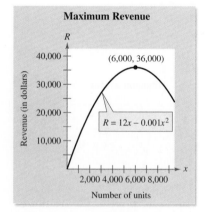

**Maximum Revenue**

(6,000, 36,000)

$R = 12x - 0.001x^2$

Revenue (in dollars)

Number of units

**FIGURE 3.38**

> **STUDY TIP** In Example 3, the revenue function was written as a function of $x$. It could also have been written as a function of $p$. That is, $R = 1000(12p - p^2)$. By finding the critical numbers of this function, you can determine that the maximum revenue occurs when $p = 6$.

## EXAMPLE 4 Finding the Maximum Profit

The marketing department for a business has determined that the demand for a product can be modeled by

$$p = \frac{50}{\sqrt{x}}.$$

The cost of producing $x$ units is given by $C = 0.5x + 500$. What price will yield a maximum profit?

### Solution

1. Let $R$ represent the revenue, $P$ the profit, $p$ the price per unit, $x$ the number of units, and $C$ the total cost of producing $x$ units.

2. Because you are maximizing the profit, the primary equation is

$$P = R - C. \qquad \text{Primary equation}$$

3. Because the revenue is $R = xp$, you can write the profit function as

$$
\begin{aligned}
P &= R - C \\
&= xp - (0.5x + 500) && \text{Substitute for } C. \\
&= x\left(\frac{50}{\sqrt{x}}\right) - 0.5x - 500 && \text{Substitute for } p. \\
&= 50\sqrt{x} - 0.5x - 500. && \text{Function of one variable}
\end{aligned}
$$

4. The feasible domain of the function is $0 < x \le 7872$. (When $x$ is greater than 7872, the profit is negative.)

5. To maximize the profit, find the critical numbers.

$$
\begin{aligned}
\frac{dP}{dx} = \frac{25}{\sqrt{x}} - 0.5 &= 0 && \text{Set derivative equal to 0.} \\
\sqrt{x} &= 50 && \text{Isolate } x\text{-term on one side.} \\
x &= 2500 && \text{Critical number}
\end{aligned}
$$

From the graph of the profit function shown in Figure 3.39, you can see that $x = 2500$ corresponds to a maximum profit. The price that corresponds to $x = 2500$ is

$$p = \frac{50}{\sqrt{x}} = \frac{50}{\sqrt{2500}} = \frac{50}{50} = \$1.00. \qquad \text{Price per unit}$$

**STUDY TIP** To find the maximum profit in Example 4, we differentiated the equation $P = R - C$ and set $dP/dx$ equal to zero. From the equation

$$\frac{dP}{dx} = \frac{dR}{dx} - \frac{dC}{dx} = 0,$$

it follows that the maximum profit occurs when the marginal revenue is equal to the marginal cost, as shown in Figure 3.40.

**TIP**

For help on the algebra in Example 4, see Example 2b in the *Chapter 3 Algebra Review,* on page 249.

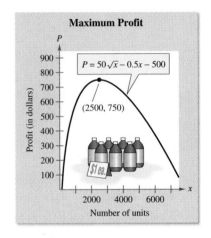

**FIGURE 3.39**

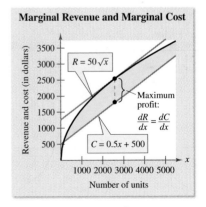

**FIGURE 3.40**

*STUDY TIP*   The following list shows some estimates of elasticities of demand for common products. *(Source: James Kearl, Principles of Economics)*

|                | Absolute Value of Elasticity |
|----------------|:----------:|
| *Item*         |            |
| Cottonseed oil | 6.92       |
| Tomatoes       | 4.60       |
| Restaurant meals | 1.63     |
| Automobiles    | 1.35       |
| Cable TV       | 1.20       |
| Beer           | 1.13       |
| Housing        | 1.00       |
| Movies         | 0.87       |
| Clothing       | 0.60       |
| Cigarettes     | 0.51       |
| Coffee         | 0.25       |
| Gasoline       | 0.15       |
| Newspapers     | 0.10       |
| Mail           | 0.05       |

Which of these items are elastic? Which are inelastic?

# Price Elasticity of Demand

One way economists measure the responsiveness of consumers to a change in the price of a product is with **price elasticity of demand.** For example, a drop in the price of vegetables might result in a much greater demand for vegetables: such a demand is called **elastic.** On the other hand, items such as milk and water are relatively unresponsive to changes in price: the demand for such items is called **inelastic.**

More formally, the elasticity of demand is the percent change of a quantity demanded $x$, divided by the percent change in its price $p$. You can develop a formula for price elasticity of demand using the approximation

$$\frac{\Delta p}{\Delta x} \approx \frac{dp}{dx},$$

which is based on the definition of the derivative. Using this approximation, you can write

$$\text{Price elasticity of demand} = \frac{\text{rate of change in demand}}{\text{rate of change in price}}$$
$$= \frac{\Delta x/x}{\Delta p/p}$$
$$= \frac{p/x}{\Delta p/\Delta x}$$
$$\approx \frac{p/x}{dp/dx}.$$

---

### Definition of Price Elasticity of Demand

If $p = f(x)$ is a differentiable function, then the **price elasticity of demand** is given by

$$\eta = \frac{p/x}{dp/dx},$$

where $\eta$ is the lowercase Greek letter eta. For a given price, the demand is **elastic** if $|\eta| > 1$, the demand is **inelastic** if $|\eta| < 1$, and the demand has **unit elasticity** if $|\eta| = 1$.

---

Price elasticity of demand is related to the total revenue function, as indicated in Figure 3.41 and the following list.

1. If the demand is *elastic,* then a decrease in price is accompanied by an increase in unit sales sufficient to increase the total revenue.
2. If the demand is *inelastic,* then a decrease in price is not accompanied by an increase in unit sales sufficient to increase the total revenue.

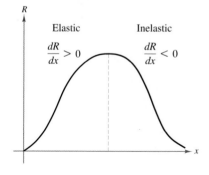

**FIGURE 3.41**   Revenue Curve

## EXAMPLE 5  Comparing Elasticity and Revenue

The demand function for a product is modeled by

$$p = \sqrt{450 - x}, \quad 0 \le x \le 450.$$

(a) Find the intervals on which the demand is elastic, inelastic, and of unit elasticity.

(b) Use the result of part (a) to describe the behavior of the revenue function.

### Solution

(a) The price elasticity of demand is given by

$$\eta = \frac{p/x}{dp/dx} \qquad \text{Formula for price elasticity of demand}$$

$$= \frac{\left( \dfrac{\sqrt{450 - x}}{x} \right)}{\left( \dfrac{-1}{2\sqrt{450 - x}} \right)} \qquad \text{Substitute for } p/x \text{ and } dp/dx.$$

$$= \frac{-900 + 2x}{x} \qquad \text{Multiply numerator and denominator by } -2\sqrt{450 - x}.$$

$$= -\frac{900}{x} + 2. \qquad \text{Rewrite as two fractions and simplify.}$$

The demand is of unit elasticity when $|\eta| = 1$. In the interval $[0, 450]$, the only solution of the equation

$$|\eta| = \left| -\frac{900}{x} + 2 \right| = 1 \qquad \text{Unit elasticity}$$

is $x = 300$. Thus, the demand is of unit elasticity when $x = 300$. For $x$-values in the interval $(0, 300)$,

$$|\eta| = \left| -\frac{900}{x} + 2 \right| > 1, \quad 0 < x < 300, \qquad \text{Elastic}$$

which implies that the demand is elastic when $0 < x < 300$. For $x$-values in the interval $(300, 450)$,

$$|\eta| = \left| -\frac{900}{x} + 2 \right| < 1, \quad 300 < x < 450, \qquad \text{Inelastic}$$

which implies that the demand is inelastic when $300 < x < 450$.

(b) From part (a), you can conclude that the revenue function $R$ is increasing on the open interval $(0, 300)$, is decreasing on the open interval $(300, 450)$, and is a maximum when $x = 300$, as indicated in Figure 3.42.

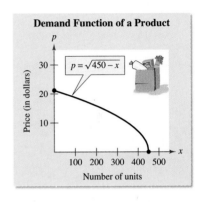

Demand Function of a Product

(a)

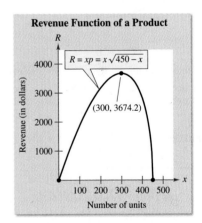

Revenue Function of a Product

(b)

**FIGURE 3.42**

---

*STUDY TIP*  In our discussion of price elasticity of demand, the price is assumed to decrease as the quantity demanded increases. Thus, the demand function $p = f(x)$ is decreasing and $dp/dx$ is negative.

## Business Terms and Formulas

We conclude with a summary of the basic business terms and formulas used in this section. A summary of the graphs of the demand, revenue, cost, and profit functions is shown in Figure 3.43.

*Summary of Business Terms and Formulas*

| | |
|---|---|
| $x$ = number of units produced (or sold) | $\eta$ = price elasticity of demand |
| $p$ = price per unit | $\quad = (p/x)/(dp/dx)$ |
| $R$ = total revenue from selling $x$ units $= xp$ | $dR/dx$ = marginal revenue |
| $C$ = total cost of producing $x$ units | $dC/dx$ = marginal cost |
| $P$ = total profit from selling $x$ units $= R - C$ | $dP/dx$ = marginal profit |
| $\overline{C}$ = average cost per unit $= \dfrac{C}{x}$ | |

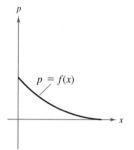

**Demand function**

Quantity demanded increases as price decreases.

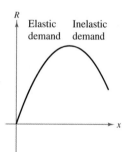

**Revenue function**

The low prices required to sell more units eventually result in a decreasing revenue.

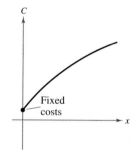

**Cost function**

The total cost to produce $x$ units includes the fixed cost.

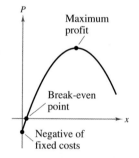

**Profit function**

The break-even point occurs when $R = C$.

**FIGURE 3.43**

---

*Group Discussion*      **Demand Function**

Throughout this text, we assume that demand functions are decreasing. Can you think of a product that has an increasing demand function? That is, can you think of a product that becomes more in demand as its price increases? Explain your reasoning, and sketch a graph of the function.

*Warm Up*   The following warm-up exercises involve skills that were covered in earlier sections. You will use these skills in the exercise set for this section.

In Exercises 1–4, evaluate the expression with $x = 150$.

**1.** $\left| -\dfrac{300}{x} + 3 \right|$

**2.** $\left| -\dfrac{600}{5x} + 2 \right|$

**3.** $\left| \dfrac{(20x^{-1/2})/x}{-10x^{-3/2}} \right|$

**4.** $\left| \dfrac{(4000/x^2)/x}{-8000x^{-3}} \right|$

In Exercises 5–10, find the marginal revenue, cost, or profit.

**5.** $C = 650 + 1.2x + 0.003x^2$

**6.** $P = 0.01x^2 + 11x$

**7.** $R = 14x - \dfrac{x^2}{2000}$

**8.** $R = 3.4x - \dfrac{x^2}{1500}$

**9.** $P = -0.7x^2 + 7x - 50$

**10.** $C = 1700 + 4.2x + 0.001x^3$

# EXERCISES 3.5

In Exercises 1–4, find the number of units $x$ that produces a maximum revenue $R$.

**1.** $R = 800x - 0.2x^2$

**2.** $R = 48x^2 - 0.02x^3$

**3.** $R = 400x - x^2$

**4.** $R = 30x^{2/3} - 2x$

In Exercises 5–8, find the number of units $x$ that produces the minimum average cost per unit $\overline{C}$.

**5.** $C = 1.25x^2 + 25x + 8000$

**6.** $C = 0.001x^3 - 5x + 250$

**7.** $C = 2x^2 + 255x + 5000$

**8.** $C = 0.02x^3 + 55x^2 + 1250$

In Exercises 9–12, find the price per unit $p$ that produces the maximum profit $P$.

| Cost Function | Demand Function |
|---|---|
| **9.** $C = 100 + 30x$ | $p = 90 - x$ |
| **10.** $C = 2400x + 5200$ | $p = 6000 - 0.4x^2$ |
| **11.** $C = 8000 + 50x + 0.03x^2$ | $p = 70 - 0.001x$ |
| **12.** $C = 35x + 500$ | $p = 50 - 0.1\sqrt{x}$ |

*Average Cost*   In Exercises 13 and 14, use the cost function to find the production level for which the average cost is a minimum. For this production level, show that the marginal cost and average cost are equal. Use a graphing utility to graph the cost function and verify your results.

**13.** $C = 2x^2 + 5x + 18$

**14.** $C = x^3 - 6x^2 + 13x$

**15.** *Profit*   A commodity has a demand function modeled by $p = 100 - 0.5x^2$ and a total cost function modeled by $C = 40x + 37.5$.

(a) What price yields a maximum profit?

(b) When the profit is maximized, what is the average cost per unit?

**16.** *Profit*   How would the answer to Exercise 15 change if the marginal cost rose from \$40 per unit to \$50 per unit? In other words, rework Exercise 15 using the cost function $C = 50x + 37.5$.

*Profit*   In Exercises 17 and 18, find the amount $s$ of advertising that maximizes the profit $P$. ($s$ and $P$ are measured in thousands of dollars.) Find the point of diminishing returns.

**17.** $P = -2s^3 + 35s^2 - 100s + 200$

**18.** $P = -0.1s^3 + 6s^2 + 400$

**19.** *Profit*   The cost per unit to produce a type of radio is $60. The manufacturer charges $90 per unit for orders of 100 or less. To encourage large orders, however, the manufacturer reduces the charge by $0.10 per radio for each order in excess of 100 units. For instance, an order of 101 radios would be $89.90 per radio, an order of 102 radios would be $89.80 per radio, and so on. Find the largest order the manufacturer should allow to obtain a maximum profit.

**20.** *Profit*   A real estate office handles 50 apartment units. When the rent is $540 per month, all units are occupied. For each $30 increase in rent, however, an average of one unit becomes vacant. Each occupied unit requires an average of $36 per month for service and repairs. What rent should be charged to obtain a maximum profit?

**21.** *Revenue*   When a wholesaler sold a certain product at $40 per unit, sales were 300 units each week. After a price increase of $5, however, the average number of units sold dropped to 275 each week. Assuming that the demand function is linear, what price per unit will yield a maximum total revenue?

**22.** *Profit*   Assume that the amount of money deposited in a bank is proportional to the square of the interest rate the bank pays on the money. Furthermore, the bank can reinvest the money at 12% simple interest. Find the interest rate the bank should pay to maximize its profit.

**23.** *Cost*   A power station is on one side of a river that is 0.5 mile wide, and a factory is 6 miles downstream on the other side of the river (see figure). It costs $6 per foot to run overland power lines and $8 per foot to run underwater power lines. Write a cost function for running the power lines from the power station to the factory. Use a graphing utility to graph your function. Estimate the value of $x$ that minimizes the cost. Explain your results.

**24.** *Cost*   An offshore oil well is 1 mile off the coast. The oil refinery is 2 miles down the coast. Laying pipe in the ocean is twice as expensive as laying it on land. Find the most economical path for the pipe from the well to the oil refinery.

**25.** *Cost*   A small business uses a minivan to make deliveries. The cost per hour for fuel is $C = v^2/600$, where $v$ is the speed of the minivan (in miles per hour). The driver is paid $10 per hour. Find the speed that minimizes the cost of a 110-mile trip. (Assume there are no costs other than fuel and wages.)

**26.** *Cost*   Repeat Exercise 25 for a fuel cost per hour of
$$C = \frac{v^2 + 360}{720}$$
and a wage of $8 per hour.

*Elasticity*   In Exercises 27–32, find the price elasticity of demand for the demand function at the indicated $x$-value. Is the demand elastic, inelastic, or of unit elasticity at the indicated $x$-value? Use a graphing utility to graph the revenue function, and identify the intervals of elasticity and inelasticity.

| Demand Function | Quantity Demanded |
|---|---|
| **27.** $p = 400 - 3x$ | $x = 20$ |
| **28.** $p = 5 - 0.03x$ | $x = 100$ |
| **29.** $p = 300 - 0.2x^2$ | $x = 30$ |
| **30.** $p = \dfrac{500}{x + 2}$ | $x = 23$ |
| **31.** $p = \dfrac{100}{x^2} + 2$ | $x = 10$ |
| **32.** $p = 100 - \sqrt{0.2x}$ | $x = 125$ |

**33.** *Elasticity*   The demand function for a product is
$$x = 20 - 2p^2.$$
(a) Consider a price of $2. If the price increases by 5%, what is the corresponding percent change in the quantity demanded?

(b) Average elasticity of demand is defined to be the percent change in quantity divided by the percent change in price. Use the percent of part (a) to find the average elasticity over the interval $[2, 2.1]$.

(c) Find the elasticity for a price of $2 and compare the result with that of part (b).

(d) Find an expression for the total revenue and find the values of $x$ and $p$ that maximize the total revenue.

**34.** *Elasticity*   The demand function for a product is
$$p^3 + x^3 = 9.$$
(a) Find the price elasticity of demand when $x = 2$.

(b) Find the values of $x$ and $p$ that maximize the total revenue.

(c) For the value of $x$ found in part (b), show that the price elasticity of demand has unit elasticity.

**35.** *Elasticity*   The demand function for a product is

$$p = (16 - x)^{1/2}, \quad 0 < x < 16.$$

(a) Find the price elasticity of demand when $x = 9$.

(b) Find the values of $x$ and $p$ that maximize the total revenue.

(c) For the value of $x$ found in part (b), show that the price elasticity of demand has unit elasticity.

**36.** *Revenue*   The demand for a car wash is

$$x = 600 - 50p,$$

where the current price is $5.00. Can revenue be increased by lowering the price and thus attracting more customers? Use price elasticity of demand to determine your answer.

**37.** *Revenue*   Repeat Exercise 36 with a demand function of

$$x = 800 - 40p.$$

**38.** A demand function is modeled by $x = a/p^m$, where $a$ is a constant and $m > 1$. Show that $\eta = -m$. In other words, show that a 1% increase in price results in an $m\%$ decrease in th quantity demanded.

**39.** *Automobile Revenues*   The annual revenue (in millions of dollars per year) for Chrysler Corporation for the years 1988–1996 can be modeled by

$$R = -146.38t^3 + 6014.19t^2 - 75{,}778.84t + 333{,}246.23,$$

where $t = 8$ corresponds to 1988. *(Source: Chrysler Corporation)*

(a) During which year, between 1988 and 1996, was Chrysler's revenue the least?

(b) During which year was the revenue the greatest?

(c) Find the revenue for each year in parts (a) and (b).

(d) Use a graphing utility to graph the revenue function. Then use the zoom and trace features to confirm the results in parts (a), (b), and (c).

**40.** *Chocolate Sales*   The revenue (in millions of dollars per year) for Hershey Foods for the years 1990–1996 can be modeled by

$$R = 6.59t^4 - 76.46t^3 + 258.22t^2 - 8.71t + 2716.3,$$

where $t = 0$ corresponds to 1990.   *(Source: Hershey Foods)*

(a) During which year, between 1990 and 1996, was Hershey's revenue increasing most rapidly?

(b) During which year was the revenue increasing at the slowest rate?

(c) Find the rate of increase or decrease for each year in parts (a) and (b).

(d) Use a graphing utility to graph the revenue function. Then use the zoom and trace features to confirm the results in parts (a), (b), and (c).

**41.** Match each graph with the function it best represents— a demand function, a revenue function, a cost function, or a profit function. Explain your reasoning. [The graphs are labeled (a)–(d).]

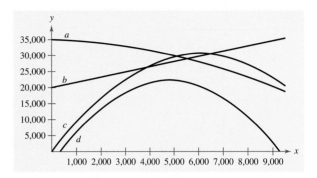

## Business Capsule

University of Michigan Economics Club

*The economics club at the University of Michigan at Flint marketed "Economist Greats" trading cards, featuring statistics on 29 famous economists, such as Milton Friedman. Profits from the cards were used to finance trips to various financial institutions.*

**42.** *Research Project*   Choose an innovative product like the one described above. Use your school's library or some other reference source to research the history of the product. Collect data about the revenue the product has generated, and find a mathematical model for the data. Summarize your findings.

# 3.6 Asymptotes

*Vertical Asymptotes and Infinite Limits  •  Horizontal Asymptotes and Limits at Infinity  •*
*Applications of Asymptotes*

## Vertical Asymptotes and Infinite Limits

In the first three sections of this chapter, you studied ways in which you can use calculus to help analyze the graph of a function. In this section, you will study another valuable aid to curve sketching—the determination of vertical and horizontal asymptotes.

Recall from Section 1.5, Example 10, that the function

$$f(x) = \frac{3}{x - 2}$$

is unbounded as $x$ approaches 2 (see Figure 3.44). This type of behavior is described by saying that the line $x = 2$ is a **vertical asymptote** of the graph of $f$. The type of limit in which $f(x)$ approaches infinity (or negative infinity) as $x$ approaches $c$ from the left or from the right is an **infinite limit.** The infinite limits for the function $f(x) = 3/(x - 2)$ can be written as

$$\lim_{x \to 2^-} \frac{3}{x - 2} = -\infty \quad \text{and} \quad \lim_{x \to 2^+} \frac{3}{x - 2} = \infty.$$

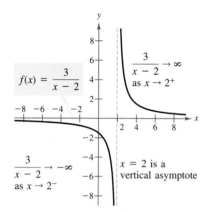

$$f(x) = \frac{3}{x - 2}$$

$\dfrac{3}{x - 2} \to \infty$ as $x \to 2^+$

$\dfrac{3}{x - 2} \to -\infty$ as $x \to 2^-$

$x = 2$ is a vertical asymptote

**FIGURE 3.44**

### Definition of Vertical Asymptote

If $f(x)$ approaches infinity (or negative infinity) as $x$ approaches $c$ from the right or from the left, then the line $x = c$ is a **vertical asymptote** of the graph of $f$.

## Technology

When you use a graphing utility to graph a function that has a vertical asymptote, the utility may try to connect separate branches of the graph. For instance, the figure at the right shows the graph of $f(x) = 3/(x - 2)$ on a graphing calculator.

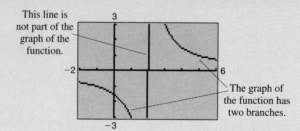

This line is not part of the graph of the function.

The graph of the function has two branches.

One of the most common instances of a vertical asymptote is the graph of a *rational function*—that is, a function of the form $f(x) = p(x)/q(x)$, where $p(x)$ and $q(x)$ are polynomials. If $c$ is a real number such that $q(c) = 0$ and $p(c) \neq 0$, the graph of $f$ has a vertical asymptote at $x = c$. Example 1 shows four cases.

## EXAMPLE 1   Finding Infinite Limits

*Limit from the Left*        *Limit from the Right*

(a) $\displaystyle\lim_{x \to 1^-} \frac{1}{x - 1} = -\infty$    $\displaystyle\lim_{x \to 1^+} \frac{1}{x - 1} = \infty$    (See Figure 3.45a.)

(b) $\displaystyle\lim_{x \to 1^-} \frac{-1}{x - 1} = \infty$    $\displaystyle\lim_{x \to 1^+} \frac{-1}{x - 1} = -\infty$    (See Figure 3.45b.)

(c) $\displaystyle\lim_{x \to 1^-} \frac{-1}{(x - 1)^2} = -\infty$    $\displaystyle\lim_{x \to 1^+} \frac{-1}{(x - 1)^2} = -\infty$    (See Figure 3.45c.)

(d) $\displaystyle\lim_{x \to 1^-} \frac{1}{(x - 1)^2} = \infty$    $\displaystyle\lim_{x \to 1^+} \frac{1}{(x - 1)^2} = \infty$    (See Figure 3.45d.)

*STUDY TIP*    Use a spreadsheet or table to convince yourself of the results shown in Example 1. For instance, in part (a), notice that the values of $f(x) = 1/(x - 1)$ decrease and increase without bound as $x$ gets closer and closer to 1 from the left and the right.

*x* Approaches 1 from the Left

| $x$ | $f(x) = 1/(x - 1)$ |
|---|---|
| 0 | $-1$ |
| 0.9 | $-10$ |
| 0.99 | $-100$ |
| 0.999 | $-1000$ |
| 0.9999 | $-10,000$ |

*x* Approaches 1 from the Right

| $x$ | $f(x) = 1/(x - 1)$ |
|---|---|
| 2 | 1 |
| 1.1 | 10 |
| 1.01 | 100 |
| 1.001 | 1000 |
| 1.0001 | 10,000 |

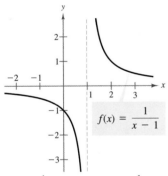

$\displaystyle\lim_{x \to 1^-} \frac{1}{x - 1} = -\infty$  $\displaystyle\lim_{x \to 1^+} \frac{1}{x - 1} = \infty$

(a)

$\displaystyle\lim_{x \to 1^-} \frac{-1}{x - 1} = \infty$  $\displaystyle\lim_{x \to 1^+} \frac{-1}{x - 1} = -\infty$

(b)

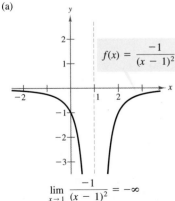

$\displaystyle\lim_{x \to 1} \frac{-1}{(x - 1)^2} = -\infty$

(c)

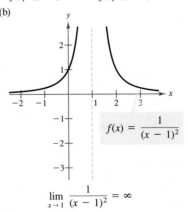

$\displaystyle\lim_{x \to 1} \frac{1}{(x - 1)^2} = \infty$

(d)

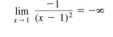

**FIGURE 3.45**

Each of the graphs in Example 1 has only one vertical asymptote. As shown in the next example, the graph of a rational function can have more than one vertical asymptote.

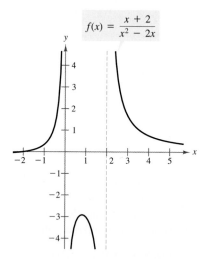

$$f(x) = \frac{x + 2}{x^2 - 2x}$$

**FIGURE 3.46** Vertical Asymptotes at $x = 0$ and $x = 2$

## EXAMPLE 2  Finding Vertical Asymptotes

Determine the vertical asymptotes of the graph of

$$f(x) = \frac{x + 2}{x^2 - 2x}.$$

*Solution*

The possible vertical asymptotes correspond to the $x$-values for which the denominator is zero.

$$x^2 - 2x = 0 \qquad \text{Set denominator equal to 0.}$$
$$x(x - 2) = 0 \qquad \text{Factor.}$$
$$x = 0, 2 \qquad \text{Zeros of denominator}$$

Because the numerator of $f(x)$ is not zero at either of these $x$-values, you can conclude that the graph of $f$ has two vertical asymptotes—one at $x = 0$ and one at $x = 2$, as shown in Figure 3.46.

## EXAMPLE 3  Finding a Vertical Asymptote

Determine the vertical asymptotes of the graph of

$$f(x) = \frac{x^2 + 2x - 8}{x^2 - 4}.$$

*Solution*

First factor the numerator and denominator. Then cancel common factors.

$$f(x) = \frac{x^2 + 2x - 8}{x^2 - 4} \qquad \text{Original function}$$

$$= \frac{(x + 4)(x - 2)}{(x + 2)(x - 2)} \qquad \text{Factor numerator and denominator.}$$

$$= \frac{(x + 4)(x - 2)}{(x + 2)(x - 2)} \qquad \text{Cancel common factors.}$$

$$= \frac{x + 4}{x + 2}, \quad x \neq 2 \qquad \text{Simplify.}$$

For all values of $x$ other than $x = 2$, the graph of this simplified function is the same as the graph of $f$. Thus, you can conclude that the graph of $f$ has only one vertical asymptote. This occurs at $x = -2$, as shown in Figure 3.47.

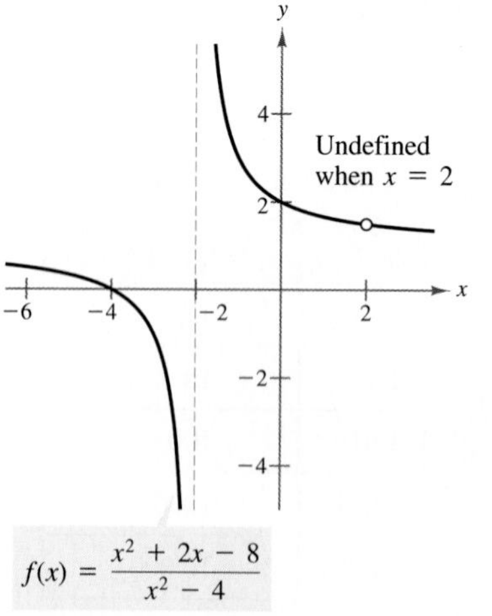

Undefined when $x = 2$

$$f(x) = \frac{x^2 + 2x - 8}{x^2 - 4}$$

**FIGURE 3.47** Vertical Asymptote at $x = -2$

From Example 3, you know that the graph of

$$f(x) = \frac{x^2 + 2x - 8}{x^2 - 4}$$

has a vertical asymptote at $x = -2$. This implies that the limit of $f(x)$ as $x \to -2$ from the right (or from the left) is either $\infty$ or $-\infty$. But without looking at the graph, how can you determine that the limit from the left is *negative* infinity and the limit from the right is *positive* infinity? That is, why is

$$\lim_{x \to 2^-} \frac{x^2 + 2x - 8}{x^2 - 4} = \cancel{\infty} \qquad \text{Limit from the left}$$

and

$$\lim_{x \to 2^+} \frac{x^2 + 2x - 8}{x^2 - 4} = \cancel{\infty}? \qquad \text{Limit from the right}$$

It is cumbersome to determine these limits analytically, and you may find the graphical method shown in Example 4 to be more efficient.

## EXAMPLE 4  Determining Infinite Limits

Find the limits.

$$\lim_{x \to 1^-} \frac{x^2 - 3x}{x - 1} \qquad \text{and} \qquad \lim_{x \to 1^+} \frac{x^2 - 3x}{x - 1}$$

### Solution

Begin by considering the function

$$f(x) = \frac{x^2 - 3x}{x - 1}.$$

Because the denominator is zero when $x = 1$ and the numerator is not zero when $x = 1$, it follows that the graph of the function has a vertical asymptote at $x = 1$. This implies that each of the given limits is either $\infty$ or $-\infty$. To determine which, use a graphing utility to graph the function, as shown in Figure 3.48. From the graph, you can see that the limit from the left is positive infinity and the limit from the right is negative infinity. That is,

$$\lim_{x \to 1^-} \frac{x^2 - 3x}{x - 1} = \infty \qquad \text{Limit from the left}$$

and

$$\lim_{x \to 1^+} \frac{x^2 - 3x}{x - 1} = -\infty. \qquad \text{Limit from the right}$$

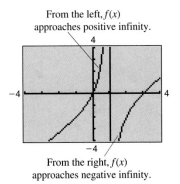

From the left, $f(x)$ approaches positive infinity.

From the right, $f(x)$ approaches negative infinity.

**FIGURE 3.48**

***STUDY TIP*** In Example 4, try evaluating $f(x)$ at $x$-values that are just barely to the left of 1. You will find that you can make the values of $f(x)$ arbitrarily large by choosing $x$ sufficiently close to 1. For instance, $f(0.99999) = 199{,}999$.

## Horizontal Asymptotes and Limits at Infinity

Another type of limit, called a **limit at infinity,** specifies a finite value approached by a function as $x$ increases (or decreases) without bound.

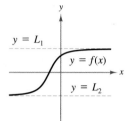

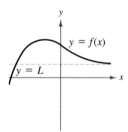

**FIGURE 3.49**

---

### Definition of Horizontal Asymptote

If $f$ is a function and $L_1$ and $L_2$ are real numbers, the statements

$$\lim_{x \to \infty} f(x) = L_1 \quad \text{and} \quad \lim_{x \to -\infty} f(x) = L_2$$

denote **limits at infinity.** The lines $y = L_1$ and $y = L_2$ are **horizontal asymptotes** of the graph of $f$.

---

Figure 3.49 shows two ways in which the graph of a function can approach one or more horizontal asymptotes. Note that it is possible for the graph of a function to cross its horizontal asymptote.

Limits at infinity share many of the properties of limits discussed in Section 1.5. When finding horizontal asymptotes, you can use the property that

$$\lim_{x \to \infty} \frac{1}{x^r} = 0, \quad r > 0 \qquad \text{and} \qquad \lim_{x \to -\infty} \frac{1}{x^r} = 0, \quad r > 0.$$

(The second limit assumes that $x^r$ is defined when $x < 0$.)

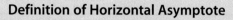

### EXAMPLE 5   Finding Limits at Infinity

Find the limit: $\displaystyle \lim_{x \to \infty} \left( 5 - \frac{2}{x^2} \right)$.

**Solution**

$$
\begin{aligned}
\lim_{x \to \infty} \left( 5 - \frac{2}{x^2} \right) &= \lim_{x \to \infty} 5 - \lim_{x \to \infty} \frac{2}{x^2} && \lim_{x \to \infty} [f(x) - g(x)] = \lim_{x \to \infty} f(x) - \lim_{x \to \infty} g(x) \\
&= \lim_{x \to \infty} 5 - 2 \left( \lim_{x \to \infty} \frac{1}{x^2} \right) && \lim_{x \to \infty} cf(x) = c \lim_{x \to \infty} f(x) \\
&= 5 - 2(0) \\
&= 5
\end{aligned}
$$

You can verify this limit by sketching the graph of

$$f(x) = 5 - \frac{2}{x^2},$$

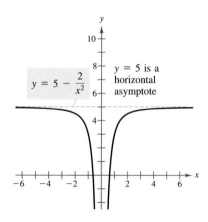

**FIGURE 3.50**

as shown in Figure 3.50. Note that the graph has $y = 5$ as a horizontal asymptote to the right. By evaluating the limit of $f(x)$ as $x \to -\infty$, you can show that this line is also a horizontal asymptote to the left.

There is an easy way to determine whether the graph of a *rational* function has a horizontal asymptote. This shortcut is based on a comparison of the degrees of the numerator and denominator of the rational function.

 **Technology**

Some functions have two horizontal asymptotes: one to the right and one to the left. For instance, try sketching the graph of

$$f(x) = \frac{x}{\sqrt{x^2 + 1}}.$$

What horizontal asymptotes does the function appear to have?

---

### Horizontal Asymptotes of Rational Functions

Let $f(x) = p(x)/q(x)$ be a rational function.

1. If the degree of the numerator is less than the degree of the denominator, then $y = 0$ is a horizontal asymptote of the graph of $f$ (to the left and to the right).

2. If the degree of the numerator is equal to the degree of the denominator, then $y = a/b$ is a horizontal asymptote of the graph of $f$ (to the left and to the right), where $a$ and $b$ are the leading coefficients of $p(x)$ and $q(x)$, respectively.

3. If the degree of the numerator is greater than the degree of the denominator, then the graph of $f$ has no horizontal asymptote.

---

## EXAMPLE 6   Finding Horizontal Asymptotes

Find the horizontal asymptotes of the graphs of the functions.

(a) $y = \dfrac{-2x + 3}{3x^2 + 1}$    (b) $y = \dfrac{-2x^2 + 3}{3x^2 + 1}$    (c) $y = \dfrac{-2x^3 + 3}{3x^2 + 1}$

### Solution

(a) Because the degree of the numerator is less than the degree of the denominator, $y = 0$ is a horizontal asymptote. (See Figure 3.51a.)

(b) Because the degree of the numerator is equal to the degree of the denominator, the line $y = -\frac{2}{3}$ is a horizontal asymptote. (See Figure 3.51b.)

(c) Because the degree of the numerator is greater than the degree of the denominator, the graph has no horizontal asymptote. (See Figure 3.51c.)

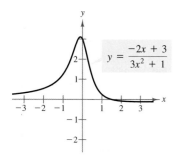

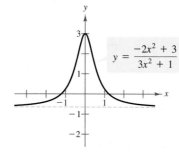

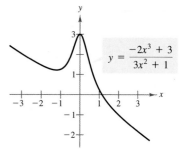

(a) $y = 0$ is a horizontal asymptote.    (b) $y = -\frac{2}{3}$ is a horizontal asymptote.    (c) No horizontal asymptote.

**FIGURE 3.51**

## Applications of Asymptotes

There are many examples of asymptotic behavior in real life. For instance, Example 7 describes the asymptotic behavior of an average cost function.

### EXAMPLE 7   Modeling Average Cost

A small business invests $5000 in a new product. In addition to this initial investment, the product will cost $0.50 per unit to produce. Find the average cost per unit if 1000 units are produced, if 10,000 units are produced, and if 100,000 units are produced. What is the limit of the average cost as the number of units produced increases?

#### Solution

From the given information, you can model the total cost $C$ (in dollars) by

$$C = 0.5x + 5000, \qquad \text{Total cost function}$$

where $x$ is the number of units produced. This implies that the average cost function is

$$\overline{C} = \frac{C}{x} = 0.5 + \frac{5000}{x}. \qquad \text{Average cost function}$$

If only 1000 units are produced, then the average cost per unit is

$$\overline{C} = 0.5 + \frac{5000}{1000} = \$5.50. \qquad \text{Average cost for 1000 units}$$

If 10,000 units are produced, then the average cost per unit is

$$\overline{C} = 0.5 + \frac{5000}{10,000} = \$1.00. \qquad \text{Average cost for 10,000 units}$$

If 100,000 units are produced, then the average cost per unit is

$$\overline{C} = 0.5 + \frac{5000}{100,000} = \$0.55. \qquad \text{Average cost for 100,000 units}$$

As $x$ approaches infinity, the limiting average cost per unit is

$$\lim_{x \to \infty} \left( 0.5 + \frac{5000}{x} \right) = \$0.50.$$

As shown in Figure 3.52, this example points out one of the major problems of small businesses. That is, it is difficult to have competitively low prices when the production level is low.

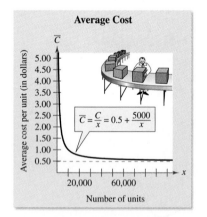

**Average Cost**

$$\overline{C} = \frac{C}{x} = 0.5 + \frac{5000}{x}$$

**FIGURE 3.52**   As $x \to \infty$, the average cost per unit approaches $0.50.

**STUDY TIP**   In Example 7, suppose that the small business had made an initial investment of $50,000. How would this change the answers to the questions? Would it change the average cost of producing $x$ units? Would it change the limiting average cost per unit?

# EXAMPLE 8  Modeling Smokestack Emission

A manufacturing plant has determined that the cost $C$ (in dollars) of removing $p\%$ of the smokestack pollutants of its main smokestack is modeled by

$$C = \frac{80,000p}{100 - p}, \qquad 0 \le p < 100.$$

What is the vertical asymptote of this function? What does the vertical asymptote mean to the plant owners?

### Solution

The graph of the cost function is shown in Figure 3.53. From the graph, you can see that $p = 100$ is the vertical asymptote. This means that as the plant attempts to remove higher and higher percents of the pollutants, the cost increases dramatically. For instance, the cost of removing 85% of the pollutants is

$$C = \frac{80,000(85)}{100 - 85} \approx \$453,333 \qquad \text{Cost for 85\% removal}$$

but the cost of removing 90% is

$$C = \frac{80,000(90)}{100 - 90} \approx \$720,000. \qquad \text{Cost for 90\% removal}$$

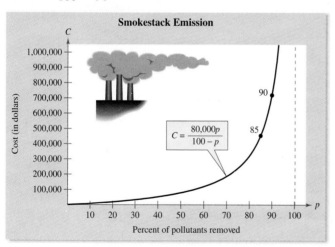

**Smokestack Emission**

$$C = \frac{80,000p}{100 - p}$$

Percent of pollutants removed

**FIGURE 3.53**

Rob Crandall/Rainbow

*During the 1980s and 1990s, industries in the United States spent billions of dollars to reduce air pollution. For instance, the estimated cost of air pollution control in 1993 was 26.6 billion dollars.* (*Source: U.S. Bureau of the Census, Current Industrial Reports*)

## *Group Discussion*     *Indirect Costs*

In Example 8, the given cost model considers only the direct cost for the manufacturing plant. Describe the possible indirect costs incurred by society as the plant attempts to remove more and more of the pollutants from its smokestack emission.

*Warm Up*

The following warm-up exercises involve skills that were covered in earlier sections. You will use these skills in the exercise set for this section.

In Exercises 1–6, find the limit.

**1.** $\lim_{x \to 2} (x + 1)$

**2.** $\lim_{x \to -1} (3x + 4)$

**3.** $\lim_{x \to -3} \dfrac{2x^2 + x - 15}{x + 3}$

**4.** $\lim_{x \to 2} \dfrac{3x^2 - 8x + 4}{x - 2}$

**5.** $\lim_{x \to 0^+} \sqrt{x}$

**6.** $\lim_{x \to 1^+} \left( x + \sqrt{x - 1} \right)$

In Exercises 7–10, find the average cost and the marginal cost.

**7.** $C = 150 + 3x$

**8.** $C = 1900 + 1.7x + 0.002x^2$

**9.** $C = 0.005x^2 + 0.5x + 1375$

**10.** $C = 760 + 0.05x$

## EXERCISES 3.6

In Exercises 1–8, find the vertical and horizontal asymptotes.

**1.** $f(x) = \dfrac{x^2 + 1}{x^2}$

**2.** $f(x) = \dfrac{4}{(x - 2)^3}$

**5.** $f(x) = \dfrac{x^3}{x^2 - 1}$

**6.** $f(x) = \dfrac{-4x}{x^2 + 4}$

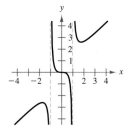

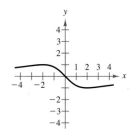

**1.** $f(x) = \dfrac{x^2 + 1}{x^2}$

**2.** $f(x) = \dfrac{4}{(x - 2)^3}$

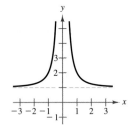

**3.** $f(x) = \dfrac{x^2 - 2}{x^2 - x - 2}$

**4.** $f(x) = \dfrac{2 + x}{1 - x}$

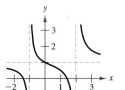

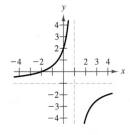

**7.** $f(x) = \dfrac{x^2 - 1}{2x^2 - 8}$

**8.** $f(x) = \dfrac{x^2 + 1}{x^3 - 8}$

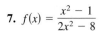

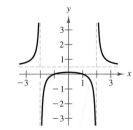

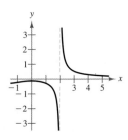

In Exercises 9–14, match the function with its graph. Use horizontal asymptotes as an aid. [The graphs are labeled (a)–(f).]

(a)
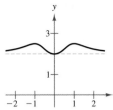

(b)
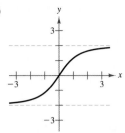

(c)
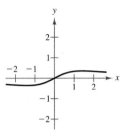

(d)
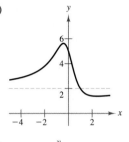

(e)
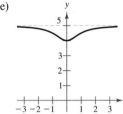

(f)
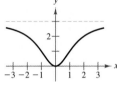

**9.** $f(x) = \dfrac{3x^2}{x^2 + 2}$

**10.** $f(x) = \dfrac{2x}{\sqrt{x^2 + 2}}$

**11.** $f(x) = \dfrac{x}{x^2 + 2}$

**12.** $f(x) = 2 + \dfrac{x^2}{x^4 + 1}$

**13.** $f(x) = 5 - \dfrac{1}{x^2 + 1}$

**14.** $f(x) = \dfrac{2x^2 - 3x + 5}{x^2 + 1}$

In Exercises 15–22, find the limit.

**15.** $\displaystyle\lim_{x \to -2^-} \dfrac{1}{(x + 2)^2}$

**16.** $\displaystyle\lim_{x \to -2^-} \dfrac{1}{x + 2}$

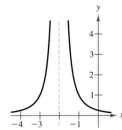

**17.** $\displaystyle\lim_{x \to 3^+} \dfrac{x - 4}{x - 3}$

**18.** $\displaystyle\lim_{x \to 1^+} \dfrac{2 + x}{1 - x}$

**19.** $\displaystyle\lim_{x \to 4^-} \dfrac{x^2}{x^2 - 16}$

**20.** $\displaystyle\lim_{x \to 4} \dfrac{x^2}{x^2 + 16}$

**21.** $\displaystyle\lim_{x \to 0^-} \left(1 + \dfrac{1}{x}\right)$

**22.** $\displaystyle\lim_{x \to 0^-} \left(x^2 - \dfrac{1}{x}\right)$

In Exercises 23–32, find the limit.

**23.** $\displaystyle\lim_{x \to \infty} \dfrac{2x - 1}{3x + 2}$

**24.** $\displaystyle\lim_{x \to \infty} \dfrac{5x^3 + 1}{10x^3 - 3x^2 + 7}$

**25.** $\displaystyle\lim_{x \to \infty} \dfrac{3x}{4x^2 - 1}$

**26.** $\displaystyle\lim_{x \to \infty} \dfrac{2x^{10} - 1}{10x^{11} - 3}$

**27.** $\displaystyle\lim_{x \to -\infty} \dfrac{5x^2}{x + 3}$

**28.** $\displaystyle\lim_{x \to \infty} \dfrac{x^3 - 2x^2 + 3x + 1}{x^2 - 3x + 2}$

**29.** $\displaystyle\lim_{x \to \infty} (2x - x^{-2})$

**30.** $\displaystyle\lim_{x \to \infty} (x + 3)^{-2}$

**31.** $\displaystyle\lim_{x \to -\infty} \left(\dfrac{2x}{x - 1} + \dfrac{3x}{x + 1}\right)$

**32.** $\displaystyle\lim_{x \to \infty} \left(\dfrac{2x^2}{x - 1} + \dfrac{3x}{x + 1}\right)$

In Exercises 33 and 34, complete the table. Then use the result to estimate the limit of $f(x)$ as $x$ approaches infinity.

**33.** $f(x) = \dfrac{x + 1}{x\sqrt{x}}$

| $x$ | $10^0$ | $10^1$ | $10^2$ | $10^3$ | $10^4$ | $10^5$ | $10^6$ |
|---|---|---|---|---|---|---|---|
| $f(x)$ | | | | | | | |

**34.** $f(x) = x^2 - x\sqrt{x(x - 1)}$

| $x$ | $10^0$ | $10^1$ | $10^2$ | $10^3$ | $10^4$ | $10^5$ | $10^6$ |
|---|---|---|---|---|---|---|---|
| $f(x)$ | | | | | | | |

In Exercises 35 and 36, use a graphing utility to complete the table and use the result to estimate the limit of $f(x)$ as $x$ approaches infinity *and* as $x$ approaches negative infinity.

**35.** $f(x) = \dfrac{2x}{\sqrt{x^2 + 4}}$

| $x$ | $-10^6$ | $-10^4$ | $-10^2$ | $10^0$ | $10^2$ | $10^4$ | $10^6$ |
|---|---|---|---|---|---|---|---|
| $f(x)$ | | | | | | | |

**36.** $f(x) = x - \sqrt{x(x - 1)}$

| $x$ | $-10^6$ | $-10^4$ | $-10^2$ | $10^0$ | $10^2$ | $10^4$ | $10^6$ |
|---|---|---|---|---|---|---|---|
| $f(x)$ | | | | | | | |

In Exercises 37–54, sketch the graph of the equation. Use intercepts, extrema, and asymptotes as sketching aids.

**37.** $y = \dfrac{2 + x}{1 - x}$

**38.** $y = \dfrac{x - 3}{x - 2}$

**39.** $f(x) = \dfrac{x^2}{x^2 + 9}$

**40.** $f(x) = \dfrac{x^2}{x^2 - 9}$

**41.** $g(x) = \dfrac{x^2}{x^2 - 16}$

**42.** $g(x) = \dfrac{x}{x^2 - 4}$

**43.** $xy^2 = 4$

**44.** $x^2 y = 4$

**45.** $y = \dfrac{2x}{1 - x}$

**46.** $y = \dfrac{2x}{1 - x^2}$

**47.** $y = 3(1 - x^{-2})$

**48.** $y = 1 + x^{-1}$

**49.** $f(x) = \dfrac{1}{x^2 - x - 2}$

**50.** $f(x) = \dfrac{x - 2}{x^2 - 4x + 3}$

**51.** $g(x) = \dfrac{x^2 - x - 2}{x - 2}$

**52.** $g(x) = \dfrac{x^2 - 9}{x + 3}$

**53.** $y = \dfrac{2x^2 - 6}{(x - 1)^2}$

**54.** $y = \dfrac{x}{(x + 1)^2}$

**55.** *Cost*   The cost $C$ (in dollars) of producing $x$ units of a product is $C = 1.35x + 4570$.

(a) Find $\overline{C}$ when $x = 100$ and when $x = 1000$.

(b) What is the limit of $\overline{C}$ as $x$ approaches infinity?

**56.** *Average Cost*   A business has a cost (in dollars) of $C = 0.5x + 500$ for producing $x$ units. Find the limit of the average cost per unit as $x$ approaches infinity.

**57.** *Cost*   The cost $C$ (in millions of dollars) for the federal government to seize $p\%$ of a type of illegal drug as it enters the country is modeled by

$$C = \dfrac{528p}{100 - p}, \qquad 0 \le p < 100.$$

(a) Find the cost of seizing 25%.

(b) Find the cost of seizing 50%.

(c) Find the cost of seizing 75%.

(d) Find the limit of $C$ as $p \to 100^-$.

**58.** *Cost*   The cost $C$ (in dollars) of removing $p\%$ of the air pollutants in the stack emission of a utility company that burns coal is modeled by

$$C = \dfrac{80{,}000p}{100 - p}, \qquad 0 \le p < 100.$$

(a) Find the cost of removing 15%.

(b) Find the cost of removing 50%.

(c) Find the cost of removing 90%.

(d) Find the limit of $C$ as $p \to 100^-$.

(e) Use a graphing utility to verify the result of part (d).

**59.** *Learning Curve*   Psychologists have developed mathematical models to predict performance $P$ (the percent of correct responses) as a function of $n$, the number of times a task is performed. One such model is

$$P = \dfrac{b + \theta a(n - 1)}{1 + \theta(n - 1)},$$

where $a$, $b$, and $\theta$ are constants that depend on the actual learning situation. Find the limit of $P$ as $n$ approaches infinity.

**60.** *Learning Curve*   Consider the learning curve given by

$$P = \dfrac{0.5 + 0.9(n - 1)}{1 + 0.9(n - 1)}, \qquad 0 < n.$$

(a) Complete the table for the model.

| $n$ | 1 | 2 | 3 | 4 | 5 | 6 | 7 | 8 | 9 | 10 |
|-----|---|---|---|---|---|---|---|---|---|----|
| $P$ |   |   |   |   |   |   |   |   |   |    |

(b) Find the limit as $n$ approaches infinity.

(c) Use a graphing utility to graph this learning curve, and interpret the graph in the context of the problem.

**61.** *Wildlife Management*   The state game commission introduces 30 elk into a new state park. The population $N$ of the herd is modeled by

$$N = \dfrac{10(3 + 4t)}{1 + 0.1t},$$

where $t$ is the time in years.

(a) Find the size of the herd after 5, 10, and 25 years.

(b) According to this model, what is the limiting size of the herd as time progresses?

**62.** *Average Profit*   The cost and revenue functions for a product are

$$C = 34.5x + 15{,}000 \quad \text{and} \quad R = 69.9x.$$

(a) Find the average profit function

$$\overline{P} = \dfrac{R - C}{x}.$$

(b) Find the average profit when $x$ is 1000, 10,000, and 100,000.

(c) What is the limit of the average profit function as $x$ approaches infinity?

**63.** Use a graphing utility to graph

$$f(x) = \dfrac{3x}{\sqrt{4x^2 + 1}}$$

and locate any horizontal asymptotes.

# Curve Sketching: A Summary

## 3.7

*Summary of Curve Sketching Techniques •*
*Summary of Simple Polynomial Graphs*

## Summary of Curve Sketching Techniques

It would be difficult to overstate the importance of using graphs in mathematics. Descarte's introduction of analytic geometry contributed significantly to the rapid advances in calculus that began during the mid-seventeenth century.

So far, you have studied several concepts that are useful in analyzing the graph of a function.

- $x$-intercepts and $y$-intercepts        (Section 1.2)

- Domain and range        (Section 1.4)

- Continuity        (Section 1.6)

- Differentiability        (Section 2.1)

- Relative extrema        (Section 3.2)

- Concavity        (Section 3.3)

- Points of inflection        (Section 3.3)

- Vertical asymptotes        (Section 3.6)

- Horizontal asymptotes        (Section 3.6)

When you are sketching the graph of a function, either by hand or with a graphing utility, remember that you cannot normally show the *entire* graph. The decision as to which part of the graph to show is crucial. For instance, which of the viewing rectangles in Figure 3.54 better represents the graph of

$$f(x) = x^3 - 25x^2 + 74x - 20?$$

The lower viewing rectangle gives a more complete view of the graph, but the context of the problem might indicate that the upper view is better. Here are some guidelines for analyzing the graph of a function.

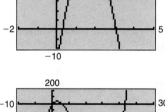

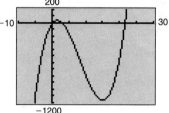

**FIGURE 3.54**

---

### Guidelines for Analyzing the Graph of a Function

1. Determine the domain and range of the function. If the function models a real-life situation, consider the context.

2. Determine the intercepts and asymptotes of the graph.

3. Locate the $x$-values where $f'(x)$ and $f''(x)$ are zero or undefined. Use the results to determine the relative extrema and points of inflection.

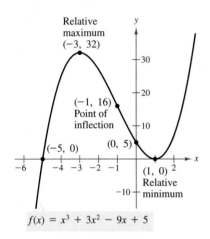

$f(x) = x^3 + 3x^2 - 9x + 5$

**FIGURE 3.55**

## EXAMPLE 1   Analyzing a Graph

Analyze the graph of

$$f(x) = x^3 + 3x^2 - 9x + 5.$$    Original function

**Solution**

Begin by finding the intercepts of the graph. This function factors as

$$f(x) = (x - 1)^2(x + 5).$$    Factored form

Thus, the x-intercepts occur when $x = 1$ and $x = -5$. The derivative is

$$f'(x) = 3x^2 + 6x - 9$$    First derivative

$$= 3(x - 1)(x + 3).$$    Factored form

Thus, the critical numbers of $f$ are $x = 1$ and $x = -3$. The second derivative of $f$ is

$$f''(x) = 6x + 6$$    Second derivative

$$= 6(x + 1),$$    Factored form

which implies that the second derivative is zero when $x = -1$. By testing the values of $f'(x)$ and $f''(x)$, as shown in the table, you can see that $f$ has one relative minimum, one relative maximum, and one point of inflection. The graph of $f$ is shown in Figure 3.55.

|  | $f(x)$ | $f'(x)$ | $f''(x)$ | Shape of graph |
|---|---|---|---|---|
| $x$ in $(-\infty, -3)$ |  | + | − | Increasing, concave downward |
| $x = -3$ | 32 | 0 | − | Relative maximum |
| $x$ in $(-3, -1)$ |  | − | − | Decreasing, concave downward |
| $x = -1$ | 16 | − | 0 | Point of inflection |
| $x$ in $(-1, 1)$ |  | − | + | Decreasing, concave upward |
| $x = 1$ | 0 | 0 | + | Relative minimum |
| $x$ in $(1, \infty)$ |  | + | + | Increasing, concave upward |

## Technology

In Example 1, you are able to find the zeros of $f$, $f'$, and $f''$ algebraically (by factoring). When this is not feasible, you can use a graphing utility to find the zeros. For instance, the function

$$g(x) = x^3 + 3x^2 - 9x + 6$$

is similar to the function in the example, but it doesn't factor with integer coefficients. Using a graphing utility, you can determine that the function has only one x-intercept, $x \approx -5.0275$.

## EXAMPLE 2 Analyzing a Graph

Analyze the graph of

$$f(x) = x^4 - 12x^3 + 48x^2 - 64x. \qquad \text{Original function}$$

### Solution

Begin by finding the intercepts of the graph. This function factors as

$$f(x) = x(x^3 - 12x^2 + 48x - 64)$$
$$= x(x - 4)^3. \qquad \text{Factored form}$$

Thus, the $x$-intercepts occur when $x = 0$ and $x = 4$. The derivative is

$$f'(x) = 4x^3 - 36x^2 + 96x - 64 \qquad \text{First derivative}$$
$$= 4(x - 1)(x - 4)^2. \qquad \text{Factored form}$$

Thus, the critical numbers of $f$ are $x = 1$ and $x = 4$. The second derivative of $f$ is

$$f''(x) = 12x^2 - 72x + 96 \qquad \text{Second derivative}$$
$$= 12(x - 4)(x - 2), \qquad \text{Factored form}$$

which implies that the second derivative is zero when $x = 2$ and $x = 4$. By testing the values of $f'(x)$ and $f''(x)$, as shown in the table, you can see that $f$ has a minimum and two points of inflection. The graph is shown in Figure 3.56.

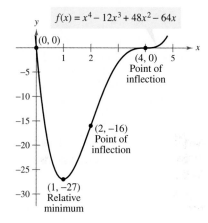

**FIGURE 3.56**

| | $f(x)$ | $f'(x)$ | $f''(x)$ | Shape of graph |
|---|---|---|---|---|
| $x$ in $(-\infty, 1)$ | | $-$ | $+$ | Decreasing, concave upward |
| $x = 1$ | $-27$ | $0$ | $+$ | Relative minimum |
| $x$ in $(1, 2)$ | | $+$ | $+$ | Increasing, concave upward |
| $x = 2$ | $-16$ | $+$ | $0$ | Point of inflection |
| $x$ in $(2, 4)$ | | $+$ | $-$ | Increasing, concave downward |
| $x = 4$ | $0$ | $0$ | $0$ | Point of inflection |
| $x$ in $(4, \infty)$ | | $+$ | $+$ | Increasing, concave upward |

**STUDY TIP**  A polynomial function of degree $n$ can have at most $n - 1$ relative extrema and at most $n - 2$ points of inflection. For instance, the third-degree polynomial in Example 1 has two relative extrema and one point of inflection. Similarly, the fourth-degree polynomial function in Example 2 has one relative extremum and two points of inflection. Is it possible for a third-degree function to have no relative extrema? Is it possible for a fourth-degree function to have no relative extrema?

**Discovery**

Show that the function in Example 3 can be rewritten as

$$f(x) = \frac{x^2 - 2x + 4}{x - 2}$$

$$= x + \frac{4}{x - 2}.$$

Use a graphing utility to graph $f$ together with the line $y = x$. How do the two graphs compare as you zoom out? Describe what is meant by a "slant asymptote." Find the slant asymptote of the function $g(x) = \dfrac{x^2 - x - 1}{x - 1}$.

## EXAMPLE 3    Analyzing a Graph

Analyze the graph of

$$f(x) = \frac{x^2 - 2x + 4}{x - 2}.$$    Original function

**Solution**

The $y$-intercept occurs at $(0, -2)$. Using the Quadratic Formula on the numerator, you can see that there are no $x$-intercepts. Because the denominator is zero when $x = 2$ (and the numerator is not zero when $x = 2$), it follows that $x = 2$ is a vertical asymptote of the graph. There are no horizontal asymptotes because the degree of the numerator is greater than the degree of the denominator. The derivative is

$$f'(x) = \frac{(x - 2)(2x - 2) - (x^2 - 2x + 4)}{(x - 2)^2}$$    First derivative

$$= \frac{x(x - 4)}{(x - 2)^2}.$$    Factored form

Thus, the critical numbers of $f$ are $x = 0$ and $x = 4$. The second derivative is

$$f''(x) = \frac{(x - 2)^2(2x - 4) - (x^2 - 4x)(2)(x - 2)}{(x - 2)^4}$$    Second derivative

$$= \frac{(x - 2)(2x^2 - 8x + 8 - 2x^2 + 8x)}{(x - 2)^4}$$

$$= \frac{8}{(x - 2)^3}.$$    Factored form

Because the second derivative has no zeros and because $x = 2$ is not in the domain of the function, you can conclude that the graph has no points of inflection. By testing the values of $f'(x)$ and $f''(x)$, as shown in the table, you can see that $f$ has one relative minimum and one relative maximum. The graph of $f$ is shown in Figure 3.57.

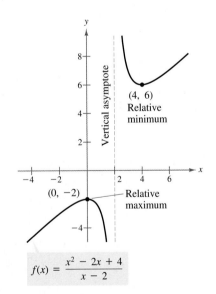

$$f(x) = \frac{x^2 - 2x + 4}{x - 2}$$

**FIGURE 3.57**

|  | $f(x)$ | $f'(x)$ | $f''(x)$ | Shape of graph |
|---|---|---|---|---|
| $x$ in $(-\infty, 0)$ |  | + | − | Increasing, concave downward |
| $x = 0$ | −2 | 0 | − | Relative maximum |
| $x$ in $(0, 2)$ |  | − | − | Decreasing, concave downward |
| $x = 2$ | Undef. | Undef. | Undef. | Vertical asymptote |
| $x$ in $(2, 4)$ |  | − | + | Decreasing, concave upward |
| $x = 4$ | 6 | 0 | + | Relative minimum |
| $x$ in $(4, \infty)$ |  | + | + | Increasing, concave upward |

## EXAMPLE 4  Analyzing a Graph

Analyze the graph of

$$f(x) = \frac{2(x^2 - 9)}{x^2 - 4}.$$  Original function

### *Solution*

Begin by writing the function in factored form.

$$f(x) = \frac{2(x - 3)(x + 3)}{(x - 2)(x + 2)}$$  Factored form

The $y$-intercept is $\left(0, \frac{9}{2}\right)$, and the $x$-intercepts are $(-3, 0)$ and $(3, 0)$. There are vertical asymptotes at $x = \pm 2$ and a horizontal asymptote at $y = 2$. The first derivative is

$$f'(x) = \frac{2[(x^2 - 4)(2x) - (x^2 - 9)(2x)]}{(x^2 - 4)^2}$$  First derivative

$$= \frac{20x}{(x^2 - 4)^2}.$$  Factored form

Thus, the critical number of $f$ is $x = 0$. The second derivative of $f$ is

$$f''(x) = \frac{(x^2 - 4)^2(20) - (20x)(2)(2x)(x^2 - 4)}{(x^2 - 4)^4}$$  Second derivative

$$= \frac{20(x^2 - 4)(x^2 - 4 - 4x^2)}{(x^2 - 4)^4}$$

$$= -\frac{20(3x^2 + 4)}{(x^2 - 4)^3}.$$  Factored form

Because the second derivative has no zeros and $x = \pm 2$ is not in the domain of the function, you can conclude that the graph has no points of inflection. By testing the values of $f'(x)$ and $f''(x)$, as shown in the table, you can see that $f$ has one relative minimum. The graph of $f$ is shown in Figure 3.58.

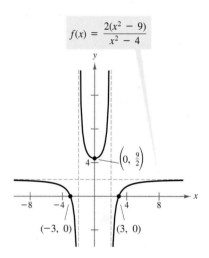

$$f(x) = \frac{2(x^2 - 9)}{x^2 - 4}$$

**FIGURE 3.58**

|  | $f(x)$ | $f'(x)$ | $f''(x)$ | Shape of graph |
|---|---|---|---|---|
| $x$ in $(-\infty, -2)$ |  | $-$ | $-$ | Decreasing, concave downward |
| $x = -2$ | Undef. | Undef. | Undef. | Vertical asymptote |
| $x$ in $(-2, 0)$ |  | $-$ | $+$ | Decreasing, concave upward |
| $x = 0$ | $\frac{9}{2}$ | $0$ | $+$ | Relative minimum |
| $x$ in $(0, 2)$ |  | $+$ | $+$ | Increasing, concave upward |
| $x = 2$ | Undef. | Undef. | Undef. | Vertical asymptote |
| $x$ in $(2, \infty)$ |  | $+$ | $-$ | Increasing, concave downward |

## Technology

Some graphing utilities will not graph the function in Example 5 properly if the function is entered as

$$f(x) = 2x^{(5/3)} - 5x^{(4/3)}.$$

To correct for this, you can enter the function as

$$f(x) = 2(\sqrt[3]{x})^5 - 5(\sqrt[3]{x})^4.$$

Try entering both functions into a graphing utility to see whether both functions produce correct graphs.

### TIP

For help on the algebra in Example 5, see Example 2a in the *Chapter 3 Algebra Review*, on page 249.

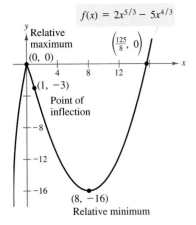

$$f(x) = 2x^{5/3} - 5x^{4/3}$$

**FIGURE 3.59**

## EXAMPLE 5 Analyzing a Graph

Analyze the graph of

$$f(x) = 2x^{5/3} - 5x^{4/3}. \qquad \text{Original function}$$

**Solution**

Begin by writing the function in factored form.

$$f(x) = x^{4/3}(2x^{1/3} - 5) \qquad \text{Factored form}$$

One of the intercepts is $(0, 0)$. A second $x$-intercept occurs when $2x^{1/3} = 5$.

$$2x^{1/3} = 5$$
$$x^{1/3} = \tfrac{5}{2}$$
$$x = \left(\tfrac{5}{2}\right)^3$$
$$x = \tfrac{125}{8}$$

The first derivative is

$$f'(x) = \tfrac{10}{3}x^{2/3} - \tfrac{20}{3}x^{1/3} \qquad \text{First derivative}$$
$$= \tfrac{10}{3}x^{1/3}(x^{1/3} - 2). \qquad \text{Factored form}$$

Thus, the critical numbers of $f$ are $x = 0$ and $x = 8$. The second derivative is

$$f''(x) = \tfrac{20}{9}x^{-1/3} - \tfrac{20}{9}x^{-2/3} \qquad \text{Second derivative}$$
$$= \tfrac{20}{9}x^{-2/3}(x^{1/3} - 1)$$
$$= \frac{20(x^{1/3} - 1)}{9x^{2/3}}. \qquad \text{Factored form}$$

Thus, possible points of inflection occur when $x = 1$ and when $x = 0$. By testing the values of $f'(x)$ and $f''(x)$, as shown in the table, you can see that $f$ has one relative maximum, one relative minimum, and one point of inflection. The graph of $f$ is shown in Figure 3.59.

| | $f(x)$ | $f'(x)$ | $f''(x)$ | Shape of graph |
|---|---|---|---|---|
| $x$ in $(-\infty, 0)$ | | $+$ | $-$ | Increasing, concave downward |
| $x = 0$ | $0$ | $0$ | Undef. | Relative maximum |
| $x$ in $(0, 1)$ | | $-$ | $-$ | Decreasing, concave downward |
| $x = 1$ | $-3$ | $-$ | $0$ | Point of inflection |
| $x$ in $(1, 8)$ | | $-$ | $+$ | Decreasing, concave upward |
| $x = 8$ | $-16$ | $0$ | $+$ | Relative minimum |
| $x$ in $(8, \infty)$ | | $+$ | $+$ | Increasing, concave upward |

## Summary of Simple Polynomial Graphs

A summary of the graphs of polynomial functions of degrees 0, 1, 2, and 3 is shown in Figure 3.60. Because of their simplicity, lower-degree polynomial functions are commonly used as mathematical models.

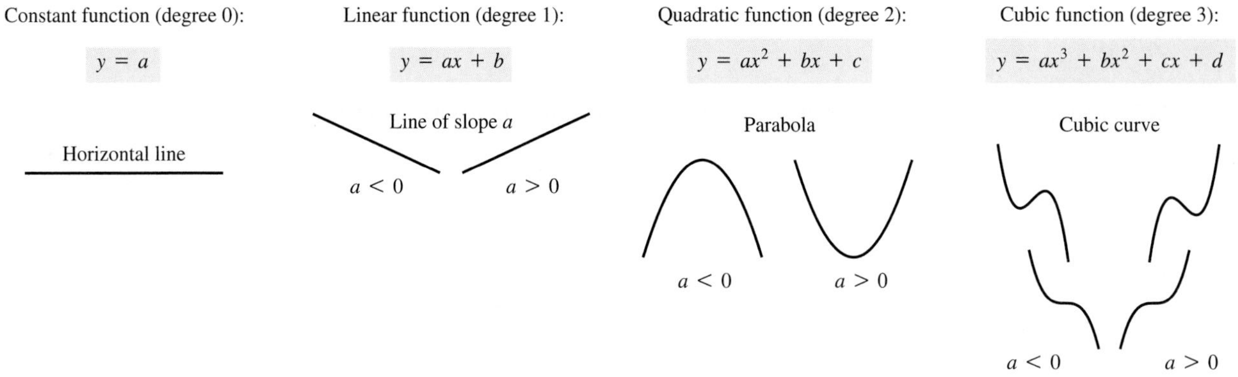

Constant function (degree 0):

$$y = a$$

Horizontal line

Linear function (degree 1):

$$y = ax + b$$

Line of slope $a$

$a < 0$    $a > 0$

Quadratic function (degree 2):

$$y = ax^2 + bx + c$$

Parabola

$a < 0$    $a > 0$

Cubic function (degree 3):

$$y = ax^3 + bx^2 + cx + d$$

Cubic curve

$a < 0$    $a > 0$

**FIGURE 3.60**

*STUDY TIP*  The graph of any cubic polynomial has one point of inflection. The slope of the graph at the point of inflection may be zero or nonzero.

---

## *Group Discussion*      **Graphs of Fourth-Degree Polynomial Functions**

In the summary presented above, the graphs of cubic functions have been classified into four basic types. How many basic types of graphs are possible with fourth-degree polynomial functions? Make a rough sketch of each type. Then use a graphing utility to classify each of the following. (In each case, use a viewing rectangle that shows all the basic characteristics of the graph.)

a. $y = x^4$

b. $y = -x^4 + 5x^2$

c. $y = x^4 - x^3 + x$

d. $y = -x^4 - 4x^3 - 3x^2 + x$

e. $y = -x^4 + 2x^2$

f. $y = -x^4 + x^3$

g. $y = x^4 - 5x$

h. $y = x^4 - 8x^3 + 8x^2$

*Warm Up*

The following warm-up exercises involve skills that were covered in earlier sections. You will use these skills in the exercise set for this section.

In Exercises 1–4, find the vertical and horizontal asymptotes of the graph.

**1.** $f(x) = \dfrac{1}{x^2}$

**2.** $f(x) = \dfrac{8}{(x - 2)^2}$

**3.** $f(x) = \dfrac{40x}{x + 3}$

**4.** $f(x) = \dfrac{x^2 - 3}{x^2 - 4x + 3}$

In Exercises 5–10, determine the open intervals on which the function is increasing or decreasing.

**5.** $f(x) = x^2 + 4x + 2$

**6.** $f(x) = -x^2 - 8x + 1$

**7.** $f(x) = x^3 - 3x + 1$

**8.** $f(x) = \dfrac{-x^3 + x^2 - 1}{x^2}$

**9.** $f(x) = \dfrac{x - 2}{x - 1}$

**10.** $f(x) = -x^3 - 4x^2 + 3x + 2$

# EXERCISES 3.7

In Exercises 1–20, sketch the graph of the function. Choose a scale that allows all relative extrema and points of inflection to be identified on the graph.

**1.** $y = -x^2 - 2x + 3$

**2.** $y = 2x^2 - 4x + 1$

**3.** $y = x^3 - 4x^2 + 6$

**4.** $y = -\frac{1}{3}(x^3 - 3x + 2)$

**5.** $y = 2 - x - x^3$

**6.** $y = x^3 + 3x^2 + 3x + 2$

**7.** $y = 3x^3 - 9x + 1$

**8.** $y = \frac{1}{3}(x - 1)^2 + 2$

**9.** $y = 3x^4 + 4x^3$

**10.** $y = 3x^4 - 6x^2$

**11.** $y = (x + 1)(x - 2)(x - 5)$

**12.** $y = -x^3 + 3x^2 + 9x - 2$

**13.** $y = x^4 - 8x^3 + 18x^2 - 16x + 5$

**14.** $y = x^4 - 4x^3 + 16x - 16$

**15.** $y = x^4 - 4x^3 + 16x$

**16.** $y = x^5 + 1$

**17.** $y = x^5 - 5x$

**18.** $y = (x - 1)^5$

**19.** $y = |2x - 3|$

**20.** $y = |x^2 - 6x + 5|$

In Exercises 21–30, use a graphing utility to graph the function. Choose a window that allows all relative extrema and points of inflection to be identified on the graph.

**21.** $y = \dfrac{x^2 + 2}{x^2 + 1}$

**22.** $y = \dfrac{x}{x^2 + 1}$

**23.** $y = 3x^{2/3} - 2x$

**24.** $y = 3x^{2/3} - x^2$

**25.** $y = 1 - x^{2/3}$

**26.** $y = (1 - x)^{2/3}$

**27.** $y = x^{1/3} + 1$

**28.** $y = x^{-1/3}$

**29.** $y = x^{5/3} - 5x^{2/3}$

**30.** $y = x^{4/3}$

In Exercises 31–40, sketch the graph of the function. Label the intercepts, relative extrema, points of inflection, and asymptotes. Then state the domain of the function.

**31.** $y = \dfrac{1}{x - 2} - 3$

**32.** $y = \dfrac{x^2 + 1}{x^2 - 2}$

**33.** $y = \dfrac{2x}{x^2 - 1}$

**34.** $y = \dfrac{x^2 - 6x + 12}{x - 4}$

**35.** $y = x\sqrt{4 - x}$

**36.** $y = x\sqrt{4 - x^2}$

**37.** $y = \dfrac{x - 3}{x}$

**38.** $y = x + \dfrac{32}{x^2}$

**39.** $y = \dfrac{x^3}{x^3 - 1}$

**40.** $y = \dfrac{x^4}{x^4 - 1}$

In Exercises 41–44, find values of $a, b, c,$ and $d$ such that the graph of

$$f(x) = ax^3 + bx^2 + cx + d$$

will resemble the given graph. Then use a graphing utility to confirm your result. (There are many correct answers.)

**41.**

**42.**

**43.**

**44.**

In Exercises 45–48, use the graph of $f'$ or $f''$ to sketch the graph of $f$. (There are many correct answers.)

**45.**
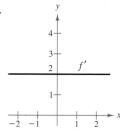

**46.**
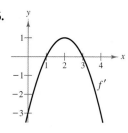

**47.**
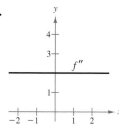

**48.**
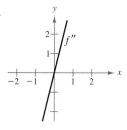

In Exercises 49 and 50, sketch a graph of a function $f$ having the given characteristics. (There are many correct answers.)

| Function | First Derivative |
|---|---|
| **49.** $f(-2) = 0$ | $f'(x) > 0, \quad -\infty < x < -1$ |
| $f(0) = 0$ | $f'(-1) = 0$ |
| | $f'(x) < 0, \quad -1 < x < 0$ |
| | $f'(0) = 0$ |
| | $f'(x) > 0, \quad 0 < x < \infty$ |

| Function | Second Derivative |
|---|---|
| **50.** $f(4) = 2$ | $f''(x) > 0, \quad x \neq 4$ |

**51.** *Cost* An employee of a delivery company earns $9 per hour driving a delivery van in an area where gasoline costs $1.20 per gallon. When the van is driven at a constant speed $s$ (in miles per hour, with $40 \leq s \leq 65$), the van gets $500/s$ miles per gallon.

(a) Find the cost $C$ as a function of $s$ for a 100-mile trip on an interstate highway.

(b) Use a graphing utility to graph the function found in part (a) and determine the most economical speed.

**52.** *Profit* The management of a company is considering three possible models for predicting the company's profit from 1995 through 2000. Model I gives the expected annual profit if the current trends continue. Models II and III give the expected annual profit for various combinations of increased labor and energy costs. In each model, $P$ is the profit (in billions of dollars) and $t = 0$ corresponds to 1995.

Model I: $P = 0.03t^2 - 0.01t + 3.39$

Model II: $P = 0.08t + 3.36$

Model III: $P = -0.07t^2 + 0.05t + 3.38$

(a) Use a graphing utility to graph all three models in the same viewing rectangle.

(b) For which models are profits increasing during the interval from 1995 through 2000?

(c) Which model is the most optimistic? Pessimistic?

**53.** *Average High Temperature* The average daily high temperature $T$ (in degrees Fahrenheit) for Pittsburgh, Pennsylvania can be modeled by

$$T = \frac{27.51 - 0.58t}{1 - 0.194t + 0.013t^2}, \quad 1 \leq t \leq 12,$$

where $t$ is the month, with $t = 1$ representing January. Use a graphing utility to graph the model and find all absolute extrema. Explain the meaning of those values. *(Source: NOAA)*

# Differentials and Marginal Analysis

## 3.8

*Differentials • Marginal Analysis • Formulas for Differentials • Error Propagation*

## Differentials

When the derivative was defined in Section 2.1 as the limit of the ratio $\Delta y/\Delta x$, it seemed natural to retain the quotient symbolism for the limit itself. Thus, the derivative of $y$ with respect to $x$ was denoted by

$$\frac{dy}{dx} = \lim_{\Delta x \to 0} \frac{\Delta y}{\Delta x}$$

*even though* we did not interpret $dy/dx$ as the quotient of two separate quantities. In this section, you will see that the quantities $dy$ and $dx$ can be assigned meanings in such a way that their quotient, when $dx \neq 0$, is equal to the derivative of $y$ with respect to $x$.

---

### Definition of Differentials

Let $y = f(x)$ represent a differentiable function. The **differential of $x$** (denoted by $dx$) is any nonzero real number. The **differential of $y$** (denoted by $dy$) is

$$dy = f'(x) \, dx.$$

---

**STUDY TIP** In this definition, $dx$ can have any nonzero value. In most applications, however, $dx$ is chosen to be small and this choice is denoted by $dx = \Delta x$.

One use of differentials is in approximating the change in $f(x)$ that corresponds to a change in $x$, as shown in Figure 3.61. This change is denoted by

$$\Delta y = f(x + \Delta x) - f(x). \qquad \text{Change in } y$$

In Figure 3.61, notice that as $\Delta x$ gets smaller and smaller, the values of $dy$ and $\Delta y$ get closer and closer. That is, when $\Delta x$ is small,

$$dy \approx \Delta y.$$

This **tangent line approximation** is the basis for most applications of differentials.

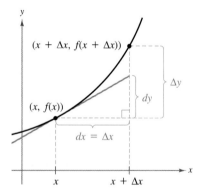

**FIGURE 3.61** Note that near the point of tangency, the graph of $f$ is very close to the tangent line. This is the essence of the approximations used in this section. In other words, near the point of tangency, $dy \approx \Delta y$.

## EXAMPLE 1   Interpreting Differentials Graphically

Consider the function

$$f(x) = x^2. \qquad \text{Original function}$$

Find the value of $dy$ when $x = 1$ and $dx = 0.01$. Compare this with the value of $\Delta y$ when $x = 1$ and $\Delta x = 0.01$. Interpret the results graphically.

### Solution

Begin by finding the derivative of $f$.

$$f'(x) = 2x \qquad \text{Derivative of } f$$

When $x = 1$ and $dx = 0.01$, the value of the differential $dy$ is

$$
\begin{aligned}
dy &= f'(x)\,dx & &\text{Differential of } y \\
&= f'(1)(0.01) & &\text{Substitute 1 for } x \text{ and 0.01 for } dx. \\
&= 2(1)(0.01) & &\text{Use } f'(x) = 2x. \\
&= 0.02. & &\text{Simplify.}
\end{aligned}
$$

When $x = 1$ and $\Delta x = 0.01$, the value of $\Delta y$ is

$$
\begin{aligned}
\Delta y &= f(x + \Delta x) - f(x) & &\text{Change in } y \\
&= f(1.01) - f(1) & &\text{Substitute 1 for } x \text{ and 0.01 for } \Delta x. \\
&= (1.01)^2 - (1)^2 & & \\
&= 0.0201. & &\text{Simplify.}
\end{aligned}
$$

Note that $dy \approx \Delta y$, as shown in Figure 3.62.

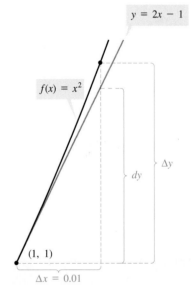

**FIGURE 3.62**

The validity of the approximation

$$dy \approx \Delta y, \qquad dx \neq 0$$

stems from the definition of the derivative. That is, the existence of the limit

$$f'(x) = \lim_{\Delta x \to 0} \frac{f(x + \Delta x) - f(x)}{\Delta x}$$

implies that when $\Delta x$ is close to zero, then $f'(x)$ is close to the difference quotient. Thus, you can write

$$\frac{f(x + \Delta x) - f(x)}{\Delta x} \approx f'(x)$$

$$f(x + \Delta x) - f(x) \approx f'(x)\,\Delta x$$

$$\Delta y \approx f'(x)\,\Delta x.$$

Substituting $dx$ for $\Delta x$ and $dy$ for $f'(x)\,dx$ produces

$$\Delta y \approx dy.$$

## Marginal Analysis

Differentials are used in economics to approximate changes in revenue, cost, and profit. Suppose that $R = f(x)$ is the total revenue for selling $x$ units of a product. When the number of units increases by 1, the change in $x$ is $\Delta x = 1$, and the change in $R$ is

$$\Delta R = f(x + \Delta x) - f(x) \approx dR = \frac{dR}{dx}\,dx.$$

In other words, you can use the differential $dR$ to approximate the change in the revenue that accompanies the sale of one additional unit. Similarly, the differentials $dC$ and $dP$ can be used to approximate the change in cost or profit that accompanies the sale (or production) of one additional unit.

### EXAMPLE 2   Using Marginal Analysis

The demand function for a product is modeled by

$$p = \sqrt{400 - x}, \qquad 0 \le x \le 400.$$

Use differentials to approximate the change in revenue as sales increase from 256 units to 257 units. Compare this with the actual change in revenue.

*Solution*

Begin by finding the marginal revenue, $dR/dx$.

### Technology

Use a graphing utility to graph the revenue function $R(x) = x\sqrt{400 - x}$ and the tangent line approximation $y = \frac{4}{3}(x - 256) + 3072$ in the viewing window $250 \le x \le 260$, $3065 \le y \le 3080$. Explain why the two curves appear to be almost identical. Where do the curves intersect? Use the graphs to verify the solution to Example 2.

$$R = xp \qquad\qquad\qquad \text{Formula for revenue}$$

$$= x\sqrt{400 - x} \qquad\qquad \text{Use } p = \sqrt{400 - x}.$$

$$\frac{dR}{dx} = x\left(\frac{1}{2}\right)(400 - x)^{-1/2}(-1) + (400 - x)^{1/2}(1) \qquad \text{Product Rule}$$

$$= \frac{800 - 3x}{2\sqrt{400 - x}} \qquad\qquad \text{Simplify.}$$

When $x = 256$ and $dx = \Delta x = 1$, you can approximate the change in the revenue to be

$$dR = \frac{dR}{dx}\,dx$$

$$= \frac{800 - 3(256)}{2\sqrt{400 - 256}}\,(1)$$

$$\approx \$1.33.$$

When $x$ increases from 256 to 257, the actual change in revenue is

$$\Delta R = 257\sqrt{400 - 257} - 256\sqrt{400 - 256}$$

$$= 3073.27 - 3072.00$$

$$\approx \$1.27.$$

## EXAMPLE 3  Using Marginal Analysis

The profit derived from selling $x$ units of an item is modeled by

$$P = 0.0002x^3 + 10x.$$

Use the differential $dP$ to approximate the change in profit when the production level changes from 50 to 51 units. Compare this with the actual gain in profit obtained by increasing the production level from 50 to 51 units.

### Solution

The marginal profit is

$$\frac{dP}{dx} = 0.0006x^2 + 10.$$

When $x = 50$ and $dx = 1$, the differential $dP$ is

$$dP = \frac{dP}{dx}\,dx$$

$$= [0.0006(50)^2 + 10](1)$$

$$= \$11.50.$$

When $x$ changes from 50 to 51 units, the actual change in profit is

$$\Delta P = [(0.0002)(51)^3 + 10(51)] - [(0.0002)(50)^3 + 10(50)]$$

$$\approx 536.53 - 525.00$$

$$= \$11.53.$$

These values are shown graphically in Figure 3.63.

> **STUDY TIP**   Example 3 uses differentials to solve the same problem that was solved in Example 5 in Section 2.3. Look back at that solution. Which approach do you prefer?

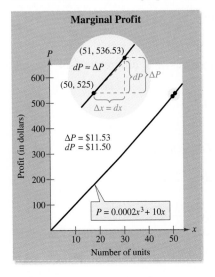

**FIGURE 3.63**

## Formulas for Differentials

You can use the definition of differentials to rewrite each differentiation rule in **differential form.** For example, if $u$ and $v$ are differentiable functions of $x$, then $du = (du/dx)\, dx$ and $dv = (dv/dx)\, dx$, which implies that you can write the Product Rule in the following differential form.

$$d[uv] = \frac{d}{dx}[uv]\, dx \qquad \text{Differential of } uv$$

$$= \left[ u\frac{dv}{dx} + v\frac{du}{dx} \right] dx \qquad \text{Product Rule}$$

$$= u\frac{dv}{dx}\, dx + v\frac{du}{dx}\, dx$$

$$= u\, dv + v\, du \qquad \text{Differential form of Product Rule}$$

The following summary gives the differential forms of the differentiation rules presented so far in the text.

| Differential Forms of Differentiation Rules | |
|---|---|
| Constant Multiple Rule: | $d[cu] = c\, du$ |
| Sum or Difference Rule: | $d[u \pm v] = du \pm dv$ |
| Product Rule: | $d[uv] = u\, dv + v\, du$ |
| Quotient Rule: | $d\left[\dfrac{u}{v}\right] = \dfrac{v\, du - u\, dv}{v^2}$ |
| Constant Rule: | $d[c] = 0$ |
| Power Rule: | $d[x^n] = nx^{n-1}\, dx$ |

The next example compares the derivatives and differentials of several simple functions.

## EXAMPLE 4   Finding Differentials

| | *Function* | *Derivative* | *Differential* |
|---|---|---|---|
| (a) | $y = x^2$ | $\dfrac{dy}{dx} = 2x$ | $dy = 2x\, dx$ |
| (b) | $y = \dfrac{3x + 2}{5}$ | $\dfrac{dy}{dx} = \dfrac{3}{5}$ | $dy = \dfrac{3}{5}\, dx$ |
| (c) | $y = 2x^2 - 3x$ | $\dfrac{dy}{dx} = 4x - 3$ | $dy = (4x - 3)\, dx$ |
| (d) | $y = \dfrac{1}{x}$ | $\dfrac{dy}{dx} = -\dfrac{1}{x^2}$ | $dy = -\dfrac{1}{x^2}\, dx$ |

## Error Propagation

A common use of differentials is the estimation of errors that are propagated by physical measuring devices. This is illustrated in Example 5.

### EXAMPLE 5  Estimating Measurement Errors

The radius of a ball bearing is measured to be 0.7 inch, as shown in Figure 3.64. This implies that the volume of the ball bearing is $\frac{4}{3}\pi(0.7)^3 \approx 1.4368$ cubic inches. You are told that the measurement of the radius is correct to within 0.01 inch. How far off could the calculation of the volume be?

**Solution**

Because the value of $r$ can be off by 0.01 inch, it follows that

$-0.01 \le \Delta r \le 0.01.$      Possible error in measuring

Using $\Delta r = dr$, you can estimate the possible error in the volume.

$V = \frac{4}{3}\pi r^3$      Formula for volume

$dV = \dfrac{dV}{dr}\,dr$      Formula for differential of $V$

$\quad = 4\pi r^2\,dr$

$\quad = 4\pi(0.7)^2(\pm 0.01)$      Substitute for $r$ and $dr$.

$\quad \approx \pm 0.0616$ cubic inch      Possible error

Thus, the volume of the ball bearing could range between $(1.4368 - 0.0616) = 1.3752$ cubic inches and $(1.4368 + 0.0616) = 1.4984$ cubic inches.

0.7 in.

**FIGURE 3.64**

In Example 5, the **relative error** in the volume is defined to be the ratio of $dV$ to $V$. This ratio is

$\dfrac{dV}{V} \approx \dfrac{\pm 0.0616}{1.4368} \approx \pm 0.0429.$

This corresponds to a **percentage error** of 4.29%.

---

### *Group Discussion*     ***Finding Propagated Errors***

a. In Example 5, if the radius of the ball bearing were measured to be 1.5 inches, correct to within 0.01 inch, would the percentage error of the volume be greater or smaller than in the example?

b. In Example 5, if the radius of the ball bearing were measured to be 0.7 inch, correct to within 0.02 inch, would the percentage error of the volume be greater or smaller than in the example?

*Warm Up*

The following warm-up exercises involve skills that were covered in earlier sections. You will use these skills in the exercise set for this section.

In Exercises 1–6, find the derivative.

**1.** $C = 44 + 0.09x^2$

**2.** $R = x(1.25 + 0.02\sqrt{x})$

**3.** $P = -0.03x^{1/3} + 1.4x - 2250$

**4.** $A = \frac{1}{4}\sqrt{3x^2}$

**5.** $C = 2\pi r$

**6.** $S = 4\pi r^2$

In Exercises 7–10, write a formula for the quantity.

**7.** Area $A$ of a circle of radius $r$

**8.** Area $A$ of a square of side $x$

**9.** Volume $V$ of a cube of edge $x$

**10.** Volume $V$ of a sphere of radius $r$

## EXERCISES 3.8

In Exercises 1–6, find the differential $dy$.

**1.** $y = 3x^2 - 4$

**2.** $y = 2x^{3/2}$

**3.** $y = (4x - 1)^3$

**4.** $y = (1 - 2x^2)^{-2}$

**5.** $y = \sqrt{x^2 + 1}$

**6.** $y = \sqrt[3]{6x^2}$

In Exercises 7–10, let $x = 1$ and $\Delta x = 0.01$. Find $\Delta y$.

**7.** $f(x) = 5x^2 - 1$

**8.** $f(x) = \sqrt{3x}$

**9.** $f(x) = \dfrac{4}{\sqrt[3]{x}}$

**10.** $f(x) = \dfrac{x}{x^2 + 1}$

In Exercises 11–14, compare the values of $dy$ and $\Delta y$.

**11.** $y = x^3$      $x = 1$      $\Delta x = dx = 0.1$

**12.** $y = 1 - 2x^2$      $x = 0$      $\Delta x = dx = -0.1$

**13.** $y = x^4 + 1$      $x = -1$      $\Delta x = dx = 0.01$

**14.** $y = 2x + 1$      $x = 2$      $\Delta x = dx = 0.01$

In Exercises 15–18, let $x = 2$ and complete the table for the function.

| $dx = \Delta x$ | $dy$ | $\Delta y$ | $\Delta y - dy$ | $dy/\Delta y$ |
|---|---|---|---|---|
| 1.000 | | | | |
| 0.500 | | | | |
| 0.100 | | | | |
| 0.010 | | | | |
| 0.001 | | | | |

**15.** $y = x^2$

**16.** $y = \dfrac{1}{x^2}$

**17.** $y = x^5$

**18.** $y = \sqrt{x}$

In Exercises 19–22, find an equation of the tangent line to the function at the indicated point. Then find the function value and the tangent line value at $f(x + \Delta x)$ and $y(x + \Delta x)$ for $\Delta x = -0.01$ and $0.01$.

| *Function* | *Point* |
|---|---|
| **19.** $f(x) = \dfrac{x}{x^2 + 1}$ | $(0, 0)$ |
| **20.** $f(x) = \sqrt{25 - x^2}$ | $(3, 4)$ |
| **21.** $f(x) = 2x^3 - x^2 + 1$ | $(-2, -19)$ |
| **22.** $f(x) = 2\sqrt[3]{x} - 1$ | $(8, 3)$ |

**23. Demand** The demand function for a product is modeled by

$$p = 75 - 0.25x.$$

(a) If $x$ changes from 7 to 8, what is the corresponding change in $p$? Compare the values of $\Delta p$ and $dp$.

(b) Repeat part (a) when $x$ changes from 70 to 71 units.

**24. Wildlife Management** A state game commission introduces 50 deer into newly acquired state game lands. The population $N$ of the herd can be modeled by

$$N = \frac{10(5 + 3t)}{1 + 0.04t},$$

where $t$ is the time in years. Use differentials to approximate the change in the herd size from $t = 5$ to $t = 6$.

**Marginal Analysis** In Exercises 25–28, use differentials to approximate the change in cost, revenue, or profit corresponding to an increase in sales of *one unit*. For instance, in Exercise 25, approximate the change in cost as $x$ increases from 12 units to 13 units. Then use a graphing utility to graph the function, and use the trace feature to verify your result.

| Function | x-Value |
|---|---|
| **25.** $C = 0.05x^2 + 4x + 10$ | $x = 12$ |
| **26.** $R = 30x - 0.15x^2$ | $x = 75$ |
| **27.** $P = -0.5x^3 + 2500x - 6000$ | $x = 50$ |
| **28.** $P = -x^2 + 60x - 100$ | $x = 25$ |

**29. Marginal Analysis** A retailer has determined that the monthly sales $x$ of a watch is 150 units when the price is $50, but decreases to 120 units when the price is $60. Assume that the demand is a linear function of the price. Find the revenue $R$ as a function of $x$ and approximate the change in revenue for a one-unit increase in sales when $x = 141$. Make a sketch showing $dR$ and $\Delta R$.

**30. Marginal Analysis** A manufacturer determines that the demand $x$ for a product is inversely proportional to the square of the price $p$. When the price is $10, the demand is 2500. Find the revenue $R$ as a function of $x$ and approximate the change in revenue for a one-unit increase in sales when $x = 3000$. Make a sketch showing $dR$ and $\Delta R$.

**31. Marginal Analysis** The demand $x$ for a radio is 30,000 units per week when the price is $25 and 40,000 units when the price is $20. The initial investment is $275,000 and the cost per unit is $17. Assume that the demand is a linear function of the price. Find the profit $P$ as a function of $x$ and approximate the change in profit for a one-unit increase in sales when $x = 28,000$. Make a sketch showing $dP$ and $\Delta P$.

**32. Marginal Analysis** The variable cost for the production of a calculator is $14.25 and the initial investment is $110,000. Find the total cost $C$ as a function of $x$, the number of units produced. Then use differentials to approximate the change in the cost for a one-unit increase in production when $x = 50,000$. Make a sketch showing $dC$ and $\Delta C$. Explain why $dC = \Delta C$ in this problem.

**33.** The area $A$ of a square of side $x$ is $A = x^2$.

(a) Compute $dA$ and $\Delta A$ in terms of $x$ and $\Delta x$.

(b) In the figure, identify the region whose area is $dA$.

(c) Identify the region whose area is $\Delta A - dA$.

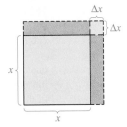

**34. Area** The side of a square is measured to be 12 inches, with a possible error of $\frac{1}{64}$ inch. Use differentials to approximate the possible error and the relative error in computing the area of the square.

**35. Area** The radius of a circle is measured to be 10 inches, with a possible error of $\frac{1}{8}$ inch. Use differentials to approximate the possible error and the relative error in computing the area of the circle.

**36. Volume and Surface Area** The edge of a cube is measured to be 12 inches, with a possible error of 0.03 inch. Use differentials to approximate the possible error and the relative error in computing (a) the volume of the cube and (b) the surface area of the cube.

**37. Volume** The radius of a sphere is measured to be 6 inches, with a possible error of 0.02 inch. Use differentials to approximate the possible error and the relative error in computing the volume of the sphere.

**38. Drug Concentration** The concentration $C$ (in milligrams per milliliter) of a drug in a patient's bloodstream $t$ hours after injection into muscle tissue is modeled by

$$C = \frac{3t}{27 + t^3}.$$

Use differentials to approximate the change in the concentration when $t$ changes from $t = 1$ to $t = 1.5$.

# Chapter 3   Algebra Review

## Solving Equations

Much of the algebra in Chapter 3 involves simplifying algebraic expressions (see pages 162 and 163) and solving algebraic equations (see page 73). The Algebra Review on page 73 illustrates some of the basic techniques for solving equations. On these two pages, you can review some of the more complicated techniques for solving equations.

When solving an equation, remember that your basic goal is to isolate the variable on one side of the equation. To do this, you use inverse operations. For instance, to get rid of the *subtract 2* in $x - 2 = 0$, you *add 2* to each side of the equation. Similarly, to get rid of the *square root* in $\sqrt{x + 3} = 2$, you *square* both sides of the equation.

### EXAMPLE 1   Solving an Equation

(a) $\dfrac{36(x^2 - 1)}{(x^2 + 3)^3} = 0$    Example 2, page 193

$36(x^2 - 1) = 0$    A fraction is zero only if its numerator is zero.

$x^2 - 1 = 0$    Divide each side by 36.

$x^2 = 1$    Add 1 to each side.

$x = \pm 1$    Take the square root of each side.

(b)  $0 = 2 - \dfrac{288}{x^2}$    Example 2, page 203

$-2 = -\dfrac{288}{x^2}$    Subtract 2 from each side.

$1 = \dfrac{144}{x^2}$    Divide each side by $-2$.

$x^2 = 144$    Multiply each side by $x^2$.

$x = \pm 12$    Take the square root of each side.

(c)     $0 = 2x(2x^2 - 3)$    Example 3, page 204

$2x = 0$  $\Longrightarrow$  $x = 0$    Set first factor equal to zero.

$2x^2 - 3 = 0$  $\Longrightarrow$  $x = \pm\sqrt{\dfrac{3}{2}}$    Set second factor equal to zero.

---

## EXAMPLE 2    Solving an Equation

(a) $\dfrac{20(x^{1/3} - 1)}{9x^{2/3}} = 0$        Example 5, page 236

$20(x^{1/3} - 1) = 0$        A fraction is zero only if its numerator is zero.

$x^{1/3} - 1 = 0$        Divide each side by 20.

$x^{1/3} = 1$        Add 1 to each side.

$x = 1$        Cube both sides.

(b) $\dfrac{25}{\sqrt{x}} - 0.5 = 0$        Example 4, page 213

$\dfrac{25}{\sqrt{x}} = 0.5$        Add 0.5 to each side.

$25 = 0.5\sqrt{x}$        Multiply each side by $\sqrt{x}$.

$50 = \sqrt{x}$        Divide each side by 0.5.

$2500 = x$        Square both sides.

(c) $x^2(4x - 3) = 0$        Example 2, page 184

$x^2 = 0 \quad \Longrightarrow \quad x = 0$        Set first factor equal to zero.

$4x - 3 = 0 \quad \Longrightarrow \quad x = \tfrac{3}{4}$        Set second factor equal to zero.

(d) $\dfrac{4x}{3(x^2 - 4)^{1/3}} = 0$        Example 4, page 176

$4x = 0$        A fraction is zero only if its numerator is zero.

$x = 0$        Divide each side by 4.

(e) $g(x) = (x - 2)(x + 1)^2$        Exercise 37, page 199

$(x - 2)(2)(x + 1) + (x + 1)^2(1) = 0$        Find derivative and set equal to zero.

$(x + 1)[2(x - 2) + (x + 1)] = 0$        Factor.

$(x + 1)(2x - 4 + x + 1) = 0$        Multiply factors.

$(x + 1)(3x - 3) = 0$        Combine like terms.

$x + 1 = 0 \quad \Longrightarrow \quad x = -1$        Set first factor equal to zero.

$3x - 3 = 0 \quad \Longrightarrow \quad x = 1$        Set second factor equal to zero.

# Chapter Summary and Study Strategies

*After studying this chapter, you should have acquired the following skills. The exercise numbers are keyed to the Review Exercises that begin on page 252. Answers to odd-numbered Review Exercises are given in the back of the text.\**

■ Find the critical numbers of a function. *(Section 3.1)*

    $c$ is a critical number of $f$ if $f'(c) = 0$ or $f'(c)$ is undefined.

■ Find the open intervals on which a function is increasing or decreasing. *(Section 3.1)*

    Increasing if $f'(x) > 0$

    Decreasing if $f'(x) < 0$

■ Find intervals on which a real-life model is increasing or decreasing, and interpret the results in context. *(Section 3.1)*

■ Use the First-Derivative Test to find the relative extrema of a function. *(Section 3.2)*

■ Find the absolute extrema of a continuous function on a closed interval. *(Section 3.2)*

■ Find minimum and maximum values of a real-life model and interpret the results in context. *(Section 3.2)*

■ Find the open intervals on which the graph of a function is concave upward or concave downward. *(Section 3.3)*

    Concave upward if $f''(x) > 0$

    Concave downward if $f''(x) < 0$

■ Find the points of inflection of the graph of a function. *(Section 3.3)*

■ Use the Second-Derivative Test to find the relative extrema of a function. *(Section 3.3)*

■ Find the point of diminishing returns of an input-output model. *(Section 3.3)*

■ Solve real-life optimization problems. *(Section 3.4)*

■ Solve business and economics optimization problems. *(Section 3.5)*

■ Find the price elasticity of demand for a demand function. *(Section 3.5)*

■ Find the vertical and horizontal asymptotes of a function and sketch its graph. *(Section 3.6)*

■ Find infinite limits and limits at infinity. *(Section 3.6)*

■ Use asymptotes to answer questions about real life. *(Section 3.6)*

■ Analyze the graph of a function. *(Section 3.7)*

■ Find the differential of a function. *(Section 3.8)*

■ Use differentials to approximate changes in a function. *(Section 3.8)*

■ Use differentials to approximate changes in real-life models. *(Section 3.8)*

\* Use a wide range of valuable study aids to help you master the material in this chapter. The *Student Solutions Guide* includes step-by-step solutions to all odd-numbered exercises to help you review and prepare. The *Algebra Review Tutorial Software* and *The Algebra of Calculus* help you brush up on your algebra skills. The *Graphing Technology Guide* offers step-by-step commands and instructions for a wide variety of graphing calculators, including the most recent models.

■ *Solve Problems Graphically, Analytically, and Numerically*   When analyzing the graph of a function, use a variety of problem-solving strategies. For instance, if you were asked to analyze the graph of $f(x) = x^3 - 4x^2 + 5x - 4$, you could begin *graphically*. That is, you could use a graphing utility to find a viewing rectangle that appears to show the important characteristics of the graph. From the graph shown below, the function appears to have one relative minimum, one relative maximum, and one point of inflection.

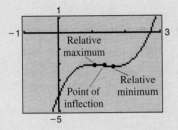

Next, you could use calculus to *analyze* the graph. Because the derivative of $f$ is $f'(x) = 3x^2 - 8x + 5 = (3x - 5)(x - 1)$, the critical numbers of $f$ are $x = \frac{5}{3}$ and $x = 1$. By the First-Derivative Test, you can conclude that $x = \frac{5}{3}$ yields a relative minimum and $x = 1$ yields a relative maximum. Because $f''(x) = 6x - 8$, you can conclude that $x = \frac{4}{3}$ yields a point of inflection. Finally, you could analyze the graph *numerically*. For instance, you could construct a table of values and observe that $f$ is increasing in the interval $(-\infty, 1)$, decreasing in the interval $\left(1, \frac{5}{3}\right)$, and increasing in the interval $\left(\frac{5}{3}, \infty\right)$.

■ *Problem-Solving Strategies*   If you get stuck when trying to solve an optimization problem, consider the following strategies.

1. *Draw a Diagram.*   If feasible, draw a diagram that represents the problem. Label all known values and unknown values on the diagram.

2. *Solve a Simpler Problem.*   Simplify the problem, or write several simple examples of the problem. For instance, if you are asked to find the dimensions that will produce a maximum area, try calculating the areas of several examples.

3. *Rewrite the Problem in Your Own Words.*   Rewriting a problem can help you understand it better.

4. *Guess and Check.*   Try guessing the answer, then check your guess in the statement of the original problem. By refining your guesses, you may be able to think of a general strategy for solving the problem.

# Chapter 3  Review Exercises

In Exercises 1–4, find the critical numbers of the function.

**1.** $f(x) = -x^2 + 2x + 4$

**2.** $g(x) = (x - 1)^2(x - 3)$

**3.** $h(x) = \sqrt{x}(x - 3)$

**4.** $f(x) = (x + 1)^3$

In Exercises 5–8, determine the open intervals on which the function is increasing or decreasing. Solve the problem analytically and graphically.

**5.** $f(x) = x^2 + x - 2$

**6.** $g(x) = -x^2 + 7x - 12$

**7.** $h(x) = \dfrac{x^2 - 3x - 4}{x - 3}$

**8.** $f(x) = -x^3 + 6x^2 - 2$

**9.** *Temperature*   The daily maximum temperature $T$ (in degrees Fahrenheit) for New York City can be modeled by

$$T = 0.036t^4 - 0.909t^3 + 5.874t^2 - 2.599t + 37.789,$$

where $0 \le t \le 12$ and $t = 0$ corresponds to January 1. *(Source: National Oceanic and Atmospheric Administration)*

(a) Find the intervals in which the model is increasing.
(b) Find the intervals in which the model is decreasing.
(c) Interpret the results of parts (a) and (b).
(d) Use a graphing utility to graph the model.

**10.** *Morning Newspapers*   The number $N$ of morning newspapers in the United States from 1970 through 1995 can be modeled by

$$N = 0.0001t^4 - 0.01t^3 + 0.25t^2 - 1.22t + 26,$$

where $0 \le t \le 25$ and $t = 0$ corresponds to 1970. *(Source: Editor and Publisher Yearbook)*

(a) Find the intervals in which the model is increasing.
(b) Find the intervals in which the model is decreasing.
(c) Interpret the results of parts (a) and (b).
(d) Use a graphing utility to graph the model.

In Exercises 11–20, use the First-Derivative Test to find the relative extrema of the function. Then use a graphing utility to confirm your result.

**11.** $f(x) = 4x^3 - 6x^2 - 2$

**12.** $f(x) = \frac{1}{4}x^4 - 8x$

**13.** $g(x) = x^2 - 16x + 12$

**14.** $h(x) = 4 + 10x - x^2$

**15.** $h(x) = 2x^2 - x^4$

**16.** $s(x) = x^4 - 8x^2 + 3$

**17.** $f(x) = \dfrac{6}{x^2 + 1}$

**18.** $f(x) = \dfrac{2}{x^2 - 1}$

**19.** $h(x) = \dfrac{x^2}{x - 2}$

**20.** $g(x) = x - 6\sqrt{x}, \quad x > 0$

In Exercises 21–28, find the absolute extrema of the function on the indicated interval. Then use a graphing utility to confirm your result.

**21.** $f(x) = x^2 + 5x + 6; \quad [-3, 0]$

**22.** $f(x) = x^4 - 2x^3; \quad [0, 2]$

**23.** $f(x) = x^3 - 12x + 1; \quad [-4, 4]$

**24.** $f(x) = \dfrac{8}{x} + x; \quad [1, 4]$

**25.** $f(x) = 3x^4 - 6x^2 + 2; \quad [0, 2]$

**26.** $f(x) = -x^4 + x^2 + 2; \quad [0, 2]$

**27.** $f(x) = \dfrac{2x}{x^2 + 1}; \quad [-1, 2]$

**28.** $f(x) = 4\sqrt{x} - x^2; \quad [0, 3]$

**29.** *Surface Area*   A right circular cylinder of radius $r$ and height $h$ has a volume of 25 cubic inches. The surface area of the cylinder is given by

$$S = 2\pi r\left(r + \frac{25}{\pi r^2}\right).$$

Use a graphing utility to graph $S$ and $S'$ and find the value of $r$ that yields the minimum surface area.

**30.** *Waste Oxidation*   When organic waste is dumped into a pond, the decomposition of the waste consumes oxygen. A model for the oxygen level $O$ (where 1 is the normal level) of a pond as waste material oxidizes is

$$O = \frac{t^2 - t + 1}{t^2 + 1}, \qquad 0 \le t,$$

where $t$ is the time in weeks.

(a) When is the oxygen level lowest? What is this level?
(b) When is the oxygen level highest? What is this level?
(c) Describe the oxygen level as $t$ increases.

In Exercises 31–34, determine the open intervals on which the graph of the function is concave upward or concave downward. Then use a graphing utility to confirm your result.

**31.** $f(x) = (x - 2)^3$

**32.** $h(x) = x^5 - 10x^2$

**33.** $g(x) = \frac{1}{4}(-x^4 + 8x^2 - 12)$

**34.** $h(x) = x^3 - 6x$

In Exercises 35–38, find the points of inflection of the graph of the function.

**35.** $f(x) = \frac{1}{2}x^4 - 4x^3$

**36.** $f(x) = (x + 2)^2(x - 4)$

**37.** $f(x) = x^3(x - 3)^2$

**38.** $f(x) = \frac{1}{4}x^4 - 2x^2 - x$

In Exercises 39–42, use the Second-Derivative Test to find the relative extrema of the function.

**39.** $f(x) = x^5 - 5x^3$

**40.** $f(x) = x(x^2 - 3x - 9)$

**41.** $f(x) = (x - 1)^3(x + 4)^2$

**42.** $f(x) = \dfrac{(x - 2)^2}{4 - x}$

In Exercises 43 and 44, identify the point of diminishing returns for the input-output function. For each function, $R$ is the revenue (in thousands of dollars) and $x$ is the amount spent on advertising (in thousands of dollars).

**43.** $R = \frac{1}{1500}(150x^2 - x^3), \qquad 0 \le x \le 100$

**44.** $R = -\frac{2}{3}(x^3 - 12x^2 - 6), \qquad 0 \le x \le 8$

**45.** *Minimum Sum*  Find two positive numbers whose product is 169 and whose sum is a minimum. Solve the problem analytically, and use a graphing utility to solve the problem graphically.

**46.** *Length*  The wall of a building is to be braced by a beam that must pass over a 5-foot fence that is parallel to the building and 4 feet from the building. Find the length of the shortest beam that can be used.

**47.** *Charitable Contributions*  The percent $P$ of income that Americans give to charities can be modeled by

$$P = 0.0014x^2 - 0.1529x + 5.855, \qquad 5 \le x \le 100,$$

where $x$ is the annual income in thousands of dollars. *(Source: Independent Sector)*

(a) What income level corresponds to the lowest percent of charitable contributions?

(b) What income level corresponds to the highest percent of charitable contributions?

(c) Use a graphing utility to verify the results of parts (a) and (b).

**48.** *Construction Costs*  A fence is to be built to enclose a rectangular region of 4800 square feet. The fencing material along three sides costs \$3 per foot. The fencing material along the fourth side costs \$4 per foot.

(a) Find the most economical dimensions of the region.

(b) How would the result of part (a) change if the fencing material costs for all sides increased by \$1 per foot?

**49.** *Tree Growth*  The growth of a red oak tree is approximated by the model

$$y = -0.003x^3 + 0.137x^2 + 0.458x - 0.839,$$
$$2 \le x \le 34,$$

where $y$ is the height of the tree in feet and $x$ is its age in years. Find the age of the tree when it is growing most rapidly. Then use a graphing utility to graph the function to confirm your result. (*Hint:* Use the viewing rectangle $-10 \le x \le 45$ and $-5 \le y \le 60$.)

**50.** *TV Usage*  The average number of hours of TV usage in the United States from 1987 to 1994 can be modeled by the equation

$$N = -0.0135t^3 + 0.457t^2 - 4.98t + 24.5,$$

where $t = 7$ corresponds to 1987.

(a) Find the intervals on which $N$ is increasing and decreasing.

(b) Find the absolute extrema on the interval $[7, 14]$.

(c) Briefly explain your results for parts (a) and (b).

**51.** *Poiseuille's Law*  The speed of blood that is $r$ centimeters from the center of an artery is

$$s(r) = c(R^2 - r^2),$$

where $c$ is a constant, $R$ is the radius of the artery, and $s$ is measured in centimeters per second. Show that the speed is a maximum at the center of an artery.

**52.** *Profit*  The demand and cost functions for a product are

$$p = 36 - 4x \quad \text{and} \quad C = 2x^2 + 6.$$

(a) What level of production will produce a maximum profit?

(b) What level of production will produce a minimum average cost per unit?

**53.** *Revenue*   For groups of 20 or more, a theater determines the ticket price $p$ according to the formula

$$p = 15 - 0.1(n - 20), \qquad 20 \le n \le N,$$

where $n$ is the number in the group. What should the value of $N$ be? Explain your reasoning.

**54.** *Cost*   The cost of fuel to run a locomotive is proportional to the $\frac{3}{2}$ power of the speed. At a speed of 25 miles per hour, the cost of fuel is $50 per hour. Other costs amount to $100 per hour. Find the speed that will minimize the cost per mile.

**55.** *Inventory Cost*   The cost $C$ of inventory depends on ordering and storage costs,

$$C = \left(\frac{Q}{x}\right)s + \left(\frac{x}{2}\right)r,$$

where $Q$ is the number of units sold per year, $r$ is the cost of storing one unit for 1 year, $s$ is the cost of placing an order, and $x$ is the number of units in the order. Determine the order size that will minimize the cost when $Q = 10{,}000$, $s = 4.5$, and $r = 5.76$.

**56.** *Profit*   The demand and cost functions for a product are

$$p = 600 - 3x \quad \text{and} \quad C = 0.3x^2 + 6x + 600,$$

where $p$ is the price per unit, $x$ is the number of units, and $C$ is the total cost. The profit for producing $x$ units is

$$P = xp - C - xt,$$

where $t$ is the excise tax per unit. Find the maximum profits for excise taxes of $t = \$5$, $t = \$10$, and $t = \$20$.

In Exercises 57 and 58, find the intervals on which the demand is elastic, inelastic, and of unit elasticity.

**57.** $p = 100 - 0.5x^2, \qquad 0 \le x \le 10\sqrt{2}$

**58.** $p = \sqrt{1800 - x}, \qquad 0 \le x \le 1800$

In Exercises 59–64, find the vertical and horizontal asymptotes of the graph. Then use a graphing utility to graph the function.

**59.** $h(x) = \dfrac{2x + 3}{x - 4}$

**60.** $g(x) = \dfrac{5x^2}{x^2 + 2}$

**61.** $f(x) = \dfrac{1 - 3x}{x}$

**62.** $h(x) = \dfrac{3x}{\sqrt{x^2 + 2}}$

**63.** $f(x) = \dfrac{3}{x^2 - 5x + 4}$

**64.** $h(x) = \dfrac{2x^2 + 3x - 5}{x - 1}$

In Exercises 65–70, find the limit, if it exists.

**65.** $\displaystyle \lim_{x \to 0^+} \left(x - \frac{1}{x^3}\right)$

**66.** $\displaystyle \lim_{x \to -1^+} \frac{x^2 - 2x + 1}{x + 1}$

**67.** $\displaystyle \lim_{x \to \infty} \frac{5x^2 + 3}{2x^2 - x + 1}$

**68.** $\displaystyle \lim_{x \to \infty} \frac{3x^2 - 2x + 3}{x + 1}$

**69.** $\displaystyle \lim_{x \to -\infty} \frac{3x^2}{x + 2}$

**70.** $\displaystyle \lim_{x \to -\infty} \left(\frac{x}{x - 2} + \frac{2x}{x + 2}\right)$

**71.** *Ultraviolet Radiation*   For a person with sensitive skin, the amount $T$ (in hours) of exposure to the sun can be modeled by

$$T = \frac{0.37s + 23.8}{s}, \qquad 0 < s \le 120,$$

where $s$ is the Sunsor Scale reading.   *(Source: Sunsor, Inc.)*

(a) Use a graphing utility to graph the model. Compare your result with the graph below.

(b) Describe the value of $T$ as $s$ increases.

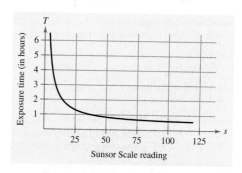

**72.** *Average Cost and Profit*   The cost and revenue functions for a product are

$$C = 10{,}000 + 48.9x \quad \text{and} \quad R = 68.5x.$$

(a) Find the average cost function.

(b) What is the limit of the average cost as $x$ approaches infinity?

(c) Find the average profit when $x$ is 1 million, 2 million, and 10 million.

(d) What is the limit of the average profit as $x$ increases without bound?

In Exercises 73–80, sketch the graph of the function. Label the intercepts, relative extrema, points of inflection, and asymptotes. State the domain of the function.

**73.** $f(x) = 4x - x^2$

**74.** $f(x) = 4x^3 - x^4$

**75.** $f(x) = x\sqrt{16 - x^2}$

**76.** $f(x) = x + \dfrac{4}{x^2}$

**77.** $f(x) = \dfrac{x + 1}{x - 1}$

**78.** $f(x) = x^2 + \dfrac{2}{x}$

**79.** $f(x) = \dfrac{2x}{1 + x^2}$

**80.** $f(x) = x^{4/5}$

In Exercises 81–84, find the differential $dy$.

**81.** $y = 6x^2 - 5$

**82.** $y = (3x^2 - 2)^3$

**83.** $y = -\dfrac{5}{\sqrt[3]{x}}$

**84.** $y = \dfrac{2 - x}{x + 5}$

In Exercises 85–88, use differentials to approximate the change in cost, revenue, or profit corresponding to an increase in sales of one unit.

**85.** $C = 40x^2 + 1225, \quad x = 10$

**86.** $C = 1.5\sqrt[3]{x} + 500, \quad x = 125$

**87.** $R = 6.25x + 0.4x^{3/2}, \quad x = 225$

**88.** $P = 0.003x^2 + 0.019x - 1200, \quad x = 750$

 **89.** *Recreational Vehicle Sales*   Sales $S$, in billions of dollars, of recreational vehicles in the United States for the years 1980 through 1994 are given in the table. *(Source: National Sporting Goods Association)*

| Year | 0 | 1 | 2 | 3 | 4 | 5 | 6 | 7 |
|------|-----|-----|-----|-----|-----|-----|-----|-----|
| Sales | 1.2 | 1.8 | 1.7 | 3.4 | 4.1 | 3.5 | 3.9 | 4.5 |

| Year | 8 | 9 | 10 | 11 | 12 | 13 | 14 |
|------|-----|-----|-----|-----|-----|-----|-----|
| Sales | 4.8 | 4.5 | 4.1 | 3.6 | 4.4 | 4.8 | 4.8 |

(a) Use the regression capabilities of a graphing utility to find a quadratic model for the data, where $t$ is the time in years, with $t = 0$ corresponding to 1980.

(b) Use a graphing utility to plot the data and graph the model.

(c) Use a graphing utility to graph $dS/dt$.

(d) The table shows that sales were down from 1989 through 1991. Does the derivative of the model show this decline? Explain. Give a possible explanation for the decline.

(e) Does the model show the full magnitude of the sales slump? Explain.

(f) Use the derivative to determine the interval of time when sales were increasing most rapidly. Explain.

 **90.** *Drug Effectiveness*   Suppose that the effectiveness $E$ of a pain-killing drug $t$ hours after entering the bloodstream is

$$E = 22.5t + 7.5t^2 - 2.5t^3, \qquad 0 \le t \le 4.5.$$

(a) Use a graphing utility to graph the equation. Choose an appropriate window.

(b) Find the maximum effectiveness the pain-killing drug attains over the interval $[0, 4.5]$.

**91.** *Surface Area and Volume*   The diameter of a sphere is measured to be 18 inches with a possible error of 0.05 inch. Use differentials to approximate the possible error in the surface area and the volume of the sphere.

**92.** *Demand*   A company finds that the demand for its product is modeled by $p = 85 - 0.125x$. If $x$ changes from 7 to 8, what is the corresponding change in $p$? Compare the values of $\Delta p$ and $dp$.

 **93.** *Economics: Revenue*   Consider the following cost and demand information about a monopoly (in dollars). Complete the table, and then use the information to answer the following questions. *(Source: Adapted from Taylor, Economics, First Edition)*

| Quantity of output | Price | Total revenue | Marginal revenue |
|------|------|------|------|
| 1 | 14.00 | | —— |
| 2 | 12.00 | | |
| 3 | 10.00 | | |
| 4 | 8.50 | | |
| 5 | 7.00 | | |
| 6 | 5.50 | | |

(a) Use the regression capabilities of a graphing utility to find a quadratic model for the total revenue data.

(b) From the total revenue model you found in part (a), use derivatives to find an equation for the marginal revenue. Now use the values for output in the table and compare the results with the values in the marginal revenue column of the table. How close was your model?

(c) What quantity maximizes total revenue for the monopoly?

# Sample Post-Graduation Exam Questions

| CPA | GMAT |
| --- | --- |
| GRE | CLAST |
| | Actuarial |

*The following questions represent the types of questions that appear on certified public accountant (CPA) exams, graduate management admission tests (GMAT), graduate records exams (GRE), actuarial exams, and college-level academic skills tests (CLAST). The answers to the questions are given in the back of the book.*

For Questions 1–3, use the data given in the graph.
*(Source: National Sporting Goods Association)*

Figure for 1–3

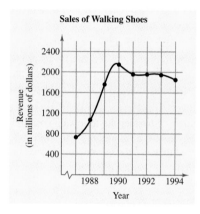

**Sales of Walking Shoes**

1. The percent increase in revenue between 1987 and 1990 was about
   (a) 40%     (b) 95%     (c) 200%     (d) 125%

2. Which of the following statements about the sales of walking shoes can be inferred from the graph?

   I. The greatest increase in revenue occurred between 1988 and 1989.

   II. In 1991, the revenue was greater than $1.8 billion.

   III. Between 1991 and 1993, the average increase in revenue per year was about $5.5 million.

   (a) I and II     (b) I and III     (c) II and III     (d) I, II, and III

3. Let $f(x)$ represent the revenue function. In 1990, $f'(x) =$
   (a) 1     (b) 1509     (c) 0     (d) undefined

4. In 1994, a company issued 75,000 shares of stock. Each share of stock was worth $85.75. Five years later, each share of stock was worth $72.21. How much less were the shares worth in 1999 than in 1994?
   (a) $1,115,500     (b) $1,051,500     (c) $1,155,000     (d) $1,015,500

5. Which of these figures most resembles the graph of $f(x) = \dfrac{x}{x^2 - 4}$?

(a) 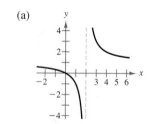   (b) 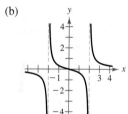   (c) 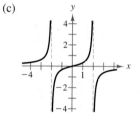   (d)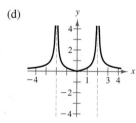

6. On April 1, 1994, Starn Corp. issued 300 of its $1000 face value bonds at $101 plus accrued interest. The bonds were dated November 1, 1993, and bear interest at an annual rate of 8% payable semiannually on November 1 and May 1. What amount did Starn receive from the bond issuance?
   (a) $313,000     (b) $327,000     (c) $303,000     (d) $300,000

# Exponential and Logarithmic Functions

# 4

**STRATEGIES** *for* **SUCCESS**

 OBJECTIVES

*When you have completed this chapter, make sure you are able to:*

❏ Graph the natural exponential function $f(x) = e^x$ and use it in applications.
❏ Calculate derivatives of exponential functions.
❏ Graph the logarithmic function $f(x) = \ln x$ and use it to solve exponential and logarithmic equations.
❏ Calculate derivatives of logarithmic functions.
❏ Solve exponential growth and decay applications.

 TOOLS

*Use these study tools to achieve the objectives above:*

**Algebra Review**
(pages 304 and 305)

**Chapter Summary and Study Strategies**
(pages 306 and 307)

**Review Exercises**
(pages 308–311)

 ADDITIONAL RESOURCES

*Use these resources to solidify your mastery of calculus:*

**Student Solutions Guide**

**Study Guide** (Additional Examples, Similar Problems, and Chapter Test)

**Algebra Review Tutorial Software**

**Graphing Technology Guide**

**Sample Post-Graduation Exam Questions**
(page 312)

**Web Exercise**
(page 285, exercise 76)

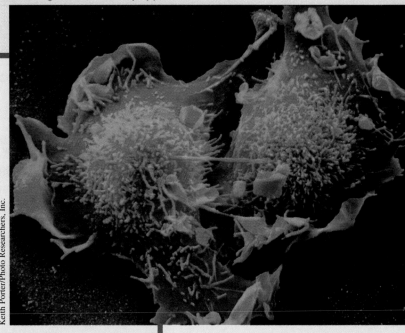

Keith Porter/Photo Researchers, Inc.

*In Exercise 58 on page 268, you will fit an exponential model to the number of cells in a population when each cell divides every 30 seconds.*

# Exponential Functions

## 4.1

*Exponential Functions • Natural Exponential Functions •
Extended Application: Compound Interest*

## Exponential Functions

You are already familiar with the behavior of algebraic functions such as

$$f(x) = x^2, \quad g(x) = \sqrt{x} = x^{1/2}, \quad \text{and} \quad h(x) = \frac{1}{x} = x^{-1},$$

each of which involves a variable raised to a constant power. By interchanging roles and raising a constant to a variable power, you obtain another important class of functions called **exponential functions.** Some simple examples are

$$f(x) = 2^x, \quad g(x) = \left(\frac{1}{10}\right)^x = \frac{1}{10^x}, \quad \text{and} \quad h(x) = 3^{2x} = 9^x.$$

In general, you can use any positive base $a \neq 1$ as the base of an exponential function.

*STUDY TIP*   In the definition at the right, the base $a = 1$ is excluded because it yields

$$f(x) = 1^x = 1.$$

This is a constant function, not an exponential function.

### Definition of Exponential Function

If $a > 0$ and $a \neq 1$, then the **exponential function** with base $a$ is given by

$$f(x) = a^x.$$

For working with exponential functions, the following properties of exponents are useful.

### Properties of Exponents

Let $a$ and $b$ be positive numbers.

1. $a^0 = 1$    2. $a^x a^y = a^{x+y}$    3. $\dfrac{a^x}{a^y} = a^{x-y}$

4. $(a^x)^y = a^{xy}$    5. $(ab)^x = a^x b^x$    6. $\left(\dfrac{a}{b}\right)^x = \dfrac{a^x}{b^x}$

7. $a^{-x} = \dfrac{1}{a^x}$

## EXAMPLE 1  Applying Properties of Exponents

(a) $(2^2)(2^3) = 2^{2+3} = 2^5 = 32$    Property 2

(b) $(2^2)(2^{-3}) = 2^{2-3} = 2^{-1} = \frac{1}{2}$    Properties 2 and 7

(c) $(3^2)^3 = 3^{2(3)} = 3^6 = 729$    Property 4

(d) $\left(\frac{1}{3}\right)^{-2} = \frac{1}{(1/3)^2} = \left(\frac{1}{1/3}\right)^2 = 3^2 = 9$    Properties 6 and 7

(e) $\dfrac{3^2}{3^3} = 3^{2-3} = 3^{-1} = \dfrac{1}{3}$    Properties 3 and 7

(f) $(2^{1/2})(3^{1/2}) = [(2)(3)]^{1/2} = 6^{1/2} = \sqrt{6}$    Property 5

Although Example 1 demonstrates the properties of exponents with integer and rational exponents, it is important to realize that the properties hold for *all* real exponents. With a calculator, you can obtain approximations of $a^x$ for any base $a$ and any real exponent $x$. Here are some examples.

$$2^{-0.6} \approx 0.660, \qquad \pi^{0.75} \approx 2.360, \qquad (1.56)^{\sqrt{2}} \approx 1.876$$

## EXAMPLE 2  Dating Organic Material

In living organic material, the ratio of radioactive carbon isotopes to the total number of carbon atoms is about 1 to $10^{12}$. When organic material dies, its radioactive carbon isotopes begin to decay, with a half-life of about 5700 years. This means that after 5700 years, the ratio of isotopes to atoms will have decreased to one-half the original ratio, after a second 5700 years the ratio will have decreased to one-fourth of the original, and so on. Figure 4.1 shows this decreasing ratio. The formula for the ratio $R$ of carbon isotopes to carbon atoms is

$$R = \left(\frac{1}{10^{12}}\right)\left(\frac{1}{2}\right)^{t/5700},$$

where $t$ is the time in years. Find the value of $R$ for the following times.

(a) 10,000 years    (b) 20,000 years    (c) 25,000 years

### Solution

(a) $R = \left(\dfrac{1}{10^{12}}\right)\left(\dfrac{1}{2}\right)^{10,000/5700} \approx 2.964 \times 10^{-13}$    Ratio for 10,000 years

(b) $R = \left(\dfrac{1}{10^{12}}\right)\left(\dfrac{1}{2}\right)^{20,000/5700} \approx 8.785 \times 10^{-14}$    Ratio for 20,000 years

(c) $R = \left(\dfrac{1}{10^{12}}\right)\left(\dfrac{1}{2}\right)^{25,000/5700} \approx 4.783 \times 10^{-14}$    Ratio for 25,000 years

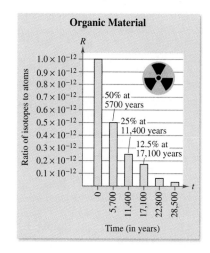

**Organic Material**

50% at 5700 years
25% at 11,400 years
12.5% at 17,100 years

Ratio of isotopes to atoms

Time (in years)

**FIGURE 4.1**

## EXAMPLE 3  Graphing Exponential Functions

Sketch the graphs of the following exponential functions.

(a) $f(x) = 2^x$     (b) $g(x) = \left(\frac{1}{2}\right)^x = 2^{-x}$     (c) $h(x) = 3^x$

### Solution

To sketch these functions by hand, you can begin by constructing a table of values, as shown below.

| $x$ | $-3$ | $-2$ | $-1$ | $0$ | $1$ | $2$ | $3$ | $4$ |
|---|---|---|---|---|---|---|---|---|
| $f(x) = 2^x$ | $\frac{1}{8}$ | $\frac{1}{4}$ | $\frac{1}{2}$ | $1$ | $2$ | $4$ | $8$ | $16$ |
| $g(x) = 2^{-x}$ | $8$ | $4$ | $2$ | $1$ | $\frac{1}{2}$ | $\frac{1}{4}$ | $\frac{1}{8}$ | $\frac{1}{16}$ |
| $h(x) = 3^x$ | $\frac{1}{27}$ | $\frac{1}{9}$ | $\frac{1}{3}$ | $1$ | $3$ | $9$ | $27$ | $81$ |

The graphs of the three functions are shown in Figure 4.2. Note that the graphs of $f(x) = 2^x$ and $h(x) = 3^x$ are increasing, whereas the graph of $g(x) = 2^{-x}$ is decreasing.

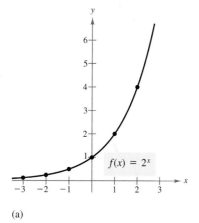

(a)

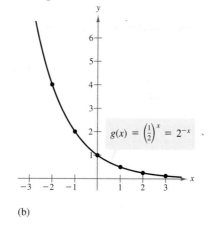

(b)

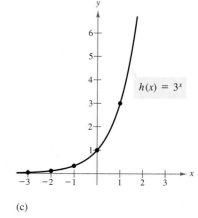

(c)

**FIGURE 4.2**

## Technology

Try graphing the functions $f(x) = 2^x$ and $h(x) = 3^x$ in the same viewing rectangle, as shown at the right. From the display, you can see that the graph of $h$ is increasing more rapidly than the graph of $f$.

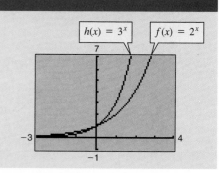

The forms of the graphs in Figure 4.2 are typical of the graphs of the exponential functions $y = a^{-x}$ and $y = a^x$, where $a > 1$. The basic characteristics of such graphs are summarized in Figure 4.3.

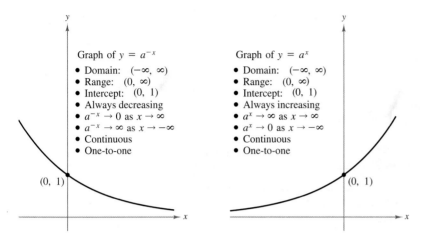

Graph of $y = a^{-x}$
- Domain: $(-\infty, \infty)$
- Range: $(0, \infty)$
- Intercept: $(0, 1)$
- Always decreasing
- $a^{-x} \to 0$ as $x \to \infty$
- $a^{-x} \to \infty$ as $x \to -\infty$
- Continuous
- One-to-one

Graph of $y = a^x$
- Domain: $(-\infty, \infty)$
- Range: $(0, \infty)$
- Intercept: $(0, 1)$
- Always increasing
- $a^x \to \infty$ as $x \to \infty$
- $a^x \to 0$ as $x \to -\infty$
- Continuous
- One-to-one

**FIGURE 4.3** Characteristics of the Exponential Functions $a^{-x}$ and $a^x (a > 1)$

## EXAMPLE 4 Graphing an Exponential Function

Sketch the graph of

$$f(x) = 3^{-x} - 1.$$

### Solution

Begin by creating a table of values, as shown below.

| $x$ | $-2$ | $-1$ | $0$ | $1$ | $2$ |
|---|---|---|---|---|---|
| $f(x)$ | $3^2 - 1 = 8$ | $3^1 - 1 = 2$ | $3^0 - 1 = 0$ | $3^{-1} - 1 = -\frac{2}{3}$ | $3^{-2} - 1 = -\frac{8}{9}$ |

From the limit

$$\lim_{x \to \infty} (3^{-x} - 1) = \lim_{x \to \infty} 3^{-x} - \lim_{x \to \infty} 1$$

$$= \lim_{x \to \infty} \frac{1}{3^x} - \lim_{x \to \infty} 1$$

$$= 0 - 1$$

$$= -1,$$

you can see that $y = -1$ is a horizontal asymptote of the graph. The graph is shown in Figure 4.4.

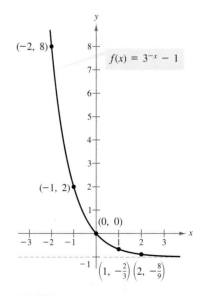

$(-2, 8)$

$f(x) = 3^{-x} - 1$

$(-1, 2)$

$(0, 0)$

$\left(1, -\frac{2}{3}\right) \left(2, -\frac{8}{9}\right)$

**FIGURE 4.4**

## Natural Exponential Functions

At the beginning of this section, exponential functions were introduced using an unspecified base $a$. In calculus, the most convenient (or natural) choice for a base is the irrational number $e$, whose decimal approximation is

$$e \approx 2.71828182846.$$

Although this choice of base may seem unusual, its convenience will become apparent as the rules for differentiating exponential functions are developed in Section 4.2. In that development, you will encounter the limit used in the following definition of $e$.

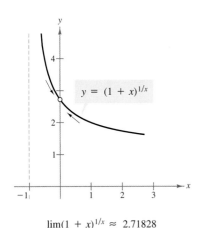

$$\lim_{x \to 0}(1 + x)^{1/x} \approx 2.71828$$

**FIGURE 4.5**

| Limit Definition of $e$ |
| --- |

The irrational number $e$ is defined to be the limit of $(1 + x)^{1/x}$ as $x \to 0$. That is,

$$\lim_{x \to 0}(1 + x)^{1/x} = e.$$

The graph of $y = (1 + x)^{1/x}$ is shown in Figure 4.5. Try reproducing this graph with a graphing utility. Then, use the zoom and trace features to find values of $y$ near $x = 0$. You will find that the $y$-values get closer and closer to the number $e \approx 2.71828$.

## EXAMPLE 5   Graphing the Natural Exponential Function

Sketch the graph of

$$f(x) = e^x.$$

**Solution**

Begin by evaluating the function for several values of $x$, as shown in the table.

| $x$ | $-2$ | $-1$ | $0$ | $1$ | $2$ |
| --- | --- | --- | --- | --- | --- |
| $f(x)$ | $e^{-2} \approx 0.135$ | $e^{-1} \approx 0.368$ | $e^0 = 1$ | $e^1 \approx 2.718$ | $e^2 \approx 7.389$ |

The graph of $f(x) = e^x$ is shown in Figure 4.6. Note that $e^x$ is positive for all values of $x$. Moreover, the graph has the $x$-axis as a horizontal asymptote to the left. That is,

$$\lim_{x \to -\infty} e^x = 0.$$

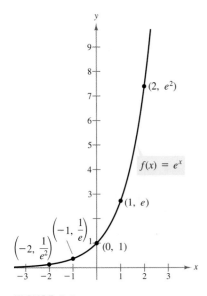

**FIGURE 4.6**

Exponential functions are often used to model the growth of a quantity or a population. When the quantity's growth is *not* restricted, an exponential model is often used. When the quantity's growth *is* restricted, the best model is often a **logistics growth function** of the form

$$f(t) = \frac{a}{1 + be^{-kt}}.$$

Graphs of both types of population growth models are shown in Figure 4.7.

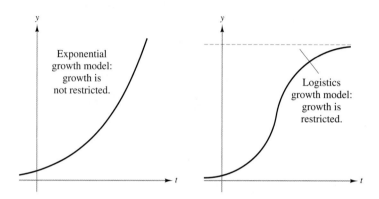

Exponential growth model: growth is not restricted.

Logistics growth model: growth is restricted.

**FIGURE 4.7**

## EXAMPLE 6 Modeling a Population

*When a culture is grown in a dish, the size of the dish and the available food limit the culture's growth.*

A bacterial culture is growing according to the *logistics growth model*

$$y = \frac{1.25}{1 + 0.25e^{-0.4t}}, \qquad 0 \le t,$$

where $y$ is the culture weight (in grams) and $t$ is the time (in hours). Find the weight of the culture after 0 hours, 1 hour, and 10 hours. What is the limit of the model as $t$ increases without bound?

**Solution**

$$y = \frac{1.25}{1 + 0.25e^{-0.4(0)}} = 1 \text{ gram} \qquad \text{Weight when } t = 0$$

$$y = \frac{1.25}{1 + 0.25e^{-0.4(1)}} \approx 1.071 \text{ grams} \qquad \text{Weight when } t = 1$$

$$y = \frac{1.25}{1 + 0.25e^{-0.4(10)}} \approx 1.244 \text{ grams} \qquad \text{Weight when } t = 10$$

As $t$ approaches infinity, the limit of $y$ is

$$\lim_{t \to \infty} \frac{1.25}{1 + 0.25e^{-0.4t}} = \lim_{t \to \infty} \frac{1.25}{1 + (0.25/e^{0.4t})} = \frac{1.25}{1 + 0} = 1.25.$$

Thus, as $t$ increases without bound, the weight of the culture approaches 1.25 grams. The graph of the model is shown in Figure 4.8.

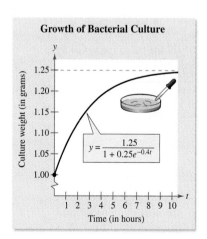

**Growth of Bacterial Culture**

$$y = \frac{1.25}{1 + 0.25e^{-0.4t}}$$

Culture weight (in grams)

Time (in hours)

**FIGURE 4.8**

## Extended Application: Compound Interest

If $P$ dollars is deposited in an account at an annual interest rate of $r$ (in decimal form), what is the balance after 1 year? The answer depends on the number of times the interest is compounded, according to the formula

$$A = P\left(1 + \frac{r}{n}\right)^n,$$

where $n$ is the number of compoundings per year. The balances for a deposit of $1000 at 8%, at various compounding periods, are shown in the table.

**Technology**

The table at the right can be reproduced using a spreadsheet or the table feature of a graphing utility. Try doing this. Do you get the same results as those shown in the table?

| Number of times compounded per year, $n$ | Balance (in dollars), $A$ |
|---|---|
| Annually, $n = 1$ | $A = P\left(1 + \frac{0.08}{1}\right)^1 = \$1080.00$ |
| Semiannually, $n = 2$ | $A = P\left(1 + \frac{0.08}{2}\right)^2 = \$1081.60$ |
| Quarterly, $n = 4$ | $A = P\left(1 + \frac{0.08}{4}\right)^4 = \$1082.43$ |
| Monthly, $n = 12$ | $A = P\left(1 + \frac{0.08}{12}\right)^{12} = \$1083.00$ |
| Daily, $n = 365$ | $A = P\left(1 + \frac{0.08}{365}\right)^{365} = \$1083.28$ |

You may be surprised to discover that as $n$ increases, the balance $A$ approaches a limit, as indicated in the following development. In this development, let $x = r/n$. Then $x \to 0$ as $n \to \infty$, and you have

$$
\begin{aligned}
A &= \lim_{n \to \infty} P\left(1 + \frac{r}{n}\right)^n \\
&= P \lim_{n \to \infty} \left[\left(1 + \frac{r}{n}\right)^{n/r}\right]^r \\
&= P\left[\lim_{x \to 0} (1 + x)^{1/x}\right]^r \quad \text{Substitute } x \text{ for } r/n. \\
&= Pe^r.
\end{aligned}
$$

*Discovery*

Use a spreadsheet or the table feature of a graphing utility to evaluate the expression

$$\left(1 + \frac{1}{n}\right)^n$$

for the following values of $n$.

| $n$ | $(1 + 1/n)^n$ |
|---|---|
| 10 | |
| 100 | |
| 1000 | |
| 10,000 | |
| 100,000 | |

What can you conclude? Try the same thing for negative values of $n$.

This limit is the balance after 1 year of **continuous compounding.** Thus, for a deposit of $1000 at 8%, compounded continuously, the balance at the end of the year would be $A = 1000e^{0.08} \approx \$1083.29$.

### Summary of Compound Interest Formulas

Let $P$ be the amount deposited, $t$ the number of years, $A$ the balance, and $r$ the annual interest rate (in decimal form).

1. Compounded $n$ times per year: $A = P[1 + (r/n)]^{nt}$

2. Compounded continuously: $A = Pe^{rt}$

The average interest rates paid by banks on savings accounts have varied greatly during the past 30 years. At times, savings accounts have earned as much as 12% annual interest and at times they have earned as little as 3%. The next example shows how the annual interest rate can affect the balance of an account.

## EXAMPLE 7  Finding Account Balances

You are creating a trust fund for your newborn nephew. You deposit $25,000 in an account, with instructions that the account be turned over to your nephew on his 25th birthday. Compare the balances in the account for the following.

(a)  6%, compounded continuously    (b)  6%, compounded quarterly

(c)  10%, compounded continuously   (d)  10%, compounded quarterly

### Solution

(a)  $25{,}000e^{0.06(25)} = \$112{,}042.23$       6%, compounded continuously

(b)  $25{,}000\left(1 + \dfrac{0.06}{4}\right)^{4(25)} = \$110{,}801.14$       6%, compounded quarterly

(c)  $25{,}000e^{0.1(25)} = \$304{,}562.35$       10%, compounded continuously

(d)  $25{,}000\left(1 + \dfrac{0.1}{4}\right)^{4(25)} = \$295{,}342.91$       10%, compounded quarterly

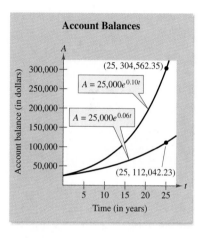

The growth of the account for parts (a) and (c) is shown in Figure 4.9. Notice the dramatic difference between the balances at 6% and at 10%.

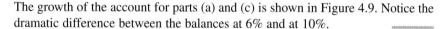

**FIGURE 4.9**

---

*Group Discussion*        *Compound Interest*

You want to invest $5000 in a certificate of deposit for 10 years. You are given the following options. Which would you choose?

a.  You can buy a 10-year certificate of deposit that earns 7%, compounded continuously. You are guaranteed a 7% rate for the entire 10 years, but you cannot withdraw the money early without paying a substantial penalty.

b.  You can buy a 5-year certificate of deposit that earns 6%, compounded continuously. After 5 years, you can reinvest your money at whatever the current interest rate is at that time.

c.  You can buy a 2-year certificate of deposit that earns 5%, compounded continuously. After 2 years, you can reinvest your money at whatever the current interest rate is at that time.

*Warm Up*

The following warm-up exercises involve skills that were covered in earlier sections. You will use these skills in the exercise set for this section.

In Exercises 1 and 2, discuss the continuity of the function.

**1.** $f(x) = \dfrac{3x^2 + 2x + 1}{x^2 + 1}$

**2.** $g(x) = \dfrac{x^2 - 9x + 20}{x - 4}$

In Exercises 3–10, find the limit.

**3.** $\displaystyle\lim_{x\to\infty} \dfrac{25}{1 + 4x}$

**4.** $\displaystyle\lim_{x\to\infty} \dfrac{16x}{3 + x^2}$

**5.** $\displaystyle\lim_{x\to\infty} \dfrac{8x^3 + 2}{2x^3 + x}$

**6.** $\displaystyle\lim_{x\to\infty} \dfrac{x}{2x}$

**7.** $\displaystyle\lim_{x\to\infty} \dfrac{3}{2 + (1/x)}$

**8.** $\displaystyle\lim_{x\to\infty} \dfrac{6}{1 + x^{-2}}$

**9.** $\displaystyle\lim_{x\to\infty} 2^{-x}$

**10.** $\displaystyle\lim_{x\to\infty} \dfrac{7}{1 + 5x}$

# EXERCISES 4.1

In Exercises 1 and 2, evaluate each expression.

**1.** (a) $5(5^3)$          (b) $27^{2/3}$
   (c) $64^{3/4}$          (d) $81^{1/2}$
   (e) $25^{3/2}$          (f) $32^{2/5}$

**2.** (a) $\left(\frac{1}{5}\right)^3$          (b) $\left(\frac{1}{8}\right)^{1/3}$

   (c) $64^{2/3}$          (d) $\left(\frac{5}{8}\right)^2$

   (e) $100^{3/2}$          (f) $4^{5/2}$

In Exercises 3–8, use the properties of exponents to simplify the expression.

**3.** (a) $(5^2)(5^3)$          (b) $(5^2)(5^{-3})$
   (c) $(5^2)^2$          (d) $5^{-3}$

**4.** (a) $\dfrac{5^3}{5^6}$          (b) $\left(\dfrac{1}{5}\right)^{-2}$

   (c) $(8^{1/2})(2^{1/2})$          (d) $(32^{3/2})\left(\frac{1}{2}\right)^{3/2}$

**5.** (a) $\dfrac{5^3}{25^2}$          (b) $(9^{2/3})(3)(3^{2/3})$

   (c) $[(25^{1/2})(5^2)]^{1/3}$          (d) $(8^2)(4^3)$

**6.** (a) $(4^3)(4^2)$          (b) $\left(\frac{1}{4}\right)^2(4^2)$
   (c) $(4^6)^{1/2}$          (d) $[(8^{-1})(8^{2/3})]^3$

**7.** (a) $(e^3)(e^4)$          (b) $(e^3)^4$
   (c) $(e^3)^{-2}$          (d) $e^0$

**8.** (a) $\left(\dfrac{1}{e}\right)^{-2}$          (b) $\left(\dfrac{e^5}{e^2}\right)^{-1}$

   (c) $\dfrac{e^5}{e^3}$          (d) $\dfrac{1}{e^{-3}}$

In Exercises 9–22, solve the equation for $x$.

**9.** $3^x = 81$          **10.** $5^{x+1} = 125$

**11.** $\left(\frac{1}{3}\right)^{x-1} = 27$          **12.** $\left(\frac{1}{5}\right)^{2x} = 625$

**13.** $4^3 = (x + 2)^3$          **14.** $4^2 = (x + 2)^2$

**15.** $x^{3/4} = 8$          **16.** $(x + 3)^{4/3} = 16$

**17.** $e^{-3x} = e$          **18.** $e^x = 1$

**19.** $e^{\sqrt{x}} = e^3$          **20.** $e^{-1/x} = \sqrt{e}$

**21.** $x^{2/3} = \sqrt[3]{e^2}$          **22.** $\dfrac{x^2}{2} = e^2$

In Exercises 23–28, match the function with its graph. [The graphs are labeled (a)–(f).]

**23.** $f(x) = 3^x$

**24.** $f(x) = 3^{-x/2}$

**25.** $f(x) = -3^x$

**26.** $f(x) = 3^{x-2}$

**27.** $f(x) = 3^{-x} - 1$

**28.** $f(x) = 3^x + 2$

(a)
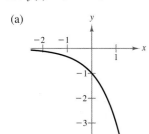

(b)
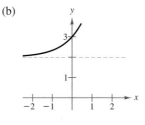

(c)
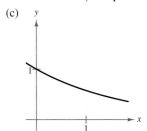

(d)
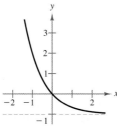

(e)
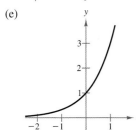

(f)
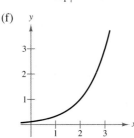

In Exercises 29–40, sketch the graph of the function.

**29.** $f(x) = 6^x$

**30.** $f(x) = 4^x$

**31.** $f(x) = \left(\frac{1}{5}\right)^x = 5^{-x}$

**32.** $f(x) = \left(\frac{1}{4}\right)^x = 4^{-x}$

**33.** $y = 3^{-x^2}$

**34.** $y = 2^{-x^2}$

**35.** $y = 3^{-|x|}$

**36.** $y = 3^{|x|}$

**37.** $s(t) = \frac{1}{4}(3^{-t})$

**38.** $s(t) = 2^{-t} + 3$

**39.** $h(x) = e^{x-2}$

**40.** $f(x) = e^{2x}$

In Exercises 41–44, use a graphing utility to sketch the graph of the function. Be sure to choose an appropriate viewing rectangle.

**41.** $N(t) = 500e^{-0.2t}$

**42.** $A(t) = 500e^{0.15t}$

**43.** $g(x) = \dfrac{2}{1 + e^{x^2}}$

**44.** $g(x) = \dfrac{10}{1 + e^{-x}}$

*Compound Interest* In Exercises 45–48, complete the table to determine the balance $A$ for $P$ dollars invested at rate $r$ for $t$ years, compounded $n$ times per year.

| $n$ | 1 | 2 | 4 | 12 | 365 | Continuous compounding |
|---|---|---|---|---|---|---|
| $A$ | | | | | | |

**45.** $P = \$1000$, $r = 3\%$, $t = 10$ years

**46.** $P = \$2500$, $r = 5\%$, $t = 20$ years

**47.** $P = \$1000$, $r = 3\%$, $t = 40$ years

**48.** $P = \$2500$, $r = 5\%$, $t = 40$ years

*Compound Interest* In Exercises 49–52, complete the table to determine the amount of money $P$ that should be invested at rate $r$ to produce a final balance of $\$100,000$ in $t$ years.

| $t$ | 1 | 10 | 20 | 30 | 40 | 50 |
|---|---|---|---|---|---|---|
| $P$ | | | | | | |

**49.** $r = 4\%$, compounded continuously

**50.** $r = 3\%$, compounded continuously

**51.** $r = 5\%$, compounded monthly

**52.** $r = 6\%$, compounded daily

**53.** *Demand* The demand function for a product is modeled by

$$p = 5000\left(1 - \frac{4}{4 + e^{-0.002x}}\right),$$

as shown in the figure. Find the price of the product if the quantity demanded is (a) $x = 100$ units and (b) $x = 500$ units. What is the limit of the price as $x$ increases without bound?

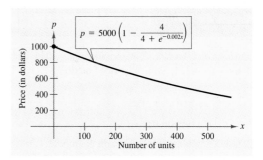

**54.** *Demand*   The demand function for a product is modeled. by $p = 500 - 0.5e^{0.004x}$, as shown in the figure. Find the price of the product if the quantity demanded is (a) $x = 1000$ units and (b) $x = 1500$ units.

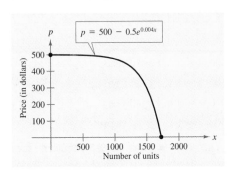

**55.** *Probability*   The average time between incoming calls at a switchboard is 3 minutes. If a call has just come in, the probability that the next call will come within the next $t$ minutes is $P(t) = 1 - e^{-t/3}$. Find the probability of the following.

(a) A call comes in within $\frac{1}{2}$ minute.
(b) A call comes in within 2 minutes.
(c) A call comes in within 5 minutes.

**56.** *Fuel Efficiency*   An automobile gets 28 miles per gallon at speeds of up to 50 miles per hour. At speeds of 50 miles per hour or greater, the number of miles per gallon drops at the rate of 12% for each 10 miles per hour. If $s$ is the speed (in miles per hour) and $y$ is the miles per gallon, then

$$y = 28e^{0.6 - 0.012s}, \quad s \geq 50.$$

Use this information to complete the table. What can you conclude?

| Speed ($s$) | 50 | 55 | 60 | 65 | 70 |
|---|---|---|---|---|---|
| Miles per gallon ($y$) | | | | | |

 **57.** *Population Growth*   The population $y$ of a bacterial culture is modeled by the logistics function

$$y = \frac{925}{1 + e^{-0.3t}},$$

where $t$ is the time in days.

(a) Use a graphing utility to graph the model.
(b) Does the population have a limit as $t$ increases without bound? Explain your answer.

(c) How would the model change if

$$y = \frac{1000}{1 + e^{-0.3t}}?$$

Explain your answer. Draw some conclusions about this type of model.

 **58.** *Biology: Cell Division*   Suppose that you have a single imaginary bacterium able to divide to form two new cells every 30 seconds. Make a table of values for the number of individuals in the population over 30-second intervals up to 5 minutes. Graph the points and use a graphing utility to fit an exponential model to the data. *(Source: Adapted from Levine/Miller,* Biology: Discovering Life, *Second Edition)*

 **59.** *Learning Theory*   In a learning theory project, the proportion $P$ of correct responses after $n$ trials can be modeled by

$$P = \frac{0.83}{1 + e^{-0.2n}}.$$

(a) Use a graphing utility to estimate the proportion of correct responses after ten trials. Check your result analytically.
(b) Use a graphing utility to estimate the number of trials required to have a proportion of correct responses of 0.75.
(c) Does the proportion of correct responses have a limit as $n$ increases without bound? Explain your answer.

**60.** *Learning Theory*   In a typing class, the average number $N$ of words per minute typed after $t$ weeks of lessons can be modeled by

$$N = \frac{95}{1 + 8.5e^{-0.12t}}.$$

(a) Use a graphing utility to estimate the average number of words per minute typed after 10 weeks. Check your result analytically.
(b) Use a graphing utility to estimate the number of weeks required to achieve an average of 70 words per minute.
(c) Does the number of words per minute have a limit as $t$ increases without bound? Explain your answer.

 In Exercises 61–64, use a graphing utility to graph the function. Determine whether the function has any horizontal asymptotes and discuss the continuity of the function.

**61.** $f(x) = \dfrac{e^x + e^{-x}}{2}$   **62.** $f(x) = \dfrac{e^x - e^{-x}}{2}$

**63.** $f(x) = \dfrac{2}{1 + e^{1/x}}$   **64.** $f(x) = \dfrac{2}{1 + 2e^{-0.2x}}$

# 4.2 | Derivatives of Exponential Functions

*Derivatives of Exponential Functions* • *Applications* •
*The Normal Probability Density Function*

## Derivatives of Exponential Functions

In Section 4.1, we stated that the most convenient base for exponential functions is the irrational number $e$. The convenience of this base stems primarily from the fact that the function $f(x) = e^x$ *is its own derivative.* You will see that this is not true of other exponential functions of the form $y = a^x$ where $a \neq e$. To verify that $f(x) = e^x$ is its own derivative, notice that the limit

$$\lim_{\Delta x \to 0} (1 + \Delta x)^{1/\Delta x} = e$$

implies that for small values of $\Delta x$, $e \approx (1 + \Delta x)^{1/\Delta x}$, or $e^{\Delta x} \approx 1 + \Delta x$. This approximation is used in the following derivation.

$$f'(x) = \lim_{\Delta x \to 0} \frac{f(x + \Delta x) - f(x)}{\Delta x} \qquad \text{Definition of derivative}$$

$$= \lim_{\Delta x \to 0} \frac{e^{x + \Delta x} - e^x}{\Delta x} \qquad \text{Use } f(x) = e^x.$$

$$= \lim_{\Delta x \to 0} \frac{e^x(e^{\Delta x} - 1)}{\Delta x} \qquad \text{Factor numerator.}$$

$$= \lim_{\Delta x \to 0} \frac{e^x[(1 + \Delta x) - 1]}{\Delta x} \qquad \text{Substitute } 1 + \Delta x \text{ for } e^{\Delta x}.$$

$$= \lim_{\Delta x \to 0} \frac{e^x(\Delta x)}{\Delta x} \qquad \text{Cancel common factor.}$$

$$= \lim_{\Delta x \to 0} e^x \qquad \text{Simplify.}$$

$$= e^x \qquad \text{Evaluate limit.}$$

If $u$ is a function of $x$, you can apply the Chain Rule to obtain the derivative of $e^u$ with respect to $x$. Both formulas are summarized as follows.

---

### Derivatives of Exponential Functions

Let $u$ be a differentiable function of $x$.

1. $\dfrac{d}{dx}[e^x] = e^x$ 　　　　　2. $\dfrac{d}{dx}[e^u] = e^u \dfrac{du}{dx}$

---

## *Discovery*

Use a spreadsheet or the table feature of a graphing utility to compare the expressions $e^{\Delta x}$ and $1 + \Delta x$ for values of $\Delta x$ near zero.

| $\Delta x$ | $e^{\Delta x}$ | $1 + \Delta x$ |
|---|---|---|
| 0.1 | | |
| 0.01 | | |
| 0.001 | | |

What can you conclude? Explain how this result is used in the development of the derivative of $f(x) = e^x$.

## EXAMPLE 1  Interpreting Derivatives Graphically

Find the slope of the tangent line of

$$f(x) = e^x \qquad \text{Original function}$$

at the points $(0, 1)$ and $(1, e)$. What conclusion can you make?

*Solution*

Because the derivative of $f$ is

$$f'(x) = e^x, \qquad \text{Derivative}$$

it follows that the slope of the tangent line to the graph of $f$ is

$$f'(0) = e^0 = 1 \qquad \text{Slope at point } (0, 1)$$

at the point $(0, 1)$, and

$$f'(1) = e^1 = e \qquad \text{Slope at point } (1, e)$$

at the point $(1, e)$, as shown in Figure 4.10. From this pattern, you can see that the slope of the tangent line to the graph of $f(x) = e^x$ at any point $(x, e^x)$ is equal to the $y$-coordinate of the point.

At the point $(1, e)$ the slope is $e \approx 2.72$.

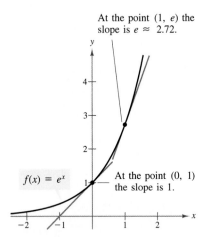

At the point $(0, 1)$ the slope is 1.

$f(x) = e^x$

**FIGURE 4.10**

## EXAMPLE 2  Differentiating Exponential Functions

Differentiate the following functions.

(a)  $f(x) = e^{2x}$  (b)  $f(x) = e^{-3x^2}$

(c)  $f(x) = 6e^{x^3}$  (d)  $f(x) = e^{-x}$

*Solution*

(a) Let $u = 2x$. Then $du/dx = 2$, and you can apply the Chain Rule.

$$f'(x) = e^u \frac{du}{dx} = e^{2x}(2) = 2e^{2x}$$

(b) Let $u = -3x^2$. Then $du/dx = -6x$, and you can apply the Chain Rule.

$$f'(x) = e^u \frac{du}{dx} = e^{-3x^2}(-6x) = -6xe^{-3x^2}$$

(c) Let $u = x^3$. Then $du/dx = 3x^2$, and you can apply the Chain Rule.

$$f'(x) = 6e^u \frac{du}{dx} = 6e^{x^3}(3x^2) = 18x^2e^{x^3}$$

(d) Let $u = -x$. Then $du/dx = -1$, and you can apply the Chain Rule.

$$f'(x) = e^u \frac{du}{dx} = e^{-x}(-1) = -e^{-x}$$

**STUDY TIP**  In Example 2, notice that when you differentiate an exponential function, the exponent does not change. For instance, the derivative of $y = e^{3x}$ is $y' = 3e^{3x}$. In both the function and its derivative, the exponent is $3x$.

The differentiation rules that you studied in Chapter 2 can be used with exponential functions, as illustrated in Example 3.

## EXAMPLE 3   Differentiating Exponential Functions

Differentiate the following functions.

(a) $f(x) = xe^x$

(b) $f(x) = \dfrac{e^x - e^{-x}}{2}$

(c) $f(x) = \dfrac{e^x}{x}$

(d) $f(x) = xe^x - e^x$

*Solution*

(a) $f(x) = xe^x$      Original function

    $f'(x) = xe^x + e^x(1)$      Product Rule

    $= xe^x + e^x$      Simplify.

(b) $f(x) = \dfrac{e^x - e^{-x}}{2}$      Original function

    $= \tfrac{1}{2}(e^x - e^{-x})$      Rewrite.

    $f'(x) = \tfrac{1}{2}(e^x + e^{-x})$      Constant Multiple Rule

(c) $f(x) = \dfrac{e^x}{x}$      Original function

    $f'(x) = \dfrac{xe^x - e^x(1)}{x^2}$      Quotient Rule

    $= \dfrac{e^x(x - 1)}{x^2}$      Simplify.

(d) $f(x) = xe^x - e^x$      Original function

    $f'(x) = [xe^x + e^x(1)] - e^x$      Product and Difference Rules

    $= xe^x + e^x - e^x$

    $= xe^x$      Simplify.

## Technology

If you have access to a symbolic differentiation utility such as *Derive, Maple, Mathematica, Mathcad,* or the *TI-92,* try using it to find the derivatives of the functions in Example 3. For example, the display at the right shows how *Derive* for Windows can be used to find the derivative of $f(x) = xe^x - e^x$.

#1: $x \cdot \text{EXP}(x) - \text{EXP}(x)$      Author.

#2: $\dfrac{d}{dx}(x \cdot \text{EXP}(x) - \text{EXP}(x))$    Differentiate.

#3: $x \cdot \hat{e}^x$      Simplify.

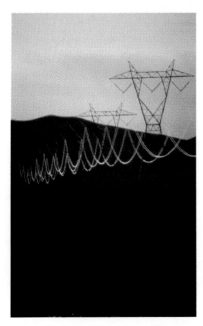

FPG International/Campbell & Boulanger

*Utility wires strung between poles have the shape of a catenary.*

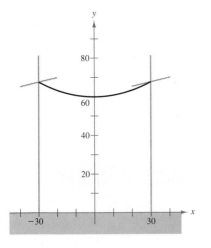

**FIGURE 4.11**

# Applications

In Chapter 3, you learned how to use derivatives to analyze the graph of a function. The next example applies these techniques to a function composed of exponential functions. In the example, notice that $e^a = e^b$ implies that $a = b$.

### EXAMPLE 4  Analyzing a Catenary

When a telephone wire is hung between two poles, the wire forms a $\cup$-shaped curve called a **catenary.** For instance, the function

$$y = 30(e^{x/60} + e^{-x/60}), \qquad -30 \le x \le 30$$

models the shape of a telephone wire strung between two poles that are 60 feet apart ($x$ and $y$ are measured in feet). Show that the lowest point on the wire is midway between the two poles. How much does the wire sag between the two poles?

### Solution

The derivative of the function is

$$y' = 30\left[e^{x/60}\left(\tfrac{1}{60}\right) + e^{-x/60}\left(-\tfrac{1}{60}\right)\right] = \tfrac{1}{2}(e^{x/60} - e^{-x/60}).$$

To find the critical numbers, set the derivative equal to zero.

| | |
|---|---|
| $\tfrac{1}{2}(e^{x/60} - e^{-x/60}) = 0$ | Set derivative equal to 0. |
| $e^{x/60} - e^{-x/60} = 0$ | Multiply both sides by 2. |
| $e^{x/60} = e^{-x/60}$ | Add $e^{-x/60}$ to both sides. |
| $\dfrac{x}{60} = -\dfrac{x}{60}$ | If $e^a = e^b$, then $a = b$. |
| $x = -x$ | Multiply both sides by 60. |
| $2x = 0$ | Add $x$ to both sides. |
| $x = 0$ | Divide both sides by 2. |

Using the First-Derivative Test, you can determine that the critical number $x = 0$ yields a relative minimum of the function. From the graph in Figure 4.11, you can see that this relative minimum is actually a minimum on the interval $[-30, 30]$. To find how much the wire sags between the two poles, you can compare its height at each pole with the height at the midpoint.

| | |
|---|---|
| $y = 30(e^{-30/60} + e^{-(-30)/60}) \approx 67.7$ feet | Height at left pole |
| $y = 30(e^{0/60} + e^{-(0)/60}) = 60$ feet | Height at midpoint |
| $y = 30(e^{30/60} + e^{-(30)/60}) \approx 67.7$ feet | Height at right pole |

From this, you can see that the wire sags about 7.7 feet.

## EXAMPLE 5   Finding a Maximum Revenue

The demand function for a product is modeled by

$$p = 56e^{-0.000012x}, \qquad \text{Demand function}$$

where $p$ is the price per unit (in dollars) and $x$ is the number of units. What price will yield a maximum revenue?

### Solution

The revenue function is

$$R = xp = 56xe^{-0.000012x}. \qquad \text{Revenue function}$$

To find the maximum revenue *analytically,* you would set the marginal revenue, $dR/dx$, equal to zero and solve for $x$. In this problem, it is easier to use a *graphical* approach. After experimenting to find a reasonable viewing rectangle, you can obtain a graph of $R$ that is similar to that shown in Figure 4.12. Using the zoom and trace features, you can conclude that the maximum revenue occurs when $x$ is about 83,300 units. To find the price that corresponds to this production level, substitute $x \approx 83{,}300$ into the demand function.

$$p \approx 56e^{-0.000012(83{,}300)}$$
$$\approx \$20.61.$$

Thus, a price of about $20.61 will yield a maximum revenue.

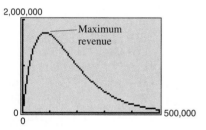

**FIGURE 4.12**   Use the zoom and trace features to approximate the *x*-value that corresponds to the maximum revenue.

*STUDY TIP*   Try solving the problem in Example 5 analytically. When you do this, you obtain

$$\frac{dR}{dx} = 56xe^{-0.000012x}(-0.000012) + e^{-0.000012x}(56) = 0.$$

Explain how you would solve this equation. What is the solution?

## The Normal Probability Density Function

If you take a course in statistics or quantitative business analysis, you will spend quite a bit of time studying the characteristics and use of the **normal probability density function** given by

$$f(x) = \frac{1}{\sigma\sqrt{2\pi}}e^{-(x-\mu)^2/2\sigma^2},$$

where $\sigma$ is the lowercase Greek letter sigma, and $\mu$ is the lowercase Greek letter mu. In this formula, $\sigma$ represents the standard deviation of the probability distribution, and $\mu$ represents the mean of the probability distribution.

### EXAMPLE 6   Exploring a Probability Density Function

Show that the graph of the normal probability density function

$$f(x) = \frac{1}{\sqrt{2\pi}}e^{-x^2/2} \qquad \text{Original function}$$

has points of inflection at $x = \pm 1$.

**Solution**

Begin by finding the second derivative of the function.

$$f'(x) = \frac{1}{\sqrt{2\pi}}(-x)e^{-x^2/2} \qquad \text{First derivative}$$

$$f''(x) = \frac{1}{\sqrt{2\pi}}[(-x)(-x)e^{-x^2/2} + (-1)e^{-x^2/2}] \qquad \text{Second derivative}$$

$$= \frac{1}{\sqrt{2\pi}}(e^{-x^2/2})(x^2 - 1) \qquad \text{Simplify.}$$

By setting the second derivative equal to zero, you can determine that $x = \pm 1$. By testing the concavity of the graph, you can then conclude that these $x$-values yield points of inflection, as shown in Figure 4.13.

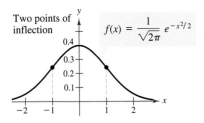

Two points of inflection

$$f(x) = \frac{1}{\sqrt{2\pi}}e^{-x^2/2}$$

**FIGURE 4.13**   The graph of the normal probability density function is bell-shaped.

---

*Group Discussion*

## The Normal Probability Density Function

Use Example 6 as a model to show that the graph of the normal probability density function with $\mu = 0$,

$$f(x) = \frac{1}{\sigma\sqrt{2\pi}}e^{-x^2/2\sigma^2},$$

has points of inflection at $x = \pm\sigma$. What is the maximum of the function? Use a graphing utility to check your answer by graphing the function for several values of $\sigma$.

*Warm Up*

The following warm-up exercises involve skills that were covered in earlier sections. You will use these skills in the exercise set for this section.

In Exercises 1–4, factor the expression.

**1.** $x^2 e^x - \frac{1}{2} e^x$

**2.** $(xe^{-x})^{-1} + e^x$

**3.** $xe^x - e^{2x}$

**4.** $e^x - xe^{-x}$

In Exercises 5–8, find the derivative of the function.

**5.** $f(x) = \dfrac{3}{7x^2}$

**6.** $g(x) = 3x^2 - \dfrac{x}{6}$

**7.** $f(x) = (4x - 3)(x^2 + 9)$

**8.** $f(t) = \dfrac{t - 2}{\sqrt{t}}$

In Exercises 9 and 10, find the relative extrema of the function.

**9.** $f(x) = \frac{1}{8}x^3 - 2x$

**10.** $f(x) = x^4 - 2x^2 + 5$

# EXERCISES 4.2

In Exercises 1–4, find the slope of the tangent line to the exponential function at the point $(0, 1)$.

**1.** $y = e^{3x}$

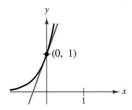

**2.** $y = e^{2x}$

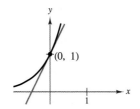

**3.** $y = e^{-x}$

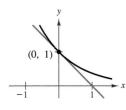

**4.** $y = e^{-2x}$

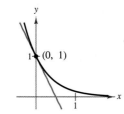

In Exercises 5–16, find the derivative of the function.

**5.** $y = e^{4x}$

**6.** $y = e^{1-x}$

**7.** $y = e^{-x^2}$

**8.** $f(x) = e^{1/x}$

**9.** $f(x) = e^{-1/x^2}$

**10.** $g(x) = e^{\sqrt{x}}$

**11.** $f(x) = (x^2 + 1)e^{4x}$

**12.** $y = (10e^{-x})^2$

**13.** $f(x) = \dfrac{2}{e^x + e^{-x}}$

**14.** $f(x) = \dfrac{e^x + e^{-x}}{2}$

**15.** $y = xe^x - 4e^{-x}$

**16.** $y = x^2 e^x - 2xe^x + 2e^x$

In Exercises 17–20, determine an equation of the tangent line to the function at the indicated point.

| *Function* | *Point* |
|---|---|
| **17.** $y = e^{-2x+x^2}$ | $(0, 1)$ |
| **18.** $g(x) = e^{x^3}$ | $\left(-1, \dfrac{1}{e}\right)$ |
| **19.** $y = x^2 e^{-x}$ | $\left(2, \dfrac{4}{e^2}\right)$ |
| **20.** $y = (e^{-x} + e^x)^3$ | $(0, 8)$ |

In Exercises 21 and 22, find $dy/dx$ implicitly.

**21.** $xe^x + 2ye^x = 0$

**22.** $e^{xy} + x^2 - y^2 = 10$

In Exercises 23–26, find the second derivative.

**23.** $f(x) = 2e^{3x} + 3e^{-2x}$

**24.** $f(x) = (1 + 2x)e^{4x}$

**25.** $f(x) = 5e^{-x} - 2e^{-5x}$

**26.** $f(x) = (3 + 2x)e^{-3x}$

In Exercises 27–30, graph and analyze the function. Include extrema, points of inflection, and asymptotes in your analysis.

**27.** $f(x) = \dfrac{1}{2 - e^{-x}}$

**28.** $f(x) = \dfrac{e^x - e^{-x}}{2}$

**29.** $f(x) = x^2 e^{-x}$

**30.** $f(x) = xe^{-x}$

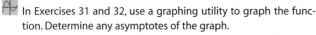 In Exercises 31 and 32, use a graphing utility to graph the function. Determine any asymptotes of the graph.

**31.** $f(x) = \dfrac{8}{1 + e^{-0.5x}}$

**32.** $g(x) = \dfrac{8}{1 + e^{-0.5/x}}$

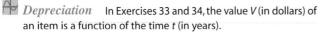 *Depreciation*    In Exercises 33 and 34, the value $V$ (in dollars) of an item is a function of the time $t$ (in years).

(a) Sketch the function over the interval $[0, 10]$. Use a graphing utility to verify your graph.

(b) Find the rate of change of $V$ when $t = 1$.

(c) Find the rate of change of $V$ when $t = 5$.

(d) Compare this depreciation model with a linear model. What are the advantages of each?

**33.** $V = 15{,}000e^{-0.6286t}$

**34.** $V = 500{,}000e^{-0.2231t}$

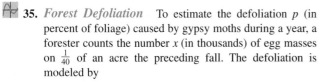 **35.** *Forest Defoliation*    To estimate the defoliation $p$ (in percent of foliage) caused by gypsy moths during a year, a forester counts the number $x$ (in thousands) of egg masses on $\frac{1}{40}$ of an acre the preceding fall. The defoliation is modeled by

$$p = \frac{300}{3 + 17e^{-1.57x}}.$$

*(Source: National Forest Service)*

(a) Use a graphing utility to graph the model.

(b) Estimate the percent of defoliation if 2000 egg masses are counted.

(c) Estimate the number of egg masses for which the amount of defoliation is increasing most rapidly.

**36.** *Learning Theory*    The average typing speed $N$ (in words per minute) after $t$ weeks of lessons is modeled by

$$N = \frac{95}{1 + 8.5e^{-0.12t}}.$$

Find the rate at which the typing speed is changing when (a) $t = 5$ weeks, (b) $t = 10$ weeks, and (c) $t = 30$ weeks.

**37.** *Compound Interest*    The balance $A$ (in dollars) in a savings account is given by $A = 5000e^{0.08t}$, where $t$ is measured in years. Find the rate at which the balance is changing when (a) $t = 1$ year, (b) $t = 10$ years, and (c) $t = 50$ years.

**38.** *Ebbinghaus Model*    The *Ebbinghaus Model* for human memory is $p = (100 - a)e^{-bt} + a$, where $p$ is the percent retained after $t$ weeks. (The constants $a$ and $b$ vary from one person to another.) If $a = 20$ and $b = 0.5$, at what rate is information being retained after 1 week? After 3 weeks?

**39.** *Orchard Yield*    The yield $V$ (in pounds per acre) for an orchard at age $t$ (in years) is modeled by

$$V = 7955.6e^{-0.0458/t}.$$

At what rate is the yield changing when $t = 5$ years? When $t = 10$ years? When $t = 25$ years?

**40.** *Self-Employed Workers*    From 1980 to 1995, the number $y$ (in thousands) of self-employed workers in the United States can be modeled by

$$y = 8622.11 + 153.97t - 0.0001e^t,$$

where $t = 0$ corresponds to 1980.    *(Source: U.S. Bureau of Labor Statistics)*

(a) Use a graphing utility to graph the model.

(b) Use the graph to estimate the rate of change in the number of self-employed people in 1980, 1990, and 1994.

(c) Confirm the result of part (b) analytically.

**41.** Use a graphing utility to graph the normal probability density function with $\mu = 0$ and $\sigma = 2, 3$, and 4 in the same viewing rectangle. What effect does the standard deviation $\sigma$ have on the function? Explain.

**42.** Use a graphing utility to graph the normal probability density function with $\sigma = 1$ and $\mu = -2, 1$, and 3 in the same viewing rectangle. What effect does the mean $\mu$ have on the function? Explain.

**43.** *Parachutist*    A parachutist jumps from a plane and opens the parachute at a height of 2000 feet. The height of the parachutist is

$$h = 1950 + 50e^{-1.6t} - 20t,$$

where $h$ is the height in feet and $t$ is the time in seconds since the parachute was opened.

(a) Find $dh/dt$ and use a graphing utility to graph $dh/dt$.

(b) Evaluate $dh/dt$ for $t = 0, 1, 5, 10$, and 20.

(c) Interpret your results from parts (a) and (b).

# 4.3 Logarithmic Functions

*The Natural Logarithmic Function • Properties of Logarithmic Functions • Solving Exponential and Logarithmic Equations • Applications*

## The Natural Logarithmic Function

From your previous algebra courses, you should be somewhat familiar with logarithms. For instance, the **common logarithm** $\log_{10} x$ is defined as

$$\log_{10} x = b \quad \text{if and only if} \quad 10^b = x.$$

The base of common logarithms is 10. In calculus, the most useful base for logarithms is the number $e$.

---

### Definition of the Natural Logarithmic Function

The **natural logarithmic function,** denoted by $\ln x$, is defined as

$$\ln x = b \quad \text{if and only if} \quad e^b = x.$$

$\ln x$ is read as "el-en of $x$" or as "the natural log of $x$."

---

This definition implies that the natural logarithmic function and the natural exponential function are inverses of each other. Thus, every logarithmic equation can be written in an equivalent exponential form and every exponential equation can be written in logarithmic form. Here are some examples.

*Logarithmic form:* $\quad \ln 1 = 0 \quad \ln e = 1 \quad \ln \dfrac{1}{e} = -1 \quad \ln 2 \approx 0.693$

*Exponential form:* $\quad e^0 = 1 \quad e^1 = e \quad e^{-1} = \dfrac{1}{e} \quad e^{0.693} \approx 2$

Because the functions $f(x) = e^x$ and $g(x) = \ln x$ are inverses of each other, their graphs are reflections of each other in the line $y = x$. This reflective property is illustrated in Figure 4.14. The figure also contains a summary of several properties of the graph of the natural logarithmic function.

Notice that the domain of the natural logarithmic function is the set of *positive real numbers*—be sure you see that $\ln x$ is not defined for zero or for negative numbers. You can test this on your calculator. If you try evaluating $\ln(-1)$ or $\ln 0$, your calculator should indicate that the value is not a real number.

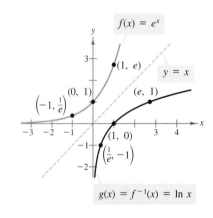

$g(x) = \ln x$

• Domain: $(0, \infty)$
• Range: $(-\infty, \infty)$
• Intercept: $(1, 0)$
• Always increasing
• $\ln x \to \infty$ as $x \to \infty$
• $\ln x \to -\infty$ as $x \to 0^+$
• Continuous
• One-to-one

**FIGURE 4.14**

### EXAMPLE 1   Graphing Logarithmic Functions

Sketch the graphs of the following functions.

(a) $f(x) = \ln(x + 1)$     (b) $f(x) = 2\ln(x - 2)$

*Solution*

(a) Because the natural logarithmic function is defined only for positive values, the domain of the function is $0 < x + 1$, or

$$-1 < x. \qquad \text{Domain}$$

To sketch the graph, begin by constructing a table of values, as shown below. Then, plot the points in the table and connect them with a smooth curve, as shown in Figure 4.15(a).

| $x$ | $-0.5$ | 0 | 0.5 | 1 | 1.5 | 2 |
|---|---|---|---|---|---|---|
| $\ln(x + 1)$ | $-0.693$ | 0 | 0.405 | 0.693 | 0.916 | 1.099 |

(b) The domain of this function is $0 < x - 2$, or

$$2 < x. \qquad \text{Domain}$$

A table of values for the function is presented below, and its graph is shown in Figure 4.15(b).

| $x$ | 2.5 | 3 | 3.5 | 4 | 4.5 | 5 |
|---|---|---|---|---|---|---|
| $2\ln(x - 2)$ | $-1.386$ | 0 | 0.811 | 1.386 | 1.833 | 2.197 |

**Technology**

What happens when you take the logarithm of a negative number? Some graphing utilities such as the *TI-85* and the *TI-86* do not give an error message for $\ln(-1)$. Instead, the graphing utility displays a complex number. For the purpose of this text, however, we will continue to say that the domain of the logarithmic function is the set of positive real numbers.

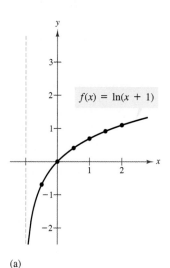

(a)

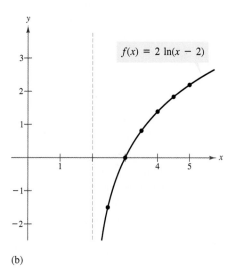

(b)

**FIGURE 4.15**

## Properties of Logarithmic Functions

Recall from Section 1.4 that inverse functions have the property that

$$f(f^{-1}(x)) = x \quad \text{and} \quad f^{-1}(f(x)) = x.$$

The properties listed below follow from the fact that the natural logarithmic function and the natural exponential function are inverses of each other.

---

### Inverse Properties of Logarithms and Exponents

1. $\ln e^x = x$     2. $e^{\ln x} = x$

---

### EXAMPLE 2   Applying Inverse Properties

Simplify the following expressions.

(a) $\ln e^{\sqrt{2}}$          (b) $e^{\ln 3x}$

*Solution*

(a) Because $\ln e^x = x$, it follows that

$$\ln e^{\sqrt{2}} = \sqrt{2}.$$

(b) Because $e^{\ln x} = x$, it follows that

$$e^{\ln 3x} = 3x.$$

Most of the properties of exponential functions can be rewritten in terms of logarithmic functions. For instance, the property

$$e^x e^y = e^{x+y}$$

states that you can multiply two exponential expressions by adding their exponents. In terms of logarithms, this property becomes

$$\ln xy = \ln x + \ln y.$$

This property and two other properties of logarithms are summarized below.

---

### Properties of Logarithms

1. $\ln xy = \ln x + \ln y$          2. $\ln \dfrac{x}{y} = \ln x - \ln y$

3. $\ln x^n = n \ln x$

---

*STUDY TIP*   There is no general property that can be used to rewrite $\ln(x + y)$. Specifically, $\ln(x + y)$ is not equal to $\ln x + \ln y$.

Rewriting a logarithm of a single quantity as the sum, difference, or multiple of logarithms is called *expanding* the logarithmic expression. The reverse procedure is called *condensing* a logarithmic expression.

### Technology

Try using a graphing utility to verify the results of Example 3b. That is, try graphing the functions

$$y = \ln \sqrt{x^2 + 1}$$

and

$$y = \frac{1}{2} \ln(x^2 + 1).$$

Because these two functions are equivalent, their graphs should coincide.

### EXAMPLE 3    Expanding Logarithmic Expressions

Use the properties of logarithms to write each expression as a sum, difference, or multiple of logarithms. (Assume $x > 0$ and $y > 0$.)

(a) $\ln \dfrac{10}{9}$     (b) $\ln \sqrt{x^2 + 1}$     (c) $\ln \dfrac{xy}{5}$     (d) $\ln \dfrac{x^2}{6y^3}$

**Solution**

(a) $\ln \frac{10}{9} = \ln 10 - \ln 9$     Property 2

(b) $\ln \sqrt{x^2 + 1} = \ln(x^2 + 1)^{1/2}$     Write with rational exponents.

$\qquad\qquad = \frac{1}{2} \ln(x^2 + 1)$     Property 3

(c) $\ln \dfrac{xy}{5} = \ln(xy) - \ln 5$     Property 2

$\qquad\quad = \ln x + \ln y - \ln 5$     Property 1

(d) $\ln \dfrac{x^2}{6y^3} = \ln x^2 - \ln 6y^3$     Property 2

$\qquad\quad = \ln x^2 - (\ln 6 + \ln y^3)$     Property 1

$\qquad\quad = \ln x^2 - \ln 6 - \ln y^3$     Simplify.

$\qquad\quad = 2 \ln x - \ln 6 - 3 \ln y$     Property 3

### EXAMPLE 4    Condensing Logarithmic Expressions

Use the properties of logarithms to rewrite each expression as the logarithm of a single quantity. (Assume $x > 0$ and $y > 0$.)

(a) $\ln x + 2 \ln y$     (b) $\ln(x + 1) + \ln(x + 2) - 3 \ln x$

**Solution**

(a) $\ln x + 2 \ln y = \ln x + \ln y^2$     Property 3

$\qquad\qquad\quad = \ln xy^2$     Property 1

(b) $\ln(x + 1) + \ln(x + 2) - 3 \ln x = \ln[(x + 1)(x + 2)] - 3 \ln x$

$\qquad\qquad\qquad\qquad\qquad = \ln(x^2 + 3x + 2) - \ln x^3$

$\qquad\qquad\qquad\qquad\qquad = \ln \dfrac{x^2 + 3x + 2}{x^3}$

# Solving Exponential and Logarithmic Equations

The inverse properties of logarithms and exponents can be used to solve exponential and logarithmic equations, as illustrated in the next two examples.

## EXAMPLE 5 Solving Exponential Equations

Solve the following equations.

(a) $e^x = 5$     (b) $10 + 3e^{0.1t} = 14$

*Solution*

(a)
| | |
|---|---|
| $e^x = 5$ | Original equation |
| $\ln e^x = \ln 5$ | Take natural log of both sides. |
| $x = \ln 5$ | Inverse property: $\ln e^x = x$ |

(b)
| | |
|---|---|
| $10 + 3e^{0.1t} = 14$ | Original equation |
| $3e^{0.1t} = 4$ | Subtract 10 from both sides. |
| $e^{0.1t} = \frac{4}{3}$ | Divide both sides by 3. |
| $\ln e^{0.1t} = \ln \frac{4}{3}$ | Take natural log of both sides. |
| $0.1t = \ln \frac{4}{3}$ | Inverse property: $\ln e^{0.1t} = 0.1t$ |
| $t = 10 \ln \frac{4}{3}$ | Multiply both sides by 10. |

> **TIP**
>
> *ALGEBRA*
>
> In the examples on this page, note that the key step in solving an exponential equation is to "take the log of both sides," and the key step in solving a logarithmic equation is to "exponentiate both sides."

## EXAMPLE 6 Solving Logarithmic Equations

Solve the following equations.

(a) $\ln x = 5$     (b) $3 + 2 \ln x^2 = 7$

*Solution*

(a)
| | |
|---|---|
| $\ln x = 5$ | Original equation |
| $e^{\ln x} = e^5$ | Exponentiate both sides. |
| $x = e^5$ | Inverse property: $e^{\ln x} = x$ |

(b)
| | |
|---|---|
| $3 + 2 \ln x^2 = 7$ | Original equation |
| $2 \ln x^2 = 4$ | Subtract 3 from both sides. |
| $\ln x^2 = 2$ | Divide both sides by 2. |
| $e^{\ln x^2} = e^2$ | Exponentiate both sides. |
| $x^2 = e^2$ | Inverse property: $e^{\ln x^2} = x^2$ |
| $x = \pm e$ | Solve for $x$. |

## Applications

### EXAMPLE 7 Finding Doubling Time

You deposit $P$ dollars in an account whose annual interest rate is $r$, compounded continuously. How long will it take for your balance to double?

*Solution*

The balance in the account after $t$ years is

$$A = Pe^{rt}.$$

Thus, the balance will have doubled when $Pe^{rt} = 2P$. To find the "doubling time," solve this equation for $t$.

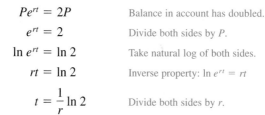

| $Pe^{rt} = 2P$ | Balance in account has doubled. |
| $e^{rt} = 2$ | Divide both sides by $P$. |
| $\ln e^{rt} = \ln 2$ | Take natural log of both sides. |
| $rt = \ln 2$ | Inverse property: $\ln e^{rt} = rt$ |
| $t = \dfrac{1}{r} \ln 2$ | Divide both sides by $r$. |

From this result, you can see that the time it takes for the balance to double is inversely proportional to the interest rate $r$. The table shows the doubling times for several interest rates. Notice that the doubling time decreases as the rate increases. The relationship between doubling time and the interest rate is shown graphically in Figure 4.16.

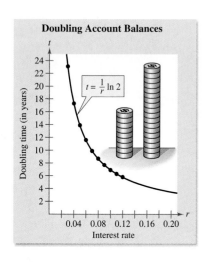

**FIGURE 4.16**

| $r$ | 3% | 4% | 5% | 6% | 7% | 8% | 9% | 10% | 11% | 12% |
|---|---|---|---|---|---|---|---|---|---|---|
| $t$ | 23.1 | 17.3 | 13.9 | 11.6 | 9.9 | 8.7 | 7.7 | 6.9 | 6.3 | 5.8 |

---

## *Group Discussion*

### *Tripling Time*

You deposit $P$ dollars in an account whose annual interest rate is $r$, compounded continuously. Use Example 7 as a model to determine how long it will take for your balance to triple. Complete the following table and use a graph to illustrate your answer.

| Rate, $r$ | 3% | 4% | 5% | 6% | 7% | 8% | 9% | 10% | 11% | 12% |
|---|---|---|---|---|---|---|---|---|---|---|
| Tripling time, $t$ | | | | | | | | | | |

*Warm Up*

The following warm-up exercises involve skills that were covered in earlier sections. You will use these skills in the exercise set for this section.

In Exercises 1–4, use the properties of exponents to simplify the expression.

**1.** $(4^2)(4^{-3})$      **2.** $(2^3)^2$

**3.** $e^0$      **4.** $(3e)^4$

In Exercises 5–8, solve for x.

**5.** $0 < x + 4$

**6.** $0 < x^2 + 1$

**7.** $0 < \sqrt{x^2 - 1}$

**8.** $0 < x - 5$

In Exercises 9 and 10, find the balance in the account after 10 years.

**9.** $P = \$1900$, $r = 6\%$, compounded continuously

**10.** $P = \$2500$, $r = 3\%$, compounded continuously

## EXERCISES 4.3

In Exercises 1–8, write the logarithmic equation as an exponential equation, or vice versa.

**1.** $\ln 2 = 0.6931\ldots$      **2.** $\ln 8.4 = 2.1282\ldots$

**3.** $\ln 0.2 = -1.6094\ldots$      **4.** $\ln 0.056 = -2.8824\ldots$

**5.** $e^0 = 1$      **6.** $e^2 = 7.3891\ldots$

**7.** $e^{-3} = 0.0498\ldots$      **8.** $e^{0.25} = 1.2840\ldots$

In Exercises 9–12, match the function with its graph. [The graphs are labeled (a)–(d).]

**9.** $f(x) = 2 + \ln x$      **10.** $f(x) = -\ln x$

**11.** $f(x) = \ln(x + 2)$      **12.** $f(x) = -\ln(x - 1)$

(a)

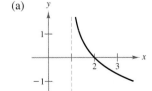

(b)

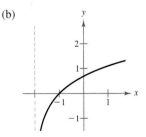

(c)

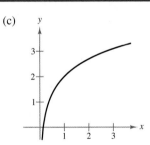

(d)

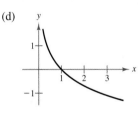

In Exercises 13–18, sketch the graph of the function.

**13.** $y = \ln(x - 1)$      **14.** $y = \ln|x|$

**15.** $y = \ln 2x$      **16.** $y = 5 + \ln x$

**17.** $y = 3 \ln x$      **18.** $y = \frac{1}{4} \ln x$

In Exercises 19–22, analytically show that the functions are inverses of each other. Then use a graphing utility to show this graphically.

**19.** $f(x) = e^{2x}$      **20.** $f(x) = e^x - 1$

    $g(x) = \ln \sqrt{x}$         $g(x) = \ln(x + 1)$

**21.** $f(x) = e^{2x-1}$      **22.** $f(x) = e^{x/3}$

    $g(x) = \frac{1}{2} + \ln \sqrt{x}$         $g(x) = \ln x^3$

In Exercises 23–28, apply the inverse properties of logarithmic and exponential functions to simplify the expression.

**23.** $\ln e^{x^2}$

**24.** $\ln e^{2x-1}$

**25.** $e^{\ln(5x+2)}$

**26.** $-1 + \ln e^{2x}$

**27.** $e^{\ln \sqrt{x}}$

**28.** $-8 + e^{\ln x^3}$

In Exercises 29 and 30, use the properties of logarithms and the fact that $\ln 2 \approx 0.6931$ and $\ln 3 \approx 1.0986$ to approximate the logarithm. Then use a calculator to confirm your approximation.

**29.** (a) $\ln 6$

(b) $\ln \frac{3}{2}$

(c) $\ln 81$

(d) $\ln \sqrt{3}$

**30.** (a) $\ln 0.25$

(b) $\ln 24$

(c) $\ln \sqrt[3]{12}$

(d) $\ln \frac{1}{72}$

In Exercises 31–40, use the properties of logarithms to write the expression as a sum, difference, or multiple of logarithms.

**31.** $\ln \frac{2}{3}$

**32.** $\ln \frac{1}{5}$

**33.** $\ln xyz$

**34.** $\ln \frac{xy}{z}$

**35.** $\ln \sqrt{x^2 + 1}$

**36.** $\ln \sqrt{2^3}$

**37.** $\ln \frac{2x}{\sqrt{x^2 - 1}}$

**38.** $\ln \frac{1}{e}$

**39.** $\ln \frac{3x(x + 1)}{(2x + 1)^2}$

**40.** $\ln z(z - 1)^2$

In Exercises 41–50, write the expression as the logarithm of a single quantity.

**41.** $\ln(x - 2) - \ln(x + 2)$

**42.** $\ln(2x + 1) + \ln(2x - 1)$

**43.** $3 \ln x + 2 \ln y - 4 \ln z$

**44.** $\frac{1}{3}[2 \ln(x + 3) + \ln x - \ln(x^2 - 1)]$

**45.** $3[\ln x + \ln(x + 3) - \ln(x + 4)]$

**46.** $2 \ln 3 - \frac{1}{2} \ln(x^2 + 1)$

**47.** $\frac{3}{2}[\ln x(x^2 + 1) - \ln(x + 1)]$

**48.** $2 \ln x + \frac{1}{2} \ln(x + 1)$

**49.** $2[\ln x - \ln(x + 1)] - 3[\ln x - \ln(x - 1)]$

**50.** $\frac{1}{2} \ln(x - 2) + \frac{3}{2} \ln(x + 2)$

In Exercises 51–64, solve for $x$ or $t$.

**51.** $e^{\ln x} = 4$

**52.** $e^{\ln x^2} - 9 = 0$

**53.** $\ln x = 0$

**54.** $2 \ln x = 4$

**55.** $e^{x+1} = 4$

**56.** $e^{-0.5x} = 0.075$

**57.** $300e^{-0.2t} = 700$

**58.** $e^{-0.0174t} = 0.5$

**59.** $5^{2x} = 15$

**60.** $2^{1-x} = 6$

**61.** $500(1.07)^t = 1000$

**62.** $400(1.06)^t = 1300$

**63.** $1000\left(1 + \frac{0.07}{12}\right)^{12t} = 3000$

**64.** $\frac{36}{1 + e^{-t}} = 20$

**65.** *Compound Interest*   A deposit of $1000 is made in an account that earns interest at an annual rate of 5%. How long will it take for the balance to double if the interest is compounded (a) annually, (b) monthly, (c) daily, and (d) continuously?

**66.** *Chemistry: Carbon Dating*   The remnants of an ancient fire in a cave in Africa showed a $^{14}_{6}\text{C}$ (carbon-14) decay rate of 3.1 counts per minute per gram of carbon. Assuming that the decay rate of $^{14}_{6}\text{C}$ in freshly cut wood (corrected for changes in the $^{14}_{6}\text{C}$ content of the atmosphere) is 13.6 counts per minute per gram of carbon, calculate the age of the remnants. The half-life of $^{14}_{6}\text{C}$ is 5730 years. Use the integrated first-order rate law,

$$\ln(N/N_0) = -kt,$$

where $N$ is the number of nuclides present at time $t$, $N_0$ is the number of nuclides present at time 0, $k = 0.693/5730$, and $t$ is the time of the fire. *(Source: Adapted from Zumdahl, Chemistry, Fourth Edition)*

**67.** *Compound Interest*   Complete the table, which shows the time $t$ necessary for $P$ dollars to triple if the interest is compounded continuously at the rate of $r$.

| $r$ | 2% | 4% | 6% | 8% | 10% | 12% | 14% |
|---|---|---|---|---|---|---|---|
| $t$ | | | | | | | |

**68.** *Demand*   The demand function for a product is

$$p = 250 - 0.8e^{0.005x},$$

where $p$ is the price per unit and $x$ is the number of units sold. Find the number of units sold for a price of (a) $p = \$200$ and (b) $p = \$125$.

**69.** *Population Growth*   The population $P$ of Pensacola, Florida from 1970 through 1994 can be modeled by

$$P = 243{,}000e^{0.01737t},$$

where $t = 0$ corresponds to 1970. According to this model, in what year will Pensacola have a population of 400,000? *(Source: U.S. Bureau of Census)*

**70.** *Population Growth* The population $P$ of Omaha, Nebraska from 1970 through 1994 can be modeled by

$$P = 556{,}000e^{0.00733t},$$

where $t = 0$ corresponds to 1970. According to this model, in what year will Omaha have a population of 700,000? *(Source: U.S. Bureau of Census)*

*Carbon Dating* In Exercises 71–74, you are given the ratio of carbon atoms in a fossil. Use the following information to estimate the age of the fossil. In living organic material, the ratio of radioactive carbon isotopes to the total number of carbon atoms is about 1 to $10^{12}$. (See Example 2 in Section 4.1.) When organic material dies, its radioactive carbon isotopes begin to decay, with a half-life of about 5700 years. Thus, the ratio $R$ of carbon isotopes to carbon-14 atoms is

$$R = 10^{-12}\left(\tfrac{1}{2}\right)^{t/5700},$$

where $t$ is the time (in years) and $t = 0$ represents the time when the organic material died.

**71.** $R = 0.32 \times 10^{-12}$  **72.** $R = 0.27 \times 10^{-12}$

**73.** $R = 0.22 \times 10^{-12}$  **74.** $R = 0.13 \times 10^{-12}$

Kenneth Garret/University of Innsbruck, Austria

*In 1991, forensic expert Rainer Henn uncovered one of the most important discoveries of modern archaeology—the body of a man that had been frozen in the Alps 5300 years ago. The age of the "iceman" was estimated by carbon dating.*

**75.** *Human Memory Model* Students in a mathematics class were given an exam and then retested monthly with equivalent exams. The average score $S$ (on a 100-point scale) for the class can be modeled by

$$S = 80 - 14\ln(t + 1), \qquad 0 \le t \le 12,$$

where $t$ is the time in months.

(a) What was the average score on the original exam?
(b) What was the average score after 4 months?
(c) After how many months was the average score 46?

**76.** *Research Project* Use a graphing utility to sketch the graph of

$$y = 10\ln\left(\frac{10 + \sqrt{100 - x^2}}{10}\right) - \sqrt{100 - x^2}$$

over the interval $(0, 10]$. This graph is called a *tractrix* or *pursuit curve*. Use your school's library or some other reference source to find information about a tractrix. Explain how such a curve can arise in a real-life setting.

**77.** Demonstrate that

$$\frac{\ln x}{\ln y} \ne \ln\frac{x}{y} = \ln x - \ln y$$

by completing the table.

| $x$ | $y$ | $\dfrac{\ln x}{\ln y}$ | $\ln\dfrac{x}{y}$ | $\ln x - \ln y$ |
|---|---|---|---|---|
| 1 | 2 | | | |
| 3 | 4 | | | |
| 10 | 5 | | | |
| 4 | 0.5 | | | |

**78.** Complete the table using $f(x) = \dfrac{\ln x}{x}$.

| $x$ | 1 | 5 | 10 | $10^2$ | $10^4$ | $10^6$ |
|---|---|---|---|---|---|---|
| $f(x)$ | | | | | | |

(a) Use the table to estimate the limit: $\lim\limits_{x\to\infty} f(x)$.
(b) Use a graphing utility to estimate the relative extrema of $f$.

In Exercises 79 and 80, use a graphing utility to verify that the functions are equivalent for $x > 0$.

**79.** $f(x) = \ln\dfrac{x^2}{4}$

$g(x) = 2\ln x - \ln 4$

**80.** $f(x) = \ln\sqrt{x(x^2 + 1)}$

$g(x) = \tfrac{1}{2}[\ln x + \ln(x^2 + 1)]$

## 4.4 Derivatives of Logarithmic Functions

*Derivatives of Logarithmic Functions* • *Applications* • *Other Bases*

## Derivatives of Logarithmic Functions

Implicit differentiation can be used to develop the derivative of the natural logarithmic function.

$$y = \ln x \qquad \text{Natural logarithmic function}$$

$$e^y = x \qquad \text{Write in exponential form.}$$

$$\frac{d}{dx}[e^y] = \frac{d}{dx}[x] \qquad \text{Differentiate with respect to } x.$$

$$e^y \frac{dy}{dx} = 1 \qquad \text{Chain Rule}$$

$$\frac{dy}{dx} = \frac{1}{e^y} \qquad \text{Divide both sides by } e^y.$$

$$\frac{dy}{dx} = \frac{1}{x} \qquad \text{Substitute } x \text{ for } e^y.$$

This result and its Chain Rule version are summarized below.

> ### Derivative of the Natural Logarithmic Function
>
> Let $u$ be a differentiable function of $x$.
>
> 1. $\dfrac{d}{dx}[\ln x] = \dfrac{1}{x}$          2. $\dfrac{d}{dx}[\ln u] = \dfrac{1}{u}\dfrac{du}{dx}$

### EXAMPLE 1   Differentiating a Logarithmic Function

Find the derivative of

$$f(x) = \ln 2x.$$

**Solution**

Let $u = 2x$. Then $du/dx = 2$, and you can apply the Chain Rule as follows.

$$f'(x) = \frac{1}{u}\frac{du}{dx} = \frac{1}{2x}(2) = \frac{1}{x}$$

---

**Discovery**

Sketch the graph of $y = \ln x$ on a piece of paper. Draw tangent lines to the graph at various points. How does the slope of these tangent lines change as you move to the right? Is the slope ever equal to zero? Use the formula for the derivative of the logarithmic function to confirm your conclusions.

## EXAMPLE 2   Differentiating Logarithmic Functions

Find the derivatives of the functions.

(a) $f(x) = \ln(2x^2 + 4)$      (b) $f(x) = x \ln x$

*Solution*

(a) Let $u = 2x^2 + 4$. Then $du/dx = 4x$, and you can apply the Chain Rule.

$$f'(x) = \frac{1}{u}\frac{du}{dx} \qquad \text{Chain Rule}$$

$$= \frac{1}{2x^2 + 4}(4x)$$

$$= \frac{2x}{x^2 + 2} \qquad \text{Simplify.}$$

(b) Using the Product Rule, you can find the derivative.

$$f'(x) = x\frac{d}{dx}[\ln x] + (\ln x)\frac{d}{dx}[x] \qquad \text{Product Rule}$$

$$= x\left(\frac{1}{x}\right) + (\ln x)(1) \qquad \text{Derivative of } \ln x$$

$$= 1 + \ln x \qquad \text{Simplify.}$$

When you are differentiating logarithmic functions, it is often helpful to use the properties of logarithms to rewrite the function *before* differentiating, as shown in the next example.

## EXAMPLE 3   Rewriting Before Differentiating

Find the derivative of $f(x) = \ln\sqrt{x + 1}$.

*Solution*

$$f(x) = \ln\sqrt{x + 1} \qquad \text{Original function}$$

$$= \ln(x + 1)^{1/2} \qquad \text{Rewrite with rational exponent.}$$

$$= \frac{1}{2}\ln(x + 1) \qquad \text{Property of logarithms}$$

$$f'(x) = \frac{1}{2}\left(\frac{1}{x + 1}\right) \qquad \text{Differentiate.}$$

$$= \frac{1}{2(x + 1)} \qquad \text{Simplify.}$$

**STUDY TIP**   To see the advantage of rewriting before differentiating, try using the Chain Rule to differentiate $f(x) = \ln\sqrt{x + 1}$ and compare your work with that shown in Example 3.

*Discovery*

What is the domain of the function $f(x) = \ln\sqrt{x + 1}$ in Example 3? What is the domain of the function $f'(x) = 1/[2(x + 1)]$? In general, you must be careful to understand the domains of functions involving logarithms. For example, are the domains of the functions $y_1 = \ln x^2$ and $y_2 = 2 \ln x$ the same? Try graphing them on your graphing utility.

The next example is an even more dramatic illustration of the benefit of rewriting a function before differentiating.

---

## EXAMPLE 4   Rewriting Before Differentiating

Find the derivative of

$$f(x) = \ln\frac{x(x^2 + 1)^2}{\sqrt{2x^3 + 1}}.$$

### Solution

$$f(x) = \ln\frac{x(x^2 + 1)^2}{\sqrt{2x^3 + 1}} \qquad\qquad \text{Original function}$$

$$= \ln x(x^2 + 1)^2 - \ln(2x^3 + 1)^{1/2} \qquad \text{Logarithmic properties}$$

$$= \ln x + \ln(x^2 + 1)^2 - \ln(2x^3 + 1)^{1/2} \qquad \text{Logarithmic properties}$$

$$= \ln x + 2\ln(x^2 + 1) - \frac{1}{2}\ln(2x^3 + 1) \qquad \text{Logarithmic properties}$$

$$f'(x) = \frac{1}{x} + 2\left(\frac{2x}{x^2 + 1}\right) - \frac{1}{2}\left(\frac{6x^2}{2x^3 + 1}\right) \qquad \text{Differentiate.}$$

$$= \frac{1}{x} + \frac{4x}{x^2 + 1} - \frac{3x^2}{2x^3 + 1} \qquad\qquad \text{Simplify.}$$

---

 **ALGEBRA TIP**   Finding the derivative of the function in Example 4 without first rewriting would be a formidable task.

$$f'(x) = \frac{1}{x(x^2 + 1)^2/\sqrt{2x^3 + 1}} \frac{d}{dx}\left[\frac{x(x^2 + 1)^2}{\sqrt{2x^3 + 1}}\right]$$

You might try showing that this yields the same result obtained in Example 4, but be careful—the algebra is messy.

---

## Technology

A symbolic differentiation utility will not generally list the derivative of the logarithmic function in the form given in Example 4. The result listed by *Derive* for Windows is shown at the right. Show that the two forms are equivalent by rewriting the answer given in Example 4.

$$\#1\colon \text{LN}\left[\frac{x \cdot (x^2 + 1)^2}{\sqrt{(2 \cdot x^3 + 1)}}\right] \qquad \text{Author.}$$

$$\#2\colon \frac{d}{dx}\,\text{LN}\left[\frac{x \cdot (x^2 + 1)^2}{\sqrt{(2 \cdot x^3 + 1)}}\right] \qquad \text{Differentiate.}$$

$$\#3\colon \frac{x \cdot (5 \cdot x^3 - 3 \cdot x + 4)}{(x^2 + 1) \cdot (2 \cdot x^3 + 1)} + \frac{1}{x} \qquad \text{Simplify.}$$

## Applications

### EXAMPLE 5  Analyzing a Graph

Analyze the graph of the function $f(x) = x^2 - \ln x$.

**Solution**

From Figure 4.17, it appears that the function has a minimum just to the left of $x = 1$. To find the minimum analytically, find the critical numbers by setting the derivative of $f$ equal to zero and solving for $x$.

$$f(x) = x^2 - \ln x \qquad \text{Original function}$$

$$f'(x) = 2x - \frac{1}{x} \qquad \text{Differentiate.}$$

$$2x - \frac{1}{x} = 0 \qquad \text{Set derivative equal to 0.}$$

$$2x = \frac{1}{x} \qquad \text{Add } 1/x \text{ to both sides.}$$

$$2x^2 = 1 \qquad \text{Multiply both sides by } x.$$

$$x^2 = \tfrac{1}{2} \qquad \text{Divide both sides by 2.}$$

$$x = \pm\sqrt{\tfrac{1}{2}} \qquad \text{Take square root of both sides.}$$

Of these two possible critical numbers, only the positive one lies in the domain of $f$. By applying the First-Derivative Test, you can confirm that the function has a relative minimum when $x = 1/\sqrt{2} \approx 0.707$.

**FIGURE 4.17**

### EXAMPLE 6  Finding a Rate of Change

A group of 200 college students were tested every 6 months over a 4-year period. The group was composed of students who took French during the fall semester of their freshman year and did not take subsequent French courses. The average test score $p$ (in percent) is modeled by

$$p = 91.6 - 15.6 \ln(t + 1), \qquad 0 \le t \le 48,$$

where $t$ is the time in months, as shown in Figure 4.18. At what rate was the average score changing after 1 year?

**Solution**

The rate of change is

$$\frac{dp}{dt} = -\frac{15.6}{t + 1}.$$

When $t = 12$, $dp/dt = -1.2$, which means that the average score was decreasing at the rate of 1.2% per month.

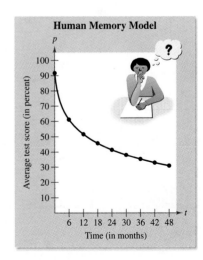

**FIGURE 4.18**

## Other Bases

This chapter began with a definition of a general exponential function

$$f(x) = a^x,$$

where $a$ is a positive number such that $a \neq 1$. The corresponding **logarithm to the base $a$** is defined by

$$\log_a x = b \quad \text{if and only if} \quad a^b = x.$$

As with the natural logarithmic function, the domain of the logarithmic function to the base $a$ is the set of positive numbers.

---

**Technology**

Use a graphing utility to graph the three functions $y_1 = \log_2 x = \ln x / \ln 2$, $y_2 = 2^x$, and $y_3 = x$ in the same viewing rectangle. Explain why the graphs of $y_1$ and $y_2$ are reflections of each other in the line $y_3 = x$.

**EXAMPLE 7  Evaluating Logarithms**

(a) $\log_2 8 = 3$                     $2^3 = 8$

(b) $\log_{10} 100 = 2$                $10^2 = 100$

(c) $\log_{10} \frac{1}{10} = -1$      $10^{-1} = \frac{1}{10}$

(d) $\log_3 81 = 4$                    $3^4 = 81$

(e) $\log_{10} 4 \approx 0.602$        $10^{0.602} \approx 4$

---

Logarithms to the base 10 are called **common logarithms.** Most calculators have only two logarithm keys—a natural logarithm key denoted by $\boxed{\text{LN}}$ and a common logarithm key denoted by $\boxed{\text{LOG}}$. Logarithms to other bases can be evaluated with the following change of base formula.

$$\log_a x = \frac{\ln x}{\ln a} \qquad \text{Change of base formula}$$

---

**EXAMPLE 8  Evaluating Logarithms**

Evaluate the following logarithms.

(a) $\log_2 3$      (b) $\log_3 6$      (c) $\log_2(-1)$

*Solution*

In each case, use the change of base formula and a calculator.

(a) $\log_2 3 = \dfrac{\ln 3}{\ln 2} \approx 1.585$ $\qquad \log_a x = \dfrac{\ln x}{\ln a}$

(b) $\log_3 6 = \dfrac{\ln 6}{\ln 3} \approx 1.631$ $\qquad \log_a x = \dfrac{\ln x}{\ln a}$

(c) $\log_2(-1)$ is not defined.

To find derivatives of exponential or logarithmic functions to bases other than $e$, you can either convert to base $e$ or use the following differentiation rules.

### Other Bases and Differentiation

Let $u$ be a differentiable function of $x$.

1. $\dfrac{d}{dx}[a^x] = (\ln a)a^x$

2. $\dfrac{d}{dx}[a^u] = (\ln a)a^u\dfrac{du}{dx}$

3. $\dfrac{d}{dx}[\log_a x] = \left(\dfrac{1}{\ln a}\right)\dfrac{1}{x}$

4. $\dfrac{d}{dx}[\log_a u] = \left(\dfrac{1}{\ln a}\right)\left(\dfrac{1}{u}\right)\dfrac{du}{dx}$

**STUDY TIP** Remember that you can convert to base $e$ using the formulas

$$a^x = e^{(\ln a)x}$$

and

$$\log_a x = \left(\dfrac{1}{\ln a}\right)\ln x.$$

## EXAMPLE 9 Finding a Rate of Change

Radioactive carbon isotopes have a half-life of 5700 years. If 1 gram of the isotopes is present in an object now, the amount $A$ (in grams) that will be present after $t$ years is

$$A = \left(\frac{1}{2}\right)^{t/5700}.$$

At what rate is the amount changing when $t = 10,000$ years?

*Solution*

The derivative of $A$ with respect to $t$ is

$$\frac{dA}{dt} = \left(\ln\frac{1}{2}\right)\left(\frac{1}{2}\right)^{t/5700}\left(\frac{1}{5700}\right).$$

When $t = 10,000$, the value of the derivative is

$$\frac{dA}{dt} = \left(\ln\frac{1}{2}\right)\left(\frac{1}{2}\right)^{10,000/5700}\left(\frac{1}{5700}\right) \approx -0.000036,$$

which implies that the amount of isotopes in the object is decreasing at the rate of 0.000036 gram per year.

---

## *Group Discussion* **Finding an Average Rate of Change**

Suppose you forgot the differentiation rules for exponential functions. Could you still answer the question in Example 9 by calculating

$$\left(\frac{1}{2}\right)^{10,001/5700} - \left(\frac{1}{2}\right)^{10,000/5700}?$$

Explain your reasoning. Illustrate each technique graphically. Which technique involves the slope of a tangent line? A secant line?

*Warm Up*

The following warm-up exercises involve skills that were covered in earlier sections. You will use these skills in the exercise set for this section.

In Exercises 1–6, expand the logarithmic expression.

**1.** $\ln(x + 1)^2$

**2.** $\ln x(x + 1)$

**3.** $\ln \dfrac{x}{x + 1}$

**4.** $\ln \left( \dfrac{x}{x - 3} \right)^3$

**5.** $\ln \dfrac{4x(x - 7)}{x^2}$

**6.** $\ln \dfrac{x^3(x + 1)}{\sqrt{x - 2}}$

In Exercises 7 and 8, find $dy/dx$ implicitly.

**7.** $y^x + xy = 7$

**8.** $x^2y - xy^2 = 3x$

In Exercises 9 and 10, find the second derivative of $f$.

**9.** $f(x) = x^2(x + 1) - 3x^3$

**10.** $f(x) = -\dfrac{1}{x^2}$

# EXERCISES 4.4

In Exercises 1–4, find the slope of the tangent line to the graph of the function at the point $(1, 0)$.

**1.** $y = \ln x^3$

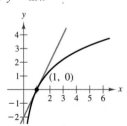

**2.** $y = \ln x^{5/2}$

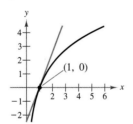

**3.** $y = \ln x^2$

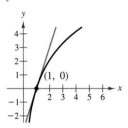

**4.** $y = \ln x^{1/2}$

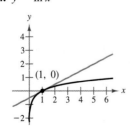

In Exercises 5–26, find the derivative of the function.

**5.** $y = \ln x^2$

**6.** $y = \ln(x^2 + 3)$

**7.** $f(x) = \ln 2x$

**8.** $f(x) = \ln(1 - x^2)$

**9.** $y = \ln\sqrt{x^4 - 4x}$

**10.** $y = \ln(1 - x)^{3/2}$

**11.** $y = \frac{1}{2}(\ln x)^6$

**12.** $y = (\ln x^2)^2$

**13.** $f(x) = x^2 \ln x$

**14.** $y = \ln\left(x\sqrt{x^2 - 1}\right)$

**15.** $y = \ln[x^2(x + 1)^3]$

**16.** $y = \ln\dfrac{x}{x^2 + 1}$

**17.** $y = \ln\dfrac{x}{x + 1}$

**18.** $y = \ln\sqrt{\dfrac{x + 1}{x - 1}}$

**19.** $y = \ln\sqrt[3]{\dfrac{x - 1}{x + 1}}$

**20.** $y = \dfrac{\ln x}{x^2}$

**21.** $y = \ln\dfrac{\sqrt{4 + x^2}}{x}$

**22.** $y = \ln\left(x + \sqrt{4 + x^2}\right)$

**23.** $g(x) = e^{-x}\ln x$

**24.** $g(x) = \ln\dfrac{e^x + e^{-x}}{2}$

**25.** $f(x) = \ln e^{x^2}$

**26.** $f(x) = \ln\dfrac{1 + e^x}{1 - e^x}$

In Exercises 27–30, write the expression with base $e$.

**27.** $2^x$

**28.** $3^x$

**29.** $\log_4 x$

**30.** $\log_3 x$

In Exercises 31–36, evaluate the logarithm.

**31.** $\log_2 4$

**32.** $\log_5 12$

**33.** $\log_3 \frac{1}{2}$

**34.** $\log_7 343$

**35.** $\log_{10} 31$

**36.** $\log_9 3$

In Exercises 37–46, find the derivative of the function.

**37.** $y = 3^x$

**38.** $y = \left(\frac{1}{4}\right)^x$

**39.** $f(x) = \log_2 x$

**40.** $g(x) = \log_5 x$

**41.** $h(x) = 4^{2x-3}$

**42.** $y = 6^{5x}$

**43.** $y = \log_{10}(x^2 + 6x)$

**44.** $f(x) = 10^{x^2}$

**45.** $y = x2^x$

**46.** $y = 3x3^x$

In Exercises 47–50, determine an equation of the tangent line to the function at the indicated point.

| Function | Point |
|---|---|
| **47.** $y = x \ln x$ | $(1, 0)$ |
| **48.** $y = \dfrac{\ln x}{x}$ | $\left(e, \dfrac{1}{e}\right)$ |
| **49.** $y = \log_3(3x + 7)$ | $\left(\dfrac{2}{3}, 2\right)$ |
| **50.** $g(x) = 25^{2x^2}$ | $\left(-\dfrac{1}{2}, 5\right)$ |

In Exercises 51–54, find $dy/dx$ implicitly.

**51.** $x^2 - 3 \ln y + y^2 = 10$

**52.** $\ln xy + 5x = 30$

**53.** $4x^3 + \ln y^2 + 2y = 2x$

**54.** $4xy + \ln(x^2 y) = 7$

In Exercises 55–58, find the second derivative of the function.

**55.** $f(x) = x \ln \sqrt{x} + 2x$

**56.** $f(x) = 3 + 2 \ln x$

**57.** $f(x) = 5^x$

**58.** $f(x) = \log_{10} x$

**59.** *Sound Intensity* The relationship between the number of decibels $\beta$ and the intensity of a sound $I$ in watts per square centimeter is given by

$$\beta = 10 \log_{10}\left(\frac{I}{10^{-16}}\right).$$

Find the rate of change in the number of decibels when the intensity is $10^{-4}$ watts per square centimeter.

**60.** *Engineering* The temperature $T$ (°F) at which water boils at selected pressures $p$ (pounds per square inch) can be modeled by

$$T = 87.97 + 34.96 \ln p + 7.91 \sqrt{p}.$$

Find the rate of change of the temperature when the pressure is 60 pounds per square inch.

In Exercises 61–66, find the slope of the graph at the indicated point. Then write an equation of the tangent line at the point.

**61.** $f(x) = 1 + 2x \ln x$

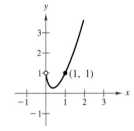

**62.** $f(x) = 2 \ln x^3$

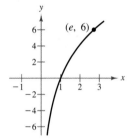

**63.** $f(x) = \ln \dfrac{5(x + 2)}{x}$

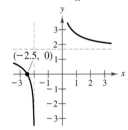

**64.** $f(x) = \ln\left(x\sqrt{x + 3}\right)$

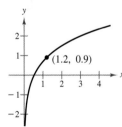

**65.** $f(x) = \log_2 x$

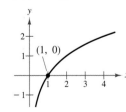

**66.** $f(x) = x^2 \log_3 x^2$

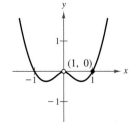

In Exercises 67–72, graph and analyze the function. Include any relative extrema and points of inflection in your analysis. Use a graphing utility to check your results.

**67.** $y = x - \ln x$

**68.** $y = \frac{1}{2}x^2 - \ln x$

**69.** $y = \dfrac{\ln x}{x}$

**70.** $y = x \ln x$

**71.** $y = x^2 \ln x$

**72.** $y = (\ln x)^2$

In Exercises 73–76, find $dx/dp$ for the demand function. Interpret this rate of change when the price is $10.

**73.** $x = \ln \dfrac{1000}{p}$

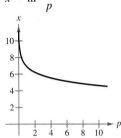

**74.** $x = 1000 - p \ln p$

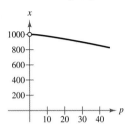

**75.** $x = \dfrac{500}{\ln(p^2 + 1)}$

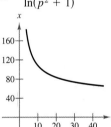

**76.** $x = 300 - 50 \ln(\ln p)$

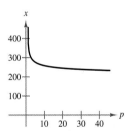

**77.** Solve the demand function in Exercise 73 for $p$. Use the result to find $dp/dx$. Then find the value of $dp/dx$ when $p = $10. What is the relationship between this derivative and $dx/dp$?

**78.** Solve the demand function in Exercise 75 for $p$. Use the result to find $dp/dx$. Then find the value of $dp/dx$ when $p = $10. What is the relationship between this derivative and $dx/dp$?

 **79.** *Minimum Average Cost* The cost of producing $x$ units of a product is

$$C = 500 + 300x - 300 \ln x, \quad x \geq 1.$$

Use a graphing utility to find the minimum average cost. Then confirm your result analytically.

**80.** *Chemistry* Radioactive einsteinium isotopes have a half-life of 276 days. If 1 gram of the isotopes is present in an object now, the amount $A$ (in grams) present after $t$ days is

$$A = \left(\frac{1}{2}\right)^{t/276}.$$

At what rate is the amount changing when $t = 500$ days?

 **81.** *Mail-Order Sales* The sales $S$ (in millions of dollars) of mail-order companies in the United States from 1985 to 1993 can be modeled by

$$S = -\frac{8,001,622 \ln t}{t^2} + \frac{9,205,046.5}{t^2} + 0.03e^t + 241,015.90,$$

where $t = 5$ corresponds to 1985. *(Source: Annual Guides to Mail Order Sales)*

(a) Use a graphing utility to graph $S$ over the interval $[5, 13]$.

(b) Estimate the amount of sales in 1991.

(c) At what rate were the sales changing in 1991?

# Business Capsule

Parrot Mountain Company

*The Parrot Mountain Company is a mail-order business started by Angel Santiago in 1990. The company offers the famous Parrot Mountain roller skates used by bird trainers and carries more than 800 bird-related items.*

**82.** *Research Project* Use your school's library or some other reference source to research information about a mail-order company, such as that mentioned above. Collect data about the company (sales or membership over a 20-year period, for example) and find a mathematical model to represent the data.

# Exponential Growth and Decay

## 4.5

*Exponential Growth and Decay • Applications*

## Exponential Growth and Decay

In this section you will learn to create models of *exponential growth and decay*. Real-life situations that involve exponential growth and decay deal with a substance or population whose *rate of change at any time t is proportional to the amount of the substance present at that time*. For example, the rate of decomposition of a radioactive substance is proportional to the amount of radioactive substance at a given instant. In its simplest form, this relationship is described by the following equation.

Rate of change of $y$ ⎤ is ⎡ proportional to $y$.

$$\frac{dy}{dt} = ky$$

In this equation, $k$ is a constant and $y$ is a function of $t$. The solution of this equation is given below.

### Law of Exponential Growth and Decay

If $y$ is a positive quantity whose rate of change with respect to time is proportional to the quantity present at any time $t$, then $y$ is of the form

$$y = Ce^{kt},$$

where $C$ is the **initial value** and $k$ is the **constant of proportionality. Exponential growth** is indicated by $k > 0$ and **exponential decay** by $k < 0$.

**Proof** Because the rate of change of $y$ is proportional to $y$, you can write

$$\frac{dy}{dt} = ky.$$

You can see that $y = Ce^{kt}$ is a solution of this equation by differentiating to obtain $dy/dt = kCe^{kt}$ and substituting

$$\frac{dy}{dt} = kCe^{kt} = k(Ce^{kt}) = ky.$$

*STUDY TIP* In the model $y = Ce^{kt}$, $C$ is called the "initial value" because when $t = 0$,

$$y = Ce^{k(0)} = C(1) = C.$$

*Discovery*

Use a graphing utility to graph $y = Ce^{2t}$ for $C = 1, 2,$ and $5$. How does the value of $C$ affect the shape of the graph? Now graph $y = 2e^{kt}$ for $k = -2, -1, 0, 1,$ and $2$. How does the value of $k$ affect the shape of the graph? Which function grows faster, $y = e^x$ or $y = x^{10}$?

*Much of the cost of nuclear energy is the cost of disposing of radioactive waste. Because of the long half-life of the waste, it must be stored in containers that will remain undisturbed for thousands of years.*

## Applications

Radioactive decay is measured in terms of **half-life,** the number of years required for half of the atoms in a sample of radioactive material to decay. The half-lives of some common radioactive isotopes are as follows.

| | |
|---|---:|
| Uranium ($U^{238}$) | 4,510,000,000  years |
| Plutonium ($Pu^{239}$) | 24,360  years |
| Carbon ($C^{14}$) | 5,730  years |
| Radium ($Ra^{226}$) | 1,620  years |
| Einsteinium ($Es^{254}$) | 270  days |
| Nobelium ($No^{257}$) | 23  seconds |

### EXAMPLE 1   Modeling Radioactive Decay

A sample contains 1 gram of radium. How much radium will remain after 1000 years?

*Solution*

Let $y$ represent the mass (in grams) of the radium in the sample. Because the rate of decay is proportional to $y$, you can apply the Law of Exponential Decay to conclude that $y$ is of the form $y = Ce^{kt}$, where $t$ is measured in years. From the given information, you know that $y = 1$ when $t = 0$. Substituting these values into the model produces

$$1 = Ce^{k(0)}, \qquad \text{Substitute 1 for } y \text{ and 0 for } t.$$

which implies that $C = 1$. Because radium has a half-life of 1620 years, you know that $y = \frac{1}{2}$ when $t = 1620$. Substituting these values into the model allows you to solve for $k$.

$$y = e^{kt} \qquad \text{Exponential decay model}$$

$$\tfrac{1}{2} = e^{k(1620)} \qquad \text{Substitute } \tfrac{1}{2} \text{ for } y \text{ and 1620 for } t.$$

$$\ln \tfrac{1}{2} = 1620k \qquad \text{Take natural log of both sides.}$$

$$\tfrac{1}{1620} \ln \tfrac{1}{2} = k \qquad \text{Divide both sides by 1620.}$$

Thus, $k \approx -0.0004279$, and the exponential decay model is

$$y = e^{-0.0004279t}.$$

To find the amount of radium remaining in the sample after 1000 years, substitute $t = 1000$ into the model. This produces

$$y = e^{-0.0004279(1000)} \approx 0.652 \text{ gram.}$$

The graph of the model is shown in Figure 4.19.

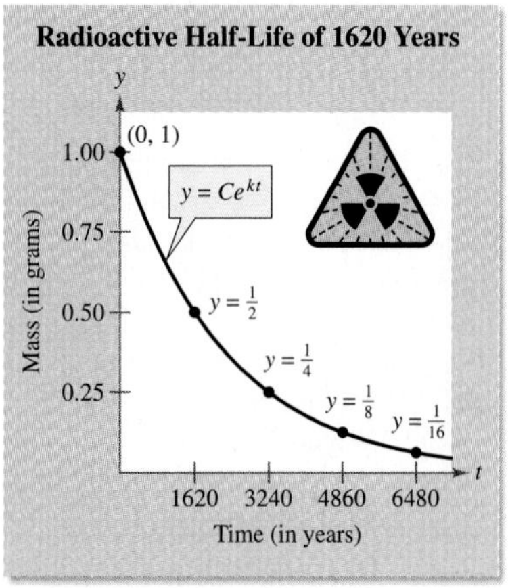

**Radioactive Half-Life of 1620 Years**

**FIGURE 4.19**

The steps used in Example 1 are summarized below.

> ## Guidelines for Modeling Exponential Growth and Decay
>
> 1. Use the given information to write *two* sets of conditions involving $y$ and $t$.
>
> 2. Substitute the given conditions into the model $y = Ce^{kt}$ and use the results to solve for the constants $C$ and $k$. (If one of the conditions involves $t = 0$, substitute that value first to solve for $C$.)
>
> 3. Use the model $y = Ce^{kt}$ to answer the question.

## EXAMPLE 2    Modeling Population Growth

In a research experiment, a population of fruit flies is increasing in accordance with the law of exponential growth. After 2 days, there are 100 flies, and after 4 days, there are 300 flies. How many flies will there be after 5 days?

### Solution

Let $y$ be the number of flies at time $t$. From the given information, you know that $y = 100$ when $t = 2$ and $y = 300$ when $t = 4$. Substituting this information into the model $y = Ce^{kt}$ produces

$$100 = Ce^{2k} \quad \text{and} \quad 300 = Ce^{4k}.$$

To solve for $k$, solve for $C$ in the first equation and substitute the result into the second equation.

$$300 = Ce^{4k} \qquad \text{Second equation}$$

$$300 = \left(\frac{100}{e^{2k}}\right)e^{4k} \qquad \text{Substitute } 100/e^{2k} \text{ for } C.$$

$$\frac{300}{100} = e^{2k} \qquad \text{Divide both sides by 100.}$$

$$\ln 3 = 2k \qquad \text{Take natural log of both sides.}$$

$$\frac{1}{2}\ln 3 = k \qquad \text{Solve for } k.$$

Using $k = \frac{1}{2}\ln 3 \approx 0.5493$, you can determine that $C \approx 100/e^{2(0.5493)} \approx 33$. Thus, the exponential growth model is

$$y = 33e^{0.5493t},$$

as shown in Figure 4.20. This implies that, after 5 days, the population is

$$y = 33e^{0.5493(5)} \approx 514 \text{ flies.}$$

**TIP**

For help on the algebra in Example 2, see Example 1b in the *Chapter 4 Algebra Review*, on page 304.

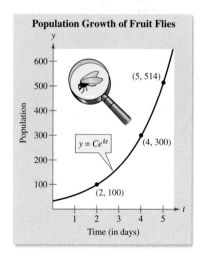

**Population Growth of Fruit Flies**

**FIGURE 4.20**

### EXAMPLE 3  Modeling Compound Interest

Money is deposited in an account for which the interest is compounded continuously. The balance in the account doubles in 6 years. What is the annual interest rate?

*Solution*

The balance $A$ in an account with continuously compounded interest is given by the exponential growth model

$$A = Pe^{rt}, \qquad\qquad \text{Exponential growth model}$$

where $P$ is the original deposit, $r$ is the annual interest rate (in decimal form), and $t$ is the time (in years). From the given information, you know that $A = 2P$ when $t = 6$, as shown in Figure 4.21. Use this information to solve for $r$.

| | |
|---|---|
| $A = Pe^{rt}$ | Exponential growth model |
| $2P = Pe^{r(6)}$ | Substitute $2P$ for $A$ and 6 for $t$. |
| $2 = e^{6r}$ | Divide both sides by $P$. |
| $\ln 2 = 6r$ | Take natural log of both sides. |
| $\frac{1}{6} \ln 2 = r$ | Divide both sides by 6. |

Thus, the annual interest rate is $r = \frac{1}{6} \ln 2 \approx 0.1155$ or about 11.55%.

Each of the examples in this section uses the exponential growth model in which the base is $e$. Exponential growth, however, can be modeled with *any* base. That is, the model

$$y = Ca^{bt}$$

also represents exponential growth. (To see this, note that the model can be written in the form $y = Ce^{(\ln a)bt}$.) In some real-life settings, bases other than $e$ are more convenient. For instance, in Example 1, knowing that the half-life of radium is 1620 years, you can immediately write the exponential decay model as

$$y = \left(\frac{1}{2}\right)^{t/1620}.$$

Using this model, the amount of radium left in the sample after 1000 years is

$$y = \left(\frac{1}{2}\right)^{1000/1620} \approx 0.652 \text{ gram,}$$

which is the same answer obtained in Example 1.

> **STUDY TIP** Can you see why you can immediately write the model $y = \left(\frac{1}{2}\right)^{t/1620}$ for the radioactive decay described in Example 1? Notice that when $t = 1620$, the value of $y$ is $\frac{1}{2}$, when $t = 3240$, the value of $y$ is $\frac{1}{4}$, and so on.

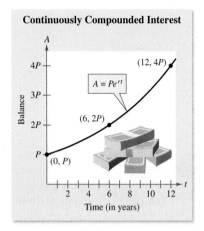

**Continuously Compounded Interest**

$A = Pe^{rt}$

$(0, P)$  $(6, 2P)$  $(12, 4P)$

Balance · Time (in years)

**FIGURE 4.21**

# Technology

## Fitting an Exponential Model to Data

Most graphing utilities have programs that allow you to find the *least squares regression exponential model* for data. Depending on the type of graphing utility, the program may fit the data to a model of the form

$$y = ab^x,$$    Exponential model with base $b$

or

$$y = ae^{bx}.$$    Exponential model with base $e$

To see how to use such a program, consider the following.

The cash flow per share $y$ for the Coca-Cola Company from 1990 through 1997 is listed in the table. *(Source: Coca-Cola Company)*

| $x$ | 0 | 1 | 2 | 3 | 4 | 5 | 6 | 7 |
|---|---|---|---|---|---|---|---|---|
| $y$ | \$0.60 | \$0.71 | \$0.84 | \$0.98 | \$1.16 | \$1.37 | \$1.60 | \$1.90 |

In the table, $x = 0$ represents 1990. To fit an exponential model to this data, enter the following coordinates into the statistical data bank of a graphing utility.

$(0, 0.6), (1, 0.71), (2, 0.84), (3, 0.98)$
$(4, 1.16), (5, 1.37), (6, 1.6), (7, 1.9)$

After running the exponential regression program with a graphing utility that uses the model $y = ab^x$, the display should read $a \approx 0.602$ and $b \approx 1.178$. (The correlation of $r \approx 0.9999$ tells you that the fit is very good.) Thus, a model for the data is

$$y = 0.602(1.178)^x.$$    Exponential model with base $b$

If you use a graphing utility that uses the model $y = ae^{bx}$, the display should read $a \approx 0.602$ and $b \approx 0.164$. The corresponding model is

$$y = 0.602e^{0.164x}.$$    Exponential model with base $e$

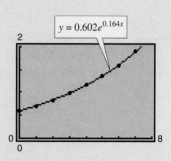

The graph of the model is shown at the right. Notice that one way to interpret the model is that the cash flow per share increased by about 16.4% each year from 1990 through 1997.

You can use either model to predict the cash flow per share in future years. For instance, in 1998 ($x = 8$), the cash flow per share is predicted to be

$$y = 0.602e^{(0.164)8} \approx \$2.23.$$

In 1996, *Value Line* predicted the cash flow for 1998 to be \$2.20.

## EXAMPLE 4   Modeling Sales

Four months after discontinuing advertising on national television, a manufacturer notices that sales have dropped from 100,000 units per month to 80,000 units. If the sales follow an exponential pattern of decline, what will they be after another 4 months?

### Solution

Let $y$ represent the number of units, let $t$ represent the time (in months), and consider the exponential decay model

$$y = Ce^{kt}.$$   Exponential decay model

From the given information, you know that $y = 100{,}000$ when $t = 0$. Using this information, you have

$$100{,}000 = Ce^0,$$

which implies that $C = 100{,}000$. To solve for $k$, use the fact that $y = 80{,}000$ when $t = 4$.

| | |
|---|---|
| $y = 100{,}000e^{kt}$ | Exponential decay model |
| $80{,}000 = 100{,}000e^{k(4)}$ | Substitute 80,000 for $y$ and 4 for $t$. |
| $0.8 = e^{4k}$ | Divide both sides by 100,000. |
| $\ln 0.8 = 4k$ | Take natural log of both sides. |
| $\frac{1}{4}\ln 0.8 = k$ | Divide both sides by 4. |

Thus, $k = \frac{1}{4}\ln 0.8 \approx -0.0558$, which means that the model is

$$y = 100{,}000e^{-0.0558t}.$$

After 4 more months ($t = 8$), you can expect sales to drop to

$$y = 100{,}000e^{-0.0558(8)} \approx 64{,}000 \text{ units},$$

as shown in Figure 4.22.

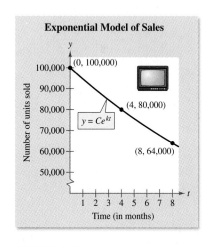

**Exponential Model of Sales**

$y = Ce^{kt}$

(0, 100,000)
(4, 80,000)
(8, 64,000)

Number of units sold — Time (in months)

**FIGURE 4.22**

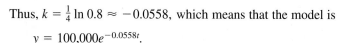

## *Group Discussion*       *Comparing Exponential and Linear Models*

In Example 4, it was assumed that the sales were following an exponential pattern of decline. This implies that the sales were dropping by the same *percent* each month. Suppose instead that the sales were following a linear pattern of decline (dropping by the same *amount* each month). How would this affect the answer to the question given in Example 4? Which of the two models do you think might be more realistic? Explain your reasoning.

## Warm Up

The following warm-up exercises involve skills that were covered in earlier sections. You will use these skills in the exercise set for this section.

In Exercises 1–4, solve the equation for $k$.

**1.** $12 = 24e^{4k}$

**2.** $10 = 3e^{5k}$

**3.** $25 = 16e^{3k}$

**4.** $22 = 32e^{20k}$

In Exercises 5–8, find the derivative of the function.

**5.** $y = 32e^{0.23t}$

**6.** $y = 24e^{-1.4t}$

**7.** $y = 18e^{0.072t}$

**8.** $y = 25e^{-0.001t}$

In Exercises 9 and 10, simplify the expression.

**9.** $e^{\ln 4}$

**10.** $4e^{\ln 3}$

## EXERCISES 4.5

In Exercises 1–6, find the exponential function $y = Ce^{kt}$ that passes through the two given points.

**1.** $y = Ce^{kt}$

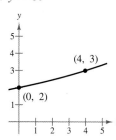

**2.** $y = Ce^{kt}$

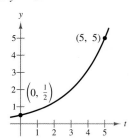

**3.** $y = Ce^{kt}$

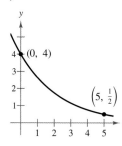

**4.** $y = Ce^{kt}$

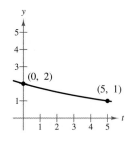

**5.** $y = Ce^{kt}$

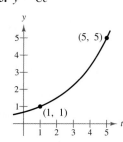

**6.** $y = Ce^{kt}$

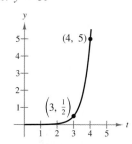

In Exercises 7–10, use the given information to write an equation for $y$. Confirm your result analytically by showing that the function satisfies the equation $dy/dt = Cy$. Does the function represent exponential growth or decay?

**7.** $\dfrac{dy}{dt} = 2y$,    $y = 10$ when $t = 0$

**8.** $\dfrac{dy}{dt} = -\dfrac{2}{3}y$,    $y = 20$ when $t = 0$

**9.** $\dfrac{dy}{dt} = -4y$,    $y = 30$ when $t = 0$

**10.** $\dfrac{dy}{dt} = 5.2y$,    $y = 18$ when $t = 0$

*Radioactive Decay* In Exercises 11–16, complete the table for the given radioactive isotope.

| Isotope | Half-life (in years) | Initial quantity | Amount after 1000 years | Amount after 10,000 years |
|---|---|---|---|---|
| 11. Ra$^{226}$ | 1,620 | 10 grams | | |
| 12. Ra$^{226}$ | 1,620 | | 1.5 grams | |
| 13. C$^{14}$ | 5,730 | | | 2 grams |
| 14. C$^{14}$ | 5,730 | 3 grams | | |
| 15. Pu$^{239}$ | 24,360 | | 2.1 grams | |
| 16. Pu$^{239}$ | 24,360 | | | 0.4 gram |

17. *Radioactive Decay* What percent of a present amount of radioactive radium (Ra$^{226}$) will remain after 900 years?

18. *Radioactive Decay* Find the half-life of a radioactive material if after 1 year 99.57% of the initial amount remains.

19. *Carbon Dating* C$^{14}$ dating assumes that the carbon dioxide on the earth today has the same radioactive content as it did centuries ago. If this is true, then the amount of C$^{14}$ absorbed by a tree that grew several centuries ago should be the same as the amount of C$^{14}$ absorbed by a similar tree today. A piece of ancient charcoal contains only 15% as much of the radioactive carbon as a piece of modern charcoal. How long ago was the tree burned to make the ancient charcoal? (The half-life of C$^{14}$ is 5730 years.)

20. *Carbon Dating* Repeat Exercise 19 for a piece of charcoal that contains 30% as much radioactive carbon as a modern piece.

21. *Population Growth* The number of a certain type of bacteria increases continuously at a rate proportional to the number present. There are 150 present at a given time and 450 present 5 hours later.

(a) How many will there be 10 hours after the initial time?
(b) How long will it take for the population to double?
(c) Does the answer to part (b) depend on the starting time? Explain.

22. *School Expenditures* In 1980, the total amount spent on schools in the United States was $311.9 billion. By 1992, it had risen to $459.3 billion. Assume the total amount spent can be modeled by exponential growth. (*Source: U.S. National Center for Education Statistics*)

(a) Estimate the total amount spent in 1990.
(b) How many years until the amount doubles?
(c) By what percent is the amount increasing each year?

*Compound Interest* In Exercises 23–28, complete the table for an account in which interest is compounded continuously.

| | Initial investment | Annual rate | Time to double | Amount after 10 years | Amount after 25 years |
|---|---|---|---|---|---|
| 23. | $1,000 | 12% | | | |
| 24. | $20,000 | 10$\frac{1}{2}$% | | | |
| 25. | $750 | | 7$\frac{3}{4}$ years | | |
| 26. | $10,000 | | 5 years | | |
| 27. | $500 | | | $1,292.85 | |
| 28. | $2,000 | | | | $6,008.33 |

29. *Effective Yield* The effective yield is the annual rate $i$ that will produce the same interest per year as the nominal rate $r$ compounded $n$ times per year.

(a) For a rate $r$ that is compounded $n$ times per year, show that the effective yield is

$$i = \left(1 + \frac{r}{n}\right)^n - 1.$$

(b) Find the effective yield for a nominal rate of 6%, compounded monthly.

30. *Effective Yield* The effective yield is the annual rate $i$ that will produce the same interest per year as the nominal rate $r$.

(a) For a rate $r$ that is compounded continuously, show that the effective yield is

$$i = e^r - 1.$$

(b) Find the effective yield for a nominal rate of 6%, compounded continuously.

*Effective Yield* In Exercises 31 and 32, use the results of Exercises 29 and 30 to complete the table showing the effective yield for a nominal rate of $r$.

| Number of compoundings per year | 4 | 12 | 365 | Continuous |
|---|---|---|---|---|
| Effective yield | | | | |

31. $r = 5\%$     32. $r = 7\frac{1}{2}\%$

**33.** *Rule of 70*   Verify that the time necessary for an investment to double its value is approximately $70/r$, where $r$ is the annual interest rate entered as a percent.

**34.** *Rule of 70*   Use the Rule of 70 given in Exercise 33 to approximate the time necessary for an investment to double in value if (a) $r = 10\%$ and (b) $r = 7\%$.

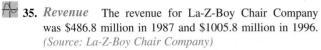 **35.** *Revenue*   The revenue for La-Z-Boy Chair Company was $486.8 million in 1987 and $1005.8 million in 1996. *(Source: La-Z-Boy Chair Company)*

(a) Use an exponential growth model to estimate the revenue in 2000.
(b) Use a linear model to estimate the 2000 revenue.
(c) Use a graphing utility to graph the models from parts (a) and (b). Which model is more accurate?

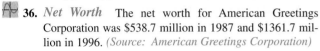 **36.** *Net Worth*   The net worth for American Greetings Corporation was $538.7 million in 1987 and $1361.7 million in 1996. *(Source: American Greetings Corporation)*

(a) Use an exponential growth model to estimate the net worth in 2000.
(b) Use a linear model to estimate the 2000 net worth.
(c) Use a graphing utility to graph the models from parts (a) and (b). Which model is more accurate?

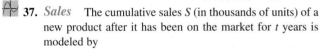 **37.** *Sales*   The cumulative sales $S$ (in thousands of units) of a new product after it has been on the market for $t$ years is modeled by

$$S = Ce^{k/t}.$$

During the first year, 5000 units were sold. The saturation point for the market is 30,000 units. That is, the limit of $S$ as $t \to \infty$ is 30,000.

(a) Solve for $C$ and $k$ in the model.
(b) How many units will be sold after 5 years?
(c) Use a graphing utility to graph the sales function.

**38.** *Sales*   The cumulative sales $S$ (in thousands of units) of a new product after it has been on the market for $t$ years is modeled by

$$S = 30(1 - 3^{kt}).$$

During the first year, 5000 units were sold.

(a) Solve for $k$ in the model.
(b) What is the saturation point for this product?
(c) How many units will be sold after 5 years?
(d) Use a graphing utility to graph the sales function.

**39.** *Learning Curve*   The management of a factory finds that the maximum number of units a worker can produce in a day is 30. The learning curve for the number of units $N$ produced per day after a new employee has worked $t$ days is modeled by $N = 30(1 - e^{kt})$. After 20 days on the job, a worker is producing 19 units in a day. How many days should pass before this worker is producing 25 units per day?

**40.** *Learning Curve*   The management in Exercise 39 requires that a new employee be producing at least 20 units per day after 30 days on the job.

(a) Find a learning curve model that describes this minimum requirement.
(b) Find the number of days before a minimal achiever is producing 25 units per day.

**41.** *Revenue*   A small business assumes that the demand function for one of its new products can be modeled by $p = Ce^{kx}$. When $p = \$45$, $x = 1000$ units, and when $p = \$40$, $x = 1200$ units.

(a) Solve for $C$ and $k$.
(b) Find the values of $x$ and $p$ that will maximize the revenue for this product.

**42.** *Revenue*   Repeat Exercise 41 given that when $p = \$5$, $x = 300$ units, and when $p = \$4$, $x = 400$ units.

**43.** *Forestry*   The value $V$ (in dollars) of a tract of timber can be modeled by

$$V = 100,000e^{0.75\sqrt{t}},$$

where $t = 0$ corresponds to 1990. If money earns interest at a rate of 4%, compounded continuously, then the present value $A$ of the timber at any time $t$ is

$$A = Ve^{-0.04t}.$$

Find the year in which the timber should be harvested to maximize the present value.

**44.** *Forestry*   Repeat Exercise 43 using the model

$$V = 100,000e^{0.6\sqrt{t}}.$$

**45.** *Earthquake Intensity*   On the Richter Scale, the magnitude $R$ of an earthquake of intensity $I$ is given by

$$R = \frac{\ln I - \ln I_0}{\ln 10},$$

where $I_0$ is the minimum intensity used for comparison. Assume $I_0 = 1$.

(a) Find the intensity of the 1906 San Francisco earthquake in which $R = 8.3$.
(b) Find the factor by which the intensity is increased when the value of $R$ is doubled.
(c) Find $dR/dI$.

# Chapter 4 Algebra Review

## Solving Exponential and Logarithmic Equations

To find the extrema or points of inflection of an exponential or logarithmic function, you must know how to solve exponential and logarithmic equations. A few examples are given on page 281. These two pages show some additional examples.

As with all equations, remember that your basic goal is to isolate the variable on one side of the equation. To do this, you use inverse operations. For instance, to get rid of an exponential expression such as $e^{2x}$, take the natural log of both sides and use the property

$$\ln e^{2x} = 2x.$$

Similarly, to get rid of a logarithmic expression such as $\log_2 3x$, exponentiate both sides and use the property

$$2^{\log_2 3x} = 3x.$$

---

### EXAMPLE 1  Solving Exponential Equations

(a)

| | |
|---|---|
| $80,000 = 100,000e^{k(4)}$ | Example 4, page 300 |
| $0.8 = e^{4k}$ | Divide each side by 100,000. |
| $\ln 0.8 = \ln e^{4k}$ | Take natural log of each side. |
| $\ln 0.8 = 4k$ | Apply the property $\ln e^a = a$. |
| $\frac{1}{4}\ln 0.8 = k$ | Divide each side by 4. |

(b)

| | |
|---|---|
| $300 = \left(\dfrac{100}{e^{2k}}\right)e^{4k}$ | Example 2, page 297 |
| $300 = (100)\dfrac{e^{4k}}{e^{2k}}$ | Rewrite product. |
| $300 = 100e^{4k-2k}$ | To divide powers, subtract exponents. |
| $300 = 100e^{2k}$ | Simplify. |
| $3 = e^{2k}$ | Divide each side by 100. |
| $\ln 3 = \ln e^{2k}$ | Take natural log of each side. |
| $\ln 3 = 2k$ | Apply the property $\ln e^a = a$. |
| $\frac{1}{2}\ln 3 = k$ | Divide each side by 2. |

## EXAMPLE 2  Solving Logarithmic Equations

(a) $\ln x = 2$                                      Original equation

$\phantom{(a) }e^{\ln x} = e^2$                      Exponentiate both sides.

$\phantom{(a) }x = e^2$                              Apply the property $e^{\ln a} = a$.

(b) $5 + 2 \ln x = 4$                                Original equation

$\phantom{(b) }2 \ln x = -1$                         Subtract 5 from each side.

$\phantom{(b) }\ln x = -\dfrac{1}{2}$                Divide each side by 2.

$\phantom{(b) }e^{\ln x} = e^{-1/2}$                 Exponentiate both sides.

$\phantom{(b) }x = e^{-1/2}$                         Apply the property $e^{\ln a} = a$.

(c) $2 \ln 3x = 4$                                   Original equation

$\phantom{(c) }\ln 3x = 2$                           Divide each side by 2.

$\phantom{(c) }e^{\ln 3x} = e^2$                     Exponentiate both sides.

$\phantom{(c) }3x = e^2$                             Apply the property $e^{\ln a} = a$.

$\phantom{(c) }x = \tfrac{1}{3}e^2$                  Divide each side by 3.

(d) $\ln x - \ln(x - 1) = 1$                         Original equation

$\phantom{(d) }\ln \dfrac{x}{x - 1} = 1$             $\ln m - \ln n = \ln(m/n)$

$\phantom{(d) }e^{\ln(x/x-1)} = e^1$                 Exponentiate both sides.

$\phantom{(d) }\dfrac{x}{x - 1} = e^1$               Apply the property $e^{\ln a} = a$.

$\phantom{(d) }x = ex - e$                           Multiply each side by $x - 1$.

$\phantom{(d) }x - ex = -e$                          Subtract $ex$ from each side.

$\phantom{(d) }x(1 - e) = -e$                        Factor.

$\phantom{(d) }x = \dfrac{-e}{1 - e}$                Divide each side by $1 - e$.

$\phantom{(d) }x = \dfrac{e}{e - 1}$                 Simplify.

# Chapter Summary and Study Strategies

*After studying this chapter, you should have acquired the following skills. The exercise numbers are keyed to the Review Exercises that begin on page 308. Answers to odd-numbered Review Exercises are given in the back of the text.\**

■ Use the properties of exponents to evaluate and simplify exponential expressions. *(Section 4.1)*     *Review Exercises 1–12*

$$a^0 = 1, \qquad a^x a^y = a^{x+y}, \qquad \frac{a^x}{a^y} = a^{x-y}, \qquad (a^x)^y = a^{xy}$$

$$(ab)^x = a^x b^x, \qquad \left(\frac{a}{b}\right)^x = \frac{a^x}{b^x}, \qquad a^{-x} = \frac{1}{a^x}$$

■ Use properties of exponents to answer questions about real life.  *(Section 4.1)*     *Review Exercises 13, 14*

■ Sketch the graphs of exponential functions.  *(Section 4.1)*     *Review Exercises 15–18*

■ Evaluate limits of exponential functions in real life.  *(Section 4.1)*     *Review Exercises 19, 20*

■ Evaluate and graph functions involving the natural exponential function.  *(Section 4.1)*     *Review Exercises 21–24*

$$\lim_{x \to 0} (1 + x)^{1/x} = e \approx 2.71828$$

■ Graph logistics growth functions.  *(Section 4.1)*     *Review Exercises 25, 26*

■ Solve compound interest problems.  *(Section 4.1)*     *Review Exercises 27–30*

$$A = P\left(1 + \frac{r}{n}\right)^{nt}, \qquad A = Pe^{rt}$$

■ Answer questions involving the natural exponential function as a real-life model. *(Section 4.1)*     *Review Exercises 31, 32*

■ Find the derivatives of natural exponential functions.  *(Section 4.2)*     *Review Exercises 33–40*

$$\frac{d}{dx}[e^x] = e^x, \qquad \frac{d}{dx}[e^u] = e^u \frac{du}{dx}$$

■ Use calculus to analyze the graphs of functions that involve the natural exponential function.  *(Section 4.2)*     *Review Exercises 41–44*

■ Use the definition of the natural logarithmic function to write exponential equations in logarithmic form, and vice versa.  *(Section 4.3)*     *Review Exercises 45–48*

$$\ln x = b \qquad \text{if and only if} \qquad e^b = x.$$

■ Sketch the graphs of natural logarithmic functions.  *(Section 4.3)*     *Review Exercises 49–52*

■ Use properties of logarithms to expand and condense logarithmic expressions. *(Section 4.3)*     *Review Exercises 53–56*

$$\ln xy = \ln x + \ln y, \qquad \ln \frac{x}{y} = \ln x - \ln y, \qquad \ln x^n = n \ln x$$

\* Use a wide range of valuable study aids to help you master the material in this chapter. The *Student Solutions Guide* includes step-by-step solutions to all odd-numbered exercises to help you review and prepare. The *Algebra Review Tutorial Software* and *The Algebra of Calculus* help you brush up on your algebra skills. The *Graphing Technology Guide* offers step-by-step commands and instructions for a wide variety of graphing calculators, including the most recent models.

■ Use inverse properties of exponential and logarithmic functions to solve exponential and logarithmic equations.   *(Section 4.3)*

*Review Exercises 57–64*

$$\ln e^x = x, \qquad e^{\ln x} = x$$

■ Use properties of natural logarithms to answer questions about real life.   *(Section 4.3)*

*Review Exercises 65, 66*

■ Find the derivatives of natural logarithmic functions.   *(Section 4.4)*

*Review Exercises 67–74*

$$\frac{d}{dx}[\ln x] = \frac{1}{x}, \qquad \frac{d}{dx}[\ln u] = \frac{1}{u}\frac{du}{dx}$$

■ Use calculus to analyze the graphs of functions that involve the natural logarithmic function.   *(Section 4.4)*

*Review Exercises 75–78*

■ Use the definition of logarithms to evaluate logarithmic expressions involving other bases. *(Section 4.4)*

*Review Exercises 79–82*

$$\log_a x = b \qquad \text{if and only if} \qquad a^b = x$$

■ Use the change of base formula to evaluate logarithmic expressions involving other bases. *(Section 4.4)*

*Review Exercises 83–86*

$$\log_a x = \frac{\ln x}{\ln a}$$

■ Find the derivatives of exponential and logarithmic functions involving other bases. *(Section 4.4)*

*Review Exercises 87–90*

$$\frac{d}{dx}[a^x] = (\ln a)a^x, \qquad \frac{d}{dx}[a^u] = (\ln a)a^u\frac{du}{dx}$$

$$\frac{d}{dx}[\log_a x] = \left(\frac{1}{\ln a}\right)\frac{1}{x}, \qquad \frac{d}{dx}[\log_a u] = \left(\frac{1}{\ln a}\right)\left(\frac{1}{u}\right)\frac{du}{dx}$$

■ Use calculus to answer questions about real-life rates of change.   *(Section 4.4)*

*Review Exercises 91, 92*

■ Use exponential growth and decay to model real-life situations.   *(Section 4.5)*

*Review Exercises 93–98*

■ *Classifying Differentiation Rules*   Differentiation rules fall into two basic classes: (1) general rules that apply to all differentiable functions and (2) specific rules that apply to special types of functions. At this point in the course, you have studied six general rules: the Constant Rule, the Constant Multiple Rule, the Sum Rule, the Difference Rule, the Product Rule, and the Quotient Rule. Although these rules were introduced in the context of algebraic functions, remember that they can also be used with exponential and logarithmic functions. You have also studied three specific rules: the Power Rule, the Exponential Rule, and the Log Rule. Each of these rules comes in two forms: the "simple" version, such as $D_x[e^x] = e^x$, and the Chain Rule version, such as $D_x[e^u] = e^u(du/dx)$.

■ *To Memorize or Not to Memorize?*   When studying mathematics, you need to memorize some formulas and rules. Much of this will come from practice—the formulas that you use most often will be committed to memory. Some formulas, however, are used only infrequently. With these, it is helpful to be able to *derive* the formula from a *known* formula. For instance, knowing the Log Rule for differentiation and the change of base formula, $\log_a x = (\ln x)/(\ln a)$, allows you to derive the formula for the derivative of a logarithmic function to base $a$.

## Chapter 4  Review Exercises

In Exercises 1–4, evaluate the expression.

**1.** $4(4^4)$             **2.** $16^{3/2}$

**3.** $\left(\frac{1}{5}\right)^4$          **4.** $\left(\frac{27}{8}\right)^{-1/3}$

In Exercises 5–8, use the properties of exponents to simplify the expression.

**5.** $\left(\frac{25}{4}\right)^0$          **6.** $(9^{1/3})(3^{1/3})$

**7.** $\dfrac{6^3}{36^2}$           **8.** $\dfrac{1}{4}\left(\dfrac{1}{2}\right)^{-3}$

In Exercises 9–12, solve the equation for x.

**9.** $5^x = 625$

**10.** $x^{3/4} = 8$

**11.** $e^{-1/2} = e^{x-1}$

**12.** $\dfrac{x^3}{3} = e^3$

**13.** *New York Stock Exchange*  The total number $y$ (in millions) of shares of stock listed on the New York Stock Exchange between 1945 and 1995 can be modeled by

$$y = 978.665(1.0954)^t, \qquad 5 \le t \le 55,$$

where $t = 5$ corresponds to 1945. Use this model to estimate the total number of shares listed in 1950, 1970, and 1990.   *(Source: New York Stock Exchange)*

**14.** *Recycling*  Between 1960 and 1994, the percent $p$ of materials in municipal waste that was recycled can be modeled by

$$p = 5.331(1.0396)^t, \qquad 0 \le t \le 34,$$

where $t = 0$ corresponds to 1960. Use the model to estimate the percent of recycled materials in 1960, 1970, and 1988.   *(Source: Franklin Associates, Ltd.)*

In Exercises 15–18, sketch the graph of the function.

**15.** $f(x) = 9^{x/2}$

**16.** $f(t) = \left(\frac{1}{6}\right)^t$

**17.** $f(x) = \left(\frac{1}{2}\right)^{2x} + 4$

**18.** $g(x) = \frac{1}{4}(2)^{-3x}$

**19.** *Demand*  The demand function for a product is

$$p = 12{,}500 - \frac{10{,}000}{2 + e^{-0.001x}},$$

where $p$ is the price per unit and $x$ is the number of units produced (see figure). What is the limit of the price as $x$ increases without bound? Explain what this means in the context of the problem.

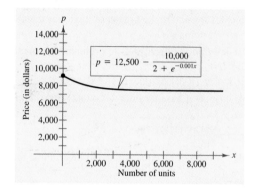

**20.** *Endangered Species*  Biologists consider a species of a plant or animal to be endangered if it is expected to become extinct in less than 20 years. The population $y$ of a certain species is modeled by

$$y = 1096e^{-0.39t}$$

(see figure). Is this species endangered? Explain your reasoning.

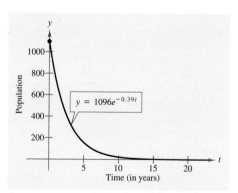

In Exercises 21–24, evaluate the function at the indicated value. Then sketch its graph.

**21.** $f(x) = 2e^{x-1}, \quad x = 2$

**22.** $f(t) = e^{4t} - 1, \quad t = 0$

**23.** $g(t) = 12e^{-0.2t}, \quad t = 17$

**24.** $g(x) = \dfrac{24}{1 + e^{-0.3x}}, \quad x = 300$

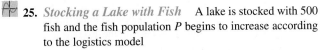 **25.** *Stocking a Lake with Fish*  A lake is stocked with 500 fish and the fish population $P$ begins to increase according to the logistics model

$$P = \frac{10,000}{1 + 19e^{-t/5}}, \quad 0 \leq t,$$

where $t$ is measured in months.

(a) Use a graphing utility to graph the function.
(b) Estimate the number of fish in the lake after 4 months.
(c) Does the population have a limit as $t$ increases without bound? Explain your reasoning.
(d) After how many months is the population increasing most rapidly? Explain your reasoning.

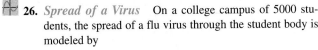 **26.** *Spread of a Virus*  On a college campus of 5000 students, the spread of a flu virus through the student body is modeled by

$$P = \frac{5000}{1 + 4999e^{-0.8t}}, \quad 0 \leq t,$$

where $P$ is the total number of infected people and $t$ is the time, measured in days.

(a) Use a graphing utility to graph the function.
(b) How many students will be infected after 5 days?
(c) According to this model, will all the students on campus become infected with the flu? Explain your reasoning.

In Exercises 27 and 28, complete the table to determine the balance $A$ when $P$ dollars is invested at an annual rate of $r$ for $t$ years, compounded $n$ times per year.

| $n$ | 1 | 2 | 4 | 12 | 365 | Continuous compounding |
|---|---|---|---|---|---|---|
| $A$ | | | | | | |

**27.** $P = \$1000$, $r = 4\%$, $t = 5$ years

**28.** $P = \$5000$, $r = 6\%$, $t = 20$ years

In Exercises 29 and 30, $2000 is deposited in an account. Decide which account, (a) or (b), will have the greater balance after 10 years.

**29.** (a) 5%, compounded continuously
     (b) 6%, compounded quarterly

**30.** (a) $6\frac{1}{2}\%$, compounded monthly
     (b) $6\frac{1}{4}\%$, compounded continuously

**31.** *Age at First Marriage*  The average age $A$ of an American man at his first marriage from 1970 to 1990 can be modeled by

$$A = 22.56 + \frac{1}{0.299 + 55.56e^{-0.4115t}}, \quad 0 \leq t \leq 20,$$

where $t = 0$ corresponds to 1970. Use this model to estimate the average age of an American man at his first marriage in 1970, 1980, and 1990.  *(Source: U.S. Center for Health Statistics)*

**32.** *Revenue*  The revenue $R$ (in millions of dollars) for Wendy's International from 1987 to 1996 can be modeled by

$$R = 979.39 + 2.82e^{0.3645t},$$

where $t = 7$ corresponds to 1987. Use this model to estimate the profit for Wendy's in 1987, 1990, and 1996.  *(Source: Wendy's International, Inc.)*

In Exercises 33–40, find the derivative of the function.

**33.** $y = 4e^{x^2}$

**34.** $y = x^2 e^x$

**35.** $y = \dfrac{x}{e^{2x}}$

**36.** $y = \dfrac{x-1}{e^x + 2x^2}$

**37.** $y = \sqrt{4e^{4x}}$

**38.** $y = \sqrt[3]{xe^{3x}}$

**39.** $y = \dfrac{5}{1 + e^{2x}}$

**40.** $y = \dfrac{e^x}{1 - xe^x}$

In Exercises 41–44, graph and analyze the function. Include any relative extrema, points of inflection, and asymptotes in your analysis.

**41.** $f(x) = 4e^{-x}$

**42.** $f(x) = x^3 e^x$

**43.** $f(x) = \dfrac{e^x}{x^2}$

**44.** $f(x) = \dfrac{1}{xe^x}$

In Exercises 45 and 46, write the logarithmic equation as an exponential equation.

**45.** $\ln 12 \approx 2.4849$      **46.** $\ln 0.6 \approx -0.5108$

In Exercises 47 and 48, write the exponential equation as a logarithmic equation.

**47.** $e^{1.5} \approx 4.4817$      **48.** $e^{-4} \approx 0.0183$

In Exercises 49–52, sketch the graph of the function.

**49.** $y = \ln(4 - x)$      **50.** $y = 5 + \ln x$

**51.** $y = \ln \dfrac{x}{3}$      **52.** $y = -\dfrac{5}{6} \ln x$

In Exercises 53–56, use the properties of logarithms to write the expression as a sum, difference, or multiple of logarithms.

**53.** $\ln \sqrt{x^2(x - 1)}$

**54.** $\ln \dfrac{x^2}{(x + 1)^3}$

**55.** $\ln\left(\dfrac{1 - x}{3x}\right)^3$

**56.** $\ln \dfrac{e^4}{5x}$

In Exercises 57-64, solve the equation for x.

**57.** $e^{\ln x} = 3$

**58.** $\ln x = 3e^{-1}$

**59.** $\ln 2x - \ln(3x - 1) = 0$

**60.** $e^{2x-1} - 6 = 0$

**61.** $\ln x + \ln(x - 3) = 0$

**62.** $e^{-1.386x} = 0.25$

**63.** $9^{6x} - 27 = 0$

**64.** $100(1.21)^x = 110$

 **65.** *Home Mortgage*   The monthly payment $M$ for a home mortgage of $P$ dollars for $t$ years at an annual interest rate of $r\%$ is given by

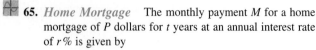

$$M = P\left\{\dfrac{\dfrac{r}{12}}{1 - \left[\dfrac{1}{(r/12) + 1}\right]^{12t}}\right\}.$$

(a) Use a graphing utility to graph the model when $P = \$100,000$ and $r = 8\%$.

(b) You are given a choice of a 20-year term or a 30-year term. Which would you choose? Explain.

**66.** *Hourly Wages*   The average hourly wage $w$ in the United States from 1970 to 1995 can be modeled by

$$w = 2.17 + \dfrac{1}{0.099 + 0.75e^{-0.171t}},$$

where $t = 0$ corresponds to 1970.   *(Source: U.S. Bureau of Labor Statistics)*

(a) Use a graphing utility to graph the model.

(b) Use the model to determine the year in which the average hourly wage was $10.

(c) For how many years past 1995 do you think this equation might be a good model for the average hourly wage? Explain your reasoning.

In Exercises 67–74, find the derivative of the function.

**67.** $f(x) = \ln 3x^2$

**68.** $y = \ln \dfrac{x(x - 1)}{x - 2}$

**69.** $y = x\sqrt{\ln x}$

**70.** $f(x) = \ln\sqrt{4x}$

**71.** $y = \dfrac{\ln x}{x^3}$

**72.** $y = \ln(x^2 - 2)^{2/3}$

**73.** $f(x) = \ln e^{-x^2}$

**74.** $y = \ln \dfrac{e^x}{1 + e^x}$

In Exercises 75–78, graph and analyze the function. Include any relative extrema and points of inflection in your analysis.

**75.** $y = \ln(x + 3)$      **76.** $y = \dfrac{8 \ln x}{x^2}$

**77.** $y = \ln \dfrac{10}{x + 2}$      **78.** $y = \ln \dfrac{x^2}{9 - x^2}$

In Exercises 79–82, evaluate the logarithm.

**79.** $\log_7 49$      **80.** $\log_2 32$

**81.** $\log_{10} 1$      **82.** $\log_4 \frac{1}{64}$

In Exercises 83–86, use the change of base formula to evaluate the logarithm. Round the result to four decimal places.

**83.** $\log_5 10$      **84.** $\log_4 12$

**85.** $\log_{16} 64$      **86.** $\log_{10} 125$

In Exercises 87–90, find the derivative of the function.

**87.** $y = \log_3(2x - 1)$

**88.** $y = \log_{10} \dfrac{3}{x}$

**89.** $y = \log_2 \dfrac{1}{x^2}$

**90.** $y = \log_{16}(x^2 - 3x)$

**91.** *Depreciation*  After $t$ years, the value $V$ of a car purchased for $20,000 is

$$V = 20,000(0.75)^t.$$

(a) Sketch a graph of the function and determine the value of the car 2 years after it was purchased.

(b) Find the rate of change of $V$ with respect to $t$ when $t = 1$ and when $t = 4$.

(c) After how many years will the car be worth $5000?

**92.** *Inflation*  If the annual rate of inflation averages 5% over the next 10 years, then the approximate cost of goods or services $C$ during any year in that decade is

$$C = P(1.05)^t,$$

where $t$ is the time in years and $P$ is the present cost.

(a) The price of an oil change is presently $24.95. Estimate the price of an oil change 10 years from now.

(b) Find the rate of change of $C$ with respect to $t$ when $t = 1$.

**93.** *Drug Decomposition*  A medical solution contains 500 milligrams of a drug per milliliter when the solution is prepared. After 40 days, it contains only 300 milligrams per milliliter. Assuming that the rate of decomposition is proportional to the concentration present, find an equation giving the concentration $A$ after $t$ days.

**94.** *Population Growth*  A population is growing continuously at the rate of $2\frac{1}{2}$% per year. Find the time necessary for the population to (a) double in size and (b) triple in size.

**95.** *Half-Life*  A sample of radioactive waste is taken from a nuclear plant. The sample contains 50 grams of strontium-90 at time $t = 0$ years and 42.031 grams after 7 years. What is the half-life of strontium-90?

**96.** *Half-Life*  The half-life of cobalt-60 is 5.2 years. Find the time it would take for a sample of 0.5 gram of cobalt-60 to decay to 0.1 gram.

**97.** *Profit*  The profit for Microsoft was $78.1 million in 1987 and $2176.0 million in 1996 (see figure). Use an exponential growth model to predict the profit in 2000. *(Source: Microsoft)*

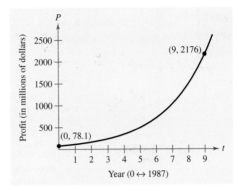

**98.** *Net Worth*  The net worth of Toys'R'Us was $1135.3 million in 1987 and $4190.6 million in 1996 (see figure). Use an exponential growth model to predict the net worth in 2000. *(Source: Toys'R'Us)*

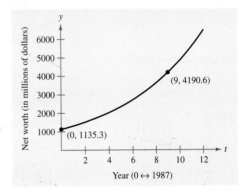

# Sample Post-Graduation Exam Questions

CPA  GMAT
GRE  CLAST
Actuarial

*The following questions represent the types of questions that appear on certified public accountant (CPA) exams, graduate management admission tests (GMAT), graduate records exams (GRE), actuarial exams, and college-level academic skills tests (CLAST). The answers to the questions are given in the back of the book.*

**1.** The rate of decay of a radioactive substance is proportional to the amount of the substance present. Three years ago there was 6 grams of substance. Now there is 5 grams. How many grams will there be 3 years from now?

   (a) 4    (b) $\frac{25}{6}$    (c) $\frac{125}{36}$    (d) $\frac{75}{36}$

**2.** In a certain town, 45% of the people have brown hair, 30% have brown eyes, and 15% have both brown hair and brown eyes. What percent of the people in the town have neither brown hair nor brown eyes?

   (a) 25%    (b) 35%    (c) 40%    (d) 50%

**3.** You have $900 and deposit it in a savings account that is compounded continuously at 4.76%. After 16 years the amount in the account will be

   (a) $1927.53    (b) $1077.81    (c) $943.88    (d) $2827.53

**4.** A bookstore orders 75 books. Each book costs the bookstore $29 and is sold for $42. The bookstore must pay a $4 service charge for each unsold book returned. If the bookstore returns seven books, how much profit will the bookstore make?

   (a) $975    (b) $947    (c) $653    (d) $681

Figure for 5–7

**Income and Expenses for Company A**

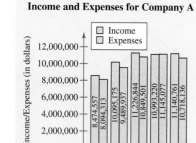

For Questions 5–7, use the data given in the graph.

**5.** In how many of the years was spending larger than in the preceding year?

   (a) 2    (b) 4    (c) 1    (d) 3

**6.** In which year was the profit the greatest?

   (a) 1991    (b) 1994    (c) 1990    (d) 1992

**7.** In 1993, profits decreased by $x$ percent from 1992 with $x$ equal to

   (a) 60%    (b) 140%    (c) 340%    (d) 40%

# Integration and Its Applications

5

## STRATEGIES *for* SUCCESS

 **OBJECTIVES**

*When you have completed this chapter, make sure you are able to:*

❐ Find the antiderivative $F$ of a function $f$; that is, $F'(x) = f(x)$.
❐ Use the General Power Rule to calculate antiderivatives.
❐ Use the Exponential Rule and the Log Rule to calculate antiderivatives.
❐ Evaluate definite integrals and apply the Fundamental Theorem of Calculus.
❐ Use the Fundamental Theorem of Calculus to find the area bounded by two graphs.
❐ Use integration to find the volume of a solid of revolution.

 **TOOLS**

*Use these study tools to achieve the objectives above:*

**Algebra Review**
(pages 374 and 375)

**Chapter Summary and Study Strategies**
(pages 376 and 377)

**Review Exercises**
(pages 378–381)

 **ADDITIONAL RESOURCES**

*Use these resources to solidify your mastery of calculus:*

**Student Solutions Guide**

**Study Guide** (Additional Examples, Similar Problems, and Chapter Test)

**Algebra Review Tutorial Software**

**Graphing Technology Guide**

**Sample Post-Graduation Exam Questions**
(page 382)

**Web Exercise**
(page 324, exercise 73)

Richard Pasley/Stock Boston

*In Exercise 72 on page 324, you will recommend how many products a company should produce based on its benefits and costs.*

## 5.1 | Antiderivatives and Indefinite Integrals

*Antiderivatives* • *Notation for Antiderivatives and Indefinite Integrals* •
*Finding Antiderivatives* • *Particular Solutions* • *Applications*

## Antiderivatives

Up to this point in the text, you have been concerned primarily with this problem: given a function, find its derivative. Many important applications of calculus involve the inverse problem: given the derivative of a function, find the function. For example, suppose you are given

$$f'(x) = 2, \quad g'(x) = 3x^2, \quad \text{and} \quad s'(t) = 4t.$$

Your goal is to determine the functions $f$, $g$, and $s$. By making educated guesses, you might come up with the following.

$$f(x) = 2x \quad \text{because} \quad \frac{d}{dx}[2x] = 2$$

$$g(x) = x^3 \quad \text{because} \quad \frac{d}{dx}[x^3] = 3x^2$$

$$s(t) = 2t^2 \quad \text{because} \quad \frac{d}{dt}[2t^2] = 4t$$

This operation of determining the original function from its derivative is the inverse operation of differentiation. It is called **antidifferentiation.**

*STUDY TIP* In this text, we use the phrase "$F(x)$ is an antiderivative of $f(x)$" synonymously with "$F$ is an antiderivative of $f$."

### Definition of Antiderivative

A function $F$ is an **antiderivative** of a function $f$ if for every $x$ in the domain of $f$, it follows that $F'(x) = f(x)$.

If $F(x)$ is an antiderivative of $f(x)$, then $F(x) + C$, where $C$ is any constant, is also an antiderivative of $f(x)$. For example,

$$F(x) = x^3, \quad G(x) = x^3 - 5, \quad \text{and} \quad H(x) = x^3 + 0.3$$

are antiderivatives of $3x^2$ because the derivative of each is $3x^2$. As it turns out, *all* antiderivatives of $3x^2$ are of the form $x^3 + C$. Thus, the process of antidifferentiation does not determine a single function, but rather a *family* of functions, each differing from the others by a constant.

## Notation for Antiderivatives and Indefinite Integrals

The antidifferentiation process is also called **integration** and is denoted by the symbol

$$\int,$$        Integral sign

called an **integral sign.** The symbol

$$\int f(x)\, dx$$        Indefinite integral

is the **indefinite integral** of $f(x)$, and it denotes the family of antiderivatives of $f(x)$. That is, if $F'(x) = f(x)$ for all $x$, then you can write

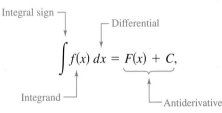

where $f(x)$ is the **integrand** and $C$ is the **constant of integration.** The differential $dx$ in the indefinite integral identifies the variable of integration. That is, the symbol $\int f(x)\, dx$ denotes the "antiderivative of $f$ *with respect to x*" just as the symbol $dy/dx$ denotes the "derivative of $y$ *with respect to x.*"

---

### Integral Notation of Antiderivatives

The notation

$$\int f(x)\, dx = F(x) + C,$$

where $C$ is an arbitrary constant, means that $F$ is an antiderivative of $f$. That is, $F'(x) = f(x)$ for all $x$ in the domain of $f$.

---

*Discovery*

Verify that $F_1(x) = x^2 - 2x$, $F_2(x) = x^2 - 2x - 1$, and $F_3(x) = (x - 1)^2$ are antiderivatives of $f(x) = 2x - 2$. Use a graphing utility to graph $F_1$, $F_2$, and $F_3$ in the same coordinate plane. How are their graphs related? What can you say about the graph of any other antiderivative of $f$?

**EXAMPLE 1**    **Notation for Antiderivatives**

Using integral notation, you can write the three antiderivatives given at the beginning of this section as follows.

(a) $\displaystyle\int 2\, dx = 2x + C$     (b) $\displaystyle\int 3x^2\, dx = x^3 + C$     (c) $\displaystyle\int 4t\, dt = 2t^2 + C$

## Finding Antiderivatives

The inverse relationship between the operations of integration and differentiation can be shown symbolically, as follows.

$$\frac{d}{dx}\left[\int f(x)\,dx\right] = f(x)$$    Differentiation is the inverse of integration.

$$\int f'(x)\,dx = f(x) + C$$    Integration is the inverse of differentiation.

This inverse relationship between integration and differentiation allows you to obtain integration formulas directly from differentiation formulas. The following summary lists the integration formulas that correspond to some of the differentiation formulas you have studied.

---

### Basic Integration Rules

1. $\displaystyle\int k\,dx = kx + C, \quad k$ is a constant    Constant Rule

2. $\displaystyle\int kf(x)\,dx = k\int f(x)\,dx$    Constant Multiple Rule

3. $\displaystyle\int [f(x) + g(x)]\,dx = \int f(x)\,dx + \int g(x)\,dx$    Sum Rule

4. $\displaystyle\int [f(x) - g(x)]\,dx = \int f(x)\,dx - \int g(x)\,dx$    Difference Rule

5. $\displaystyle\int x^n\,dx = \frac{x^{n+1}}{n+1} + C, \quad n \neq -1$    Simple Power Rule

---

*STUDY TIP*    You will study the General Power Rule for integration in Section 5.2 and the Exponential and Log Rules in Section 5.3.

Be sure you see that the Simple Power Rule has the restriction that $n$ cannot be $-1$. Thus, you *cannot* use the Simple Power Rule to evaluate the integral

$$\int \frac{1}{x}\,dx.$$

To evaluate this integral, you need the Log Rule (Section 5.3).

*STUDY TIP*    In Example 2b, the integral $\int 1\,dx$ is usually shortened to the form $\int dx$.

## EXAMPLE 2    Finding Indefinite Integrals

(a) $\displaystyle\int 2\,dx = 2x + C$    (b) $\displaystyle\int 1\,dx = x + C$    (c) $\displaystyle\int -5\,dt = -5t + C$

## EXAMPLE 3  Finding an Indefinite Integral

Find the indefinite integral $\int 3x\,dx$.

### Solution

$$\int 3x\,dx = 3\int x\,dx \qquad \text{Constant Multiple Rule}$$

$$= 3\int x^1\,dx \qquad \text{Rewrite } x \text{ as } x^1.$$

$$= 3\left(\frac{x^2}{2}\right) + C \qquad \text{Power Rule with } n = 1$$

$$= \frac{3}{2}x^2 + C \qquad \text{Simplify.}$$

In finding indefinite integrals, a strict application of the basic integration rules tends to produce cumbersome constants of integration. For instance, in Example 3, you could have written

$$\int 3x\,dx = 3\int x\,dx = 3\left(\frac{x^2}{2} + C\right) = \frac{3}{2}x^2 + 3C.$$

However, because $C$ represents *any* constant, it is unnecessary to write $3C$ as the constant of integration. You can simply write $\frac{3}{2}x^2 + C$.

In Example 3, note that the general pattern of integration is similar to that of differentiation.

| Given: | Rewrite: | Integrate: | Simplify: |
|---|---|---|---|
| $\int 3x\,dx$ | $3\int x^1\,dx$ | $3\left(\frac{x^2}{2}\right) + C$ | $\frac{3}{2}x^2 + C$ |

## EXAMPLE 4  Rewriting Before Integrating

| | Given Integral | Rewrite | Integrate | Simplify |
|---|---|---|---|---|
| (a) | $\int \dfrac{1}{x^3}\,dx$ | $\int x^{-3}\,dx$ | $\dfrac{x^{-2}}{-2} + C$ | $-\dfrac{1}{2x^2} + C$ |
| (b) | $\int \sqrt{x}\,dx$ | $\int x^{1/2}\,dx$ | $\dfrac{x^{3/2}}{3/2} + C$ | $\dfrac{2}{3}x^{3/2} + C$ |

**Technology**

If you have access to a symbolic integration program such as *Derive, Maple, Mathcad, Mathematica,* or the *TI-92,* try using it to find antiderivatives. For instance, the following steps show how *Derive* for Windows can be used to find the antiderivative in Example 3. Note that the program does not list a constant of integration.

#1: $3 \cdot x$      Author.

#2: $\int 3 \cdot x\,dx$      Integrate.

#3: $\dfrac{3 \cdot x^2}{2}$      Simplify.

***STUDY TIP*** Remember that you can check your answer to an antidifferentiation problem by differentiating. For instance, in Example 4b, you can check that $\frac{2}{3}x^{3/2}$ is the correct antiderivative by differentiating to obtain

$$\frac{d}{dx}\left[\frac{2}{3}x^{3/2}\right] = \left(\frac{2}{3}\right)\left(\frac{3}{2}\right)x^{1/2}$$

$$= \sqrt{x}.$$

With the five basic integration rules, you can integrate *any* polynomial function, as demonstrated in the next example.

### EXAMPLE 5   Integrating Polynomial Functions

Find the following indefinite integrals.

(a) $\displaystyle\int (x + 2)\, dx$     (b) $\displaystyle\int (3x^4 - 5x^2 + x)\, dx$

**Solution**

(a)  Use the Sum Rule to integrate each part separately.

$$\int (x + 2)\, dx = \int x\, dx + \int 2\, dx$$

$$= \frac{x^2}{2} + 2x + C$$

(b)  Try to identify each basic integration rule used to evaluate this integral.

$$\int (3x^4 - 5x^2 + x)\, dx = 3\left(\frac{x^5}{5}\right) - 5\left(\frac{x^3}{3}\right) + \frac{x^2}{2} + C$$

$$= \frac{3}{5}x^5 - \frac{5}{3}x^3 + \frac{1}{2}x^2 + C$$

**TIP**

For help on the algebra in Example 6, see Example 1a in the *Chapter 5 Algebra Review*, on page 374.

**ALGEBRA**

### EXAMPLE 6   Rewriting Before Integrating

Find the indefinite integral

$$\int \frac{x + 1}{\sqrt{x}}\, dx.$$

**Solution**

Begin by rewriting the quotient in the integrand as a sum. Then rewrite each term using rational exponents.

$$\int \frac{x + 1}{\sqrt{x}}\, dx = \int \left(\frac{x}{\sqrt{x}} + \frac{1}{\sqrt{x}}\right) dx \qquad \text{Rewrite as a sum.}$$

$$= \int (x^{1/2} + x^{-1/2})\, dx \qquad \text{Use rational exponents.}$$

$$= \frac{x^{3/2}}{3/2} + \frac{x^{1/2}}{1/2} + C \qquad \text{Apply Power Rule.}$$

$$= \frac{2}{3}x^{3/2} + 2x^{1/2} + C \qquad \text{Simplify.}$$

**STUDY TIP**   When integrating quotients, don't make the mistake of integrating the numerator and denominator separately. For instance, in Example 6, be sure you see that

$$\int \frac{x + 1}{\sqrt{x}}\, dx \neq \frac{\int (x + 1)\, dx}{\int \sqrt{x}\, dx}.$$

If you have access to a symbolic integration program, try using it for this example. When we tried it, the program listed the antiderivative as $2\sqrt{x}(x + 3)/3$, which is equivalent to the result listed above.

## Particular Solutions

You have already seen that the equation $y = \int f(x) \, dx$ has many solutions, each differing from the others by a constant. This means that the graphs of any two antiderivatives of $f$ are vertical translations of each other. For instance, Figure 5.1 shows the graphs of several antiderivatives of the form

$$y = F(x) = \int (3x^2 - 1) \, dx = x^3 - x + C.$$

Each of these antiderivatives is a solution of $dy/dx = 3x^2 - 1$.

In many applications of integration, you are given enough information to determine a **particular solution.** To do this, you need only know the value of $F(x)$ for one value of $x$. (This information is called an **initial condition.**) For example, in Figure 5.1, there is only one curve that passes through the point $(2, 4)$. To find this curve, use the following information.

$F(x) = x^3 - x + C$    General solution

$F(2) = 4$    Initial condition

By using the initial condition in the general solution, you can determine that $F(2) = 2^3 - 2 + C = 4$, which implies that $C = -2$. Thus, the particular solution is

$F(x) = x^3 - x - 2.$    Particular solution

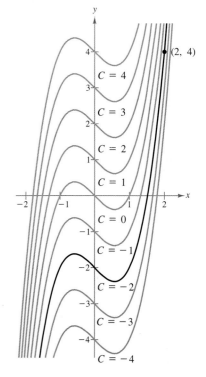

$F(x) = x^3 - x + C$

**FIGURE 5.1**

---

## EXAMPLE 7    Finding a Particular Solution

Find the general solution of $F'(x) = 2x - 2$, and find the particular solution that satisfies the initial condition $F(1) = 2$.

### Solution

Begin by integrating to find the general solution.

$$F(x) = \int (2x - 2) \, dx \qquad \text{Integrate } F'(x) \text{ to obtain } F(x).$$

$$= x^2 - 2x + C \qquad \text{General solution}$$

Using the initial condition $F(1) = 2$, you can write

$$F(1) = 1^2 - 2(1) + C = 2,$$

which implies that $C = 3$. Thus, the particular solution is

$$F(x) = x^2 - 2x + 3. \qquad \text{Particular solution}$$

This solution is shown graphically in Figure 5.2. Note that each of the gray curves represents a solution of the equation $F'(x) = 2x - 2$. The black curve, however, is the only solution that passes through the point $(1, 2)$, which means that $F(x) = x^2 - 2x + 3$ is the only solution that satisfies the initial condition.

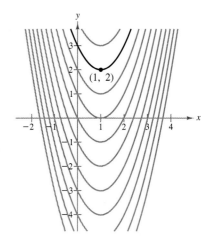

**FIGURE 5.2**

## Applications

In Chapter 2, you used the general position function for a falling object, $s(t) = -16t^2 + v_0 t + s_0$, where $s(t)$ is the height (in feet) and $t$ is the time (in seconds). In the next example, integration is used to *derive* this function.

### EXAMPLE 8   Deriving a Position Function

A ball is thrown upward with an initial velocity of 64 feet per second from an initial height of 80 feet, as shown in Figure 5.3. Derive the position function giving the height $s$ (in feet) as a function of the time $t$ (in seconds). When does the ball hit the ground?

*Solution*

Let $t = 0$ represent the initial time. Then the two given conditions can be written as follows.

$$s(0) = 80 \qquad \text{Initial height is 80 feet.}$$
$$s'(0) = 64 \qquad \text{Initial velocity is 64 feet per second.}$$

Because the acceleration due to gravity is $-32$ feet per second per second, you have the following.

$$s''(t) = -32 \qquad \text{Acceleration due to gravity}$$
$$s'(t) = \int -32\, dt \qquad \text{Integrate } s''(t) \text{ to obtain } s'(t).$$
$$= -32t + C_1 \qquad \text{Velocity function}$$

Using the initial velocity, you can conclude that $C_1 = 64$.

$$s'(t) = -32t + 64 \qquad \text{Velocity function}$$
$$s(t) = \int (-32t + 64)\, dt \qquad \text{Integrate } s'(t) \text{ to obtain } s(t).$$
$$= -16t^2 + 64t + C_2 \qquad \text{Position function}$$

Using the initial height, it follows that $C_2 = 80$. Thus, the position function is

$$s(t) = -16t^2 + 64t + 80. \qquad \text{Position function}$$

To find the time when the ball hits the ground, set the position function equal to zero and solve for $t$.

$$-16t^2 + 64t + 80 = 0 \qquad \text{Set } s(t) \text{ equal to zero.}$$
$$-16(t + 1)(t - 5) = 0 \qquad \text{Factor.}$$
$$t = -1, \quad t = 5 \qquad \text{Solve for } t.$$

Because the time must be positive, you can conclude that the ball hits the ground 5 seconds after it is thrown.

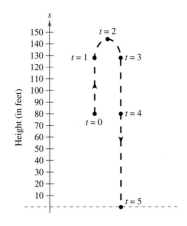

**FIGURE 5.3**

## EXAMPLE 9   Finding a Cost Function

The marginal cost for producing $x$ units of a product is modeled by

$$\frac{dC}{dx} = 32 - 0.04x.$$                    Marginal cost

It costs $50 to produce one unit. Find the total cost of producing 200 units.

### Solution

To find the cost function, integrate the marginal cost function.

$$C = \int (32 - 0.04x)\, dx$$                    Integrate $dC/dx$ to obtain $C$.

$$= 32x - 0.04\left(\frac{x^2}{2}\right) + K$$

$$= 32x - 0.02x^2 + K$$                    Cost function

To solve for $K$, use the initial condition that $C = 50$ when $x = 1$.

$$50 = 32(1) - 0.02(1)^2 + K$$                    Substitute 50 for $C$ and 1 for $x$.

$$18.02 = K$$                    Solve for $K$.

Thus, the total cost function is

$$C = 32x - 0.02x^2 + 18.02,$$                    Cost function

which implies that the cost of producing 200 units is

$$C = 32(200) - 0.02(200)^2 + 18.02$$

$$= \$5618.02.$$

## *Group Discussion*     *Investigating Marginal Cost*

In Example 9, you were given a marginal cost function of

$$\frac{dC}{dx} = 32 - 0.04x.$$

This means that the cost of making each additional unit decreases by about $0.04. You can confirm this by finding the cost of making different amounts of the product.

| $x$ | 1 | 2 | 3 | 4 | 5 | 6 |
|---|---|---|---|---|---|---|
| $C$ | $50.00 | $81.94 | $113.84 | $145.70 | $177.52 | $209.30 |

From this, you can see that the first unit cost $50, the second cost $31.94, the third cost $31.90, the fourth cost $31.86, the fifth cost $31.82, and so on. At what production level would this costing scheme cease to make sense? Explain your reasoning.

*Warm Up*

The following warm-up exercises involve skills that were covered in earlier sections. You will use these skills in the exercise set for this section.

In Exercises 1–6, rewrite the expression using rational exponents.

**1.** $\dfrac{\sqrt{x}}{x}$

**2.** $\sqrt[3]{2x}\,(2x)$

**3.** $\sqrt{5x^3} + \sqrt{x^5}$

**4.** $\dfrac{1}{\sqrt{x}} + \dfrac{1}{\sqrt[3]{x^2}}$

**5.** $\dfrac{(x+1)^3}{\sqrt{x+1}}$

**6.** $\dfrac{\sqrt{x}}{\sqrt[3]{x}}$

In Exercises 7–10, let $(x, y) = (2, 2)$, and solve the equation for $C$.

**7.** $y = x^2 + 5x + C$

**8.** $y = 3x^3 - 6x + C$

**9.** $y = -16x^2 + 26x + C$

**10.** $y = -\frac{1}{4}x^4 - 2x^2 + C$

## EXERCISES 5.1

In Exercises 1–8, verify the statement by showing that the derivative of the right side is equal to the integrand of the left side.

**1.** $\displaystyle \int \left( -\frac{9}{x^4} \right) dx = \frac{3}{x^3} + C$

**2.** $\displaystyle \int \frac{4}{\sqrt{x}}\,dx = 8\sqrt{x} + C$

**3.** $\displaystyle \int \left( 4x^3 - \frac{1}{x^2} \right) dx = x^4 + \frac{1}{x} + C$

**4.** $\displaystyle \int \left( 1 - \frac{1}{\sqrt[3]{x^2}} \right) dx = x - 3\sqrt[3]{x} + C$

**5.** $\displaystyle \int 2x^3 \sqrt{x}\,dx = \frac{4}{9}x^{9/2} + C$

**6.** $\displaystyle \int (x-2)(x+2)\,dx = \frac{x^3}{3} - 4x + C$

**7.** $\displaystyle \int \frac{x^2 - 1}{x^{3/2}}\,dx = \frac{2(x^2 + 3)}{3\sqrt{x}} + C$

**8.** $\displaystyle \int \frac{2x - 1}{x^{4/3}}\,dx = \frac{3(x + 1)}{\sqrt[3]{x}} + C$

In Exercises 9–18, find the indefinite integral and check your result by differentiation.

**9.** $\displaystyle \int 6\,dx$

**10.** $\displaystyle \int -4\,dx$

**11.** $\displaystyle \int 3t^2\,dt$

**12.** $\displaystyle \int t^4\,dt$

**13.** $\displaystyle \int 5x^{-3}\,dx$

**14.** $\displaystyle \int 4y^{-3}\,dy$

**15.** $\displaystyle \int du$

**16.** $\displaystyle \int e\,dt$

**17.** $\displaystyle \int x^{3/2}\,dx$

**18.** $\displaystyle \int v^{-1/2}\,dv$

In Exercises 19–24, complete a table using Example 4 as a model. Use a symbolic integration program to verify your results.

**19.** $\displaystyle \int \sqrt[3]{x}\,dx$

**20.** $\displaystyle \int \frac{1}{x^2}\,dx$

**21.** $\displaystyle \int \frac{1}{x\sqrt{x}}\,dx$

**22.** $\displaystyle \int x(x^2 + 3)\,dx$

**23.** $\displaystyle \int \frac{1}{2x^3}\,dx$

**24.** $\displaystyle \int \frac{1}{(2x)^3}\,dx$

In Exercises 25 and 26, find two functions that have the given derivative, and sketch the graph of each. (There is more than one correct answer.)

**25.**

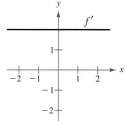

**26.**

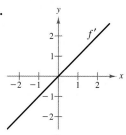

In Exercises 27–36, find the indefinite integral and check your result by differentiation.

**27.** $\displaystyle\int (x^3 + 2)\, dx$

**28.** $\displaystyle\int (x^2 - 2x + 3)\, dx$

**29.** $\displaystyle\int (2x^{4/3} + 3x - 1)\, dx$

**30.** $\displaystyle\int \left(\sqrt{x} + \frac{1}{2\sqrt{x}}\right) dx$

**31.** $\displaystyle\int \sqrt[3]{x^2}\, dx$

**32.** $\displaystyle\int (\sqrt[4]{x^3} + 1)\, dx$

**33.** $\displaystyle\int \frac{1}{x^4}\, dx$

**34.** $\displaystyle\int \frac{1}{4x^2}\, dx$

**35.** $\displaystyle\int (2x + x^{-1/2})\, dx$

**36.** $\displaystyle\int \frac{t^2 + 2}{t^2}\, dt$

In Exercises 37–42, use a symbolic integration program to find the indefinite integral.

**37.** $\displaystyle\int u(3u^2 + 1)\, du$

**38.** $\displaystyle\int \sqrt{x}(x + 1)\, dx$

**39.** $\displaystyle\int (x - 1)(6x - 5)\, dx$

**40.** $\displaystyle\int (2t^2 - 1)^2\, dt$

**41.** $\displaystyle\int y^2 \sqrt{y}\, dy$

**42.** $\displaystyle\int (1 + 3t)t^2\, dt$

In Exercises 43–48, find the particular solution $y = f(x)$ that satisfies the differential equation and initial condition.

**43.** $f'(x) = 3\sqrt{x} + 3;\quad f(1) = 4$

**44.** $f'(x) = \frac{1}{5}x - 2;\quad f(10) = -10$

**45.** $f'(x) = 6x(x - 1);\quad f(1) = -1$

**46.** $f'(x) = (2x - 3)(2x + 3);\quad f(3) = 0$

**47.** $f'(x) = \dfrac{2 - x}{x^3},\ x > 0;\quad f(2) = \dfrac{3}{4}$

**48.** $f'(x) = -\dfrac{5}{x^2},\ x > 0;\quad f(1) = 2$

In Exercises 49 and 50, you are shown a family of graphs, each of which is a solution of the given differential equation. Find the equation of the particular solution that passes through the indicated point.

**49.** $\dfrac{dy}{dx} = -5x - 2$

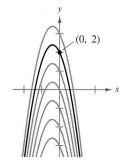

**50.** $\dfrac{dy}{dx} = 2(x - 1)$

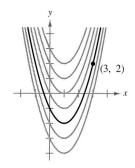

In Exercises 51 and 52, find the equation of the function $f$ whose graph passes through the point.

| *Derivative* | *Point* |
|---|---|
| **51.** $f'(x) = 6\sqrt{x} - 10$ | $(4, 2)$ |
| **52.** $f'(x) = \dfrac{6}{x^2}$ | $(2, 5)$ |

In Exercises 53–56, find a function $f$ that satisfies the given conditions.

**53.** $f''(x) = 2,\quad f'(2) = 5,\quad f(2) = 10$

**54.** $f''(x) = x^2,\quad f'(0) = 6,\quad f(0) = 3$

**55.** $f''(x) = x^{-2/3},\quad f'(8) = 6,\quad f(0) = 0$

**56.** $f''(x) = x^{-3/2},\quad f'(1) = 2,\quad f(9) = -4$

*Cost*   In Exercises 57–60, find the cost function for the marginal cost and fixed cost.

| *Marginal Cost* | *Fixed Cost $(x = 0)$* |
|---|---|
| **57.** $\dfrac{dC}{dx} = 85$ | $\$5500$ |
| **58.** $\dfrac{dC}{dx} = \dfrac{1}{50}x + 10$ | $\$1000$ |
| **59.** $\dfrac{dC}{dx} = \dfrac{1}{20\sqrt{x}} + 4$ | $\$750$ |
| **60.** $\dfrac{dC}{dx} = \dfrac{\sqrt[4]{x}}{10} + 10$ | $\$2300$ |

*Demand Function*   In Exercises 61 and 62, find the revenue and demand functions for the given marginal revenue. (Use the fact that $R = 0$ when $x = 0$.)

**61.** $\dfrac{dR}{dx} = 225 - 3x$     **62.** $\dfrac{dR}{dx} = 100 - 6x - 2x^2$

*Profit*   In Exercises 63 and 64, find the profit function for the given marginal profit and initial condition.

| Marginal Profit | Initial Condition |
|---|---|

**63.** $\dfrac{dP}{dx} = -18x + 1650$     $P(15) = \$22{,}725$

**64.** $\dfrac{dP}{dx} = -40x + 250$     $P(5) = \$650$

*Vertical Motion*   In Exercises 65–68, use $a(t) = -32$ feet per second per second as the acceleration due to gravity. (Neglect air resistance.)

**65.** A ball is thrown vertically upward with an initial velocity of 60 feet per second. How high will the ball go?

**66.** *Grand Canyon*   The Grand Canyon is 6000 feet deep at the deepest part. A rock is dropped from this height. Express the height of the rock as a function of the time $t$ (in seconds). How long will it take the rock to hit the canyon floor?

**67.** With what initial velocity must an object be thrown upward from the ground to reach the height of the Washington Monument (550 feet)?

**68.** A balloon, rising vertically with a velocity of 16 feet per second, releases a sandbag at an instant when the balloon is 64 feet above the ground.

  (a) How many seconds after its release will the bag strike the ground?

  (b) With what velocity will the bag strike the ground?

**69.** *Cost*   A company produces a product for which the marginal cost of producing $x$ units is

$$\frac{dC}{dx} = 2x - 12$$

and the fixed costs are $125.

  (a) Find the total cost function and the average cost function.

  (b) Find the total cost of producing 50 units.

  (c) In part (b), how much of the total cost is fixed? How much is variable? Give examples of fixed costs associated with the manufacturing of a product. Give examples of variable costs.

**70.** *Population Growth*   The growth rate of a city's population is modeled by

$$\frac{dP}{dt} = 500t^{1.06},$$

where $t$ is the time in years. The city's population is now 50,000. What will the population be in 10 years?

**71.** *Natural Gas Consumption*   The consumption $S$ (in quadrillion Btu's) of natural gas in the United States increased steadily between 1985 and 1994. The rate of increase can be modeled by

$$\frac{dS}{dt} = 0.012t^2 - 0.182t + 1.045, \quad 5 \le t \le 14,$$

where $t = 5$ represents 1985. In 1985, the consumption was 17.8 quadrillion Btu's. Find a model for the consumption from 1985 through 1994 and determine the consumption in 1994.   *(Source: U.S. Energy Information Administration)*

   **72.** *Economics: Marginal Benefits and Costs*   The table gives the marginal benefit and marginal cost of producing $x$ products for a given company. Plot the points of each column and use the regression capabilities of a graphing utility to find a linear model for marginal benefit and a quadratic model for marginal cost. Then use integration to find the benefit $B$ and cost $C$ equations. Assume $B(0) = 0$ and $C(0) = 425$. Finally, find the intervals where the benefit exceeds the cost of producing $x$ products, and make a recommendation for how many products the company should produce based on your findings.   *(Source: Adapted from Taylor,* Economics, *First Edition)*

|  | 1 | 2 | 3 | 4 | 5 |
|---|---|---|---|---|---|
| Marginal benefit | 330 | 320 | 290 | 270 | 250 |
| Marginal cost | 150 | 120 | 100 | 110 | 120 |

|  | 6 | 7 | 8 | 9 | 10 |
|---|---|---|---|---|---|
| Marginal benefit | 230 | 210 | 190 | 170 | 160 |
| Marginal cost | 140 | 160 | 190 | 250 | 320 |

**73.** *Research Project*   Use your school's library or some other reference source to research a company that markets a natural resource. Find data on the revenue of the company and on the consumption of the resource. Then find a model for each. Is the company's revenue related to the consumption of the resource? Explain.

<table>
<tr><td>**5.2**</td><td># The General Power Rule</td></tr>
</table>

*The General Power Rule • Substitution •*
*Extended Application: Propensity to Consume*

## The General Power Rule

In Section 5.1, you used the Simple Power Rule

$$\int x^n \, dx = \frac{x^{n+1}}{n+1} + C, \qquad n \neq -1$$

to find antiderivatives of functions expressed as powers of $x$ alone. In this section, you will study a technique for finding antiderivatives of more complicated functions.

To begin, consider how you might find the antiderivative of $2x(x^2 + 1)^3$. Because you are hunting for a function whose derivative is $2x(x^2 + 1)^3$, you might discover the following.

$$\frac{d}{dx}[(x^2 + 1)^4] = 4(x^2 + 1)^3(2x) \qquad \text{Use Chain Rule.}$$

$$\frac{d}{dx}\left[\frac{(x^2 + 1)^4}{4}\right] = (x^2 + 1)^3(2x) \qquad \text{Divide both sides by 4.}$$

$$\frac{(x^2 + 1)^4}{4} + C = \int 2x(x^2 + 1)^3 \, dx \qquad \text{Write in integral form.}$$

The key to this solution is the presence of the factor $2x$ in the integrand. In other words, this solution works because $2x$ is precisely the derivative of $(x^2 + 1)$. Letting $u = x^2 + 1$, you can write

$$\int \overbrace{(x^2 + 1)^3}^{u^3} \underbrace{2x \, dx}_{du} = \int u^3 \, du = \frac{u^4}{4} + C.$$

This is an example of the **General Power Rule** for integration.

---

### General Power Rule for Integration

If $u$ is a differentiable function of $x$, then

$$\int u^n \frac{du}{dx} \, dx = \int u^n \, du = \frac{u^{n+1}}{n+1} + C, \qquad n \neq -1.$$

When using the General Power Rule, you must first identify a factor $u$ of the integrand that is raised to a power. Then, you must show that its derivative $du/dx$ is also a factor of the integrand. This is demonstrated in Example 1.

### EXAMPLE 1   Applying the General Power Rule

Find the following indefinite integrals.

(a) $\displaystyle \int 3(3x - 1)^4 \, dx$    (b) $\displaystyle \int (2x + 1)(x^2 + x) \, dx$

(c) $\displaystyle \int 3x^2 \sqrt{x^3 - 2} \, dx$    (d) $\displaystyle \int \frac{-4x}{(1 - 2x^2)^2} \, dx$

**Solution**

(a) $\displaystyle \int 3(3x - 1)^4 \, dx = \int \overbrace{(3x - 1)^4}^{u^n} \overbrace{(3)}^{\frac{du}{dx}} \, dx$         Let $u = 3x - 1$.

$\displaystyle \qquad\qquad = \frac{(3x - 1)^5}{5} + C$         General Power Rule

> **STUDY TIP**   Example 1b illustrates a case of the General Power Rule that is sometimes overlooked—when the power is $n = 1$. In this case, the rule takes the form
>
> $$\int u \frac{du}{dx} \, dx = \frac{u^2}{2} + C.$$

(b) $\displaystyle \int (2x + 1)(x^2 + x) \, dx = \int \overbrace{(x^2 + x)}^{u^n} \overbrace{(2x + 1)}^{\frac{du}{dx}} \, dx$         Let $u = x^2 + x$.

$\displaystyle \qquad\qquad = \frac{(x^2 + x)^2}{2} + C$         General Power Rule

(c) $\displaystyle \int 3x^2 \sqrt{x^3 - 2} \, dx = \int \overbrace{(x^3 - 2)^{1/2}}^{u^n} \overbrace{(3x^2)}^{\frac{du}{dx}} \, dx$         Let $u = x^3 - 2$.

$\displaystyle \qquad\qquad = \frac{(x^3 - 2)^{3/2}}{3/2} + C$         General Power Rule

$\displaystyle \qquad\qquad = \frac{2}{3}(x^3 - 2)^{3/2} + C$         Simplify.

(d) $\displaystyle \int \frac{-4x}{(1 - 2x^2)^2} \, dx = \int \overbrace{(1 - 2x^2)^{-2}}^{u^n} \overbrace{(-4x)}^{\frac{du}{dx}} \, dx$         Let $u = 1 - 2x^2$.

$\displaystyle \qquad\qquad = \frac{(1 - 2x^2)^{-1}}{-1} + C$         General Power Rule

$\displaystyle \qquad\qquad = -\frac{1}{1 - 2x^2} + C$         Simplify.

Many times, part of the derivative $du/dx$ is missing from the integrand, and in *some* cases you can make the necessary adjustments to apply the General Power Rule.

## EXAMPLE 2   Multiplying and Dividing by a Constant

Find the indefinite integral $\int x(3 - 4x^2)^2 \, dx$.

**TIP**

For help on the algebra in Example 2, see Example 1b in the *Chapter 5 Algebra Review,* on page 374.

**Solution**

Let $u = 3 - 4x^2$. To apply the General Power Rule, you need to create $du/dx = -8x$ as a factor of the integrand. You can accomplish this by multiplying and dividing by the constant $-8$.

$$\int x(3 - 4x^2)^2 \, dx = \int \left(-\frac{1}{8}\right) \overbrace{(3 - 4x^2)^2}^{u^n} \overbrace{(-8x)}^{\frac{du}{dx}} \, dx \qquad \text{Multiply and divide by } -8.$$

$$= -\frac{1}{8} \int (3 - 4x^2)^2(-8x) \, dx \qquad \text{Factor } -\frac{1}{8} \text{ out of integrand.}$$

$$= \left(-\frac{1}{8}\right)\frac{(3 - 4x^2)^3}{3} + C \qquad \text{General Power Rule}$$

$$= -\frac{(3 - 4x^2)^3}{24} + C \qquad \text{Simplify.}$$

*STUDY TIP*   Try using the Chain Rule to check the result of Example 2. After differentiating $-\frac{1}{24}(3 - 4x^2)^3$ and simplifying, you should obtain the original integrand.

## EXAMPLE 3   A Failure of the General Power Rule

Find the indefinite integral $\int -8(3 - 4x^2)^2 \, dx$.

**Solution**

Let $u = 3 - 4x^2$. As in Example 2, to apply the General Power Rule you must create $du/dx = -8x$ as a factor of the integrand. In Example 2, you could do this by multiplying and dividing by a constant, and then factoring that constant out of the integrand. This strategy doesn't work with variables. That is,

$$\int -8(3 - 4x^2)^2 \, dx \neq \frac{1}{x}\int (3 - 4x^2)^2(-8x) \, dx.$$

To find this indefinite integral, you can expand the integrand and use the Simple Power Rule.

$$\int -8(3 - 4x^2)^2 \, dx = \int (-72 + 192x^2 - 128x^4) \, dx$$

$$= -72x + 64x^3 - \frac{128}{5}x^5 + C$$

*STUDY TIP*   In Example 3, be sure you see that you cannot factor variable quantities outside the integral sign. After all, if this were permissible, then you could move the entire integrand outside the integral sign and eliminate the need for all integration rules except the rule $\int dx = x + C$.

When an integrand contains an extra constant factor that is not needed as part of $du/dx$, you can simply move the factor outside the integral sign, as illustrated in the next example.

### TIP

For help on the algebra in Example 4, see Example 1c in the *Chapter 5 Algebra Review*, on page 374.

ALGEBRA

---

## EXAMPLE 4   Applying the General Power Rule

Find the indefinite integral $\displaystyle\int 7x^2\sqrt{x^3+1}\,dx$.

### Solution

Let $u = x^3 + 1$. Then you need to create $du/dx = 3x^2$ by multiplying and dividing by 3. The constant factor $\frac{7}{3}$ is not needed as part of $du/dx$, and can be moved outside the integral sign.

$$\int 7x^2\sqrt{x^3+1}\,dx = \int 7x^2(x^3+1)^{1/2}\,dx \qquad \text{Write with rational exponents.}$$

$$= \int \frac{7}{3}(x^3+1)^{1/2}(3x^2)\,dx \qquad \text{Multiply and divide by 3.}$$

$$= \frac{7}{3}\int (x^3+1)^{1/2}(3x^2)\,dx \qquad \text{Factor } \frac{7}{3} \text{ outside integral.}$$

$$= \frac{7}{3}\frac{(x^3+1)^{3/2}}{3/2} + C \qquad \text{General Power Rule}$$

$$= \frac{14}{9}(x^3+1)^{3/2} + C \qquad \text{Simplify.}$$

## Technology

If you use a symbolic integration utility, such as *Derive*, *Maple*, *Mathcad*, *Mathematica*, or the *TI-92*, to find indefinite integrals, you should be in for some surprises. This is true because integration is not nearly as straightforward as differentiation. By trying different integrands, you should be able to find several that the program cannot solve: in such situations, it may list a new indefinite integral. You should also be able to find several that have horrendous antiderivatives, some with functions that you may not recognize. Two examples, using *Derive* for Windows, are shown at the right.

| | | |
|---|---|---|
| #1: | $\sqrt{(x^3+1)}$ | Author. |
| #2: | $\int \sqrt{(x^3+1)}\,dx$ | Integrate. |
| #3: | $\dfrac{3\cdot\int\dfrac{1}{\sqrt{(x^3+1)}}\,dx}{5} + \dfrac{2\cdot x\cdot\sqrt{(x^3+1)}}{5}$ | Simplify. |
| #4: | $\sqrt{(x^2+1)}$ | Author. |
| #5: | $\int \sqrt{(x^2+1)}\,dx$ | Integrate. |
| #6: | $\dfrac{\mathrm{LN}(\sqrt{(x^2+1)}+x)}{2} + \dfrac{x\cdot\sqrt{(x^2+1)}}{2}$ | Simplify. |

## Substitution

The integration technique used in Examples 1, 2, and 4 depends on your ability to recognize or create an integrand of the form $u^n \, du/dx$. With more complicated integrands, it is difficult to recognize the steps needed to fit the integrand to a basic integration formula. When this occurs, an alternative procedure called **substitution** or **change of variables** can be helpful. With this procedure, you completely rewrite the integral in terms of $u$ and $du$. That is, if $u = f(x)$, then $du = f'(x) \, dx$, and the General Power Rule takes the form

$$\int u^n \frac{du}{dx} \, dx = \int u^n \, du. \qquad \text{General Power Rule}$$

*Discovery*

Calculate the derivative of each of the following functions. Which one is the antiderivative of $f(x) = \sqrt{1 - 3x}$?

$$F(x) = (1 - 3x)^{3/2} + C$$
$$F(x) = \tfrac{2}{3}(1 - 3x)^{3/2} + C$$
$$F(x) = -\tfrac{2}{9}(1 - 3x)^{3/2} + C$$

### EXAMPLE 5   Integrating by Substitution

Find the indefinite integral $\displaystyle \int \sqrt{1 - 3x} \, dx$.

#### Solution

Begin by letting $u = 1 - 3x$. Then, $du/dx = -3$ and $du = -3 \, dx$. This implies that $dx = -\tfrac{1}{3} du$, and you can find the indefinite integral as follows.

$$\int \sqrt{1 - 3x} \, dx = \int (1 - 3x)^{1/2} \, dx \qquad \text{Write with rational exponent.}$$

$$= \int u^{1/2} \left( -\frac{1}{3} du \right) \qquad \text{Substitute } u \text{ and } du.$$

$$= -\frac{1}{3} \int u^{1/2} \, du \qquad \text{Factor } -\tfrac{1}{3} \text{ out of integrand.}$$

$$= -\frac{1}{3} \frac{u^{3/2}}{3/2} + C \qquad \text{Apply Power Rule.}$$

$$= -\frac{2}{9} u^{3/2} + C \qquad \text{Simplify.}$$

$$= -\frac{2}{9}(1 - 3x)^{3/2} + C \qquad \text{Substitute } 1 - 3x \text{ for } u.$$

To become efficient at integration, you should learn to use *both* techniques discussed in this section. For simpler integrals, you should use pattern recognition and create $du/dx$ by multiplying and dividing by an appropriate constant. For more complicated integrals, you should use a formal change of variables, as illustrated in Example 5. (You will learn more about this technique in Chapter 6.) For the integrals in this section's exercise set, try working several of the problems twice—once with pattern recognition and once using formal substitution.

## Extended Application: Propensity to Consume

In 1994, the U.S. poverty level for a family of four was about $15,200. Families at or below the poverty level tend to consume 100% of their income—that is, they use all their income to purchase necessities such as food, clothing, and shelter. As income level increases, the average consumption tends to drop below 100%. For instance, a family earning $17,000 may be able to save $510 and thus consume only $16,490 (97%) of their income. As the income increases, the ratio of consumption to savings tends to decrease. The rate of change of consumption with respect to income is called the **marginal propensity to consume.** *(Source: U.S. Bureau of Census)*

### EXAMPLE 6    Analyzing Consumption

For a family of four in 1994, the marginal propensity to consume income $x$ can be modeled by

$$\frac{dQ}{dx} = \frac{0.97}{(x - 15{,}199)^{0.03}}, \qquad x \geq 15{,}200,$$

where $Q$ represents the income consumed. Use the model to estimate the amount consumed by a family of four whose 1994 income was $26,000.

### Solution

Begin by integrating $dQ/dx$ to find a model for the consumption $Q$. Use the initial condition that $Q = 15{,}200$ when $x = 15{,}200$.

$$\frac{dQ}{dx} = \frac{0.97}{(x - 15{,}199)^{0.03}} \qquad \text{Marginal propensity to consume}$$

$$Q = \int \frac{0.97}{(x - 15{,}199)^{0.03}}\, dx \qquad \text{Integrate to obtain } Q.$$

$$= \int 0.97(x - 15{,}199)^{-0.03}\, dx \qquad \text{Rewrite.}$$

$$= (x - 15{,}199)^{0.97} + C \qquad \text{General Power Rule}$$

$$= (x - 15{,}199)^{0.97} + 15{,}199 \qquad \text{Use initial condition to find } C.$$

Using this model, you can estimate that a family of four with an income of $x = 26{,}000$ consumed about $23,373. The graph of $Q$ is shown in Figure 5.4.

---

**TIP**

When you use the initial condition to find the value of $C$ in Example 6, you substitute 15,200 for $Q$ and 15,200 for $x$.

$$Q = (x - 15{,}199)^{0.97} + C$$
$$15{,}200 = (15{,}200 - 15{,}199)^{0.97} + C$$
$$15{,}200 = 1 + C$$
$$15{,}199 = C$$

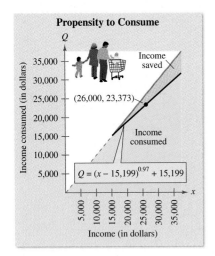

**Propensity to Consume**

$Q = (x - 15{,}199)^{0.97} + 15{,}199$

**FIGURE 5.4**

---

*Group Discussion*     ## Extending the Example

According to the model in Example 6, at what income level would a family of four consume $30,000? Use the graph to verify your results.

## Warm Up

The following warm-up exercises involve skills that were covered in earlier sections. You will use these skills in the exercise set for this section.

In Exercises 1–6, find the indefinite integral.

**1.** $\displaystyle \int (2x^3 + 1)\, dx$

**2.** $\displaystyle \int (x^{1/2} + 3x - 4)\, dx$

**3.** $\displaystyle \int \frac{1}{x^2}\, dx$

**4.** $\displaystyle \int \frac{1}{3t^3}\, dt$

**5.** $\displaystyle \int (1 + 2t)t^{3/2}\, dt$

**6.** $\displaystyle \int \sqrt{x}(2x - 1)\, dx$

In Exercises 7–10, simplify the expression.

**7.** $\displaystyle \left(-\frac{5}{4}\right)\frac{(x - 2)^4}{4}$

**8.** $\displaystyle \left(\frac{1}{6}\right)\frac{(x - 1)^{-2}}{-2}$

**9.** $\displaystyle (6)\frac{(x^2 + 3)^{2/3}}{2/3}$

**10.** $\displaystyle \left(\frac{5}{2}\right)\frac{(1 - x^3)^{-1/2}}{-1/2}$

# EXERCISES 5.2

In Exercises 1–6, identify $u$ and $du/dx$ for the integral $\int u^n (du/dx)\, dx$.

**1.** $\displaystyle \int (5x^2 + 1)^2 (10x)\, dx$

**2.** $\displaystyle \int (3 - 4x^2)^3 (-8x)\, dx$

**3.** $\displaystyle \int \sqrt{1 - x^2}\,(-2x)\, dx$

**4.** $\displaystyle \int 3x^2 \sqrt{x^3 + 1}\, dx$

**5.** $\displaystyle \int \left(4 + \frac{1}{x^2}\right)\left(\frac{-2}{x^3}\right) dx$

**6.** $\displaystyle \int \frac{1}{(1 + 2x)^2}(2)\, dx$

In Exercises 7–26, find the indefinite integral and check the result by differentiation.

**7.** $\displaystyle \int (1 + 2x)^4 (2)\, dx$

**8.** $\displaystyle \int (x^2 - 1)^3 (2x)\, dx$

**9.** $\displaystyle \int \sqrt{5x^2 - 4}\,(10x)\, dx$

**10.** $\displaystyle \int \sqrt{3 - x^3}\,(3x^2)\, dx$

**11.** $\displaystyle \int (x - 1)^4\, dx$

**12.** $\displaystyle \int (x - 3)^{5/2}\, dx$

**13.** $\displaystyle \int x(x^2 - 1)^7\, dx$

**14.** $\displaystyle \int x(1 - 2x^2)^3\, dx$

**15.** $\displaystyle \int \frac{x^2}{(1 + x^3)^2}\, dx$

**16.** $\displaystyle \int \frac{x^2}{(x^3 - 1)^2}\, dx$

**17.** $\displaystyle \int \frac{x + 1}{(x^2 + 2x - 3)^2}\, dx$

**18.** $\displaystyle \int \frac{6x}{(1 + x^2)^3}\, dx$

**19.** $\displaystyle \int \frac{x - 2}{\sqrt{x^2 - 4x + 3}}\, dx$

**20.** $\displaystyle \int \frac{4x + 6}{(x^2 + 3x + 7)^3}\, dx$

**21.** $\displaystyle \int 5x \sqrt[3]{1 - x^2}\, dx$

**22.** $\displaystyle \int u^3 \sqrt{u^4 + 2}\, du$

**23.** $\displaystyle \int \frac{4x}{\sqrt{1 + x^2}}\, dx$

**24.** $\displaystyle \int \frac{x^2}{\sqrt{1 - x^3}}\, dx$

**25.** $\displaystyle \int \frac{-3}{\sqrt{2x + 3}}\, dx$

**26.** $\displaystyle \int \frac{t + 2t^2}{\sqrt{t}}\, dt$

In Exercises 27–32, use a symbolic integration utility to find the indefinite integral.

**27.** $\displaystyle \int \frac{x^3}{\sqrt{1 - x^4}}\, dx$

**28.** $\displaystyle \int \left(1 + \frac{1}{t}\right)^3 \left(\frac{1}{t^2}\right) dt$

**29.** $\displaystyle \int \frac{1}{\sqrt{2x}}\, dx$

**30.** $\displaystyle \int \frac{1}{3x^2}\, dx$

**31.** $\displaystyle \int (x^3 + 3x)(x^2 + 1)\, dx$

**32.** $\displaystyle \int (3 - 2x - 4x^2)(1 + 4x)\, dx$

In Exercises 33–40, use formal substitution (as illustrated in Example 5) to find the indefinite integral.

**33.** $\int x(6x^2 - 1)^3 \, dx$

**34.** $\int x^2(1 - x^3)^2 \, dx$

**35.** $\int x^2(2 - 3x^3)^{3/2} \, dx$

**36.** $\int t\sqrt{t^2 + 1} \, dt$

**37.** $\int \dfrac{x}{\sqrt{x^2 + 25}} \, dx$

**38.** $\int \dfrac{3}{\sqrt{2x + 1}} \, dx$

**39.** $\int \dfrac{x^2 + 1}{\sqrt{x^3 + 3x + 4}} \, dx$

**40.** $\int \sqrt{x}\,(4 + x^{3/2})^2 \, dx$

In Exercises 41–44, perform the integration in two ways: once using the Simple Power Rule and once using the General Power Rule. Explain the difference in the results.

**41.** $\int (2x - 1)^2 \, dx$

**42.** $\int (3 - 2x)^2 \, dx$

**43.** $\int x(x^2 - 1)^2 \, dx$

**44.** $\int x(2x^2 + 1)^2 \, dx$

**45.** Find the equation of the function $f$ whose graph passes through the point $\left(0, \frac{4}{3}\right)$ and whose derivative is
$$f'(x) = x\sqrt{1 - x^2}.$$

**46.** Find the equation of the function $f$ whose graph passes through the point $\left(0, \frac{7}{3}\right)$ and whose derivative is
$$f'(x) = x\sqrt{1 - x^2}.$$

 **47.** *Cost*   The marginal cost of a product is modeled by $dC/dx = 4/\sqrt{x + 1}$. When $x = 15$, $C = 50$.

(a) Find the cost function.
(b) Use a graphing utility to graph $dC/dx$ and $C$ in the same viewing rectangle.

 **48.** *Cost*   The marginal cost of a product is modeled by
$$\frac{dC}{dx} = \frac{12}{\sqrt[3]{12x + 1}}.$$
When $x = 13$, $C = 100$.

(a) Find the cost function.
(b) Use a graphing utility to graph $dC/dx$ and $C$ in the same viewing rectangle.

*Supply Function*   In Exercises 49 and 50, find the supply function $x = f(p)$ that satisfies the given conditions.

**49.** $\dfrac{dx}{dp} = p\sqrt{p^2 - 25}, \quad x = 600$ when $p = \$13$

**50.** $\dfrac{dx}{dp} = \dfrac{10}{\sqrt{p - 3}}, \quad x = 100$ when $p = \$3$

*Demand Function*   In Exercises 51 and 52, find the demand function $x = f(p)$ that satisfies the given conditions.

**51.** $\dfrac{dx}{dp} = -\dfrac{6000p}{(p^2 - 16)^{3/2}}, \quad x = 5000$ when $p = \$5$

**52.** $\dfrac{dx}{dp} = -\dfrac{400}{(0.02p - 1)^3}, \quad x = 10{,}000$ when $p = \$100$

**53.** *Tree Growth*   An evergreen nursery usually sells a certain type of shrub after 5 years of growth and shaping. The growth rate during those 5 years is approximated by
$$\frac{dh}{dt} = \frac{17.6t}{\sqrt{17.6t^2 + 1}},$$
where $t$ is time in years and $h$ is height in inches. The seedlings are 6 inches tall when planted ($t = 0$).

(a) Find the height function.
(b) How tall are the shrubs when they are sold?

**54.** *Cash Flow*   The rate of disbursement $dQ/dt$ of a $2 million federal grant is proportional to the square of $100 - t$, where $t$ is the time (in days, $0 \le t \le 100$) and $Q$ is the amount that remains to be disbursed. Find the amount that remains to be disbursed after 50 days. Assume that the entire grant will be disbursed after 100 days.

 In Exercises 55 and 56, (a) use the marginal propensity to consume, $dQ/dx$, to write $Q$ as a function of $x$, where $x$ is the income and $Q$ is the income consumed. Assume that 100% of the income is consumed for families that have an annual income of $20,000 or less. (b) Use the result of part (a) to complete the table showing the income consumed and the income saved, $x - Q$, for various incomes. (c) Use a graphing utility to graphically represent the income consumed and saved.

| $x$ | 20,000 | 50,000 | 100,000 | 150,000 |
|---|---|---|---|---|
| $Q$ |  |  |  |  |
| $x - Q$ |  |  |  |  |

**55.** $\dfrac{dQ}{dx} = \dfrac{0.95}{(x - 19{,}999)^{0.05}}, \quad 20{,}000 \le x$

**56.** $\dfrac{dQ}{dx} = \dfrac{0.93}{(x - 19{,}999)^{0.07}}, \quad 20{,}000 \le x$

 In Exercises 57 and 58, use a symbolic integration utility to find the indefinite integral. Verify the result by differentiating.

**57.** $\int \dfrac{1}{\sqrt{x} + \sqrt{x + 1}} \, dx$

**58.** $\int \dfrac{x}{\sqrt{3x + 2}} \, dx$

## 5.3 Exponential and Logarithmic Integrals

*Using the Exponential Rule • Using the Log Rule*

## Using the Exponential Rule

Each of the differentiation rules for exponential functions has its corresponding integration rule.

---

### Integrals of Exponential Functions

Let $u$ be a differentiable function of $x$.

$$\int e^x \, dx = e^x + C \qquad \text{Simple Exponential Rule}$$

$$\int e^u \frac{du}{dx} \, dx = \int e^u \, du = e^u + C \qquad \text{General Exponential Rule}$$

---

### EXAMPLE 1 Integrating Exponential Functions

(a) $\displaystyle \int 2e^x \, dx = 2 \int e^x \, dx$        Constant Multiple Rule

$\qquad\qquad\quad = 2e^x + C$        Simple Exponential Rule

(b) $\displaystyle \int 2e^{2x} \, dx = \int e^{2x}(2) \, dx$        Let $u = 2x$, then $\dfrac{du}{dx} = 2.$

$\qquad\qquad\quad = \displaystyle \int e^u \frac{du}{dx} \, dx$

$\qquad\qquad\quad = e^{2x} + C$        General Exponential Rule

(c) $\displaystyle \int (e^x + x) \, dx = \int e^x \, dx + \int x \, dx$        Sum Rule

$\qquad\qquad\qquad = e^x + \dfrac{x^2}{2} + C$        Simple Exponential and Power Rules

You can check each of these results by differentiating.

## Technology

If you use a symbolic integration utility to find antiderivatives of exponential or logarithmic functions, you can easily obtain results that are beyond the scope of this course. For instance, the following antiderivative of $e^{x^2}$ involves the imaginary unit $i$ and the probability function called "ERF." In this course, you are not expected to interpret or use such results. You can simply state that the function cannot be integrated using elementary functions.

**#1:** $\text{EXP}(\text{x}^2)$

**#2:** $\displaystyle\int \text{EXP}(\text{x}^2)\, \text{dx}$

**#3:** $-\dfrac{\sqrt{\pi} \cdot \hat{\imath} \cdot \text{ERF}(\hat{\imath} \cdot \text{x})}{2}$

**EXAMPLE 2** **Integrating an Exponential Function**

Find the indefinite integral $\displaystyle\int e^{3x+1}\, dx$.

**Solution**

Let $u = 3x + 1$, then $du/dx = 3$. You can introduce the missing factor of 3 in the integrand by multiplying and dividing by 3.

$$\int e^{3x+1}\, dx = \frac{1}{3}\int e^{3x+1}(3)\, dx \qquad \text{Multiply and divide by 3.}$$

$$= \frac{1}{3}\int e^{u}\frac{du}{dx}\, dx \qquad \text{Substitute } u \text{ and } du/dx.$$

$$= \frac{1}{3}e^{u} + C \qquad \text{General Exponential Rule}$$

$$= \frac{1}{3}e^{3x+1} + C \qquad \text{Substitute for } u.$$

**EXAMPLE 3** **Integrating an Exponential Function**

Find the indefinite integral $\displaystyle\int 5xe^{-x^2}\, dx$.

**Solution**

Let $u = -x^2$, then $du/dx = -2x$. You can create the factor of $(-2x)$ in the integrand by multiplying and dividing by $-2$.

$$\int 5xe^{-x^2}\, dx = \int \left(-\frac{5}{2}\right)e^{-x^2}(-2x)\, dx \qquad \text{Multiply and divide by } -2.$$

$$= -\frac{5}{2}\int e^{-x^2}(-2x)\, dx \qquad \text{Factor the constant } -\tfrac{5}{2} \text{ out of the integrand.}$$

$$= -\frac{5}{2}\int e^{u}\frac{du}{dx}\, dx \qquad \text{Substitute } u \text{ and } du/dx.$$

$$= -\frac{5}{2}e^{u} + C \qquad \text{General Exponential Rule}$$

$$= -\frac{5}{2}e^{-x^2} + C \qquad \text{Substitute for } u.$$

**TIP**

**ALGEBRA**

For help on the algebra in Example 3, see Example 1d in the *Chapter 5 Algebra Review,* on page 374.

**STUDY TIP** Remember that you cannot introduce a missing *variable* in the integrand. For instance, you cannot find $\int e^{x^2}\, dx$ by multiplying and dividing by $2x$ and then factoring $1/(2x)$ out of the integral. That is,

$$\int e^{x^2}\, dx \neq \frac{1}{2x}\int e^{x^2}(2x)\, dx.$$

## Using the Log Rule

When the Power Rules for integration were introduced in Sections 5.1 and 5.2, you saw that they work for powers other than $n = -1$.

$$\int x^n \, dx = \frac{x^{n+1}}{n+1} + C, \qquad n \neq -1 \qquad \text{Simple Power Rule}$$

$$\int u^n \frac{du}{dx} \, dx = \int u^n \, du = \frac{u^{n+1}}{n+1} + C, \qquad n \neq -1 \qquad \text{General Power Rule}$$

The Log Rules for integration allow you to integrate functions of the form $\int x^{-1} \, dx$ and $\int u^{-1} \, du$.

### Integrals of Logarithmic Functions

Let $u$ be a differentiable function of $x$.

$$\int \frac{1}{x} \, dx = \ln|x| + C \qquad \text{Simple Logarithmic Rule}$$

$$\int \frac{du/dx}{u} \, dx = \int \frac{1}{u} \, du = \ln|u| + C \qquad \text{General Logarithmic Rule}$$

You can verify each of these rules by differentiating. For instance, to verify that $d/dx[\ln|x|] = 1/x$, notice that

$$\frac{d}{dx}[\ln x] = \frac{1}{x} \qquad \text{and} \qquad \frac{d}{dx}[\ln(-x)] = \frac{-1}{-x} = \frac{1}{x}.$$

## EXAMPLE 4  Integrating Logarithmic Functions

(a) $\displaystyle \int \frac{4}{x} \, dx = 4 \int \frac{1}{x} \, dx$  Constant Multiple Rule

$\displaystyle \qquad = 4 \ln|x| + C$  Simple Logarithmic Rule

(b) $\displaystyle \int \frac{2x}{x^2} \, dx = \int \frac{du/dx}{u} \, dx$  Let $u = x^2$, then $\dfrac{du}{dx} = 2x$.

$\displaystyle \qquad = \ln|u| + C$  General Logarithmic Rule

$\displaystyle \qquad = \ln x^2 + C$  Substitute for $u$.

(c) $\displaystyle \int \frac{3}{3x+1} \, dx = \int \frac{du/dx}{u} \, dx$  Let $u = 3x + 1$, then $\dfrac{du}{dx} = 3$.

$\displaystyle \qquad = \ln|u| + C$  General Logarithmic Rule

$\displaystyle \qquad = \ln|3x + 1| + C$  Substitute for $u$.

**STUDY TIP**   Notice the absolute values in the Log Rules. For those special cases in which $u$ or $x$ cannot be negative, you can omit the absolute value. For instance, in Example 4b, it is not necessary to write the anti-derivative as $\ln|x^2| + C$ because $x^2$ cannot be negative.

---

### EXAMPLE 5   Using the Log Rule

Find the indefinite integral $\displaystyle\int \frac{1}{2x - 1}\, dx$.

**Solution**

Let $u = 2x - 1$, then $du/dx = 2$. You can create the necessary factor of 2 in the integrand by multiplying and dividing by 2.

$$\int \frac{1}{2x - 1}\, dx = \frac{1}{2}\int \frac{2}{2x - 1}\, dx \qquad \text{Multiply and divide by 2.}$$

$$= \frac{1}{2}\int \frac{du/dx}{u}\, dx \qquad \text{Substitute } u \text{ and } \frac{du}{dx}.$$

$$= \frac{1}{2}\ln|u| + C \qquad \text{General Log Rule}$$

$$= \frac{1}{2}\ln|2x - 1| + C \qquad \text{Substitute for } u.$$

---

### EXAMPLE 6   Using the Log Rule

Find the indefinite integral $\displaystyle\int \frac{6x}{x^2 + 1}\, dx$.

**Solution**

Let $u = x^2 + 1$, then $du/dx = 2x$. You can create the necessary factor of $2x$ in the integrand by factoring a 3 out of the integrand.

$$\int \frac{6x}{x^2 + 1}\, dx = 3\int \frac{2x}{x^2 + 1}\, dx \qquad \text{Factor 3 out of integrand.}$$

$$= 3\int \frac{du/dx}{u}\, dx \qquad \text{Substitute } u \text{ and } \frac{du}{dx}.$$

$$= 3\ln|u| + C \qquad \text{General Log Rule}$$

$$= 3\ln(x^2 + 1) + C \qquad \text{Substitute for } u.$$

Integrals to which the Log Rule can be applied are often given in disguised form. For instance, if a rational function has a numerator of degree greater than or equal to that of the denominator, you should use long division to rewrite the integrand. Here is an example.

$$\int \frac{x^2 + 6x + 1}{x^2 + 1}\, dx = \int \left(1 + \frac{6x}{x^2 + 1}\right) dx$$

$$= x + 3\ln(x^2 + 1) + C.$$

**ALGEBRA**

**TIP**

For help on the algebra in the integral at the right, see Example 2d in the *Chapter 5 Algebra Review,* on page 375.

The next example summarizes some additional situations in which it is helpful to rewrite the integrand in order to recognize the antiderivative.

## EXAMPLE 7  Rewriting Before Integrating

Find the following indefinite integrals.

(a) $\displaystyle\int \frac{3x^2 + 2x - 1}{x^2}\, dx$ 　　(b) $\displaystyle\int \frac{1}{1 + e^{-x}}\, dx$ 　　(c) $\displaystyle\int \frac{x^2 + x + 1}{x - 1}\, dx$

**TIP**
For help on the algebra in Example 7, see Example 2a–c in the *Chapter 5 Algebra Review,* on page 375.

ALGEBRA

*Solution*

(a) Begin by rewriting the integrand as the sum of three fractions.

$$\int \frac{3x^2 + 2x - 1}{x^2}\, dx = \int \left(\frac{3x^2}{x^2} + \frac{2x}{x^2} - \frac{1}{x^2}\right) dx$$

$$= \int \left(3 + \frac{2}{x} - \frac{1}{x^2}\right) dx$$

$$= 3x + 2 \ln |x| + \frac{1}{x} + C$$

(b) Begin by rewriting the integrand by multiplying and dividing by $e^x$.

$$\int \frac{1}{1 + e^{-x}}\, dx = \int \left(\frac{e^x}{e^x}\right) \frac{1}{1 + e^{-x}}\, dx$$

$$= \int \frac{e^x}{e^x + 1}\, dx$$

$$= \ln (e^x + 1) + C$$

(c) Begin by dividing the numerator by the denominator.

$$\int \frac{x^2 + x + 1}{x - 1}\, dx = \int \left(x + 2 + \frac{3}{x - 1}\right) dx$$

$$= \frac{x^2}{2} + 2x + 3 \ln |x - 1| + C$$

## *Group Discussion*　　***Using the General Log Rule***

One of the most common applications of the Log Rule is to find indefinite integrals of the form

$$\int \frac{a}{bx + c}\, dx.$$

Describe a quick way to find this indefinite integral. Then apply your technique to the following.

a. $\displaystyle\int \frac{1}{2x - 5}\, dx$ 　　b. $\displaystyle\int \frac{4}{3x + 2}\, dx$ 　　c. $\displaystyle\int \frac{7}{8x - 3}\, dx$

Use a symbolic integration utility to check your results.

## Warm Up

The following warm-up exercises involve skills that were covered in earlier sections. You will use these skills in the exercise set for this section.

In Exercises 1 and 2, find the domain of the function.

**1.** $y = \ln(2x - 5)$

**2.** $y = \ln(x^2 - 5x + 6)$

In Exercises 3–6, use long division to rewrite the quotient.

**3.** $\dfrac{x^2 + 4x + 2}{x + 2}$

**4.** $\dfrac{x^2 - 6x + 9}{x - 4}$

**5.** $\dfrac{x^3 + 4x^2 - 30x - 4}{x^2 - 4x}$

**6.** $\dfrac{x^4 - x^3 + x^2 + 15x + 2}{x^2 + 5}$

In Exercises 7–10, evaluate the integral.

**7.** $\displaystyle\int \left( x^3 + \frac{1}{x^2} \right) dx$

**8.** $\displaystyle\int \frac{x^2 + 2x}{x} \, dx$

**9.** $\displaystyle\int \frac{x^3 + 4}{x^2} \, dx$

**10.** $\displaystyle\int \frac{x + 3}{x^3} \, dx$

## EXERCISES 5.3

In Exercises 1–12, use the Exponential Rule to find the indefinite integral.

**1.** $\displaystyle\int 2e^{2x} \, dx$

**2.** $\displaystyle\int -3e^{-3x} \, dx$

**3.** $\displaystyle\int e^{4x} \, dx$

**4.** $\displaystyle\int e^{-0.25x} \, dx$

**5.** $\displaystyle\int 9xe^{-x^2} \, dx$

**6.** $\displaystyle\int 3xe^{0.5x^2} \, dx$

**7.** $\displaystyle\int 5x^2 \, e^{x^3} \, dx$

**8.** $\displaystyle\int (2x + 1)e^{x^2 + x} \, dx$

**9.** $\displaystyle\int (x^2 + 2x)e^{x^3 + 3x^2 - 1} \, dx$

**10.** $\displaystyle\int 3(x - 4)e^{x^2 - 8x} \, dx$

**11.** $\displaystyle\int 5e^{2-x} \, dx$

**12.** $\displaystyle\int 3e^{-(x+1)/2} \, dx$

In Exercises 13–24, use the Log Rule to find the indefinite integral.

**13.** $\displaystyle\int \frac{1}{x + 1} \, dx$

**14.** $\displaystyle\int \frac{1}{x - 5} \, dx$

**15.** $\displaystyle\int \frac{1}{3 - 2x} \, dx$

**16.** $\displaystyle\int \frac{1}{6x - 5} \, dx$

**17.** $\displaystyle\int \frac{x}{x^2 + 1} \, dx$

**18.** $\displaystyle\int \frac{x^2}{3 - x^3} \, dx$

**19.** $\displaystyle\int \frac{x^2}{x^3 + 1} \, dx$

**20.** $\displaystyle\int \frac{x}{x^2 + 4} \, dx$

**21.** $\displaystyle\int \frac{x + 3}{x^2 + 6x + 7} \, dx$

**22.** $\displaystyle\int \frac{x^2 + 2x + 3}{x^3 + 3x^2 + 9x + 1} \, dx$

**23.** $\displaystyle\int \frac{1}{x \ln x} \, dx$

**24.** $\displaystyle\int \frac{e^x}{1 + e^x} \, dx$

In Exercises 25–32, use a symbolic integration utility to find the indefinite integral.

**25.** $\displaystyle\int \frac{1}{x^2} e^{2/x} \, dx$

**26.** $\displaystyle\int \frac{1}{x^3} e^{1/4x^2} \, dx$

**27.** $\displaystyle\int \frac{1}{\sqrt{x}} e^{\sqrt{x}} \, dx$

**28.** $\displaystyle\int (e^x - e^{-x})^2 \, dx$

**29.** $\displaystyle\int \frac{e^{-x}}{1 + e^{-x}} \, dx$

**30.** $\displaystyle\int \frac{3e^x}{2 + e^x} \, dx$

**31.** $\displaystyle\int \frac{4e^{2x}}{5 - e^{2x}} \, dx$

**32.** $\displaystyle\int \frac{-e^{3x}}{2 - e^{3x}} \, dx$

In Exercises 33–48, use any basic integration formula or formulas to find the indefinite integral.

**33.** $\int \dfrac{e^{2x} + 2e^x + 1}{e^x}\, dx$

**34.** $\int (6x + e^x)\sqrt{3x^2 + e^x}\, dx$

**35.** $\int e^x \sqrt{1 - e^x}\, dx$

**36.** $\int \dfrac{2(e^x - e^{-x})}{(e^x + e^{-x})^2}\, dx$

**37.** $\int \dfrac{1}{(x - 1)^2}\, dx$

**38.** $\int \dfrac{1}{\sqrt{x + 1}}\, dx$

**39.** $\int \dfrac{x^2 - 4}{x}\, dx$

**40.** $\int \dfrac{x + 5}{x}\, dx$

**41.** $\int \dfrac{x^3 - 8x}{2x^2}\, dx$

**42.** $\int \dfrac{x - 1}{4x}\, dx$

**43.** $\int \dfrac{2}{1 + e^{-x}}\, dx$

**44.** $\int \dfrac{3}{1 + e^{-3x}}\, dx$

**45.** $\int \dfrac{x^2 + 2x + 5}{x - 1}\, dx$

**46.** $\int \dfrac{x - 3}{x + 3}\, dx$

**47.** $\int \dfrac{1 + e^{-x}}{1 + xe^{-x}}\, dx$

**48.** $\int \dfrac{5}{e^{-5x} + 7}\, dx$

In Exercises 49 and 50, find the equation of the function $f$ whose graph passes through the point.

| Derivative | Point |
| --- | --- |
| **49.** $f'(x) = \dfrac{x^2 + 4x + 3}{x - 1}$ | $(2, 4)$ |
| **50.** $f'(x) = \dfrac{x^3 - 4x^2 + 3}{x - 3}$ | $(4, -1)$ |

**51. Bacteria Growth**  A population of bacteria is growing at the rate of

$$\frac{dP}{dt} = \frac{3000}{1 + 0.25t},$$

where $t$ is the time in days. When $t = 0$, the population is 1000.

(a) Write an equation that models the population $P$ in terms of the time $t$.

(b) What is the population after 3 days?

(c) After how many days will the population be 12,000?

**52. Population Decline**  Because of an insufficient oxygen supply, the trout population in a lake is dying. The population's rate of change can be modeled by

$$\frac{dP}{dt} = -125e^{-t/20},$$

where $t$ is the time in days. When $t = 0$, the population is 2500.

(a) Write an equation that models the population $P$ in terms of the time $t$.

(b) What is the population after 15 days?

(c) According to this model, how long will it take for the entire trout population to die?

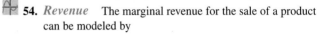 **53. Demand**  The marginal price for the demand of a product can be modeled by $dp/dx = 0.1e^{-x/500}$, where $x$ is the quantity demanded. When the demand is 600 units, the price is $30.

(a) Find the demand function, $p = f(x)$.

(b) Use a graphing utility to graph the demand function. Does price increase or decrease as demand increases?

(c) Use the zoom and trace features of the graphing utility to find the quantity demanded when the price is $22.

**54. Revenue**  The marginal revenue for the sale of a product can be modeled by

$$\frac{dR}{dx} = 50 - 0.02x + \frac{100}{x + 1},$$

where $x$ is the quantity demanded.

(a) Find the revenue function.

(b) Use a graphing utility to graph the revenue function.

(c) Find the revenue when 1500 units are sold.

(d) Use the zoom and trace features of the graphing utility to find the number of units sold when the revenue is $60,230.

**55. ATM Transactions**  From 1985 through 1995, the number of automatic teller machine (ATM) transactions $T$ (in millions) in the United States changed at the rate of

$$\frac{dT}{dt} = 189.167e^{0.11t},$$

where $t = 0$ corresponds to 1985. In 1993, there were 7705 million transactions.  *(Source: Bank Network News)*

(a) Write a model that gives the total number of ATM transactions per year.

(b) Use the model to find the number of ATM transactions in 1990.

**56. Sales**  The rate of change in sales for Wal-Mart from 1987 through 1996 can be modeled by

$$\frac{ds}{dt} = 58.17t^2 + \frac{26{,}272.58}{t},$$

where $s$ is the sales (in millions) and $t = 7$ corresponds to 1987. In 1992, the sales for Wal-Mart were $55,484 million.  *(Source: Wal-Mart)*

(a) Find a model for sales from 1987 through 1996.

(b) Find Wal-Mart's sales in 1994.

<table>
<tr><td>**5.4**</td><td># Area and the Fundamental Theorem of Calculus</td></tr>
</table>

*Area and Definite Integrals • The Fundamental Theorem of Calculus •*
*Marginal Analysis • Average Value • Even and Odd Functions*

## Area and Definite Integrals

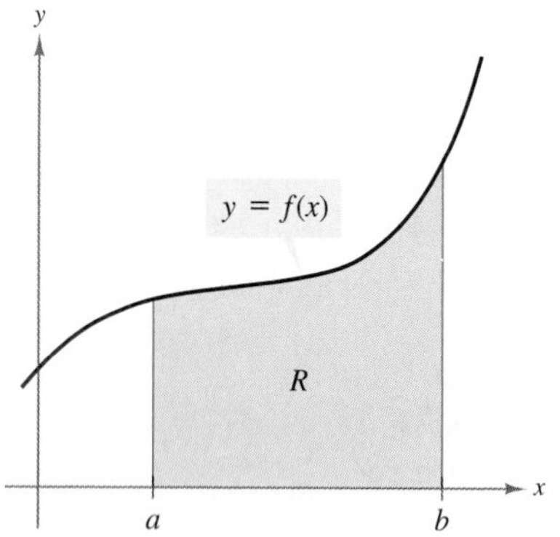

From your study of geometry, you know that area is a number that defines the size of a bounded region. For simple regions, such as rectangles, triangles, and circles, area can be found using geometric formulas.

In this section, you will learn how to use calculus to find the areas of non-standard regions, such as the region $R$ shown in Figure 5.5.

**FIGURE 5.5**   $\displaystyle\int_a^b f(x)\,dx = $ Area

---

### Definition of a Definite Integral

Let $f$ be nonnegative and continuous on the closed interval $[a, b]$. The area of the region bounded by the graph of $f$, the $x$-axis, and the lines $x = a$ and $x = b$ is denoted by

$$\text{Area} = \int_a^b f(x)\,dx.$$

The expression $\int_a^b f(x)\,dx$ is called the **definite integral** from $a$ to $b$, where $a$ is the **lower limit of integration** and $b$ is the **upper limit of integration.**

---

### EXAMPLE 1   Evaluating a Definite Integral

Evaluate the definite integral $\displaystyle\int_0^2 2x\,dx$.

*Solution*

This definite integral represents the area of the region bounded by the graph of $f(x) = 2x$, the $x$-axis, and the line $x = 2$, as shown in Figure 5.6. The region is triangular, with a height of four units and a base of two units.

$$\int_0^2 2x\,dx = \frac{1}{2}(\text{base})(\text{height}) \qquad \text{Formula for area of triangle}$$

$$= \frac{1}{2}(2)(4)$$

$$= 4 \qquad\qquad\qquad\qquad \text{Simplify.}$$

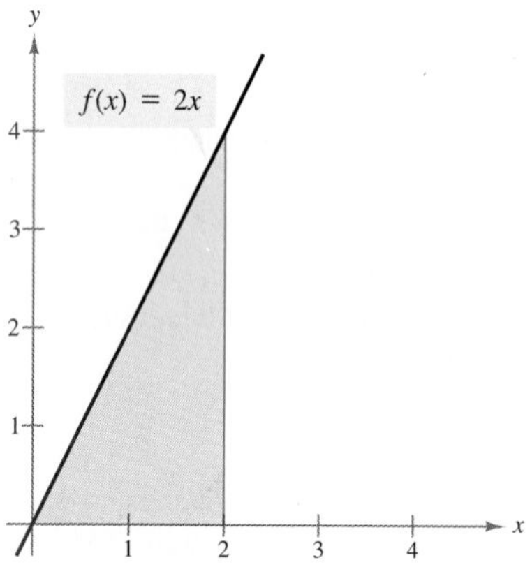

**FIGURE 5.6**

## The Fundamental Theorem of Calculus

Consider the function $A(x)$, which denotes the area of the region shown in Figure 5.7. To discover the relationship between $A(x)$ and $f(x)$, let $x$ increase by an amount $\Delta x$. This increases the area by $\Delta A$. Let $f(m)$ and $f(M)$ denote the minimum and maximum values of $f$ on the interval $[x, x + \Delta x]$.

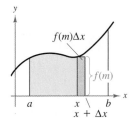 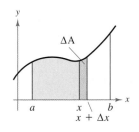 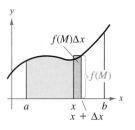

**FIGURE 5.8**

**FIGURE 5.7**  $A(x) = $ Area from $a$ to $x$

As indicated in Figure 5.8, you can write the following inequality.

$$f(m)\Delta x \leq \Delta A \leq f(M)\Delta x \qquad \text{See Figure 5.8.}$$

$$f(m) \leq \frac{\Delta A}{\Delta x} \leq f(M) \qquad \text{Divide each term by } \Delta x.$$

$$\lim_{\Delta x \to 0} f(m) \leq \lim_{\Delta x \to 0} \frac{\Delta A}{\Delta x} \leq \lim_{\Delta x \to 0} f(M) \qquad \text{Take limit of each term.}$$

$$f(x) \leq A'(x) \leq f(x) \qquad \text{Definition of derivative of } A(x)$$

Thus, $f(x) = A'(x)$, and $A(x) = F(x) + C$, where $F'(x) = f(x)$. Because $A(a) = 0$, it follows that $C = -F(a)$. Thus, $A(x) = F(x) - F(a)$, which implies that

$$A(b) = \int_a^b f(x)\, dx = F(b) - F(a).$$

This equation tells you that *if you can find an antiderivative for $f$, then you can use the antiderivative to evaluate the definite integral $\int_a^b f(x)\, dx$.* This result is called the **Fundamental Theorem of Calculus.**

---

### The Fundamental Theorem of Calculus

If $f$ is nonnegative and continuous on the closed interval $[a, b]$, then

$$\int_a^b f(x)\, dx = F(b) - F(a),$$

where $F$ is any function such that $F'(x) = f(x)$ for all $x$ in $[a, b]$.

---

**STUDY TIP**   There are two basic ways to introduce the Fundamental Theorem of Calculus. One way uses an area function, as shown here. The other uses a summation process, as shown in the appendix.

Three comments about the Fundamental Theorem of Calculus are in order.

1. The Fundamental Theorem of Calculus describes a way of *evaluating* a definite integral, not a procedure for finding antiderivatives.

2. In applying the Fundamental Theorem, it is helpful to use the notation

$$\int_a^b f(x)dx = F(x) \Big]_a^b = F(b) - F(a).$$

3. The constant of integration $C$ can be dropped because

$$\int_a^b f(x)dx = \left[ F(x) + C \right]_a^b$$
$$= [F(b) + C] - [F(a) + C]$$
$$= F(b) - F(a) + C - C$$
$$= F(b) - F(a).$$

In the development of the Fundamental Theorem of Calculus, $f$ was assumed to be nonnegative on the closed interval $[a, b]$. As such, the definite integral was defined as an area. Now, with the Fundamental Theorem, the definition can be extended to include functions that are negative on all or part of the closed interval $[a, b]$. Specifically, if $f$ is *any* function that is continuous on a closed interval $[a, b]$, then the **definite integral** of $f(x)$ from $a$ to $b$ is defined to be

$$\int_a^b f(x)dx = F(b) - F(a),$$

where $F$ is an antiderivative of $f$. Remember that definite integrals do not necessarily represent areas and can be negative, zero, or positive.

**STUDY TIP**   Be sure you see the distinction between indefinite and definite integrals. The *indefinite integral*

$$\int f(x) \, dx$$

denotes a family of *functions*, each of which is an antiderivative of $f$, whereas the *definite integral*

$$\int_a^b f(x) \, dx$$

is a *number*.

---

### Properties of Definite Integrals

Let $f$ and $g$ be continuous on the closed interval $[a, b]$.

1. $\displaystyle\int_a^b kf(x) \, dx = k \int_a^b f(x) \, dx, \quad k$ is a constant.

2. $\displaystyle\int_a^b [f(x) \pm g(x)]dx = \int_a^b f(x) \, dx \pm \int_a^b g(x) \, dx$

3. $\displaystyle\int_a^b f(x) \, dx = \int_a^c f(x) \, dx + \int_c^b f(x) \, dx, \quad a < c < b$

4. $\displaystyle\int_a^a f(x) \, dx = 0$

5. $\displaystyle\int_a^b f(x) \, dx = -\int_b^a f(x) \, dx$

## EXAMPLE 2  Finding Area by the Fundamental Theorem

Find the area of the region bounded by the x-axis and the graph of

$$f(x) = x^2 - 1, \quad 1 \le x \le 2.$$

### Solution

Note that $f(x) \ge 0$ on the interval $1 \le x \le 2$, as shown in Figure 5.9. Therefore, you can represent the area of the region by a definite integral. To find the area, use the Fundamental Theorem of Calculus.

$$\begin{aligned} \text{Area} &= \int_1^2 (x^2 - 1)\, dx & \text{Definition of definite integral} \\ &= \left[ \frac{x^3}{3} - x \right]_1^2 & \text{Find antiderivative.} \\ &= \left[ \frac{(2)^3}{3} - 2 \right] - \left[ \frac{(1)^3}{3} - 1 \right] & \text{Apply Fundamental Theorem.} \\ &= \frac{4}{3} & \text{Simplify.} \end{aligned}$$

Thus, the area of the region is $\frac{4}{3}$ square units.

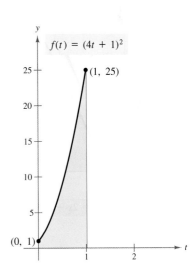

**FIGURE 5.9**  Area $= \int_1^2 (x^2 - 1)dx$

*STUDY TIP*  It is easy to make errors in signs when evaluating definite integrals. To avoid such errors, enclose the values of the antiderivative at the upper and lower limits of integration in separate sets of parentheses, as shown above.

## EXAMPLE 3  Evaluating a Definite Integral

Evaluate the definite integral

$$\int_0^1 (4t + 1)^2 \, dt,$$

and sketch the region whose area is represented by the integral.

### Solution

$$\begin{aligned} \int_0^1 (4t + 1)^2 \, dt &= \frac{1}{4} \int_0^1 (4t + 1)^2 (4) \, dt & \text{Multiply and divide by 4.} \\ &= \frac{1}{4} \left[ \frac{(4t + 1)^3}{3} \right]_0^1 & \text{Find antiderivative.} \\ &= \frac{1}{4} \left[ \left( \frac{5^3}{3} \right) - \left( \frac{1}{3} \right) \right] & \text{Apply Fundamental Theorem.} \\ &= \frac{31}{3} & \text{Simplify.} \end{aligned}$$

The region is shown in Figure 5.10.

**FIGURE 5.10**

## Average Value

The *average value* of a function on a closed interval is defined as follows.

> ### Definition of the Average Value of a Function
>
> If $f$ is continuous on $[a, b]$, then the **average value** of $f$ on $[a, b]$ is
>
> $$\text{Average value of } f \text{ on } [a, b] = \frac{1}{b - a}\int_a^b f(x)\, dx.$$

### EXAMPLE 7  Finding the Average Cost

The cost per unit $c$ of producing a product over a 2-year period is modeled by

$$c = 0.005t^2 + 0.01t + 13.15, \qquad 0 \le t \le 24,$$

where $t$ is the time in months. Approximate the average cost per unit over the 2-year period.

### Solution

The average cost can be found by integrating $c$ over the interval $[0, 24]$.

$$
\begin{aligned}
\text{Average cost per unit} &= \frac{1}{24}\int_0^{24} (0.005t^2 + 0.01t + 13.15)\, dt \\
&= \frac{1}{24}\left[\frac{0.005t^3}{3} + \frac{0.01t^2}{2} + 13.15t\right]_0^{24} \\
&= \frac{1}{24}(341.52) \\
&= \$14.23 \qquad \text{(See Figure 5.12.)}
\end{aligned}
$$

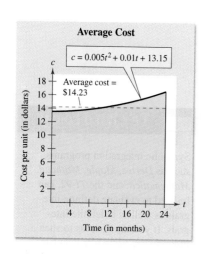

**Average Cost**

$c = 0.005t^2 + 0.01t + 13.15$

Average cost = $14.23

Cost per unit (in dollars)

Time (in months)

**FIGURE 5.12**

To check the reasonableness of the average value found in Example 7, assume that one unit is produced each month, beginning with $t = 0$ and ending with $t = 24$. When $t = 0$, the cost is

$$c = 0.005(0)^2 + 0.01(0) + 13.15 = \$13.15.$$

Similarly, when $t = 1$, the cost is

$$c = 0.005(1)^2 + 0.01(1) + 13.15 \approx \$13.17.$$

Each month, the cost increases, and the average of the 25 costs is

$$\frac{13.15 + 13.17 + 13.19 + 13.23 + \cdots + 16.27}{25} \approx \$14.25.$$

## Even and Odd Functions

Several common functions have graphs that are symmetric to the $y$-axis or the origin, as shown in Figure 5.13. If the graph of $f$ is symmetric to the $y$-axis, as in Figure 5.13(a), then

$$f(-x) = f(x), \qquad \text{Even function}$$

and $f$ is called an **even** function. If the graph of $f$ is symmetric to the origin, as in Figure 5.13(b), then

$$f(-x) = -f(x), \qquad \text{Odd function}$$

and $f$ is called an **odd** function.

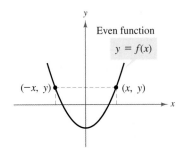

(a) $y$-Axis symmetry

### Integration of Even and Odd Functions

1. If $f$ is an *even* function, then $\displaystyle\int_{-a}^{a} f(x)\,dx = 2\int_{0}^{a} f(x)\,dx.$

2. If $f$ is an *odd* function, then $\displaystyle\int_{-a}^{a} f(x)\,dx = 0.$

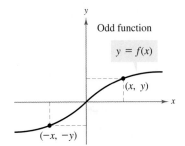

(b) Origin symmetry

**FIGURE 5.13**

**EXAMPLE 8** **Integrating Even and Odd Functions**

(a) Because $f(x) = x^2$ is even,

$$\int_{-2}^{2} x^2\,dx = 2\int_{0}^{2} x^2\,dx = 2\left[\frac{x^3}{3}\right]_{0}^{2} = 2\left(\frac{8}{3} - 0\right) = \frac{16}{3}.$$

(b) Because $f(x) = x^3$ is odd,

$$\int_{-2}^{2} x^3\,dx = 0.$$

## Group Discussion     Using Geometry to Evaluate Definite Integrals

When using the Fundamental Theorem of Calculus to evaluate $\int_a^b f(x)\,dx$, remember that you must first be able to find an antiderivative of $f(x)$. If you are unable to find an antiderivative, you cannot use the Fundamental Theorem. In some cases, you can still evaluate the definite integral. For instance, explain how you can use geometry to evaluate

$$\int_{-2}^{2} \sqrt{4 - x^2}\,dx.$$

Use a symbolic integration utility to verify your answer.

## Warm Up

The following warm-up exercises involve skills that were covered in earlier sections. You will use these skills in the exercise set for this section.

In Exercises 1–4, find the indefinite integral.

**1.** $\displaystyle \int (3x + 7)\, dx$

**2.** $\displaystyle \int \left(x^{3/2} + 2\sqrt{x}\right) dx$

**3.** $\displaystyle \int \frac{1}{5x}\, dx$

**4.** $\displaystyle \int e^{-6x}\, dx$

In Exercises 5 and 6, evaluate the expression when $a = 5$ and $b = 3$.

**5.** $\left(\dfrac{a}{5} - a\right) - \left(\dfrac{b}{5} - b\right)$

**6.** $\left(6a - \dfrac{a^3}{3}\right) - \left(6b - \dfrac{b^3}{3}\right)$

In Exercises 7–10, integrate the marginal function.

**7.** $\dfrac{dC}{dx} = 0.02x^{3/2} + 29{,}500$

**8.** $\dfrac{dR}{dx} = 9000 + 2x$

**9.** $\dfrac{dP}{dx} = 25{,}000 - 0.01x$

**10.** $\dfrac{dC}{dx} = 0.03x^2 + 4600$

## EXERCISES 5.4

In Exercises 1–4, sketch the region whose area is represented by the definite integral. Then use a geometric formula to evaluate the integral.

**1.** $\displaystyle \int_0^2 3\, dx$

**2.** $\displaystyle \int_0^3 2x\, dx$

**3.** $\displaystyle \int_0^5 (x + 1)\, dx$

**4.** $\displaystyle \int_{-3}^3 \sqrt{9 - x^2}\, dx$

In Exercises 5–10, find the area of the region.

**5.** $y = x - x^2$

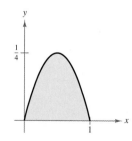

**6.** $y = \dfrac{1}{x^2}$

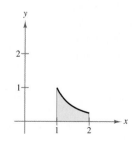

**7.** $y = 1 - x^4$

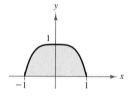

**8.** $y = 3e^{-x/2}$

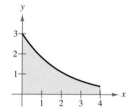

**9.** $y = \sqrt[3]{2x}$

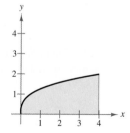

**10.** $y = \dfrac{x^2 + 4}{x}$

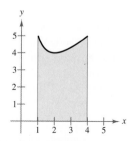

In Exercises 11–34, evaluate the definite integral.

**11.** $\int_0^1 2x\,dx$

**12.** $\int_2^7 3\,dv$

**13.** $\int_{-1}^0 (2x + 1)\,dx$

**14.** $\int_2^5 (-3x + 4)\,dx$

**15.** $\int_{-1}^1 (2t - 1)^2\,dt$

**16.** $\int_0^3 (3x^2 + x - 2)\,dx$

**17.** $\int_0^1 (2t - 1)^2\,dt$

**18.** $\int_2^2 (x - 3)^4\,dx$

**19.** $\int_{-1}^1 \left(\sqrt[3]{t} - 2\right) dt$

**20.** $\int_1^4 \sqrt{\dfrac{2}{x}}\,dx$

**21.** $\int_1^4 \dfrac{2u - 1}{\sqrt{u}}\,du$

**22.** $\int_0^1 \dfrac{x - \sqrt{x}}{3}\,dx$

**23.** $\int_{-1}^0 \left(t^{1/3} - t^{2/3}\right) dt$

**24.** $\int_0^4 \left(x^{1/2} + x^{1/4}\right) dx$

**25.** $\int_0^4 \dfrac{1}{\sqrt{2x + 1}}\,dx$

**26.** $\int_0^2 \dfrac{x}{\sqrt{1 + 2x^2}}\,dx$

**27.** $\int_0^1 e^{-2x}\,dx$

**28.** $\int_1^2 e^{1-x}\,dx$

**29.** $\int_1^3 \dfrac{e^{3/x}}{x^2}\,dx$

**30.** $\int_{-1}^1 (e^x - e^{-x})\,dx$

**31.** $\int_{-1}^1 |4x|\,dx$

**32.** $\int_0^3 |2x - 3|\,dx$

**33.** $\int_0^4 \left(2 - |x - 2|\right) dx$

**34.** $\int_{-4}^4 \left(4 - |x|\right) dx$

In Exercises 35–38, evaluate the definite integral by hand. Then use a symbolic integration utility to evaluate the definite integral. Briefly explain any differences in your results.

**35.** $\int_{-1}^2 \dfrac{x}{x^2 - 9}\,dx$

**36.** $\int_2^3 \dfrac{x + 1}{x^2 + 2x - 3}\,dx$

**37.** $\int_0^3 \dfrac{2e^x}{2 + e^x}\,dx$

**38.** $\int_1^2 \dfrac{(2 + \ln x)^3}{x}\,dx$

In Exercises 39–44, evaluate the definite integral. Then sketch the region whose area is represented by the integral.

**39.** $\int_1^3 (4x - 3)\,dx$

**40.** $\int_0^2 (x + 4)\,dx$

**41.** $\int_0^1 (x - x^3)\,dx$

**42.** $\int_0^1 \sqrt{x}(1 - x)\,dx$

**43.** $\int_2^4 \dfrac{3x^2}{x^3 - 1}\,dx$

**44.** $\int_0^{\ln 6} \dfrac{e^x}{2}\,dx$

In Exercises 45–48, find the area of the region bounded by the graphs.

**45.** $y = 3x^2 + 1$, $y = 0$, $x = 0$, and $x = 2$

**46.** $y = 1 + \sqrt{x}$, $y = 0$, $x = 0$, and $x = 4$

**47.** $y = (x + 5)/x$, $y = 0$, $x = 1$, and $x = 5$

**48.** $y = 3e^x$, $y = 0$, $x = -2$, and $x = 1$

In Exercises 49 and 50, use the values $\int_0^5 f(x)\,dx = 8$ and $\int_0^5 g(x)\,dx = 3$ to evaluate the definite integral.

**49.** (a) $\int_0^5 [f(x) + g(x)]\,dx$ (b) $\int_0^5 [f(x) - g(x)]\,dx$

(c) $\int_0^5 -4f(x)\,dx$ (d) $\int_0^5 [f(x) - 3g(x)]\,dx$

**50.** (a) $\int_0^5 2g(x)\,dx$ (b) $\int_5^0 f(x)\,dx$

(c) $\int_5^5 f(x)\,dx$ (d) $\int_0^5 [f(x) - f(x)]\,dx$

In Exercises 51–56, use a graphing utility to graph the function over the interval. Find the average value of the function over the interval. Then find all $x$-values in the interval for which the function is equal to its average value.

| Function | Interval |
|---|---|
| **51.** $f(x) = 6 - x^2$ | $[-2, 2]$ |
| **52.** $f(x) = 5e^{0.2(x - 10)}$ | $[0, 10]$ |
| **53.** $f(x) = x\sqrt{4 - x^2}$ | $[0, 2]$ |
| **54.** $f(x) = \dfrac{5x}{x^2 + 1}$ | $[0, 7]$ |
| **55.** $f(x) = x - 2\sqrt{x}$ | $[0, 4]$ |
| **56.** $f(x) = \dfrac{1}{(x - 3)^2}$ | $[0, 2]$ |

In Exercises 57–60, state whether the function is even, odd, or neither.

**57.** $f(x) = 3x^4$

**58.** $g(x) = x^3 - 2x$

**59.** $g(t) = 2t^5 - 3t^2$

**60.** $f(t) = 5t^4 + 1$

**61.** Use the value $\int_0^2 x^2\,dx = \dfrac{8}{3}$ to evaluate the definite integral. Explain your reasoning.

(a) $\int_{-2}^0 x^2\,dx$ (b) $\int_{-2}^2 x^2\,dx$ (c) $\int_0^2 -x^2\,dx$

**62.** Use the value $\int_0^2 x^3\,dx = 4$ to evaluate the definite integral. Explain your reasoning.

(a) $\int_{-2}^0 x^3\,dx$ (b) $\int_{-2}^2 x^3\,dx$ (c) $\int_0^2 3x^3\,dx$

*Marginal Analysis*   In Exercises 63–68, find the change in cost C, revenue R, or profit P, for the given marginal. In each case, assume the number of units x increases by 3 from the specified value of x.

| Marginal | Number of Units, x |
|---|---|
| **63.** $\dfrac{dC}{dx} = 2.25$ | $x = 100$ |
| **64.** $\dfrac{dC}{dx} = \dfrac{20,000}{x^2}$ | $x = 10$ |
| **65.** $\dfrac{dR}{dx} = 48 - 3x$ | $x = 12$ |
| **66.** $\dfrac{dR}{dx} = 75\left(20 + \dfrac{900}{x}\right)$ | $x = 500$ |
| **67.** $\dfrac{dP}{dx} = \dfrac{400 - x}{150}$ | $x = 200$ |
| **68.** $\dfrac{dP}{dx} = 12.5\left(40 - 3\sqrt{x}\right)$ | $x = 125$ |

*Capital Accumulation*   In Exercises 69–72, you are given the rate of investment $dI/dt$. Find the capital accumulation over a 5-year period by evaluating the definite integral

$$\text{Capital accumulation} = \int_0^5 \frac{dI}{dt}\, dt,$$

where t is the time in years.

**69.** $\dfrac{dI}{dt} = 500$       **70.** $\dfrac{dI}{dt} = 500\sqrt{t + 1}$

**71.** $\dfrac{dI}{dt} = 100t$       **72.** $\dfrac{dI}{dt} = \dfrac{12,000t}{(t^2 + 2)^2}$

**73.** *Cost*   The total cost of purchasing and maintaining a piece of equipment for x years can be modeled by

$$C = 5000\left(25 + 3\int_0^x t^{1/4}dt\right).$$

Find the total cost after (a) 1 year, (b) 5 years, and (c) 10 years.

**74.** *Depreciation*   A company purchases a new machine for which the rate of depreciation can be modeled by

$$\frac{dV}{dt} = 10,000(t - 6), \quad 0 \le t \le 5,$$

where V is the value of the machine after t years. Set up and evaluate the definite integral that yields the total loss of value of the machine over the first 3 years.

**75.** *Compound Interest*   A deposit of \$2250 is made in a savings account at an annual interest rate of 12%, compounded continuously. Find the average balance in the account during the first 5 years.

**76.** *Blood Flow*   The velocity v of blood at a distance r from the center of an artery of radius R can be modeled by

$$v = k(R^2 - r^2),$$

where k is a constant. Find the average velocity along a radius of the artery. (Use 0 and R as the limits of integration.)

**77.** *Computer Industry*   The rate of change in revenue for the computer and data processing industry in the United States from 1985 through 1994 can be modeled by

$$\frac{dR}{dt} = 1.62t + 21.14e^{-t}, \quad 0 \le t \le 9,$$

where R is the revenue (in billions of dollars) and $t = 0$ represents 1985. In 1985, the revenue was \$45.1 billion. (*Source: U.S. Bureau of Census*)

(a) Write a model for the revenue as a function of t.

(b) What was the average revenue for 1985 through 1994?

**78.** *Wildlife Population*   The rate of change of the number of coyotes $N(t)$ in a population is directly proportional to $650 - N(t)$, where t is time in years.

$$\frac{dN}{dt} = k[650 - N(t)]$$

When $t = 0$, the population is 300, and when $t = 2$, the population has increased to 500.

(a) Find the population function.

(b) Find the average number of coyotes over the first 5 years.

In Exercises 79–82, use a symbolic integration utility to evaluate the definite integral.

**79.** $\displaystyle\int_3^6 \frac{x}{3\sqrt{x^2 - 8}}\, dx$

**80.** $\displaystyle\int_{1/2}^1 (x + 1)\sqrt{1 - x}\, dx$

**81.** $\displaystyle\int_2^5 \left(\frac{1}{x^2} - \frac{1}{x^3}\right) dx$

**82.** $\displaystyle\int_0^1 x^3(x^3 + 1)^3\, dx$

# 5.5 — The Area of a Region Bounded by Two Graphs

*Area of a Region Bounded by Two Graphs • Consumer Surplus and Producer Surplus • Applications*

## Area of a Region Bounded by Two Graphs

With a few modifications, you can extend the use of definite integrals from finding the area of a region *under a graph* to finding the area of a region *bounded by two graphs*. To see how this is done, consider the region bounded by the graphs of $f$, $g$, $x = a$, and $x = b$, as shown in Figure 5.14. If the graphs of both $f$ and $g$ lie above the $x$-axis, then you can interpret the area of the region between the graphs as the area of the region under the graph of $g$ subtracted from the area of the region under the graph of $f$, as shown in Figure 5.14.

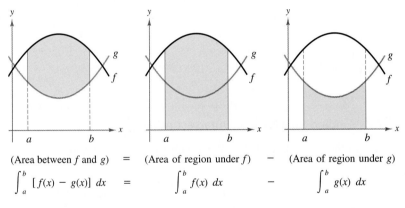

(Area between $f$ and $g$)  =  (Area of region under $f$)  −  (Area of region under $g$)

$$\int_a^b [f(x) - g(x)]\, dx \quad = \quad \int_a^b f(x)\, dx \quad - \quad \int_a^b g(x)\, dx$$

**FIGURE 5.14**

Although Figure 5.14 depicts the graphs of $f$ and $g$ lying above the $x$-axis, this is not necessary, and the same integrand $[f(x) - g(x)]$ can be used as long as both functions are continuous and $g(x) \le f(x)$ on the interval $[a, b]$.

---

### Area of a Region Bounded by Two Graphs

If $f$ and $g$ are continuous on $[a, b]$ and $g(x) \le f(x)$ for all $x$ in the interval, then the area of the region bounded by the graphs of $f$, $g$, $x = a$, and $x = b$ is given by

$$A = \int_a^b [f(x) - g(x)]\, dx.$$

EXAMPLE 1   **Finding the Area Bounded by Two Graphs**

Find the area of the region bounded by the graphs of $y = x^2 + 2$ and $y = x$, for $0 \leq x \leq 1$.

**Solution**

Begin by sketching the graphs of both functions, as shown in Figure 5.15. From the figure, you can see that $x \leq x^2 + 2$ for all $x$ in $[0, 1]$. Thus, you can let $f(x) = x^2 + 2$ and $g(x) = x$. Then compute the area as follows.

$$\text{Area} = \int_a^b [f(x) - g(x)]\, dx \qquad \text{Area between } f \text{ and } g$$

$$= \int_0^1 [(x^2 + 2) - (x)]\, dx \qquad \text{Substitute for } f \text{ and } g.$$

$$= \int_0^1 (x^2 - x + 2)\, dx$$

$$= \left[ \frac{x^3}{3} - \frac{x^2}{2} + 2x \right]_0^1 \qquad \text{Find antiderivative.}$$

$$= \frac{11}{6} \text{ square units} \qquad \text{Apply Fundamental Theorem.}$$

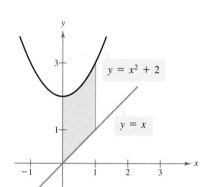

**FIGURE 5.15**

EXAMPLE 2   **Finding Area Between Intersecting Graphs**

Find the area of the region bounded by the graphs of $y = 2 - x^2$ and $y = x$.

**Solution**

In this problem, the values of $a$ and $b$ are not given and you must compute them by finding the points of intersection of the two graphs. To do this, equate the two functions and solve for $x$. When you do this, you will obtain $x = -2$ and $x = 1$. From Figure 5.16, you can see that the graph of $f(x) = 2 - x^2$ lies above the graph of $g(x) = x$ for all $x$ in the interval $[-2, 1]$.

$$\text{Area} = \int_a^b [f(x) - g(x)]\, dx \qquad \text{Area between } f \text{ and } g$$

$$= \int_{-2}^1 [(2 - x^2) - (x)]\, dx \qquad \text{Substitute for } f \text{ and } g.$$

$$= \int_{-2}^1 (-x^2 - x + 2)\, dx$$

$$= \left[ -\frac{x^3}{3} - \frac{x^2}{2} + 2x \right]_{-2}^1 \qquad \text{Find antiderivative.}$$

$$= \frac{9}{2} \text{ square units} \qquad \text{Apply Fundamental Theorem.}$$

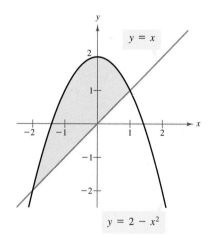

**FIGURE 5.16**

## EXAMPLE 3  Finding Area Below the x-Axis

Find the area of the region bounded by the graph of $y = x^2 - 3x - 4$ and the
x-axis.

### Solution

Begin by finding the intercepts of the graph. To do this, set the function equal to
zero and solve for x.

$$x^2 - 3x - 4 = 0 \qquad \text{Set function equal to 0.}$$
$$(x - 4)(x + 1) = 0 \qquad \text{Factor.}$$
$$x = 4, -1 \qquad \text{Solve for } x.$$

From Figure 5.17, you can see that $x^2 - 3x - 4 \leq 0$ for all $x$ in the interval
$[-1, 4]$. Therefore, you can let $f(x) = 0$ and $g(x) = x^2 - 3x - 4$, and compute
the area as follows.

$$\text{Area} = \int_a^b [f(x) - g(x)] \, dx \qquad \text{Area between } f \text{ and } g$$

$$= \int_{-1}^4 [(0) - (x^2 - 3x - 4)] \, dx \qquad \text{Substitute for } f \text{ and } g.$$

$$= \int_{-1}^4 (-x^2 + 3x + 4) \, dx$$

$$= \left[ -\frac{x^3}{3} + \frac{3x^2}{2} + 4x \right]_{-1}^4 \qquad \text{Find antiderivative.}$$

$$= \frac{125}{6} \qquad \text{Apply Fundamental Theorem.}$$

Therefore, the area of the region is $\frac{125}{6}$ square units.

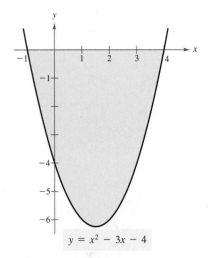

$y = x^2 - 3x - 4$

**FIGURE 5.17**

## Technology

Most graphing utilities can display regions
that are bounded by two graphs. For
instance, to graph the region in Example 3
on a *TI-82* or *TI-83*, set the viewing rectangle to $-1 \leq x \leq 4$ and $-7 \leq y \leq 1$. Then
enter

$$\text{Shade}(X^2 - 3X - 4, 0)$$

and press ENTER. You should obtain the
graph shown at the right.

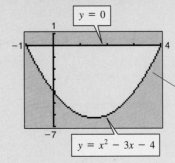

$y = 0$

Region lying below
the line $y = 0$
and above the graph of
$y = x^2 - 3x - 4$

$y = x^2 - 3x - 4$

Sometimes two graphs intersect at more than two points. To determine the area of the region bounded by two such graphs, you must find *all* points of intersection and check to see which graph is above the other in each interval determined by the points.

## EXAMPLE 4  Using Multiple Points of Intersection

Find the area of the region bounded by the graphs of

$$f(x) = 3x^3 - x^2 - 10x \qquad \text{and} \qquad g(x) = -x^2 + 2x.$$

### Solution

To find the points of intersection of the two graphs, set the functions equal to each other and solve for $x$.

$$f(x) = g(x) \qquad \text{Set } f(x) \text{ equal to } g(x).$$
$$3x^3 - x^2 - 10x = -x^2 + 2x \qquad \text{Substitute for } f(x) \text{ and } g(x).$$
$$3x^3 - 12x = 0 \qquad \text{Write in standard form.}$$
$$3x(x^2 - 4) = 0$$
$$3x(x - 2)(x + 2) = 0 \qquad \text{Factor.}$$
$$x = 0, 2, -2 \qquad \text{Solve for } x.$$

These three points of intersection determine two intervals of integration: $[-2, 0]$ and $[0, 2]$. In Figure 5.18, you can see that $g(x) \le f(x)$ in the interval $[-2, 0]$, and that $f(x) \le g(x)$ in the interval $[0, 2]$. Therefore, you must use two integrals to determine the area of the region bounded by the graphs of $f$ and $g$—one for the interval $[-2, 0]$ and one for the interval $[0, 2]$.

$$\text{Area} = \int_{-2}^{0} [f(x) - g(x)] \, dx + \int_{0}^{2} [g(x) - f(x)] \, dx$$

$$= \int_{-2}^{0} (3x^3 - 12x) \, dx + \int_{0}^{2} (-3x^3 + 12x) \, dx$$

$$= \left[ \frac{3x^4}{4} - 6x^2 \right]_{-2}^{0} + \left[ -\frac{3x^4}{4} + 6x^2 \right]_{0}^{2}$$

$$= (0 - 0) - (12 - 24) + (-12 + 24) - (0 + 0)$$

$$= 24$$

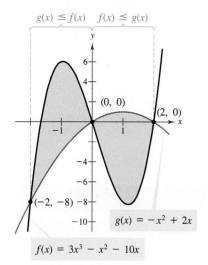

$g(x) \le f(x)$    $f(x) \le g(x)$

(0, 0)

(2, 0)

(−2, −8)

$g(x) = -x^2 + 2x$

$f(x) = 3x^3 - x^2 - 10x$

**FIGURE 5.18**

Thus, the region has an area of 24 square units.

**STUDY TIP**  It is easy to make an error when calculating areas such as that in Example 4. To give yourself some idea about the reasonableness of your solution, you could make a careful sketch of the region on graph paper and then use the grid on the graph paper to approximate the area. Try doing this with the graph shown in Figure 5.18. Is your approximation close to 24 square units?

## Consumer Surplus and Producer Surplus

You already know that a demand function relates the price of a product to the consumer demand. A **supply function** relates the price of a product to producers' willingness to supply the product. Whereas a typical demand function is decreasing, a typical supply function is increasing. That is, as the price increases, more and more producers become willing to supply the product. The point $(x_0, p_0)$ at which a demand function $p = D(x)$ and a supply function $p = S(x)$ intersect is called the **point of equilibrium.**

Economists call the area of the region bounded by the graph of the demand function, the horizontal line $p = p_0$, and the vertical line $x = 0$ the **consumer surplus.** Similarly, the area of the region bounded by the graph of the supply function, the horizontal line $p = p_0$, and the vertical line $x = 0$ is called the **producer surplus,** as shown in Figure 5.19.

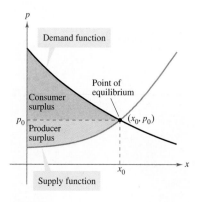

**FIGURE 5.19**

## EXAMPLE 5   Finding Surpluses

The demand and supply functions for a product are modeled by

*Demand:*  $p = -0.36x + 9$    and    *Supply:*  $p = 0.14x + 2,$

where $x$ is the number of units (in millions). Find the consumer and producer surpluses for this product.

### Solution

By equating the demand and supply functions, you can determine that the point of equilibrium occurs when $x = 14$ (million) and the price is \$3.96 per unit. The consumer surplus and producer surplus are shown in Figure 5.20.

$$\text{Consumer surplus} = \int_0^{14} (\text{demand function} - \text{price}) \, dx$$

$$= \int_0^{14} [(-0.36x + 9) - 3.96] \, dx$$

$$= \left[ -0.18x^2 + 5.04x \right]_0^{14}$$

$$= 35.28$$

$$\text{Producer surplus} = \int_0^{14} (\text{price} - \text{supply function}) \, dx$$

$$= \int_0^{14} [3.96 - (0.14x + 2)] \, dx$$

$$= \left[ -0.07x^2 + 1.96x \right]_0^{14}$$

$$= 13.72$$

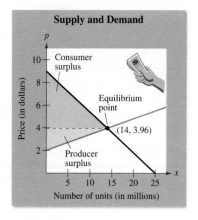

**FIGURE 5.20**

Martin Rogers/Tony Stone Images

*In 1995, the United States consumed about 6.5 billion barrels of petroleum.*

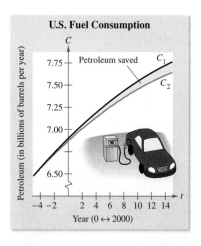

**FIGURE 5.21**

## Applications

In addition to consumer and producer surpluses, there are many other types of applications involving the area of a region bounded by two graphs. Example 6 illustrates one of these applications.

### EXAMPLE 6   Modeling Petroleum Consumption

In the annual *Energy Outlook* for 1996, the U.S. Energy Information Administration projected the consumption $C$ (in billions of barrels per year) of petroleum to follow the model

$$C_1 = -0.00137t^2 + 0.0781t + 6.90, \qquad -5 \leq t \leq 15,$$

where $t = 0$ represents January 1, 2000. If the actual consumption more closely followed the model

$$C_2 = -0.00137t^2 + 0.0721t + 6.87, \qquad -5 \leq t \leq 15,$$

how much petroleum would be saved?

#### Solution

The petroleum saved can be represented as the area of the region between the graphs of $C_1$ and $C_2$, as shown in Figure 5.21.

$$\text{Petroleum saved} = \int_{-5}^{15} [C_1 - C_2]\, dt$$

$$= \int_{-5}^{15} (0.006t + 0.03)\, dt$$

$$= (0.003t^2 + 0.03t) \Big]_{-5}^{15}$$

$$= 1.2 \text{ billion barrels}$$

So, there would be about 1,200,000,000 barrels of petroleum saved.

---

## Group Discussion    *Finding Units for Area*

In Example 6, the vertical axis is measured in billions of barrels per year and the horizontal axis is measured in years. Explain why the area of the region shown in Figure 5.21 is measured in billions of barrels.

## Warm Up

The following warm-up exercises involve skills that were covered in earlier sections. You will use these skills in the exercise set for this section.

In Exercises 1–4, simplify the expression.

**1.** $(-x^2 + 4x + 3) - (x + 1)$

**2.** $(-2x^2 + 3x + 9) - (-x + 5)$

**3.** $(-x^3 + 3x^2 - 1) - (x^2 - 4x + 4)$

**4.** $(3x + 1) - (-x^3 + 9x + 2)$

In Exercises 5–10, find the points of intersection of the graphs.

**5.** $f(x) = x^2 - 4x + 4, \ g(x) = 4$

**6.** $f(x) = -3x^2, \ g(x) = 6 - 9x$

**7.** $f(x) = x^2, \ g(x) = -x + 6$

**8.** $f(x) = \frac{1}{2}x^3, \ g(x) = 2x$

**9.** $f(x) = x^2 - 3x, \ g(x) = 3x - 5$

**10.** $f(x) = e^x, \ g(x) = e$

# EXERCISES 5.5

In Exercises 1–8, find the area of the region.

**1.** $f(x) = x^2 - 6x$
$g(x) = 0$

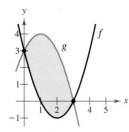

**2.** $f(x) = x^2 + 2x + 1$
$g(x) = 2x + 5$

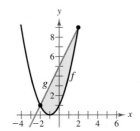

**5.** $f(x) = 3(x^3 - x)$
$g(x) = 0$

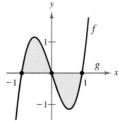

**6.** $f(x) = (x - 1)^3$
$g(x) = x - 1$

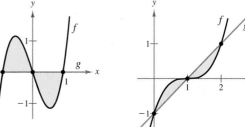

**3.** $f(x) = x^2 - 4x + 3$
$g(x) = -x^2 + 2x + 3$

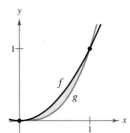

**4.** $f(x) = x^2$
$g(x) = x^3$

**7.** $f(x) = e^x - 1$
$g(x) = 0$

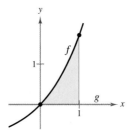

**8.** $f(x) = -x + 3$
$g(x) = 2x^{-1}$

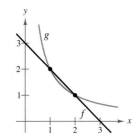

In Exercises 9–14, sketch the region whose area is represented by the definite integral.

**9.** $\int_0^4 [(x + 1) - \frac{1}{2}x]\, dx$

**10.** $\int_{-1}^1 [(1 - x^2) - (x^2 - 1)]\, dx$

**11.** $\int_{-2}^2 [2x^2 - (x^4 - 2x^2)]\, dx$

**12.** $\int_{-4}^0 [(x - 6) - (x^2 + 5x - 6)]\, dx$

**13.** $\int_{-1}^2 [(y^2 + 2) - 1]\, dy$

**14.** $\int_{-2}^3 [(y + 6) - y^2]\, dy$

In Exercises 15–30, sketch the region bounded by the graphs of the functions and find the area of the region.

**15.** $y = 2x,\ y = 4 - 2x,\ y = 0$

**16.** $y = \dfrac{1}{x^2},\ y = 0,\ x = 1,\ x = 5$

**17.** $f(x) = x(x^2 - 3x + 3),\ g(x) = x^2$

**18.** $y = x^3 - 2x + 1,\ y = -2x,\ x = 1$

**19.** $f(x) = \sqrt[3]{x},\ g(x) = x$

**20.** $f(x) = \sqrt{3x + 1},\ g(x) = x + 1$

**21.** $y = x^2 - 4x + 3,\ y = 3 + 4x - x^2$

**22.** $f(y) = y^2,\ g(y) = y + 2$

**23.** $f(y) = y(2 - y),\ g(y) = -y$

**24.** $f(y) = \sqrt{y},\ y = 9,\ x = 0$

**25.** $f(x) = e^{0.5x},\ g(x) = -\dfrac{1}{x},\ x = 1,\ x = 2$

**26.** $f(x) = \dfrac{1}{x},\ g(x) = -e^x,\ x = \dfrac{1}{2},\ x = 1$

**27.** $y = \dfrac{4}{x},\ y = x,\ x = 1,\ x = 4$

**28.** $y = \dfrac{8}{x},\ y = x^2,\ x = 1,\ x = 4$

**29.** $y = xe^{-x^2},\ y = 0,\ x = 0,\ x = 1$

**30.** $y = \dfrac{e^{1/x}}{x^2},\ y = 0,\ x = 1,\ x = 3$

In Exercises 31–34, use a graphing utility to graph the region bounded by the graphs of the functions, and find the area of the region.

**31.** $f(x) = x^2 - 4x,\ g(x) = 0$

**32.** $f(x) = 3 - 2x - x^2,\ g(x) = 0$

**33.** $f(x) = x^2 + 2x + 1,\ g(x) = x + 1$

**34.** $f(x) = -x^2 + 4x + 2,\ g(x) = x + 2$

In Exercises 35 and 36, use integration to find the area of the triangular region having the given vertices.

**35.** $(0, 0),\ (4, 0),\ (4, 4)$     **36.** $(0, 0),\ (4, 0),\ (6, 4)$

*Consumer and Producer Surpluses*   In Exercises 37–42, find the consumer and producer surpluses.

| Demand Function | Supply Function |
|---|---|
| **37.** $p_1(x) = 50 - 0.5x$ | $p_2(x) = 0.125x$ |
| **38.** $p_1(x) = 300 - x$ | $p_2(x) = 100 + x$ |
| **39.** $p_1(x) = 200 - 0.02x^2$ | $p_2(x) = 100 + x$ |
| **40.** $p_1(x) = 1000 - 0.4x^2$ | $p_2(x) = 42x$ |
| **41.** $p_1(x) = \dfrac{10{,}000}{\sqrt{x + 100}}$ | $p_2(x) = 100\sqrt{0.05x + 10}$ |
| **42.** $p_1(x) = \sqrt{25 - 0.1x}$ | $p_2(x) = \sqrt{9 + 0.1x} - 2$ |

**43.** *Writing*   Describe the characteristics of typical demand and supply functions.

**44.** *Writing*   Suppose that the demand and supply functions for a product did not intersect. What could you conclude?

*Revenue*   In Exercises 45 and 46, two models, $R_1$ and $R_2$, are given for revenue (in billions of dollars per year) for a large corporation. Both models are estimates of revenues for 1996–2000, with $t = 6$ representing 1996. Which model is projecting the greater revenue? How much more total revenue does that model project over the 5-year period?

**45.** $R_1 = 7.21 + 0.58t,\ R_2 = 7.21 + 0.45t$

**46.** $R_1 = 7.21 + 0.26t + 0.02t^2,\ R_2 = 7.21 + 0.1t + 0.01t^2$

**47.** *Fuel Cost*   The projected fuel cost $C$ (in millions of dollars per year) for an airline company from 1995 through 2005 is $C_1 = 568.5 + 7.15t$, where $t = 5$ represents 1995. If the company purchases more efficient airplane engines, fuel costs are expected to decrease and follow the model $C_2 = 525.6 + 6.43t$. How much can the company save with the more efficient engines? Explain your reasoning.

**48.** *Epidemic* An epidemic was spreading such that $t$ weeks from its outbreak it had infected

$$N_1(t) = 0.1t^2 + 0.5t + 150, \quad 0 \le t \le 50,$$

people. Twenty-five weeks after the outbreak, a vaccine was developed and administered to the public. At that point, the number of people infected was governed by the model

$$N_2(t) = -0.2t^2 + 6t + 200.$$

Approximate the number of people that the vaccine prevented from becoming ill during the epidemic.

**49.** *Beef Consumption* For the years 1980–1993, the per capita consumption of beef (in pounds per year) in the United States can be modeled by

$$B(t) = \begin{cases} 72.1 + 0.98\sqrt{t}, & 0 \le t \le 5 \\ 94.6 - 4.71t + 0.17t^2, & 5 < t \le 13, \end{cases}$$

where $t = 0$ represents 1980.

(a) Use a graphing utility to graph this model.
(b) Suppose the beef consumption from 1986 to 1993 had continued to follow the model for 1980 through 1985. How much additional beef would have been consumed from 1986 through 1993?

**50.** *Consumer and Producer Surpluses* Factory orders for an air conditioner are about 6000 units per week when the price is $331 and about 8000 units per week when the price is $303. The supply function is given by $p = 0.0275x$. Find the consumer and producer surpluses. (Assume the demand function is linear.)

**51.** *Consumer and Producer Surpluses* Repeat Exercise 50 with a demand of about 6000 units per week when the price is $325 and about 8000 units per week when the price is $300. Find the consumer and producer surpluses. (Assume the demand function is linear.)

**52.** *Profit* The revenue from a manufacturing process (in millions of dollars per year) is projected to follow the model $R = 100$ for 10 years. Over the same period of time, the cost (in millions of dollars per year) is projected to follow the model $C = 60 + 0.2t^2$, where $t$ is the time (in years). Approximate the profit over the 10-year period.

**53.** *Profit* Repeat Exercise 52 for revenue and cost models given by $R = 100 + 0.08t$ and $C = 60 + 0.2t^2$.

**54.** *Lorenz Curve* Economists use *Lorenz curves* to illustrate the distribution of income in a country. Letting $x$ represent the percent of families in a country and $y$ the percent of total income, the model $y = x$ would represent a country in which each family had the same income. The Lorenz curve, $y = f(x)$, represents the actual income distribution. The area between these two models, for $0 \le x \le 100$, indicates the "income inequality" of a country. In 1993, the Lorenz curve for the United States could be modeled by

$$y = 0.0614x^{3/2} + 0.3867x - 0.0387, \quad 0 \le x \le 100,$$

where $x$ is measured from the poorest to the wealthiest families. Find the income inequality for the United States in 1993. *(Source: U.S. Bureau of Census)*

**55.** *Income Distribution* Using the Lorenz curve in Exercise 54, complete the table, which lists the percent of total income earned by each quintile in the United States in 1993.

| Quintile | Lowest | 2nd | 3rd | 4th | Highest |
|----------|--------|-----|-----|-----|---------|
| Percent  |        |     |     |     |         |

## Business Capsule

Alternative Design

*In 1991, interior designer Courtney Sloane started the company Alternative Design in Jersey City, NJ. The company, whose 1994 sales were $350,000, renovates interiors and designs furniture for residential and commercial clients. Sloane started the business part-time while working as a design consultant.*

**56.** *Research Project* Use your school's library or some other reference source to research a small company similar to that described above. Describe the costs, market conditions, and competition that affect the company's success.

# 5.6 The Definite Integral as the Limit of a Sum

*The Midpoint Rule • The Definite Integral as the Limit of a Sum*

## The Midpoint Rule

In Section 5.4, you learned that you cannot use the Fundamental Theorem of Calculus to evaluate a definite integral unless you can find an antiderivative of the integrand. In cases when this cannot be done, you can approximate the value of the integral using an approximation technique. One such technique is called the **Midpoint Rule.** (Two other techniques are discussed in Section 6.5.)

### EXAMPLE 1   Approximating the Area of a Plane Region

Use the five rectangles in Figure 5.22 to approximate the area of the region bounded by the graph of $f(x) = -x^2 + 5$, the $x$-axis, and the lines $x = 0$ and $x = 2$.

### Solution

You can find the heights of the five rectangles by evaluating $f$ at the midpoint of each of the following intervals.

$$\left[0, \frac{2}{5}\right], \quad \left[\frac{2}{5}, \frac{4}{5}\right], \quad \left[\frac{4}{5}, \frac{6}{5}\right], \quad \left[\frac{6}{5}, \frac{8}{5}\right], \quad \left[\frac{8}{5}, \frac{10}{5}\right]$$

Evaluate $f(x)$ at the midpoints of these intervals.

The width of each rectangle is $\frac{2}{5}$. Hence, the sum of the five areas is

$$
\begin{aligned}
\text{Area} &\approx \frac{2}{5}f\left(\frac{1}{5}\right) + \frac{2}{5}f\left(\frac{3}{5}\right) + \frac{2}{5}f\left(\frac{5}{5}\right) + \frac{2}{5}f\left(\frac{7}{5}\right) + \frac{2}{5}f\left(\frac{9}{5}\right) \\
&= \frac{2}{5}\left[f\left(\frac{1}{5}\right) + f\left(\frac{3}{5}\right) + f\left(\frac{5}{5}\right) + f\left(\frac{7}{5}\right) + f\left(\frac{9}{5}\right)\right] \\
&= \frac{2}{5}\left(\frac{124}{25} + \frac{116}{25} + \frac{100}{25} + \frac{76}{25} + \frac{44}{25}\right) \\
&= \frac{920}{125} \\
&= 7.36.
\end{aligned}
$$

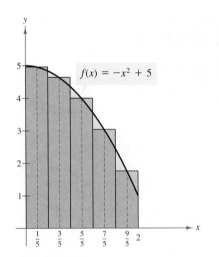

$f(x) = -x^2 + 5$

**FIGURE 5.22**

For the region in Example 1, you can find the exact area with a definite integral. That is,

$$\text{Area} = \int_0^2 (-x^2 + 5)\, dx$$

$$= \frac{22}{3}$$

$$\approx 7.33.$$

The approximation procedure used in Example 1 is the **Midpoint Rule.** You can use the Midpoint Rule to approximate *any* definite integral—not just those representing area. The basic steps are summarized below.

---

### Guidelines for Using the Midpoint Rule

To approximate the definite integral $\int_a^b f(x)\, dx$ with the Midpoint Rule, use the following steps.

1. Divide the interval $[a, b]$ into $n$ subintervals, each of width

$$\Delta x = \frac{b - a}{n}.$$

2. Find the midpoint of each subinterval.

$$\text{Midpoints} = \{x_1, x_2, x_3, \ldots, x_n\}$$

3. Evaluate $f$ at each midpoint and form the following sum.

$$\int_a^b f(x)\, dx \approx \frac{b - a}{n}[f(x_1) + f(x_2) + f(x_3) + \cdots + f(x_n)]$$

---

**STUDY TIP** In Example 1, the Midpoint Rule is used to approximate an integral whose exact value can be found with the Fundamental Theorem of Calculus. This was done to illustrate the accuracy of the rule. In practice, of course, you would use the Midpoint Rule to approximate the values of definite integrals for which you cannot find an antiderivative. Examples 2 and 3 illustrate such integrals.

An important characteristic of the Midpoint Rule is that the approximation tends to improve as $n$ increases. The table below shows the approximations for the area of the region described in Example 1 for various values of $n$. For example, for $n = 10$, the Midpoint Rule yields

$$\int_0^2 (-x^2 + 5)\, dx = \frac{2}{10}\left[f\left(\frac{1}{10}\right) + f\left(\frac{3}{10}\right) + \cdots + f\left(\frac{19}{10}\right)\right]$$

$$\approx 7.3400.$$

| $n$ | 5 | 10 | 15 | 20 | 25 | 30 |
|---|---|---|---|---|---|---|
| Approximation | 7.3600 | 7.3400 | 7.3363 | 7.3350 | 7.3344 | 7.3341 |

Note that as $n$ increases, the approximation gets closer and closer to the exact value of the integral, which was found to be $\frac{22}{3} \approx 7.3333$.

# Technology

## Programming the Midpoint Rule

The easiest way to use the Midpoint Rule to approximate the definite integral $\int_a^b f(x)\,dx$ is to program it into a computer or programmable calculator. For instance, the following program will evaluate the Midpoint Rule on a *TI-82* or *TI-83* calculator. See the appendix for other models.

```
PROGRAM: MIDPOINT
:Disp "LOWER LIMIT"              Prompt for value of a.
:Input A                         Input value of a.
:Disp "UPPER LIMIT"              Prompt for value of b.
:Input B                         Input value of b.
:Disp "N DIVISIONS"             Prompt for value of n.
:Input N                         Input value of n.
:0 → S                           Initialize sum of areas.
:(B−A)/N → W                     Calculate width of subinterval.
:1 → J                           Initialize counter.
:Lbl 1                           Begin loop.
:A + (J−1)W → L                  Calculate left endpoint.
:A + JW → R                      Calculate right endpoint.
:(L + R)/2 → X                   Calculate midpoint of subinterval.
:S + WY₁ → S                     Add area to sum.
:IS > (J, N)                     Test counter.
:Goto 1                          End loop.
:Disp "APPROXIMATION"
:Disp S                          Display approximation.
```

Before executing the program, enter the function as $Y_1$. When the program is executed, you will be prompted to enter the lower and upper limits of integration, and the number of subintervals you want to use.

With most integrals, you can determine the accuracy of the approximation by using increasingly larger values of $n$. For example, if you use the program to approximate the value of $\int_0^2 \sqrt{x^3 + 1}\,dx$, you will obtain the following approximations.

| $n$ | 10 | 20 | 30 | 40 | 50 | 60 | 70 |
|---|---|---|---|---|---|---|---|
| Approximation | 3.238 | 3.240 | 3.241 | 3.241 | 3.241 | 3.241 | 3.241 |

Thus, it seems clear that the approximation of 3.241 is accurate to three decimal places.

## EXAMPLE 2  Using the Midpoint Rule

Use the Midpoint Rule with $n = 5$ to approximate

$$\int_0^1 \frac{1}{x^2 + 1}\,dx.$$

### Solution

With $n = 5$, the interval $[0, 1]$ is divided into five subintervals.

$$\left[0, \frac{1}{5}\right], \quad \left[\frac{1}{5}, \frac{2}{5}\right], \quad \left[\frac{2}{5}, \frac{3}{5}\right], \quad \left[\frac{3}{5}, \frac{4}{5}\right], \quad \left[\frac{4}{5}, 1\right]$$

The midpoints of these intervals are $\frac{1}{10}, \frac{3}{10}, \frac{5}{10}, \frac{7}{10}$, and $\frac{9}{10}$. Because each subinterval has a width of $\Delta x = (1 - 0)/5 = \frac{1}{5}$, you can approximate the value of the definite integral as follows.

$$\int_0^1 \frac{1}{x^2 + 1}\,dx = \frac{1}{5}\left(\frac{1}{1.01} + \frac{1}{1.09} + \frac{1}{1.25} + \frac{1}{1.49} + \frac{1}{1.81}\right)$$

$$\approx 0.786$$

The region whose area is represented by the definite integral is shown in Figure 5.23. The actual area of this region is $\pi/4 \approx 0.785$. Thus, the approximation is off by only 0.001.

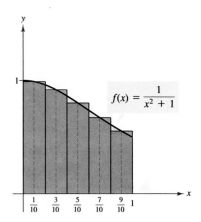

**FIGURE 5.23**

## EXAMPLE 3  Using the Midpoint Rule

Use the Midpoint Rule with $n = 10$ to approximate $\displaystyle\int_1^3 \sqrt{x^2 + 1}\,dx$.

### Solution

Begin by dividing the interval $[1, 3]$ into ten subintervals. The midpoints of these intervals are

$$\frac{11}{10}, \quad \frac{13}{10}, \quad \frac{3}{2}, \quad \frac{17}{10}, \quad \frac{19}{10}, \quad \frac{21}{10}, \quad \frac{23}{10}, \quad \frac{5}{2}, \quad \frac{27}{10}, \quad \text{and} \quad \frac{29}{10}.$$

Because each subinterval has a width of $\Delta x = (3 - 1)/10 = \frac{1}{5}$, you can approximate the value of the definite integral as follows.

$$\int_1^3 \sqrt{x^2 + 1}\,dx = \frac{1}{5}\left[\sqrt{(1.1)^2 + 1} + \sqrt{(1.3)^2 + 1} + \cdots + \sqrt{(2.9)^2 + 1}\right]$$

$$\approx 4.504$$

The region whose area is represented by the definite integral is shown in Figure 5.24. Using techniques that are not within the scope of this course, it can be shown that the actual area is

$$\frac{1}{2}\left[3\sqrt{10} + \ln\left(3 + \sqrt{10}\right) - \sqrt{2} - \ln\left(1 + \sqrt{2}\right)\right] \approx 4.505.$$

Thus, the approximation is off by only 0.001.

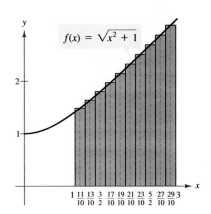

**FIGURE 5.24**

## The Definite Integral as the Limit of a Sum

Consider the closed interval $[a, b]$, divided into $n$ subintervals whose midpoints are $x_i$ and whose widths are $\Delta x = (b - a)/n$. In this section you have seen that the midpoint approximation

$$\int_a^b f(x)\, dx \approx f(x_1)\Delta x + f(x_2)\Delta x + f(x_3)\Delta x + \cdots + f(x_n)\Delta x$$

$$= [f(x_1) + f(x_2) + f(x_3) + \cdots + f(x_n)]\Delta x$$

becomes better and better as $n$ increases. In fact, the limit of this sum as $n$ approaches infinity is exactly equal to the definite integral. That is,

$$\int_a^b f(x)\, dx = \lim_{n\to\infty} [f(x_1) + f(x_2) + f(x_3) + \cdots + f(x_n)]\Delta x.$$

It can be shown that this limit is valid as long as $x_i$ is *any* point in the *i*th interval.

### EXAMPLE 4   Approximating a Definite Integral

Use a computer, programmable calculator, or symbolic integration utility to approximate the definite integral

$$\int_0^1 e^{-x^2}\, dx.$$

#### Solution

Using the program on page 362, with $n = 10, 20, 30, 40,$ and $50$, it appears that the value of the integral is approximately 0.7468. If you have access to a computer or calculator with a built-in program to approximate definite integrals, try using it to approximate this integral. When we used *Derive* for Windows to approximate the integral, we obtained 0.746824.

---

## *Group Discussion*    A Failure of the Midpoint Rule

Suppose you use the Midpoint Rule to approximate the definite integral

$$\int_0^1 \frac{1}{x^2}\, dx.$$

Use a program similar to the one on page 362 to complete the table for $n = 10, 20, 30, 40, 50,$ and $60$.

| $n$ | 10 | 20 | 30 | 40 | 50 | 60 |
|---|---|---|---|---|---|---|
| Approximation | | | | | | |

Why are the approximations getting larger? Why isn't the Midpoint Rule working?

## Warm Up

The following warm-up exercises involve skills that were covered in earlier sections. You will use these skills in the exercise set for this section.

In Exercises 1–6, find the midpoint of the interval.

**1.** $\left[0, \frac{1}{3}\right]$  **2.** $\left[\frac{1}{10}, \frac{2}{10}\right]$

**3.** $\left[\frac{3}{20}, \frac{4}{20}\right]$  **4.** $\left[1, \frac{7}{6}\right]$

**5.** $\left[2, \frac{31}{15}\right]$  **6.** $\left[\frac{26}{9}, 3\right]$

In Exercises 7–10, find the limit.

**7.** $\lim\limits_{x \to \infty} \dfrac{2x^2 + 4x - 1}{3x^2 - 2x}$  **8.** $\lim\limits_{x \to \infty} \dfrac{4x + 5}{7x - 5}$

**9.** $\lim\limits_{x \to \infty} \dfrac{x - 7}{x^2 + 1}$  **10.** $\lim\limits_{x \to \infty} \dfrac{5x^3 + 1}{x^3 + x^2 + 4}$

## EXERCISES 5.6

In Exercises 1–4, use the Midpoint Rule with $n = 4$ to approximate the area of the region. Compare your result with the exact area obtained with a definite integral.

**1.** $f(x) = -2x + 3, [0, 1]$  **2.** $f(x) = \sqrt{x} + 1, [0, 2]$

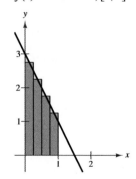

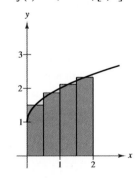

**3.** $f(x) = \sqrt{x}, [0, 1]$  **4.** $f(x) = 1 - x^2, [-1, 1]$

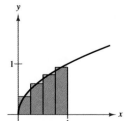

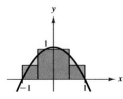

In Exercises 5–12, use the Midpoint Rule with $n = 4$ to approximate the area of the region bounded by the graph of $f$ and the $x$-axis over the interval. Compare your result with the exact area. Sketch the region.

| | Function | Interval |
|---|---|---|
| **5.** | $f(x) = x^2 + 2$ | $[-1, 1]$ |
| **6.** | $f(x) = 3x - 4$ | $[2, 5]$ |
| **7.** | $f(x) = 2x^2$ | $[1, 3]$ |
| **8.** | $f(x) = 2x - x^3$ | $[0, 1]$ |
| **9.** | $f(x) = x^2 - x^3$ | $[0, 1]$ |
| **10.** | $f(x) = x^2 - x^3$ | $[-1, 0]$ |
| **11.** | $f(x) = x(1 - x)^2$ | $[0, 1]$ |
| **12.** | $f(x) = x^2(3 - x)$ | $[0, 3]$ |

In Exercises 13–16, use a program similar to that on page 362 to approximate the area of the region. How large must $n$ be to obtain an approximation that is correct to within 0.01?

**13.** $\displaystyle\int_0^4 (2x^2 + 3)\, dx$  **14.** $\displaystyle\int_0^4 (2x^3 + 3)\, dx$

**15.** $\displaystyle\int_1^2 (2x^2 - x + 1)\, dx$  **16.** $\displaystyle\int_1^2 (x^3 - 1)\, dx$

In Exercises 17 and 18, use the Midpoint Rule with $n = 4$ to approximate the area of the region. Compare your result with the exact area obtained with a definite integral.

**17.** $f(y) = \frac{1}{4}y, [2, 4]$

**18.** $f(y) = 4y - y^2, [0, 4]$

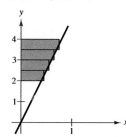
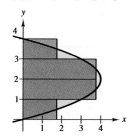

*Trapezoidal Rule* In Exercises 19 and 20, use the Trapezoidal Rule with $n = 8$ to approximate the definite integral. Compare the result with the exact value and the approximation obtained with $n = 8$ and the Midpoint Rule. Which approximation technique appears to be better? Let $f$ be continuous on $[a, b]$ and let $n$ be the number of equal subintervals (see figure). Then the Trapezoidal Rule for approximating $\int_a^b f(x)\, dx$ is

$$\frac{b - a}{2n}[f(x_0) + 2f(x_1) + \cdots + 2f(x_{n-1}) + f(x_n)].$$

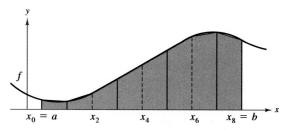

**19.** $\int_0^2 x^3\, dx$

**20.** $\int_1^3 \frac{1}{x^2}\, dx$

In Exercises 21–24, use the Trapezoidal Rule to approximate the definite integral with $n = 4$.

**21.** $\int_0^2 \frac{1}{x + 1}\, dx$

**22.** $\int_0^4 \sqrt{1 + x^2}\, dx$

**23.** $\int_{-1}^1 \frac{1}{x^2 + 1}\, dx$

**24.** $\int_1^5 \frac{\sqrt{x - 1}}{x}\, dx$

 In Exercises 25 and 26, use a computer or programmable calculator to approximate the definite integral using the Midpoint Rule and Trapezoidal Rule for $n = 4, 8, 12, 16,$ and $20$.

**25.** $\int_0^4 \sqrt{2 + 3x^2}\, dx$

**26.** $\int_0^2 \frac{5}{x^3 + 1}\, dx$

In Exercises 27 and 28, use the Trapezoidal Rule with $n = 10$ to approximate the area of the region bounded by the graphs of the equations.

**27.** $y = \sqrt{\dfrac{x^3}{4 - x}}, \quad y = 0, \quad x = 3$

**28.** $y = x\sqrt{\dfrac{4 - x}{4 + x}}, \quad y = 0, \quad x = 4$

**29.** *Velocity and Acceleration* The table lists the velocity $v$ (in feet per second) of an accelerating car over a 20-second interval. Use the Trapezoidal Rule to approximate the distance in feet that the car travels during the 20 seconds. (The distance is given by $s = \int_0^{20} v\, dt$.)

| Time, $t$ | 0 | 5 | 10 | 15 | 20 |
|---|---|---|---|---|---|
| Velocity, $v$ | 0.0 | 29.3 | 51.3 | 66.0 | 73.3 |

**30.** *Area* To estimate the surface area of a pond, a surveyor takes several measurements, as shown in the figure. Estimate the surface area of the pond using (a) the Midpoint Rule and (b) the Trapezoidal Rule.

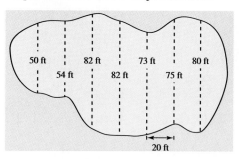

 **31.** *Approximation of Pi* Use the Midpoint and the Trapezoidal Rules with $n = 4$ to approximate $\pi$ where

$$\pi = \int_0^1 \frac{4}{1 + x^2}\, dx.$$

Then use a graphing utility to evaluate the definite integral. Compare all of your results.

# Volumes of Solids of Revolution

## 5.7

*The Disc Method • The Washer Method • Applications*

## The Disc Method

As shown in Figure 5.25, a **solid of revolution** is formed by revolving a plane region about a line. The line is called the **axis of revolution.**

To develop a formula for finding the volume of a solid of revolution, consider a continuous function $f$ that is nonnegative on the interval $[a, b]$. Suppose that the area of the region is approximated by $n$ rectangles, each of width $\Delta x$, as shown in Figure 5.26. By revolving the rectangles about the $x$-axis, you obtain $n$ circular discs, each with a volume of $\pi[f(x_i)]^2\Delta x$. The volume of the solid formed by revolving the region about the $x$-axis is approximately equal to the sum of the volumes of the $n$ discs. Moreover, by taking the limit as $n$ approaches infinity, you can see that the exact volume is given by a definite integral. This result is called the **Disc Method.**

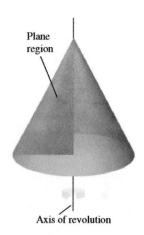

**FIGURE 5.25**

### The Disc Method

The volume of the solid formed by revolving the region bounded by the graph of $f$ and the $x$-axis $(a \le x \le b)$ about the $x$-axis is

$$\text{Volume} = \pi \int_a^b [f(x)]^2 \, dx.$$

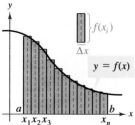

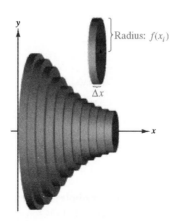

Approximation by *n* rectangles

Approximation by *n* discs

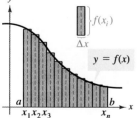

**FIGURE 5.26**

## EXAMPLE 1   Finding the Volume of a Solid of Revolution

Find the volume of the solid formed by revolving the region bounded by the graph of $f(x) = -x^2 + x$ and the $x$-axis about the $x$-axis.

### Solution

Begin by sketching the region bounded by the graph of $f$ and the $x$-axis. As shown in Figure 5.27(a), sketch a representative rectangle whose height is $f(x)$ and whose width is $\Delta x$. From this rectangle, you can see that the radius of the solid is

$$\text{Radius} = f(x) = -x^2 + x.$$

Using the Disc Method, you can find the volume of the solid of revolution.

$$\text{Volume} = \pi \int_0^1 [f(x)]^2 \, dx \qquad \text{Disc Method}$$

$$= \pi \int_0^1 (-x^2 + x)^2 \, dx \qquad \text{Substitute for } f(x).$$

$$= \pi \int_0^1 (x^4 - 2x^3 + x^2) \, dx \qquad \text{Expand integrand.}$$

$$= \pi \left[ \frac{x^5}{5} - \frac{x^4}{2} + \frac{x^3}{3} \right]_0^1 \qquad \text{Find antiderivative.}$$

$$= \frac{\pi}{30} \qquad \text{Apply Fundamental Theorem.}$$

$$\approx 0.105 \qquad \text{Round to three decimal places.}$$

Thus, the volume of the solid is about 0.105 cubic unit.

## Technology

Try using the integration capabilities of a graphing utility to verify the solution in Example 1. The following command on a *TI-82* or *TI-83* graphing calculator should yield a volume of approximately 0.10472.

fnInt($\pi$(−X∧2+X)∧2,X,0,1)

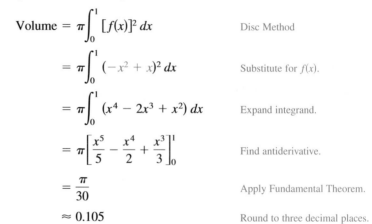

(a)  Plane region

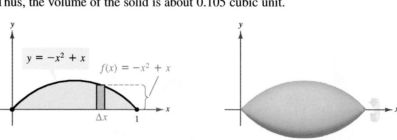

(b)  Solid of revolution

**FIGURE 5.27**

**STUDY TIP**   In Example 1, the entire problem was solved *without* referring to the three-dimensional sketch given in Figure 5.27(b). In general, to set up the integral for calculating the volume of a solid of revolution, a sketch of the plane region is more useful than a sketch of the solid, because the radius is more readily visualized in the plane region.

## The Washer Method

You can extend the Disc Method to find the volume of a solid of revolution with a *hole*. Consider a region that is bounded by the graphs of $f$ and $g$, as shown in Figure 5.28(a). If the region is revolved about the $x$-axis, then the volume of the resulting solid can be found by applying the Disc Method to $f$ and $g$ and subtracting the results.

$$\text{Volume} = \pi \int_a^b [f(x)]^2 \, dx - \pi \int_a^b [g(x)]^2 \, dx$$

Writing this as a single integral produces the **Washer Method.**

---

### The Washer Method

Let $f$ and $g$ be continuous and nonnegative on the closed interval $[a, b]$, as shown in Figure 5.28(a). If $g(x) \le f(x)$ for all $x$ in the interval, then the volume of the solid formed by revolving the region bounded by the graphs of $f$ and $g$ $(a \le x \le b)$ about the $x$-axis is

$$\text{Volume} = \pi \int_a^b \{[f(x)]^2 - [g(x)]^2\} \, dx.$$

$f(x)$ is the **outer radius** and $g(x)$ is the **inner radius.**

---

In Figure 5.28(b), note that the solid of revolution has a hole. Moreover, the radius of the hole is $g(x)$, the inner radius.

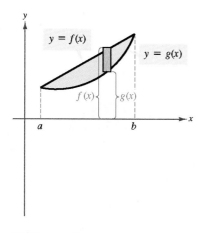

(a) Plane region          (b) Solid of revolution with hole

**FIGURE 5.28**

## EXAMPLE 2   Using the Washer Method

Find the volume of the solid formed by revolving the region bounded by the graphs of $f(x) = \sqrt{25 - x^2}$ and $g(x) = 3$ about the $x$-axis (see Figure 5.29).

### Solution

First find the points of intersection of $f$ and $g$ by setting $f(x)$ equal to $g(x)$ and solving for $x$.

$$f(x) = g(x) \qquad \text{Set } f(x) \text{ equal to } g(x).$$
$$\sqrt{25 - x^2} = 3 \qquad \text{Substitute for } f(x) \text{ and } g(x).$$
$$25 - x^2 = 9 \qquad \text{Square both sides.}$$
$$16 = x^2$$
$$\pm 4 = x \qquad \text{Solve for } x.$$

Using $f(x)$ as the outer radius and $g(x)$ as the inner radius, you can find the volume of the solid as follows.

$$\text{Volume} = \pi \int_{-4}^{4} \{[f(x)]^2 - [g(x)]^2\}\, dx \qquad \text{Washer Method}$$

$$= \pi \int_{-4}^{4} \left[\left(\sqrt{25 - x^2}\right)^2 - (3)^2\right] dx \qquad \text{Substitute for } f(x) \text{ and } g(x).$$

$$= \pi \int_{-4}^{4} (16 - x^2)\, dx \qquad \text{Simplify.}$$

$$= \pi \left[ 16x - \frac{x^3}{3} \right]_{-4}^{4} \qquad \text{Find antiderivative.}$$

$$= \frac{256\pi}{3} \qquad \text{Apply Fundamental Theorem.}$$

$$\approx 268.08 \qquad \text{Round to two decimal places.}$$

Thus, the volume of the solid is about 268.08 cubic inches.

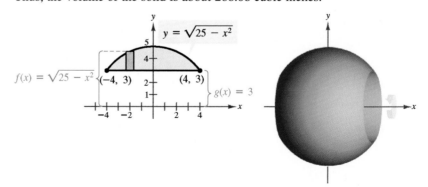

(a)  Plane region

(b)  Solid of revolution

**FIGURE 5.29**

# Applications

## EXAMPLE 3   Finding a Football's Volume

A regulation-size football can be modeled as a solid of revolution formed by revolving the graph of

$$f(x) = -0.0944x^2 + 3.4, \qquad -5.5 \le x \le 5.5$$

about the $x$-axis, as shown in Figure 5.30. Use this model to find the volume of a football. (In the model, $x$ and $y$ are measured in inches.)

### Solution

To find the volume of the solid of revolution, use the Disc Method.

$$\text{Volume} = \pi \int_{-5.5}^{5.5} [f(x)]^2 \, dx \qquad \text{Disc Method}$$

$$= \pi \int_{-5.5}^{5.5} (-0.0944x^2 + 3.4)^2 \, dx \qquad \text{Substitute for } f(x).$$

$$\approx 232 \text{ cubic inches} \qquad \text{Volume}$$

Thus, the volume of the football is about 232 cubic inches.

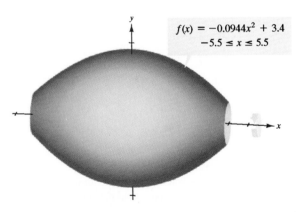

$$f(x) = -0.0944x^2 + 3.4$$
$$-5.5 \le x \le 5.5$$

**FIGURE 5.30**   A football-shaped solid is formed by revolving a parabolic segment about the $x$-axis.

North Wind Picture Archives

*American football, in its modern form, is a twentieth-century invention. In the 1800s a rough, soccer-like game was played with a "round football." In 1905, at the request of President Theodore Roosevelt, the Intercollegiate Athletic Association (which became the NCAA in 1910) was formed. With the introduction of the forward pass in 1906, the shape of the ball was altered to make it easier to grip.*

---

## Group Discussion     Testing the Reasonableness of an Answer

A football is about 11 inches long and has a diameter of about 7 inches. In Example 3, the volume of a football was approximated to be 232 cubic inches. Explain how you can determine whether this answer is reasonable.

## Warm Up

The following warm-up exercises involve skills that were covered in earlier sections. You will use these skills in the exercise set for this section.

In Exercises 1–6, solve for x.

**1.** $x^2 = 2x$

**2.** $-x^2 + 4x = x^2$

**3.** $x = -x^3 + 5x$

**4.** $x^2 + 1 = x + 3$

**5.** $-x + 4 = \sqrt{4x - x^2}$

**6.** $\sqrt{x - 1} = \frac{1}{2}(x - 1)$

In Exercises 7–10, evaluate the integral.

**7.** $\displaystyle\int_0^2 2e^{2x}\, dx$

**8.** $\displaystyle\int_{-1}^3 \frac{2x + 1}{x^2 + x + 2}\, dx$

**9.** $\displaystyle\int_0^2 x\sqrt{x^2 + 1}\, dx$

**10.** $\displaystyle\int_1^5 \frac{(\ln x)^2}{x}\, dx$

## EXERCISES 5.7

In Exercises 1–16, find the volume of the solid formed by revolving the region bounded by the graph(s) of the equation(s) about the x-axis.

**1.** $y = \sqrt{4 - x^2}$

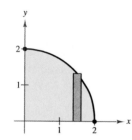

**2.** $y = x^2$

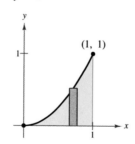

**3.** $y = \sqrt{x}$

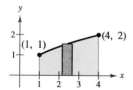

**4.** $y = \sqrt{4 - x^2}$

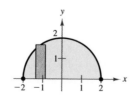

**5.** $y = 4 - x^2$, $y = 0$

**6.** $y = x$, $y = 0$, $x = 4$

**7.** $y = 1 - \frac{1}{4}x^2$, $y = 0$

**8.** $y = x^2 + 1$, $y = 5$

**9.** $y = -x + 1$, $y = 0$, $x = 0$

**10.** $y = x$, $y = e^{x-1}$, $x = 0$

**11.** $y = \dfrac{1}{x} - \dfrac{1}{2}$, $y = -\dfrac{1}{2}x + 1$

**12.** $y = \sqrt{x}$, $y = 0$, $x = 4$

**13.** $y = 2x^2$, $y = 0$, $x = 2$

**14.** $y = \dfrac{1}{x}$, $y = 0$, $x = 1$, $x = 3$

**15.** $y = e^x$, $y = 0$, $x = 0$, $x = 1$

**16.** $y = x^2$, $y = 4x - x^2$

In Exercises 17–24, find the volume of the solid formed by revolving the region bounded by the graph(s) of the equation(s) about the y-axis.

**17.** $y = x^2$, $y = 4$, $0 \le x \le 2$

**18.** $y = \sqrt{16 - x^2}$, $y = 0$, $0 \le x \le 4$

**19.** $x = 1 - \frac{1}{2}y$, $x = 0$, $y = 0$

**20.** $x = y(y - 1)$, $x = 0$

**21.** $y = x^{2/3}$

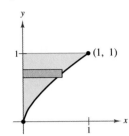

**22.** $x = -y^2 + 4y$

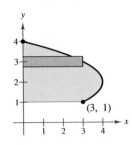

**23.** $y = \sqrt{4 - x}$, $y = 0$, $x = 0$

**24.** $y = 4$, $y = 0$, $x = 2$, $x = 0$

**25.** The line segment from $(0, 0)$ to $(6, 3)$ is revolved about the $x$-axis to form a cone. What is the volume of the cone?

**26.** Use the Disc Method to verify that the volume of a right circular cone is $\frac{1}{3}\pi r^2 h$, where $r$ is the radius of the base and $h$ is the height.

**27.** Use the Disc Method to verify that the volume of a sphere of radius $r$ is $\frac{4}{3}\pi r^3$.

**28.** The right half of the ellipse $9x^2 + 25y^2 = 225$ is revolved about the $y$-axis to form an oblate spheroid (shaped like an M&M candy). Find the volume of the spheroid.

**29.** The upper half of the ellipse $9x^2 + 16y^2 = 144$ is revolved about the $x$-axis to form a prolate spheroid (shaped like a football). Find the volume of the spheroid.

**30.** *Fuel Tank*   A tank on the wing of a jet is modeled by revolving the region bounded by the graph of $y = \frac{1}{8}x^2\sqrt{2 - x}$ and the $x$-axis about the $x$-axis, where $x$ and $y$ are measured in meters (see figure). Find the volume of the tank.

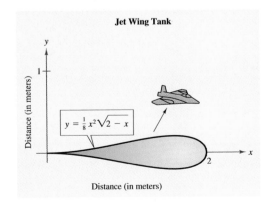

Jet Wing Tank

$y = \frac{1}{8}x^2\sqrt{2 - x}$

Distance (in meters)

Distance (in meters)

**31.** *Fish Population*   A pond is to be stocked with a species of fish. The food supply in 500 cubic feet of pond water can adequately support one fish. The pond is nearly circular, is 20 feet deep at its center, and has a radius of 200 feet. The bottom of the pond can be modeled by

$$y = 20[(0.005x)^2 - 1].$$

(a) How much water is in the pond?

(b) How many fish can the pond support?

**32.** *Modeling a Pond*   A pond is approximately circular, with a diameter of 400 feet (see figure). Starting at the center, the depth of the water is measured every 25 feet and recorded in the table.

| x | 0 | 25 | 50 | 75 | 100 |
|---|---|----|----|----|-----|
| Depth | 20 | 19 | 19 | 17 | 15 |

| x | 125 | 150 | 175 | 200 |
|---|-----|-----|-----|-----|
| Depth | 14 | 10 | 6 | 0 |

(a) Use a graphing utility to plot the depths and graph the model of the pond's depth, $y = 20 - 0.00045x^2$.

(b) Use the model in part (a) to find the pond's volume.

(c) Use the result of part (b) to approximate the number of gallons of water in the pond ($1 \text{ ft}^3 \approx 7.48$ gal).

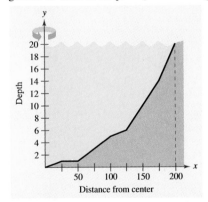

Depth

Distance from center

In Exercises 33 and 34, use a program similar to the one on page 362 to approximate the volume of a solid generated by revolving the region bounded by the graphs of the equations about the $x$-axis.

**33.** $y = \sqrt[3]{x + 1}$, $y = 0$, $x = 0$, $x = 7$

**34.** $y = \dfrac{10}{x^2 + 1}$, $y = 0$, $x = 0$, $x = 3$

## Chapter 5 Algebra Review

### "Unsimplifying an Algebraic Expression"

In algebra it is often helpful to write an expression in simplest form. In this chapter, you have seen that the reverse is often true in integration. That is, to fit an integrand to an integration formula, it often helps to "unsimplify" the expression. To do this, you use the same algebraic rules, but your goal is different. Here are some examples.

### EXAMPLE 1  Rewriting an Algebraic Expression

(a) $\dfrac{x+1}{\sqrt{x}} = \dfrac{x}{\sqrt{x}} + \dfrac{1}{\sqrt{x}}$

　　Example 6, page 318
　　Rewrite as two fractions.

　　$= \dfrac{x^1}{x^{1/2}} + \dfrac{1}{x^{1/2}}$

　　Rewrite with rational exponents.

　　$= x^{1-1/2} + x^{-1/2}$

　　Properties of exponents

　　$= x^{1/2} + x^{-1/2}$

　　Simplify exponent.

(b) $x(3 - 4x^2)^2 = \dfrac{-8}{-8}x(3 - 4x^2)^2$

　　Example 2, page 327
　　Multiply and divide by $-8$.

　　$= \left(-\dfrac{1}{8}\right)(-8)x(3 - 4x^2)^2$

　　Regroup.

　　$= \left(-\dfrac{1}{8}\right)(3 - 4x^2)^2(-8x)$

　　Regroup.

(c) $7x^2\sqrt{x^3 + 1} = 7x^2(x^3 + 1)^{1/2}$

　　Example 4, page 328
　　Rewrite with rational exponents.

　　$= \dfrac{3}{3}(7x^2)(x^3 + 1)^{1/2}$

　　Multiply and divide by 3.

　　$= \dfrac{7}{3}(3x^2)(x^3 + 1)^{1/2}$

　　Regroup.

　　$= \dfrac{7}{3}(x^3 + 1)^{1/2}(3x^2)$

　　Regroup.

(d) $5xe^{-x^2} = \dfrac{-2}{-2}(5x)e^{-x^2}$

　　Example 3, page 334
　　Multiply and divide by $-2$.

　　$= \left(-\dfrac{5}{2}\right)(-2x)e^{-x^2}$

　　Regroup.

　　$= \left(-\dfrac{5}{2}\right)e^{-x^2}(-2x)$

　　Regroup.

## EXAMPLE 2  Rewriting an Algebraic Expression

(a) $\dfrac{3x^2 + 2x - 1}{x^2} = \dfrac{3x^2}{x^2} + \dfrac{2x}{x^2} - \dfrac{1}{x^2}$

Example 7(a), page 337
Rewrite as separate fractions.

$\qquad\qquad = 3 + \dfrac{2}{x} - x^{-2}$

Properties of exponents.

$\qquad\qquad = 3 + 2\left(\dfrac{1}{x}\right) - x^{-2}$

Regroup.

(b) $\dfrac{1}{1 + e^{-x}} = \left(\dfrac{e^x}{e^x}\right)\dfrac{1}{1 + e^{-x}}$

Example 7(b), page 337
Multiply and divide by $e^x$.

$\qquad = \dfrac{e^x}{e^x + e^x(e^{-x})}$

Multiply.

$\qquad = \dfrac{e^x}{e^x + e^{x-x}}$

Property of exponents

$\qquad = \dfrac{e^x}{e^x + e^0}$

Simplify exponent.

$\qquad = \dfrac{e^x}{e^x + 1}$

$e^0 = 1$

(c) $\dfrac{x^2 + x + 1}{x - 1} = x + 2 + \dfrac{3}{x - 1}$

Example 7(c), page 337
Use long division as shown below.

$$
\begin{array}{r}
x + 2 \\
x - 1 \overline{\smash{)}\, x^2 + x + 1} \\
\underline{x^2 - x} \\
2x + 1 \\
\underline{2x - 2} \\
3
\end{array}
$$

(d) $\dfrac{x^2 + 6x + 1}{x^2 + 1} = 1 + \dfrac{6x}{x^2 + 1}$

Bottom of page 336
Use long division as shown below.

$$
\begin{array}{r}
1 \\
x^2 + 1 \overline{\smash{)}\, x^2 + 6x + 1} \\
\underline{x^2 \qquad + 1} \\
6x
\end{array}
$$

# Chapter Summary and Study Strategies

*After studying this chapter, you should have acquired the following skills. The exercise numbers are keyed to the Review Exercises that begin on page 378. Answers to odd-numbered Review Exercises are given in the back of the text.* *

■ Use basic integration formulas to find indefinite integrals.  *(Section 5.1)*        *Review Exercises 1–8*

$$\int k\, dx = kx + C \qquad\qquad \int [f(x) - g(x)]\, dx = \int f(x)\, dx - \int g(x)\, dx$$

$$\int kf(x)\, dx = k \int f(x)\, dx \qquad\qquad \int x^n\, dx = \frac{x^{n+1}}{n+1} + C, \quad n \neq -1$$

$$\int [f(x) + g(x)]\, dx = \int f(x)\, dx + \int g(x)\, dx$$

■ Use initial conditions to find particular solutions of indefinite integrals.  *(Section 5.1)*        *Review Exercises 9–12*

■ Use antiderivatives to solve real-life problems.  *(Section 5.1)*        *Review Exercises 13, 14*

■ Use the General Power Rule to find indefinite integrals.  *(Section 5.2)*        *Review Exercises 15–22*

$$\int u^n \frac{du}{dx}\, dx = \int u^n\, du = \frac{u^{n+1}}{n+1} + C, \quad n \neq -1$$

■ Use the General Power Rule to solve real-life problems.  *(Section 5.2)*        *Review Exercises 23, 24*

■ Use the Exponential and Log Rules to find indefinite integrals.  *(Section 5.3)*        *Review Exercises 25–30*

$$\int e^x\, dx = e^x + C \qquad\qquad \int \frac{1}{x}\, dx = \ln|x| + C$$

$$\int e^u \frac{du}{dx}\, dx = \int e^u\, du = e^u + C \qquad\qquad \int \frac{du/dx}{u}\, dx = \int \frac{1}{u}\, du = \ln|u| + C$$

■ Use a symbolic integration utility to find indefinite integrals.  *(Section 5.3)*        *Review Exercises 31, 32*

■ Find the area of a region bounded by the graph of a function and the *x*-axis.  *(Section 5.4)*        *Review Exercises 33–38*

■ Use the Fundamental Theorem of Calculus to evaluate a definite integral.  *(Section 5.4)*        *Review Exercises 39–50*

$$\int_a^b f(x)\, dx = F(x) \Big]_a^b = F(b) - F(a), \qquad \text{where } F'(x) = f(x)$$

■ Use definite integrals to solve marginal analysis problems.  *(Section 5.4)*        *Review Exercises 51, 52*

■ Find the average value of a function over a closed interval.  *(Section 5.4)*        *Review Exercises 53–56*

$$\text{Average value} = \frac{1}{b-a} \int_a^b f(x)\, dx$$

■ Use average values to solve real-life problems.  *(Section 5.4)*        *Review Exercises 57–60*

* Use a wide range of valuable study aids to help you master the material in this chapter. The *Student Solutions Guide* includes step-by-step solutions to all odd-numbered exercises to help you review and prepare. The *Algebra Review Tutorial Software* and *The Algebra of Calculus* help you brush up on your algebra skills. The *Graphing Technology Guide* offers step-by-step commands and instructions for a wide variety of graphing calculators, including the most recent models.

■ Use properties of even and odd functions to help evaluate definite integrals.                    *Review Exercises 61–64*
  *(Section 5.4)*

  Even function: $f(-x) = f(x)$          Odd function: $f(-x) = -f(x)$

■ Find the area of a region bounded by two graphs.   *(Section 5.5)*                                *Review Exercises 65–72*

  $$A = \int_a^b [f(x) - g(x)]\, dx$$

■ Find consumer and producer surpluses.   *(Section 5.5)*                                           *Review Exercises 73, 74*

■ Use the area of a region bounded by two graphs to solve real-life problems.                       *Review Exercises 75–78*
  *(Section 5.5)*

■ Use the Midpoint Rule to approximate the value of a definite integral.   *(Section 5.6)*          *Review Exercises 79–82*

  $$\int_a^b f(x)\, dx \approx \frac{b - a}{n}[f(x_1) + f(x_2) + f(x_3) + \cdots + f(x_n)]$$

■ Use the Disc Method to find the volume of a solid of revolution.   *(Section 5.7)*                *Review Exercises 83–86*

  $$\text{Volume} = \pi \int_a^b [f(x)]^2\, dx$$

■ Use the Washer Method to find the volume of a solid of revolution with a hole.                    *Review Exercises 87–90*
  *(Section 5.7)*

  $$\text{Volume} = \pi \int_a^b \{[f(x)]^2 - [g(x)]^2\}\, dx$$

■ Use solids of revolution to solve real-life problems.   *(Section 5.7)*                           *Review Exercises 91, 92*

■ *Indefinite and Definite Integrals*   When evaluating integrals, remember that
  an indefinite integral is a *family of antiderivatives*, each differing by a constant $C$,
  whereas a definite integral is a number.

■ *Checking Antiderivatives by Differentiating*   When finding an antiderivative,
  remember that you can check your result by differentiating. For example, you can
  check that the antiderivative

  $$\int (3x^3 - 4x)\, dx = \frac{3}{4}x^4 - 2x^2 + C$$

  is correct by differentiating to obtain

  $$\frac{d}{dx}\left[\frac{3}{4}x^4 - 2x^2 + C\right] = 3x^3 - 4x.$$

  Because the derivative is equal to the original integrand, you know that the
  antiderivative is correct.

■ *Grouping Symbols and the Fundamental Theorem*   When using the
  Fundamental Theorem of Calculus to evaluate a definite integral, you can avoid
  sign errors by using grouping symbols. Here is an example.

  $$\int_1^3 (x^3 - 9x)\, dx = \left[\frac{x^4}{4} - \frac{9x^2}{2}\right]_1^3 = \left[\frac{3^4}{4} - \frac{9(3^2)}{2}\right] - \left[\frac{1^4}{4} - \frac{9(1^2)}{2}\right] = \frac{81}{4} - \frac{81}{2} - \frac{1}{4} + \frac{9}{2} = -16$$

# Chapter 5  Review Exercises

In Exercises 1–8, find the indefinite integral.

**1.** $\displaystyle\int 16\,dx$

**2.** $\displaystyle\int \tfrac{3}{5}x\,dx$

**3.** $\displaystyle\int (2x^2 + 5x)\,dx$

**4.** $\displaystyle\int (5 - 6x^2)\,dx$

**5.** $\displaystyle\int \frac{2}{3\sqrt[3]{x}}\,dx$

**6.** $\displaystyle\int 6x^2\sqrt{x}\,dx$

**7.** $\displaystyle\int \left(\sqrt[3]{x^4} + 3x\right)dx$

**8.** $\displaystyle\int \left(\frac{4}{\sqrt{x}} + \sqrt{x}\right)dx$

In Exercises 9–12, find the particular solution, $y = f(x)$, that satisfies the given conditions.

**9.** $f'(x) = 3x + 1,\ \ f(2) = 6$

**10.** $f'(x) = x^{-1/3} - 1,\ \ f(8) = 4$

**11.** $f''(x) = 2x^2,\ \ f'(3) = 10,\ \ f(3) = 6$

**12.** $f''(x) = \dfrac{6}{\sqrt{x}} + 3,\ \ f'(1) = 12,\ \ f(4) = 56$

**13.** *Vertical Motion*   An object is projected upward from the ground with an initial velocity of 80 feet per second.

(a) How long does it take the object to rise to its maximum height?

(b) What is the maximum height?

(c) When is the velocity of the object half of its initial velocity?

(d) What is the height of the object when its velocity is one-half the initial velocity?

**14.** *Forecasting*   The weekly revenue for a new product has been increasing. The rate of change of the revenue can be modeled by $dR/dt = 0.675t^{3/2}$ for $0 \le t \le 225$, where $t$ is the time (in weeks). When $t = 0$, $R = 0$.

(a) Find a model for the revenue function.

(b) When will the weekly revenue be $27,000?

In Exercises 15–22, find the indefinite integral.

**15.** $\displaystyle\int (1 + 5x)^2\,dx$

**16.** $\displaystyle\int (x - 6)^{4/3}\,dx$

**17.** $\displaystyle\int \frac{1}{\sqrt{5x - 1}}\,dx$

**18.** $\displaystyle\int \frac{4x}{\sqrt{1 - 3x^2}}\,dx$

**19.** $\displaystyle\int x(1 - 4x^2)\,dx$

**20.** $\displaystyle\int \frac{x^2}{(x^3 - 4)^2}\,dx$

**21.** $\displaystyle\int (x^4 - 2x)(2x^3 - 1)\,dx$

**22.** $\displaystyle\int \frac{\sqrt{x}}{(1 - x^{3/2})^3}\,dx$

**23.** *Production*   The output $P$ (in board feet) of a small sawmill changes according to the model $dP/dt = 2t(0.001t^2 + 0.5)^{1/4}$ for $0 \le t \le 40$, where $t$ is measured in hours. Find the number of board feet produced in (a) 6 hours and (b) 12 hours.

**24.** *Cost*   The marginal cost for a catering service to cater to $x$ people can be modeled by

$$\frac{dC}{dx} = \frac{5x}{\sqrt{x^2 + 1000}}.$$

When $x = 225$, the cost is $1136.06. Find the cost of catering to (a) 500 people and (b) 1000 people.

In Exercises 25–30, find the indefinite integral.

**25.** $\displaystyle\int 3e^{-3x}\,dx$

**26.** $\displaystyle\int (2t - 1)e^{t^2 - t}\,dt$

**27.** $\displaystyle\int (x - 1)e^{x^2 - 2x}\,dx$

**28.** $\displaystyle\int \frac{4}{6x - 1}\,dx$

**29.** $\displaystyle\int \frac{x^2}{1 - x^3}\,dx$

**30.** $\displaystyle\int \frac{x - 4}{x^2 - 8x}\,dx$

In Exercises 31 and 32, use a symbolic integration utility to find the indefinite integral.

**31.** $\displaystyle\int \frac{(\sqrt{x} + 1)^2}{\sqrt{x}}\,dx$

**32.** $\displaystyle\int \frac{e^{5x}}{5 + e^{5x}}\,dx$

In Exercises 33–38, find the area of the region.

**33.** $f(x) = 4 - 2x$

**34.** $f(x) = 4 - x^2$

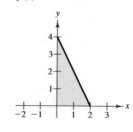

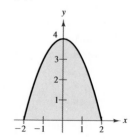

**35.** $f(y) = (y - 2)^2$

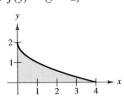

**36.** $f(x) = \sqrt{9 - x^2}$

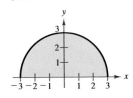

**37.** $f(x) = 2/(x + 1)$

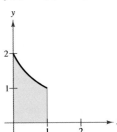

**38.** $f(x) = 2xe^{x^2 - 4}$

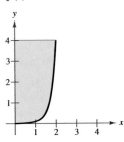

In Exercises 39–50, use the Fundamental Theorem of Calculus to evaluate the definite integral.

**39.** $\displaystyle\int_0^4 (2 + x)\, dx$

**40.** $\displaystyle\int_{-1}^1 (t^2 + 2)\, dt$

**41.** $\displaystyle\int_{-1}^1 (4t^3 - 2t)\, dt$

**42.** $\displaystyle\int_1^4 2x\sqrt{x}\, dx$

**43.** $\displaystyle\int_0^3 \frac{1}{\sqrt{1 + x}}\, dx$

**44.** $\displaystyle\int_3^6 \frac{x}{3\sqrt{x^2 - 8}}\, dx$

**45.** $\displaystyle\int_1^2 \left(\frac{1}{x^2} - \frac{1}{x^3}\right) dx$

**46.** $\displaystyle\int_0^1 x^2(x^3 + 1)^3\, dx$

**47.** $\displaystyle\int_1^3 \frac{(3 + \ln x)}{x}\, dx$

**48.** $\displaystyle\int_0^{\ln 5} e^{x/5}\, dx$

**49.** $\displaystyle\int_{-1}^1 3xe^{x^2 - 1}\, dx$

**50.** $\displaystyle\int_1^3 \frac{1}{x(\ln x + 2)^2}\, dx$

**51.** *Cost*  The marginal cost of serving a typical additional client at a law firm can be modeled by $dC/dx = 675 + 0.5x$, where $x$ is the number of clients. How does the cost $C$ change when $x$ increases from 50 to 51 clients?

**52.** *Profit*  The marginal profit obtained by selling $x$ dollars of automobile insurance can be modeled by

$$\frac{dP}{dx} = 0.4\left(1 - \frac{5000}{x}\right), \qquad x \geq 5000.$$

Find the change in the profit when $x$ increases from $75,000 to $100,000.

In Exercises 53–56, find the average value of the function on the closed interval. Then find all $x$-values in the interval for which the function is equal to its average value.

**53.** $f(x) = \dfrac{4}{\sqrt{x - 1}}, \quad [5, 10]$

**54.** $f(x) = \dfrac{20 \ln x}{x}, \quad [2, 10]$

**55.** $f(x) = e^{5 - x}, \quad [2, 5]$

**56.** $f(x) = x^3, \quad [0, 2]$

**57.** *Checking Account*  An interest-bearing checking account yields 4% interest compounded continuously. If you deposit $500 in such an account, and never write checks, what will the average value of the account be over a period of 2 years? Explain your reasoning.

**58.** *Fuel Cost*  Suppose that the price $p$ of gasoline can be modeled by

$$p = 1.00 + 0.1t + 0.02t^2,$$

where $t = 0$ represents January 1, 1990. Find the cost of gasoline for an automobile that is driven 15,000 miles per year and gets 33 miles per gallon from 1990 through 1994.

**59.** *Major League Baseball*  The average salary $S$ (in thousands of dollars) of a Major League Baseball player has been increasing. The rate of change of the salary can be modeled by

$$\frac{dS}{dt} = 145.3 - 79.8t + 13.8t^2 - 0.4t^3,$$

where $t$ is the year, with $t = 0$ representing 1980. *(Source: Major League Baseball Players Association)*

(a) Find the salary function in terms of the year if the average salary in 1980 was $144,000.

(b) If the average salary continues to increase at this rate, in what year would the average salary surpass $10,000,000?

**60.** *Respiratory Cycle*  The volume $V$ (in liters) of air in the lungs during a 5-second respiratory cycle is approximated by the model

$$V = 0.1729t + 0.1522t^2 - 0.0374t^3,$$

where $t$ is time in seconds.

(a) Use a graphing utility to sketch the graph of the equation on the interval $[0, 5]$.

(b) Determine the intervals on which the function is increasing and decreasing.

(c) Determine the maximum volume during the respiratory cycle.

(d) Determine the average volume of air in the lungs during one cycle.

(e) Briefly explain your results for parts (a) through (d).

In Exercises 61–64, explain how the given value can be used to evaluate the second integral.

**61.** $\int_0^2 6x^5 \, dx = 64, \quad \int_{-2}^2 6x^5 \, dx$

**62.** $\int_0^3 (x^4 + x^2) \, dx = 57.6, \quad \int_{-3}^3 (x^4 + x^2) \, dx$

**63.** $\int_1^2 \frac{4}{x^2} \, dx = 2, \quad \int_{-2}^{-1} \frac{4}{x^2} \, dx$

**64.** $\int_0^1 (x^3 - x) \, dx = -\frac{1}{4}, \quad \int_{-1}^0 (x^3 - x) \, dx$

In Exercises 65–70, sketch the region bounded by the graphs of the equations. Then find the area of the region.

**65.** $y = \frac{1}{x^2}, \ y = 4, \ x = 3$

**66.** $y = 1 - \frac{1}{2}x, \ y = x - 2, \ y = 1$

**67.** $y = \frac{4}{\sqrt{x+1}}, \ y = 0, \ x = 0, \ x = 8$

**68.** $y = \sqrt{x}(x - 1), \ y = 0$

**69.** $y = (x - 3)^2, \ y = 8 - (x - 3)^2$

**70.** $y = 4 - x, \ y = x^2 - 5x + 8, \ x = 0$

In Exercises 71 and 72, use a graphing utility to graph the region bounded by the graphs of the equations. Then find the area of the region.

**71.** $y = x, \ y = 2 - x^2$      **72.** $y = x, \ y = x^5$

*Consumer and Producer Surpluses*   In Exercises 73 and 74, find the consumer surplus and producer surplus for the demand and supply functions.

**73.** Supply function: $p_2(x) = 500 - x$

Demand function: $p_1(x) = 1.25x + 162.5$

**74.** Supply function: $p_2(x) = \sqrt{100,000 - 0.15x^2}$

Demand function: $p_1(x) = \sqrt{0.01x^2 + 36,000}$

**75.** *Beer Consumption*   The per capita adult consumption of beer $y$ (in gallons per year) in the United States from 1970 through 1981 can be modeled by

$$y = -0.015t^2 + 0.74t + 30.59, \quad 0 \le t \le 11,$$

where $t = 0$ represents 1970. The consumption from 1981 through 1994 can be modeled by

$$y = -0.06t^2 + 2.75t + 292/t - 12.78, \ 11 < t < 24.$$

If the consumption had continued to follow the first model from 1981 through 1994, how much more or less beer would have been consumed?   *(Source: U.S. Department of Agriculture)*

**76.** *Credit Card Sales*   The annual sales $s$ (in millions of dollars per year) for VISA, Mastercard, and American Express from 1993 through 1996 can be modeled by

$s = 25.18t + 81.79$      VISA

$s = 14.23t + 72.14$      Mastercard

$s = 1.49t + 19.77,$      American Express

where $3 \le t \le 6$ represents the 4-year period from 1993 through 1996.   *(Source: Credit Card News)*

(a) From 1993 through 1996, how much more were VISA's sales than Mastercard's sales?

(b) From 1993 through 1996, how much more were VISA's sales than American Express's sales?

**77.** *Ice Cream Sales*   The sales $S$ (in millions of dollars per year) for Ben and Jerry's Ice Cream from 1987 through 1992 can be modeled by

$$S = -509.32 + 173.92t - 19.06t^2 + 0.75t^3, \\ 7 \le t \le 12,$$

where $t = 7$ represents 1987. From 1992 through 1996, the sales can be modeled by

$$S = -1067.22 + 251.47t - 17.83t^2 + 0.43t^3, \\ 12 < t \le 16.$$

Use a graphing utility to graph both models for $7 \le t \le 16$. Using the graphs, predict whether the sales would have increased or decreased if the first model had continued to be valid from 1992 through 1996. Analytically determine how much more or less Ben and Jerry's sales would have been.   *(Source: Ben and Jerry's Ice Cream)*

**78.** *Psychology: Sleep Patterns*   The graph shows three areas, representing Awake time, REM sleep time, and non-REM sleep time, over a typical individual's lifetime. Make generalizations about the amount of total sleep, non-REM sleep, and REM sleep an individual gets as he or she gets older. If you wanted to estimate mathematically the amount of non-REM sleep an individual gets between birth and 50 years, how would you do so? How would you mathematically estimate the amount of REM sleep an individual gets during this interval?   *(Source: Adapted from Bernstein/Clarke-Stewart/Roy/Wickens, Psychology, Fourth Edition)*

Figure for 78

**Sleep Patterns**

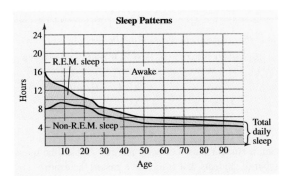

In Exercises 87–90, find the volume of the solid of revolution formed by revolving the region about the *x*-axis.

**87.** $y = 2x + 1$, $y = 1$

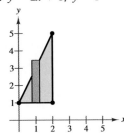

**88.** $y = \sqrt{x}$, $y = 2$

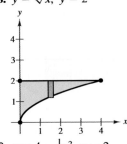

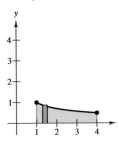

In Exercises 79–82, use the Midpoint Rule with $n = 4$ to approximate the definite integral. Then use a programmable calculator or computer with $n = 20$. Compare the two approximations.

**79.** $\int_0^2 (x^2 + 1)^2 \, dx$

**80.** $\int_{-1}^1 \sqrt{1 - x^2} \, dx$

**81.** $\int_0^1 \frac{1}{x^2 + 1} \, dx$

**82.** $\int_{-1}^1 e^{3-x^2} \, dx$

In Exercises 83–86, use the Disc Method to find the volume of the solid of revolution formed by revolving the region about the *x*-axis.

**83.** $y = \dfrac{1}{\sqrt{x}}$

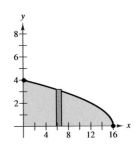

**84.** $y = \sqrt{16 - x}$

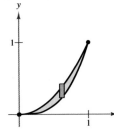

**89.** $y = x^2$, $y = x^3$

**90.** $y = 4 - \frac{1}{2}x^2$, $y = 2$

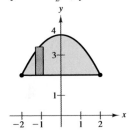

**91.** *Manufacturing*  A manufacturer drills a hole with radius 0.25 inch through the center of a metal sphere whose radius is 1 inch. What is the volume of the resulting ring?

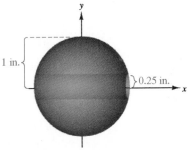

**92.** *Design*  To create a computer design for a funnel, an engineer revolves the region bounded by the lines $y = 3 - x$, $y = 0$, and $x = 0$ about the *y*-axis, where *x* and *y* are measured in feet. Find the volume of the funnel.

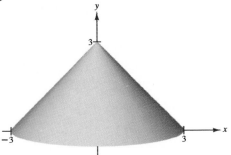

**85.** $y = e^{1-x}$

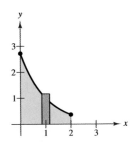

**86.** $y = \dfrac{1}{x}$

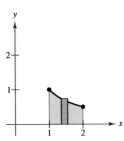

# Sample Post-Graduation Exam Questions

CPA   GMAT
GRE   CLAST
    Actuarial

*The following questions represent the types of questions that appear on certified public accountant (CPA) exams, graduate management admission tests (GMAT), graduate records exams (GRE), actuarial exams, and college-level academic skills tests (CLAST). The answers to the questions are given in the back of the book.*

For Questions 1–3, use the data given in the table.

| Number of students | Number of correct answers |
|---|---|
| 12 | 45 to 50 |
| 11 | 40 to 44 |
| 14 | 35 to 39 |
| 10 | 30 to 34 |
| 7 | 0 to 29 |

**1.** To pass a 50-question exam, a student must correctly answer 75% of the questions. What is the maximum number of students that could have passed the exam?

  (a) 23      (b) 47      (c) 30      (d) 37

**2.** What percent of the class answered 40 or more questions correctly?

  (a) 69%      (b) 43%      (c) 22%      (d) 31%

**3.** The number of students who answered 35 to 39 questions correctly is $y$ times the number who answered 29 or fewer correctly, where $y$ is

  (a) $\frac{1}{2}$      (b) 1      (c) 2      (d) $\frac{7}{5}$

**4.** For which of the following statements is $x = -2$ a solution?

  I. $x^2 + 4x + 4 \le 0$

  II. $|3x + 6| = 0$

  III. $x^2 + 7x + 10 > 0$

  (a) I, II, and III      (b) I and II      (c) II and III      (d) I and III

**Figure for 5**

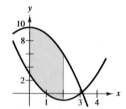

**5.** The area of the shaded region in the graph is $A$ square units, with $A$ equal to

  (a) $\frac{50}{3}$      (b) $\frac{25}{3}$      (c) 50      (d) $\frac{4}{3}$

For Questions 6–8, use the following data.

  56, 58, 54, 54, 59, 56, 55, 57, 56, 62

**6.** The mean of the data is

  (a) 55.5      (b) 56.7      (c) 56.5      (d) 56

**7.** The median of the data is

  (a) 55.5      (b) 56.7      (c) 56.5      (d) 56

**8.** The mode of the data is

  (a) 55.5      (b) 56.7      (c) 56.5      (d) 56

# Techniques of Integration 6

## STRATEGIES *for* SUCCESS

 **OBJECTIVES**

*When you have completed this chapter, make sure you are able to:*

❑ Find indefinite and definite integrals using integration by substitution.
❑ Evaluate integrals using integration by parts and apply integration to present value applications.
❑ Evaluate integrals using partial fractions and apply integration to the logistics growth model.
❑ Use tables of integrals to evaluate indefinite and definite integrals.
❑ Use the Trapezoidal Rule and Simpson's Rule to approximate definite integrals.
❑ Evaluate improper integrals with infinite limits of integration and with infinite integrands.

 **TOOLS**

*Use these study tools to achieve the objectives above:*

**Algebra Review**
(pages 442 and 443)

**Chapter Summary and Study Strategies**
(pages 444 and 445)

**Review Exercises**
(pages 446–449)

 **ADDITIONAL RESOURCES**

*Use these resources to solidify your mastery of calculus:*

**Student Solutions Guide**

**Study Guide** (Additional Examples, Similar Problems, and Chapter Test)

**Algebra Review Tutorial Software**

**Graphing Technology Guide**

**Sample Post-Graduation Exam Questions**
(page 450)

**Web Exercise**
(page 411, exercise 59)

Henryk T. Kaiser/The Picture Cube

*In Exercise 69 on page 401, you will determine the future value of an investment.*

<table>
<tr><td>

## 6.1

</td><td>

# Integration by Substitution

</td></tr>
</table>

*Review of Basic Integration Formulas • Integration by Substitution • Substitution and Definite Integrals • Applications*

## Review of Basic Integration Formulas

Each of the basic integration rules you studied in Chapter 5 was derived from a corresponding differentiation rule. It may surprise you to learn that, although you now have all the necessary tools for *differentiating* algebraic, exponential, and logarithmic functions, your set of tools for *integrating* these functions is by no means complete. The primary objective of this chapter is to develop several techniques that greatly expand the set of integrals to which the basic integration formulas can be applied.

### Basic Integration Formulas

1. Constant Rule:
$$\int k \, dx = kx + C$$

2. Simple Power Rule $(n \neq -1)$:
$$\int x^n \, dx = \frac{x^{n+1}}{n+1} + C$$

3. General Power Rule $(n \neq -1)$:
$$\int u^n \frac{du}{dx} \, dx = \int u^n \, du = \frac{u^{n+1}}{n+1} + C$$

4. Simple Exponential Rule:
$$\int e^x \, dx = e^x + C$$

5. General Exponential Rule:
$$\int e^u \frac{du}{dx} \, dx = \int e^u \, du = e^u + C$$

6. Simple Log Rule:
$$\int \frac{1}{x} \, dx = \ln|x| + C$$

7. General Log Rule:
$$\int \frac{du/dx}{u} \, dx = \int \frac{1}{u} \, du = \ln|u| + C$$

You don't have to work many integration problems before you realize that integration is not nearly as straightforward as differentiation. A major part of any integration problem is determining which basic integration formula (or formulas) to use to solve the problem. This requires remembering the basic formulas, familiarity with various procedures for rewriting integrands in the basic forms, and lots of practice.

# Integration by Substitution

There are several techniques for rewriting an integral so that it fits one or more of the basic formulas. One of the most powerful techniques is **integration by substitution.** With this technique, you choose part of the integrand to be $u$ and then rewrite the entire integral in terms of $u$.

## EXAMPLE 1   Integration by Substitution

Use the substitution $u = x + 1$ to find the indefinite integral

$$\int \frac{x}{(x + 1)^2}\, dx.$$

### Solution

From the substitution $u = x + 1$,

$$x = u - 1, \quad \frac{du}{dx} = 1, \quad \text{and} \quad dx = du.$$

Replacing *all* instances of $x$ and $dx$ with the appropriate $u$-variable forms produces the following.

$$\int \frac{x}{(x + 1)^2}\, dx = \int \frac{u - 1}{u^2}\, du \qquad \text{Substitute for } x \text{ and } dx.$$

$$= \int \left( \frac{u}{u^2} - \frac{1}{u^2} \right) du \qquad \text{Write as separate fractions.}$$

$$= \int \left( \frac{1}{u} - \frac{1}{u^2} \right) du \qquad \text{Simplify.}$$

$$= \ln |u| + \frac{1}{u} + C \qquad \text{Find antiderivative.}$$

$$= \ln |x + 1| + \frac{1}{x + 1} + C \qquad \text{Substitute for } u.$$

*STUDY TIP*    When you use integration by substitution, you need to realize that your integral should contain just one variable. For instance, the integrals

$$\int \frac{x}{(x + 1)^2}\, dx$$

and

$$\int \frac{u - 1}{u^2}\, du$$

are in the correct form, but the integral

$$\int \frac{x}{u^2}\, dx$$

is not.

The basic steps for integration by substitution are outlined in the following guidelines.

## Guidelines for Integration by Substitution

1. Let $u$ be a function of $x$ (usually part of the integrand).

2. Solve for $x$ and $dx$ in terms of $u$ and $du$.

3. Convert the entire integral to $u$-variable form and try to fit it to one or more of the basic integration formulas. If none fits, try a different substitution.

4. After integrating, rewrite the antiderivative as a function of $x$.

## EXAMPLE 2  Integration by Substitution

Find the indefinite integral $\int x\sqrt{x^2 - 1}\,dx$.

**Solution**

Consider the substitution $u = x^2 - 1$, which produces $du = 2x\,dx$. To create $2x\,dx$ as part of the integral, multiply and divide by 2.

$$\int x\sqrt{x^2 - 1}\,dx = \frac{1}{2}\int \overbrace{(x^2 - 1)^{1/2}}^{u^{1/2}}\overbrace{2x\,dx}^{du} \qquad \text{Multiply and divide by 2.}$$

$$= \frac{1}{2}\int u^{1/2}\,du \qquad \text{Substitute for } x \text{ and } dx.$$

$$= \frac{1}{2}\frac{u^{3/2}}{3/2} + C \qquad \text{Power Rule}$$

$$= \frac{1}{3}u^{3/2} + C \qquad \text{Simplify.}$$

$$= \frac{1}{3}(x^2 - 1)^{3/2} + C \qquad \text{Substitute for } u.$$

You can check this result by differentiating.

## EXAMPLE 3  Integration by Substitution

Find the indefinite integral $\int \dfrac{e^{3x}}{1 + e^{3x}}\,dx$.

**Solution**

Consider the substitution $u = 1 + e^{3x}$, which produces $du = 3e^{3x}\,dx$. To create $3e^{3x}\,dx$ as part of the integral, multiply and divide by 3.

$$\int \frac{e^{3x}}{1 + e^{3x}}\,dx = \frac{1}{3}\int \overbrace{\frac{1}{1 + e^{3x}}}^{1/u}\overbrace{3e^{3x}\,dx}^{du} \qquad \text{Multiply and divide by 3.}$$

$$= \frac{1}{3}\int \frac{1}{u}\,du \qquad \text{Substitute for } x \text{ and } dx.$$

$$= \frac{1}{3}\ln|u| + C \qquad \text{Log Rule}$$

$$= \frac{1}{3}\ln(1 + e^{3x}) + C \qquad \text{Substitute for } u.$$

Note that the absolute value is not necessary in the final answer because the quantity $(1 + e^{3x})$ is positive for all values of $x$.

# EXAMPLE 4  Integration by Substitution

Find the indefinite integral

$$\int x\sqrt{x-1}\,dx.$$

### Solution

Consider the substitution $u = x - 1$, which produces $du = dx$ and $x = u + 1$.

$$\int x\sqrt{x-1}\,dx = \int (u+1)(u^{1/2})\,du \qquad \text{Substitute for } x \text{ and } dx.$$

$$= \int (u^{3/2} + u^{1/2})\,du$$

$$= \frac{u^{5/2}}{5/2} + \frac{u^{3/2}}{3/2} + C \qquad \text{Power Rule}$$

$$= \frac{2}{5}(x-1)^{5/2} + \frac{2}{3}(x-1)^{3/2} + C \qquad \text{Substitute for } u.$$

This form of the antiderivative can be further simplified.

$$\frac{2}{5}(x-1)^{5/2} + \frac{2}{3}(x-1)^{3/2} + C = \frac{6}{15}(x-1)^{5/2} + \frac{10}{15}(x-1)^{3/2} + C$$

$$= \frac{2}{15}(x-1)^{3/2}[3(x-1) + 5] + C$$

$$= \frac{2}{15}(x-1)^{3/2}(3x+2) + C$$

You can check this answer by differentiating.

Example 4 demonstrates one of the characteristics of integration by substitution. That is, the form of the antiderivative as it exists immediately after resubstitution into $x$-variable form can often be simplified. Thus, when working the exercises in this section, don't assume that your answer is incorrect just because it doesn't look exactly like the answer given in the back of the text. You may be able to reconcile the two answers by algebraic simplification.

## Technology

If you have access to a symbolic integration utility, such as *Derive, Maple, Mathcad, Mathematica,* or the *TI-92,* try using it to solve several of the exercises in this section. For instance, the display at the right shows how *Derive* for Windows can be used to find the indefinite integral in Example 4.

#1: $x \cdot \sqrt{(x-1)}$ 　　　　Author.

#2: $\int x \cdot \sqrt{(x-1)}\,dx$ 　　　　Integrate.

#3: $\dfrac{2 \cdot (x-1)^{3/2} \cdot (3 \cdot x + 2)}{15}$ 　　　　Simplify.

## Substitution and Definite Integrals

The fourth step outlined in the guidelines for integration by substitution suggests that you convert back to the variable $x$. To evaluate *definite* integrals, however, it is often more convenient to determine the limits of integration for the variable $u$. This is often easier than converting back to the variable $x$ and evaluating the antiderivative at the original limits.

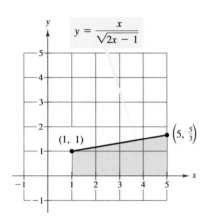

**FIGURE 6.1**   Region Before Substitution

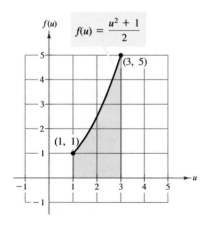

**FIGURE 6.2**   Region After Substitution

---

### EXAMPLE 5   Using Substitution with a Definite Integral

Evaluate the definite integral $\displaystyle\int_1^5 \frac{x}{\sqrt{2x - 1}} \, dx$.

#### Solution

Use the substitution $u = \sqrt{2x - 1}$, which implies that $u^2 = 2x - 1$, $x = \frac{1}{2}(u^2 + 1)$, and $dx = u \, du$. Before substituting, determine the new upper and lower limits of integration.

> *Lower limit:* When $x = 1$, $u = \sqrt{2(1) - 1} = 1$.
> *Upper limit:* When $x = 5$, $u = \sqrt{2(5) - 1} = 3$.

Now, substitute and integrate, as follows.

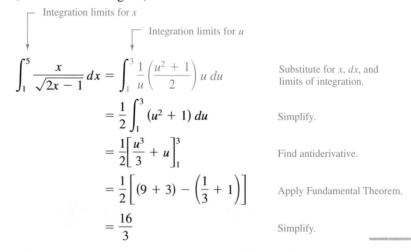

$$\int_1^5 \frac{x}{\sqrt{2x - 1}} \, dx = \int_1^3 \frac{1}{u}\left(\frac{u^2 + 1}{2}\right) u \, du \qquad \text{Substitute for } x, dx, \text{ and limits of integration.}$$

$$= \frac{1}{2}\int_1^3 (u^2 + 1) \, du \qquad \text{Simplify.}$$

$$= \frac{1}{2}\left[\frac{u^3}{3} + u\right]_1^3 \qquad \text{Find antiderivative.}$$

$$= \frac{1}{2}\left[(9 + 3) - \left(\frac{1}{3} + 1\right)\right] \qquad \text{Apply Fundamental Theorem.}$$

$$= \frac{16}{3} \qquad \text{Simplify.}$$

**STUDY TIP**   In Example 5, you can interpret the equation

$$\int_1^5 \frac{x}{\sqrt{2x - 1}} \, dx = \int_1^3 \frac{1}{u}\left(\frac{u^2 + 1}{2}\right) u \, du$$

graphically to mean that the two different regions shown in Figures 6.1 and 6.2 have the same area.

## Applications

Integration can be used to find the probability that an event will occur. In such an application, the real-life situation is modeled by a *probability density function f*, and the probability that *x* will lie between *a* and *b* is represented by

$$P(a \le x \le b) = \int_a^b f(x)\, dx.$$

The probability $P(a \le x \le b)$ must be a number between 0 and 1.

*Psychologists use definite integrals to represent the probability that an event will occur. For instance, a probability of 0.5 means that an event will occur about 50% of the time.*

### EXAMPLE 6  Finding a Probability

A psychologist finds that the probability that a participant in a memory experiment will recall between *a* and *b* percent (in decimal form) of the material is

$$P(a \le x \le b) = \int_a^b \frac{28}{9} x \sqrt[3]{1 - x}\, dx, \quad 0 \le a \le b \le 1.$$

Find the probability that a randomly chosen participant will recall between 0% and 87.5% of the material.

**Solution**

Let $u = \sqrt[3]{1 - x}$. Then $u^3 = 1 - x$, $x = 1 - u^3$, and $dx = -3u^2\, du$.

    *Lower limit:* When $x = 0$, $u = \sqrt[3]{1 - 0} = 1$.
    *Upper limit:* When $x = 0.875$, $u = \sqrt[3]{1 - 0.875} = 0.5$.

To find the probability, substitute and integrate, as follows.

$$\int_0^{0.875} \frac{28}{9} x \sqrt[3]{1 - x}\, dx = \int_1^{1/2} \left[ \frac{28}{9} (1 - u^3)(u)(-3u^2) \right] du$$

$$= \left( \frac{28}{3} \right) \int_1^{1/2} (u^6 - u^3)\, du$$

$$= \frac{28}{3} \left[ \frac{u^7}{7} - \frac{u^4}{4} \right]_1^{1/2}$$

$$\approx 0.865$$

Thus, the probability is about 86.5%, as indicated in Figure 6.3.

**TIP**

*ALGEBRA*

For help on the algebra in Example 6, see Example 1a in the *Chapter 6 Algebra Review,* on page 442.

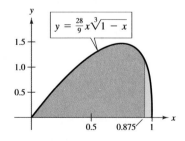

**FIGURE 6.3**

*Group Discussion*       **Extending the Example**

In Example 6, explain how you could find a value of *b* such that $P(0 \le x \le b) = 0.5$.

*Warm Up*

The following warm-up exercises involve skills that were covered in earlier sections. You will use these skills in the exercise set for this section.

In Exercises 1–6, evaluate the indefinite integral.

**1.** $\int 5 \, dx$

**2.** $\int x^{3/2} \, dx$

**3.** $\int 2x(x^2 + 1)^3 \, dx$

**4.** $\int 6e^{6x} \, dx$

**5.** $\int \frac{2}{2x + 1} \, dx$

**6.** $\int 2xe^{-x^2} \, dx$

In Exercises 7–10, simplify the expression.

**7.** $2x(x - 1)^2 + x(x - 1)$

**8.** $6x(x + 4)^3 - 3x^2(x + 4)^2$

**9.** $3(x + 7)^{1/2} - 2x(x + 7)^{-1/2}$

**10.** $(x + 5)^{1/3} - 5(x + 5)^{-2/3}$

## EXERCISES 6.1

In Exercises 1–28, evaluate the indefinite integral.

**1.** $\int (x - 2)^4 \, dx$

**2.** $\int (x + 5)^{3/2} \, dx$

**3.** $\int \frac{2}{(t - 9)^2} \, dt$

**4.** $\int \frac{2t - 1}{t^2 - t + 2} \, dt$

**5.** $\int \sqrt{1 + x} \, dx$

**6.** $\int (3 + x)^5 \, dx$

**7.** $\int \frac{12x + 2}{3x^2 + x} \, dx$

**8.** $\int \frac{6x^2 + 2}{x^3 + x} \, dx$

**9.** $\int \frac{1}{(5x + 1)^3} \, dx$

**10.** $\int \frac{1}{(3x + 1)^2} \, dx$

**11.** $\int \frac{1}{\sqrt{x + 1}} \, dx$

**12.** $\int \frac{1}{\sqrt{5x + 1}} \, dx$

**13.** $\int \frac{e^{3x}}{1 - e^{3x}} \, dx$

**14.** $\int 2xe^{3x^2} \, dx$

**15.** $\int \frac{x^2}{x - 1} \, dx$

**16.** $\int \frac{2x}{x - 4} \, dx$

**17.** $\int x\sqrt{x^2 + 4} \, dx$

**18.** $\int \frac{t}{\sqrt{1 - t^2}} \, dt$

**19.** $\int e^{5x} \, dx$

**20.** $\int \frac{e^x}{1 + e^x} \, dx$

**21.** $\int \frac{x}{(x + 1)^4} \, dx$

**22.** $\int \frac{x^2}{(x + 1)^3} \, dx$

**23.** $\int \frac{x}{(3x - 1)^2} \, dx$

**24.** $\int \frac{5x}{(x - 4)^3} \, dx$

**25.** $\int \frac{1}{\sqrt{t - 1}} \, dt$

**26.** $\int \frac{1}{\sqrt{x} + 1} \, dx$

**27.** $\int \frac{2\sqrt{t} + 1}{t} \, dt$

**28.** $\int \frac{6x + \sqrt{2x}}{x} \, dx$

In Exercises 29–36, evaluate the definite integral.

**29.** $\int_0^4 \sqrt{2x + 1} \, dx$

**30.** $\int_2^4 \sqrt{4x + 1} \, dx$

**31.** $\int_0^1 3xe^{x^2} \, dx$

**32.** $\int_0^2 e^{-2x} \, dx$

**33.** $\int_0^4 \frac{x}{(x + 4)^2} \, dx$

**34.** $\int_0^1 x(x + 5)^4 \, dx$

**35.** $\int_0^{0.5} x(1 - x)^3 \, dx$

**36.** $\int_0^{0.5} x^2(1 - x)^3 \, dx$

In Exercises 37–44, find the area of the region bounded by the graphs of the equations. Then use a graphing utility to graph the region and verify your answer.

**37.** $y = x\sqrt{x - 3}$, $y = 0$, $x = 7$

**38.** $y = x\sqrt{2x + 1}$, $y = 0$, $x = 4$

**39.** $y = x^2\sqrt{1 - x}$, $y = 0$, $x = -3$

**40.** $y = x^2\sqrt{x + 2}$ $y = 0$, $x = 7$

**41.** $y = \dfrac{x^2 - 1}{\sqrt{2x - 1}}$, $y = 0$, $x = 1$, $x = 5$

**42.** $y = \dfrac{2x - 1}{\sqrt{x + 3}}$, $y = 0$, $x = \dfrac{1}{2}$, $x = 6$

**43.** $y = x\sqrt[3]{x + 1}$, $y = 0$, $x = 0$, $x = 7$

**44.** $y = x\sqrt[3]{x - 2}$, $y = 0$, $x = 2$, $x = 10$

In Exercises 45–48, find the area of the region bounded by the graphs of the equations.

**45.** $y = -x\sqrt{x + 2}$, $y = 0$

**46.** $y = x\sqrt[3]{1 - x}$, $y = 0$

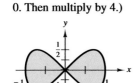

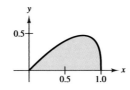

**47.** $y^2 = x^2(1 - x^2)$

(*Hint:* Find the area of the region bounded by $y = x\sqrt{1 - x^2}$ and $y = 0$. Then multiply by 4.)

**48.** $y = 1/(1 + \sqrt{x})$, $y = 0$, $x = 0$, $x = 4$

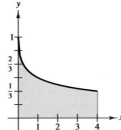

In Exercises 49 and 50, find the volume of the solid generated by revolving the region bounded by the graphs of the equations about the $x$-axis.

**49.** $y = x\sqrt{1 - x^2}$

**50.** $y = \sqrt{x}(1 - x)^2$, $y = 0$

In Exercises 51 and 52, find the average amount by which the function $f$ exceeds the function $g$ on the interval.

**51.** $f(x) = \dfrac{1}{x + 1}$, $g(x) = \dfrac{x}{(x + 1)^2}$, $[0, 1]$

**52.** $f(x) = x\sqrt{4x + 1}$, $g(x) = 2\sqrt{x^3}$, $[0, 2]$

**53.** *Probability* The probability of recall in an experiment is modeled by

$$P(a \leq x \leq b) = \int_a^b \frac{15}{4}x\sqrt{1 - x}\,dx,$$

where $x$ is the percent of recall (see figure).

(a) What is the probability of recalling between 40% and 80%?

(b) What is the median percent recall? That is, for what value of $b$ is $P(0 \leq x \leq b) = 0.5$?

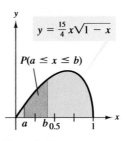

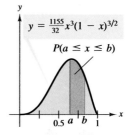

Figure for 53          Figure for 54

**54.** *Probability* The probability of finding between $a$ and $b$ percent of iron in ore samples is modeled by

$$P(a \leq x \leq b) = \int_a^b \frac{1155}{32}x^3(1 - x)^{3/2}\,dx$$

(see figure). Find the probability that a sample will contain between (a) 0% and 25% and (b) 50% and 100% of iron.

**55.** *Weather* During a 2-week period in March in a small town near Lake Erie, the measurable snowfall (in inches) on the ground can be modeled by

$$S(t) = t\sqrt{14 - t}, \quad 0 \leq t \leq 14.$$

Use a graphing utility to graph the equation. Then find the average amount of snow on the ground during the 2-week period.

**56.** *Revenue* A company sells a seasonal product that has a daily revenue modeled by

$$R = 0.06t^2(365 - t)^{1/2} + 1250, \quad 0 \leq t \leq 365.$$

Find the average daily revenue over the period of 1 year.

**57.** *Revenue* Describe a product whose seasonal sales pattern resembles the model in Exercise 56. Explain your reasoning.

In Exercises 58 and 59, use a program similar to the Midpoint Rule program on page 362 with $n = 10$ to approximate the area of the region bounded by the graph(s) of the equation(s).

**58.** $y = \sqrt[3]{x}\sqrt{4 - x}$, $y = 0$          **59.** $y^2 = x^2(1 - x^2)$

<table>
<tr><td>

**6.2**

</td><td>

# Integration by Parts and Present Value

*Integration by Parts • Present Value*

</td></tr>
</table>

## Integration by Parts

In this section, you will study an integration technique called **integration by parts.** This technique is particularly useful for integrands involving the products of algebraic and exponential or logarithmic functions, such as $\int x^2 e^x \, dx$ and $\int x \ln x \, dx$. Integration by parts is based on the Product Rule.

$$\frac{d}{dx}[uv] = u\frac{dv}{dx} + v\frac{du}{dx} \qquad\qquad \text{Product Rule}$$

$$uv = \int u\frac{dv}{dx}\,dx + \int v\frac{du}{dx}\,dx \qquad\qquad \text{Integrate both sides.}$$

$$uv = \int u \, dv + \int v \, du \qquad\qquad \text{Write in differential form.}$$

$$\int u \, dv = uv - \int v \, du \qquad\qquad \text{Rewrite.}$$

---

### Integration by Parts

Let $u$ and $v$ be differentiable functions of $x$.

$$\int u \, dv = uv - \int v \, du$$

---

Note that the formula for integration by parts expresses the original integral in terms of another integral. Depending on the choices for $u$ and $dv$, it may be easier to evaluate the second integral than the original one.

**STUDY TIP**   When using integration by parts, note that you can first choose $dv$ or first choose $u$. After you do, however, the choice of the other factor is determined— it must be the remaining portion of the integrand. Also note that $dv$ *must* contain the differential $dx$ of the original integral.

---

### Guidelines for Integration by Parts

1. Let $dv$ be the most complicated portion of the integrand that fits a basic integration formula. Let $u$ be the remaining factor.

2. Let $u$ be the portion of the integrand whose derivative is a simpler function than $u$ itself. Let $dv$ be the remaining factor.

---

## EXAMPLE 1   Integration by Parts

Find the indefinite integral $\int xe^x \, dx$.

### Solution

To apply integration by parts, you must rewrite the original integral in the form $\int u \, dv$. That is, you must break $xe^x \, dx$ into two factors—one "part" representing $u$ and the other "part" representing $dv$. There are several ways to do this.

$$\int \underbrace{(x)}_{u}\underbrace{(e^x \, dx)}_{dv} \qquad \int \underbrace{(e^x)}_{u}\underbrace{(x \, dx)}_{dv} \qquad \int \underbrace{(1)}_{u}\underbrace{(xe^x \, dx)}_{dv} \qquad \int \underbrace{(xe^x)}_{u}\underbrace{(dx)}_{dv}$$

Following the guidelines, you should choose the first option because $dv = e^x \, dx$ is the most complicated portion of the integrand that fits a basic integration formula *and* because the derivative of $u = x$ is simpler than $x$.

$$dv = e^x \, dx \quad \Longrightarrow \quad v = \int dv = \int e^x \, dx = e^x$$

$$u = x \quad \Longrightarrow \quad du = dx$$

With these substitutions, you can apply the integration by parts formula as follows.

$$\int xe^x \, dx = xe^x - \int e^x \, dx \qquad \text{\small $\int u \, dv = uv - \int v \, du$}$$

$$= xe^x - e^x + C \qquad \text{\small Integrate $\int e^x \, dx$.}$$

**STUDY TIP**   In Example 1, notice that you do not need to include a constant of integration when solving $v = \int e^x \, dx = e^x$. To see why this is true, try replacing $e^x$ by $e^x + C_1$ in the solution.

$$\int xe^x \, dx = x(e^x + C_1) - \int (e^x + C_1) \, dx$$

After integrating, you can see that the terms involving $C_1$ cancel.

| #1: $x \cdot \hat{e}^x$ | Author. |
|---|---|
| #2: $\int x \cdot \hat{e}^x \, dx$ | Integrate. |
| #3: $\hat{e}^x \cdot (x - 1)$ | Simplify. |

### Technology

If you have access to a symbolic integration utility, such as *Derive*, *Maple*, *Mathcad*, *Mathematica*, or the *TI-92*, try using it to solve several of the exercises in this section. For instance, the display at the left shows how *Derive* for Windows can be used to find the indefinite integral in Example 1. Note that the form of the integral is slightly different from that obtained in Example 1.

*STUDY TIP*   To remember the integration by parts formula, you might like to use the following "Z" pattern. The top row represents the original integral, the diagonal row represents $uv$, and the bottom row represents the new integral.

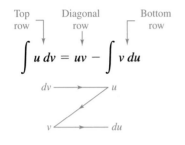

Top row     Diagonal row     Bottom row

$$\int u\, dv = uv - \int v\, du$$

$dv$ — $u$

$v$ — $du$

## EXAMPLE 2   Integration by Parts

Find the indefinite integral $\int x^2 \ln x\, dx$.

### Solution

For this integral, $x^2$ is more easily integrated than $\ln x$. Furthermore, the derivative of $\ln x$ is simpler than $\ln x$. Therefore, you should choose $dv = x^2\, dx$.

$$dv = x^2\, dx \quad\Longrightarrow\quad v = \int dv = \int x^2\, dx = \frac{x^3}{3}$$

$$u = \ln x \quad\Longrightarrow\quad du = \frac{1}{x}\, dx$$

Using these substitutions, apply the integration by parts formula as follows.

$$\int x^2 \ln x\, dx = \frac{x^3}{3} \ln x - \int \left(\frac{x^3}{3}\right)\!\left(\frac{1}{x}\right) dx \qquad \int u\, dv = uv - \int v\, du$$

$$= \frac{x^3}{3} \ln x - \frac{1}{3}\int x^2\, dx \qquad\qquad \text{Simplify.}$$

$$= \frac{x^3}{3} \ln x - \frac{x^3}{9} + C \qquad\qquad \text{Integrate.}$$

## EXAMPLE 3   Integrating by Parts with a Single Factor

Find the indefinite integral $\int \ln x\, dx$.

### Solution

This integral is unusual because it has only one factor. In such cases, you should choose $dv = dx$ and choose $u$ to be the single factor.

$$dv = dx \quad\Longrightarrow\quad v = \int dv = \int dx = x$$

$$u = \ln x \quad\Longrightarrow\quad du = \frac{1}{x}\, dx$$

Using these substitutions, apply the integration by parts formula as follows.

$$\int \ln x\, dx = x \ln x - \int \left(\frac{1}{x}\right)(x)\, dx \qquad \int u\, dv = uv - \int v\, du$$

$$= x \ln x - \int dx \qquad\qquad \text{Simplify.}$$

$$= x \ln x - x + C \qquad\qquad \text{Integrate.}$$

## EXAMPLE 4   Using Integration by Parts Repeatedly

Find the indefinite integral $\int x^2 e^x \, dx$.

### Solution

Using the guidelines, notice that the derivative of $x^2$ becomes simpler, whereas the derivative of $e^x$ does not. Therefore, you should let $u = x^2$ and let $dv = e^x \, dx$.

$$dv = e^x \, dx \implies v = \int dv = \int e^x \, dx = e^x$$

$$u = x^2 \implies du = 2x \, dx$$

Using these substitutions, apply the integration by parts formula as follows.

$$\int x^2 e^x \, dx = x^2 e^x - \int 2xe^x \, dx \qquad \text{First application of integration by parts}$$

To evaluate the new integral on the right, apply integration by parts a second time, using the following substitutions.

$$dv = e^x \, dx \implies v = \int dv = \int e^x \, dx = e^x$$

$$u = 2x \implies du = 2 \, dx$$

Using these substitutions, apply the integration by parts formula as follows.

$$\int x^2 e^x \, dx = x^2 e^x - \int 2xe^x \, dx \qquad \text{First application of integration by parts}$$

$$= x^2 e^x - \left( 2xe^x - \int 2e^x \, dx \right) \qquad \text{Second application of integration by parts}$$

$$= x^2 e^x - 2xe^x + 2e^x + C \qquad \text{Integrate.}$$

$$= e^x(x^2 - 2x + 2) + C \qquad \text{Simplify.}$$

You can confirm this result by differentiating.

> **STUDY TIP**   Remember that you can check an indefinite integral by differentiating. For instance, in Example 4, try differentiating the antiderivative
>
> $$e^x(x^2 - 2x + 2) + C$$
>
> to check that you obtain the original integrand, $x^2 e^x$.

When making repeated applications of integration by parts, be careful not to interchange the substitutions in successive applications. For instance, in Example 4, the first substitutions were $dv = e^x \, dx$ and $u = x^2$. If in the second application you had switched to $dv = 2x \, dx$ and $u = e^x$, you would have reversed the previous integration and returned to the *original* integral.

$$\int x^2 e^x \, dx = x^2 e^x - \left( x^2 e^x - \int x^2 e^x \, dx \right)$$

$$= \int x^2 e^x \, dx$$

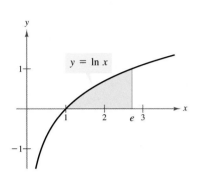

**FIGURE 6.4**

## EXAMPLE 5 Evaluating a Definite Integral

Evaluate the definite integral

$$\int_{1}^{e} \ln x \, dx.$$

### Solution

Integration by parts was used to find the antiderivative of $\ln x$ in Example 3. Using this result, you can evaluate the definite integral as follows.

$$\int_{1}^{e} \ln x \, dx = \left[ x \ln x - x \right]_{1}^{e} \qquad \text{Use result of Example 3.}$$

$$= (e \ln e - e) - (1 \ln 1 - 1) \qquad \text{Apply Fundamental Theorem.}$$

$$= (e - e) - (0 - 1)$$

$$= 1 \qquad \text{Simplify.}$$

The area represented by this definite integral is shown in Figure 6.4.

Before starting the exercises in this section, remember that it is not enough to know *how* to use the various integration techniques. You also must know *when* to use them. Integration is first and foremost a problem of recognition—recognizing which formula or technique to apply to obtain an antiderivative. Often, a slight alteration of an integrand will necessitate the use of a different integration technique. Here are some examples.

| *Integral* | *Technique* | *Antiderivative* |
|---|---|---|
| $\displaystyle\int x \ln x \, dx$ | Integration by parts | $\dfrac{x^2}{2} \ln x - \dfrac{x^2}{4} + C$ |
| $\displaystyle\int \dfrac{\ln x}{x} \, dx$ | Power Rule: $\displaystyle\int u^n \dfrac{du}{dx} \, dx$ | $\dfrac{(\ln x)^2}{2} + C$ |
| $\displaystyle\int \dfrac{1}{x \ln x} \, dx$ | Log Rule: $\displaystyle\int \dfrac{1}{u} \dfrac{du}{dx} \, dx$ | $\ln|\ln x| + C$ |

As you gain experience with integration by parts, your skill in determining $u$ and $dv$ will improve. The summary below gives suggestions for choosing $u$ and $dv$.

---

### Summary of Common Uses of Integration by Parts

1. $\displaystyle\int x^n e^{ax} \, dx$     Let $u = x^n$ and $dv = e^{ax} \, dx$. (Examples 1 and 4)

2. $\displaystyle\int x^n \ln x \, dx$     Let $u = \ln x$ and $dv = x^n \, dx$. (Examples 2 and 3)

## Present Value

The **present value** of a future payment is the amount that would have to be deposited today to produce the future payment. What is the present value of a future payment of $1000 one year from now? Because of inflation, $1000 today buys more than $1000 a year from now. The following definition considers only the effect of inflation.

---

### Present Value

If $c(t)$ represents a continuous income function in dollars per year and the annual rate of inflation is $r$, then the actual total income over $t_1$ years is

$$\text{Actual income over } t_1 \text{ years} = \int_0^{t_1} c(t)\, dt$$

and its **present value** is

$$\text{Present value} = \int_0^{t_1} c(t)e^{-rt}\, dt.$$

---

**TIP**

*ALGEBRA*

According to this definition, if the rate of inflation were 4%, then the present value of $1000 a year from now is just $980.26.

Ignoring inflation, the equation for present value also applies to an interest-bearing account where the annual interest rate $r$ is compounded continuously and $c(t)$ is an income function in dollars per year.

## EXAMPLE 6    Finding Present Value

You have just won a state lottery for $1,000,000. You will be paid $50,000 a year for 20 years. Assuming an annual inflation rate of 6%, what is the present value of this income?

### Solution

The income function for your winnings is $c(t) = 50,000$. Thus,

$$\text{Actual income} = \int_0^{20} 50,000\, dt = \Big[ 50,000t \Big]_0^{20} = \$1,000,000.$$

Because you do not receive this entire amount now, its present value is

$$\text{Present value} = \int_0^{20} 50,000e^{-0.06t}\, dt = \left[ \frac{50,000}{-0.06}e^{-0.06t} \right]_0^{20} \approx \$582,338.$$

This present value represents the amount that the state must deposit now to cover your payments over the next 20 years. This shows why state lotteries are so profitable—for the states!

Michael Dwyer/Stock Boston

*In 1994, 36 states had state lotteries, and they totaled about $29 billion in revenue. Of this, 57% was paid out in prizes and 6% was used to cover the state lotteries' administration costs. These costs include the cost of making the lottery tickets. Chris and Chuck Harter, who own American Games, Inc., sell lottery tickets and bingo cards to states and charitable organizations.*

## EXAMPLE 7 Finding Present Value

A company expects its income during the next 5 years to be given by

$$c(t) = 100{,}000t, \qquad 0 \le t \le 5.$$

Assuming an annual inflation rate of 10%, what is the present value of this income?

### Solution

The present value is

$$\text{Present value} = \int_0^5 100{,}000te^{-0.1t}\, dt = 100{,}000\int_0^5 te^{-0.1t}\, dt.$$

Using integration by parts, let $dv = e^{-0.1t}\, dt$ and obtain the following.

$$dv = e^{-0.1t}\, dt \quad \Longrightarrow \quad v = \int dv = \int e^{-0.1t}\, dt = -10e^{-0.1t}$$

$$u = t \quad \Longrightarrow \quad du = dt$$

This implies that

$$\int te^{-0.1t}\, dt = -10te^{-0.1t} + 10\int e^{-0.1t}\, dt$$

$$= -10te^{-0.1t} - 100e^{-0.1t}$$

$$= -10e^{-0.1t}(t + 10).$$

Therefore, the present value is

$$\text{Present value} = 100{,}000\int_0^5 te^{-0.1t}\, dt$$

$$= 100{,}000\left[ -10e^{-0.1t}(t + 10) \right]_0^5$$

$$\approx \$902{,}040.$$

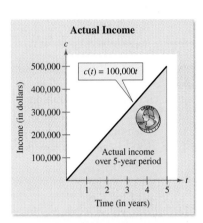

**Actual Income**

$c(t) = 100{,}000t$

Actual income over 5-year period

Income (in dollars) — Time (in years)

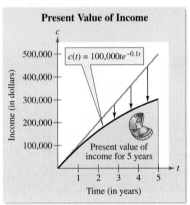

**Present Value of Income**

$c(t) = 100{,}000te^{-0.1t}$

Present value of income for 5 years

Income (in dollars) — Time (in years)

**FIGURE 6.5**

## Group Discussion

### Present Value

As shown in Figure 6.5, the actual income for the business in Example 7, over the next 5 years, is

$$\text{Actual income} = \int_0^5 100{,}000t\, dt = \$1{,}250{,}000.$$

If you were selling this company for its present value over a 5-year period, would it be to your advantage to argue that the inflation rate was lower than 10%? Explain your reasoning.

*Warm Up*

The following warm-up exercises involve skills that were covered in earlier sections. You will use these skills in the exercise set for this section.

In Exercises 1–6, find $f'(x)$.

**1.** $f(x) = 3x^4 + 2x - 1$

**2.** $f(x) = \frac{1}{4}x^2 + \frac{1}{8}x + 4$

**3.** $f(x) = \ln(x^4 + x^2)$

**4.** $f(x) = e^{x^3}$

**5.** $f(x) = x^2 e^x$

**6.** $f(x) = \dfrac{x + 1}{x - 3}$

In Exercises 7–10, find the area between the graphs of $f$ and $g$.

**7.** $f(x) = -x^2 + 4, \ g(x) = x^2 - 4$

**8.** $f(x) = -x^2 + 2, \ g(x) = 1$

**9.** $f(x) = 4x, \ g(x) = x^2$

**10.** $f(x) = x^3 - 3x^2 + 2, \ g(x) = x - 1$

## EXERCISES 6.2

In Exercises 1–6, use integration by parts to find the indefinite integral.

**1.** $\displaystyle\int xe^{3x}\,dx$

**2.** $\displaystyle\int xe^{-x}\,dx$

**3.** $\displaystyle\int x^2 e^{-x}\,dx$

**4.** $\displaystyle\int x^2 e^{2x}\,dx$

**5.** $\displaystyle\int \ln 2x\,dx$

**6.** $\displaystyle\int \ln x^2\,dx$

In Exercises 7–26, find the indefinite integral. (*Hint:* Integration by parts is not required for all the integrals.)

**7.** $\displaystyle\int e^{4x}\,dx$

**8.** $\displaystyle\int e^{-2x}\,dx$

**9.** $\displaystyle\int xe^{4x}\,dx$

**10.** $\displaystyle\int xe^{-2x}\,dx$

**11.** $\displaystyle\int xe^{x^2}\,dx$

**12.** $\displaystyle\int x^2 e^{x^3}\,dx$

**13.** $\displaystyle\int x^2 e^x\,dx$

**14.** $\displaystyle\int \dfrac{x}{e^x}\,dx$

**15.** $\displaystyle\int \dfrac{e^{1/t}}{t^2}\,dt$

**16.** $\displaystyle\int x^3 \ln x\,dx$

**17.** $\displaystyle\int t \ln(t + 1)\,dt$

**18.** $\displaystyle\int \dfrac{1}{x(\ln x)^3}\,dx$

**19.** $\displaystyle\int x(\ln x)^2\,dx$

**20.** $\displaystyle\int \ln 3x\,dx$

**21.** $\displaystyle\int \dfrac{(\ln x)^2}{x}\,dx$

**22.** $\displaystyle\int x\sqrt{x - 1}\,dx$

**23.** $\displaystyle\int x(x + 1)^2\,dx$

**24.** $\displaystyle\int \dfrac{x}{\sqrt{2 + 3x}}\,dx$

**25.** $\displaystyle\int \dfrac{xe^{2x}}{(2x + 1)^2}\,dx$

**26.** $\displaystyle\int \dfrac{x^3 e^{x^2}}{(x^2 + 1)^2}\,dx$

In Exercises 27–30, find the area of the region bounded by the graphs of the equations. Then use a graphing utility to graph the region and verify your answer.

**27.** $y = x^3 e^x, \ y = 0, \ x = 0, \ x = 2$

**28.** $y = (x^2 - 1)e^x, \ y = 0, \ x = -1, \ x = 1$

**29.** $y = x^2 \ln x, \ y = 0, \ x = 1, \ x = e$

**30.** $y = \dfrac{\ln x}{x^2}, \ y = 0, \ x = 1, \ x = e$

In Exercises 31–34, evaluate the definite integral.

**31.** $\displaystyle\int_0^1 x^2 e^x\,dx$

**32.** $\displaystyle\int_0^2 \dfrac{x^2}{e^x}\,dx$

**33.** $\displaystyle\int_1^e x^5 \ln x\,dx$

**34.** $\displaystyle\int_0^1 \ln(1 + 2x)\,dx$

In Exercises 35–38, find the indefinite integral using the specified method.

**35.** $\int 2x\sqrt{2x - 3}\, dx$

   (a) By parts, letting $dv = \sqrt{2x - 3}\, dx$
   (b) By substitution, letting $u = \sqrt{2x - 3}$

**36.** $\int x\sqrt{4 + x}\, dx$

   (a) By parts, letting $dv = \sqrt{4 + x}\, dx$
   (b) By substitution, letting $u = \sqrt{4 + x}$

**37.** $\int \dfrac{x}{\sqrt{4 + 5x}}\, dx$

   (a) By parts, letting $dv = \dfrac{1}{\sqrt{4 + 5x}}\, dx$
   (b) By substitution, letting $u = \sqrt{4 + 5x}$

**38.** $\int x\sqrt{4 - x}\, dx$

   (a) By parts, letting $dv = \sqrt{4 - x}\, dx$
   (b) By substitution, letting $u = \sqrt{4 - x}$

In Exercises 39 and 40, use integration by parts to verify the formula.

**39.** $\int x^n \ln x\, dx = \dfrac{x^{n+1}}{(n+1)^2}[-1 + (n+1)\ln x] + C,$

   $n \neq -1$

**40.** $\int x^n e^{ax}\, dx = \dfrac{x^n e^{ax}}{a} - \dfrac{n}{a}\int x^{n-1} e^{ax}\, dx$

In Exercises 41–44, use the results of Exercises 39 and 40 to find the indefinite integral.

**41.** $\int x^2 e^{5x}\, dx$

**42.** $\int x e^{-3x}\, dx$

**43.** $\int x^{-2} \ln x\, dx$

**44.** $\int x^{1/2} \ln x\, dx$

In Exercises 45 and 46, find the area of the region bounded by the graphs of the given equations.

**45.** $y = xe^{-x},\ y = 0,\ x = 4$

**46.** $y = \frac{1}{9}xe^{-x/3},\ y = 0,\ x = 0,\ x = 3$

**47.** Given the region bounded by the graphs of $y = 2\ln x$, $y = 0$, and $x = e$, find

   (a) the area of the region.
   (b) the volume of the solid generated by revolving the region about the x-axis.

**48.** Find the volume of the solid generated by revolving the region bounded by the graphs of $y = xe^x$, $y = 0$, $x = 0$, and $x = 1$ about the x-axis (see figure).

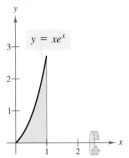

In Exercises 49–52, use a symbolic integration utility to evaluate the integral.

**49.** $\displaystyle\int_0^2 t^3 e^{-4t}\, dt$

**50.** $\displaystyle\int_1^4 \ln x(x^2 + 4)\, dx$

**51.** $\displaystyle\int_0^5 x^4(25 - x^2)^{3/2}\, dx$

**52.** $\displaystyle\int_1^e x^9 \ln x\, dx$

**53.** *Product Demand*  A manufacturing company forecasts that the demand $x$ (in units) for its product over the next 10 years can be modeled by

   $$x = 500(20 + te^{-0.1t}), \qquad 0 \le t \le 10,$$

   where $t$ is the time in years.

   (a) Use a graphing utility to decide whether the company is forecasting an increase or a decrease in demand over the decade.
   (b) According to the model, what is the total demand over the next 10 years?
   (c) Find the average annual demand during the 10-year period.

**54.** *Capital Campaign*  The board of trustees of a college is planning a 5-year capital gifts campaign to raise money for the college. The goal is to have an annual gift income $I$ that is modeled by

   $$I = 2000(375 + 68te^{-0.2t}), \qquad 0 \le t \le 5,$$

   where $t$ is the time in years.

   (a) Use a graphing utility to decide whether the board of trustees expects the gift income to increase or decrease over the 5-year period.
   (b) Find the expected total gift income over the 5-year period.
   (c) Determine the average annual gift income over the 5-year period. Compare the result with the income given when $t = 3$.

**55.** *Learning Theory*  A model for the ability $M$ of a child to memorize, measured on a scale from 0 to 10, is

$$M = 1 + 1.6t \ln t, \qquad 0 < t \le 4,$$

where $t$ is the child's age in years. Find the average value of this model between

(a) the child's first and second birthdays.
(b) the child's third and fourth birthdays.

**56.** *Revenue*  A company sells a seasonal product. The revenue from the product can be modeled by

$$R = 410.5t^2 e^{-t/30} + 25,000, \qquad 0 \le t \le 365,$$

where $t$ is the time in days.

(a) Find the average daily receipts during the first quarter, which is given by $0 \le t \le 91$.
(b) Find the average daily receipts during the fourth quarter, which is given by $274 \le t \le 365$.
(c) Find the total daily receipts during the year.

*Present Value*  In Exercises 57–62, find the present value of the income $c$ (measured in dollars) over $t_1$ years at the given annual inflation rate $r$.

**57.** $c = 5000$, $r = 5\%$, $t_1 = 4$ years

**58.** $c = 450$, $r = 4\%$, $t_1 = 10$ years

**59.** $c = 150,000 + 2500t$, $r = 4\%$, $t_1 = 10$ years

**60.** $c = 30,000 + 500t$, $r = 7\%$, $t_1 = 6$ years

**61.** $c = 1000 + 50e^{t/2}$, $r = 6\%$, $t_1 = 4$ years

**62.** $c = 5000 + 25te^{t/10}$, $r = 6\%$, $t_1 = 10$ years

**63.** *Present Value*  A company expects its income $c$ during the next 4 years to be modeled by

$$c = 150,000 + 75,000t.$$

(a) Find the actual income for the business over the 4 years.
(b) Assuming an annual inflation rate of 4%, what is the present value of this income?

**64.** *Contract Value*  A professional athlete signs a 3-year contract in which the earnings can be modeled by

$$c = 300,000 + 125,000t.$$

(a) Find the actual value of the athlete's contract.
(b) Assuming an annual inflation rate of 5%, what is the present value of the contract?

**65.** Use a program similar to the Midpoint Rule program on page 362 with $n = 10$ to approximate

$$\int_1^4 \frac{4}{\sqrt{x} + \sqrt[3]{x}} \, dx.$$

**66.** Use a program similar to the Midpoint Rule program on page 362 with $n = 12$ to approximate the volume of the solid generated by revolving the region bounded by the graphs of

$$y = \frac{10}{\sqrt{xe^x}}, \quad y = 0, \quad x = 1, \quad \text{and} \quad x = 4$$

about the $x$-axis.

*Future Value*  In Exercises 67 and 68, find the future value of the income (in dollars) given by $f(t)$ over $t_1$ years at the annual interest rate of $r$. If the function $f$ represents a continuous investment over a period of $t_1$ years at an annual interest rate of $r$ (compounded continuously), then the **future value** of the investment is given by

$$\text{Future value} = e^{rt_1} \int_0^{t_1} f(t) e^{-rt} \, dt.$$

**67.** $f(t) = 3000$, $r = 8\%$, $t_1 = 10$ years

**68.** $f(t) = 3000e^{0.05t}$, $r = 10\%$, $t_1 = 5$ years

**69.** *Finance: Future Value*  Use the equation from Exercises 67 and 68 to calculate the following.  *(Source: Adapted from Garman/Forgue, Personal Finance, Fifth Edition)*

(a) The future value of $1200 saved each year for 10 years earning 7 percent interest.
(b) A person who wishes to invest $1200 each year finds one choice expected to pay 9 percent interest per year and another riskier choice that may pay 10 percent interest per year. What is the difference in return (future value) if the investment is made for 15 years?

**70.** *College Expenses*  In 1997, the total cost to attend Notre Dame University for 1 year was estimated to be $23,660. If your grandparents had continuously invested in a college fund according to the model

$$f(t) = 300t$$

for 18 years, at an annual interest rate of 10%, would the fund have grown enough to allow you to cover 4 years of expenses at Notre Dame?

# Partial Fractions and Logistics Growth

*Partial Fractions  •  Logistics Growth Function*

## Partial Fractions

In Sections 6.1 and 6.2, you studied integration by substitution and by parts. In this section you will study a third technique called **partial fractions.** This technique involves the decomposition of a rational function into the sum of two or more simple rational functions. For instance, suppose you know that

$$\frac{x + 7}{x^2 - x - 6} = \frac{2}{x - 3} - \frac{1}{x + 2}.$$

Knowing the "partial fractions" on the right side would allow you to integrate the left side as follows.

$$\int \frac{x + 7}{x^2 - x - 6}\, dx = \int \left( \frac{2}{x - 3} - \frac{1}{x + 2} \right) dx$$

$$= 2 \int \frac{1}{x - 3}\, dx - \int \frac{1}{x + 2}\, dx$$

$$= 2 \ln|x - 3| - \ln|x + 2| + C$$

To use this method, you must be able to factor the denominator of the original rational function *and* find the partial fraction decomposition of the function.

*STUDY TIP*    A rational function $p(x)/q(x)$ is *proper* if the degree of the numerator is less than the degree of the denominator.

### Partial Fractions

To find the partial fraction decomposition of the *proper* rational function $p(x)/q(x)$, factor $q(x)$ and write an equation that has the form

$$\frac{p(x)}{q(x)} = \text{(sum of partial fractions)}.$$

For each *distinct* linear factor $(ax + b)$, the right side should include a term of the form

$$\frac{A}{ax + b}.$$

For each *repeated* linear factor $(ax + b)^n$, the right side should include $n$ terms of the form

$$\frac{A_1}{ax + b} + \frac{A_2}{(ax + b)^2} + \cdots + \frac{A_n}{(ax + b)^n}.$$

## EXAMPLE 1   Finding a Partial Fraction Decomposition

Write the partial fraction decomposition for

$$\frac{x + 7}{x^2 - x - 6}.$$

### Solution

Begin by factoring the denominator as $x^2 - x - 6 = (x - 3)(x + 2)$. Then, write the partial fraction decomposition as

$$\frac{x + 7}{x^2 - x - 6} = \frac{A}{x - 3} + \frac{B}{x + 2}.$$

To solve this equation for $A$ and $B$, multiply both sides of the equation by the least common denominator $(x - 3)(x + 2)$. This produces the following **basic equation.**

$$x + 7 = A(x + 2) + B(x - 3) \qquad \text{Basic equation}$$

Because this equation is true for all $x$, you can substitute any convenient values of $x$ into the equation. The $x$-values that are especially convenient are the ones that make a factor of the least common denominator zero: $x = -2$ and $x = 3$.

*Substitute $x = -2$:*

$$\begin{aligned}
x + 7 &= A(x + 2) + B(x - 3) && \text{Basic equation} \\
-2 + 7 &= A(-2 + 2) + B(-2 - 3) && \text{Substitute } -2 \text{ for } x. \\
5 &= A(0) + B(-5) && \text{Simplify.} \\
-1 &= B && \text{Solve for } B.
\end{aligned}$$

*Substitute $x = 3$:*

$$\begin{aligned}
x + 7 &= A(x + 2) + B(x - 3) && \text{Basic equation} \\
3 + 7 &= A(3 + 2) + B(3 - 3) && \text{Substitute 3 for } x. \\
10 &= A(5) + B(0) && \text{Simplify.} \\
2 &= A && \text{Solve for } A.
\end{aligned}$$

Now that you have solved the basic equation for $A$ and $B$, you can write the partial fraction decomposition as

$$\frac{x + 7}{x^2 - x - 6} = \frac{2}{x - 3} - \frac{1}{x + 2},$$

as indicated at the beginning of this section.

**TIP**

ALGEBRA

You can check the result in Example 1 by subtracting the partial fractions to obtain the original fraction, as shown in Example 1b in the *Chapter 6 Algebra Review,* on page 442.

---

**STUDY TIP**   Be sure you see that the substitutions for $x$ in Example 1 are chosen for their convenience in solving for $A$ and $B$. The value $x = -2$ is selected because it eliminates the term $A(x + 2)$, and the value $x = 3$ is chosen because it eliminates the term $B(x - 3)$.

## Technology

The use of partial fractions depends on the ability to factor the denominator. If this cannot be easily done, then partial fractions should not be used. For instance, consider the integral

$$\int \frac{5x^2 + 20x + 6}{x^3 + 2x^2 + x + 1} \, dx.$$

This integral is only slightly different from that in Example 2, yet it is immensely more difficult to solve. When we tried using a symbolic integration utility on this integral, it was unable to solve it. Of course, if the integral is a definite integral (as is true in many applied problems), then you can use an approximation technique such as the Midpoint Rule.

### TIP

You can check the partial fraction decomposition in Example 2 by combining the partial fractions to obtain the original fraction, as shown in Example 1c in the *Chapter 6 Algebra Review*, on page 442. Also, for help with the algebra used to simplify the answer, see Example 2c on page 443.

## EXAMPLE 2   Integrating with Repeated Factors

Find the indefinite integral $\int \dfrac{5x^2 + 20x + 6}{x^3 + 2x^2 + x} \, dx.$

**Solution**

Begin by factoring the denominator as $x(x + 1)^2$. Then, write the partial fraction decomposition as

$$\frac{5x^2 + 20x + 6}{x(x + 1)^2} = \frac{A}{x} + \frac{B}{x + 1} + \frac{C}{(x + 1)^2}.$$

To solve this equation for $A$, $B$, and $C$, multiply both sides of the equation by the least common denominator $x(x + 1)^2$.

$$5x^2 + 20x + 6 = A(x + 1)^2 + Bx(x + 1) + Cx \qquad \text{Basic equation}$$

Now, solve for $A$ and $C$ by substituting $x = -1$ and $x = 0$ into the basic equation.

*Substitute $x = -1$:*

$$5(-1)^2 + 20(-1) + 6 = A(-1 + 1)^2 + B(-1)(-1 + 1) + C(-1)$$
$$-9 = A(0) + B(0) - C$$
$$9 = C \qquad \text{Solve for } C.$$

*Substitute $x = 0$:*

$$5(0)^2 + 20(0) + 6 = A(0 + 1)^2 + B(0)(0 + 1) + C(0)$$
$$6 = A(1) + B(0) + C(0)$$
$$6 = A \qquad \text{Solve for } A.$$

At this point, you have exhausted the convenient choices for $x$ and have yet to solve for $B$. When this happens, you can use *any* other $x$-value along with the known values of $A$ and $C$.

*Substitute $x = 1$, $A = 6$, and $C = 9$:*

$$5(1)^2 + 20(1) + 6 = (6)(1 + 1)^2 + B(1)(1 + 1) + (9)(1)$$
$$31 = 6(4) + B(2) + 9(1)$$
$$-1 = B \qquad \text{Solve for } B.$$

Now that you have solved for $A$, $B$, and $C$, you can use the partial fraction decomposition to integrate.

$$\int \frac{5x^2 + 20x + 6}{x^3 + 2x^2 + x} \, dx = \int \left( \frac{6}{x} - \frac{1}{x + 1} + \frac{9}{(x + 1)^2} \right) dx$$

$$= 6 \ln|x| - \ln|x + 1| + 9\frac{(x + 1)^{-1}}{-1} + C$$

$$= \ln \left| \frac{x^6}{x + 1} \right| - \frac{9}{x + 1} + C$$

You can use the partial fraction decomposition technique outlined in Examples 1 and 2 only with a *proper* rational function—that is, a rational function whose numerator is of lower degree than its denominator. If the numerator is of equal or greater degree, you must divide first. For instance, the rational function

$$\frac{x^3}{x^2 + 1}$$

is improper because the degree of the numerator is greater than the degree of the denominator. Before applying partial fractions to this function, you should divide the denominator into the numerator to obtain

$$\frac{x^3}{x^2 + 1} = x - \frac{x}{x^2 + 1}.$$

## EXAMPLE 3 Integrating an Improper Rational Function

Find the indefinite integral $\displaystyle\int \frac{x^5 + x - 1}{x^4 - x^3}\, dx.$

### Solution

This rational function is improper—its numerator has a degree greater than that of its denominator. Thus, you should begin by dividing the denominator into the numerator to obtain

$$\frac{x^5 + x - 1}{x^4 - x^3} = x + 1 + \frac{x^3 + x - 1}{x^4 - x^3}.$$

Now, applying partial fraction decomposition produces

$$\frac{x^3 + x - 1}{x^3(x - 1)} = \frac{A}{x} + \frac{B}{x^2} + \frac{C}{x^3} + \frac{D}{x - 1}.$$

Multiplying both sides by the least common denominator $x^3(x - 1)$ produces the basic equation.

$$x^3 + x - 1 = Ax^2(x - 1) + Bx(x - 1) + C(x - 1) + Dx^3 \qquad \text{Basic equation}$$

Using techniques similar to those in the first two examples, you can solve for $A$, $B$, $C$, and $D$ to obtain

$$A = 0, \quad B = 0, \quad C = 1, \quad \text{and} \quad D = 1.$$

Thus, you can integrate as follows.

$$\begin{aligned}
\int \frac{x^5 + x - 1}{x^4 - x^3}\, dx &= \int \left( x + 1 + \frac{x^3 + x - 1}{x^4 - x^3} \right) dx \\
&= \int \left( x + 1 + \frac{1}{x^3} + \frac{1}{x - 1} \right) dx \\
&= \frac{x^2}{2} + x - \frac{1}{2x^2} + \ln|x - 1| + C
\end{aligned}$$

**TIP**

You can check the partial fraction decomposition in Example 3 by combining the partial fractions to obtain the original fraction, as shown in Example 2a in the *Chapter 6 Algebra Review*, on page 443.

ALGEBRA

# Logistics Growth Function

In Section 4.5, you saw that exponential growth occurs in situations for which the rate of growth is proportional to the quantity present at any given time. That is, if $y$ is the quantity at time $t$, then

$$\frac{dy}{dt} = ky \qquad \frac{dy}{dt} \text{ is proportional to } y.$$

$$y = Ce^{kt}. \qquad \text{Exponential growth function}$$

Exponential growth is unlimited. As long as $C$ and $k$ are positive, the value of $Ce^{kt}$ can be made arbitrarily large by choosing sufficiently large values of $t$.

In many real-life situations, however, the growth of a quantity is limited and cannot increase beyond a certain size $L$, as shown in Figure 6.6. The **logistics growth** model assumes that the rate of growth is proportional to both the quantity $y$ and the difference between the quantity and the limit $L$. That is,

$$\frac{dy}{dt} = ky(L - y). \qquad \frac{dy}{dt} \text{ is proportional to } y \text{ and } (L - y).$$

The solution of this *differential equation* is given in Example 4.

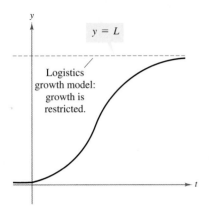

**FIGURE 6.6**

Logistics growth model: growth is restricted.

$y = L$

*STUDY TIP* The logistics growth model in Example 4 was simplified by assuming that the limit of the quantity $y$ is 1. If the limit had been $L$, then the solution would have been

$$y = \frac{L}{1 + be^{-kt}}.$$

In the fourth step of the solution, notice that partial fractions were used to integrate the left side of the equation.

---

## EXAMPLE 4   Deriving the Logistics Growth Function

Solve the equation $\dfrac{dy}{dt} = ky(1 - y)$. (Assume $y > 0$ and $1 - y > 0$.)

*Solution*

$$\frac{dy}{dt} = ky(1 - y) \qquad \text{Differential equation}$$

$$\frac{1}{y(1 - y)} \, dy = k \, dt \qquad \text{Write in differential form.}$$

$$\int \frac{1}{y(1 - y)} \, dy = \int k \, dt \qquad \text{Integrate both sides.}$$

$$\int \left( \frac{1}{y} + \frac{1}{1 - y} \right) dy = \int k \, dt \qquad \text{Rewrite using partial fractions.}$$

$$\ln y - \ln(1 - y) = kt + C_1 \qquad \text{Find antiderivative.}$$

$$\ln \frac{y}{1 - y} = kt + C_1 \qquad \text{Simplify.}$$

$$\frac{y}{1 - y} = Ce^{kt} \qquad \text{Exponentiate and let } e^{C_1} = C.$$

Solving this equation for $y$ produces

$$y = \frac{1}{1 + be^{-kt}}, \qquad \text{Logistics growth function}$$

where $b = 1/C$.

## EXAMPLE 5  Comparing Logistics Growth Functions

Use a graphing utility to investigate the effects of the values of $L$, $b$, and $k$ on the graph of

$$y = \frac{L}{1 + be^{-kt}}.$$   Logistics growth function $(L > 0, b > 0, k > 0)$

### Solution

The value of $L$ determines the horizontal asymptote of the graph to the right. In other words, as $t$ increases without bound, the graph approaches a limit of $L$.

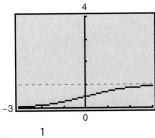
$$y = \frac{1}{1 + e^{-t}}$$

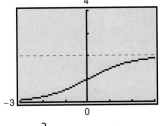
$$y = \frac{2}{1 + e^{-t}}$$

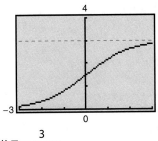
$$y = \frac{3}{1 + e^{-t}}$$

The value of $b$ determines the point of inflection of the graph. When $b = 1$, the point of inflection occurs when $t = 0$. If $b > 1$, the point of inflection is to the right of the $y$-axis. If $0 < b < 1$, the point of inflection is to the left of the $y$-axis.

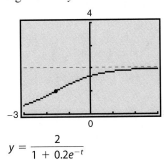
$$y = \frac{2}{1 + 0.2e^{-t}}$$

$$y = \frac{2}{1 + e^{-t}}$$

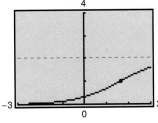
$$y = \frac{2}{1 + 5e^{-t}}$$

The value of $k$ determines the rate of growth of the graph. For fixed values of $b$ and $L$, larger values of $k$ correspond to more rapid rates of growth.

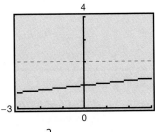
$$y = \frac{2}{1 + e^{-0.2t}}$$

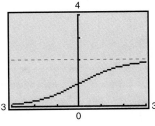
$$y = \frac{2}{1 + e^{-t}}$$

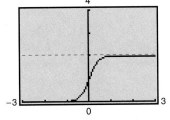
$$y = \frac{2}{1 + e^{-5t}}$$

## EXAMPLE 6 Modeling a Population

The state game commission releases 100 deer into a game refuge. During the first 5 years, the population increases to 432 deer. The commission believes that the population can be modeled by logistics growth with a limit of 2000 deer. Write the logistics growth model for this population. Then use the model to create a table showing the size of the deer population over the next 30 years.

### Solution

Let $y$ represent the number of deer in year $t$. Assuming a logistics growth model means that the rate of change in the population is proportional to both $y$ and $(2000 - y)$. That is,

$$\frac{dy}{dt} = ky(2000 - y), \quad 100 \le y \le 2000.$$

The solution of this equation is

$$y = \frac{2000}{1 + be^{-kt}}.$$

Using the fact that $y = 100$ when $t = 0$, you can solve for $b$.

$$100 = \frac{2000}{1 + be^{-k(0)}} \quad \Longrightarrow \quad b = 19$$

Then, using the fact that $y = 432$ when $t = 5$, you can solve for $k$.

$$432 = \frac{2000}{1 + 19e^{-k(5)}} \quad \Longrightarrow \quad k \approx 0.33106$$

Thus, the logistics growth model for the population is

$$y = \frac{2000}{1 + 19e^{-0.33106t}}. \qquad \text{Logistics growth model}$$

The population, in 5-year intervals, is shown in the table.

| Time, $t$ | 0 | 5 | 10 | 15 | 20 | 25 | 30 |
|---|---|---|---|---|---|---|---|
| Population, $y$ | 100 | 432 | 1181 | 1766 | 1951 | 1990 | 1998 |

PhotoDisc

*In the United States, the National Park System has more than 330 protected areas for wildlife, and the National Wildlife Refuge System includes more than 400 refuges. The deer shown in this photo is a white-tailed buck.*

## Group Discussion    *Logistics Growth*

Analyze the graph of the logistics growth function in Example 6. During which years is the *rate of growth* of the herd increasing? During which years is the *rate of growth* of the herd decreasing? How would these answers change if, instead of a limit of 2000 deer, the game refuge could contain a limit of 3000 deer?

## Warm Up

The following warm-up exercises involve skills that were covered in earlier sections. You will use these skills in the exercise set for this section.

In Exercises 1–6, factor the expression.

**1.** $x^2 - x - 12$

**2.** $x^2 - 25$

**3.** $x^3 - x^2 - 2x$

**4.** $x^3 - 4x^2 + 4x$

**5.** $x^3 - 4x^2 + 5x - 2$

**6.** $x^3 - 5x^2 + 7x - 3$

In Exercises 7–10, rewrite the improper rational function as the sum of a proper rational function and a polynomial.

**7.** $\dfrac{x^2 - 2x + 1}{x - 2}$

**8.** $\dfrac{2x^2 - 4x + 1}{x - 1}$

**9.** $\dfrac{x^3 - 3x^2 + 2}{x - 2}$

**10.** $\dfrac{x^3 + 4x^2 + 5x + 2}{x^2 - 1}$

# EXERCISES 6.3

In Exercises 1–12, write the expression as a sum of partial fractions.

**1.** $\dfrac{2(x + 20)}{x^2 - 25}$

**2.** $\dfrac{10x + 3}{x^2 + x}$

**3.** $\dfrac{8x + 3}{x^2 - 3x}$

**4.** $\dfrac{3x + 11}{x^2 - 2x - 3}$

**5.** $\dfrac{4x - 13}{x^2 - 3x - 10}$

**6.** $\dfrac{7x + 5}{6(2x^2 + 3x + 1)}$

**7.** $\dfrac{3x^2 - 2x - 5}{x^3 + x^2}$

**8.** $\dfrac{3x^2 - x + 1}{x(x + 1)^2}$

**9.** $\dfrac{x + 1}{3(x - 2)^2}$

**10.** $\dfrac{3x - 4}{(x - 5)^2}$

**11.** $\dfrac{8x^2 + 15x + 9}{(x + 1)^3}$

**12.** $\dfrac{6x^2 - 5x}{x^3 + 6x^2 + 12x + 8}$

In Exercises 13–32, find the indefinite integral.

**13.** $\displaystyle\int \dfrac{1}{x^2 - 1}\, dx$

**14.** $\displaystyle\int \dfrac{9}{x^2 - 9}\, dx$

**15.** $\displaystyle\int \dfrac{-2}{x^2 - 16}\, dx$

**16.** $\displaystyle\int \dfrac{-4}{x^2 - 4}\, dx$

**17.** $\displaystyle\int \dfrac{1}{3x^2 - x}\, dx$

**18.** $\displaystyle\int \dfrac{3}{x^2 - 3x}\, dx$

**19.** $\displaystyle\int \dfrac{1}{2x^2 + x}\, dx$

**20.** $\displaystyle\int \dfrac{5}{x^2 + x - 6}\, dx$

**21.** $\displaystyle\int \dfrac{3}{x^2 + x - 2}\, dx$

**22.** $\displaystyle\int \dfrac{1}{4x^2 - 9}\, dx$

**23.** $\displaystyle\int \dfrac{5 - x}{2x^2 + x - 1}\, dx$

**24.** $\displaystyle\int \dfrac{x + 1}{x^2 + 4x + 3}\, dx$

**25.** $\displaystyle\int \dfrac{x^2 + 12x + 12}{x^3 - 4x}\, dx$

**26.** $\displaystyle\int \dfrac{3x^2 - 7x - 2}{x^3 - x}\, dx$

**27.** $\displaystyle\int \dfrac{x + 2}{x^2 - 4x}\, dx$

**28.** $\displaystyle\int \dfrac{4x^2 + 2x - 1}{x^3 + x^2}\, dx$

**29.** $\displaystyle\int \dfrac{4 - 3x}{(x - 1)^2}\, dx$

**30.** $\displaystyle\int \dfrac{x^4}{(x - 1)^3}\, dx$

**31.** $\displaystyle\int \dfrac{3x^2 + 3x + 1}{x(x^2 + 2x + 1)}\, dx$

**32.** $\displaystyle\int \dfrac{3x}{x^2 - 6x + 9}\, dx$

In Exercises 33–40, evaluate the definite integral.

**33.** $\displaystyle\int_4^5 \dfrac{1}{9 - x^2}\, dx$

**34.** $\displaystyle\int_0^1 \dfrac{3}{2x^2 + 5x + 2}\, dx$

**35.** $\displaystyle\int_1^5 \dfrac{x - 1}{x^2(x + 1)}\, dx$

**36.** $\displaystyle\int_0^1 \dfrac{x^2 - x}{x^2 + x + 1}\, dx$

**37.** $\displaystyle\int_0^1 \frac{x^3}{x^2 - 2}\,dx$

**38.** $\displaystyle\int_0^1 \frac{x^3 - 1}{x^2 - 4}\,dx$

**39.** $\displaystyle\int_1^2 \frac{x^3 - 4x^2 - 3x + 3}{x^2 - 3x}\,dx$

**40.** $\displaystyle\int_2^4 \frac{x^4 - 4}{x^2 - 1}\,dx$

In Exercises 41–44, find the area of the shaded region.

**41.** $y = \dfrac{14}{16 - x^2}$

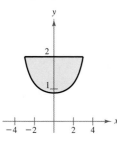

**42.** $y = \dfrac{-4}{(x + 2)(x - 3)}$

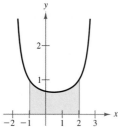

**43.** $y = \dfrac{x + 1}{x^2 - x}$

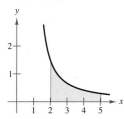

**44.** $y = \dfrac{x^2 + 2x - 1}{x^2 - 4}$

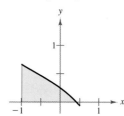

In Exercises 45–48, write the partial fraction decomposition for the rational expression. Check your result algebraically. Then assign a value to the constant $a$ and use a graphing utility to check the result graphically.

**45.** $\dfrac{1}{a^2 - x^2}$

**46.** $\dfrac{1}{x(x + a)}$

**47.** $\dfrac{1}{x(a - x)}$

**48.** $\dfrac{1}{(x + 1)(a - x)}$

In Exercises 49–52, use a graphing utility to graph the function. Then find the volume of the solid generated by revolving the region bounded by the graphs of the given equations about the x-axis by

(a) using the integration capabilities of a graphing utility.

(b) integrating by hand using partial fraction decomposition.

**49.** $y = \dfrac{10}{x(x + 10)}$, $y = 0$, $x = 1$, $x = 5$

**50.** $y = \dfrac{-4}{(x + 1)(x - 4)}$, $y = 0$, $x = 0$, $x = 3$

**51.** $y = \dfrac{2}{x^2 - 4}$, $x = 1$, $x = -1$, $y = 0$

**52.** $y = \dfrac{25x}{x^2 + x - 6}$, $x = -2$, $x = 0$, $y = 0$

**53.** *Population Growth* A conservation organization releases 100 animals of an endangered species into a game preserve. The organization believes that the preserve has a capacity of 1000 animals and that the growth of the herd will be logistic. That is, the size $y$ of the herd will follow the equation

$$\int \frac{1}{y(1000 - y)}\,dy = \int k\,dt,$$

where $t$ is measured in years. Find this logistic curve. (To solve for the constant of integration $C$ and the proportionality constant $k$, assume $y = 100$ when $t = 0$ and $y = 134$ when $t = 2$.) Use a graphing utility to graph your solution.

**54.** *Epidemic Model* A single infected individual enters a community of 500 individuals susceptible to the disease. The disease spreads at a rate proportional to the product of the total number infected and the number of susceptible individuals not yet infected. A model for the time it takes for the disease to spread to $x$ individuals is

$$t = 5010 \int \frac{1}{(x + 1)(500 - x)}\,dx,$$

where $t$ is the time in hours.

(a) Find the time it takes for 75% of the population to become infected (when $t = 0$, $x = 1$).

(b) Find the number of people infected after 100 hours.

**55.** *Marketing* After test-marketing a new menu item, a fast-food restaurant predicts that sales of a new item will grow according to the model

$$\frac{dS}{dt} = \frac{2t}{(t + 4)^2},$$

where $t$ is the time in weeks and $S$ is the sales in thousands of dollars. Find the sales of the menu item at 10 weeks.

**56.** *Population Growth* One gram of a bacterial culture is present at time $t = 0$, and 10 grams is the upper limit of the culture's weight. The time required for the culture to grow to $y$ grams is modeled by

$$kt = \int \frac{1}{y(10 - y)}\,dy,$$

where $y$ is the weight of the culture in grams and $t$ is the time in hours.

(a) Verify that the weight of the culture at time $t$ is modeled by

$$y = \frac{10}{1 + 9e^{-10kt}}.$$

Use the fact that $y = 1$ when $t = 0$.

(b) Use the graph to determine the constant $k$.

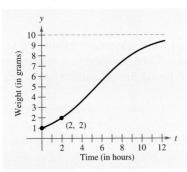

**57.** *Doctorate Degrees*   The number of doctorate degrees awarded each year from 1985 to 1993 can be modeled by

$$N = \frac{73t^2 - 2877t - 2950}{4t^2 - 96t - 100},$$

where $t = 5$ represents 1985 and $N$ is the number of degrees awarded in thousands. Find the total number of degrees awarded between 1985 and 1993. Then find the average number of degrees awarded during this time period. *(Source: U.S. National Center for Educational Statistics)*

**58.** *Biology: Population Growth*   The graph shows the logistic growth of two species of the single-celled *Paramecium* in a laboratory culture. During which time intervals is the rate of growth of each species increasing? During which time intervals is the rate of growth of each species decreasing? Which species has a higher limiting population under these conditions?   *(Source: Adapted from Levine/Miller, Biology: Discovering Life, Second Edition)*

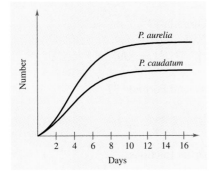

# Business Capsule

Baye Limousines

*Lolita Sweet received an undergraduate degree in business management, and for several years worked in accounting and managerial positions at several different companies. In 1993, Sweet started her own limousine company in the San Francisco area. In its first year, the company had only one limousine and one driver, and grossed only $24,000. By 1996, the company had grown to 16 limousines and 18 drivers, with revenues of $312,000.*

**59.** *Research Project*   Use your school's library or some other reference source to research the opportunity cost of attending graduate school for 2 years to receive a Masters of Business Administration (MBA) degree rather than working for 2 years with a bachelor's degree. Write a short paper describing these costs.

# 6.4

# Integration Tables and Completing the Square

*Integration Tables • Reduction Formulas • Completing the Square*

## Integration Tables

So far in this chapter, you have studied three integration techniques to be used along with the basic integration formulas. Certainly these techniques and formulas do not cover every possible method for finding an antiderivative, but they do cover most of the important ones.

In this section, you will expand the list of integration formulas to form a table of integrals. As you add new integration formulas to the basic list, two things occur. On one hand, it becomes increasingly difficult to memorize, or even become familiar with, the entire list of formulas. On the other hand, with a longer list you need fewer techniques for fitting an integral to one of the formulas on the list. The procedure of integrating by means of a long list of formulas is called **integration by tables.** (The tables in this section constitute only a partial listing of integration formulas. Much longer lists exist, some having several hundred formulas.)

Integration by tables should not be considered a trivial task. It requires considerable thought and insight, and it often requires substitution. Many people find a table of integrals to be a valuable supplement to the integration techniques discussed in the first three sections of this chapter. We encourage you to gain competence in the use of integration tables, as well as to continue to improve in the use of the various integration techniques. In doing so, you should find that a combination of techniques and tables is the most versatile approach to integration.

Each integration formula in the table on the following three pages can be developed using one or more of the techniques you have studied. You should try to verify several of the formulas. For instance, Formula 4

$$\int \frac{u}{(a + bu)^2} \, du = \frac{1}{b^2} \left( \frac{a}{a + bu} + \ln|a + bu| \right) + C \qquad \text{Formula 4}$$

can be verified using partial fractions, Formula 17

$$\int \frac{\sqrt{a + bu}}{u} \, du = 2\sqrt{a + bu} + a \int \frac{1}{u\sqrt{a + bu}} \, du \qquad \text{Formula 17}$$

can be verified using integration by parts, and Formula 37

$$\int \frac{1}{1 + e^u} \, du = u - \ln(1 + e^u) + C \qquad \text{Formula 37}$$

can be verified using substitution and partial fractions.

*STUDY TIP*   A symbolic integration utility, such as *Derive*, *Maple*, *Mathcad*, *Mathematica*, or the *TI-92*, consists, in part, of a database of integration tables. The primary difference between using a symbolic integrator and using a table of integrals is that with a symbolic integrator the computer searches through the database to find a fit. With a table of integrals, *you* must do the searching.

In the following table of integrals, the formulas have been grouped into eight different types according to the form of the integrand.

Forms involving $u^n$        Forms involving $a + bu$

Forms involving $\sqrt{a + bu}$        Forms involving $\sqrt{u^2 \pm a^2}$

Forms involving $u^2 - a^2$        Forms involving $\sqrt{a^2 - u^2}$

Forms involving $e^u$        Forms involving $\ln u$

## Table of Integrals

*Forms involving $u^n$*

1. $\displaystyle\int u^n \, du = \frac{u^{n+1}}{n+1} + C, \quad n \neq -1$

2. $\displaystyle\int \frac{1}{u} \, du = \ln|u| + C$

*Forms involving $a + bu$*

3. $\displaystyle\int \frac{u}{a + bu} \, du = \frac{1}{b^2}(bu - a \ln|a + bu|) + C$

4. $\displaystyle\int \frac{u}{(a + bu)^2} \, du = \frac{1}{b^2}\left(\frac{a}{a + bu} + \ln|a + bu|\right) + C$

5. $\displaystyle\int \frac{u}{(a + bu)^n} \, du = \frac{1}{b^2}\left[\frac{-1}{(n-2)(a + bu)^{n-2}} + \frac{a}{(n-1)(a + bu)^{n-1}}\right] + C, \quad n \neq 1, 2$

6. $\displaystyle\int \frac{u^2}{a + bu} \, du = \frac{1}{b^3}\left[-\frac{bu}{2}(2a - bu) + a^2 \ln|a + bu|\right] + C$

7. $\displaystyle\int \frac{u^2}{(a + bu)^2} \, du = \frac{1}{b^3}\left(bu - \frac{a^2}{a + bu} - 2a \ln|a + bu|\right) + C$

8. $\displaystyle\int \frac{u^2}{(a + bu)^3} \, du = \frac{1}{b^3}\left[\frac{2a}{a + bu} - \frac{a^2}{2(a + bu)^2} + \ln|a + bu|\right] + C$

9. $\displaystyle\int \frac{u^2}{(a + bu)^n} \, du = \frac{1}{b^3}\left[\frac{-1}{(n-3)(a + bu)^{n-3}} + \frac{2a}{(n-2)(a + bu)^{n-2}} - \frac{a^2}{(n-1)(a + bu)^{n-1}}\right] + C, \quad n \neq 1, 2, 3$

10. $\displaystyle\int \frac{1}{u(a + bu)} \, du = \frac{1}{a} \ln\left|\frac{u}{a + bu}\right| + C$

11. $\displaystyle\int \frac{1}{u(a + bu)^2} \, du = \frac{1}{a}\left(\frac{1}{a + bu} + \frac{1}{a} \ln\left|\frac{u}{a + bu}\right|\right) + C$

12. $\displaystyle\int \frac{1}{u^2(a + bu)} \, du = -\frac{1}{a}\left(\frac{1}{u} + \frac{b}{a} \ln\left|\frac{u}{a + bu}\right|\right) + C$

13. $\displaystyle\int \frac{1}{u^2(a + bu)^2} \, du = -\frac{1}{a^2}\left[\frac{a + 2bu}{u(a + bu)} + \frac{2b}{a} \ln\left|\frac{u}{a + bu}\right|\right] + C$

## Table of Integrals (*continued*)

*Forms involving* $\sqrt{a + bu}$

14. $\displaystyle \int u^n \sqrt{a + bu}\, du = \frac{2}{b(2n + 3)}\left[ u^n(a + bu)^{3/2} - na \int u^{n-1}\sqrt{a + bu}\, du \right]$

15. $\displaystyle \int \frac{1}{u\sqrt{a + bu}}\, du = \frac{1}{\sqrt{a}} \ln\left| \frac{\sqrt{a + bu} - \sqrt{a}}{\sqrt{a + bu} + \sqrt{a}} \right| + C, \quad a > 0$

16. $\displaystyle \int \frac{1}{u^n \sqrt{a + bu}}\, du = \frac{-1}{a(n-1)}\left[ \frac{\sqrt{a + bu}}{u^{n-1}} + \frac{(2n - 3)b}{2} \int \frac{1}{u^{n-1}\sqrt{a + bu}}\, du \right], \quad n \neq 1$

17. $\displaystyle \int \frac{\sqrt{a + bu}}{u}\, du = 2\sqrt{a + bu} + a \int \frac{1}{u\sqrt{a + bu}}\, du$

18. $\displaystyle \int \frac{\sqrt{a + bu}}{u^n}\, du = \frac{-1}{a(n-1)}\left[ \frac{(a + bu)^{3/2}}{u^{n-1}} + \frac{(2n - 5)b}{2} \int \frac{\sqrt{a + bu}}{u^{n-1}}\, du \right], \quad n \neq 1$

19. $\displaystyle \int \frac{u}{\sqrt{a + bu}}\, du = -\frac{2(2a - bu)}{3b^2}\sqrt{a + bu} + C$

20. $\displaystyle \int \frac{u^n}{\sqrt{a + bu}}\, du = \frac{2}{(2n + 1)b}\left( u^n\sqrt{a + bu} - na \int \frac{u^{n-1}}{\sqrt{a + bu}}\, du \right)$

*Forms involving* $\sqrt{u^2 \pm a^2}, \quad a > 0$

21. $\displaystyle \int \sqrt{u^2 \pm a^2}\, du = \frac{1}{2}\left( u\sqrt{u^2 \pm a^2} \pm a^2 \ln\left| u + \sqrt{u^2 \pm a^2} \right| \right) + C$

22. $\displaystyle \int u^2 \sqrt{u^2 \pm a^2}\, du = \frac{1}{8}\left[ u(2u^2 \pm a^2)\sqrt{u^2 \pm a^2} - a^4 \ln\left| u + \sqrt{u^2 \pm a^2} \right| \right] + C$

23. $\displaystyle \int \frac{\sqrt{u^2 + a^2}}{u}\, du = \sqrt{u^2 + a^2} - a \ln\left| \frac{a + \sqrt{u^2 + a^2}}{u} \right| + C$

24. $\displaystyle \int \frac{\sqrt{u^2 \pm a^2}}{u^2}\, du = \frac{-\sqrt{u^2 \pm a^2}}{u} + \ln\left| u + \sqrt{u^2 \pm a^2} \right| + C$

25. $\displaystyle \int \frac{1}{\sqrt{u^2 \pm a^2}}\, du = \ln\left| u + \sqrt{u^2 \pm a^2} \right| + C$

26. $\displaystyle \int \frac{1}{u\sqrt{u^2 + a^2}}\, du = -\frac{1}{a} \ln\left| \frac{a + \sqrt{u^2 + a^2}}{u} \right| + C$

27. $\displaystyle \int \frac{u^2}{\sqrt{u^2 \pm a^2}}\, du = \frac{1}{2}\left( u\sqrt{u^2 \pm a^2} \mp a^2 \ln\left| u + \sqrt{u^2 \pm a^2} \right| \right) + C$

28. $\displaystyle \int \frac{1}{u^2\sqrt{u^2 \pm a^2}}\, du = \mp \frac{\sqrt{u^2 \pm a^2}}{a^2 u} + C$

## Table of Integrals (*continued*)

*Forms involving $u^2 - a^2$,   $a > 0$*

29. $\displaystyle\int \frac{1}{u^2 - a^2} \, du = -\int \frac{1}{a^2 - u^2} \, du = \frac{1}{2a} \ln\left|\frac{u - a}{u + a}\right| + C$

30. $\displaystyle\int \frac{1}{(u^2 - a^2)^n} \, du = \frac{-1}{2a^2(n - 1)}\left[\frac{u}{(u^2 - a^2)^{n-1}} + (2n - 3)\int \frac{1}{(u^2 - a^2)^{n-1}} \, du\right], \quad n \neq 1$

*Forms involving $\sqrt{a^2 - u^2}$,   $a > 0$*

31. $\displaystyle\int \frac{\sqrt{a^2 - u^2}}{u} \, du = \sqrt{a^2 - u^2} - a \ln\left|\frac{a + \sqrt{a^2 - u^2}}{u}\right| + C$

32. $\displaystyle\int \frac{1}{u\sqrt{a^2 - u^2}} \, du = -\frac{1}{a} \ln\left|\frac{a + \sqrt{a^2 - u^2}}{u}\right| + C$

33. $\displaystyle\int \frac{1}{u^2\sqrt{a^2 - u^2}} \, du = \frac{-\sqrt{a^2 - u^2}}{a^2 u} + C$

*Forms involving $e^u$*

34. $\displaystyle\int e^u \, du = e^u + C$

35. $\displaystyle\int u e^u \, du = (u - 1)e^u + C$

36. $\displaystyle\int u^n e^u \, du = u^n e^u - n\int u^{n-1} e^u \, du$

37. $\displaystyle\int \frac{1}{1 + e^u} \, du = u - \ln(1 + e^u) + C$

38. $\displaystyle\int \frac{1}{1 + e^{nu}} \, du = u - \frac{1}{n} \ln(1 + e^{nu}) + C$

*Forms involving $\ln u$*

39. $\displaystyle\int \ln u \, du = u(-1 + \ln u) + C$

40. $\displaystyle\int u \ln u \, du = \frac{u^2}{4}(-1 + 2 \ln u) + C$

41. $\displaystyle\int u^n \ln u \, du = \frac{u^{n+1}}{(n + 1)^2}[-1 + (n + 1) \ln u] + C, \quad n \neq -1$

42. $\displaystyle\int (\ln u)^2 \, du = u[2 - 2 \ln u + (\ln u)^2] + C$

43. $\displaystyle\int (\ln u)^n \, du = u(\ln u)^n - n\int (\ln u)^{n-1} \, du$

**Technology**

Throughout this section, remember that a symbolic integration utility, such as *Derive*, *Mathcad*, *Mathematica*, *Maple*, or the *TI-92*, can be used instead of integration tables. If you have access to such a utility, try using it to find the indefinite integrals in Examples 1 and 2.

**EXAMPLE 1**   **Using Integration Tables**

Find the indefinite integral $\displaystyle \int \frac{x}{\sqrt{x-1}}\,dx$.

**Solution**

Because the expression inside the radical is linear, you should consider forms involving $\sqrt{a+bu}$, as in Formula 19.

$$\int \frac{u}{\sqrt{a+bu}}\,du = -\frac{2(2a-bu)}{3b^2}\sqrt{a+bu} + C \qquad \text{Formula 19}$$

Using this formula, let $a = -1$, $b = 1$, and $u = x$. Then $du = dx$, and you obtain the following.

$$\int \frac{x}{\sqrt{x-1}}\,dx = -\frac{2(-2-x)}{3}\sqrt{x-1} + C \qquad \begin{array}{l}\text{Substitute values of}\\ a, b, \text{ and } u.\end{array}$$

$$= \frac{2}{3}(2+x)\sqrt{x-1} + C \qquad \text{Simplify.}$$

**EXAMPLE 2**   **Using Integration Tables**

Find the indefinite integral $\displaystyle \int x\sqrt{x^4-9}\,dx$.

**Solution**

Because it is not clear which formula to use, you can begin by letting $u = x^2$ and $du = 2x\,dx$. With this substitution, you can write the integral as follows.

$$\int x\sqrt{x^4-9}\,dx = \frac{1}{2}\int \sqrt{(x^2)^2-9}\,(2x)\,dx \qquad \text{Multiply and divide by 2.}$$

$$= \frac{1}{2}\int \sqrt{u^2-9}\,du \qquad \text{Substitute } u \text{ and } du.$$

Now, it appears that you can use Formula 21.

$$\int \sqrt{u^2-a^2}\,du = \frac{1}{2}\left(u\sqrt{u^2-a^2} - a^2\ln\left|u+\sqrt{u^2-a^2}\right|\right) + C$$

Letting $a = 3$, you obtain

$$\int x\sqrt{x^4-9}\,dx = \frac{1}{2}\int \sqrt{u^2-a^2}\,du$$

$$= \frac{1}{2}\left[\frac{1}{2}\left(u\sqrt{u^2-a^2} - a^2\ln\left|u+\sqrt{u^2-a^2}\right|\right)\right] + C$$

$$= \frac{1}{4}\left(x^2\sqrt{x^4-9} - 9\ln\left|x^2+\sqrt{x^4-9}\right|\right) + C.$$

## EXAMPLE 3  Using Integration Tables

Find the indefinite integral $\displaystyle\int \frac{1}{x\sqrt{x+1}}\,dx$.

**Solution**

Considering forms involving $\sqrt{a+bu}$, where $a = 1$, $b = 1$, and $u = x$, you can use Formula 15.

$$\int \frac{1}{u\sqrt{a+bu}}\,du = \frac{1}{\sqrt{a}}\ln\left|\frac{\sqrt{a+bu}-\sqrt{a}}{\sqrt{a+bu}+\sqrt{a}}\right| + C, \qquad a > 0.$$

Therefore,

$$\int \frac{1}{x\sqrt{x+1}}\,dx = \int \frac{1}{u\sqrt{a+bu}}\,du = \frac{1}{\sqrt{a}}\ln\left|\frac{\sqrt{a+bu}-\sqrt{a}}{\sqrt{a+bu}+\sqrt{a}}\right| + C$$

$$= \ln\left|\frac{\sqrt{x+1}-1}{\sqrt{x+1}+1}\right| + C.$$

## EXAMPLE 4  Using Integration Tables

Evaluate the definite integral $\displaystyle\int_0^2 \frac{x}{1+e^{-x^2}}\,dx$.

**Solution**

Of the forms involving $e^u$, Formula 37

$$\int \frac{1}{1+e^u}\,du = u - \ln(1+e^u) + C$$

seems most appropriate. To use this formula, let $u = -x^2$ and $du = -2x\,dx$.

$$\int \frac{x}{1+e^{-x^2}}\,dx = -\frac{1}{2}\int \frac{1}{1+e^{-x^2}}(-2x)\,dx = -\frac{1}{2}\int \frac{1}{1+e^u}\,du$$

$$= -\frac{1}{2}\left[u - \ln(1+e^u)\right] + C$$

$$= -\frac{1}{2}\left[-x^2 - \ln(1+e^{-x^2})\right] + C$$

$$= \frac{1}{2}\left[x^2 + \ln(1+e^{-x^2})\right] + C$$

So, the value of the definite integral is

$$\int_0^2 \frac{x}{1+e^{-x^2}}\,dx = \frac{1}{2}\left[x^2 + \ln(1+e^{-x^2})\right]_0^2 \approx 1.66,$$

as shown in Figure 6.7.

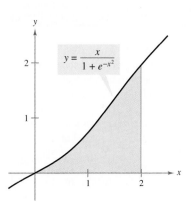

**FIGURE 6.7**

# Reduction Formulas

Several of the formulas in the integration table have the form

$$\int f(x)\, dx = g(x) + \int h(x)\, dx,$$

where the right side contains an integral. Such integration formulas are called **reduction formulas** because they reduce the original integral to the sum of a function and a simpler integral.

**TIP**

For help on the algebra in Example 5, see Example 2b in the *Chapter 6 Algebra Review*, on page 443.

ALGEBRA

## EXAMPLE 5   Using a Reduction Formula

Find the indefinite integral $\int x^2 e^x\, dx$.

### Solution

Using Formula 36,

$$\int u^n e^u\, du = u^n e^u - n \int u^{n-1} e^u\, du,$$

you can let $u = x$ and $n = 2$. Then $du = dx$, and you can write

$$\int x^2 e^x\, dx = x^2 e^x - 2 \int x e^x\, dx.$$

Then, using Formula 35,

$$\int u e^u\, du = (u - 1)e^u + C,$$

you can write

$$\begin{aligned}
\int x^2 e^x\, dx &= x^2 e^x - 2 \int x e^x\, dx \\
&= x^2 e^x - 2(x - 1)e^x + C \\
&= x^2 e^x - 2x e^x + 2e^x + C \\
&= e^x(x^2 - 2x + 2) + C
\end{aligned}$$

## Technology

You have now studied two ways to find the indefinite integral in Example 5. Example 5 uses an integration table, and Example 4 in Section 6.2 uses integration by parts. A third way would be to use a symbolic integrator. For instance, the screen at the right shows how to use *Derive* for Windows to solve the integral.

| | | |
|---|---|---|
| #1: $x^2 \cdot \hat{e}^x$ | | Author. |
| #2: $\int x^2 \cdot \hat{e}^x\, dx$ | | Integrate. |
| #3: $\hat{e}^x \cdot (x^2 - 2 \cdot x + 2)$ | | Simplify. |

## Completing the Square

Many integration formulas involve the sum or difference of two squares. You can extend the use of these formulas by an algebraic procedure called **completing the square**. This procedure is demonstrated in Example 6.

### EXAMPLE 6 Completing the Square

Find the indefinite integral

$$\int \frac{1}{x^2 - 4x + 1} \, dx.$$

*Solution*

Begin by writing the denominator as the difference of two squares.

$$x^2 - 4x + 1 = (x^2 - 4x + 4) - 4 + 1 = (x - 2)^2 - 3$$

Therefore, you can rewrite the original integral as

$$\int \frac{1}{x^2 - 4x + 1} \, dx = \int \frac{1}{(x - 2)^2 - 3} \, dx.$$

Considering $u = x - 2$ and $a = \sqrt{3}$, you can apply Formula 29

$$\int \frac{1}{u^2 - a^2} \, du = \frac{1}{2a} \ln \left| \frac{u - a}{u + a} \right| + C$$

to conclude that

$$\int \frac{1}{x^2 - 4x + 1} \, dx = \int \frac{1}{(x - 2)^2 - 3} \, dx$$

$$= \int \frac{1}{u^2 - a^2} \, du$$

$$= \frac{1}{2a} \ln \left| \frac{u - a}{u + a} \right| + C$$

$$= \frac{1}{2\sqrt{3}} \ln \left| \frac{x - 2 - \sqrt{3}}{x - 2 + \sqrt{3}} \right| + C.$$

---

*Group Discussion* **Using Integration Tables**

Which integration formulas on pages 413–415 would you use to integrate the following? Explain your choice of $u$ for each integral.

a. $\displaystyle\int \frac{e^x}{e^x + 1} \, dx$    b. $\displaystyle\int \frac{1}{e^x + 1} \, dx$    c. $\displaystyle\int \frac{1}{e^{2x} + 1} \, dx$

Use a symbolic integration utility to check that your choice of $u$ was correct.

*Warm Up*

The following warm-up exercises involve skills that were covered in earlier sections. You will use these skills in the exercise set for this section.

In Exercises 1–4, expand the expression.

**1.** $(x + 4)^2$

**2.** $\left(x + \frac{1}{2}\right)^2$

**3.** $(x - 1)^2$

**4.** $\left(x - \frac{1}{3}\right)^2$

In Exercises 5–8, write the partial fraction decomposition for the function.

**5.** $\dfrac{4}{x(x + 2)}$

**6.** $\dfrac{3}{x(x - 4)}$

**7.** $\dfrac{x + 4}{x^2(x - 2)}$

**8.** $\dfrac{3x^2 + 4x - 8}{x(x - 2)(x + 1)}$

In Exercises 9 and 10, use integration by parts to find the indefinite integral.

**9.** $\displaystyle\int 2xe^x \, dx$

**10.** $\displaystyle\int 3x^2 \ln x \, dx$

## EXERCISES 6.4

In Exercises 1–8, use the indicated formula from the table of integrals in this section to find the indefinite integral.

**1.** $\displaystyle\int \dfrac{x}{(2 + 3x)^2} \, dx$, Formula 4

**2.** $\displaystyle\int \dfrac{1}{x(2 + 3x)^2} \, dx$, Formula 11

**3.** $\displaystyle\int \dfrac{x}{\sqrt{2 + 3x}} \, dx$, Formula 19

**4.** $\displaystyle\int \dfrac{4}{x^2 - 9} \, dx$, Formula 29

**5.** $\displaystyle\int \dfrac{2x}{\sqrt{x^4 - 9}} \, dx$, Formula 25

**6.** $\displaystyle\int x^2 \sqrt{x^2 + 9} \, dx$, Formula 22

**7.** $\displaystyle\int x^3 e^{x^2} \, dx$, Formula 35

**8.** $\displaystyle\int \dfrac{x}{1 + e^{x^2}} \, dx$, Formula 37

In Exercises 9–34, use the table of integrals in this section to find the indefinite integral.

**9.** $\displaystyle\int \dfrac{1}{x(1 + x)} \, dx$

**10.** $\displaystyle\int \dfrac{1}{x(1 + x)^2} \, dx$

**11.** $\displaystyle\int \dfrac{1}{x\sqrt{x^2 + 9}} \, dx$

**12.** $\displaystyle\int \dfrac{1}{\sqrt{x^2 - 1}} \, dx$

**13.** $\displaystyle\int \dfrac{1}{x\sqrt{4 - x^2}} \, dx$

**14.** $\displaystyle\int \dfrac{\sqrt{x^2 - 9}}{x^2} \, dx$

**15.** $\displaystyle\int x \ln x \, dx$

**16.** $\displaystyle\int x^2 (\ln x^3)^2 \, dx$

**17.** $\displaystyle\int \dfrac{6x}{1 + e^{3x^2}} \, dx$

**18.** $\displaystyle\int \dfrac{1}{1 + e^x} \, dx$

**19.** $\displaystyle\int x\sqrt{x^4 - 4} \, dx$

**20.** $\displaystyle\int \dfrac{x}{x^4 - 9} \, dx$

**21.** $\displaystyle\int \dfrac{t^2}{(2 + 3t)^3} \, dt$

**22.** $\displaystyle\int \dfrac{\sqrt{3 + 4t}}{t} \, dt$

**23.** $\displaystyle\int \dfrac{s}{s^2\sqrt{3 + s}} \, ds$

**24.** $\displaystyle\int \sqrt{3 + x^2} \, dx$

**25.** $\displaystyle\int \frac{x^2}{(3 + 2x)^5}\, dx$

**26.** $\displaystyle\int \frac{1}{x^2\sqrt{x^2 - 4}}\, dx$

**27.** $\displaystyle\int \frac{1}{x^2\sqrt{1 - x^2}}\, dx$

**28.** $\displaystyle\int \frac{2x}{(1 - 3x)^2}\, dx$

**29.** $\displaystyle\int x^2 \ln x \, dx$

**30.** $\displaystyle\int xe^{x^2}\, dx$

**31.** $\displaystyle\int \frac{x^2}{(3x - 5)^2}\, dx$

**32.** $\displaystyle\int \frac{1}{2x^2(2x - 1)^2}\, dx$

**33.** $\displaystyle\int \frac{\ln x}{x(4 + 3 \ln x)}\, dx$

**34.** $\displaystyle\int (\ln x)^3 \, dx$

In Exercises 35–40, use integration tables to find the exact area of the region bounded by the graphs of the equations. Then use a graphing utility to graph the region and approximate the area.

**35.** $y = \dfrac{x}{\sqrt{x + 1}}$, $y = 0$, $x = 8$

**36.** $y = \dfrac{2}{1 + e^{4x}}$, $y = 0$, $x = 0$, $x = 1$

**37.** $y = \dfrac{x}{1 + e^{x^2}}$, $y = 0$, $x = 2$

**38.** $y = \dfrac{e^x}{1 - e^{2x}}$, $y = 0$, $x = 1$, $x = 2$

**39.** $y = x^2\sqrt{x^2 + 4}$, $y = 0$, $x = \sqrt{5}$

**40.** $y = \dfrac{1}{\sqrt{x}(1 + 2\sqrt{x})}$, $y = 0$, $x = 1$, $x = 4$

In Exercises 41–44, evaluate the definite integral.

**41.** $\displaystyle\int_0^5 \frac{x}{\sqrt{5 + 2x}}\, dx$

**42.** $\displaystyle\int_0^5 \frac{x}{(4 + x)^2}\, dx$

**43.** $\displaystyle\int_0^4 \frac{6}{1 + e^{0.5x}}\, dx$

**44.** $\displaystyle\int_1^4 x \ln x \, dx$

In Exercises 45–48, find the indefinite integral (a) using integration tables and (b) using the specified method.

| Integral | Method |
|---|---|
| **45.** $\displaystyle\int x^2 e^x\, dx$ | Integration by parts |
| **46.** $\displaystyle\int x^4 \ln x\, dx$ | Integration by parts |
| **47.** $\displaystyle\int \frac{1}{x^2(x + 1)}\, dx$ | Partial fractions |
| **48.** $\displaystyle\int \frac{1}{x^2 - 75}\, dx$ | Partial fractions |

In Exercises 49 and 50, complete the square to express each polynomial as the sum or difference of squares.

**49.** (a) $x^2 + 6x$      (b) $x^2 - 8x + 9$
    (c) $x^4 + 2x^2 - 5$      (d) $3 - 2x - x^2$

**50.** (a) $2x^2 + 12x + 14$      (b) $3x^2 - 12x - 9$
    (c) $x^2 - 2x$      (d) $9 + 8x - x^2$

In Exercises 51–58, complete the square and then use integration tables to find the indefinite integral.

**51.** $\displaystyle\int \frac{1}{x^2 + 6x - 8}\, dx$

**52.** $\displaystyle\int \frac{1}{x^2 + 4x - 5}\, dx$

**53.** $\displaystyle\int \frac{1}{(x - 1)\sqrt{x^2 - 2x + 2}}\, dx$

**54.** $\displaystyle\int \sqrt{x^2 - 6x}\, dx$

**55.** $\displaystyle\int \frac{1}{2x^2 - 4x - 6}\, dx$

**56.** $\displaystyle\int \frac{\sqrt{7 - 6x - x^2}}{x + 3}\, dx$

**57.** $\displaystyle\int \frac{x}{\sqrt{x^4 + 2x^2 + 2}}\, dx$

**58.** $\displaystyle\int \frac{x\sqrt{x^4 + 4x^2 + 5}}{x^2 + 2}\, dx$

***Population Growth*** In Exercises 59 and 60, use a graphing utility to graph the growth function. Use the table of integrals to find the average value of the growth function over the interval, where $N$ is the size of a population and $t$ is the time in days.

**59.** $N = \dfrac{50}{1 + e^{4.8 - 1.9t}}$,   $[3, 4]$

**60.** $N = \dfrac{375}{1 + e^{4.20 - 0.25t}}$,   $[21, 28]$

**61.** ***Revenue*** The marginal revenue (in dollars per year) for a new product is modeled by
$$\frac{dR}{dt} = 10,000\left[1 - \frac{1}{(1 + 0.1t^2)^{1/2}}\right],$$
where $t$ is the time in years. Estimate the total revenue from sales of the product over its first 2 years on the market.

**62.** ***Consumer and Producer Surpluses*** Find the consumer surplus and the producer surplus for a product with the given demand and supply functions.
$$Demand:\ p = \frac{60}{\sqrt{x^2 + 81}}, \qquad Supply:\ p = \frac{x}{3}$$

**63.** ***Profit*** The net profit $P$ (in millions of dollars per year) for Hershey Foods from 1988 through 1996 can be modeled by
$$P = \sqrt{375.67t^2 - 715.86}, \qquad 8 \leq t \leq 16,$$
where $t$ is the time in years, with $t = 8$ corresponding to 1988. Find the average net profit over that time period. (*Source: Hershey Foods*)

6.5

# Numerical Integration

*Trapezoidal Rule  •  Simpson's Rule  •  Error Analysis*

## Trapezoidal Rule

In Section 5.6, you studied one technique for approximating the value of a *definite* integral—the Midpoint Rule. In this section, you will study two other approximation techniques: the **Trapezoidal Rule** and **Simpson's Rule.**

To develop the Trapezoidal Rule, consider a function $f$ that is nonnegative and continuous on the closed interval $[a, b]$. To approximate the area represented by $\int_a^b f(x)\,dx$, partition the interval into $n$ subintervals, each of width

$$\Delta x = \frac{b - a}{n}. \qquad \text{Width of each subinterval}$$

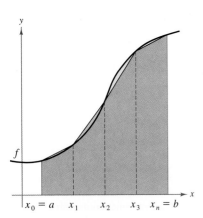

Next, form $n$ trapezoids, as shown in Figure 6.8. As you can see in Figure 6.9, the area of the first trapezoid is

$$\text{Area of first trapezoid} = \frac{b - a}{n}\left[\frac{f(x_0) + f(x_1)}{2}\right].$$

The areas of the other trapezoids follow a similar pattern, and the sum of the $n$ areas is

$$\frac{b - a}{n}\left[\frac{f(x_0) + f(x_1)}{2} + \frac{f(x_1) + f(x_2)}{2} + \cdots + \frac{f(x_{n-1}) + f(x_n)}{2}\right]$$

$$= \frac{b - a}{2n}[f(x_0) + f(x_1) + f(x_1) + f(x_2) + \cdots + f(x_{n-1}) + f(x_n)]$$

$$= \frac{b - a}{2n}[f(x_0) + 2f(x_1) + 2f(x_2) + \cdots + 2f(x_{n-1}) + f(x_n)].$$

Although this development assumes $f$ to be continuous *and* nonnegative on $[a, b]$, the resulting formula is valid as long as $f$ is continuous on $[a, b]$.

**FIGURE 6.8**

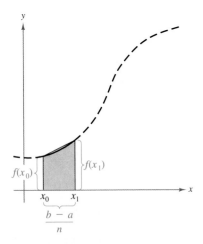

**FIGURE 6.9**

---

### The Trapezoidal Rule

If $f$ is continuous on $[a, b]$, then

$$\int_a^b f(x)\,dx \approx \frac{b - a}{2n}[f(x_0) + 2f(x_1) + \cdots + 2f(x_{n-1}) + f(x_n)].$$

## EXAMPLE 1   Using the Trapezoidal Rule

Use the Trapezoidal Rule to approximate the definite integral

$$\int_0^1 e^x\,dx.$$

Compare the results for $n = 4$ and $n = 8$.

### Solution

When $n = 4$, the width of each subinterval is $(1 - 0)/4 = \frac{1}{4}$ and the endpoints of the subintervals are

$$x_0 = 0, \quad x_1 = \frac{1}{4}, \quad x_2 = \frac{1}{2}, \quad x_3 = \frac{3}{4}, \quad \text{and} \quad x_4 = 1,$$

as indicated in Figure 6.10. Therefore, by the Trapezoidal Rule,

$$\int_0^1 e^x\,dx = \frac{1}{8}(e^0 + 2e^{0.25} + 2e^{0.5} + 2e^{0.75} + e^1)$$

$$\approx 1.7272. \qquad \text{Approximation using } n = 4$$

When $n = 8$, the width of each subinterval is $(1 - 0)/8 = \frac{1}{8}$ and the endpoints of the subintervals are

$$x_0 = 0, \quad x_1 = \frac{1}{8}, \quad x_2 = \frac{1}{4}, \quad x_3 = \frac{3}{8}, \quad x_4 = \frac{1}{2}$$

$$x_5 = \frac{5}{8}, \quad x_6 = \frac{3}{4}, \quad x_7 = \frac{7}{8}, \quad \text{and} \quad x_8 = 1,$$

as indicated in Figure 6.11. Therefore, by the Trapezoidal Rule,

$$\int_0^1 e^x\,dx = \frac{1}{16}(e^0 + 2e^{0.125} + 2e^{0.25} + \cdots + 2e^{0.875} + e^1)$$

$$\approx 1.7205. \qquad \text{Approximation using } n = 8$$

Of course, for *this particular* integral, you could have found an antiderivative and used the Fundamental Theorem of Calculus to find the exact value of the definite integral. The exact value is

$$\int_0^1 e^x\,dx = e - 1 \approx 1.718282. \qquad \text{Exact value}$$

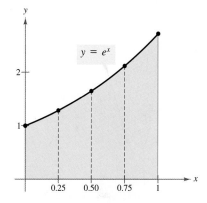

**FIGURE 6.10**   Four Subintervals

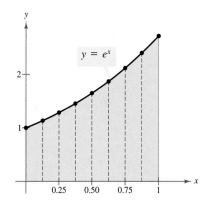

**FIGURE 6.11**   Eight Subintervals

There are two important points that should be made concerning the Trapezoidal Rule. First, the approximation tends to become more accurate as $n$ increases. For instance, in Example 1, if $n = 16$, the Trapezoidal Rule yields an approximation of 1.7188. Second, although you could have used the Fundamental Theorem of Calculus to evaluate the integral in Example 1, this theorem cannot be used to evaluate an integral as simple as $\int_0^1 e^{x^2}\,dx$, because $e^{x^2}$ has no elementary function as an antiderivative. Yet, the Trapezoidal Rule can be easily applied to this integral.

## Simpson's Rule

One way to view the Trapezoidal Rule is to say that on each subinterval, $f$ is approximated by a first-degree polynomial. In Simpson's Rule, $f$ is approximated by a second-degree polynomial on each subinterval.

To develop Simpson's Rule, partition the interval $[a, b]$ into an *even number* $n$ of subintervals, each of width $(b - a)/n$. On the subinterval $[x_0, x_2]$, approximate the function $f$ by the second-degree polynomial $p(x)$ that passes through the points

$$(x_0, f(x_0)), \qquad (x_1, f(x_1)), \qquad \text{and} \qquad (x_2, f(x_2)),$$

as shown in Figure 6.12. The Fundamental Theorem of Calculus can be used to show that

$$\int_{x_0}^{x_2} f(x)\,dx \approx \int_{x_0}^{x_2} p(x)\,dx$$

$$= \frac{x_2 - x_0}{6}\left[p(x_0) + 4p\left(\frac{x_0 + x_2}{2}\right) + p(x_2)\right]$$

$$= \frac{2[(b - a)/n]}{6}[p(x_0) + 4p(x_1) + p(x_2)]$$

$$= \frac{b - a}{3n}[f(x_0) + 4f(x_1) + f(x_2)].$$

Repeating this process on the subintervals $[x_{i-2}, x_i]$ produces

$$\int_a^b f(x)\,dx \approx \frac{b - a}{3n}[f(x_0) + 4f(x_1) + f(x_2) + f(x_2) + 4f(x_3) +$$

$$f(x_4) + \cdots + f(x_{n-2}) + 4f(x_{n-1}) + f(x_n)].$$

By grouping like terms, you can obtain the following approximation, which is known as Simpson's Rule. This rule is named after the English mathematician Thomas Simpson (1710–1761).

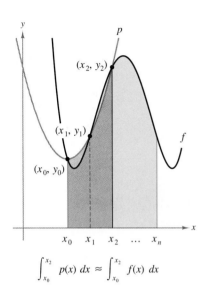

$$\int_{x_0}^{x_2} p(x)\,dx \approx \int_{x_0}^{x_2} f(x)\,dx$$

**FIGURE 6.12**

---

### Simpson's Rule ($n$ is Even)

If $f$ is continuous on $[a, b]$, then

$$\int_a^b f(x)\,dx \approx \frac{b - a}{3n}[f(x_0) + 4f(x_1) + 2f(x_2) + 4f(x_3) +$$

$$\cdots + 4f(x_{n-1}) + f(x_n)].$$

---

*STUDY TIP*    The coefficients in Simpson's Rule have the pattern

$$1 \quad 4 \quad 2 \quad 4 \quad 2 \quad 4 \quad \cdots \quad 4 \quad 2 \quad 4 \quad 1.$$

In Example 1, the Trapezoidal Rule was used to estimate the value of

$$\int_0^1 e^x \, dx.$$

The next example uses Simpson's Rule to approximate the same integral.

## EXAMPLE 2   Using Simpson's Rule

Use Simpson's Rule to approximate the definite integral

$$\int_0^1 e^x \, dx.$$

Compare the results for $n = 4$ and $n = 8$.

### Solution

When $n = 4$, the width of each subinterval is $(1 - 0)/4 = \frac{1}{4}$ and the endpoints of the subintervals are

$$x_0 = 0, \quad x_1 = \frac{1}{4}, \quad x_2 = \frac{1}{2}, \quad x_3 = \frac{3}{4}, \quad \text{and} \quad x_4 = 1,$$

as indicated in Figure 6.13. Therefore, by Simpson's Rule,

$$\int_0^1 e^x dx = \frac{1}{12}(e^0 + 4e^{0.25} + 2e^{0.5} + 4e^{0.75} + e^1)$$

$$\approx 1.718319. \qquad \text{Approximation using } n = 4$$

When $n = 8$, the width of each subinterval is $(1 - 0)/8 = \frac{1}{8}$ and the endpoints of the subintervals are

$$x_0 = 0, \quad x_1 = \frac{1}{8}, \quad x_2 = \frac{1}{4}, \quad x_3 = \frac{3}{8}, \quad x_4 = \frac{1}{2}$$

$$x_5 = \frac{5}{8}, \quad x_6 = \frac{3}{4}, \quad x_7 = \frac{7}{8}, \quad \text{and} \quad x_8 = 1,$$

as indicated in Figure 6.14. Therefore, by Simpson's Rule,

$$\int_0^1 e^x dx = \frac{1}{24}(e^0 + 4e^{0.125} + 2e^{0.25} + \cdots + 4e^{0.875} + e^1)$$

$$\approx 1.718284. \qquad \text{Approximation using } n = 8$$

Recall that the exact value of this integral is

$$\int_0^1 e^x dx = e - 1 \approx 1.718282.$$

Thus, with only eight subintervals, you obtained an approximation that is correct to the nearest 0.000002—an impressive result.

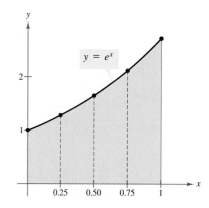

**FIGURE 6.13**

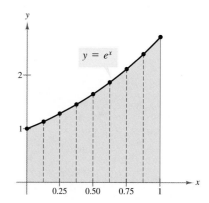

**FIGURE 6.14**

> **STUDY TIP**   Comparing the results of Examples 1 and 2, you can see that for a given value of $n$, Simpson's Rule tends to be more accurate than the Trapezoidal Rule.

## Technology

### Programming Simpson's Rule

In Section 5.6, you saw how to program the Midpoint Rule into a computer or programmable calculator. The program below will evaluate Simpson's Rule on a *TI-82* or *TI-83* calculator. See the appendix for other models.

| | |
|---|---|
| PROGRAM: SIMPSONS | |
| :Disp "LOWER LIMIT" | Prompt for value of $a$. |
| :Input A | Input value of $a$. |
| :Disp "UPPER LIMIT" | Prompt for value of $b$. |
| :Input B | Input value of $b$. |
| :Disp "N/2 DIVISIONS" | Prompt for value of $n/2$. |
| :Input D | Input value of $n/2$. |
| :0 → S | Initialize sum of areas. |
| :(B−A)/(2D) → W | Calculate width of subinterval. |
| :1 → J | Initialize counter. |
| :Lbl 1 | Begin loop. |
| :A+2(J−1)W → L | Calculate left endpoint. |
| :A+2JW → R | Calculate right endpoint. |
| :(L+R)/2 → M | Calculate midpoint of subinterval. |
| :L → X | |
| :Y₁ → L | Evaluate $f(x)$ at left endpoint. |
| :M → X | |
| :Y₁ → M | Evaluate $f(x)$ at midpoint. |
| :R → X | |
| :Y₁ → R | Evaluate $f(x)$ at right endpoint. |
| :W(L+4M+R)/3+S → S | Simpson's Rule |
| :IS > (J, D) | Check value of index, $J$. |
| :Goto 1 | End loop. |
| :Disp "APPROXIMATION" | |
| :Disp S | Display approximation. |

Before executing the program, enter the function as $Y_1$. When the program is executed, you will be prompted to enter the lower and upper limits of integration, and *half* the number of subintervals you want to use.

To check your steps, try running the program with the integral in Example 2. When $n = 8$ (input 4), you should obtain an approximation of

$$\int_0^1 e^x \, dx \approx 1.718284155.$$

# Error Analysis

In Examples 1 and 2, you were able to calculate the exact value of the integral and compare that value with the approximations to see how good they were. In practice, you need to have a different way of telling how good an approximation is: such a way is provided in the next result.

## Error in Trapezoidal and Simpson's Rules

The error $E$ in approximating $\int_a^b f(x)\,dx$ is as follows.

*Trapezoidal Rule:* $|E| \leq \dfrac{(b-a)^3}{12n^2}\left[\max|f''(x)|\right], \qquad a \leq x \leq b$

*Simpson's Rule:* $|E| \leq \dfrac{(b-a)^5}{180n^4}\left[\max|f^{(4)}(x)|\right], \qquad a \leq x \leq b$

This result indicates that the errors generated by the Trapezoidal Rule and Simpson's Rule have upper bounds dependent on the extreme values of $f''(x)$ and $f^{(4)}(x)$ in the interval $[a, b]$. Furthermore, the bounds for the errors can be made arbitrarily small by *increasing n*. To determine what value of $n$ to choose, consider the following steps.

*Trapezoidal Rule*

1. Find $f''(x)$.

2. Find the maximum of $|f''(x)|$ on the interval $[a, b]$.

3. Set up the inequality $|E| \leq \dfrac{(b-a)^3}{12n^2}\left[\max|f''(x)|\right]$.

4. For an error less than $\epsilon$, solve for $n$ in the inequality
$$\dfrac{(b-a)^3}{12n^2}\left[\max|f''(x)|\right] < \epsilon.$$

5. Partition $[a, b]$ into $n$ subintervals and apply the Trapezoidal Rule.

*Simpson's Rule*

1. Find $f^{(4)}(x)$.

2. Find the maximum of $|f^{(4)}(x)|$ on the interval $[a, b]$.

3. Set up the inequality $|E| \leq \dfrac{(b-a)^5}{180n^4}\left[\max|f^{(4)}(x)|\right]$.

4. For an error less than $\epsilon$, solve for $n$ in the inequality
$$\dfrac{(b-a)^5}{180n^4}\left[\max|f^{(4)}(x)|\right] < \epsilon.$$

5. Partition $[a, b]$ into $n$ subintervals and apply Simpson's Rule.

*Discovery*

How does the error in the Trapezoidal Rule decrease as you increase $n$, the number of subintervals? In Example 1 on page 423, you approximated the integral $\int_0^1 e^x\,dx$ using the Trapezoidal Rule with four and eight subintervals. Because the exact value of this integral is known to be $e - 1$, the errors are

$\quad n = 4 \qquad$ error: $0.008918$

and

$\quad n = 8 \qquad$ error: $0.002218$.

Notice that the error diminishes by a factor of approximately 4 when the number of subintervals is doubled. Explain how this is consistent with the formula for the Trapezoidal Rule error. What do you think the error would be if you used 16 subintervals in Example 1? (The actual error is 0.0005593, which is approximately 0.002218/4.) How does the error in Simpson's Rule diminish as you increase the number of subintervals?

## EXAMPLE 3   Using the Trapezoidal Rule

Use the Trapezoidal Rule to estimate the value of the integral

$$\int_0^1 e^{-x^2}\,dx$$

such that the approximation error is less than 0.01.

**Solution**

1.  Begin by finding the second derivative of $f(x) = e^{-x^2}$.

$$\begin{aligned}
f(x) &= e^{-x^2} \\
f'(x) &= -2xe^{-x^2} \\
f''(x) &= 4x^2e^{-x^2} - 2e^{-x^2} \\
&= 2e^{-x^2}(2x^2 - 1)
\end{aligned}$$

2.  $f''$ has only one critical number in the interval $[0, 1]$, and the maximum value of $|f''(x)|$ on this interval is $|f''(0)| = 2$.

3.  The error $E$ using the Trapezoidal Rule is bounded by

$$|E| \le \frac{(b-a)^3}{12n^2}(2) = \frac{1}{12n^2}(2) = \frac{1}{6n^2}.$$

4.  To ensure that the approximation has an error of less than 0.01, you should choose $n$ such that

$$\frac{1}{6n^2} < 0.01.$$

Solving for $n$, you can determine that $n$ must be 5 or more.

5.  Partition $[0, 1]$ into five subintervals, as shown in Figure 6.15. Then apply the Trapezoidal Rule to obtain

$$\int_0^1 e^{-x^2}\,dx = \frac{1}{10}\left(\frac{1}{e^0} + \frac{2}{e^{0.04}} + \frac{2}{e^{0.16}} + \frac{2}{e^{0.36}} + \frac{2}{e^{0.64}} + \frac{1}{e^1}\right)$$
$$\approx 0.744,$$

with an error of less than 0.01.

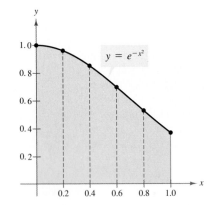

$y = e^{-x^2}$

**FIGURE 6.15**

---

*Group Discussion*     **Using the Trapezoidal Rule**

In the second step of the solution to Example 3, it is stated that the maximum value of $|f''(x)|$ is 2. Explain why this is true. Verify this result by using a graphing utility to graph $f''$.

## Warm Up

The following warm-up exercises involve skills that were covered in earlier sections. You will use these skills in the exercise set for this section.

In Exercises 1–6, find the indicated derivative.

**1.** $f(x) = \dfrac{1}{x}$, $f''(x)$

**2.** $f(x) = \ln(2x + 1)$, $f^{(4)}(x)$

**3.** $f(x) = 2 \ln x$, $f^{(4)}(x)$

**4.** $f(x) = x^3 - 2x^2 + 7x - 12$, $f''(x)$

**5.** $f(x) = e^{2x}$, $f^{(4)}(x)$

**6.** $f(x) = e^{x^2}$, $f''(x)$

In Exercises 7 and 8, find the absolute maximum of $f$ on the interval.

**7.** $f(x) = -x^2 + 6x + 9$, $[0, 4]$

**8.** $f(x) = \dfrac{8}{x^3}$, $[1, 2]$

In Exercises 9 and 10, solve for $n$.

**9.** $\dfrac{1}{4n^2} < 0.001$

**10.** $\dfrac{1}{16n^4} < 0.0001$

## EXERCISES 6.5

In Exercises 1–10, use the Trapezoidal Rule and Simpson's Rule to approximate the value of the definite integral for the indicated value of $n$. Compare these results with the exact value of the definite integral. Round your answers to four decimal places.

**1.** $\displaystyle\int_0^2 x^2 \, dx$, $n = 4$

**2.** $\displaystyle\int_0^1 \left(\dfrac{x^2}{2} + 1\right) dx$, $n = 4$

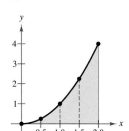

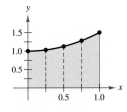

**3.** $\displaystyle\int_0^2 (x^4 + 1) \, dx$, $n = 4$

**4.** $\displaystyle\int_1^2 \dfrac{1}{x} \, dx$, $n = 4$

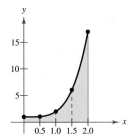

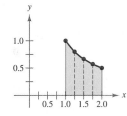

**5.** $\displaystyle\int_0^2 x^3 \, dx$, $n = 8$

**6.** $\displaystyle\int_1^2 \dfrac{1}{x} \, dx$, $n = 8$

**7.** $\displaystyle\int_1^2 \dfrac{1}{x^2} \, dx$, $n = 4$

**8.** $\displaystyle\int_0^4 \sqrt{x} \, dx$, $n = 8$

**9.** $\displaystyle\int_0^1 \dfrac{1}{1 + x} \, dx$, $n = 4$

**10.** $\displaystyle\int_0^2 x\sqrt{x^2 + 1} \, dx$, $n = 4$

In Exercises 11–18, approximate the integral using (a) the Trapezoidal Rule and (b) Simpson's Rule. (Round your answers to three significant digits.)

| Definite Integral | Subdivisions |
|---|---|

**11.** $\displaystyle\int_0^1 \frac{1}{1+x^2}\,dx$     $n = 4$

**12.** $\displaystyle\int_0^2 \frac{1}{\sqrt{1+x^3}}\,dx$     $n = 4$

**13.** $\displaystyle\int_0^1 \sqrt{1-x^2}\,dx$     $n = 4$

**14.** $\displaystyle\int_0^1 \sqrt{1-x^2}\,dx$     $n = 8$

**15.** $\displaystyle\int_0^2 e^{-x^2}\,dx$     $n = 2$

**16.** $\displaystyle\int_0^2 e^{-x^2}\,dx$     $n = 4$

**17.** $\displaystyle\int_0^3 \frac{1}{2-2x+x^2}\,dx$     $n = 6$

**18.** $\displaystyle\int_0^3 \frac{x}{2+x+x^2}\,dx$     $n = 6$

*Present Value*   In Exercises 19 and 20, use a program similar to the Simpson's Rule program on page 426 with $n = 8$ to approximate the present value of the income $c(t)$ over $t_1$ years at the given annual interest rate $r$. Then use the integration capabilities of a graphing utility to approximate the present value. Compare the results. (Present value is defined in Section 6.2.)

**19.** $c(t) = 6000 + 200\sqrt{t}$, $r = 7\%$, $t_1 = 4$

**20.** $c(t) = 200{,}000 + 15{,}000\sqrt[3]{t}$, $r = 10\%$, $t_1 = 8$

*Marginal Analysis*   In Exercises 21 and 22, use a program similar to the Simpson's Rule program on page 426 with $n = 4$ to approximate the change in revenue from the marginal revenue function $dR/dx$. In each case, assume that the number of units sold $x$ increases from 14 to 16.

**21.** $\dfrac{dR}{dx} = 5\sqrt{8000 - x^3}$     **22.** $\dfrac{dR}{dx} = 50\sqrt{x}\sqrt{20 - x}$

*Probability*   In Exercises 23 and 24, use a program similar to the Simpson's Rule program on page 426 with $n = 6$ to approximate the indicated normal probability. The standard normal probability density function is

$$f(x) = \frac{1}{\sqrt{2\pi}}e^{-x^2/2}.$$

If $x$ is chosen at random from a population with this density, then the probability that $x$ lies in the interval $[a, b]$ is $P(a \le x \le b) = \int_a^b f(x)\,dx$.

**23.** $P(0 \le x \le 1)$

**24.** $P(0 \le x \le 2)$

*Surveying*   In Exercises 25 and 26, use a program similar to the Simpson's Rule program on page 426 to estimate the number of square feet of land in the lot, where $x$ and $y$ are measured in feet, as shown in the figures. In each case, the land is bounded by a stream and two straight roads.

**25.**

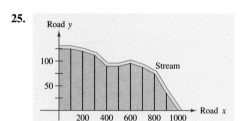

| $x$ | 0 | 100 | 200 | 300 | 400 | 500 |
|---|---|---|---|---|---|---|
| $y$ | 125 | 125 | 120 | 112 | 90 | 90 |

| $x$ | 600 | 700 | 800 | 900 | 1000 |
|---|---|---|---|---|---|
| $y$ | 95 | 88 | 75 | 35 | 0 |

**26.**

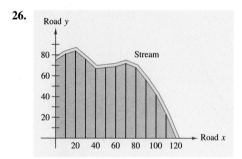

| $x$ | 0 | 10 | 20 | 30 | 40 | 50 | 60 |
|---|---|---|---|---|---|---|---|
| $y$ | 75 | 81 | 84 | 76 | 67 | 68 | 69 |

| $x$ | 70 | 80 | 90 | 100 | 110 | 120 |
|---|---|---|---|---|---|---|
| $y$ | 72 | 68 | 56 | 42 | 23 | 0 |

In Exercises 27–30, use the error formula to find bounds for the error in approximating the integral using (a) the Trapezoidal Rule and (b) Simpson's Rule. (Let $n = 4$.)

**27.** $\displaystyle\int_0^2 x^4 \, dx$

**28.** $\displaystyle\int_0^1 \frac{1}{x + 1} \, dx$

**29.** $\displaystyle\int_0^1 e^{x^3} \, dx$

**30.** $\displaystyle\int_0^1 e^{-x^2} \, dx$

In Exercises 31–34, use the error formula to find $n$ such that the error in the approximation of the definite integral is less than 0.0001 using (a) the Trapezoidal Rule and (b) Simpson's Rule.

**31.** $\displaystyle\int_0^1 x^4 \, dx$

**32.** $\displaystyle\int_1^3 \frac{1}{x} \, dx$

**33.** $\displaystyle\int_1^3 e^{2x} \, dx$

**34.** $\displaystyle\int_3^5 \ln x \, dx$

In Exercises 35–38, use the program for Simpson's Rule given on page 426 to approximate the integral.

**35.** $\displaystyle\int_1^4 x\sqrt{x + 4} \, dx$

**36.** $\displaystyle\int_1^4 x^2\sqrt{x + 4} \, dx$

**37.** $\displaystyle\int_2^5 10xe^{-x} \, dx$

**38.** $\displaystyle\int_2^5 10x^2e^{-x} \, dx$

**39.** Prove that Simpson's Rule is exact when used to approximate the integral of a cubic polynomial function, and demonstrate the result for

$$\int_0^1 x^3 \, dx, \qquad n = 2.$$

**40.** Use a program similar to the Simpson's Rule program on page 426 with $n = 4$ to find the volume of the solid generated by revolving the region bounded by the graphs of
$$y = x\sqrt[3]{x + 4}, \qquad y = 0, \qquad \text{and} \qquad x = 4$$
about the $x$-axis.

In Exercises 41 and 42, use the following definite integral to find the required arc length. If $f$ has a continuous derivative, then the arc length of $f$ between the points $(a, f(a))$ and $(b, f(b))$ is

$$\int_a^b \sqrt{1 + [f'(x)]^2} \, dx.$$

**41. Arc Length** The suspension cable on a bridge that is 400 feet long is in the shape of a parabola whose equation is $y = x^2/800$ (see figure). Use a program similar to the Simpson's Rule program on page 426 with $n = 12$ to approximate the length of the cable. Compare this result with the length obtained by using the table in Section 6.4 to perform the integration.

Figure for 41

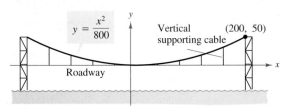

**42. Arc Length** A fleeing object leaves the origin and moves up the $y$-axis (see figure). At the same time, a pursuer leaves the point $(1, 0)$ and always moves toward the fleeing object. If the pursuer's speed is twice that of the fleeing object, the equation of the pursuer's path is

$$y = \frac{1}{3}(x^{3/2} - 3x^{1/2} + 2).$$

Find the distance traveled by the pursuer by integrating over the interval $[0, 1]$.

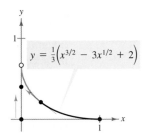

**43. Medicine** A body assimilates a 12-hour cold tablet at a rate modeled by

$$\frac{dC}{dt} = 8 - \ln(t^2 - 2t + 4), \qquad 0 \le t \le 12,$$

where $dC/dt$ is measured in grams per hour and $t$ is the time (in hours). Find the total amount of the drug absorbed into the body during the 12 hours.

**44. Medicine** The concentration $M$ (in grams per liter) of a 6-hour allergy medicine in a body is modeled by

$$M = 12 - 4\ln(t^2 - 4t + 6), \qquad 0 \le t \le 6,$$

where $t$ is the time in hours since the allergy medication was taken. Find the average level of concentration in the body over the 6-hour period.

**45. Magazine Subscribers** The rate of change $S$ of subscribers to a newly introduced magazine is modeled by

$$\frac{dS}{dt} = 1000t^2e^{-t}, \qquad 0 \le t \le 6,$$

where $t$ is the time in years. Find the total increase in the number of subscribers during the first 6 years.

# Improper Integrals

## Improper Integrals

The definition of the definite integral

$$\int_a^b f(x)\,dx$$

includes the requirements that the interval $[a, b]$ be finite and that $f$ be bounded on $[a, b]$. In this section, you will study integrals that do not satisfy these requirements because of one of the following.

1. One or both of the limits of integration are infinite.

2. $f$ has an infinite discontinuity on the interval $[a, b]$.

Integrals having either of these characteristics are called **improper integrals.** For instance, the integrals

$$\int_0^\infty e^{-x}\,dx \quad \text{and} \quad \int_{-\infty}^\infty \frac{1}{x^2 + 1}\,dx$$

are improper because one or both limits of integration are infinite, as indicated in Figure 6.16. Similarly, the integrals

$$\int_1^5 \frac{1}{\sqrt{x - 1}}\,dx \quad \text{and} \quad \int_{-2}^2 \frac{1}{(x + 1)^2}\,dx$$

are improper because their integrands approach infinity somewhere in the interval of integration, as indicated in Figure 6.17.

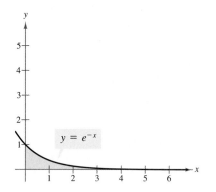

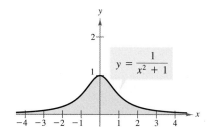

**FIGURE 6.16**

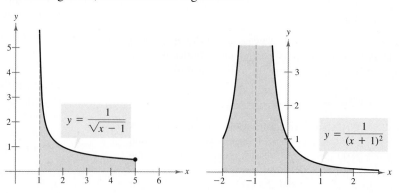

**FIGURE 6.17**

## Integrals with Infinite Limits of Integration

To see how to evaluate an improper integral, consider the integral shown in Figure 6.18. As long as $b$ is a real number that is greater than 1 (no matter how large), this is a definite integral whose value is

$$\int_1^b \frac{1}{x^2}\, dx = \left[-\frac{1}{x}\right]_1^b = -\frac{1}{b} + 1 = 1 - \frac{1}{b}.$$

The table shows the values of this integral for several values of $b$.

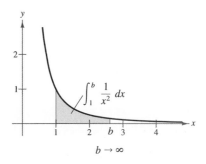

**FIGURE 6.18**

| $b$ | 2 | 5 | 10 | 100 | 1000 | 10,000 |
|---|---|---|---|---|---|---|
| $\int_1^b \frac{1}{x^2}\, dx = 1 - \frac{1}{b}$ | 0.5000 | 0.8000 | 0.9000 | 0.9900 | 0.9990 | 0.9999 |

From this table, it appears that the value of the integral is approaching a limit as $b$ increases without bound. This limit is denoted by the following *improper integral*.

$$\int_1^\infty \frac{1}{x^2}\, dx = \lim_{b \to \infty} \int_1^b \frac{1}{x^2}\, dx$$

$$= \lim_{b \to \infty} \left(1 - \frac{1}{b}\right)$$

$$= 1$$

### Improper Integrals (Infinite Limits of Integration)

1. If $f$ is continuous on the interval $[a, \infty)$, then
$$\int_a^\infty f(x)\, dx = \lim_{b \to \infty} \int_a^b f(x)\, dx.$$

2. If $f$ is continuous on the interval $(-\infty, b]$, then
$$\int_{-\infty}^b f(x)\, dx = \lim_{a \to -\infty} \int_a^b f(x)\, dx.$$

3. If $f$ is continuous on the interval $(-\infty, \infty)$, then
$$\int_{-\infty}^\infty f(x)\, dx = \int_{-\infty}^c f(x)\, dx + \int_c^\infty f(x)\, dx,$$
where $c$ is any real number.

In each case, if the limit exists, then the improper integral **converges;** otherwise the improper integral **diverges.** Thus, in the third case, the integral will diverge if either one of the integrals on the right diverges.

## EXAMPLE 1   Evaluating an Improper Integral

Determine the convergence or divergence of the improper integral

$$\int_1^\infty \frac{1}{x}\,dx.$$

### Solution

Begin by applying the definition of an improper integral.

$$\int_1^\infty \frac{1}{x}\,dx = \lim_{b\to\infty} \int_1^b \frac{1}{x}\,dx \qquad \text{Definition of improper integral}$$

$$= \lim_{b\to\infty} \left[\ln x\right]_1^b \qquad \text{Find antiderivative.}$$

$$= \lim_{b\to\infty} (\ln b - 0) \qquad \text{Apply Fundamental Theorem.}$$

$$= \infty \qquad \text{Evaluate limit.}$$

Because the limit is infinite, the improper integral diverges.

As you begin to work with improper integrals, you will find that integrals that appear to be similar can have very different values. For instance, consider the two improper integrals

$$\int_1^\infty \frac{1}{x}\,dx = \infty \qquad \text{Divergent integral}$$

and

$$\int_1^\infty \frac{1}{x^2}\,dx = 1. \qquad \text{Convergent integral}$$

The first integral diverges and the second converges to 1. Graphically, this means that the areas shown in Figure 6.19 are very different. The region lying between the graph of $y = 1/x$ and the $x$-axis (for $x \geq 1$) has an *infinite* area, and the region lying between the graph of $y = 1/x^2$ and the $x$-axis (for $x \geq 1$) has a *finite* area.

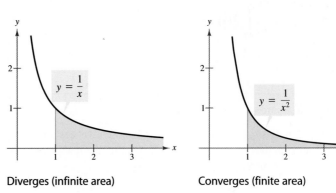

Diverges (infinite area)          Converges (finite area)

**FIGURE 6.19**

---

### Technology

Symbolic integration utilities, such as *Derive*, *Maple*, *Mathcad*, *Mathematica*, or the *TI-92*, evaluate improper integrals in much the same way they evaluate definite integrals. For instance, the steps below show how to use *Derive* for Windows to evaluate the improper integral $\int_1^\infty (1/x^2)\,dx$.

#1: $\dfrac{1}{x^2}$

#2: $\displaystyle\int_1^\infty \dfrac{1}{x^2}\,dx$

#3: 1

## EXAMPLE 2 Evaluating an Improper Integral

Evaluate the improper integral

$$\int_{-\infty}^{0} \frac{1}{(1-2x)^{3/2}} \, dx.$$

*Solution*

Begin by applying the definition of an improper integral.

$$\int_{-\infty}^{0} \frac{1}{(1-2x)^{3/2}} \, dx = \lim_{a \to -\infty} \int_{a}^{0} \frac{1}{(1-2x)^{3/2}} \, dx \qquad \text{Definition of improper integral}$$

$$= \lim_{a \to -\infty} \left[ \frac{1}{\sqrt{1-2x}} \right]_{a}^{0} \qquad \text{Find antiderivative.}$$

$$= \lim_{a \to -\infty} \left( 1 - \frac{1}{\sqrt{1-2a}} \right) \qquad \text{Apply Fundamental Theorem.}$$

$$= 1 - 0 \qquad \text{Evaluate limit.}$$

$$= 1 \qquad \text{Simplify.}$$

Thus, the improper integral converges to 1. As shown in Figure 6.20, this implies that the region lying between the graph of $y = 1/(1-2x)^{3/2}$ and the *x*-axis (for $x \le 0$) has an area of 1.

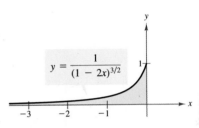

$$y = \frac{1}{(1-2x)^{3/2}}$$

**FIGURE 6.20**

## EXAMPLE 3 Evaluating an Improper Integral

Evaluate the improper integral

$$\int_{0}^{\infty} 2xe^{-x^2} \, dx.$$

*Solution*

Begin by applying the definition of an improper integral.

$$\int_{0}^{\infty} 2xe^{-x^2} \, dx = \lim_{b \to \infty} \int_{0}^{b} 2xe^{-x^2} \, dx \qquad \text{Definition of improper integral}$$

$$= \lim_{b \to \infty} \left[ -e^{-x^2} \right]_{0}^{b} \qquad \text{Find antiderivative.}$$

$$= \lim_{b \to \infty} \left( -e^{-b^2} + 1 \right) \qquad \text{Apply Fundamental Theorem.}$$

$$= 0 + 1 \qquad \text{Evaluate limit.}$$

$$= 1 \qquad \text{Simplify.}$$

Thus, the improper integral converges to 1. As shown in Figure 6.21, this implies that the region lying between the graph of $y = 2xe^{-x^2}$ and the *x*-axis (for $x \ge 0$) has an area of 1.

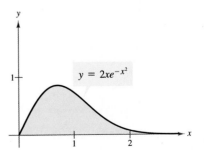

$$y = 2xe^{-x^2}$$

**FIGURE 6.21**

## Integrals with Infinite Integrands

### Improper Integrals (Infinite Integrands)

1. If $f$ is continuous on the interval $[a, b)$ and approaches infinity at $b$, then

$$\int_a^b f(x)\, dx = \lim_{c \to b^-} \int_a^c f(x)\, dx.$$

2. If $f$ is continuous on the interval $(a, b]$ and approaches infinity at $a$, then

$$\int_a^b f(x)\, dx = \lim_{c \to a^+} \int_c^b f(x)\, dx.$$

3. If $f$ is continuous on the interval $[a, b]$, except for some $c$ in $(a, b)$ at which $f$ approaches infinity, then

$$\int_a^b f(x)\, dx = \int_a^c f(x)\, dx + \int_c^b f(x)\, dx.$$

In each case, if the limit exists, then the improper integral **converges;** otherwise the improper integral **diverges.**

### EXAMPLE 4   Evaluating an Improper Integral

Evaluate the improper integral

$$\int_1^2 \frac{1}{\sqrt[3]{x - 1}}\, dx.$$

**Solution**

$$
\begin{aligned}
\int_1^2 \frac{1}{\sqrt[3]{x - 1}}\, dx &= \lim_{b \to 1^+} \int_b^2 \frac{1}{\sqrt[3]{x - 1}}\, dx && \text{Definition of improper integral} \\[2mm]
&= \lim_{b \to 1^+} \left[ \frac{3}{2}(x - 1)^{2/3} \right]_b^2 && \text{Find antiderivative.} \\[2mm]
&= \lim_{b \to 1^+} \left[ \frac{3}{2} - \frac{3}{2}(b - 1)^{2/3} \right] && \text{Apply Fundamental Theorem.} \\[2mm]
&= \frac{3}{2} - 0 && \text{Evaluate limit.} \\[2mm]
&= \frac{3}{2} && \text{Simplify.}
\end{aligned}
$$

Thus, the integral converges to $\frac{3}{2}$. This implies that the region shown in Figure 6.22 has an area of $\frac{3}{2}$.

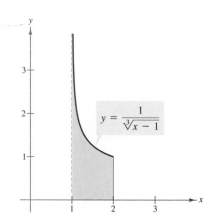

$$y = \frac{1}{\sqrt[3]{x - 1}}$$

**FIGURE 6.22**

## EXAMPLE 5 Evaluating an Improper Integral

Evaluate the improper integral

$$\int_1^2 \frac{2}{x^2 - 2x}\, dx.$$

**Solution**

$$\int_1^2 \frac{2}{x^2 - 2x}\, dx = \int_1^2 \left(\frac{1}{x-2} - \frac{1}{x}\right) dx \qquad \text{Use partial fractions.}$$

$$= \lim_{b \to 2^-} \int_1^b \left(\frac{1}{x-2} - \frac{1}{x}\right) dx \qquad \text{Definition of improper integral}$$

$$= \lim_{b \to 2^-} \left[\ln|x-2| - \ln|x|\right]_1^b \qquad \text{Find antiderivative.}$$

$$= -\infty \qquad \text{Evaluate limit.}$$

Thus, you can conclude that the integral diverges. This implies that the region shown in Figure 6.23 has an infinite area.

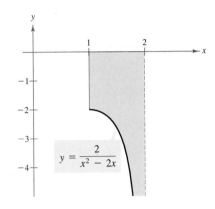

**FIGURE 6.23**

## EXAMPLE 6 Evaluating an Improper Integral

Evaluate the improper integral

$$\int_{-1}^2 \frac{1}{x^3}\, dx.$$

**Solution**

This integral is improper because the integrand has an infinite discontinuity at the interior value $x = 0$, as shown in Figure 6.24. Thus, you can write

$$\int_{-1}^2 \frac{1}{x^3}\, dx = \int_{-1}^0 \frac{1}{x^3}\, dx + \int_0^2 \frac{1}{x^3}\, dx.$$

By applying the definition of an improper integral, you can show that each of these integrals diverges. Therefore, the original improper integral also diverges.

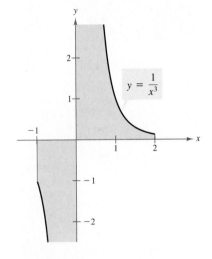

**FIGURE 6.24**

**STUDY TIP** Had you not recognized that the integral in Example 6 was improper, you would have obtained the incorrect result

$$\int_{-1}^2 \frac{1}{x^3}\, dx = \left[-\frac{1}{2x^2}\right]_{-1}^2 = -\frac{1}{8} + \frac{1}{2} = \frac{3}{8}. \qquad \text{Incorrect}$$

Improper integrals in which the integrand has an infinite discontinuity *between* the limits of integration are often overlooked, so keep alert for such possibilities. Even symbolic integrators can have trouble with this type of integral: when we used *Derive* for Windows to evaluate this integral, it gave us the same incorrect result.

## Applications

In Section 4.2, you studied the graph of the *normal probability density function*

$$f(x) = \frac{1}{\sigma\sqrt{2\pi}}e^{-(x-\mu)^2/2\sigma^2}.$$

This function is used in statistics to represent a population that is normally distributed with a mean of $\mu$ and a standard deviation of $\sigma$. Specifically, if an outcome $x$ is chosen at random from the population, the probability that $x$ will have a value between $a$ and $b$ is

$$P(a \leq x \leq b) = \int_a^b \frac{1}{\sigma\sqrt{2\pi}}e^{-(x-\mu)^2/2\sigma^2}\,dx.$$

As shown in Figure 6.25, the probability $P(-\infty < x < \infty)$ is

$$P(-\infty < x < \infty) = \int_{-\infty}^{\infty} \frac{1}{\sigma\sqrt{2\pi}}e^{-(x-\mu)^2/2\sigma^2}\,dx = 1.$$

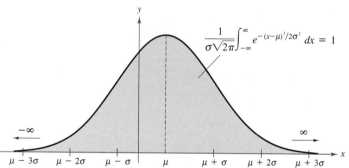

**FIGURE 6.25**

Bob Daemmrich/Stock Boston

*Many professional basketball players are over $6\frac{1}{2}$ feet tall. If a man is chosen at random from the population, the probability that he is $6\frac{1}{2}$ feet tall or taller is less than half of one percent.*

### EXAMPLE 7  Finding a Probability

In 1996, the mean height of American men (between 18 and 24 years old) was 70 inches, and the standard deviation was 3 inches. If an 18- to 24-year-old man was chosen at random from the population, what is the probability that he was 6 feet tall or taller?  *(Source: National Center for Health Statistics)*

**Solution**

Using a mean of $\mu = 70$ and a standard deviation of $\sigma = 3$, the probability $P(72 \leq x < \infty)$ is given by the improper integral

$$P(72 \leq x < \infty) = \int_{72}^{\infty} \frac{1}{3\sqrt{2\pi}}e^{-(x-70)^2/18}\,dx.$$

Using a symbolic integrator, you can approximate the value of this integral to be 0.252. Thus, the probability that the man was 6 feet tall or taller is about 25.2%.

## EXAMPLE 8    Finding Volume

The solid formed by revolving the graph of

$$f(x) = \frac{1}{x}, \quad 1 \le x < \infty$$

about the $x$-axis is called **Gabriel's horn.** (See Figure 6.26.) Find the volume of Gabriel's horn.

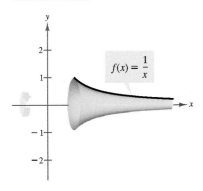

**FIGURE 6.26**

### Discovery

Example 8 shows that the volume of Gabriel's horn is finite. What is the *area* under the curve $f(x) = 1/x$ and above the $x$-axis for $x \ge 1$? It can be shown that the surface area of Gabriel's horn is given by the integral

$$2\pi \int_1^\infty \frac{1}{x}\sqrt{1 + \frac{1}{x^4}}\, dx.$$

Use a symbolic integration utility to show that this improper integral diverges. Hence, Gabriel's horn has finite volume but infinite surface area! How could you paint Gabriel's horn?

### Solution

You can find the volume of the horn using the Disc Method (see Section 5.7).

$$\text{Volume} = \int_1^\infty \pi\left(\frac{1}{x}\right)^2 dx \qquad \text{Disc Method}$$

$$= \lim_{b\to\infty} \int_1^b \frac{\pi}{x^2}\, dx \qquad \text{Definition of improper integral}$$

$$= \lim_{b\to\infty} \left[-\frac{\pi}{x}\right]_1^b \qquad \text{Find antiderivative.}$$

$$= \lim_{b\to\infty} \left(\pi - \frac{\pi}{b}\right) \qquad \text{Apply Fundamental Theorem.}$$

$$= \pi \qquad \text{Evaluate limit.}$$

Thus, Gabriel's horn has a volume of $\pi$ cubic units.

## Group Discussion    Finite or Infinite?

Use a symbolic integration utility to help you decide which of the following improper integrals converge. What can you conclude?

a. $\displaystyle\int_1^\infty \frac{1}{x^{0.9}}\, dx$     b. $\displaystyle\int_1^\infty \frac{1}{x^{1.0}}\, dx$     c. $\displaystyle\int_1^\infty \frac{1}{x^{1.1}}\, dx$

## Warm Up

The following warm-up exercises involve skills that were covered in earlier sections. You will use these skills in the exercise set for this section.

In Exercises 1–6, find the limit.

**1.** $\lim\limits_{x \to 2} (2x + 5)$

**2.** $\lim\limits_{x \to 1} \left( \dfrac{1}{x} + 2x^2 \right)$

**3.** $\lim\limits_{x \to -4} \dfrac{x + 4}{x^2 - 16}$

**4.** $\lim\limits_{x \to 0} \dfrac{x^2 - 2x}{x^3 + 3x^2}$

**5.** $\lim\limits_{x \to 1} \dfrac{1}{\sqrt{x - 1}}$

**6.** $\lim\limits_{x \to -3} \dfrac{x^2 + 2x - 3}{x + 3}$

In Exercises 7–10, evaluate the expression when $x = b$ and when $x = 0$.

**7.** $\dfrac{4}{3}(2x - 1)^3$

**8.** $\dfrac{1}{x - 5} + \dfrac{3}{(x - 2)^2}$

**9.** $\ln(5 - 3x^2) - \ln(x + 1)$

**10.** $e^{3x^2} + e^{-3x^2}$

# EXERCISES 6.6

In Exercises 1–14, determine whether the improper integral converges. If it does, evaluate the integral.

**1.** $\displaystyle\int_0^\infty e^{-x}\, dx$

**2.** $\displaystyle\int_{-\infty}^0 e^{2x}\, dx$

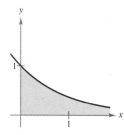

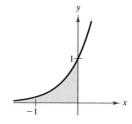

**3.** $\displaystyle\int_1^\infty \dfrac{1}{x^2}\, dx$

**4.** $\displaystyle\int_1^\infty \dfrac{1}{\sqrt{x}}\, dx$

**5.** $\displaystyle\int_0^\infty e^{x/3}\, dx$

**6.** $\displaystyle\int_0^\infty \dfrac{5}{e^{2x}}\, dx$

**7.** $\displaystyle\int_5^\infty \dfrac{x}{\sqrt{x^2 - 16}}\, dx$

**8.** $\displaystyle\int_{1/2}^\infty \dfrac{1}{\sqrt{2x - 1}}\, dx$

**9.** $\displaystyle\int_{-\infty}^0 e^{-x}\, dx$

**10.** $\displaystyle\int_{-\infty}^{-1} \dfrac{1}{x^2}\, dx$

**11.** $\displaystyle\int_{-\infty}^\infty e^{|x|}\, dx$

**12.** $\displaystyle\int_{-\infty}^\infty \dfrac{|x|}{x^2 + 1}\, dx$

**13.** $\displaystyle\int_{-\infty}^\infty 2xe^{-3x^2}\, dx$

**14.** $\displaystyle\int_{-\infty}^\infty x^2 e^{-x^3}\, dx$

In Exercises 15–18, determine the divergence or convergence of the given improper integral. Evaluate the integral if it converges.

**15.** $\displaystyle\int_0^4 \dfrac{1}{\sqrt{x}}\, dx$

**16.** $\displaystyle\int_3^4 \dfrac{1}{\sqrt{x - 3}}\, dx$

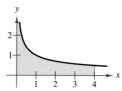

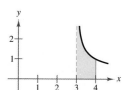

**17.** $\int_0^2 \dfrac{1}{(x-1)^{2/3}}\, dx$

**18.** $\int_0^2 \dfrac{1}{(x-1)^2}\, dx$

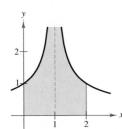

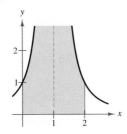

In Exercises 19–28, evaluate the improper integral.

**19.** $\int_0^1 \dfrac{1}{1-x}\, dx$

**20.** $\int_0^{27} \dfrac{5}{\sqrt[3]{x}}\, dx$

**21.** $\int_0^9 \dfrac{1}{\sqrt{9-x}}\, dx$

**22.** $\int_0^2 \dfrac{x}{\sqrt{4-x^2}}\, dx$

**23.** $\int_0^1 \dfrac{1}{x^2}\, dx$

**24.** $\int_0^1 \dfrac{1}{x}\, dx$

**25.** $\int_0^2 \dfrac{1}{\sqrt[3]{x-1}}\, dx$

**26.** $\int_0^2 \dfrac{1}{(x-1)^{4/3}}\, dx$

**27.** $\int_3^4 \dfrac{1}{\sqrt{x^2-9}}\, dx$

**28.** $\int_3^5 \dfrac{1}{x^2\sqrt{x^2-9}}\, dx$

In Exercises 29–32, complete the table for the specified values of $a$ and $n$ to demonstrate that

$$\lim_{x \to \infty} x^n e^{-ax} = 0, \quad a > 0, n > 0.$$

| $x$ | 1 | 10 | 25 | 50 |
|-----|---|----|----|----|
| $x^n e^{-ax}$ | | | | |

**29.** $a = 1, n = 1$

**30.** $a = 2, n = 4$

**31.** $a = \frac{1}{2}, n = 2$

**32.** $a = \frac{1}{2}, n = 5$

In Exercises 33–36, use the results of Exercises 29–32 to evaluate the improper integral.

**33.** $\int_0^\infty x^2 e^{-x}\, dx$

**34.** $\int_0^\infty (x-1)e^{-x}\, dx$

**35.** $\int_0^\infty x e^{-2x}\, dx$

**36.** $\int_0^\infty x e^{-x}\, dx$

**37.** *Present Value*  You are asked to calculate the pay for a business that is forecast to yield a continuous flow of profit at the rate of $500,000 per year. If money will earn interest at the nominal rate of 9% per year compounded continuously, what is the present value of the business (a) for 20 years and (b) forever?

**38.** *Present Value*  Repeat Exercise 37 for a farm that is expected to produce a profit of $75,000 per year. Assume that money will earn interest at the nominal rate of 8% compounded continuously.

*Capitalized Cost*  In Exercises 39 and 40, find the capitalized cost $C$ of an asset (a) for $n = 5$ years, (b) for $n = 10$ years, and (c) forever. The **capitalized cost** is given by

$$C = C_0 + \int_0^n c(t)e^{-rt}dt,$$

where $C_0$ is the original investment, $t$ is the time in years, $r$ is the annual interest rate compounded continuously, and $c(t)$ is the annual cost of maintenance. [*Hint:* For part (c), see Exercises 29–32.]

**39.** $C_0 = \$650{,}000$, $c(t) = 25{,}000$, $r = 10\%$

**40.** $C_0 = \$650{,}000$, $c(t) = \$25{,}000(1 + 0.08t)$, $r = 12\%$

In Exercises 41 and 42, (a) find the area of the region bounded by the graphs of the given equations and (b) find the volume of the solid generated by revolving the region about the $x$-axis.

**41.** $y = \dfrac{1}{x^2}, y = 0, x \geq 1$

**42.** $y = e^{-x}, y = 0, x \geq 0$

**43.** *Women's Height*  The mean height of American women between the ages of 25 and 34 is 64.5 inches, and the standard deviation is 2.4 inches. Find the probability that a 25- to 34-year-old woman chosen at random is

(a) between 5 and 6 feet tall.

(b) 5 feet 8 inches or taller.

(c) 6 feet or taller.

*(Source: U.S. National Center for Health Statistics)*

**44.** *Quality Control*  A company manufactures wooden yardsticks. The lengths of the yardsticks are normally distributed, with a mean of 36 inches and a standard deviation of 0.2 inch. Find the probability that a yardstick is (a) longer than 35.5 inches and (b) longer than 35.9 inches.

## Chapter 6  *Algebra Review*

### Algebra and Integration Techniques

Integration techniques involve many different algebraic skills. Study the examples on these two pages. Be sure that you understand the algebra used in each step.

---

### EXAMPLE 1   Algebra and Integration Techniques

(a) $\dfrac{28}{9}(1 - u^3)(u)(-3u^2)$                 Example 6, page 389

$= \dfrac{(3)(28)}{9}(1 - u^3)(-u^3)$                 Regroup factors.

$= \dfrac{28}{3}(1 - u^3)(-u^3)$                 Simplify fraction.

$= \dfrac{28}{3}(u^6 - u^3)$                 Multiply.

(b) $\dfrac{2}{x - 3} - \dfrac{1}{x + 2}$                 Example 1, page 403

$= \dfrac{2(x + 2)}{(x - 3)(x + 2)} - \dfrac{(x - 3)}{(x - 3)(x + 2)}$                 Rewrite with common denominator.

$= \dfrac{2(x + 2) - (x - 3)}{(x - 3)(x + 2)}$                 Rewrite as single fraction.

$= \dfrac{2x + 4 - x + 3}{x^2 - x - 6}$                 Multiply factors.

$= \dfrac{x + 7}{x^2 - x - 6}$                 Combine like terms.

(c) $\dfrac{6}{x} - \dfrac{1}{x + 1} + \dfrac{9}{(x + 1)^2}$                 Example 2, page 404

$= \dfrac{6(x + 1)^2}{x(x + 1)^2} - \dfrac{x(x + 1)}{x(x + 1)^2} + \dfrac{9x}{x(x + 1)^2}$                 Rewrite with common denominator.

$= \dfrac{6(x + 1)^2 - x(x + 1) + 9x}{x(x + 1)^2}$                 Rewrite as single fraction.

$= \dfrac{6x^2 + 12x + 6 - x^2 - x + 9x}{x^3 + 2x^2 + x}$                 Multiply factors.

$= \dfrac{5x^2 + 20x + 6}{x^3 + 2x^2 + x}$                 Combine like terms.

## EXAMPLE 2 Algebra and Integration Techniques

(a) $x + 1 + \dfrac{1}{x^3} + \dfrac{1}{x - 1}$

Example 3, page 405

$= \dfrac{(x + 1)(x^3)(x - 1)}{x^3(x - 1)} + \dfrac{x - 1}{x^3(x - 1)} + \dfrac{x^3}{x^3(x - 1)}$

Rewrite with common denominator.

$= \dfrac{(x + 1)(x^3)(x - 1) + (x - 1) + x^3}{x^3(x - 1)}$

Rewrite as single fraction.

$= \dfrac{(x^2 - 1)(x^3) + x - 1 + x^3}{x^3(x - 1)}$

$(x + 1)(x - 1) = x^2 - 1$

$= \dfrac{x^5 - x^3 + x - 1 + x^3}{x^4 - x^3}$

Multiply factors.

$= \dfrac{x^5 + x - 1}{x^4 - x^3}$

Combine like terms.

(b) $x^2 e^x - 2(x - 1)e^x$

Example 5, page 418

$= x^2 e^x - 2(xe^x - e^x)$

Multiply factors.

$= x^2 e^x - 2xe^x + 2e^x$

Multiply factors.

$= e^x(x^2 - 2x + 2)$

Factor.

(c) $6 \ln|x| - \ln|x + 1| + 9\dfrac{(x + 1)^{-1}}{-1}$

Example 2, page 404

$= \ln|x|^6 - \ln|x + 1| + 9\dfrac{(x + 1)^{-1}}{-1}$

$m \ln n = \ln n^m$

$= \ln|x^6| - \ln|x + 1| + 9\dfrac{(x + 1)^{-1}}{-1}$

Property of absolute value

$= \ln\dfrac{|x^6|}{|x + 1|} + 9\dfrac{(x + 1)^{-1}}{-1}$

$\ln m - \ln n = \ln \dfrac{m}{n}$

$= \ln\left|\dfrac{x^6}{x + 1}\right| + 9\dfrac{(x + 1)^{-1}}{-1}$

$\left|\dfrac{a}{b}\right| = \dfrac{|a|}{|b|}$

$= \ln\left|\dfrac{x^6}{x + 1}\right| - 9(x + 1)^{-1}$

Rewrite sum as difference.

$= \ln\left|\dfrac{x^6}{x + 1}\right| - \dfrac{9}{x + 1}$

Rewrite with positive exponent.

# Chapter Summary and Study Strategies

*After studying this chapter, you should have acquired the following skills. The exercise numbers are keyed to the Review Exercises that begin on page 446. Answers to odd-numbered Review Exercises are given in the back of the text.\**

■ Use the basic integration formulas to find indefinite integrals.  *(Section 6.1)*                    *Review Exercises 1–12*

Constant Rule: $\int k\,dx = kx + C$

Power Rules: $\int x^n dx = \dfrac{x^{n+1}}{n+1} + C, \quad \int u^n \dfrac{du}{dx} dx = \dfrac{u^{n+1}}{n+1} + C, \quad n \neq -1$

Exponential Rules: $\int e^x dx = e^x + C, \quad \int e^u \dfrac{du}{dx} dx = \int e^u\,du = e^u + C$

Log Rules: $\int \dfrac{1}{x} dx = \ln|x| + C, \quad \int \dfrac{du/dx}{u} dx = \int \dfrac{1}{u} du = \ln|u| + C$

■ Use substitution to find indefinite integrals.  *(Section 6.1)*                                        *Review Exercises 13–20*

■ Use substitution to evaluate definite integrals.  *(Section 6.1)*                                      *Review Exercises 21–24*

■ Use integration to solve real-life problems.  *(Section 6.1)*                                          *Review Exercises 25–28*

■ Use integration by parts to find indefinite integrals.  *(Section 6.2)*                                *Review Exercises 29–32*

$\int u\,dv = uv - \int v\,du$

■ Use integration by parts repeatedly to find indefinite integrals.  *(Section 6.2)*                     *Review Exercises 33, 34*

■ Find the present value of future income.  *(Section 6.2)*                                              *Review Exercises 35–42*

■ Use partial fractions to find indefinite integrals.  *(Section 6.3)*                                   *Review Exercises 43–48*

■ Use logistics growth functions to model real-life situations.  *(Section 6.3)*                         *Review Exercises 49, 50*

■ Use integration tables to find indefinite integrals.  *(Section 6.4)*                                  *Review Exercises 51–56*

■ Use reduction formulas to find indefinite integrals.  *(Section 6.4)*                                  *Review Exercises 57–60*

■ Use completing the square to find indefinite integrals.  *(Section 6.4)*                               *Review Exercises 61–64*

■ Use the Trapezoidal Rule to approximate definite integrals.  *(Section 6.5)*                           *Review Exercises 65–68*

$\int_a^b f(x)dx \approx \dfrac{b-a}{2n}[f(x_0) + 2f(x_1) + \cdots + 2f(x_{n-1}) + f(x_n)]$

■ Use Simpson's Rule to approximate definite integrals.  *(Section 6.5)*                                 *Review Exercises 69–72*

$\int_a^b f(x)dx \approx \dfrac{b-a}{3n}[f(x_0) + 4f(x_1) + 2f(x_2) + 4f(x_3) + \cdots + 4f(x_{n-1}) + f(x_n)]$

■ Analyze the sizes of the errors when approximating definite integrals with the                         *Review Exercises 73, 74*
Trapezoidal Rule.  *(Section 6.5)*

$|E| \leq \dfrac{(b-a)^3}{12n^2}[\max|f''(x)|], \quad a \leq x \leq b$

---

\* Use a wide range of valuable study aids to help you master the material in this chapter. The *Student Solutions Guide* includes step-by-step solutions to all odd-numbered exercises to help you review and prepare. The *Algebra Review Tutorial Software* and *The Algebra of Calculus* help you brush up on your algebra skills. The *Graphing Technology Guide* offers step-by-step commands and instructions for a wide variety of graphing calculators, including the most recent models.

■ Analyze the sizes of the errors when approximating definite integrals with Simpson's Rule. *(Section 6.5)*

*Review Exercises 75, 76*

$$|E| \le \frac{(b-a)^5}{180n^4}[\max|f^{(4)}(x)|], \quad a \le x \le b$$

■ Evaluate improper integrals with infinite limits of integration. *(Section 6.6)*

*Review Exercises 77–80*

$$\int_a^\infty f(x)dx = \lim_{b\to\infty} \int_a^b f(x)dx, \quad \int_{-\infty}^b f(x)dx = \lim_{a\to-\infty} \int_a^b f(x)dx$$

■ Evaluate improper integrals with infinite integrands. *(Section 6.6)*

*Review Exercises 81–84*

$$\int_a^b f(x)dx = \lim_{c\to b^-} \int_a^c f(x)dx, \quad \int_a^b f(x)dx = \lim_{c\to a^+} \int_c^b f(x)dx$$

■ Use improper integrals to solve real-life problems. *(Section 6.6)*

*Review Exercises 85–87*

■ *Use a Variety of Approaches* To be efficient at finding antiderivatives, you need to use a variety of approaches.

1. Check to see whether the integral fits one of the basic integration formulas—you should have these formulas memorized.

2. Try an integration technique such as substitution, integration by parts, partial fractions, or completing the square to rewrite the integral in a form that fits one of the basic integration formulas.

3. Use a table of integrals.

4. Use a symbolic integration utility such as *Derive, Maple, Mathcad, Mathematica,* or the *TI-92.*

■ *Use Numerical Integration* When solving a definite integral, remember that you cannot apply the Fundamental Theorem of Calculus unless you can find an antiderivative of the integrand. This is not always possible—even with a symbolic integration utility. In such cases, you can use a numerical technique such as the Midpoint Rule, the Trapezoidal Rule, or Simpson's Rule to approximate the value of the integral.

■ *Improper Integrals* When solving integration problems, remember that the symbols used to denote definite integrals are the same as those used to denote improper integrals. Evaluating an improper integral as a definite integral can lead to an incorrect value. For instance, if you evaluated the integral

$$\int_{-2}^1 \frac{1}{x^2}dx$$

as though it were a definite integral, you would obtain a value of $-\frac{3}{2}$. This is not, however, correct. This integral is actually a divergent improper integral. If you have access to a symbolic integration utility, try using it to evaluate this integral—it will probably make the same mistake.

# Chapter 6   Review Exercises

In Exercises 1–12, use a basic integration formula to find the indefinite integral.

**1.** $\displaystyle\int dt$

**2.** $\displaystyle\int (x^2 + 2x - 1)dx$

**3.** $\displaystyle\int (x + 5)^3 dx$

**4.** $\displaystyle\int \frac{2}{(x - 1)^2} dx$

**5.** $\displaystyle\int e^{10x} dx$

**6.** $\displaystyle\int 3xe^{-x^2} dx$

**7.** $\displaystyle\int \frac{1}{5x} dx$

**8.** $\displaystyle\int \frac{2x^3 - x}{x^4 - x^2 + 1} dx$

**9.** $\displaystyle\int x\sqrt{x^2 + 4}\, dx$

**10.** $\displaystyle\int \frac{1}{\sqrt{2x - 9}} dx$

**11.** $\displaystyle\int \frac{2e^x}{3 + e^x} dx$

**12.** $\displaystyle\int (x^2 - 1)e^{x^3 - 3x} dx$

In Exercises 13–20, use substitution to find the indefinite integral.

**13.** $\displaystyle\int x(x - 2)^3 dx$

**14.** $\displaystyle\int x(1 - x)^2 dx$

**15.** $\displaystyle\int x\sqrt{x + 1}\, dx$

**16.** $\displaystyle\int x^2 \sqrt{x + 1}\, dx$

**17.** $\displaystyle\int 2x\sqrt{x - 3}\, dx$

**18.** $\displaystyle\int \frac{\sqrt{x}}{1 + \sqrt{x}} dx$

**19.** $\displaystyle\int (x + 1)\sqrt{1 - x}\, dx$

**20.** $\displaystyle\int \frac{x}{x - 1} dx$

In Exercises 21–24, use substitution to evaluate the definite integral. Use a symbolic integration utility to check your answer.

**21.** $\displaystyle\int_2^3 x\sqrt{x - 2}\, dx$

**22.** $\displaystyle\int_2^3 x^2 \sqrt{x - 2}\, dx$

**23.** $\displaystyle\int_1^3 x^2(x - 1)^3 dx$

**24.** $\displaystyle\int_{-3}^0 x(x + 3)^4 dx$

**25.** *Probability*   The probability of recall in an experiment is found to be

$$P(a \le x \le b) = \int_a^b \frac{105}{16}x^2 \sqrt{1 - x}\, dx,$$

where $x$ represents the percent of recall (see figure).

(a) Find the probability that a randomly chosen individual will recall 80% of the material.

(b) What is the median percent recall? That is, for what value of $b$ is it true that $P(0 \le x \le b) = 0.5$?

Figure for 25

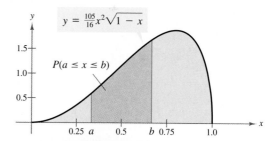

**26.** *Probability*   The probability of locating between $a$ and $b$ percent of oil and gas deposits in a region is

$$P(a \le x \le b) = \int_a^b \frac{140}{27}x^2(1 - x)^{1/3}\, dx$$

(see figure).

(a) Find the probability that between 40% and 60% of the deposits will be found.

(b) Find the probability that between 0% and 50% of the deposits will be found.

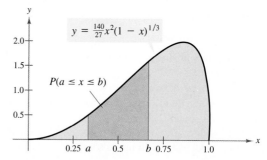

**27.** *Profit*   The net profit for Coca-Cola in millions of dollars per year from 1987 through 1996 can be modeled by

$$P = 449.63 + 4.15t^2 \ln t,$$

where $t$ is the time in years, with $t = 7$ corresponding to 1987.   *(Source: Coca-Cola)*

(a) Find the average net profit for the years 1987 through 1996.

(b) Find the total net profit for the years 1987 through 1996.

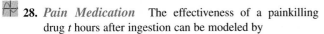

**28.** *Pain Medication*    The effectiveness of a painkilling drug $t$ hours after ingestion can be modeled by

$$E(t) = te^{-0.4t},$$

where the effectiveness $E$ is measured as a percent.

(a) Use a graphing utility to graph the model over the interval $0 \le t \le 24$.

(b) Over what intervals is the effectiveness increasing and decreasing?

(c) At what time does the medication reach maximum effectiveness?

(d) Determine the average effectiveness of the medication over the 24-hour period.

(e) At what time does the medication have the same effectiveness as the average effectiveness for the entire interval?

In Exercises 29–32, use integration by parts to find the indefinite integral.

**29.** $\displaystyle\int \frac{\ln x}{\sqrt{x}}\,dx$

**30.** $\displaystyle\int \sqrt{x}\,\ln x\,dx$

**31.** $\displaystyle\int (x-1)e^x\,dx$

**32.** $\displaystyle\int \ln\!\left(\frac{x}{x+1}\right)dx$

In Exercises 33 and 34, use integration by parts repeatedly to find the indefinite integral. Use a symbolic integration utility to check your answer.

**33.** $\displaystyle\int 2x^2e^{2x}\,dx$

**34.** $\displaystyle\int (\ln x)^3\,dx$

In Exercises 35–38, find the present value of the income given by $c(t)$ (measured in dollars) over $t_1$ years at the given annual inflation rate.

**35.** $c(t) = 10{,}000,\ r = 4\%,\ t_1 = 5$ years

**36.** $c(t) = 20{,}000 + 1500t,\ r = 6\%,\ t_1 = 10$ years

**37.** $c(t) = 12{,}000t,\ r = 5\%,\ t_1 = 10$ years

**38.** $c(t) = 10{,}000 + 100e^{t/2},\ r = 5\%,\ t_1 = 5$ years

**39.** *Economics: Present Value*    Calculate the present values of the following:

(a) $1000 per year for 5 years at interest rates of 5, 10, and 15%

(b) A lottery ticket that pays $100,000 per year after taxes over 20 years, assuming an inflation rate of 8%

*(Source: Adapted from Boyes/Melvin,* Economics, *Third Edition)*

**40.** *Finance: Present Value*    You receive $1000 at the end of each year for the next 3 years to help with college expenses. Assuming an annual interest rate of 6%, what is the present value of that stream of payments? *(Source: Adapted from Garman/Forgue,* Personal Finance, *Fifth Edition)*

**41.** *Finance: Present Value*    Determine the amount a person planning for retirement would need to deposit today to be able to withdraw $6000 each year for the next 10 years from an account earning 6%. *(Source: Adapted from Garman/Forgue,* Personal Finance, *Fifth Edition)*

**42.** *Finance: Present Value*    A person invests $50,000 earning 6%. If $6000 is withdrawn each year, use present value to determine how many years it will take for the fund to run out. *(Source: Adapted from Garman/Forgue,* Personal Finance, *Fifth Edition)*

In Exercises 43–48, use partial fractions to find the indefinite integral.

**43.** $\displaystyle\int \frac{1}{x(x+5)}\,dx$

**44.** $\displaystyle\int \frac{4x-2}{3(x-1)^2}\,dx$

**45.** $\displaystyle\int \frac{4x-13}{x^2-3x-10}\,dx$

**46.** $\displaystyle\int \frac{4x^2-x-5}{x^2(x+5)}\,dx$

**47.** $\displaystyle\int \frac{x^2}{x^2+2x-15}\,dx$

**48.** $\displaystyle\int \frac{x^2+2x-12}{x(x+3)}\,dx$

**49.** *Sales Growth*    A new product initially sells 1250 units per week. After 6 months, the number of sales increases to 6500. The sales can be modeled by logistics growth with a limit of 10,000 units per week.

(a) Find a logistics growth model for the number of units.

(b) Use the model to complete the table.

| Time, $t$ | 0 | 3 | 6 | 12 | 24 |
|-----------|---|---|---|----|----|
| Sales, $y$ |   |   |   |    |    |

(c) Use the graph below to approximate the time $t$ that sales will be 7500.

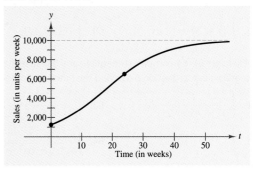

**50.** *Population Growth* A conservation society has introduced a population of 300 ring-neck pheasants into a new area. After 5 years the population has increased to 966. The population can be modeled by logistics growth with a limit of 2700 pheasants.

(a) Find a logistics growth model for the population of ring-neck pheasants.

(b) How many pheasants were present after 4 years?

(c) How long will it take to establish a population of 1750 pheasants?

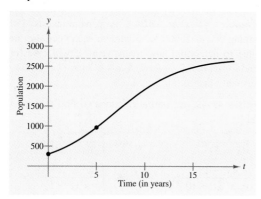

In Exercises 51–56, use the table of integrals in Section 6.4 to evaluate the integral.

**51.** $\int \dfrac{\sqrt{x^2 + 25}}{x}\, dx$

**52.** $\int \dfrac{1}{x(4 + 3x)}\, dx$

**53.** $\int \dfrac{1}{(x^2 - 4)}\, dx$

**54.** $\int x(\ln x^2)^2\, dx$

**55.** $\int_0^3 \dfrac{x}{\sqrt{1 + x}}\, dx$

**56.** $\int_1^3 \dfrac{1}{x^2\sqrt{16 - x^2}}\, dx$

In Exercises 57–60, use a reduction formula from the table of integrals in Section 6.4 to find the indefinite integral.

**57.** $\int \dfrac{\sqrt{1 + x}}{x}\, dx$

**58.** $\int \dfrac{1}{(x^2 - 9)^2}\, dx$

**59.** $\int (x - 5)^3 e^{x - 5}\, dx$

**60.** $\int (\ln x)^4\, dx$

In Exercises 61–64, complete the square and then use the table of integrals given in Section 6.4 to find the indefinite integral.

**61.** $\int \dfrac{1}{x^2 + 4x - 21}\, dx$

**62.** $\int \dfrac{1}{x^2 - 8x - 52}\, dx$

**63.** $\int \sqrt{x^2 - 10x}\, dx$

**64.** $\int \dfrac{x}{\sqrt{x^4 + 6x^2 + 10}}\, dx$

In Exercises 65–68, use the Trapezoidal Rule to approximate the definite integral.

**65.** $\int_1^3 \dfrac{1}{x^2}\, dx, \ n = 4$

**66.** $\int_0^2 (x^2 + 1)\, dx, \ n = 4$

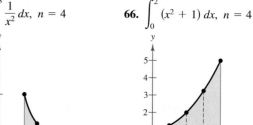

**67.** $\int_1^2 \dfrac{1}{1 + \ln x}\, dx, \ n = 4$

**68.** $\int_0^2 \dfrac{1}{\sqrt{1 + x^3}}\, dx, \ n = 8$

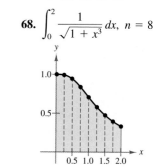

In Exercises 69–72, use Simpson's Rule to approximate the definite integral.

**69.** $\int_1^2 \dfrac{1}{x^3}\, dx, \ n = 4$

**70.** $\int_1^2 x^3\, dx, \ n = 4$

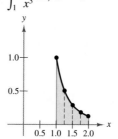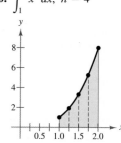

**71.** $\int_0^1 \dfrac{x^{3/2}}{2 - x^2}\, dx, \ n = 4$

**72.** $\int_0^1 e^{x^2}\, dx, \ n = 6$

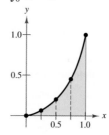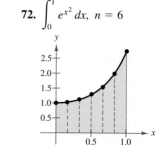

In Exercises 73 and 74, use the error formula to find bounds for the error in approximating the integral using the Trapezoidal Rule.

**73.** $\int_0^2 e^{2x}\, dx, \quad n = 4$

**74.** $\int_0^2 e^{2x}\, dx, \quad n = 8$

In Exercises 75 and 76, use the error formula to find bounds for the error in approximating the integral using Simpson's Rule.

**75.** $\int_2^4 \frac{1}{x-1}\, dx, \quad n = 4$

**76.** $\int_2^4 \frac{1}{x-1}\, dx, \quad n = 8$

In Exercises 77–84, evaluate the improper integral.

**77.** $\int_0^\infty 4xe^{-2x^2}\, dx$

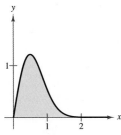

**78.** $\int_{-\infty}^0 \frac{3}{(1-3x)^{2/3}}\, dx$

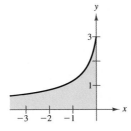

**79.** $\int_{-\infty}^0 \frac{1}{3x^2}\, dx$

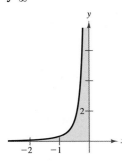

**80.** $\int_0^\infty 2x^2 e^{-x^3}\, dx$

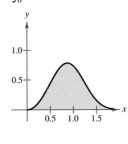

**81.** $\int_0^4 \frac{1}{\sqrt{4x}}\, dx$

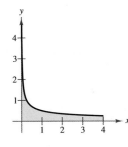

**82.** $\int_1^2 \frac{x}{16(x-1)^2}\, dx$

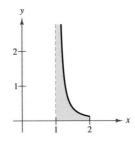

**83.** $\int_2^3 \frac{1}{\sqrt{x-2}}\, dx$

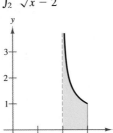

**84.** $\int_0^2 \frac{x+2}{(x-1)^2}\, dx$

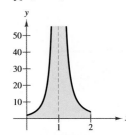

**85.** *Present Value*   You are considering buying a franchise that yields a continuous income stream of $50,000 per year. Find the present value of the franchise (a) for 15 years and (b) forever. Assume that money earns 6% interest per year, compounded continuously.

**86.** *Capitalized Cost*   A company invests $1.5 million in a new manufacturing plant that will cost $50,000 per year in maintenance. Find the capitalized cost for (a) 20 years and (b) forever. Assume that money earns 6% interest, compounded continuously.

**87.** *Per Capita Income*   In 1995, the per capita income per state was approximately normally distributed with a mean of $21,875.30 and a standard deviation of $2988.40 (see figure). Find the probability that a state has a per capita income of (a) $25,000 or greater and (b) $30,000 or greater. *(Source: U.S. Bureau of Economic Analysis)*

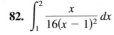

$$y = \frac{1}{2988.40\sqrt{2\pi}}e^{-(x-21,875.3)^2/2(2988.4)^2}$$

# Sample Post-Graduation Exam Questions

*The following questions represent the types of questions that appear on certified public accountant (CPA) exams, graduate management admission tests (GMAT), graduate records exams (GRE), actuarial exams, and college-level academic skills tests (CLAST). The answers to the questions are given in the back of the book.*

Figure for 1–4

Resident Population by Age
State of Florida, 1992

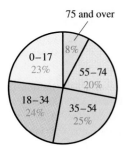

Resident Population by Age
State of Florida, 1995

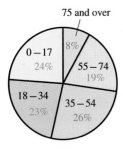

Figure for 7

For Questions 1–4, the total 1992 population was 13,488,000; the total 1995 population was 14,166,000. Also use the data given in the graphs.    *(Source: U.S. Bureau of the Census)*

**1.** Find the number of people aged 75 and over for the year 1992.

    (a) 986,800        (b) 944,160        (c) 1,079,840        (d) 863,450

**2.** Find the increase in population of 35- to 54-year-olds from 1992 to 1995.

    (a) 311,160        (b) 134,880        (c) 674,400        (d) 622,760

**3.** In 1992, how many people were 54 years old or younger?

    (a) 3,776,640        (b) 6,339,360        (c) 9,711,360        (d) 10,341,180

**4.** In what age group did the population and the population percent decrease between 1992 and 1995?

    (a) 0–17        (b) 75 and over    (c) 18–34        (d) 55–74

**5.** $\displaystyle\int_{1}^{6} \frac{x}{\sqrt{x+3}} =$

    (a) $\frac{10}{3}$        (b) $-\frac{20}{3}$        (c) $-4$        (d) $\frac{20}{3}$

**6.** $\displaystyle\lim_{x \to \infty} \frac{2x^2 + 5x - 16}{5x^2 - 15x + 48} =$

    (a) 0        (b) $\frac{2}{5}$        (c) $\infty$        (d) $\frac{1}{3}$

**7.** The city council is planning to construct a new street, as shown in the figure. Construction costs of the new street are $110 per linear foot. What is the projected cost for constructing the new street?

    (a) $2,904,000    (b) $4,065,600    (c) $1,904,000    (d) $3,864,000

**8.** The following information pertains to Varn Co.:

    Sales $1,000,000,    Variable Costs $200,000,    Fixed Costs $50,000

    What is Varn's break-even point in sales dollars?

    (a) $40,000        (b) $250,000        (c) $62,500        (d) $200,000

# Functions of Several Variables

## 7

### STRATEGIES *for* SUCCESS

 **OBJECTIVES**

*When you have completed this chapter, make sure you are able to:*

❑ Analyze surfaces and graph functions of two variables on the three-dimensional coordinate system.
❑ Calculate partial derivatives of functions of several variables.
❑ Find extrema of functions of several variables using the First- and Second-Partials Tests.
❑ Use Lagrange multipliers to solve constrained optimization problems.
❑ Use the least squares regression line for mathematical modeling.
❑ Evaluate double integrals and use them to find areas and volumes.

 **TOOLS**

*Use these study tools to achieve the objectives above:*

**Algebra Review**
(pages 536 and 537)

**Chapter Summary and Study Strategies**
(pages 538 and 539)

**Review Exercises**
(pages 540–543)

**ADDITIONAL RESOURCES**

*Use these resources to solidify your mastery of calculus:*

**Student Solutions Guide**

**Study Guide** (Additional Examples, Similar Problems, and Chapter Test)

**Algebra Review Tutorial Software**

**Graphing Technology Guide**

**Sample Post-Graduation Exam Questions**
(page 544)

**Web Exercise**
(page 509, exercise 46)

Spencer Grant/Photo Researchers, Inc.

*In Exercise 48 on page 478, you will discuss how a contour map representing seismic amplitudes of a fault horizon is used in earthquake studies.*

# The Three-Dimensional Coordinate System

*The Three-Dimensional Coordinate System  •  The Distance and Midpoint Formulas  •
The Equation of a Sphere  •  Traces of Surfaces*

## The Three-Dimensional Coordinate System

Recall that the Cartesian plane is determined by two perpendicular number lines called the $x$-axis and the $y$-axis. These axes together with their point of intersection (the origin) allow you to develop a two-dimensional coordinate system for identifying points in a plane. To identify a point in space, you must introduce a third dimension to the model. The geometry of this three-dimensional model is called **solid analytic geometry.**

*Discovery*

Describe the location of a point $(x, y, z)$ if $x = 0$. Describe the location of a point $(x, y, z)$ if $x = 0$ and $y = 0$. What can you say about the ordered triple $(x, y, z)$ if the point is located on the $y$-axis? What can you say about the ordered triple $(x, y, z)$ if the point is located in the $xz$-plane?

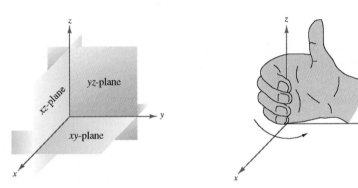

**FIGURE 7.1**          **FIGURE 7.2**

You can construct a **three-dimensional coordinate system** by passing a $z$-axis perpendicular to both the $x$- and $y$-axes at the origin. Figure 7.1 shows the positive portion of each coordinate axis. Taken as pairs, the axes determine three **coordinate planes: the $xy$-plane, the $xz$-plane, and the $yz$-plane.** These three coordinate planes separate the three-dimensional coordinate system into eight **octants.** The first octant is the one for which all three coordinates are positive. In this three-dimensional system, a point $P$ in space is determined by an ordered triple $(x, y, z)$, where $x$, $y$, and $z$ are as follows.

$x$ = directed distance from $yz$-plane to $P$

$y$ = directed distance from $xz$-plane to $P$

$z$ = directed distance from $xy$-plane to $P$

A three-dimensional coordinate system can have either a **left-handed** or a **right-handed** orientation. In this text, we work exclusively with right-handed systems, as shown in Figure 7.2.

## EXAMPLE 1   Plotting Points in Space

Plot the following points in space.

(a) $(2, -3, 3)$      (b) $(-2, 6, 2)$      (c) $(1, 4, 0)$      (d) $(2, 2, -3)$

### Solution

To plot the point $(2, -3, 3)$, notice that $x = 2$, $y = -3$, and $z = 3$. To help visualize the point (see Figure 7.3), locate the point $(2, -3)$ in the $xy$-plane (denoted by a cross). The point $(2, -3, 3)$ lies three units above the cross. The other three points are also shown in the figure.

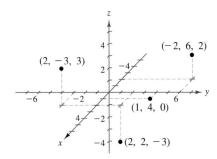

**FIGURE 7.3**

# The Distance and Midpoint Formulas

Many of the formulas established for the two-dimensional coordinate system can be extended to three dimensions. For example, to find the distance between two points in space, you can use the Pythagorean Theorem twice, as shown in Figure 7.4. By doing this, you will obtain the formula for the distance between two points in space.

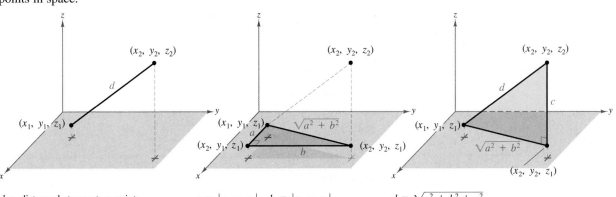

**FIGURE 7.4**

## Distance Formula in Space

The distance between the points $(x_1, y_1, z_1)$ and $(x_2, y_2, z_2)$ is

$$d = \sqrt{(x_2 - x_1)^2 + (y_2 - y_1)^2 + (z_2 - z_1)^2}.$$

### EXAMPLE 2   Finding the Distance Between Two Points

Find the distance between $(1, 0, 2)$ and $(2, 4, -3)$.

**Solution**

$$
\begin{aligned}
d &= \sqrt{(x_2 - x_1)^2 + (y_2 - y_1)^2 + (z_2 - z_1)^2} && \text{Distance Formula} \\
&= \sqrt{(2 - 1)^2 + (4 - 0)^2 + (-3 - 2)^2} && \text{Substitute.} \\
&= \sqrt{1 + 16 + 25} && \text{Simplify.} \\
&= \sqrt{42}. && \text{Simplify.}
\end{aligned}
$$

Notice the similarity between the Distance Formula in the plane and the Distance Formula in space. The Midpoint Formulas in the plane and in space are also similar.

## Midpoint Formula in Space

The midpoint of the line segment joining the points $(x_1, y_1, z_1)$ and $(x_2, y_2, z_2)$ is

$$\text{Midpoint} = \left( \frac{x_1 + x_2}{2}, \frac{y_1 + y_2}{2}, \frac{z_1 + z_2}{2} \right).$$

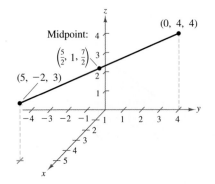

**FIGURE 7.5**

### EXAMPLE 3   Using the Midpoint Formula

Find the midpoint of the line segment joining $(5, -2, 3)$ and $(0, 4, 4)$.

**Solution**

Using the Midpoint Formula, the midpoint is

$$\left( \frac{5 + 0}{2}, \frac{-2 + 4}{2}, \frac{3 + 4}{2} \right) = \left( \frac{5}{2}, 1, \frac{7}{2} \right),$$

as shown in Figure 7.5.

## The Equation of a Sphere

A **sphere** with center at $(h, k, l)$ and radius $r$ is defined to be the set of all points $(x, y, z)$ such that the distance between $(x, y, z)$ and $(h, k, l)$ is $r$, as shown in Figure 7.6. Using the Distance Formula, this condition can be written as

$$\sqrt{(x - h)^2 + (y - k)^2 + (z - l)^2} = r.$$

By squaring both sides of this equation, you obtain the standard equation of a sphere.

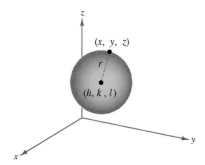

**FIGURE 7.6** Sphere: Radius $r$, Center $(h, k, l)$

### Standard Equation of a Sphere

The **standard equation of a sphere** whose center is $(h, k, l)$ and whose radius is $r$ is

$$(x - h)^2 + (y - k)^2 + (z - l)^2 = r^2.$$

## EXAMPLE 4   Finding the Equation of a Sphere

Find the standard equation for the sphere whose center is $(2, 4, 3)$ and whose radius is 3. Does this sphere intersect the $xy$-plane?

### Solution

$$(x - h)^2 + (y - k)^2 + (z - l)^2 = r^2 \qquad \text{Standard equation}$$
$$(x - 2)^2 + (y - 4)^2 + (z - 3)^2 = 3^2 \qquad \text{Substitute.}$$

From the graph shown in Figure 7.7, you can see that the center of the sphere lies three units above the $xy$-plane. Because the sphere has a radius of 3, you can conclude that it does intersect the $xy$-plane—at the point $(2, 4, 0)$.

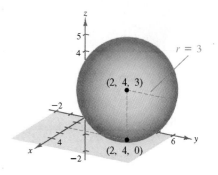

**FIGURE 7.7**

## EXAMPLE 5   Finding the Equation of a Sphere

Find the equation of the sphere that has the points $(3, -2, 6)$ and $(-1, 4, 2)$ as endpoints of a diameter.

### Solution

By the Midpoint Formula, the center of the sphere is

$$(h, k, l) = \left( \frac{3 + (-1)}{2}, \frac{-2 + 4}{2}, \frac{6 + 2}{2} \right) \qquad \text{Midpoint Formula}$$

$$= (1, 1, 4). \qquad \text{Simplify.}$$

By the Distance Formula, the radius is

$$r = \sqrt{(3 - 1)^2 + (-2 - 1)^2 + (6 - 4)^2} \qquad \begin{array}{l}\text{Distance Formula using}\\ (3, -2, 6) \text{ and } (1, 1, 4)\end{array}$$

$$= \sqrt{17}. \qquad \text{Simplify.}$$

Therefore, the standard equation of the sphere is

$$(x - h)^2 + (y - k)^2 + (z - l)^2 = r^2 \qquad \text{Formula for a sphere}$$

$$(x - 1)^2 + (y - 1)^2 + (z - 4)^2 = 17. \qquad \text{Substitute.}$$

## EXAMPLE 6   Finding the Center and Radius of a Sphere

Find the center and radius of the sphere whose equation is

$$x^2 + y^2 + z^2 - 2x + 4y - 6z + 8 = 0.$$

### Solution

You can obtain the standard equation of the sphere by completing the square. To do this, begin by grouping terms with the same variable. Then add "the square of half the coefficient of each linear term" to both sides of the equation. For instance, to complete the square of $(x^2 - 2x)$, add $\left[ \frac{1}{2}(-2) \right]^2 = 1$ to both sides.

$$x^2 + y^2 + z^2 - 2x + 4y - 6z + 8 = 0$$

$$(x^2 - 2x + \quad) + (y^2 + 4y + \quad) + (z^2 - 6z + \quad) = -8$$

$$(x^2 - 2x + 1) + (y^2 + 4y + 4) + (z^2 - 6z + 9) = -8 + 1 + 4 + 9$$

$$(x - 1)^2 + (y + 2)^2 + (z - 3)^2 = 6$$

Therefore, the center of the sphere is $(1, -2, 3)$, and its radius is $\sqrt{6}$, as shown in Figure 7.8.

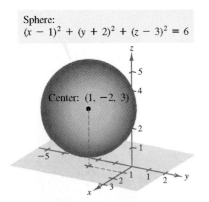

Sphere:
$(x - 1)^2 + (y + 2)^2 + (z - 3)^2 = 6$

Center: $(1, -2, 3)$

**FIGURE 7.8**

Note in Example 6 that the points satisfying the equation of the sphere are "surface points," not "interior points." In general, the collection of points satisfying an equation involving $x$, $y$, and $z$ is called a **surface in space.**

# Traces of Surfaces

Finding the intersection of a surface with one of the three coordinate planes (or with a plane parallel to one of the three coordinate planes) helps visualize the surface. Such an intersection is called a **trace** of the surface. For example, the *xy*-trace of a surface consists of all points that are common to both the surface *and* the *xy*-plane. Similarly, the *xz*-trace of a surface consists of all points that are common to both the surface and the *xz*-plane.

---

### EXAMPLE 7   Finding a Trace of a Surface

Sketch the *xy*-trace of the sphere whose equation is

$$(x - 3)^2 + (y - 2)^2 + (z + 4)^2 = 5^2.$$

#### Solution

To find the *xy*-trace of this surface, use the fact that every point in the *xy*-plane has a *z*-coordinate of zero. This means that if you substitute $z = 0$ into the given equation, the resulting equation will represent the intersection of the surface with the *xy*-plane.

$$(x - 3)^2 + (y - 2)^2 + (z + 4)^2 = 5^2 \qquad \text{Equation of sphere}$$
$$(x - 3)^2 + (y - 2)^2 + (0 + 4)^2 = 25 \qquad \text{Let } z = 0 \text{ to find } xy\text{-trace.}$$
$$(x - 3)^2 + (y - 2)^2 + 16 = 25$$
$$(x - 3)^2 + (y - 2)^2 = 9$$
$$(x - 3)^2 + (y - 2)^2 = 3^2 \qquad \text{Equation of circle}$$

From this equation, you can see that the *xy*-trace is a circle of radius 3, as shown in Figure 7.9.

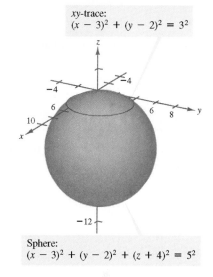

*xy*-trace:
$(x - 3)^2 + (y - 2)^2 = 3^2$

Sphere:
$(x - 3)^2 + (y - 2)^2 + (z + 4)^2 = 5^2$

**FIGURE 7.9**

---

## *Group Discussion*   *Comparing Two and Three Dimensions*

In this section, you saw similarities between formulas in two-dimensional coordinate geometry and formulas in three-dimensional coordinate geometry. In two-dimensional coordinate geometry, the graph of the equation

$$ax + by + c = 0$$

is a line. In three-dimensional coordinate geometry, what is the graph of the equation

$$ax + by + c = 0?$$

Is it a line? Explain your reasoning. Use a three-dimensional graphing utility to verify your result.

*Warm Up*

The following warm-up exercises involve skills that were covered in earlier sections. You will use these skills in the exercise set for this section.

In Exercises 1–4, find the distance between the points.

**1.** $(5, 1), (3, 5)$       **2.** $(2, 3), (-1, -1)$

**3.** $(-5, 4), (-5, -4)$       **4.** $(-3, 6), (-3, -2)$

In Exercises 5–8, find the midpoint of the line segment connecting the points.

**5.** $(2, 5), (6, 9)$       **6.** $(-1, -2), (3, 2)$

**7.** $(-6, 0), (6, 6)$       **8.** $(-4, 3), (2, -1)$

In Exercises 9 and 10, write the standard equation of the circle.

**9.** Center: $(2, 3)$; Radius: 2       **10.** Diameter endpoints: $(4, 0), (-2, 8)$

## EXERCISES 7.1

In Exercises 1–4, plot the points on the same three-dimensional coordinate system.

**1.** (a) $(2, 1, 3)$

   (b) $(-1, 2, 1)$

**2.** (a) $(3, -2, 5)$

   (b) $\left(\frac{3}{2}, 4, -2\right)$

**3.** (a) $(5, -2, 2)$

   (b) $(5, -2, -2)$

**4.** (a) $(0, 4, -5)$

   (b) $(4, 0, 5)$

In Exercises 5–8, find the distance between the two points.

**5.** $(4, 1, 5), (8, 2, 6)$

**6.** $(-4, -1, 1), (2, -1, 5)$

**7.** $(-1, -5, 7), (-3, 4, -4)$

**8.** $(8, -2, 2), (8, -2, 4)$

In Exercises 9–12, find the coordinates of the midpoint of the line segment joining the given points.

**9.** $(6, -9, 1), (-2, -1, 5)$

**10.** $(4, 0, -6), (8, 8, 20)$

**11.** $(-5, -2, 5), (6, 3, -7)$

**12.** $(0, -2, 5), (4, 2, 7)$

In Exercises 13–16, find $(x, y, z)$.

**13.**

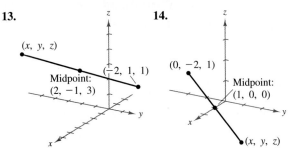

**14.**

**15.**

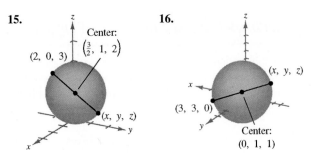

**16.**

In Exercises 17–20, find the lengths of the sides of the triangle with the indicated vertices, and determine whether the triangle is a right triangle, an isosceles triangle, or neither of these.

**17.** $(0, 0, 0), (2, 2, 1), (2, -4, 4)$

**18.** $(5, 3, 4), (7, 1, 3), (3, 5, 3)$

**19.** $(-2, 2, 4), (-2, 2, 6), (-2, 4, 8)$

**20.** $(5, 0, 0), (0, 2, 0), (0, 0, -3)$

In Exercises 21–28, find the standard form of the equation of the sphere.

**21.**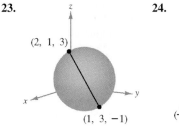
$(0, 2, 2)$, $r = 2$

**22.**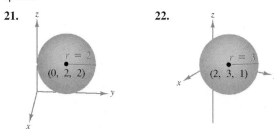
$(2, 3, 1)$, $r = 3$

**23.**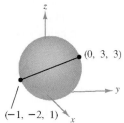
$(2, 1, 3)$
$(1, 3, -1)$

**24.**
$(0, 3, 3)$
$(-1, -2, 1)$

**25.** Center: $(1, 1, 5)$; Radius: 3

**26.** Center: $(4, -1, 1)$; Radius: 5

**27.** Diameter endpoints: $(2, 0, 0), (0, 6, 0)$

**28.** Center: $(-2, 1, 1)$; Tangent to the $xy$-coordinate plane

In Exercises 29–34, find the sphere's center and radius.

**29.** $x^2 + y^2 + z^2 - 2x + 6y + 8z + 1 = 0$

**30.** $x^2 + y^2 + z^2 - 5x = 0$

**31.** $x^2 + y^2 + z^2 - 8y = 0$

**32.** $x^2 + y^2 + z^2 - 4y + 6z + 4 = 0$

**33.** $2x^2 + 2y^2 + 2z^2 - 4x - 12y - 8z + 3 = 0$

**34.** $x^2 + y^2 + z^2 = 36$

In Exercises 35 and 36, sketch the $xy$-trace of the sphere.

**35.** $(x - 1)^2 + (y - 3)^2 + (z - 2)^2 = 25$

**36.** $x^2 + y^2 + z^2 - 6x - 10y + 6z + 30 = 0$

In Exercises 37 and 38, sketch the $yz$-trace of the sphere.

**37.** $x^2 + y^2 + z^2 - 4x - 4y - 6z - 12 = 0$

**38.** $x^2 + y^2 + z^2 - 6x - 10y + 6z + 30 = 0$

In Exercises 39–42, sketch the trace of the intersection of each plane with the given sphere.

**39.** $x^2 + y^2 + z^2 = 25$; (a) $z = 3$, (b) $x = 4$

**40.** $x^2 + y^2 + z^2 = 169$; (a) $x = 5$, (b) $y = 12$

**41.** $x^2 + y^2 + z^2 - 4x - 6y + 9 = 0$; (a) $x = 2$, (b) $y = 3$

**42.** $x^2 + y^2 + z^2 - 8x - 6z + 16 = 0$; (a) $x = 4$, (b) $z = 3$

**43.** *Crystals*  Crystals are classified according to their symmetry. Crystals shaped like cubes are classified as isometric. Suppose you have mapped the vertices of a crystal onto a three-dimensional coordinate system. Determine $(x, y, z)$ if the crystal is isometric.

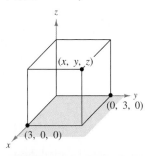
$(x, y, z)$
$(0, 3, 0)$
$(3, 0, 0)$

Dr. E.R. Degginger

*Halite crystals (rock salt) are classified as isometric.*

**44.** *Earth*  Assume that the earth is a sphere with a radius of 3963 miles. If the center of the earth is placed at the origin of a three-dimensional coordinate system, what is the equation of the sphere? Lines of longitude that run north-south could be represented by what trace(s)? What shape would each of these traces form? Why? Lines of latitude that run east-west could be represented by what trace(s)? Why? What shape would each of these traces form? Why?

## 7.2 Surfaces in Space

*Equations of Planes in Space • Drawing Planes in Space • Quadric Surfaces*

### Equations of Planes in Space

In Section 7.1, you studied one type of surface in space—a sphere. In this section, you will study a second type—a plane in space. The **general equation of a plane** in space is

$$ax + by + cz = d. \qquad \text{General equation of a plane}$$

Note the similarity of this equation to the general equation of a line in the plane. In fact, if you intersect the plane represented by this equation with each of the three coordinate planes, you will obtain traces that are lines, as shown in Figure 7.10.

In Figure 7.10, the points where the plane intersects the three coordinate axes are the $x$-, $y$-, and $z$-intercepts of the plane. By connecting these three points, you can form a triangular region, which helps you visualize the plane in space.

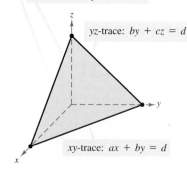

$xz$-trace: $ax + cz = d$
Plane: $ax + by + cz = d$
$yz$-trace: $by + cz = d$
$xy$-trace: $ax + by = d$

**FIGURE 7.10**

### EXAMPLE 1   Sketching a Plane in Space

Find the $x$-, $y$-, and $z$-intercepts of the plane given by

$$3x + 2y + 4z = 12.$$

Then sketch the plane.

**Solution**

To find the $x$-intercept, let both $y$ and $z$ be zero.

$$3x + 2(0) + 4(0) = 12 \qquad \text{Substitute 0 for } y \text{ and } z.$$
$$3x = 12 \qquad \text{Simplify.}$$
$$x = 4 \qquad \text{Solve for } x.$$

Thus, the $x$-intercept is $(4, 0, 0)$. To find the $y$-intercept, let $x$ and $z$ be zero and conclude that $y = 6$. Thus, the $y$-intercept is $(0, 6, 0)$. Similarly, by letting $x$ and $y$ be zero, you can determine that $z = 3$ and that the $z$-intercept is $(0, 0, 3)$. Figure 7.11 shows the triangular portion of the plane formed by connecting the three intercepts.

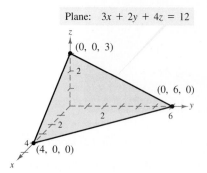

Plane: $3x + 2y + 4z = 12$

$(0, 0, 3)$
$(0, 6, 0)$
$(4, 0, 0)$

**FIGURE 7.11**   Sketch Made by Connecting Intercepts:
$(4, 0, 0), (0, 6, 0), (0, 0, 3)$

## Drawing Planes in Space

The planes shown in Figures 7.10 and 7.11 each have three intercepts. When this occurs, we suggest that you draw the plane by sketching the triangular region formed by connecting the three intercepts.

It is possible for a plane in space to have fewer than three intercepts. This occurs when one or more of the coefficients in the equation $ax + by + cz = d$ is zero. Figure 7.12 shows some planes in space that have only one intercept, and Figure 7.13 shows some that have only two intercepts. In each figure, note the use of dashed lines and shading to give the illusion of three dimensions.

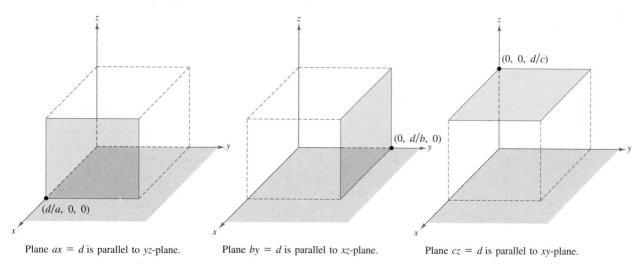

Plane $ax = d$ is parallel to $yz$-plane.     Plane $by = d$ is parallel to $xz$-plane.     Plane $cz = d$ is parallel to $xy$-plane.

**FIGURE 7.12**   Planes Parallel to Coordinate Planes

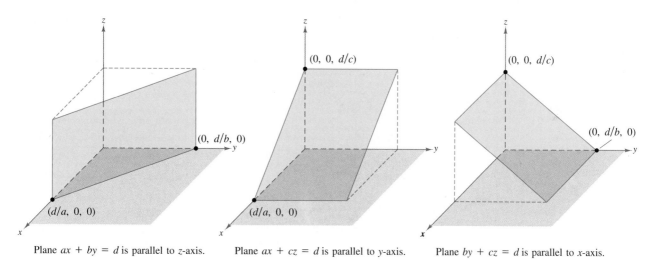

Plane $ax + by = d$ is parallel to $z$-axis.     Plane $ax + cz = d$ is parallel to $y$-axis.     Plane $by + cz = d$ is parallel to $x$-axis.

**FIGURE 7.13**   Planes Parallel to Coordinate Axes

## Quadric Surfaces

A third common surface is a **quadric surface.** Every quadric surface has an equation of the form

$$Ax^2 + By^2 + Cz^2 + Dx + Ey + Fz + G = 0. \qquad \text{Second-degree equation}$$

There are six basic quadric surfaces.

1. Elliptic cone
2. Elliptic paraboloid
3. Hyperbolic paraboloid
4. Ellipsoid
5. Hyperboloid of one sheet
6. Hyperboloid of two sheets

The six types are summarized on pages 464 and 465. Notice that each surface is pictured with two types of three-dimensional sketches. The computer-generated sketches use traces with hidden lines to give the illusion of three dimensions. The artist-rendered sketches use shading to create the same illusion.

All of the quadric surfaces on pages 464 and 465 are centered at the origin and have axes along the coordinate axes. Moreover, only one of several possible orientations of each surface is shown. If the surface has a different center or is oriented along a different axis, then its standard equation will change accordingly. For instance, the ellipsoid

$$\frac{x^2}{1^2} + \frac{y^2}{3^2} + \frac{z^2}{2^2} = 1$$

has $(0, 0, 0)$ as its center, but the ellipsoid

$$\frac{(x - 2)^2}{1^2} + \frac{(y + 1)^2}{3^2} + \frac{(z - 4)^2}{2^2} = 1$$

has $(2, -1, 4)$ as its center. A computer-generated graph of the first ellipsoid is shown in Figure 7.14.

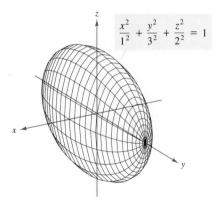

$$\frac{x^2}{1^2} + \frac{y^2}{3^2} + \frac{z^2}{2^2} = 1$$

**FIGURE 7.14**

# Technology

## Using a Three-Dimensional Graphing Utility

Most three-dimensional graphing utilities represent surfaces by sketching several traces of the surface. The traces are usually taken in equally spaced parallel planes. Depending on the graphing utility, the sketch can be made with one set, two sets, or three sets of traces. For instance, the two sketches shown below use two sets of traces: in one set the traces are parallel to the $xz$-plane and in the other set the traces are parallel to the $yz$-plane.

To sketch the graph of an equation involving $x$, $y$, and $z$ with a three-dimensional "function grapher," you must first solve the equation for $z$. After entering the equation, you need to specify a rectangular viewing box (the three-dimensional analog of a viewing rectangle).

The two sketches shown below were generated by *Derive* for Windows. If you have access to a three-dimensional graphing utility, try using it to duplicate these graphs.

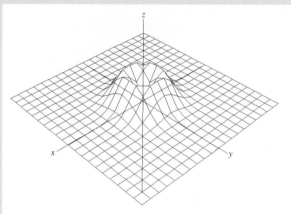

Generated by Derive

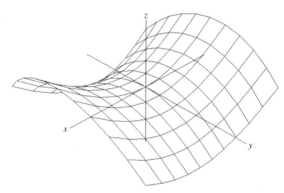

Generated by Derive

Equation: $z = (x^2 + y^2)e^{1-x^2-y^2}$

Grid: 20 traces by 20 traces

Viewing Box: $-5 \le x \le 5$
$-5 \le y \le 5$
$-5 \le z \le 5$

Equation: $z = \frac{1}{4}x^2 - \frac{1}{4}y^2$

Grid: 10 traces by 10 traces

Viewing Box: $-4 \le x \le 4$
$-4 \le y \le 4$
$-4 \le z \le 4$

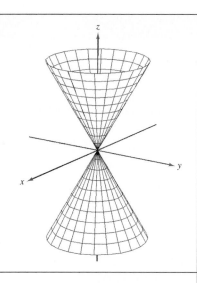

### Elliptic Cone

$$\frac{x^2}{a^2} + \frac{y^2}{b^2} - \frac{z^2}{c^2} = 0$$

| *Trace* | *Plane* |
|---------|---------|
| Ellipse | Parallel to $xy$-plane |
| Hyperbola | Parallel to $xz$-plane |
| Hyperbola | Parallel to $yz$-plane |

The axis of the cone corresponds to the variable whose coefficient is negative. The traces in the coordinate planes parallel to this axis are intersecting lines.

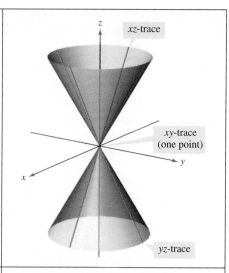

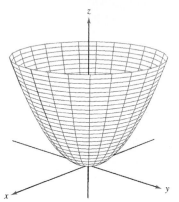

### Elliptic Paraboloid

$$z = \frac{x^2}{a^2} + \frac{y^2}{b^2}$$

| *Trace* | *Plane* |
|---------|---------|
| Ellipse | Parallel to $xy$-plane |
| Parabola | Parallel to $xz$-plane |
| Parabola | Parallel to $yz$-plane |

The axis of the paraboloid corresponds to the variable raised to the first power.

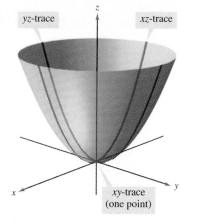

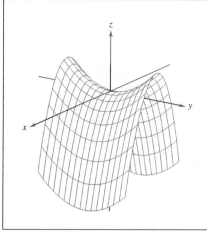

### Hyperbolic Paraboloid

$$z = \frac{y^2}{b^2} - \frac{x^2}{a^2}$$

| *Trace* | *Plane* |
|---------|---------|
| Hyperbola | Parallel to $xy$-plane |
| Parabola | Parallel to $xz$-plane |
| Parabola | Parallel to $yz$-plane |

The axis of the paraboloid corresponds to the variable raised to the first power.

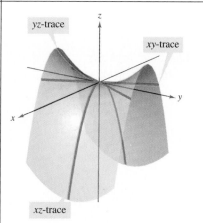

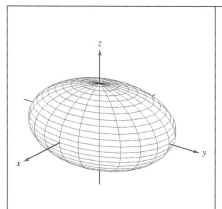

### Ellipsoid

$$\frac{x^2}{a^2} + \frac{y^2}{b^2} + \frac{z^2}{c^2} = 1$$

| Trace | Plane |
|-------|-------|
| Ellipse | Parallel to $xy$-plane |
| Ellipse | Parallel to $xz$-plane |
| Ellipse | Parallel to $yz$-plane |

The surface is a sphere if the coefficients $a$, $b$, and $c$ are equal and nonzero.

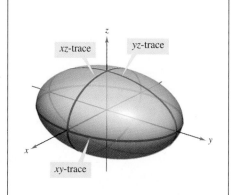

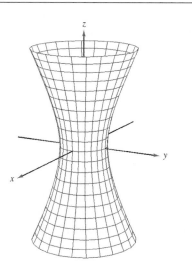

### Hyperboloid of One Sheet

$$\frac{x^2}{a^2} + \frac{y^2}{b^2} - \frac{z^2}{c^2} = 1$$

| Trace | Plane |
|-------|-------|
| Ellipse | Parallel to $xy$-plane |
| Hyperbola | Parallel to $xz$-plane |
| Hyperbola | Parallel to $yz$-plane |

The axis of the hyperboloid corresponds to the variable whose coefficient is negative.

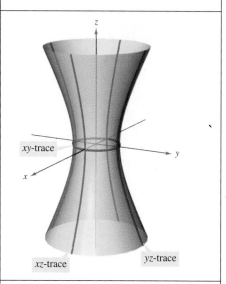

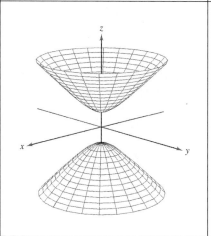

### Hyperboloid of Two Sheets

$$\frac{z^2}{c^2} - \frac{x^2}{a^2} - \frac{y^2}{b^2} = 1$$

| Trace | Plane |
|-------|-------|
| Ellipse | Parallel to $xy$-plane |
| Hyperbola | Parallel to $xz$-plane |
| Hyperbola | Parallel to $yz$-plane |

The axis of the hyperboloid corresponds to the variable whose coefficient is positive. There is no trace in the coordinate plane perpendicular to this axis.

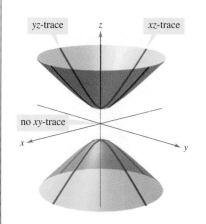

When classifying quadric surfaces, note that the two types of paraboloids have one variable raised to the first power. The other four types of quadric surfaces have equations that are of second degree in *all* three variables.

## EXAMPLE 2    Classifying a Quadric Surface

Classify the surface given by

$$x - y^2 - z^2 = 0.$$

Describe the traces of the surface in the $xy$-plane, the $xz$-plane, and the plane given by $x = 1$.

### Solution

Because $x$ is raised only to the first power, the surface is a paraboloid whose axis is the $x$-axis, as shown in Figure 7.15. In standard form, the equation is

$$x = y^2 + z^2.$$

The traces in the $xy$-plane, the $xz$-plane, and the plane given by $x = 1$ are as follows.

| Trace in $xy$-plane $(z = 0)$: | $x = y^2$ | Parabola |
|---|---|---|
| Trace in $xz$-plane $(y = 0)$: | $x = z^2$ | Parabola |
| Trace in plane $x = 1$: | $y^2 + z^2 = 1$ | Circle |

These three traces are shown in Figure 7.16. From the traces, you can see that the surface is an elliptic (or circular) paraboloid. If you have access to a three-dimensional graphing utility, try using it to graph this surface. If you do this, you will discover that sketching surfaces in space is not a simple task—even with a graphing utility.

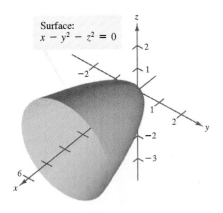

Surface:
$x - y^2 - z^2 = 0$

**FIGURE 7.15**    Elliptic Paraboloid

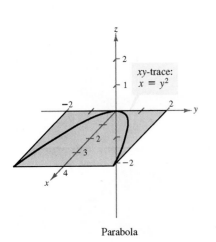

$xy$-trace:
$x = y^2$

Parabola

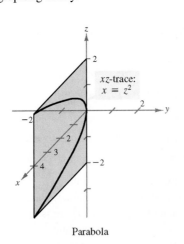

$xz$-trace:
$x = z^2$

Parabola

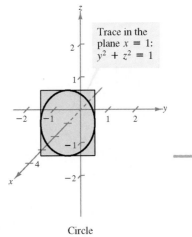

Trace in the plane $x = 1$:
$y^2 + z^2 = 1$

Circle

**FIGURE 7.16**

## EXAMPLE 3 Classifying Quadric Surfaces

(a) The equation

$$x^2 - 4y^2 - 4z^2 - 4 = 0 \qquad \text{Original equation}$$

can be written in standard form as

$$\frac{x^2}{4} - y^2 - z^2 = 1. \qquad \text{Standard form}$$

From the standard form, you can see that the graph is a hyperboloid of two sheets, with the $x$-axis as its axis, as shown in Figure 7.17(a).

(b) The equation

$$x^2 + 4y^2 + z^2 - 4 = 0 \qquad \text{Original equation}$$

can be written in standard form as

$$\frac{x^2}{4} + y^2 + \frac{z^2}{4} = 1. \qquad \text{Standard form}$$

From the standard form, you can see that the graph is an ellipsoid, as shown in Figure 7.17(b).

Surface:
$x^2 - 4y^2 - 4z^2 - 4 = 0$

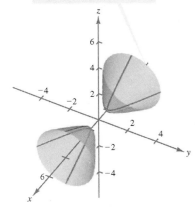

(a)

Surface:
$x^2 + 4y^2 + z^2 - 4 = 0$

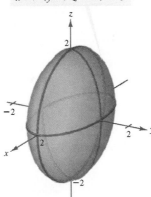

(b)

**FIGURE 7.17**

*Group Discussion*     **Classifying Quadric Surfaces**

Classify the following quadric surfaces. Use a three-dimensional graphing utility to verify your results.

a. $\dfrac{x^2}{2^2} - \dfrac{y^2}{4^2} + \dfrac{z^2}{3^2} = 0$     b. $\dfrac{x^2}{2^2} + \dfrac{y^2}{4^2} + \dfrac{z^2}{3^2} = 1$     c. $\dfrac{x^2}{2^2} - \dfrac{y^2}{4^2} + \dfrac{z^2}{3^2} = 1$

*Warm Up*

The following warm-up exercises involve skills that were covered in earlier sections. You will use these skills in the exercise set for this section.

In Exercises 1–4, find the $x$- and $y$-intercepts of the function.

**1.** $3x + 4y = 12$      **2.** $6x + y = -8$      **3.** $-2x + y = -2$      **4.** $-x - y = 5$

In Exercises 5–8, rewrite the expression by completing the square.

**5.** $x^2 + y^2 + z^2 - 2x - 4y - 6z + 15 = 0$      **6.** $x^2 + y^2 - z^2 - 8x + 4y - 6z + 11 = 0$

**7.** $z - 2 = x^2 + y^2 + 2x - 2y$      **8.** $x^2 + y^2 + z^2 - 6x + 10y + 26z = -202$

In Exercises 9 and 10, write the expression in standard form.

**9.** $16x^2 - 16y^2 + 16z^2 = 4$      **10.** $9x^2 - 9y^2 + 9z^2 = 36$

# EXERCISES 7.2

In Exercises 1–12, find the intercepts and sketch the graph of the plane.

**1.** $4x + 2y + 6z = 12$      **2.** $3x + 6y + 2z = 6$

**3.** $3x + 3y + 5z = 15$      **4.** $x + y + z = 3$

**5.** $2x - y + 3z = 8$      **6.** $2x - y + z = 4$

**7.** $z = 3$      **8.** $y = -4$

**9.** $y + z = 5$      **10.** $x + 2y = 8$

**11.** $x + y - z = 0$      **12.** $x - 3z = 3$

In Exercises 13–20, determine whether the planes $a_1x + b_1y + c_1z = d_1$ and $a_2x + b_2y + c_2z = d_2$ are parallel, perpendicular, or neither. The planes are parallel if there exists a nonzero constant $k$ such that $a_1 = ka_2$, $b_1 = kb_2$, $c_1 = kc_2$, and perpendicular if $a_1a_2 + b_1b_2 + c_1c_2 = 0$.

**13.** $5x - 3y + z = 4$, $x + 4y + 7z = 1$

**14.** $3x + y - 4z = 3$, $-9x - 3y + 12z = 4$

**15.** $x - 5y - z = 1$, $5x - 25y - 5z = -3$

**16.** $x = 6$, $y = -1$

**17.** $x + 2y = 3$, $4x + 8y = 5$

**18.** $x + 3y + z = 7$, $x - 5z = 0$

**19.** $2x + y = 3$, $3x - 5z = 0$

**20.** $2x - z = 1$, $4x + y + 8z = 10$

In Exercises 21–26, find the distance between the point and the plane (see figure). The distance $D$ between a point $(x_0, y_0, z_0)$ and the plane $ax + by + cz + d = 0$ is

$$D = \frac{|ax_0 + by_0 + cz_0 + d|}{\sqrt{a^2 + b^2 + c^2}}.$$

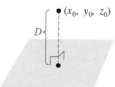

Plane:
$ax + by + cz + d = 0$

**21.** $(0, 0, 0)$, $2x + 3y + z = 12$

**22.** $(1, 5, -4)$, $3x - y + 2z = 6$

**23.** $(1, 2, 3)$, $2x - y + z = 4$

**24.** $(10, 0, 0)$, $x - 3y + 4z = 6$

**25.** $(1, 0, -1)$, $2x - 4y + 3z = 12$

**26.** $(2, -1, 0)$, $3x + 3y + 2z = 6$

In Exercises 27–34, match the given equation with the correct graph. [The graphs are labeled (a)–(h).]

**27.** $\dfrac{x^2}{9} + \dfrac{y^2}{16} + \dfrac{z^2}{9} = 1$

**28.** $x^2 - \dfrac{4y^2}{15} + z^2 = -\dfrac{4}{15}$

**29.** $4x^2 + 4y^2 - z^2 = 4$

**30.** $y^2 = 4x^2 + 9z^2$

**31.** $4x^2 - 4y + z^2 = 0$

**32.** $12z = -3y^2 + 4x^2$

**33.** $4x^2 - y^2 + 4z = 0$

**34.** $x^2 + y^2 + z^2 = 9$

(a)

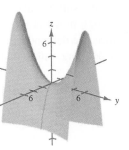

(b)

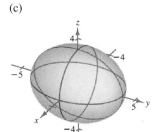

(c)

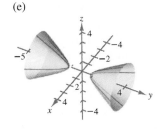

(d)

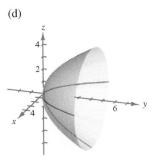

(e)

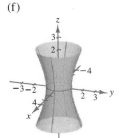

(f)

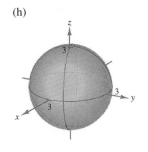

(g)

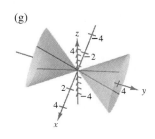

(h)

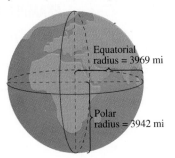

In Exercises 35–38, describe the trace of the surface in the indicated planes.

| Surface | Planes |
|---|---|
| **35.** $x^2 - y - z^2 = 0$ | $xy$-plane, $y = 1$, $yz$-plane |
| **36.** $y = x^2 + z^2$ | $xy$-plane, $y = 1$, $yz$-plane |
| **37.** $\dfrac{x^2}{4} + y^2 + z^2 = 1$ | $xy$-plane, $xz$-plane, $yz$-plane |
| **38.** $y^2 + z^2 - x^2 = 1$ | $xy$-plane, $xz$-plane, $yz$-plane |

In Exercises 39–52, identify the quadric surface.

**39.** $x^2 + \dfrac{y^2}{4} + z^2 = 1$

**40.** $\dfrac{x^2}{9} + \dfrac{y^2}{16} + \dfrac{z^2}{16} = 1$

**41.** $25x^2 + 25y^2 - z^2 = 5$

**42.** $9x^2 + 4y^2 - 8z^2 = 72$

**43.** $x^2 - y + z^2 = 0$

**44.** $z = 4x^2 + y^2$

**45.** $x^2 - y^2 + z = 0$

**46.** $z^2 - x^2 - \dfrac{y^2}{4} = 1$

**47.** $4x^2 - y^2 + 4z^2 = -16$

**48.** $z^2 = x^2 + \dfrac{y^2}{4}$

**49.** $z^2 = 9x^2 + y^2$

**50.** $4y = x^2 + z^2$

**51.** $3z = -y^2 + x^2$

**52.** $z^2 = 2x^2 + 2y^2$

In Exercises 53–56, use a three-dimensional graphing utility to graph the function.

**53.** $z = y^2 - x^2 + 1$

**54.** $z = x^2 + y^2 + 1$

**55.** $z = \dfrac{x^2}{2} + \dfrac{y^2}{4}$

**56.** $z = \dfrac{1}{12}\sqrt{144 - 16x^2 - 9y^2}$

**57.** *Shape of the Earth*   Because of the forces caused by its rotation, the earth is actually an oblate ellipsoid rather than a sphere. The equatorial radius is 3969 miles and the polar radius is 3942 miles. Find an equation of the ellipsoid. Assume that the center of the earth is at the origin and the $xy$-trace ($z = 0$) corresponds to the equator.

<table>
<tr><td>**7.3**</td><td>

# Functions of Several Variables

*Functions of Several Variables  •  The Graph of a Function of Two Variables  •*
*Level Curves and Contour Maps  •  Applications*
</td></tr>
</table>

## Functions of Several Variables

In the first six chapters of this text, you studied functions of a single independent variable. Many quantities in science, business, and technology, however, are functions not of one, but of two or more variables. For instance, the demand function for a product is often dependent on the price *and* the advertising, rather than on the price alone.

The notation for functions of two or more variables is similar to that used for functions of a single variable. For example,

$$f(x, y) = x^2 + xy \qquad \text{and} \qquad g(x, y) = e^{x+y}$$

$$\underbrace{\qquad\qquad}_{\text{2 variables}} \qquad\qquad\qquad \underbrace{\qquad\qquad}_{\text{2 variables}}$$

are functions of two variables, and

$$f(x, y, z) = x + 2y - 3z$$

$$\underbrace{\qquad\qquad}_{\text{3 variables}}$$

is a function of three variables.

---

### Definition of a Function of Two Variables

If to each ordered pair $(x, y)$ in some set $D$ there corresponds a unique real number $z = f(x, y)$, then $f$ is called a **function of $x$ and $y$.** The set $D$ is the **domain** of $f$, and the corresponding set of $z$-values is the **range** of $f$. Functions of three, four, or more variables are defined similarly.

---

### EXAMPLE 1   Evaluating Functions of Several Variables

(a)  For $f(x, y) = 2x^2 - y^2$, you can evaluate $f(2, 3)$ as follows.

$$f(2, 3) = 2(2)^2 - (3)^2 = 8 - 9 = -1$$

(b)  For $f(x, y, z) = e^x(y + z)$, you can evaluate $f(0, -1, 4)$ as follows.

$$f(0, -1, 4) = e^0(-1 + 4) = (1)(3) = 3$$

# The Graph of a Function of Two Variables

A function of two variables can be represented graphically as a surface in space by letting $z = f(x, y)$. When sketching the graph of a function of $x$ and $y$, remember that even though the graph is three-dimensional, the domain of the function is two-dimensional—it consists of the points in the $xy$-plane for which the function is defined. As with functions of a single variable, unless specifically restricted, the domain of a function of two variables is assumed to be the set of all points $(x, y)$ for which the defining equation has meaning.

---

### EXAMPLE 2   Finding the Domain and Range of a Function

Find the domain and range of the function

$$f(x, y) = \sqrt{64 - x^2 - y^2}.$$

*Solution*

Because no restrictions are given, the domain is assumed to be the set of all points for which the defining equation makes sense.

$$64 - x^2 - y^2 \geq 0 \qquad \text{Quantity inside radical must be nonnegative.}$$
$$x^2 + y^2 \leq 64 \qquad \text{Domain of the function}$$

Thus, the domain is the set of all points that lie on or inside the circle given by $x^2 + y^2 = 8^2$. The range of $f$ is the set

$$0 \leq z \leq 8. \qquad \text{Range of the function}$$

As shown in Figure 7.18, the graph of the function is a hemisphere.

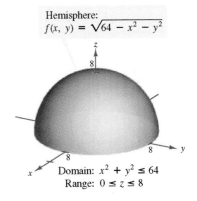

Hemisphere:
$f(x, y) = \sqrt{64 - x^2 - y^2}$

Domain: $x^2 + y^2 \leq 64$
Range: $0 \leq z \leq 8$

**FIGURE 7.18**

## Technology

Some three-dimensional graphing utilities can graph equations in $x$, $y$, and $z$. Others are programmed to graph only functions of $x$ and $y$. A surface in space represents the graph of a function of $x$ and $y$ only if each vertical line intersects the surface at most once. For instance, the surface shown in Figure 7.18 passes this vertical line test, but the surface at the right (drawn by *Mathematica*) does not represent the graph of a function of $x$ and $y$.

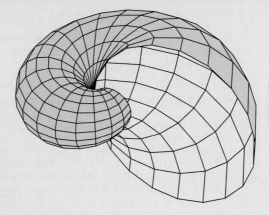

Some vertical lines intersect this surface more than once. Thus, the surface does not pass the vertical line test and is not a function of $x$ and $y$.

## Level Curves and Contour Maps

A **contour map** of a surface is created by *projecting* traces, taken in evenly spaced planes that are parallel to the *xy*-plane, onto the *xy*-plane. Each projection is a **level curve** of the surface.

Contour maps are used to create weather maps, topographical maps, and population density maps. For instance, Figure 7.19(a) shows a graph of a "mountain and valley" surface given by $z = f(x, y)$. Each of the level curves in Figure 7.19(b) represents the intersection of the surface $z = f(x, y)$ with a plane $z = c$, where $c = 828, 830, \ldots, 854$.

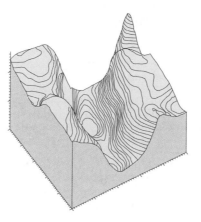

(a) Surface                    (b) Contour Map

**FIGURE 7.19**

## EXAMPLE 3   Reading a Contour Map

The "contour map" in Figure 7.20 was computer generated using data collected by satellite instrumentation. The map uses color to represent levels of chlorine nitrate in the atmosphere. Chlorine nitrate contributes to the ozone depletion in the earth's atmosphere. The red areas represent the highest level of chlorine nitrate and the dark blue areas represent the lowest level. Describe the areas that have the highest levels of chlorine nitrate.   *(Source: Lockheed Missiles and Space Company)*

### Solution

The highest levels of chlorine nitrate are in the Antarctic Ocean, surrounding Antarctica. Although chlorine nitrate is not itself harmful to ozone, it has a tendency to convert to chlorine monoxide, which *is* harmful to ozone. Once the chlorine nitrate is converted to chlorine monoxide, it no longer shows on the contour map. Thus, Antarctica itself shows little chlorine nitrate—the nitrate has been converted to monoxide. If you have seen maps showing the "ozone hole" in the earth's atmosphere, you know that the hole occurs over Antarctica.

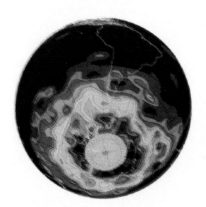

Lockheed Missiles and Space Company

**FIGURE 7.20**

# EXAMPLE 4   Reading a Contour Map

The contour map shown in Figure 7.21 represents the population density of the United States in 1990. Discuss the use of color to represent the level curves. *(Source: U.S. Bureau of Census)*

## Solution

You can see from the key that the light yellow regions have population densities of less than ten people per square mile. As the color darkens, the population density increases, so that the brown regions have population densities of more than 500 people per square mile. The most densely populated regions are urban areas and are represented by three different types of dots.

One advantage of such a map is that it allows you to "see" the population density of the country at a glance. From the map, it is clear that the Rocky Mountain region of the country has a population density that is much less than the region between Chicago and New York City.

*STUDY TIP*   In Figure 7.21, the level curves do not correspond to equally spaced population density levels. If, for example, you wanted to represent population density levels of 0–50, 50–100, . . . , and 450–500, how would the map change? Do you think that the change would better illustrate the population density levels in the United States?

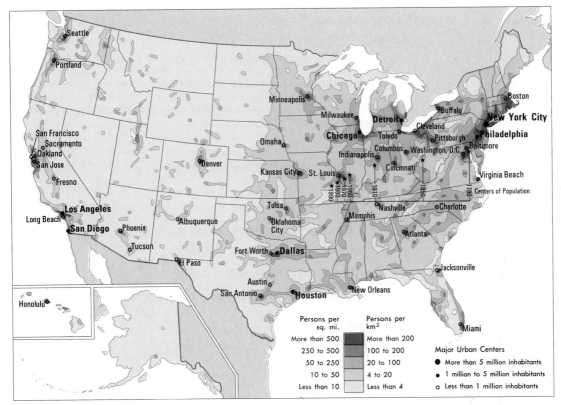

From *The World Book Encyclopedia* © World Book, Inc. By permission of the publisher.

**FIGURE 7.21**

## Applications

The **Cobb-Douglas production function** is used in economics to represent the number of units produced by varying amounts of labor and capital. Let $x$ represent the number of units of labor and let $y$ represent the number of units of capital. Then, the number of units produced is modeled by

$$f(x, y) = Cx^a y^{1-a},$$

where $C$ is a constant and $0 < a < 1$.

### EXAMPLE 5   Using a Production Function

A manufacturer estimates that its production (measured in units of a product) can be modeled by

$$f(x, y) = 100x^{0.6}y^{0.4},$$

where the labor $x$ is measured in person-hours and the capital $y$ is measured in thousands of dollars.

(a) What is the production level when $x = 1000$ and $y = 500$?

(b) What is the production level when $x = 2000$ and $y = 1000$?

(c) How does doubling the amounts of labor and capital from part a to part b affect the production?

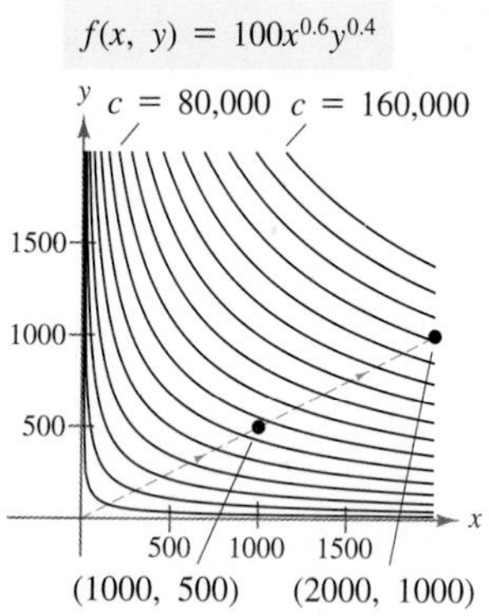

*Solution*

(a) When $x = 1000$ and $y = 500$, the production level is
$$f(1000, 500) = 100(1000)^{0.6}(500)^{0.4}$$
$$\approx 75{,}786 \text{ units.}$$

(b) When $x = 2000$ and $y = 1000$, the production level is
$$f(2000, 1000) = 100(2000)^{0.6}(1000)^{0.4}$$
$$\approx 151{,}572 \text{ units.}$$

(c) When the amounts of labor and capital are doubled, the production level also doubles. In Exercise 42, you are asked to show that this is characteristic of the Cobb-Douglas production function.

A contour graph of this function is shown in Figure 7.22.

$f(x, y) = 100x^{0.6}y^{0.4}$

$c = 80{,}000 \quad c = 160{,}000$

**FIGURE 7.22**   Level Curves (at Increments of 10,000)

> **STUDY TIP**   In Figure 7.22, note that the level curves of the function
> $$f(x, y) = 100x^{0.6}y^{0.4}$$
> occur at increments of 10,000.

## EXAMPLE 6   Finding Monthly Payments

The monthly payment $M$ for an installment loan of $P$ dollars taken out over $t$ years at an annual rate of $r$ is

$$M = f(P, r, t) = \frac{\dfrac{Pr}{12}}{1 - \left[\dfrac{1}{1 + (r/12)}\right]^{12t}}.$$

(a) Find the monthly payment for a home mortgage of $95,000 taken out for 30 years at an annual interest rate of 9%.

(b) Find the monthly payment for a car loan of $14,000 taken out for 5 years at an annual interest rate of 11%.

### Solution

(a) If $P = \$95{,}000$, $r = 0.09$, and $t = 30$, then the monthly payment is

$$M = f(95{,}000, 0.09, 30)$$

$$= \frac{\dfrac{(95{,}000)(0.09)}{12}}{1 - \left[\dfrac{1}{1 + (0.09/12)}\right]^{12(30)}}$$

$$= \$764.39.$$

(b) If $P = \$14{,}000$, $r = 0.11$, and $t = 5$, then the monthly payment is

$$M = f(14{,}000, 0.11, 5)$$

$$= \frac{\dfrac{(14{,}000)(0.11)}{12}}{1 - \left[\dfrac{1}{1 + (0.11/12)}\right]^{12(5)}}$$

$$= \$304.39.$$

*For many Americans, buying a house is the largest single purchase they will ever make. During the 1970s, 1980s, and 1990s, the annual interest rate on home mortgages varied drastically. It was as high as 18% and as low as 5%. Such variations can change monthly payments by hundreds of dollars.*

---

## *Group Discussion*      **Monthly Payments**

You are taking out a home mortgage for $100,000, and you are given the following options. Which option would you choose? Explain your reasoning.

a. A fixed annual rate of 10%, over a term of 20 years.

b. A fixed annual rate of 9%, over a term of 30 years.

c. An adjustable annual rate of 9%, over a term of 20 years. The annual rate can fluctuate—each year it is set at 1% above the prime rate.

d. A fixed annual rate of 9%, over a term of 15 years.

*Warm Up*

The following warm-up exercises involve skills that were covered in earlier sections. You will use these skills in the exercise set for this section.

In Exercises 1–4, evaluate the function when $x = -3$.

**1.** $f(x) = 5 - 2x$

**2.** $f(x) = -x^2 + 4x + 5$

**3.** $y = \sqrt{4x^2 - 3x + 4}$

**4.** $y = \sqrt[3]{34 - 4x + 2x^2}$

In Exercises 5–8, find the domain of the function.

**5.** $f(x) = 5x^2 + 3x - 2$

**6.** $g(x) = \dfrac{1}{2x} - \dfrac{2}{x + 3}$

**7.** $h(y) = \sqrt{y - 5}$

**8.** $f(y) = \sqrt{y^2 - 5}$

In Exercises 9 and 10, evaluate the expression.

**9.** $(476)^{0.65}$

**10.** $(251)^{0.35}$

## EXERCISES 7.3

In Exercises 1–14, find the function values.

**1.** $f(x, y) = \dfrac{x}{y}$

(a) $f(3, 2)$    (b) $f(-1, 4)$    (c) $f(30, 5)$
(d) $f(5, y)$    (e) $f(x, 2)$    (f) $f(5, t)$

**2.** $f(x, y) = 4 - x^2 - 4y^2$

(a) $f(0, 0)$    (b) $f(0, 1)$    (c) $f(2, 3)$
(d) $f(1, y)$    (e) $f(x, 0)$    (f) $f(t, 1)$

**3.** $f(x, y) = xe^y$

(a) $f(5, 0)$    (b) $f(3, 2)$    (c) $f(2, -1)$
(d) $f(5, y)$    (e) $f(x, 2)$    (f) $f(t, t)$

**4.** $g(x, y) = \ln|x + y|$

(a) $g(2, 3)$    (b) $g(5, 6)$    (c) $g(e, 0)$
(d) $g(0, 1)$    (e) $g(2, -3)$    (f) $g(e, e)$

**5.** $h(x, y, z) = \dfrac{xy}{z}$

(a) $h(2, 3, 9)$    (b) $h(1, 0, 1)$

**6.** $f(x, y, z) = \sqrt{x + y + z}$

(a) $f(0, 5, 4)$    (b) $f(6, 8, -3)$

**7.** $V(r, h) = \pi r^2 h$

(a) $V(3, 10)$    (b) $V(5, 2)$

**8.** $F(r, N) = 500\left(1 + \dfrac{r}{12}\right)^N$

(a) $F(0.09, 60)$    (b) $F(0.14, 240)$

**9.** $A(P, r, t) = P\left[\left(1 + \dfrac{r}{12}\right)^{12t} - 1\right]\left(1 + \dfrac{12}{r}\right)$

(a) $A(100, 0.10, 10)$    (b) $A(275, 0.0925, 40)$

**10.** $A(P, r, t) = Pe^{rt}$

(a) $A(500, 0.10, 5)$    (b) $A(1500, 0.12, 20)$

**11.** $f(x, y) = \displaystyle\int_x^y (2t - 3)\,dt$

(a) $f(1, 2)$    (b) $f(1, 4)$

**12.** $g(x, y) = \displaystyle\int_x^y \dfrac{1}{t}\,dt$

(a) $g(4, 1)$    (b) $g(6, 3)$

**13.** $f(x, y) = x^2 - 2y$

(a) $f(x + \Delta x, y)$    (b) $\dfrac{f(x, y + \Delta y) - f(x, y)}{\Delta y}$

**14.** $f(x, y) = 3xy + y^2$

(a) $f(x + \Delta x, y)$    (b) $\dfrac{f(x, y + \Delta y) - f(x, y)}{\Delta y}$

In Exercises 15–18, describe the region $R$ in the $xy$-coordinate plane that corresponds to the domain of the function, and find the range of the function.

**15.** $f(x, y) = \sqrt{16 - x^2 - y^2}$

**16.** $f(x, y) = x^2 + y^2 - 1$

**17.** $f(x, y) = e^{x/y}$

**18.** $f(x, y) = \ln(x + y)$

In Exercises 19–28, describe the region $R$ in the $xy$-coordinate plane that corresponds to the domain of the function.

**19.** $f(x, y) = \sqrt{9 - 9x^2 - y^2}$

**20.** $f(x, y) = \sqrt{x^2 + y^2 - 1}$

**21.** $f(x, y) = \dfrac{x}{y}$

**22.** $f(x, y) = \dfrac{4y}{x - 1}$

**23.** $f(x, y) = \dfrac{1}{xy}$

**24.** $g(x, y) = \dfrac{1}{x - y}$

**25.** $h(x, y) = x\sqrt{y}$

**26.** $f(x, y) = \sqrt{xy}$

**27.** $g(x, y) = \ln(4 - x - y)$

**28.** $f(x, y) = ye^{1/x}$

In Exercises 29–32, match the graph of the surface with one of the contour maps. [The contour maps are labeled (a)–(d).]

**29.** $f(x, y) = x^2 + \dfrac{y^2}{4}$

**30.** $f(x, y) = e^{1 - x^2 + y^2}$

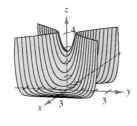

**31.** $f(x, y) = e^{1 - x^2 - y^2}$

**32.** $f(x, y) = \ln|y - x^2|$

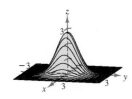

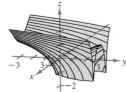

(a)

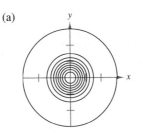

(b)

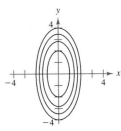

(c)

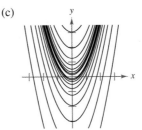

(d)

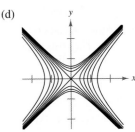

In Exercises 33–40, describe the level curves of the function. Sketch the level curves for the given $c$-values.

| Function | $c$-Values |
|---|---|
| **33.** $z = x + y$ | $c = -1, 0, 2, 4$ |
| **34.** $z = 6 - 2x - 3y$ | $c = 0, 2, 4, 6, 8, 10$ |
| **35.** $z = \sqrt{16 - x^2 - y^2}$ | $c = 0, 1, 2, 3, 4$ |
| **36.** $f(x, y) = x^2 + y^2$ | $c = 0, 2, 4, 6, 8$ |
| **37.** $f(x, y) = xy$ | $c = \pm 1, \pm 2, \ldots, \pm 6$ |
| **38.** $z = e^{xy}$ | $c = 1, 2, 3, 4, \frac{1}{2}, \frac{1}{3}, \frac{1}{4}$ |
| **39.** $f(x, y) = \dfrac{x}{x^2 + y^2}$ | $c = \pm\frac{1}{2}, \pm 1, \pm\frac{3}{2}, \pm 2$ |
| **40.** $f(x, y) = \ln(x - y)$ | $c = 0, \pm\frac{1}{2}, \pm 1, \pm\frac{3}{2}, \pm 2$ |

**41.** *Cobb-Douglas Production Function* A manufacturer estimates the Cobb-Douglas production function to be
$$f(x, y) = 100x^{0.75}y^{0.25}.$$
Estimate production level when $x = 1500$ and $y = 1000$.

**42.** *Cobb-Douglas Production Function* Use the Cobb-Douglas production function (Example 5) to show that if both the number of units of labor and the number of units of capital are doubled, the production level is also doubled.

**43.** *Cost* A company manufactures two types of woodburning stoves: a freestanding model and a fireplace-insert model. The cost function for producing $x$ freestanding stoves and $y$ fireplace-insert stoves is
$$C(x, y) = 27\sqrt{xy} + 195x + 215y + 980.$$
Find the cost when $x = 80$ and $y = 20$.

**44.** *Forestry*   The **Doyle Log Rule** is one of several methods used to determine the lumber yield of a log in board feet in terms of its diameter $d$ in inches and its length $L$ in feet. The number of board feet is given by

$$N(d, L) = \left(\frac{d - 4}{4}\right)^2 L.$$

(a) Find the number of board feet of lumber in a log with a diameter of 22 inches and a length of 12 feet.

(b) Find $N(30, 12)$.

**45.** *Profit*   A corporation manufactures a product at two locations. The costs of producing $x_1$ units at location 1 and $x_2$ units at location 2 are

$$C_1(x_1) = 0.02x_1^2 + 4x_1 + 500$$

and

$$C_2(x_2) = 0.05x_2^2 + 4x_2 + 275,$$

respectively. If the product sells for \$15 per unit, then the profit function for the product is

$$P(x_1, x_2) = 15(x_1 + x_2) - C_1(x_1) - C_2(x_2).$$

Find (a) $P(250, 150)$ and (b) $P(300, 200)$.

**46.** *Queuing Model*   The average amount of time that a customer waits in line for service is given by

$$W(x, y) = \frac{1}{x - y}, \qquad y < x,$$

where $y$ is the average arrival rate and $x$ is the average service rate ($x$ and $y$ are measured in the number of customers per hour). Evaluate $W$ at the following points.

(a) $(15, 10)$    (b) $(12, 9)$    (c) $(12, 6)$    (d) $(4, 2)$

**47.** *Investment*   In 1994, an investment of \$1000 was made in a bond earning 10% compounded annually. The investor pays tax at rate $R$, and the annual rate of inflation is $I$. In the year 2004, the value $V$ of the bond in constant 1994 dollars is

$$V(I, R) = 1000\left[\frac{1 + 0.10(1 - R)}{1 + I}\right]^{10}.$$

Use this function of two variables to complete the table.

| | Inflation Rate, $I$ | | |
|---|---|---|---|
| Tax Rate, $R$ | | 0.00 | 0.03 | 0.05 |
| | 0.00 | | | |
| | 0.28 | | | |
| | 0.35 | | | |

**48.** *Geology: A Contour Map*   The contour map below represents color-coded seismic amplitudes of a fault horizon and a projected contour map, which is used in earthquake studies. *(Source: Adapted from Shipman/Wilson/Todd,* An Introduction to Physical Science, *Eighth Edition)*

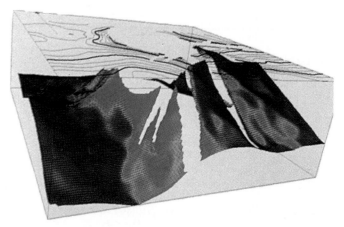

Shipman, *An Introduction to Physical Science* 8/e © 1997, Houghton Mifflin Company

(a) Discuss the use of color to represent the level curves.

(b) Do the level curves correspond to equally spaced amplitudes? Explain.

**49.** *Earnings*   The earnings per share for McDonald's Corporation from 1987 through 1996 can be modeled by

$$z = 0.186x + 0.071y - 0.396,$$

where $x$ is total revenue (in billions of dollars) and $y$ is the total net worth (in billions of dollars). *(Source: McDonald's Corporation)*

(a) Find the earnings per share when $x = 8$ and $y = 6$.

(b) Which of the two variables in this model has the greater influence on the earnings per share? Explain.

# 7.4

# Partial Derivatives

*Functions of Two Variables • Graphical Interpretation of Partial Derivatives •*
*Functions of Three Variables • Higher-Order Partial Derivatives*

## Functions of Two Variables

Real-life applications of functions of several variables are often concerned with how changes in one of the variables will affect the values of the functions. For instance, an economist who wants to determine the effect of a tax increase on the economy might make calculations using different tax rates while holding all other variables, such as unemployment, constant.

You can follow a similar procedure to find the rate of change of a function $f$ with respect to one of its independent variables. That is, you find the derivative of $f$ with respect to one independent variable, while holding the other variables constant. This process is called **partial differentiation,** and each derivative is called a **partial derivative.** A function of several variables has as many partial derivatives as it has independent variables.

---

### Partial Derivatives of a Function of Two Variables

If $z = f(x, y)$, then the **first partial derivatives of $f$ with respect to $x$ and $y$** are the functions $\partial z/\partial x$ and $\partial z/\partial y$, defined as follows.

$$\frac{\partial z}{\partial x} = \lim_{\Delta x \to 0} \frac{f(x + \Delta x, y) - f(x, y)}{\Delta x} \qquad \text{\textit{y} is held constant.}$$

$$\frac{\partial z}{\partial y} = \lim_{\Delta y \to 0} \frac{f(x, y + \Delta y) - f(x, y)}{\Delta y} \qquad \text{\textit{x} is held constant.}$$

---

*STUDY TIP* Note that this definition indicates that partial derivatives of a function of two variables are determined by temporarily considering one variable to be fixed. For instance, if $z = f(x, y)$, then to find $\partial z/\partial x$, you consider $y$ to be constant and differentiate with respect to $x$. Similarly, to find $\partial z/\partial y$, you consider $x$ to be constant and differentiate with respect to $y$.

## EXAMPLE 1   Finding Partial Derivatives

Find $\partial z/\partial x$ and $\partial z/\partial y$ for the function $z = 3x - x^2y^2 + 2x^3y$.

*Solution*

$$\frac{\partial z}{\partial x} = 3 - 2xy^2 + 6x^2y \qquad \text{Hold \textit{y} constant and differentiate with respect to \textit{x}.}$$

$$\frac{\partial z}{\partial y} = -2x^2y + 2x^3 \qquad \text{Hold \textit{x} constant and differentiate with respect to \textit{y}.}$$

---

**Notation for First Partial Derivatives**

The first partial derivatives of $z = f(x, y)$ are denoted by

$$\frac{\partial z}{\partial x} = f_x(x, y) = z_x = \frac{\partial}{\partial x}[f(x, y)]$$

and

$$\frac{\partial z}{\partial y} = f_y(x, y) = z_y = \frac{\partial}{\partial y}[f(x, y)].$$

The values of the first partial derivatives at the point $(a, b)$ are denoted by

$$\frac{\partial z}{\partial x}\bigg|_{(a, b)} = f_x(a, b) \quad \text{and} \quad \frac{\partial z}{\partial y}\bigg|_{(a, b)} = f_y(a, b).$$

---

## EXAMPLE 2  Finding and Evaluating Partial Derivatives

Find the first partial derivatives of $f(x, y) = xe^{x^2y}$ and evaluate each at the point $(1, \ln 2)$.

### Solution

To find the first partial derivative with respect to $x$, hold $y$ constant and differentiate using the Product Rule.

$$f_x(x, y) = x\frac{\partial}{\partial x}[e^{x^2y}] + e^{x^2y}\frac{\partial}{\partial x}[x] \qquad \text{Product Rule}$$

$$= x(2xy)e^{x^2y} + e^{x^2y} \qquad \text{$y$ is held constant.}$$

$$= e^{x^2y}(2x^2y + 1) \qquad \text{Simplify.}$$

At the point $(1, \ln 2)$, the value of this derivative is

$$f_x(1, \ln 2) = e^{(1)^2(\ln 2)}[2(1)^2(\ln 2) + 1] \qquad \text{Substitute for $x$ and $y$.}$$

$$= 2(2 \ln 2 + 1) \qquad \text{Simplify.}$$

$$\approx 4.773. \qquad \text{Use a calculator.}$$

To find the first partial derivative with respect to $y$, hold $x$ constant and differentiate to obtain

$$f_y(x, y) = x(x^2)e^{x^2y} \qquad \text{Constant Multiple Rule}$$

$$= x^3e^{x^2y}. \qquad \text{Simplify.}$$

At the point $(1, \ln 2)$, the value of this derivative is

$$f_y(1, \ln 2) = (1)^3e^{(1)^2(\ln 2)} \qquad \text{Substitute for $x$ and $y$.}$$

$$= 2 \qquad \text{Simplify.}$$

---

## Technology

Symbolic differentiation utilities, such as *Derive, Maple, Mathcad, Mathematica*, and the *TI-92*, can be used to find partial derivatives of a function of two variables. For instance, when *Derive* for Windows is used to find $f_x(x, y)$ for the function in Example 2, you obtain the following.

#1: $x \cdot \hat{e}^{x^2 \cdot y}$

#2: $\dfrac{d}{dx}[x \cdot \hat{e}^{x^2 \cdot y}]$

#3: $\hat{e}^{x^2 \cdot y} \cdot (2 \cdot x^2 \cdot y + 1)$

## Graphical Interpretation of Partial Derivatives

At the beginning of this course, you spent a lot of time studying graphical inter-pretations of the derivative of a function of a single variable. There, you found that $f'(x_0)$ represents the slope of the tangent line to the graph of $y = f(x)$ at the point $(x_0, y_0)$. The partial derivatives of a function of two variables also have use-ful graphical interpretations. Consider the function

$$z = f(x, y). \qquad \text{Function of two variables}$$

As shown in Figure 7.23(a), the graph of this function is a surface in space. If the variable $y$ is fixed, say at $y = y_0$, then

$$z = f(x, y_0) \qquad \text{Function of one variable}$$

is a function of one variable. The graph of this function is the curve that is the intersection of the plane $y = y_0$ and the surface $z = f(x, y)$. On this curve, the par-tial derivative

$$f_x(x, y_0) \qquad \text{Slope in } x\text{-direction}$$

represents the slope in the plane $y = y_0$, as shown in Figure 7.23(a). In a similar way, if the variable $x$ is fixed, say at $x = x_0$, then

$$z = f(x_0, y) \qquad \text{Function of one variable}$$

is a function of one variable. Its graph is the intersection of the plane $x = x_0$ and the surface $z = f(x, y)$. On this curve, the partial derivative

$$f_y(x_0, y) \qquad \text{Slope in } y\text{-direction}$$

represents the slope in the plane $x = x_0$, as shown in Figure 7.23(b).

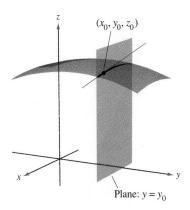

(a) $f_x(x, y_0)$ = slope in $x$-direction

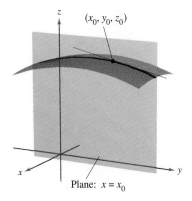

(b) $f_y(x_0, y)$ = slope in $y$-direction

**FIGURE 7.23**

**EXAMPLE 3    Finding Slopes in the *x*- and *y*- Directions**

Find the slope of the surface given by

$$f(x, y) = -\frac{x^2}{2} - y^2 + \frac{25}{8}$$

at the point $\left(\frac{1}{2}, 1, 2\right)$ in (a) the *x*-direction and (b) the *y*-direction.

*Solution*

(a) To find the slope in the *x*-direction, hold *y* constant and differentiate with respect to *x* to obtain

$$f_x(x, y) = -x. \qquad \text{Partial derivative with respect to } x$$

At the point $\left(\frac{1}{2}, 1, 2\right)$, the slope in the *x*-direction is

$$f_x\left(\tfrac{1}{2}, 1\right) = -\tfrac{1}{2}, \qquad \text{Slope in } x\text{-direction}$$

as shown in Figure 7.24(a).

(b) To find the slope in the *y*-direction, hold *x* constant and differentiate with respect to *y* to obtain

$$f_y(x, y) = -2y. \qquad \text{Partial derivative with respect to } y$$

At the point $\left(\frac{1}{2}, 1, 2\right)$, the slope in the *y*-direction is

$$f_y\left(\tfrac{1}{2}, 1\right) = -2, \qquad \text{Slope in } y\text{-direction}$$

as shown in Figure 7.24(b).

*Discovery*

Find the partial derivatives $f_x$ and $f_y$ at $(0, 0)$ for the function in Example 3. What are the slopes of *f* in the *x*- and *y*-directions at $(0, 0)$? Describe the shape of the graph of *f* at this point.

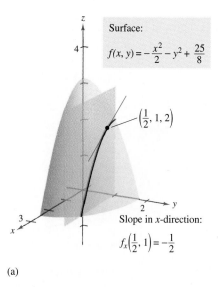

Surface:

$$f(x, y) = -\frac{x^2}{2} - y^2 + \frac{25}{8}$$

$\left(\frac{1}{2}, 1, 2\right)$

Slope in *x*-direction:

$$f_x\left(\tfrac{1}{2}, 1\right) = -\tfrac{1}{2}$$

(a)

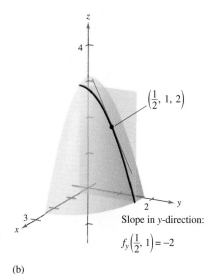

$\left(\frac{1}{2}, 1, 2\right)$

Slope in *y*-direction:

$$f_y\left(\tfrac{1}{2}, 1\right) = -2$$

(b)

**FIGURE 7.24**

Consumer products in the same market or in related markets can be classified as **complementary** or **substitute products.** If two products have a complementary relationship, an increase in the sale of one product will be accompanied
by an increase in the sale of the other product. For instance, videocassette
recorders and videocassettes have a complementary relationship.

If two products have a substitute relationship, an increase in the sale of one
product will be accompanied by a decrease in the sale of the other product. For
instance, videocassette recorders and videodisc recorders both compete in the
same home entertainment market and you would expect a drop in the price of one
to be a deterrent to the sale of the other.

## EXAMPLE 4 Examining a Demand Function

The demand functions for two products are represented by

$$x_1 = f(p_1, p_2) \quad \text{and} \quad x_2 = g(p_1, p_2),$$

where $p_1$ and $p_2$ are the prices per unit for the two products, and $x_1$ and $x_2$ are the
numbers of units sold. The graphs of two different demand functions for $x_1$ are
shown below. Use them to classify the products as complementary or substitute.

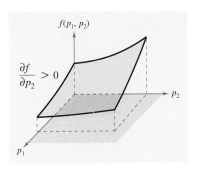

(a)

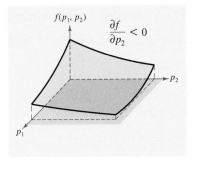

(b)

### Solution

(a) Notice that graph (a) represents the demand for the *first product*. From the
graph of this function, you can see that for a fixed price $p_1$, an increase in $p_2$
results in an increase in the demand for the first product. Remember that an
increase in $p_2$ will also result in a decrease in the demand for the second product. So, if $\partial f/\partial p_2 > 0$, the two products have a substitute relationship.

(b) Notice that graph (b) represents a different demand for the *first product*. From
the graph of this function, you can see that for a fixed price $p_1$, an increase in
$p_2$ results in a decrease in the demand for the first product. Remember that an
increase in $p_2$ will also result in a decrease in the demand for the second product. So, if $\partial f/\partial p_2 < 0$, the two products have a complementary relationship.

Phil Degginger

*As of 1997, Subway had been
chosen as the number one franchise
by Entrepreneur magazine eight
times. Between 1987 and 1997,
Subway opened 10,000 new
stores. What type of product is
complementary to a Subway
sandwich? What type of product
is a substitute?*

## Functions of Three Variables

The concept of a partial derivative can be extended in a natural way to functions of three or more variables. For instance, the function $w = f(x, y, z)$ has three partial derivatives, each of which is formed by considering two of the variables to be constant. That is, to define the partial derivative of $w$ with respect to $x$, consider $y$ *and* $z$ to be constant and write

$$\frac{\partial w}{\partial x} = f_x(x, y, z) = \lim_{\Delta x \to 0} \frac{f(x + \Delta x, y, z) - f(x, y, z)}{\Delta x}.$$

To define the partial derivative of $w$ with respect to $y$, consider $x$ *and* $z$ to be constant and write

$$\frac{\partial w}{\partial y} = f_y(x, y, z) = \lim_{\Delta y 0} \frac{f(x, y + \Delta y, z) - f(x, y, z)}{\Delta y}.$$

To define the partial derivative of $w$ with respect to $z$, consider $x$ *and* $y$ to be constant and write

$$\frac{\partial w}{\partial z} = f_z(x, y, z) = \lim_{\Delta z 0} \frac{f(x, y, z + \Delta z) - f(x, y, z)}{\Delta z}.$$

### Technology

Symbolic differentiation utilities, such as *Derive*, *Maple*, *Mathcad*, *Mathematica*, and the *TI-92*, can be used to find partial derivatives of a function of three or more variables. For instance, when *Derive* for Windows is used to find $f_x(x, y, z)$ for the function in Example 5, you obtain the following.

#1: $x \cdot \hat{e}^{x \cdot y + 2 \cdot z}$

#2: $\dfrac{d}{dx}[x \cdot \hat{e}^{x \cdot y + 2 \cdot z}]$

#3: $\hat{e}^{x \cdot y + 2 \cdot z} \cdot (x \cdot y + 1)$

### EXAMPLE 5  Finding Partial Derivatives of a Function

Find the three partial derivatives of the function

$$w = xe^{xy + 2z}.$$

**Solution**

Holding $y$ and $z$ constant, you obtain

$$\frac{\partial w}{\partial x} = x\frac{\partial}{\partial x}[e^{xy+2z}] + e^{xy+2z}\frac{\partial}{\partial x}[x] \qquad \text{Product Rule}$$

$$= x(ye^{xy+2z}) + e^{xy+2z}(1) \qquad \text{Hold } y \text{ and } z \text{ constant.}$$

$$= (xy + 1)e^{xy+2z}. \qquad \text{Simplify.}$$

Holding $x$ and $z$ constant, you obtain

$$\frac{\partial w}{\partial y} = x(x)e^{xy+2z} \qquad \text{Hold } x \text{ and } z \text{ constant.}$$

$$= x^2 e^{xy+2z}. \qquad \text{Simplify.}$$

Holding $x$ and $y$ constant, you obtain

$$\frac{\partial w}{\partial z} = x(2)e^{xy+2z} \qquad \text{Hold } x \text{ and } y \text{ constant.}$$

$$= 2xe^{xy+2z}. \qquad \text{Simplify.}$$

Note that the Product Rule was used only when finding the partial derivative with respect to $x$. Can you see why?

## Higher-Order Partial Derivatives

As with ordinary derivatives, it is possible to find partial derivatives of second, third, or higher order. For instance, there are four different ways to find a second partial derivative of $z = f(x, y)$.

$$\frac{\partial}{\partial x}\left(\frac{\partial f}{\partial x}\right) = \frac{\partial^2 f}{\partial x^2} = f_{xx} \qquad \text{Differentiate twice with respect to } x.$$

$$\frac{\partial}{\partial y}\left(\frac{\partial f}{\partial y}\right) = \frac{\partial^2 f}{\partial y^2} = f_{yy} \qquad \text{Differentiate twice with respect to } y.$$

$$\frac{\partial}{\partial y}\left(\frac{\partial f}{\partial x}\right) = \frac{\partial^2 f}{\partial y \partial x} = f_{xy} \qquad \begin{array}{l}\text{Differentiate first with respect to } x \\ \text{and then with respect to } y.\end{array}$$

$$\frac{\partial}{\partial x}\left(\frac{\partial f}{\partial y}\right) = \frac{\partial^2 f}{\partial x \partial y} = f_{yx} \qquad \begin{array}{l}\text{Differentiate first with respect to } y \\ \text{and then with respect to } x.\end{array}$$

The third and fourth cases are *mixed* partial derivatives. Notice that with the two types of notation for mixed partials, different conventions are used for indicating the order of differentiation. For instance, the partial derivative

$$\frac{\partial}{\partial y}\left(\frac{\partial f}{\partial x}\right) = \frac{\partial^2 f}{\partial y \partial x} \qquad \text{Right-to-left order}$$

indicates differentiation with respect to $x$ first, but the partial

$$(f_y)_x = f_{yx} \qquad \text{Left-to-right order}$$

indicates differentiation with respect to $y$ first. To remember this, note that in each case you differentiate first with respect to the variable "nearest" $f$.

---

## EXAMPLE 6  Finding Second Partial Derivatives

Find the second partial derivatives of

$$f(x, y) = 3xy^2 - 2y + 5x^2y^2$$

and determine the value of $f_{xy}(-1, 2)$.

### Solution

Begin by finding the first partial derivatives

$$f_x(x, y) = 3y^2 + 10xy^2 \quad \text{and} \quad f_y(x, y) = 6xy - 2 + 10x^2y.$$

Then, differentiating with respect to $x$ and $y$ produces

$$f_{xx}(x, y) = 10y^2, \qquad f_{yy}(x, y) = 6x + 10x^2$$
$$f_{xy}(x, y) = 6y + 20xy, \qquad f_{yx}(x, y) = 6y + 20xy.$$

Finally, the value of $f_{xy}(x, y)$ at the point $(-1, 2)$ is

$$f_{xy}(-1, 2) = 6(2) + 20(-1)(2) = 12 - 40 = -28.$$

**STUDY TIP**   Notice in Example 6 that the two mixed partials are equal. This is often the case. In fact, it can be shown that if a function has continuous second partial derivatives, then the order in which the partial derivatives are taken is irrelevant.

A function of two variables has two first partial derivatives and four second partial derivatives. For a function of three variables, there are three first partials,

$$f_x, f_y, \text{ and } f_z$$

and nine second partials,

$$f_{xx}, f_{xy}, f_{xz}, f_{yx}, f_{yy}, f_{yz}, f_{zx}, f_{zy}, \text{ and } f_{zz},$$

six of which are mixed partials. To find partial derivatives of order three and higher, follow the same pattern used to find second partial derivatives. For instance, if $z = f(x, y)$, then

$$z_{xxx} = \frac{\partial}{\partial x}\left(\frac{\partial^2 f}{\partial x^2}\right) = \frac{\partial^3 f}{\partial x^3} \quad \text{and} \quad z_{xxy} = \frac{\partial}{\partial y}\left(\frac{\partial^2 f}{\partial x^2}\right) = \frac{\partial^3 f}{\partial y \partial x^2}.$$

## EXAMPLE 7   Finding Second Partial Derivatives

Find the second partial derivatives of

$$f(x, y, z) = ye^x + x \ln z.$$

### Solution

Begin by finding the first partial derivatives.

$$f_x(x, y, z) = ye^x + \ln z, \quad f_y(x, y, z) = e^x, \quad f_z(x, y, z) = \frac{x}{z}$$

Then, differentiate with respect to $x$, $y$, and $z$ to find the nine second partial derivatives.

$$f_{xx}(x, y, z) = ye^x, \quad f_{xy}(x, y, z) = e^x, \quad f_{xz}(x, y, z) = \frac{1}{z}$$

$$f_{yx}(x, y, z) = e^x, \quad f_{yy}(x, y, z) = 0, \quad f_{yz}(x, y, z) = 0$$

$$f_{zx}(x, y, z) = \frac{1}{z}, \quad f_{zy}(x, y, z) = 0, \quad f_{zz}(x, y, z) = -\frac{x}{z^2}$$

---

## *Group Discussion*     *Interpreting a Partial Derivative*

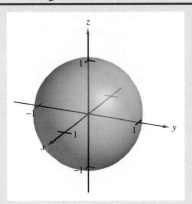

The upper half of the sphere shown at the left is given by

$$f(x, y) = \sqrt{1 - x^2 - y^2}.$$

a. Describe the values of $f_x(x, 0)$. Can you find a value of $x$ for which this partial derivative is 1? Use a three-dimensional graphing utility to explain your answer graphically.

b. Describe the values of $f_y(0, y)$. Can you find a value of $y$ for which this partial derivative is 1? Use a three-dimensional graphing utility to explain your answer graphically.

*Warm Up*

The following warm-up exercises involve skills that were covered in earlier sections. You will use these skills in the exercise set for this section.

In Exercises 1–8, find the derivative of the function.

**1.** $f(x) = \sqrt{x^2 + 3}$

**2.** $g(x) = (3 - x^2)^3$

**3.** $g(t) = te^{2t+1}$

**4.** $f(x) = e^{2x}\sqrt{1 - e^{2x}}$

**5.** $f(x) = \ln(3 - 2x)$

**6.** $u(t) = \ln\sqrt{t^3 - 6t}$

**7.** $g(x) = \dfrac{5x^2}{(4x - 1)^2}$

**8.** $f(x) = \dfrac{(x + 2)^3}{(x^2 - 9)^2}$

In Exercises 9 and 10, evaluate the derivative at the point $(2, 6)$.

**9.** $f(x) = x^2e^{x-2}$

**10.** $g(x) = x\sqrt{x^2 - x + 2}$

## EXERCISES 7.4

In Exercises 1–14, find the first partial derivatives with respect to x and with respect to y.

**1.** $f(x, y) = 2x - 3y + 5$

**2.** $f(x, y) = x^2 - 3y^2 + 7$

**3.** $f(x, y) = 5\sqrt{x} - 6y^2$

**4.** $f(x, y) = x^{-1/2} + 4y^{3/2}$

**5.** $f(x, y) = \dfrac{x}{y}$

**6.** $z = x\sqrt{y}$

**7.** $f(x, y) = \sqrt{x^2 + y^2}$

**8.** $f(x, y) = \dfrac{xy}{x^2 + y^2}$

**9.** $z = x^2e^{2y}$

**10.** $z = xe^{x+y}$

**11.** $h(x, y) = e^{-(x^2+y^2)}$

**12.** $g(x, y) = e^{x/y}$

**13.** $z = \ln\dfrac{x - y}{(x + y)^2}$

**14.** $g(x, y) = \ln\sqrt{x^2 + y^2}$

In Exercises 15–20, let $f(x, y) = 3x^2ye^{x-y}$ and $g(x, y) = 3xy^2e^{y-x}$. Find each of the following.

**15.** $f_x(x, y)$

**16.** $f_y(x, y)$

**17.** $g_x(x, y)$

**18.** $g_y(x, y)$

**19.** $f_x(1, 1)$

**20.** $g_x(-2, -2)$

In Exercises 21–28, evaluate $f_x$ and $f_y$ at the point.

| Function | Point |
|---|---|
| **21.** $f(x, y) = 3x^2 + xy - y^2$ | $(2, 1)$ |
| **22.** $f(x, y) = x^2 - 3xy + y^2$ | $(1, -1)$ |
| **23.** $f(x, y) = e^{3xy}$ | $(0, 4)$ |
| **24.** $f(x, y) = e^xy^2$ | $(0, 2)$ |

| Function | Point |
|---|---|
| **25.** $f(x, y) = \dfrac{xy}{x - y}$ | $(2, -2)$ |
| **26.** $f(x, y) = \dfrac{4xy}{\sqrt{x^2 + y^2}}$ | $(1, 0)$ |
| **27.** $f(x, y) = \ln(x^2 + y^2)$ | $(1, 0)$ |
| **28.** $f(x, y) = \ln\sqrt{xy}$ | $(-1, -1)$ |

In Exercises 29–34, evaluate $w_x$, $w_y$, and $w_z$ at the point.

| Function | Point |
|---|---|
| **29.** $w = \sqrt{x^2 + y^2 + z^2}$ | $(2, -1, 2)$ |
| **30.** $w = \dfrac{xy}{x + y + z}$ | $(1, 2, 0)$ |
| **31.** $w = \ln\sqrt{x^2 + y^2 + z^2}$ | $(3, 0, 4)$ |
| **32.** $w = \dfrac{1}{\sqrt{1 - x^2 - y^2 - z^2}}$ | $(0, 0, 0)$ |
| **33.** $w = 2xz^2 + 3xyz - 6y^2z$ | $(1, -1, 2)$ |
| **34.** $w = xye^{z^2}$ | $(2, 1, 0)$ |

In Exercises 35–38, find values of x and y such that $f_x(x, y) = 0$ and $f_y(x, y) = 0$ simultaneously.

**35.** $f(x, y) = x^2 + 4xy + y^2 - 4x + 16y + 3$

**36.** $f(x, y) = 3x^3 - 12xy + y^3$

**37.** $f(x, y) = \dfrac{1}{x} + \dfrac{1}{y} + xy$

**38.** $f(x, y) = \ln(x^2 + y^2 + 1)$

In Exercises 39–46, find the slope of the surface at the indicated point in (a) the x-direction and (b) the y-direction.

| *Function* | *Point* |
|---|---|
| **39.** $z = 2x - 3y + 5$ | $(2, 1, 6)$ |
| **40.** $z = xy$ | $(1, 2, 2)$ |
| **41.** $z = x^2 - 9y^2$ | $(3, 1, 0)$ |
| **42.** $z = x^2 + 4y^2$ | $(2, 1, 8)$ |
| **43.** $z = \sqrt{25 - x^2 - y^2}$ | $(3, 0, 4)$ |
| **44.** $z = \dfrac{x}{y}$ | $(3, 1, 3)$ |
| **45.** $z = 4 - x^2 - y^2$ | $(1, 1, 2)$ |

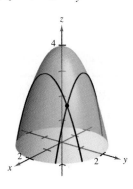

**46.** $z = x^2 - y^2$      $(-2, 1, 3)$

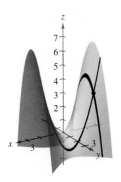

In Exercises 47–50, show that $\partial^2 z / \partial x \partial y = \partial^2 z / \partial y \partial x$.

**47.** $z = x^2 - 2xy + 3y^2$

**48.** $z = x^4 - 3x^2y^2 + y^4$

**49.** $z = \dfrac{e^{2xy}}{4x}$

**50.** $z = \dfrac{x^2 - y^2}{2xy}$

In Exercises 51–58, find the second partial derivatives

$$\frac{\partial^2 z}{\partial x^2}, \quad \frac{\partial^2 z}{\partial y^2}, \quad \frac{\partial^2 z}{\partial y \partial x}, \quad \text{and} \quad \frac{\partial^2 z}{\partial x \partial y}.$$

**51.** $z = x^3 - 4y^2$

**52.** $z = 3x^2 - xy + 2y^3$

**53.** $z = 4x^3 + 3xy^2 - 4y^3$

**54.** $z = \sqrt{9 - x^2 - y^2}$

**55.** $z = \dfrac{xy}{x - y}$

**56.** $z = \dfrac{x}{x + y}$

**57.** $z = xe^{-y^2}$

**58.** $z = xe^y + ye^x$

In Exercises 59–62, evaluate the second partial derivatives $f_{xx}$, $f_{xy}$, $f_{yy}$, and $f_{yx}$ at the point.

| *Function* | *Point* |
|---|---|
| **59.** $f(x, y) = x^4 - 3x^2y^2 + y^2$ | $(1, 0)$ |
| **60.** $f(x, y) = \sqrt{x^2 + y^2}$ | $(0, 2)$ |
| **61.** $f(x, y) = \ln(x - y)$ | $(2, 1)$ |
| **62.** $f(x, y) = x^2e^y$ | $(-1, 0)$ |

In Exercises 63–66, find the first partial derivatives with respect to $x$, $y$, and $z$.

**63.** $w = 3x^2y - 5xyz + 10yz^2$

**64.** $w = \sqrt{x^2 + y^2 + z^2}$

**65.** $w = \dfrac{xy}{x + y + z}$

**66.** $w = \dfrac{1}{\sqrt{1 - x^2 - y^2 - z^2}}$

**67.** *Marginal Cost*   A company manufactures two models of bicycles: a mountain bike and a racing bike. The cost function for producing $x$ mountain bikes and $y$ racing bikes is

$$C = 10\sqrt{xy} + 149x + 189y + 675.$$

Find the marginal costs ($\partial C / \partial x$ and $\partial C / \partial y$) when $x = 120$ and $y = 160$.

**68.** *Marginal Revenue*   A corporation has two plants that produce the same product. If $x_1$ and $x_2$ are the numbers of units produced at plant 1 and plant 2, respectively, then the total revenue for the product is given by

$$R = 200x_1 + 200x_2 - 4x_1^2 - 8x_1x_2 - 4x_2^2.$$

If $x_1 = 4$ and $x_2 = 12$, find the following.

(a) The marginal revenue for plant 1, $\partial R / \partial x_1$

(b) The marginal revenue for plant 2, $\partial R / \partial x_2$

**69.** *Marginal Productivity*   Let $x = 1000$ and $y = 500$ in the Cobb-Douglas production function

$$f(x, y) = 100x^{0.6}y^{0.4}.$$

(a) Find the marginal productivity of labor, $\partial f/\partial x$.

(b) Find the marginal productivity of capital, $\partial f/\partial y$.

**70.** *Marginal Productivity*   Repeat Exercise 69 for the production function $f(x, y) = 100x^{0.75}y^{0.25}$.

**71.** *Complementary and Substitute Products*   Using the notation of Example 4 in this section, we let $x_1$ and $x_2$ be the demands for products 1 and 2, respectively, and $p_1$ and $p_2$ the prices of products 1 and 2, respectively . Determine whether the following demand functions describe complementary or substitute product relationships.

(a) $x_1 = 150 - 2p_1 - \frac{5}{2}p_2,$   $x_2 = 350 - \frac{3}{2}p_1 - 3p_2$

(b) $x_1 = 150 - 2p_1 + 1.8p_2,$   $x_2 = 350 + \frac{3}{4}p_1 - 1.9p_2$

(c) $x_1 = \dfrac{1000}{\sqrt{p_1 p_2}},$   $x_2 = \dfrac{750}{p_2\sqrt{p_1}}$

**72.** *Psychology*   Early in the twentieth century, an intelligence test called the *Stanford-Binet Test* (more commonly know as the *IQ test*) was developed. In this test, an individual's mental age $M$ is divided by the individual's chronological age $C$ and the quotient is multiplied by 100. The result is the individual's *IQ*.

$$IQ(M, C) = \frac{M}{C} \times 100$$

Find the partial derivatives of *IQ* with respect to $M$ and with respect to $C$. Evaluate the partial derivatives at the point $(12, 10)$ and interpret the result.   *(Source: Adapted from Bernstein/Clark-Stewart/Roy/Wickens, Psychology, Fourth Edition)*

**73.** *University Admissions*   Let $N$ be the number of applicants to a university, $p$ the charge for food and housing at the university, and $t$ the tuition. Suppose that $N$ is a function of $p$ and $t$ such that $\partial N/\partial p < 0$ and $\partial N/\partial t < 0$. How would you interpret the fact that both partials are negative?

**74.** *Apparent Temperature*   A measure of what hot weather feels like to two average persons is the Apparent Temperature index. A model for this index is

$$A = 0.885t - 78.7h + 1.20th + 2.70,$$

where $A$ is the apparent temperature, $t$ is the air temperature, and $h$ is the relative humidity in decimal form. *(Source: The UMAP Journal)*

(a) Find $\partial A/\partial t$ and $\partial A/\partial h$ when $t = 90°F$ and $h = 0.80$.

(b) Which has a greater effect on $A$, air temperature or humidity? Explain.

**75.** *Marginal Utility*   The utility function $U = f(x, y)$ is a measure of the utility (or satisfaction) derived by a person from the consumption of two goods $x$ and $y$. Suppose the utility function is $U = -5x^2 + xy - 3y^2$.

(a) Determine the marginal utility of good $x$.

(b) Determine the marginal utility of good $y$.

(c) When $x = 2$ and $y = 3$, should a person consume one more unit of good $x$ or one more unit of good $y$? Explain your reasoning.

(d) Use a three-dimensional graphing utility to graph the function. Interpret the marginal utilities of goods $x$ and $y$ graphically.

## Business Capsule

The Volant Ski Corporation

*Hank Kashiwa, co-founder of Volant Ski Corp., first used a video brochure to educate customers about their stainless steel skis. In 1997, Kashiwa took a team of seven key staff members from coast to coast in a series of "Town Meetings" to demonstrate the benefits that Volant's technology brings to skiing and snowboarding.*

**76.** *Research Project*   Use your school's library or other reference source to research a company that increased the demand for its product by creative advertising. Write a paper about the company. Use graphs to show how a change in demand is a change in a product's marginal utility.

# Extrema of Functions of Two Variables

*Relative Extrema • The First-Partials Test for Relative Extrema •
The Second-Partials Test for Relative Extrema • Applications of Extrema*

## Relative Extrema

Earlier in the text, you learned how to use derivatives to find the relative minimum and relative maximum values of a function of a single variable. In this section, you will learn how to use partial derivatives to find the relative minimum and relative maximum values of a function of two variables.

---

### Relative Extrema of a Function of Two Variables

Let $f$ be a function defined on a region containing $(x_0, y_0)$. The number $f(x_0, y_0)$ is a **relative maximum** of $f$ if there is a circular region $R$ centered at $(x_0, y_0)$ such that

$$f(x, y) \leq f(x_0, y_0) \qquad \text{\small $f(x_0, y_0)$ is a relative maximum.}$$

for all $(x, y)$ in $R$. The number $f(x_0, y_0)$ is a **relative minimum** of $f$ if there is a circular region $R$ centered at $(x_0, y_0)$ such that

$$f(x, y) \geq f(x_0, y_0) \qquad \text{\small $f(x_0, y_0)$ is a relative minimum.}$$

for all $(x, y)$ in $R$.

---

Graphically, you can think of a point on a surface as a relative maximum if it is at least as high as all "nearby" points on the surface, and you can think of a point on a surface as a relative minimum if it is at least as low as all "nearby" points on the surface. Several relative maxima and relative minima are shown in Figure 7.25.

As in single-variable calculus, you need to distinguish between relative extrema and absolute extrema of a function of two variables. The number $f(x_0, y_0)$ is an absolute maximum of $f$ in the region $R$ if it is greater than or equal to all other function values in the region. For instance, the function

$$f(x, y) = -(x^2 + y^2)$$

graphs as a paraboloid, opening downward, with vertex at $(0, 0, 0)$. The number $f(0, 0) = 0$ is an absolute maximum of the function over the entire $xy$-plane. An absolute minimum of $f$ in a region is defined similarly.

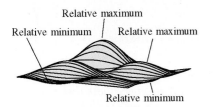

Relative maximum

Relative minimum   Relative maximum

Relative minimum

**FIGURE 7.25** Relative Extrema

## The First-Partials Test for Relative Extrema

To locate the relative extrema of a function of two variables, you can use a procedure that is similar to the First-Derivative Test used for functions of a single variable.

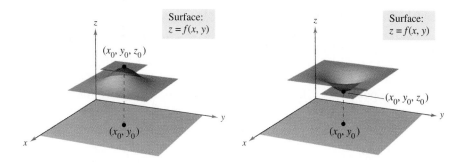

Relative Maximum                Relative Minimum

**FIGURE 7.26**

### First-Partials Test for Relative Extrema

If $f(x_0, y_0)$ is a relative extremum of $f$ on an open region $R$ in the $xy$-plane, and the first partial derivatives of $f$ exist in $R$, then

$$f_x(x_0, y_0) = 0 \quad \text{and} \quad f_y(x_0, y_0) = 0,$$

as shown in Figure 7.26.

***STUDY TIP*** An *open* region in the $xy$-plane is similar to an open interval on the real number line. For instance, the region $R$ consisting of the interior of the circle $x^2 + y^2 = 1$ is an open region. If the region $R$ consists of the interior of the circle *and* the points on the circle, then it is a *closed* region.

A point $(x_0, y_0)$ is a **critical point** of $f$ if $f_x(x_0, y_0)$ or $f_y(x_0, y_0)$ is undefined or if

$$f_x(x_0, y_0) = 0 \quad \text{and} \quad f_y(x_0, y_0) = 0. \qquad \text{Critical point}$$

The First-Partials Test states that if the first partial derivatives exist, then you need only examine values of $f(x, y)$ at critical points to find the relative extrema. As is true for a function of a single variable, however, the critical points of a function of two variables do not always yield relative extrema. For instance, the point $(0, 0)$ is a critical point of the surface shown in Figure 7.27, but $f(0, 0)$ is not a relative extremum of the function. Such points are called **saddle points** of the function.

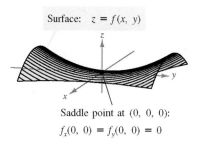

Saddle point at $(0, 0, 0)$:
$f_x(0, 0) = f_y(0, 0) = 0$

**FIGURE 7.27**

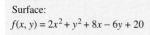

Surface:
$$f(x, y) = 2x^2 + y^2 + 8x - 6y + 20$$

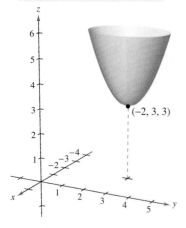

**FIGURE 7.28**

## EXAMPLE 1   Finding Relative Extrema

Find the relative extrema of

$$f(x, y) = 2x^2 + y^2 + 8x - 6y + 20.$$

**Solution**

Begin by finding the first partial derivatives of $f$.

$$f_x(x, y) = 4x + 8 \quad\text{and}\quad f_y(x, y) = 2y - 6$$

Because these partial derivatives are defined for all points in the $xy$-plane, the only critical points are those for which both first partial derivatives are zero. To locate these points, set $f_x(x, y)$ and $f_y(x, y)$ equal to 0, and solve the resulting system of equations.

$$4x + 8 = 0 \qquad \text{Set } f_x(x, y) \text{ equal to } 0.$$
$$2y - 6 = 0 \qquad \text{Set } f_y(x, y) \text{ equal to } 0.$$

The solution of this system is $x = -2$ and $y = 3$. Therefore, the point $(-2, 3)$ is the only critical number of $f$. From the graph of the function, as shown in Figure 7.28, you can see that this critical point yields a relative minimum of the function. Thus, the function has only one relative extremum, which is

$$f(-2, 3) = 3. \qquad \text{Relative minimum}$$

Example 1 shows a relative minimum occurring at one type of critical point—the type for which both $f_x(x, y)$ and $f_y(x, y)$ are zero. The next example shows a relative maximum that occurs at the other type of critical point—the type for which either $f_x(x, y)$ or $f_y(x, y)$ is undefined.

## EXAMPLE 2   Finding Relative Extrema

Find the relative extrema of

$$f(x, y) = 1 - (x^2 + y^2)^{1/3}.$$

**Solution**

Begin by finding the first partial derivatives of $f$.

$$f_x(x, y) = -\frac{2x}{3(x^2 + y^2)^{2/3}} \quad\text{and}\quad f_y(x, y) = -\frac{2y}{3(x^2 + y^2)^{2/3}}$$

These partial derivatives are defined for all points in the $xy$-plane *except* the point $(0, 0)$. Thus, $(0, 0)$ is a critical point of $f$. Moreover, this is the only critical point because there are no other values of $x$ and $y$ for which either partial is undefined or for which both partials are zero. From the graph of the function, as shown in Figure 7.29, you can see that this critical point yields a relative maximum of the function. Thus, the function has only one relative extremum, which is

$$f(0, 0) = 1. \qquad \text{Relative maximum}$$

Surface:
$$f(x, y) = 1 - (x^2 + y^2)^{1/3}$$

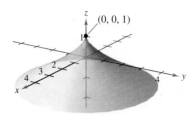

**FIGURE 7.29**   $f_x(x, y)$ and $f_y(x, y)$ are undefined at $(0, 0)$.

## Technology

### Classifying Extrema with a Graphing Utility

The First-Partials Test tells you that the relative extrema of a function $z = f(x, y)$ can occur only at the critical points of the function. The test does not, however, give you a method for determining whether a critical point actually yields a relative minimum, a relative maximum, or neither. Once you have located the critical points of a function, you can use technology to find the relative extrema graphically. Four examples of surfaces are shown below. Try using the graphs and the given critical numbers to classify the relative extrema.

*Surface:* $f(x, y) = x^2 - 2x + y^2 + 4y + 5$

*Critical point:* $(1, -2)$

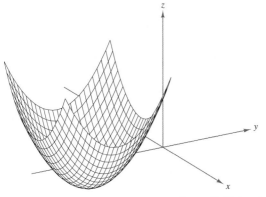

Generated by Mathematica

*Surface:* $f(x, y) = \sqrt{x^2 + y^2}$

*Critical point:* $(0, 0)$

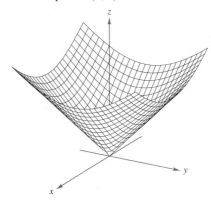

Generated by Mathematica

*Surface:* $f(x, y) = \dfrac{1}{1 + x^2 + y^2}$

*Critical point:* $(0, 0)$

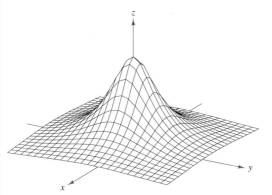

Generated by Mathematica

*Surface:* $f(x, y) = 8xye^{-x^2 - y^2}$

*Critical points:* $(0, 0), \left(\pm\dfrac{1}{\sqrt{2}}, \pm\dfrac{1}{\sqrt{2}}\right), \left(\pm\dfrac{1}{\sqrt{2}}, \mp\dfrac{1}{\sqrt{2}}\right)$

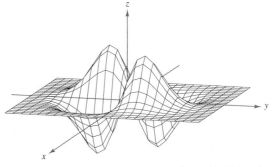

Generated by Mathematica

## The Second-Partials Test for Relative Extrema

For functions such as those in Examples 1 and 2, you can determine the *type* of extrema at the critical points by sketching the graph of the function. For more complicated functions, a graphical approach is not so easy to use. The **Second-Partials Test** is an analytical test that can be used to determine whether a critical number yields a relative minimum, a relative maximum, or neither.

**STUDY TIP**   If $d > 0$, then $f_{xx}(a, b)$ and $f_{yy}(a, b)$ must have the same sign. Thus, you can replace $f_{xx}(a, b)$ with $f_{yy}(a, b)$ in the first two parts of the test.

**TIP**

ALGEBRA

For help in solving the system of equations

$$y - x^3 = 0$$
$$x - y^3 = 0$$

in Example 3, see Example 1a in the *Chapter 7 Algebra Review*, on page 536.

---

### Second-Partials Test

If $f$ has continuous first and second partial derivatives in an open region and there exists a point $(a, b)$ in the region such that $f_x(a, b) = 0$ and $f_y(a, b) = 0$, then the quantity

$$d = f_{xx}(a, b) f_{yy}(a, b) - [f_{xy}(a, b)]^2$$

can be used as follows.

1. $f(a, b)$ is a relative minimum if $d > 0$ and $f_{xx}(a, b) > 0$.

2. $f(a, b)$ is a relative maximum if $d > 0$ and $f_{xx}(a, b) < 0$.

3. $(a, b, f(a, b))$ is a saddle point if $d < 0$.

4. The test gives no information if $d = 0$.

---

### EXAMPLE 3   Applying the Second-Partials Test

Find the relative extrema and saddle points of $f(x, y) = xy - \frac{1}{4}x^4 - \frac{1}{4}y^4$.

**Solution**

Begin by finding the critical points of $f$. Because

$$f_x(x, y) = y - x^3 \qquad \text{and} \qquad f_y(x, y) = x - y^3$$

are defined for all points in the $xy$-plane, the only critical points are those for which both first partial derivatives are zero. By solving the equations $y - x^3 = 0$ and $x - y^3 = 0$ simultaneously, you can determine that the critical points are $(1, 1)$, $(-1, -1)$, and $(0, 0)$. Furthermore, because

$$f_{xx}(x, y) = -3x^2, \quad f_{yy}(x, y) = -3y^2, \quad \text{and} \quad f_{xy}(x, y) = 1,$$

you can use the quantity $d = f_{xx}(a, b) f_{yy}(a, b) - [f_{xy}(a, b)]^2$ to classify the critical points as follows.

| Critical Point | $d$ | $f_{xx}(x, y)$ | Conclusion |
|---|---|---|---|
| $(1, 1)$ | $(-3)(-3) - 1 = 8$ | $-3$ | Relative maximum |
| $(-1, -1)$ | $(-3)(-3) - 1 = 8$ | $-3$ | Relative maximum |
| $(0, 0)$ | $(0)(0) - 1 = -1$ | $0$ | Saddle point |

The graph of $f$ is shown in Figure 7.30.

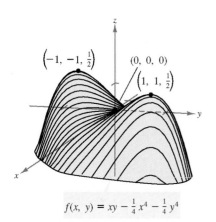

$$f(x, y) = xy - \tfrac{1}{4}x^4 - \tfrac{1}{4}y^4$$

**FIGURE 7.30**

# Applications of Extrema

## EXAMPLE 4  Finding a Maximum Profit

A company makes two substitute products whose demand functions are

$$x_1 = 200(p_2 - p_1) \qquad \text{Demand for product 1}$$
$$x_2 = 500 + 100p_1 - 180p_2, \qquad \text{Demand for product 2}$$

where $p_1$ and $p_2$ are the prices per unit (in dollars) and $x_1$ and $x_2$ are the numbers of units sold. The costs of producing the two products are \$0.50 and \$0.75 per unit, respectively. Find the prices that will yield a maximum profit.

### Solution

The cost and revenue functions are as follows.

$$
\begin{aligned}
C &= 0.5x_1 + 0.75x_2 &&\text{Cost}\\
&= 0.5(200)(p_2 - p_1) + 0.75(500 + 100p_1 - 180p_2) &&\text{Substitute.}\\
&= 375 - 25p_1 - 35p_2 &&\text{Simplify.}\\
R &= p_1 x_1 + p_2 x_2 &&\text{Revenue}\\
&= p_1(200)(p_2 - p_1) + p_2(500 + 100p_1 - 180p_2) &&\text{Substitute.}\\
&= -200p_1^2 - 180p_2^2 + 300p_1 p_2 + 500p_2 &&\text{Simplify.}
\end{aligned}
$$

This implies that the profit function is

$$
\begin{aligned}
P &= R - C &&\text{Profit}\\
&= -200p_1^2 - 180p_2^2 + 300p_1 p_2 + 500p_2 - (375 - 25p_1 - 35p_2)\\
&= -200p_1^2 - 180p_2^2 + 300p_1 p_2 + 25p_1 + 535p_2 - 375.
\end{aligned}
$$

The maximum profit occurs when the two first partial derivatives are zero.

$$\frac{\partial P}{\partial p_1} = -400p_1 + 300p_2 + 25 = 0$$

$$\frac{\partial P}{\partial p_2} = 300p_1 - 360p_2 + 535 = 0$$

By solving this system simultaneously, you can conclude that the solution is $p_1 = \$3.14$ and $p_2 = \$4.10$. From the graph of the function shown in Figure 7.31, you can see that this critical number yields a maximum. Thus, the maximum profit is

$$P = P(3.14, 4.10) = \$761.48.$$

*STUDY TIP*   In Example 4, you can check that the two products are substitutes by observing that $x_1$ increases as $p_2$ increases and $x_2$ increases as $p_1$ increases.

**TIP**

For help in solving the system of equations in Example 4, see Example 1b in the *Chapter 7 Algebra Review,* on page 536.

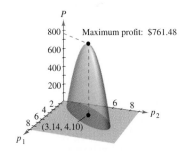

**FIGURE 7.31**

*STUDY TIP*   In Example 4, to convince yourself that the maximum profit is \$761.48, try substituting other prices into the profit function. For each pair of prices, you will obtain a profit that is less than \$761.48. For instance, if $p_1 = \$2$ and $p_2 = \$3$, then the profit is $P = P(2, 3) = \$660.00.$

ALGEBRA

**TIP**

For help in solving the system of equations

$$y(24 - 12x - 4y) = 0$$
$$x(24 - 6x - 8y) = 0$$

in Example 5, see Example 2a in the *Chapter 7 Algebra Review,* on page 537.

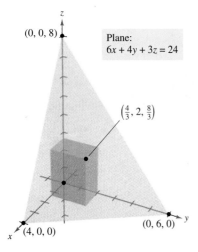

Plane:
$6x + 4y + 3z = 24$

$(0, 0, 8)$

$\left(\frac{4}{3}, 2, \frac{8}{3}\right)$

$(4, 0, 0)$

$(0, 6, 0)$

**FIGURE 7.32**

## EXAMPLE 5 Finding a Maximum Volume

Consider all possible rectangular boxes that are resting on the $xy$-plane with one vertex at the origin and the opposite vertex in the plane $6x + 4y + 3z = 24$, as shown in Figure 7.32. Of all such boxes, which has the greatest volume?

### Solution

Because one vertex of the box lies in the plane given by $6x + 4y + 3z = 24$ or $z = \frac{1}{3}(24 - 6x - 4y)$, you can write the volume of the box as

$$
\begin{aligned}
V &= xyz && \text{Volume = (width)(length)(height)} \\
&= xy\left(\tfrac{1}{3}\right)(24 - 6x - 4y) && \text{Substitute for } z. \\
&= \tfrac{1}{3}(24xy - 6x^2y - 4xy^2). && \text{Simplify.}
\end{aligned}
$$

To find the critical numbers, set the first partial derivatives equal to zero.

$$
\begin{aligned}
V_x &= \tfrac{1}{3}(24y - 12xy - 4y^2) && \text{Partial with respect to } x \\
&= \tfrac{1}{3}y(24 - 12x - 4y) && \text{Factor.} \\
&= 0 && \text{Set equal to 0.} \\
V_y &= \tfrac{1}{3}(24x - 6x^2 - 8xy) && \text{Partial with respect to } y \\
&= \tfrac{1}{3}x(24 - 6x - 8y) && \text{Factor.} \\
&= 0 && \text{Set equal to 0.}
\end{aligned}
$$

The four solutions of this system are $(0, 0)$, $(0, 6)$, $(4, 0)$, and $\left(\frac{4}{3}, 2\right)$. Using the Second-Partials Test, you can determine that the maximum volume occurs when the width is $x = \frac{4}{3}$ and the length is $y = 2$. For these values, the height of the box is

$$z = \tfrac{1}{3}\left[24 - 6\left(\tfrac{4}{3}\right) - 4(2)\right] = \tfrac{8}{3}.$$

Thus, the maximum volume is

$$V = xyz = \left(\tfrac{4}{3}\right)(2)\left(\tfrac{8}{3}\right) = \tfrac{64}{9} \text{ cubic units.}$$

---

## *Group Discussion*

## *Using the Second-Partials Test*

In Example 5, explain why you can disregard the three critical numbers $(0, 0)$, $(0, 6)$, and $(4, 0)$. Then use the second partial derivatives

$$
\begin{aligned}
V_{xx}(x, y) &= -4y \\
V_{yy}(x, y) &= -\tfrac{8}{3}x \\
V_{xy}(x, y) &= \tfrac{1}{3}(24 - 12x - 8y)
\end{aligned}
$$

to explain why the critical number $\left(\frac{4}{3}, 2\right)$ yields a maximum.

*Warm Up*

The following warm-up exercises involve skills that were covered in earlier sections. You will use these skills in the exercise set for this section.

In Exercises 1–6, solve the system of equations.

1. $\begin{cases} 5x = 15 \\ 3x - 2y = 5 \end{cases}$

2. $\begin{cases} \frac{1}{2}y = 3 \\ -x + 5y = 19 \end{cases}$

3. $\begin{cases} x + y = 5 \\ x - y = -3 \end{cases}$

4. $\begin{cases} 2x - 4y = 14 \\ 3x + y = 7 \end{cases}$

5. $\begin{cases} x^2 + x = 0 \\ 2yx + y = 0 \end{cases}$

6. $\begin{cases} 3y^2 + 6y = 0 \\ xy + x + 2 = 0 \end{cases}$

In Exercises 7–10, find all first and second partial derivatives of the function.

7. $z = 4x^3 - 3y^2$

8. $z = 2x^2 - 3xy + y^2$

9. $z = x^4 - \sqrt{xy} + 2y$

10. $z = xe^{xy}$

## EXERCISES 7.5

In Exercises 1–4, find any critical points and relative extrema of the function.

1. $f(x, y) = x^2 - y^2 + 4x - 8y - 11$

2. $f(x, y) = x^2 + y^2 + 2x - 6y + 6$

3. $f(x, y) = \sqrt{x^2 + y^2 + 1}$

4. $f(x, y) = \sqrt{25 - (x - 2)^2 - y^2}$

In Exercises 5–20, examine each function for relative extrema and saddle points.

5. $f(x, y) = (x - 1)^2 + (y - 3)^2$

6. $f(x, y) = 9 - (x - 3)^2 - (y + 2)^2$

7. $f(x, y) = 2x^2 + 2xy + y^2 + 2x - 3$

8. $f(x, y) = -x^2 - 5y^2 + 8x - 10y - 13$

9. $f(x, y) = -5x^2 + 4xy - y^2 + 16x + 10$

10. $f(x, y) = x^2 + 6xy + 10y^2 - 4y + 4$

11. $f(x, y) = 3x^2 + 2y^2 - 12x - 4y + 7$

12. $f(x, y) = -3x^2 - 2y^2 + 3x - 4y + 5$

13. $f(x, y) = x^2 - y^2 + 4x - 4y - 8$

14. $f(x, y) = x^2 - 3xy - y^2$

15. $f(x, y) = xy$

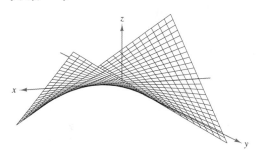

16. $f(x, y) = 12x + 12y - xy - x^2 - y^2$

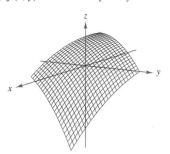

**17.** $f(x, y) = (x^2 + 4y^2)e^{1-x^2-y^2}$

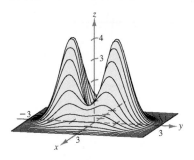

**18.** $f(x, y) = e^{-(x^2+y^2)}$

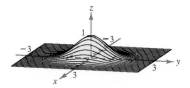

**19.** $f(x, y) = e^{xy}$

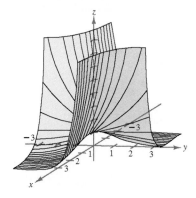

**20.** $f(x, y) = -\dfrac{4x}{x^2 + y^2 + 1}$

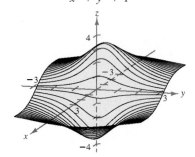

In Exercises 21–24, determine whether there is a relative maximum, a relative minimum, a saddle point, or insufficient information to determine the nature of the function $f(x, y)$ at the critical point $(x_0, y_0)$.

**21.** $f_{xx}(x_0, y_0) = 16$  
  $f_{yy}(x_0, y_0) = 4$  
  $f_{xy}(x_0, y_0) = 8$

**22.** $f_{xx}(x_0, y_0) = -4$  
  $f_{yy}(x_0, y_0) = -6$  
  $f_{xy}(x_0, y_0) = 3$

**23.** $f_{xx}(x_0, y_0) = -7$  
  $f_{yy}(x_0, y_0) = 4$  
  $f_{xy}(x_0, y_0) = 9$

**24.** $f_{xx}(x_0, y_0) = 20$  
  $f_{yy}(x_0, y_0) = 8$  
  $f_{xy}(x_0, y_0) = 9$

In Exercises 25–30, find the critical points and test for relative extrema. List the critical points for which the Second-Partials Test fails.

**25.** $f(x, y) = (xy)^2$   **26.** $f(x, y) = \sqrt{x^2 + y^2}$

**27.** $f(x, y) = x^3 + y^3$

**28.** $f(x, y) = x^3 + y^3 - 3x^2 + 6y^2 + 3x + 12y + 7$

**29.** $f(x, y) = x^{2/3} + y^{2/3}$   **30.** $f(x, y) = (x^2 + y^2)^{2/3}$

In Exercises 31 and 32, find the critical points of the function and, from the form of the function, determine whether each critical point is a relative maximum or a relative minimum.

**31.** $f(x, y, z) = (x - 1)^2 + (y + 3)^2 + z^2$

**32.** $f(x, y, z) = 6 - [x(y + 2)(z - 1)]^2$

*True or False*   In Exercises 33–36, determine whether the statement is true or false. If it is false, explain why or give an example that shows it is false.

**33.** If $d > 0$ and $f_x(a, b) < 0$, then $f(a, b)$ is a relative minimum.

**34.** A saddle point always occurs at a critical point.

**35.** If $f(x, y)$ has a relative maximum $(x_0, y_0, z_0)$, then $f_x(x_0, y_0) = f_y(x_0, y_0) = 0$.

**36.** The function $f(x, y) = \sqrt[3]{x^2 + y^2}$ has a relative maximum at the origin.

In Exercises 37–40, find three positive numbers $x$, $y$, and $z$ that satisfy the given conditions.

**37.** The sum is 30 and the product is maximum.

**38.** The sum is 32 and $P = xy^2z$ is maximum.

**39.** The sum is 30 and the sum of the squares is minimum.

**40.** The sum is 1 and the sum of the squares is minimum.

**41.** *Revenue*   A company manufactures two products. The total revenue from $x_1$ units of product 1 and $x_2$ units of product 2 is

$$R = -5x_1{}^2 - 8x_2{}^2 - 2x_1x_2 + 42x_1 + 102x_2.$$

Find $x_1$ and $x_2$ so as to maximize the revenue.

**42.** *Revenue*   A retail outlet sells two competitive products, the prices of which are $p_1$ and $p_2$. Find $p_1$ and $p_2$ so as to maximize the total revenue

$$R = 500p_1 + 800p_2 + 1.5p_1p_2 - 1.5p_1{}^2 - p_2{}^2.$$

*Revenue*   In Exercises 43 and 44, find $p_1$ and $p_2$ so as to maximize the total revenue $R = x_1p_1 + x_2p_2$ for a retail outlet that sells two competitive products with the given demand functions.

**43.** $x_1 = 1000 - 2p_1 + p_2$, $x_2 = 1500 + 2p_1 - 1.5p_2$

**44.** $x_1 = 1000 - 4p_1 + 2p_2$, $x_2 = 900 + 4p_1 - 3p_2$

**45.** *Profit*   A corporation manufactures a product at two locations. The cost functions for producing $x_1$ units at location 1 and $x_2$ units at location 2 are

$$C_1 = 0.05x_1{}^2 + 15x_1 + 5400$$

and

$$C_2 = 0.03x_2{}^2 + 15x_2 + 6100,$$

respectively. The demand function for the product is given by

$$p = 225 - 0.4(x_1 + x_2),$$

and therefore the total revenue function is

$$R = [225 - 0.4(x_1 + x_2)](x_1 + x_2).$$

Find the production levels at the two locations that will maximize the profit

$$P = R - C_1 - C_2.$$

**46.** *Cost*   The material for constructing the base of an open box costs 1.5 times as much as the material for constructing the sides. Find the dimensions of the box of largest volume that can be made for a fixed amount of money $C$.

**47.** *Volume*   Find the dimensions of a rectangular package of largest volume that may be sent by parcel post assuming that the sum of the length and the girth (perimeter of a cross section) cannot exceed 108 inches.

**48.** *Volume*   Show that the rectangular box of given volume and minimum surface area is a cube.

**49.** *Hardy-Weinberg Law*   Common blood types are determined genetically by the three alleles A, B, and O. (An allele is any of a group of possible mutational forms of a gene.) A person whose blood type is AA, BB, or OO is homozygous. A person whose blood type is AB, AO, or BO is heterozygous. The Hardy-Weinberg Law states that the proportion $P$ of heterozygous individuals in any given population is

$$P(p, q, r) = 2pq + 2pr + 2qr,$$

where $p$ represents the percent of allele A in the population, $q$ represents the percent of allele B in the population, and $r$ represents the percent of allele O in the population. Use the fact that $p + q + r = 1$ (the sum of the three must equal 100%) to show that the maximum proportion of heterozygous individuals in any population is $\frac{2}{3}$.

**50.** *Stocking a Lake*   A lake is to be stocked with smallmouth and largemouth bass. Let $x$ represent the number of smallmouth bass and let $y$ represent the number of largemouth bass in the lake. The weight of each fish is dependent on the population densities. After a 6-month period, the weight of a single smallmouth bass is

$$W_1 = 3 - 0.002x - 0.005y,$$

and the weight of a single largemouth bass is

$$W_2 = 4.5 - 0.003x - 0.004y.$$

Assuming that no fish die during the 6-month period, how many smallmouth and largemouth bass should be stocked in the lake so that the *total* weight $T$ of bass in the lake is a maximum?

**51.** *Medicine*   In order to treat a certain bacterial infection, a combination of two drugs is being tested. Studies have shown that the duration of the infection in laboratory tests can be modeled by

$$D(x, y) = x^2 + 2y^2 - 18x - 24y + 2xy + 120,$$

where $x$ is the dosage in hundreds of milligrams of the first drug and $y$ is the dosage in hundreds of milligrams of the second drug. Determine the partial derivatives of $D$ with respect to $x$ and with respect to $y$. Find the amount of each drug necessary to minimize the duration of the infection.

# Lagrange Multipliers

## 7.6

*Lagrange Multipliers with One Constraint •*
*Lagrange Multipliers with Two Constraints*

(0, 0, 8)

Plane:
$6x + 4y + 3z = 24$

$\left(\frac{4}{3}, 2, \frac{8}{3}\right)$

(0, 6, 0)    $y$

$x$    (4, 0, 0)

**FIGURE 7.33**

## Lagrange Multipliers with One Constraint

In Example 5 in Section 7.5, you were asked to find the dimensions of the rectangular box of maximum volume that would fit in the first octant beneath the plane $6x + 4y + 3z = 24$, as shown again in Figure 7.33. Another way of stating this problem is to say that you are asked to find the maximum of

$$V = xyz \qquad \text{Objective function}$$

subject to the constraint

$$6x + 4y + 3z - 24 = 0. \qquad \text{Constraint}$$

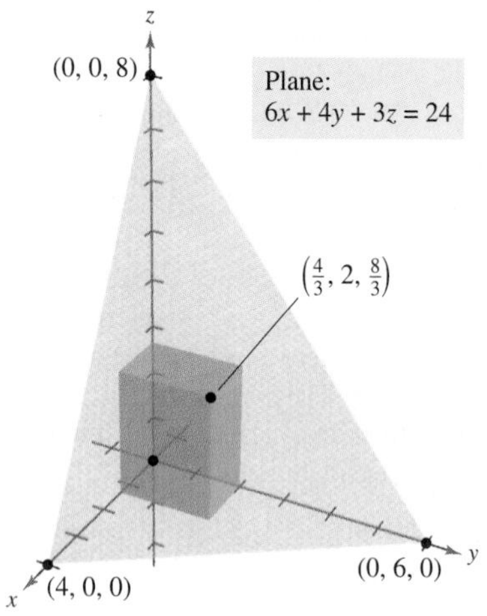

This type of problem is called a **constrained optimization** problem. In Section 7.5, you answered this question by solving for $z$ in the constraint equation and then rewriting $V$ as a function of two variables.

In this section, you will study a different (and often better) way to solve constrained optimization problems. This method involves the use of variables called **Lagrange multipliers,** named after the French mathematician Joseph Louis Lagrange (1736–1813).

### Method of Lagrange Multipliers

If $f(x, y)$ has a maximum or minimum subject to the constraint $g(x, y) = 0$, then it will occur at one of the critical numbers of the function $F$ defined by

$$F(x, y, \lambda) = f(x, y) - \lambda g(x, y).$$

The variable $\lambda$ (the lowercase Greek letter lambda) is called a **Lagrange multiplier.** For functions of three variables, $F$ has the form

$$F(x, y, z, \lambda) = f(x, y, z) - \lambda g(x, y, z).$$

The method of Lagrange multipliers gives you a way of finding critical points but does not tell you whether these points yield minima, maxima, or neither. To make this distinction, you must rely on the context of the problem.

## EXAMPLE 1 Using Lagrange Multipliers: One Constraint

Find the maximum of

$$V = xyz \qquad \text{Objective function}$$

subject to the constraint

$$6x + 4y + 3z - 24 = 0. \qquad \text{Constraint}$$

### Solution

First, let $f(x, y, z) = xyz$ and $g(x, y, z) = 6x + 4y + 3z - 24$. Then, define a new function $F$ by

$$F(x, y, z, \lambda) = f(x, y, z) - \lambda g(x, y, z)$$
$$= xyz - \lambda(6x + 4y + 3z - 24).$$

To find the critical numbers of $F$, set the partial derivatives of $F$ with respect to $x$, $y$, $z$, and $\lambda$ equal to zero and obtain

$$F_x(x, y, z, \lambda) = yz - 6\lambda = 0$$
$$F_y(x, y, z, \lambda) = xz - 4\lambda = 0$$
$$F_z(x, y, z, \lambda) = xy - 3\lambda = 0$$
$$F_\lambda(x, y, z, \lambda) = -6x - 4y - 3z + 24 = 0.$$

Solving for $\lambda$ in the first equation and substituting into the second and third equations produces the following.

$$xz - 4\left(\frac{yz}{6}\right) = 0 \implies y = \frac{3}{2}x$$

$$xy - 3\left(\frac{yz}{6}\right) = 0 \implies z = 2x$$

Next, substitute these values for $y$ and $z$ into the equation $F_\lambda(x, y, z, \lambda) = 0$ and solve for $x$.

$$F_\lambda(x, y, z, \lambda) = 0$$
$$-6x - 4\left(\tfrac{3}{2}x\right) - 3(2x) + 24 = 0$$
$$-18x = -24$$
$$x = \tfrac{4}{3}$$

Using this $x$-value, you can conclude that the critical values are $x = \tfrac{4}{3}$, $y = 2$, and $z = \tfrac{8}{3}$, which implies that the maximum is

$$V = xyz \qquad \text{Objective function}$$

$$= \left(\frac{4}{3}\right)(2)\left(\frac{8}{3}\right) \qquad \text{Substitute values of } x, y, \text{ and } z.$$

$$= \frac{64}{9} \text{ cubic units.} \qquad \text{Maximum volume}$$

*STUDY TIP* Example 1 shows how Lagrange multipliers can be used to solve the same problem that was solved in Example 5 in Section 7.5.

TIP

ALGEBRA

The most difficult aspect of many Lagrange Multiplier problems is the complicated algebra needed to solve the system of equations arising from $F(x, y, \lambda) = f(x, y) - \lambda g(x, y)$. There is no general way to proceed in every case, so you should study the examples carefully, and refer to the *Chapter 7 Algebra Review,* on pages 536 and 537.

Joseph Nettis/Stock Boston

*For many industrial applications, a simple robot costs as much as a year's wages and benefits for three workers. Thus, manufacturers must carefully balance the amount of money spent on labor and capital.*

## EXAMPLE 2 Finding a Maximum Production

A manufacturer's production is modeled by the Cobb-Douglas function

$$f(x, y) = 100x^{3/4}y^{1/4}, \qquad \text{Objective function}$$

where $x$ represents the units of labor and $y$ represents the units of capital. Each labor unit costs \$150 and each capital unit costs \$250. The total expenses for labor and capital cannot exceed \$50,000. Find the maximum production level.

### Solution

Because total labor and capital expenses cannot exceed \$50,000, the constraint is

$$150x + 250y = 50{,}000 \qquad \text{Constraint}$$
$$150x + 250y - 50{,}000 = 0. \qquad \text{Standard form}$$

To find the maximum production level, begin by writing the function

$$F(x, y, \lambda) = 100x^{3/4}y^{1/4} - \lambda(150x + 250y - 50{,}000).$$

Next, set the partial derivatives of this function equal to zero.

$$F_x(x, y, \lambda) = 75x^{-1/4}y^{1/4} - 150\lambda = 0 \qquad \text{Equation 1}$$
$$F_y(x, y, \lambda) = 25x^{3/4}y^{-3/4} - 250\lambda = 0 \qquad \text{Equation 2}$$
$$F_\lambda(x, y, \lambda) = -150x - 250y + 50{,}000 = 0 \qquad \text{Equation 3}$$

The strategy for solving such a system must be customized to the particular system. In this case, you can solve for $\lambda$ in the first equation, substitute into the second equation, solve for $x$, substitute into the third equation, and solve for $y$.

$$75x^{-1/4}y^{1/4} - 150\lambda = 0 \qquad \text{Equation 1}$$
$$\lambda = \tfrac{1}{2}x^{-1/4}y^{1/4} \qquad \text{Solve for } \lambda.$$
$$25x^{3/4}y^{-3/4} - 250\left(\tfrac{1}{2}\right)x^{-1/4}y^{1/4} = 0 \qquad \text{Substitute in Equation 2.}$$
$$25x - 125y = 0 \qquad \text{Multiply by } x^{1/4}y^{3/4}.$$
$$x = 5y \qquad \text{Solve for } x.$$
$$-150(5y) - 250y + 50{,}000 = 0 \qquad \text{Substitute in Equation 3.}$$
$$-1000y = -50{,}000 \qquad \text{Simplify.}$$
$$y = 50 \qquad \text{Solve for } y.$$

Using this value for $y$, it follows that $x = 5(50) = 250$. Thus, the maximum production level of

$$f(250, 50) = 100(250)^{3/4}(50)^{1/4} \qquad \text{Substitute for } x \text{ and } y.$$
$$\approx 16{,}719 \text{ units} \qquad \text{Maximum production}$$

occurs when $x = 250$ units of labor and $y = 50$ units of capital.

Economists call the Lagrange multiplier obtained in a production function the **marginal productivity of money.** For instance, in Example 2, the marginal productivity of money when $x = 250$ and $y = 50$ is

$$\lambda = \tfrac{1}{2}x^{-1/4}y^{1/4} = \tfrac{1}{2}(250)^{-1/4}(50)^{1/4} \approx 0.334.$$

This means that if one additional dollar is spent on production, approximately 0.334 additional unit of the product can be produced.

## EXAMPLE 3   Finding a Maximum Production

In Example 2, suppose that $70,000 is available for labor and capital. What is the maximum number of units that can be produced?

### Solution

You could rework the entire problem, as demonstrated in Example 2. However, because the only change in the problem is the availability of additional money to spend on labor and capital, you can use the fact that the marginal productivity of money is

$$\lambda \approx 0.334.$$

Because an additional $20,000 is available and the maximum production in Example 2 was 16,719 units, you can conclude that the maximum production is now

$$16,719 + (0.334)(20,000) \approx 23,400 \text{ units.}$$

Try using the procedure demonstrated in Example 2 to confirm this result.

## Technology

You can use a three-dimensional graphing utility to confirm graphically the results of Examples 2 and 3. Begin by graphing the surface $f(x, y) = 100x^{3/4}y^{1/4}$. Then graph the vertical plane given by $150x + 250y = 50,000$. As shown at the right, the maximum production level corresponds to the highest point on the intersection of the surface and the plane.

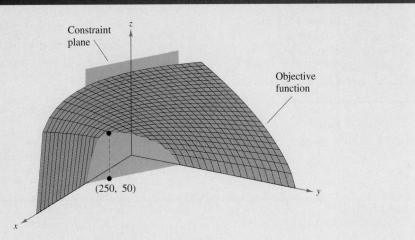

Intersection of the surface $z = 100x^{3/4}y^{1/4}$ and the plane $150x + 250y = 50,000$

**EXAMPLE 4** **Optimizing with Four Variables**

Find the minimum of the function

$$f(x, y, z, w) = x^2 + y^2 + z^2 + w^2 \qquad \text{Objective function}$$

subject to the constraint

$$3x + 2y - 4z + w - 3 = 0. \qquad \text{Constraint}$$

**Solution**

Begin by forming the function

$$F(x, y, z, w, \lambda) = x^2 + y^2 + z^2 + w^2 - \lambda(3x + 2y - 4z + w - 3).$$

Next, set the five partial derivatives of $F$ equal to zero.

$$F_x(x, y, z, w, \lambda) = 2x - 3\lambda = 0 \qquad \text{Equation 1}$$
$$F_y(x, y, z, w, \lambda) = 2y - 2\lambda = 0 \qquad \text{Equation 2}$$
$$F_z(x, y, z, w, \lambda) = 2z + 4\lambda = 0 \qquad \text{Equation 3}$$
$$F_w(x, y, z, w, \lambda) = 2w - \lambda = 0 \qquad \text{Equation 4}$$
$$F_\lambda(x, y, z, w, \lambda) = -3x - 2y + 4z - w + 3 = 0 \qquad \text{Equation 5}$$

For this system, you can solve the first four equations in terms of $\lambda$.

$$2x - 3\lambda = 0 \implies x = \tfrac{3}{2}\lambda$$
$$2y - 2\lambda = 0 \implies y = \lambda$$
$$2z + 4\lambda = 0 \implies z = -2\lambda$$
$$2w - \lambda = 0 \implies w = \tfrac{1}{2}\lambda$$

Substituting these values into the fifth equation allows you to solve for $\lambda$.

$$-3x - 2y + 4z - w + 3 = 0$$
$$-3\left(\tfrac{3}{2}\lambda\right) - 2(\lambda) + 4(-2\lambda) - \left(\tfrac{1}{2}\lambda\right) = -3$$
$$-15\lambda = -3$$
$$\lambda = \tfrac{1}{5}$$

Therefore, the critical numbers are

$$x = \tfrac{3}{10}, \quad y = \tfrac{1}{5}, \quad z = -\tfrac{2}{5}, \quad \text{and} \quad w = \tfrac{1}{10},$$

and the minimum value of $f$ subject to the given constraint is

$$f\left(\tfrac{3}{10}, \tfrac{1}{5}, -\tfrac{2}{5}, \tfrac{1}{10}\right) = \left(\tfrac{3}{10}\right)^2 + \left(\tfrac{1}{5}\right)^2 + \left(-\tfrac{2}{5}\right)^2 + \left(\tfrac{1}{10}\right)^2 \qquad \text{Substitute.}$$
$$= \tfrac{9}{100} + \tfrac{1}{25} + \tfrac{4}{25} + \tfrac{1}{100} \qquad \text{Simplify.}$$
$$= \tfrac{3}{10}. \qquad \text{Minimum value}$$

Thus, the minimum value of $f(x, y, z, w)$ is $\tfrac{3}{10}$.

*Discovery*

In Example 4, what would be the minimum value of the objective function $f(x, y, z, w) = x^2 + y^2 + z^2 + w^2$ if there were no constraint? What would be the maximum value of $f$ in this case? How do you know that the answer to Example 4 is the minimum value of $f$ subject to the given constraint, and not the maximum value?

In Example 4 in Section 7.5, you found the maximum profit for two substitute products whose demand functions are

$$x_1 = 200(p_2 - p_1)$$ Demand for product 1

$$x_2 = 500 + 100p_1 - 180p_2.$$ Demand for product 2

With this model, the total demand, $x_1 + x_2$, is completely determined by the prices $p_1$ and $p_2$. In many real-life situations, this assumption is too simplistic: regardless of the prices of the substitute brands, the annual total demands for some products, such as toothpaste, are relatively constant. In such situations, the total demand is **limited,** and variations in price do not affect the total demand as much as they affect the market share of the substitute brands.

## EXAMPLE 5   Finding the Maximum Profit

A company makes two substitute products whose demand functions are

$$x_1 = 200(p_2 - p_1)$$ Demand for product 1

$$x_2 = 500 + 100p_1 - 180p_2,$$ Demand for product 2

where $p_1$ and $p_2$ are the prices per unit (in dollars) and $x_1$ and $x_2$ are the numbers of units sold. The costs of producing the two products are \$0.50 and \$0.75 per unit, respectively. The total demand is limited to 200 units per year. Find the prices that will yield a maximum profit.

### Solution

From Example 4 in Section 7.5, the profit function is modeled by

$$P = -200p_1{}^2 - 180p_2{}^2 + 300p_1p_2 + 25p_1 + 535p_2 - 375.$$

The total demand for the two products is

$$x_1 + x_2 = 200(p_2 - p_1) + 500 + 100p_1 - 180p_2$$
$$= -100p_1 + 20p_2 + 500.$$

Because the total demand is limited to 200 units,

$$-100p_1 + 20p_2 + 500 = 200.$$ Constraint

Using Lagrange multipliers, you can determine that the maximum profit occurs when $p_1 = \$3.94$ and $p_2 = \$4.69$. This corresponds to an annual profit of \$712.21.

> **STUDY TIP**   The constrained optimization problem in Example 5 is represented graphically in Figure 7.34. The graph of the objective function is a paraboloid and the graph of the constraint is a vertical plane. In the "unconstrained" optimization problem on page 495, the maximum profit occurred at the vertex of the paraboloid. In this "constrained" problem, however, the maximum profit corresponds to the highest point on the curve that is the intersection of the paraboloid and the vertical "constraint" plane.

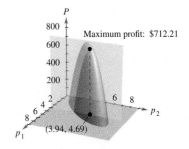

**FIGURE 7.34**

## Lagrange Multipliers with Two Constraints

In Examples 1 through 5, each of the optimization problems contained only one constraint. When an optimization problem has two constraints, you need to introduce a second Lagrange multiplier. The customary symbol for this second multiplier is $\mu$, the Greek letter mu.

### EXAMPLE 6  Using Lagrange Multipliers: Two Constraints

Find the minimum value of

$$f(x, y, z) = x^2 + y^2 + z^2 \qquad \text{Objective function}$$

subject to the constraints

$$x + y - 3 = 0 \qquad \text{Constraint 1}$$
$$x + z - 5 = 0. \qquad \text{Constraint 2}$$

**Solution**

Begin by forming the function

$$F(x, y, z, \lambda, \mu) = x^2 + y^2 + z^2 - \lambda(x + y - 3) - \mu(x + z - 5).$$

Next, set the five partial derivatives equal to zero, and solve the resulting system of equations for $x$, $y$, and $z$.

$$F_x(x, y, z, \lambda, \mu) = 2x - \lambda - \mu = 0 \qquad \text{Equation 1}$$
$$F_y(x, y, z, \lambda, \mu) = 2y - \lambda = 0 \qquad \text{Equation 2}$$
$$F_z(x, y, z, \lambda, \mu) = 2z - \mu = 0 \qquad \text{Equation 3}$$
$$F_\lambda(x, y, z, \lambda, \mu) = -x - y + 3 = 0 \qquad \text{Equation 4}$$
$$F_\mu(x, y, z, \lambda, \mu) = -x - z + 5 = 0 \qquad \text{Equation 5}$$

Solving this system of equations produces $x = \frac{8}{3}$, $y = \frac{1}{3}$, and $z = \frac{7}{3}$. Thus, the minimum value of $f(x, y, z)$ is

$$f\left(\frac{8}{3}, \frac{1}{3}, \frac{7}{3}\right) = \left(\frac{8}{3}\right)^2 + \left(\frac{1}{3}\right)^2 + \left(\frac{7}{3}\right)^2 = \frac{38}{3}.$$

## *Group Discussion*    Solving a System of Equations

Explain how you would solve the system of equations shown in Example 6. Illustrate your description by actually solving the system.

Suppose that in Example 6 you had been asked to find the maximum value of $f(x, y, z)$ subject to the two constraints. How would your answer have been different? Use a three-dimensional graphing utility to verify your conclusion.

*Warm Up*

The following warm-up exercises involve skills that were covered in earlier sections. You will use these skills in the exercise set for this section.

In Exercises 1–6, solve the system of linear equations.

**1.** $\begin{cases} 4x - 6y = 3 \\ 2x + 3y = 2 \end{cases}$

**2.** $\begin{cases} 6x - 6y = 5 \\ -3x - y = 1 \end{cases}$

**3.** $\begin{cases} 5x - y = 25 \\ x - 5y = 15 \end{cases}$

**4.** $\begin{cases} 4x - 9y = 5 \\ -x + 8y = -2 \end{cases}$

**5.** $\begin{cases} 2x - y + z = 3 \\ 2x + 2y + z = 4 \\ -x + 2y + 3z = -1 \end{cases}$

**6.** $\begin{cases} -x - 4y + 6z = -2 \\ x - 3y - 3z = 4 \\ 3x + y + 3z = 0 \end{cases}$

In Exercises 7–10, find all first partial derivatives.

**7.** $f(x, y) = x^2y + xy^2$

**8.** $f(x, y) = 25(xy + y^2)^2$

**9.** $f(x, y, z) = x(x^2 - 2xy + yz)$

**10.** $f(x, y, z) = z(xy + xz + yz)$

## EXERCISES 7.6

In Exercises 1–12, use Lagrange multipliers to find the indicated extremum. In each case, assume that x and y are positive.

| Objective Function | Constraint |
|---|---|
| **1.** Maximize $f(x, y) = xy$ | $x + y = 10$ |
| **2.** Maximize $f(x, y) = xy$ | $2x + y = 4$ |
| **3.** Minimize $f(x, y) = x^2 + y^2$ | $x + y - 4 = 0$ |
| **4.** Minimize $f(x, y) = x^2 + y^2$ | $-2x - 4y + 5 = 0$ |
| **5.** Maximize $f(x, y) = x^2 - y^2$ | $y - x^2 = 0$ |
| **6.** Maximize $f(x, y) = x^2 - y^2$ | $x - 2y + 6 = 0$ |
| **7.** Maximize $f(x, y) = 3x + xy + 3y$ | $x + y = 25$ |
| **8.** Maximize $f(x, y) = 3x + y + 10$ | $x^2y = 6$ |
| **9.** Maximize $f(x, y) = \sqrt{6 - x^2 - y^2}$ | $x + y - 2 = 0$ |
| **10.** Minimize $f(x, y) = \sqrt{x^2 + y^2}$ | $2x + 4y - 15 = 0$ |
| **11.** Maximize $f(x, y) = e^{xy}$ | $x^2 + y^2 - 8 = 0$ |
| **12.** Minimize $f(x, y) = 2x + y$ | $xy = 32$ |

In Exercises 13–18, use Lagrange multipliers to find the indicated extremum. In each case, assume that x, y, and z are positive.

**13.** Minimize $f(x, y, z) = 2x^2 + 3y^2 + 2z^2$

Constraint: $x + y + z - 24 = 0$

**14.** Maximize $f(x, y, z) = xyz$

Constraint: $x + y + z - 6 = 0$

**15.** Minimize $f(x, y, z) = x^2 + y^2 + z^2$

Constraint: $x + y + z = 1$

**16.** Minimize $f(x, y) = x^2 - 8x + y^2 - 12y + 48$

Constraint: $x + y = 8$

**17.** Maximize $f(x, y, z) = x + y + z$

Constraint: $x^2 + y^2 + z^2 = 1$

**18.** Maximize $f(x, y, z) = x^2y^2z^2$

Constraint: $x^2 + y^2 + z^2 = 1$

In Exercises 19 and 20, use Lagrange multipliers with the objective function

$$f(x, y, z, w) = 2x^2 + y^2 + z^2 + 2w^2$$

with the given constraints to find the indicated extremum. In each case, assume that $x, y, z,$ and $w$ are nonnegative.

**19.** Maximize $f(x, y, z, w)$

Constraint: $2x + 2y + z + w = 2$

**20.** Maximize $f(x, y, z, w)$

Constraint: $x + y + 2z + 2w = 4$

In Exercises 21–24, use Lagrange multipliers to find the indicated extremum of $f$ subject to two constraints. In each case, assume that $x, y,$ and $z$ are nonnegative.

**21.** Maximize $f(x, y, z) = xyz$

Constraints: $x + y + z = 24$, $x - y + z = 12$

**22.** Minimize $f(x, y, z) = x^2 + y^2 + z^2$

Constraints: $x + 2z = 4$, $x + y = 8$

**23.** Maximize $f(x, y, z) = xyz$

Constraints: $x^2 + z^2 = 5$, $x - 2y = 0$

**24.** Maximize $f(x, y, z) = xy + yz$

Constraints: $x + 2y = 6$, $x - 3z = 0$

In Exercises 25–28, find three positive numbers $x, y,$ and $z$ that satisfy the given conditions.

**25.** The sum is 120 and the product is maximum.

**26.** The sum is 120 and the sum of the squares is minimum.

**27.** The sum is $S$ and the product is maximum.

**28.** The sum is $S$ and the sum of the squares is minimum.

In Exercises 29–32, find the minimum distance from the curve or surface to the specified point. (*Hint:* Start by minimizing the square of the distance.)

**29.** Line: $x + 2y = 5$, $(0, 0)$

Minimize $d^2 = x^2 + y^2$.

**30.** Circle: $(x - 4)^2 + y^2 = 4$, $(0, 10)$

Minimize $d^2 = x^2 + (y - 10)^2$.

**31.** Plane: $x + y + z = 1$, $(2, 1, 1)$

Minimize $d^2 = (x - 2)^2 + (y - 1)^2 + (z - 1)^2$.

**32.** Cone: $z = \sqrt{x^2 + y^2}$, $(4, 0, 0)$

Minimize $d^2 = (x - 4)^2 + y^2 + z^2$.

**33.** *Volume* Find the dimensions of the rectangular package of largest volume subject to the constraint that the sum of the length and the girth cannot exceed 108 inches (see figure). (*Hint:* Maximize $V = xyz$ subject to the constraint $x + 2y + 2z = 108$.)

Figure for 33       Figure for 34

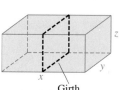

Girth

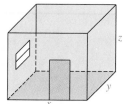

**34.** *Cost* In redecorating an office, the cost for new carpeting is five times the cost of wallpapering a wall. Find the dimensions of the largest office that can be redecorated for a fixed cost $C$ (see figure). (*Hint:* Maximize $V = xyz$ subject to $5xy + 2xz + 2yz = C$.)

**35.** *Cost* A cargo container (in the shape of a rectangular solid) must have a volume of 480 cubic feet. Use Lagrange multipliers to find the dimensions of the container of this size that has a minimum cost, if the bottom will cost $5 per square foot to construct and the sides and top will cost $3 per square foot to construct.

**36.** *Cost* A manufacturer has an order for 1000 units that can be produced at two locations. Let $x_1$ and $x_2$ be the numbers of units produced at the two plants. Find the number of units that should be produced at each plant to minimize the cost if the cost function is given by

$$C = 0.25x_1^2 + 25x_1 + 0.05x_2^2 + 12x_2.$$

**37.** *Cost* A manufacturer has an order for 2000 units that can be produced at two locations. Let $x_1$ and $x_2$ be the numbers of units produced at the two plants. The cost function is modeled by

$$C = 0.25x_1^2 + 10x_1 + 0.15x_2^2 + 12x_2.$$

Find the number of units that should be produced at each plant to minimize the cost.

**38.** *Hardy-Weinberg Law* Repeat Exercise 49 in Section 7.5 using Lagrange multipliers—that is, maximize

$$P(p, q, r) = 2pq + 2pr + 2qr$$

subject to the constraint

$$p + q + r = 1.$$

**39.** *Least-Cost Rule* The production function for a company is

$$f(x, y) = 100x^{0.25}y^{0.75},$$

where $x$ is the number of units of labor and $y$ is the number of units of capital. Suppose that labor costs $48 per unit, capital costs $36 per unit, and management sets a production goal of 20,000 units.

(a) Find the numbers of units of labor and capital needed to meet the production goal while minimizing the cost.

(b) Show that the conditions of part (a) are met when

$$\frac{\text{Marginal productivity of labor}}{\text{Marginal productivity of capital}} = \frac{\text{unit price of labor}}{\text{unit price of capital}}.$$

This proportion is called the **Least-Cost Rule** (or Equimarginal Rule).

**40.** *Least-Cost Rule* Repeat Exercise 39 for the production function

$$f(x, y) = 100x^{0.6}y^{0.4}.$$

**41.** *Production* The production function for a company is

$$f(x, y) = 100x^{0.25}y^{0.75},$$

where $x$ is the number of units of labor and $y$ is the number of units of capital. Suppose that labor costs $48 per unit and capital costs $36 per unit. The total cost of labor and capital is limited to $100,000.

(a) Find the maximum production level for this manufacturer.

(b) Find the marginal productivity of money.

(c) Use the marginal productivity of money to find the maximum number of units that can be produced if $125,000 is available for labor and capital.

**42.** *Production* Repeat Exercise 41 for the production function

$$f(x, y) = 100x^{0.06}y^{0.04}.$$

**43.** *Bacteria Culture* A microbiologist must prepare a culture medium in which to grow a certain type of bacteria. The percent of salt contained in this medium is $S = 12xyz$, where $x$, $y$, and $z$ are the nutrient solutions to be mixed in the medium. For the bacteria to grow, the medium must be 13% salt. The nutrient solutions cost $1, $2, and $3 per liter. How much of each nutrient solution should be used to minimize the cost of the culture medium?

**44.** *Bacteria Culture* Repeat Exercise 43 for a salt-content model of

$$S = 0.01x^2y^2z^2.$$

**45.** *Investment Strategy* An investor is considering three different stocks in which to invest $300,000. The average annual dividends for the stocks are

| | |
|---|---|
| Dow Chemicals ($D$) | 3.9% |
| CIGNA Corp. ($C$) | 4.5% |
| Pennzoil Company ($P$) | 4.0%. |

The amount invested in Pennzoil must follow the equation

$$3000D - 3000C + P^2 = 0.$$

How much should be invested in each stock to yield a maximum of dividends?

**46.** *Research Project* Use your school's library or some other reference source to write a paper about two different types of available investment options. Find examples of each type and find the data about their dividends for the past 10 years. What are the similarities and differences between the two types?

**47.** Use Lagrange multipliers to show that the maximum area of a rectangle with dimensions $x$ and $y$ and a given perimeter $P$ is $\frac{1}{16}P^2$.

**48.** An investor is considering three different stocks in which to invest $20,000. Average annual dividends for the stocks are

| | |
|---|---|
| Anheuser-Busch ($A$) | 1.9% |
| General Electric ($G$) | 2.1% |
| Kellogg Company ($K$) | 2.4%. |

The amount invested in General Electric must follow the equation

$$2000A - 2000K + G^2 = 0.$$

How much should be invested in each stock to yield a maximum of dividends?

**49.** A rancher plans to use an existing stone wall and the side of a barn as a boundary for two adjacent rectangular corrals. Fencing for the perimeter costs $10 per foot. To separate the corrals, a fence that costs $4 per foot will divide the region. The total area of the two corrals is to be 6000 square feet. Use Lagrange multipliers to find the dimensions that will minimize the cost of the fencing. What is the minimum cost?

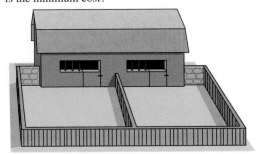

## 7.7 | Least Squares Regression Analysis

*Measuring the Accuracy of a Mathematical Model* •
*Least Squares Regression Line* • *Least Squares Regression Quadratic*

## Measuring the Accuracy of a Mathematical Model

When seeking a mathematical model to fit real-life data, you should try to find a model that is both as *simple* and as *accurate* as possible. For instance, a simple linear model for the points shown in Figure 7.35 is

$$f(x) = 1.8566x - 5.0246. \qquad \text{Linear model}$$

However, Figure 7.36 shows that by choosing a slightly more complicated quadratic model,

$$g(x) = 0.1996x^2 - 0.7281x + 1.3749, \qquad \text{Quadratic model}$$

you can obtain significantly greater accuracy.

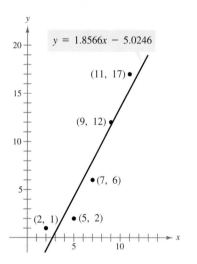

**FIGURE 7.35**

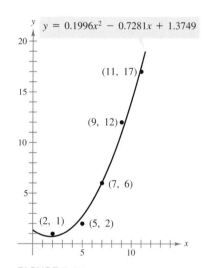

**FIGURE 7.36**

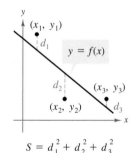

$$S = d_1^2 + d_2^2 + d_3^2$$

**FIGURE 7.37**

To measure how well the model $y = f(x)$ fits a collection of points, sum the squares of the differences between the actual $y$-values and the model's $y$-values. This sum is called the **sum of the squared errors** and is denoted by $S$. Graphically, $S$ is the sum of the squares of the vertical distances between the graph of $f$ and the given points in the plane, as shown in Figure 7.37.

### Definition of the Sum of the Squared Errors

The **sum of the squared errors** for the model $y = f(x)$ with respect to the points $(x_1, y_1), (x_2, y_2), \ldots, (x_n, y_n)$ is given by

$$S = [f(x_1) - y_1]^2 + [f(x_2) - y_2]^2 + \cdots + [f(x_n) - y_n]^2.$$

## EXAMPLE 1   Finding the Sum of the Squared Errors

Find the sum of the squared errors for the linear and quadratic models

$$f(x) = 1.8566x - 5.0246 \qquad \text{Linear model}$$
$$g(x) = 0.1996x^2 - 0.7281x + 1.3749 \qquad \text{Quadratic model}$$

(see Figures 7.35 and 7.36) with respect to the points

$$(2, 1), (5, 2), (7, 6), (9, 12), (11, 17).$$

### Solution

Begin by evaluating each model at the given $x$-values, as shown in the table.

| $x$ | 2 | 5 | 7 | 9 | 11 |
|---|---|---|---|---|---|
| Actual $y$-values | 1 | 2 | 6 | 12 | 17 |
| Linear model, $f(x)$ | $-1.3114$ | 4.2584 | 7.9716 | 11.6848 | 15.3980 |
| Quadratic model, $g(x)$ | 0.7171 | 2.7244 | 6.0586 | 10.9896 | 17.5174 |

For the linear model $f$, the sum of the squared errors is

$$S = (-1.3114 - 1)^2 + (4.2584 - 2)^2 + (7.9716 - 6)^2$$
$$+ (11.6848 - 12)^2 + (15.3980 - 17)^2$$
$$\approx 16.9959.$$

Similarly, the sum of the squared errors for the quadratic model $g$ is

$$S = (0.7171 - 1)^2 + (2.7244 - 2)^2 + (6.0586 - 6)^2$$
$$+ (10.9896 - 12)^2 + (17.5174 - 17)^2$$
$$\approx 1.8968.$$

*STUDY TIP*   In Example 1, note that the sum of the squared errors for the quadratic model is less than the sum of the squared errors for the linear model, which confirms that the quadratic model is a better fit.

## Least Squares Regression Line

The sum of the squared errors can be used to determine which of several models is the best fit for a collection of data. In general, if the sum of the squared errors of $f$ is less than the sum of the squared errors of $g$, then $f$ is said to be a better fit for the data than $g$. In regression analysis, you consider all possible models of a certain type. The one that is defined to be the best-fitting model is the one with the least sum of the squared errors. Example 2 shows how to use the optimization techniques described in Section 7.5 to find the best-fitting linear model for a collection of data.

**TIP**

For help in solving the system of equations in Example 2, see Example 2b in the *Chapter 7 Algebra Review,* on page 537.

ALGEBRA

### EXAMPLE 2   Finding the Best Linear Model

Find the values of $a$ and $b$ such that the linear model

$$f(x) = ax + b$$

has a minimum sum of the squared errors for the points

$$(-3, 0), (-1, 1), (0, 2), (2, 3).$$

*Solution*

The sum of the squared errors is

$$
\begin{aligned}
S &= [f(x_1) - y_1]^2 + [f(x_2) - y_2]^2 + [f(x_3) - y_3]^2 + [f(x_4) - y_4]^2 \\
&= (-3a + b - 0)^2 + (-a + b - 1)^2 + (b - 2)^2 + (2a + b - 3)^2 \\
&= 14a^2 - 4ab + 4b^2 - 10a - 12b + 14.
\end{aligned}
$$

To find the values of $a$ and $b$ for which $S$ is a minimum, you can use the techniques described in Section 7.5. That is, find the partial derivatives of $S$.

$$\frac{\partial S}{\partial a} = 28a - 4b - 10 \qquad \text{Differentiate with respect to } a.$$

$$\frac{\partial S}{\partial b} = -4a + 8b - 12 \qquad \text{Differentiate with respect to } b.$$

Next, set each partial derivative equal to zero.

$$28a - 4b - 10 = 0 \qquad \text{Set } \partial S/\partial a \text{ equal to } 0.$$
$$-4a + 8b - 12 = 0 \qquad \text{Set } \partial S/\partial b \text{ equal to } 0.$$

The solution of this system of linear equations is

$$a = \frac{8}{13} \quad \text{and} \quad b = \frac{47}{26}.$$

Thus, the best-fitting linear model for the given points is

$$f(x) = \frac{8}{13}x + \frac{47}{26}.$$

The graph of this model is shown in Figure 7.38.

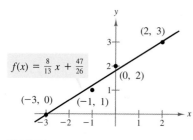

$$f(x) = \frac{8}{13}x + \frac{47}{26}$$

**FIGURE 7.38**

The line in Example 2 is called the **least squares regression line** for the given data. The solution shown in Example 2 can be generalized to find a formula for the least squares regression line, as follows. Consider the linear model

$$f(x) = ax + b$$

and the points $(x_1, y_1), (x_2, y_2), \ldots, (x_n, y_n)$. The sum of the squared errors is

$$S = \sum_{i=1}^{n} [f(x_i) - y_i]^2 = \sum_{i=1}^{n} (ax_i + b - y_i)^2.$$

To minimize $S$, set the partial derivatives $\partial S / \partial a$ and $\partial S / \partial b$ equal to zero and solve for $a$ and $b$. The results are summarized below.

---

### The Least Squares Regression Line

The **least squares regression line** for the points

$$(x_1, y_1), (x_2, y_2), \ldots, (x_n, y_n)$$

is $y = ax + b$, where

$$a = \frac{n \sum_{i=1}^{n} x_i y_i - \sum_{i=1}^{n} x_i \sum_{i=1}^{n} y_i}{n \sum_{i=1}^{n} x_i^2 - \left(\sum_{i=1}^{n} x_i\right)^2} \quad \text{and} \quad b = \frac{1}{n} \left(\sum_{i=1}^{n} y_i - a \sum_{i=1}^{n} x_i\right).$$

---

In the formula for the least squares regression line, note that if the $x$-values are symmetrically spaced about zero, then

$$\sum_{i=1}^{n} x_i = 0$$

and the formulas for $a$ and $b$ simplify to

$$a = \frac{n \sum_{i=1}^{n} x_i y_i}{n \sum_{i=1}^{n} x_i^2} \quad \text{and} \quad b = \frac{1}{n} \sum_{i=1}^{n} y_i.$$

Note also that only the *development* of the least squares regression line involves partial derivatives. The *application* of this formula is simply a matter of computing the values of $a$ and $b$—a task that is performed much more simply on a calculator or a computer than by hand.

*Discovery*

Graph the three points $(2, 2)$, $(2, 1)$, and $(2.1, 1.5)$ and visually estimate the least squares regression line for this data. Now use the formulas on this page or a graphing utility to show that the equation of the line is actually $y = 1.5$. In general, the least squares regression line for "nearly vertical" data can be quite unusual. Show that by interchanging the roles of $x$ and $y$, you can obtain a better linear approximation.

### EXAMPLE 3 Modeling Hourly Wage

The average hourly wages $y$ (in dollars per hour) for production workers in manufacturing industries from 1987 through 1995 are listed in the table. Find the least squares regression line for the data and use the result to estimate the average hourly wage in 1998. *(Source: U.S. Bureau of Labor Statistics)*

| Year | 1987 | 1988 | 1989 | 1990 | 1991 | 1992 | 1993 | 1994 | 1995 |
|------|------|------|------|------|------|------|------|------|------|
| $y$ | 9.91 | 10.19 | 10.48 | 10.83 | 11.18 | 11.46 | 11.74 | 12.06 | 12.35 |

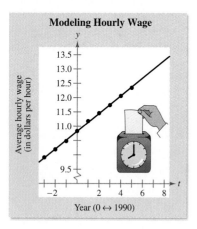

**Modeling Hourly Wage**

**FIGURE 7.39**

*Solution*

Let $t$ represent the year, with $t = 0$ corresponding to 1990. Then, you need to find the linear model that best fits the points

$(-3, 9.91), (-2, 10.19), (-1, 10.48), (0, 10.83), (1, 11.18),$
$(2, 11.46), (3, 11.74), (4, 12.06), (5, 12.35).$

Using a calculator with a built-in least squares regression program, you can determine that the best-fitting line is

$$y = 10.82 + 0.31t.$$

With this model, you can estimate the 1998 average hourly wage to be

$$y = 10.82 + 0.31(8) = \$13.30 \text{ per hour.}$$

This result is shown graphically in Figure 7.39.

## Technology

Most graphing utilities have a built-in linear regression program. For instance, the steps for finding the linear regression model in Example 3 on a *TI-82* are shown at the right. For the *TI-83* use the keystrokes ⎡STAT⎤, ⎡CALC⎤, 4, ⎡ENTER⎤ after inputting the data. Steps for other calculators are given in the appendix. When you run such a program, the "*r*-value" gives a measure of how well the model fits the data. The closer the value of $|r|$ is to 1, the better the fit. For the data in Example 3, $r \approx 0.9996$, which implies that the model is a very good fit.

⎡STAT⎤
⎡EDIT⎤

| L1 | L2 |
|-----|-------|
| −3 | 9.91 |
| −2 | 10.19 |
| −1 | 10.48 |
| 0 | 10.83 |
| 1 | 11.18 |
| 2 | 11.46 |
| 3 | 11.74 |
| 4 | 12.06 |
| 5 | 12.35 |

⎡STAT⎤
⎡CALC⎤
5 ⎡ENTER⎤

Enter data.

Cursor to CALC.
Select LinReg option.

LinReg
$y = ax + b$
$a = .3086666667$    Slope
$b = 10.82466667$    $y$-Intercept
$r = .9995897455$    Correlation coefficient

## Least Squares Regression Quadratic

When using regression analysis to model data, remember that the least squares regression line provides only the best *linear* model for a set of data. It does not necessarily provide the best possible model. For instance, in Example 1, you saw that the quadratic model was a better fit than the linear model.

Regression analysis can be performed with many different types of models, such as exponential or logarithmic models. The following development shows how to find the best-fitting quadratic model for a collection of data points. Consider a quadratic model of the form

$$f(x) = ax^2 + bx + c.$$

The sum of the squared errors for this model is

$$S = \sum_{i=1}^{n} [f(x_i) - y_i]^2 = \sum_{i=1}^{n} (ax_i^2 + bx_i + c - y_i)^2.$$

To find the values of $a$, $b$, and $c$ that minimize $S$, set the three partial derivatives, $\partial S / \partial a$, $\partial S / \partial b$, and $\partial S / \partial c$, equal to zero.

$$\frac{\partial S}{\partial a} = \sum_{i=1}^{n} 2x_i^2 (ax_i^2 + bx_i + c - y_i) = 0$$

$$\frac{\partial S}{\partial b} = \sum_{i=1}^{n} 2x_i (ax_i^2 + bx_i + c - y_i) = 0$$

$$\frac{\partial S}{\partial c} = \sum_{i=1}^{n} 2(ax_i^2 + bx_i + c - y_i) = 0$$

By expanding this system, you obtain the result given in the following summary.

 **Technology**

Most graphing utilities have built-in programs for finding the least squares regression quadratic. This program works just like the program for the least squares line. You should use this program to check the exercises you do.

---

### Least Squares Regression Quadratic

The **least squares regression quadratic** for the points

$$(x_1, y_1), (x_2, y_2), \ldots, (x_n, y_n)$$

is $y = ax^2 + bx + c$, where $a$, $b$, and $c$ are the solutions of the following system of equations.

$$a\sum_{i=1}^{n} x_i^4 + b\sum_{i=1}^{n} x_i^3 + c\sum_{i=1}^{n} x_i^2 = \sum_{i=1}^{n} x_i^2 y_i$$

$$a\sum_{i=1}^{n} x_i^3 + b\sum_{i=1}^{n} x_i^2 + c\sum_{i=1}^{n} x_i = \sum_{i=1}^{n} x_i y_i$$

$$a\sum_{i=1}^{n} x_i^2 + b\sum_{i=1}^{n} x_i + cn = \sum_{i=1}^{n} y_i$$

## EXAMPLE 4   Modeling the Number of CD's Sold

The numbers $y$ of music CDs in millions sold from 1984 through 1994 are listed in the table. Find the least squares regression quadratic for the data and use the result to estimate the number of CDs sold in 1998.   *(Source: Recording Industry Association of America)*

| Year | 1984 | 1985 | 1986 | 1987 | 1988 | 1989 | 1990 | 1991 | 1992 | 1993 | 1994 |
|---|---|---|---|---|---|---|---|---|---|---|---|
| $y$ | 5.8 | 22.6 | 53.0 | 102.1 | 149.7 | 207.2 | 286.5 | 333.3 | 407.5 | 495.4 | 662.1 |

### Solution

Let $t$ represent the year, with $t = 0$ corresponding to 1990. Then, you need to find the quadratic model that best fits the points

$$(-6, 5.8), (-5, 22.6), (-4, 53.0), (-3, 102.1), (-2, 149.7), (-1, 207.2),$$
$$(0, 286.5), (1, 333.3), (2, 407.5), (3, 495.4), (4, 662.1).$$

Using a calculator with a built-in least squares regression program, you can determine that the best-fitting quadratic is

$$y = 4.727t^2 + 71.59t + 267.3.$$

With this model, you can estimate the number of CDs sold in 1998 to be

$$y = 4.727(8)^2 + 71.59(8) + 267.3 \approx 1140.$$

This result is shown graphically in Figure 7.40.

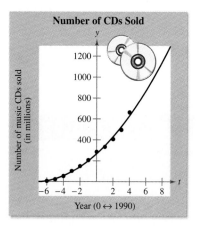

**Number of CDs Sold**

Number of music CDs sold (in millions)

Year (0 ↔ 1990)

**FIGURE 7.40**

---

## *Group Discussion*

## *Least Squares Regression Analysis*

Most graphing utilities, such as a *TI-82* or *TI-83*, have several built-in least squares regression programs. Try entering the points

$$(-6, 5.8), (-5, 53.0), (-4, 53.0), (-3, 102.1), (-2, 149.7), (-1, 207.2),$$
$$(0, 286.5), (1, 333.3), (2, 407.5), (3, 495.4), (4, 662.1)$$

from Example 4 in a graphing utility. Then run one of the regression programs for the points. The quadratic model in Example 4 has a correlation coefficient of $r \approx 0.997$. Can you find a model, such as an exponential model or power model

$$y = ab^x \qquad \text{Exponential model}$$
$$y = ax^b, \qquad \text{Power model}$$

that has a better correlation coefficient?

*Warm Up*

The following warm-up exercises involve skills that were covered in earlier sections. You will use these skills in the exercise set for this section.

In Exercises 1 and 2, evaluate the expression.

**1.** $(2.5 - 1)^2 + (3.25 - 2)^2 + (4.1 - 3)^2$

**2.** $(1.1 - 1)^2 + (2.08 - 2)^2 + (2.95 - 3)^2$

In Exercises 3 and 4, find the partial derivatives of $S$.

**3.** $S = a^2 + 6b^2 - 4a - 8b - 4ab + 6$

**4.** $S = 4a^2 + 9b^2 - 6a - 4b - 2ab + 8$

In Exercises 5–10, evaluate the summation.

**5.** $\displaystyle\sum_{i=1}^{5} i$  **6.** $\displaystyle\sum_{i=1}^{6} 2i$  **7.** $\displaystyle\sum_{i=1}^{4} \frac{1}{i}$  **8.** $\displaystyle\sum_{i=1}^{3} i^2$  **9.** $\displaystyle\sum_{i=1}^{6} (2 - i)^2$  **10.** $\displaystyle\sum_{i=1}^{5} (30 - i^2)$

## EXERCISES 7.7

In Exercises 1–4, (a) use the method of least squares to find the least squares regression line, and (b) calculate the sum of the squared errors.

**1.**

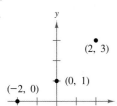

**2.**

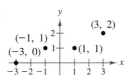

**3.**

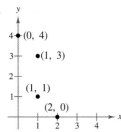

**4.**

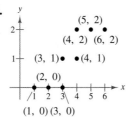

In Exercises 5–14, use a graphing utility with a least squares regression program to find the least squares regression line for the given points.

**5.** $(-2, 0), (-1, 1), (0, 1), (1, 2), (2, 3)$

**6.** $(-4, -1), (-2, 0), (2, 4), (4, 5)$

**7.** $(-2, 2), (2, 6), (3, 7)$

**8.** $(-5, 1), (1, 3), (2, 3), (2, 5)$

**9.** $(-3, 4), (-1, 2), (1, 1), (3, 0)$

**10.** $(-10, 10), (-5, 8), (3, 6), (7, 4), (5, 0)$

**11.** $(0, 0), (1, 1), (3, 4), (4, 2), (5, 5)$

**12.** $(1, 0), (3, 3), (5, 6)$

**13.** $(0, 6), (4, 3), (5, 0), (8, -4), (10, -5)$

**14.** $(5, 2), (0, 0), (2, 1), (7, 4), (10, 6), (12, 6)$

In Exercises 15–18, use partial derivatives to find the values of $a$ and $b$ such that the linear model $f(x) = ax + b$ has a minimum sum of the squared errors for the given points.

**15.** $(-2, -1), (0, 0), (2, 3)$

**16.** $(-3, 0), (-1, 1), (1, 1), (3, 2)$

**17.** $(-2, 4), (-1, 1), (0, -1), (1, -3)$

**18.** $(-5, -3), (-4, -2), (-2, -1), (-1, 1)$

In Exercises 19–22, use a graphing utility with a least squares regression program to find the least squares regression quadratic for the given points. Then plot the points and graph the least squares quadratic.

**19.** $(-2, 0), (-1, 0), (0, 1), (1, 2), (2, 5)$

**20.** $(-4, 5), (-2, 6), (2, 6), (4, 2)$

**21.** $(1, 0), (2, 1), (3, 7), (4, 13)$

**22.** $(0, 10), (1, 9), (2, 6), (3, 0)$

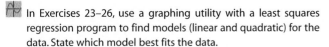

In Exercises 23–26, use a graphing utility with a least squares regression program to find models (linear and quadratic) for the data. State which model best fits the data.

**23.** $(-4, 1), (-3, 2), (-2, 2), (-1, 4), (0, 6), (1, 8), (2, 9)$

**24.** $(-1, -4), (0, -3), (1, -3), (2, 0), (4, 5), (6, 9), (9, 3)$

**25.** $(0, 769), (1, 677), (2, 601), (3, 543), (4, 489), (5, 411)$

**26.** $(1, 10.3), (2, 14.2), (3, 18.9), (4, 23.7), (5, 29.1), (6, 35)$

**27.** *Demand*   A store manager wants to know the demand for a product as a function of price. The daily sales for three different prices of the product are listed in the table.

| Price, $x$ | $1.00 | $1.25 | $1.50 |
|---|---|---|---|
| Demand, $y$ | 450 | 375 | 330 |

(a) Use a graphing utility with a least squares regression program to find the least squares regression line for the data.
(b) Estimate the demand when the price is $1.40.
(c) What price will create a demand of 500 products?

**28.** *Demand*   A hardware retailer wants to know the demand for a tool as a function of price. The monthly sales for four different prices of the tool are listed in the table.

| Price, $x$ | $25 | $30 | $35 | $40 |
|---|---|---|---|---|
| Demand, $y$ | 82 | 75 | 67 | 55 |

(a) Use a graphing utility with a least squares regression program to find the least squares regression line for the data.
(b) Estimate the demand when the price is $32.95.
(c) What price will create a demand of 83 tools?

**29.** *Crop Yield*   A farmer used four test plots to determine the relationship between wheat yield in bushels per acre and the amount of fertilizer in hundreds of pounds per acre. The results are given in the table.

| Fertilizer, $x$ | 1.0 | 1.5 | 2.0 | 2.5 |
|---|---|---|---|---|
| Yield, $y$ | 35 | 44 | 50 | 56 |

(a) Use a graphing utility with a least squares regression program to find the least squares regression line for the data.
(b) Estimate the yield for a fertilizer application of 160 pounds per acre.

**30.** *Carcinogen Contamination*   After contamination by a carcinogen, people in different geographic regions were assigned an exposure index that represented the degree of contamination. Use the following data and a graphing utility with a least squares regression program to find a least squares regression line to estimate the mortality per 100,000 people for a given exposure.

| Exposure, $x$ | 1.35 | 2.67 | 3.93 | 5.14 | 7.43 |
|---|---|---|---|---|---|
| Mortality, $y$ | 118.5 | 135.2 | 167.3 | 197.6 | 204.7 |

**31.** *Infant Mortality*   To study the number of infant deaths per 1000 live births in the United States, a medical researcher obtains the following data. *(Source: Department of Health and Human Services)*

| Year | 1950 | 1960 | 1970 | 1980 | 1990 | 1993 |
|---|---|---|---|---|---|---|
| Deaths, $y$ | 29.2 | 26.0 | 20.0 | 12.6 | 9.2 | 8.5 |

(a) Use a graphing utility with a least squares regression program to find the least squares regression line for the data and use this line to estimate the number of infant deaths in 2000. Let $t = 0$ represent 1970.
(b) Find the least squares regression quadratic for the data and use the model to estimate the number of infant deaths in 2000.

**32.** *Population Growth*   The table gives the approximate world population (in billions) for six different years.

| Year | 1800 | 1850 | 1900 | 1950 | 1990 | 1996 |
|---|---|---|---|---|---|---|
| Time, $t$ | $-2$ | $-1$ | 0 | 1 | 1.8 | 1.92 |
| Population, $y$ | 0.9 | 1.1 | 1.5 | 2.4 | 5.2 | 5.8 |

(a) During the 1800s, population growth was almost linear. Use a graphing utility with a least squares regression program to find a least squares regression line for those years and use the line to estimate the population in 1875.
(b) Use a graphing utility with a least squares regression program to find a least squares regression quadratic for the data from 1850 through 1996 and use the model to estimate the population in the year 2000.
(c) Even though the rate of growth of the population has begun to decline, most demographers believe the population size will pass the eight billion mark sometime in the next 50 years. What do you think?

**33.** *Engine Design*  After developing a new turbocharger for an automobile engine, the following experimental data was obtained for speed in miles per hour at 2-second intervals.

| Time, $x$ | 0 | 2 | 4 | 6 | 8 | 10 |
|-----------|---|----|----|----|----|----|
| Speed, $y$ | 0 | 15 | 30 | 50 | 65 | 70 |

(a) Use a graphing utility with a least squares regression program to find a least squares regression quadratic for the data.

(b) Use the model to estimate the speed after 5 seconds.

**34.** *New York Stock Exchange*  The daily averages of shares traded (in thousands) on the NYSE for selected years are listed in the table. *(Source: New York Stock Exchange)*

| Year | 1970 | 1975 | 1980 |
|------|------|------|------|
| Shares | 11,564 | 18,551 | 44,871 |

| Year | 1985 | 1990 | 1995 |
|------|------|------|------|
| Shares | 109,169 | 156,777 | 346,101 |

(a) Use a graphing utility with a least squares regression program to find a least squares regression quadratic that models the data.

(b) Use the model to estimate the number of shares that will be traded in 2000.

In Exercises 35–38, use a graphing utility with a least squares regression program to find any model that best fits the data points.

**35.** $(1, 13), (2, 16.5), (4, 24), (5, 28), (8, 39), (11, 50.25),$ $(17, 72), (20, 85)$

**36.** $(1, 5.5), (3, 7.75), (6, 15.2), (8, 23.5), (11, 46), (15, 110)$

**37.** $(1, 1.5), (2.5, 8.5), (5, 13.5), (8, 16.7), (9, 18), (20, 22)$

**38.** $(0, 0.5), (1, 7.6), (3, 60), (4.2, 117), (5, 170), (7.9, 380)$

In Exercises 39–42, plot the points and determine whether the data has positive, negative, or no correlation (see graphs at top of next column). Then use a graphing utility to find the value of $r$ and confirm your result. The number $r$ is called the **correlation coefficient.** It is a measure of how well the model fits the data. Correlation coefficients vary between $-1$ and $1$, and the closer $|r|$ is to 1, the better the model.

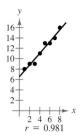
$r = 0.981$

$r = -0.866$

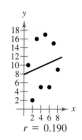
$r = 0.190$

Positive Correlation   Negative Correlation   No Correlation

**39.** $(1, 4), (2, 6), (3, 8), (4, 11), (5, 13), (6, 15)$

**40.** $(1, 7.5), (2, 7), (3, 7), (4, 6), (5, 5), (6, 4.9)$

**41.** $(1, 3), (2, 6), (3, 2), (4, 3), (5, 9), (6, 1)$

**42.** $(0.5, 2), (0.75, 1.75), (1, 3), (1.5, 3.2), (2, 3.7), (2.6, 4)$

*True or False*  In Exercises 43–46, determine whether the statement is true or false. Explain your reasoning.

**43.** Data that is modeled by $y = 3.29x - 4.17$ has a negative correlation.

**44.** Data that is modeled by $y = -0.238x + 25$ has a negative correlation.

**45.** If the correlation coefficient $r \approx -0.98781$, the model is a good fit.

**46.** A correlation coefficient of $r \approx 0.201$ implies that the data has no correlation.

**47.** *Finance: Median Income*  In the table below are the median income levels for various age levels in the United States. Use the data and a graphing utility with a least squares regression program to find the least squares regression quadratic for the data and use the resulting model to estimate the median income for someone who is 28 years old. *(Source: Adapted from Garman/Forgue,* Personal Finance, *Fifth Edition)*

| Age Level | Median Income |
|-----------|---------------|
| 20 | $20,700 |
| 30 | 33,500 |
| 40 | 43,700 |
| 50 | 49,400 |
| 60 | 35,800 |
| 70 | 19,000 |

| **7.8** | # Double Integrals and Area in the Plane |
|---|---|

*Double Integrals • Finding Area with a Double Integral*

## Double Integrals

In Section 7.4, you learned that it is meaningful to differentiate functions of several variables by differentiating with respect to one variable at a time while holding the other variables constant. It should not be surprising to learn that you can *integrate* functions of two or more variables using a similar procedure. For instance, if you are given the partial derivative

$$f_x(x, y) = 2xy, \qquad \text{Partial with respect to } x$$

then, by holding $y$ constant, you can integrate with respect to $x$ to obtain

$$\int f_x(x, y)\, dx = f(x, y) = x^2 y + C(y).$$

This procedure is called **partial integration with respect to $x$.** Note that the "constant of integration" $C(y)$ is assumed to be a function of $y$, because $y$ is fixed during integration with respect to $x$. Similarly, if you are given the partial derivative

$$f_y(x, y) = x^2 + 2, \qquad \text{Partial with respect to } y$$

then, by holding $x$ constant, you can integrate with respect to $y$ to obtain

$$\int f_y(x, y)\, dy = f(x, y) = x^2 y + 2y + C(x).$$

In this case, the "constant of integration" $C(x)$ is assumed to be a function of $x$, because $x$ is fixed during integration with respect to $y$.

To evaluate a definite integral of a function of two or more variables, you can apply the Fundamental Theorem of Calculus to one variable while holding the others constant, as follows.

$$\int_{1}^{2y} 2xy\, dx = x^2 y \Big]_{1}^{2y} = (2y)^2 y - (1)^2 y = 4y^3 - y.$$

$x$ is the variable of integration and $y$ is fixed.     Replace $x$ by the limits of integration.     The result is a function of $y$.

Note that you omit the constant of integration, just as you do for a definite integral of a function of one variable.

## EXAMPLE 1   Finding Partial Integrals

(a) $\displaystyle\int_{1}^{x} (2x^2y^{-2} + 2y)\, dy = \left[\dfrac{-2x^2}{y} + y^2\right]_{1}^{x}$    Hold $x$ constant.

$$= \left(\dfrac{-2x^2}{x} + x^2\right) - \left(\dfrac{-2x^2}{1} + 1\right)$$

$$= 3x^2 - 2x - 1$$

(b) $\displaystyle\int_{y}^{5y} \sqrt{x - y}\, dx = \left[\dfrac{2}{3}(x - y)^{3/2}\right]_{y}^{5y}$    Hold $y$ constant.

$$= \dfrac{2}{3}[(5y - y)^{3/2} - (y - y)^{3/2}]$$

$$= \dfrac{16}{3}y^{3/2}$$

In Example 1a, note that the definite integral defines a function of $x$ and can *itself* be integrated. An "integral of an integral" is called a **double integral.** With a function of two variables, there are two types of double integrals.

$$\int_{a}^{b}\int_{g_1(x)}^{g_2(x)} f(x, y)\, dy\, dx = \int_{a}^{b}\left[\int_{g_1(x)}^{g_2(x)} f(x, y)\, dy\right] dx$$

$$\int_{a}^{b}\int_{g_1(y)}^{g_2(y)} f(x, y)\, dx\, dy = \int_{a}^{b}\left[\int_{g_1(y)}^{g_2(y)} f(x, y)\, dx\right] dy$$

Notice that the difference between these two types is the order in which the integration is performed, *dy dx* or *dx dy*.

## EXAMPLE 2   Evaluating a Double Integral

$$\int_{1}^{2}\int_{0}^{x} (2xy + 3)\, dy\, dx = \int_{1}^{2}\left[\int_{0}^{x} (2xy + 3)\, dy\right] dx$$

$$= \int_{1}^{2} \left[xy^2 + 3y\right]_{0}^{x} dx$$

$$= \int_{1}^{2} (x^3 + 3x)\, dx$$

$$= \left[\dfrac{x^4}{4} + \dfrac{3x^2}{2}\right]_{1}^{2}$$

$$= \left(\dfrac{2^4}{4} + \dfrac{3(2^2)}{2}\right) - \left(\dfrac{1^4}{4} + \dfrac{3(1^2)}{2}\right)$$

$$= \dfrac{33}{4}$$

### Technology

A symbolic integration utility, such as *Derive*, *Maple*, *Mathcad*, *Mathematica*, or the *TI-92*, can be used to evaluate double integrals. To do this, you need to enter the integrand, then integrate twice— once with respect to one of the variables and then with respect to the other variable. For instance, when *Derive* for Windows is used to evaluate the integral in Example 2, you obtain the following.

$\#1:\ 2 \cdot x \cdot y + 3$

$\#2:\ \displaystyle\int_{0}^{x} (2 \cdot x \cdot y + 3)\, dy$

$\#3:\ \displaystyle\int_{1}^{2}\int_{0}^{x} (2 \cdot x \cdot y + 3)\, dy\, dx$

$\#4:\ \dfrac{33}{4}$

## Finding Area with a Double Integral

One of the simplest applications of a double integral is finding the area of a plane region. For instance, consider the region $R$ that is bounded by

$$a \leq x \leq b \qquad \text{and} \qquad g_1(x) \leq y \leq g_2(x).$$

Using the techniques described in Section 5.5, you know that the area of $R$ is

$$\int_a^b [g_2(x) - g_1(x)]\, dx.$$

This same area is also given by the double integral

$$\int_a^b \int_{g_1(x)}^{g_2(x)} dy\, dx$$

because

$$\int_a^b \int_{g_1(x)}^{g_2(x)} dy\, dx = \int_a^b \Big[ y \Big]_{g_1(x)}^{g_2(x)} dx$$

$$= \int_a^b [g_2(x) - g_1(x)]\, dx.$$

Figure 7.41 shows the two basic types of plane regions whose areas can be determined by a double integral.

*STUDY TIP*    In Figure 7.41, note that the horizontal or vertical orientation of the narrow rectangle indicates the order of integration. The "outer" variable of integration always corresponds to the width of the rectangle. Notice also that the outer limits of integration for a double integral are constant, whereas the inner limits may be functions of the outer variable.

**Determining Area in the Plane by Double Integrals**

Region is bounded by
$$a \leq x \leq b$$
$$g_1(x) \leq y \leq g_2(x)$$

Region is bounded by
$$c \leq y \leq d$$
$$h_1(y) \leq x \leq h_2(y)$$

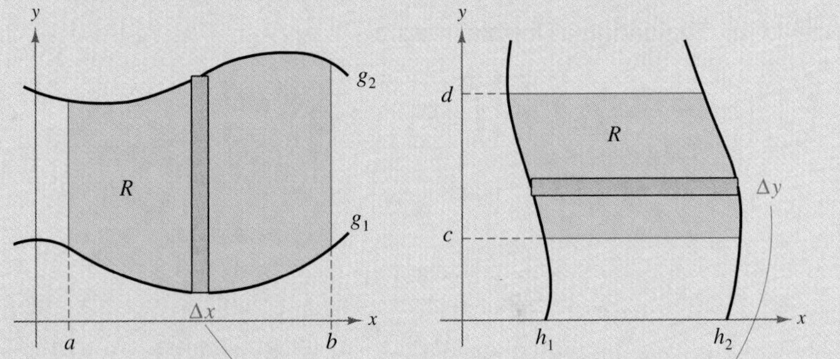

$$\text{Area} = \int_a^b \int_{g_1(x)}^{g_2(x)} dy\, dx$$

$$\text{Area} = \int_c^d \int_{h_1(y)}^{h_2(y)} dx\, dy$$

**FIGURE 7.41**

## EXAMPLE 3   Finding Area by a Double Integral

Use a double integral to find the area of the rectangular region shown in Figure 7.42.

### Solution

The bounds for $x$ are $1 \le x \le 5$ and the bounds for $y$ are $2 \le y \le 4$. Therefore, the area of the region is

$$\int_1^5 \int_2^4 dy \, dx = \int_1^5 \left[ y \right]_2^4 dx \qquad \text{Integrate with respect to } y.$$

$$= \int_1^5 (4 - 2) \, dx \qquad \text{Apply Fundamental Theorem.}$$

$$= \int_1^5 2 \, dx \qquad \text{Simplify.}$$

$$= \left[ 2x \right]_1^5 \qquad \text{Integrate with respect to } x.$$

$$= 10 - 2 \qquad \text{Apply Fundamental Theorem.}$$

$$= 8. \qquad \text{Simplify.}$$

You can confirm this by noting that the rectangle is two units by four units.

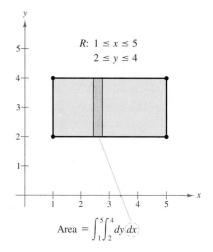

$$\text{Area} = \int_1^5 \int_2^4 dy \, dx$$

**FIGURE 7.42**

## EXAMPLE 4   Finding Area by a Double Integral

Use a double integral to find the area of the region bounded by the graphs of $y = x^2$ and $y = x^3$.

### Solution

As shown in Figure 7.43, the two graphs intersect when $x = 0$ and $x = 1$. Choosing $x$ to be the outer variable, the bounds for $x$ are $0 \le x \le 1$ and the bounds for $y$ are $x^3 \le y \le x^2$. This implies that the area of the region is

$$\int_0^1 \int_{x^3}^{x^2} dy \, dx = \int_0^1 \left[ y \right]_{x^3}^{x^2} dx \qquad \text{Integrate with respect to } y.$$

$$= \int_0^1 (x^2 - x^3) \, dx \qquad \text{Apply Fundamental Theorem.}$$

$$= \left[ \frac{x^3}{3} - \frac{x^4}{4} \right]_0^1 \qquad \text{Integrate with respect to } x.$$

$$= \frac{1}{3} - \frac{1}{4} \qquad \text{Apply Fundamental Theorem.}$$

$$= \frac{1}{12}. \qquad \text{Simplify.}$$

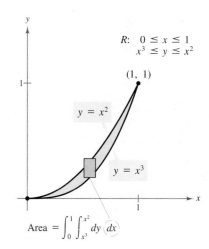

$$\text{Area} = \int_0^1 \int_{x^3}^{x^2} dy \, dx$$

**FIGURE 7.43**

In setting up double integrals, the most difficult task is likely to be determining the correct limits of integration. This can be simplified by making a sketch of the region $R$ and identifying the appropriate bounds for $x$ and $y$.

### EXAMPLE 5  Changing the Order of Integration

For the double integral

$$\int_0^2 \int_{y^2}^4 dx \, dy,$$

(a) sketch the region $R$ whose area is represented by the integral,

(b) rewrite the integral so that $x$ is the outer variable, and

(c) show that both orders of integration yield the same value.

### Solution

(a) From the limits of integration, you know that

$$y^2 \le x \le 4, \qquad \text{Variable bounds for } x$$

which means that the region $R$ is bounded on the left by the parabola $x = y^2$ and on the right by the line $x = 4$. Furthermore, because

$$0 \le y \le 2, \qquad \text{Constant bounds for } y$$

you know that the region lies above the $x$-axis, as shown in Figure 7.44.

(b) If you interchange the order of integration so that $x$ is the outer variable, then $x$ will have constant bounds of integration given by $0 \le x \le 4$. Solving for $y$ in the equation $x = y^2$ implies that the bounds for $y$ are $0 \le y \le \sqrt{x}$, as shown in Figure 7.45. Thus, with $x$ as the outer variable, the integral can be written as

$$\int_0^4 \int_0^{\sqrt{x}} dy \, dx.$$

(c) Both integrals yield the same value.

$$\int_0^2 \int_{y^2}^4 dx \, dy = \int_0^2 \left[ x \right]_{y^2}^4 dy = \int_0^2 (4 - y^2) \, dy = \left[ 4y - \frac{y^3}{3} \right]_0^2 = \frac{16}{3}$$

$$\int_0^4 \int_0^{\sqrt{x}} dy \, dx = \int_0^4 \left[ y \right]_0^{\sqrt{x}} dx = \int_0^4 \sqrt{x} \, dx = \left[ \frac{2}{3} x^{3/2} \right]_0^4 = \frac{16}{3}$$

To designate a double integral or an area of a region without specifying a particular order of integration, you can use the symbol

$$\iint_R dA,$$

where $dA = dx \, dy$ or $dA = dy \, dx$.

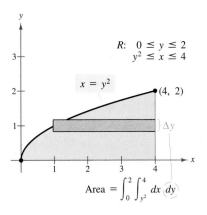

R:  $0 \le y \le 2$,
    $y^2 \le x \le 4$

$x = y^2$

(4, 2)

$\} \Delta y$

Area $= \int_0^2 \int_{y^2}^4 dx \, dy$

**FIGURE 7.44**

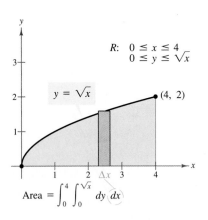

R:  $0 \le x \le 4$
    $0 \le y \le \sqrt{x}$

$y = \sqrt{x}$

(4, 2)

$\Delta x$

Area $= \int_0^4 \int_0^{\sqrt{x}} dy \, dx$

**FIGURE 7.45**

## EXAMPLE 6   Finding Area by a Double Integral

Use a double integral to calculate the area denoted by

$$\int_{R}\int dA,$$

where $R$ is the region bounded by $y = x$ and $y = x^2 - x$.

### Solution

Begin by sketching the region $R$, as shown in Figure 7.46. From the sketch, you can see that vertical rectangles of width $dx$ are more convenient than horizontal ones. Therefore, $x$ is the outer variable of integration and its constant bounds are $0 \le x \le 2$. This implies that the bounds for $y$ are $x^2 - x \le y \le x$, and the area is given by

$$\int_{R}\int dA = \int_{0}^{2}\int_{x^2-x}^{x} dy\, dx \qquad \text{Substitute bounds for region.}$$

$$= \int_{0}^{2}\left[y\right]_{x^2-x}^{x} dx \qquad \text{Integrate with respect to } y.$$

$$= \int_{0}^{2}\left[x - (x^2 - x)\right] dx \qquad \text{Apply Fundamental Theorem.}$$

$$= \int_{0}^{2}(2x - x^2)\, dx \qquad \text{Simplify.}$$

$$= \left[x^2 - \frac{x^3}{3}\right]_{0}^{2} \qquad \text{Integrate with respect to } x.$$

$$= 4 - \frac{8}{3} \qquad \text{Apply Fundamental Theorem.}$$

$$= \frac{4}{3}. \qquad \text{Simplify.}$$

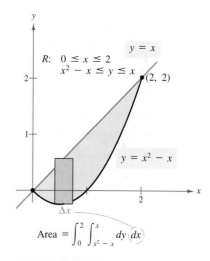

**FIGURE 7.46**

As you are working the exercises for this section, you should be aware that the primary uses of double integrals will be discussed in Section 7.9. We have introduced double integrals by way of areas in the plane so that you can gain practice in finding the limits of integration. When setting up a double integral, remember that your first step should be to sketch the region $R$. After doing this, you have two choices of integration orders: $dx\, dy$ or $dy\, dx$.

## *Group Discussion*    **Sketching Regions in the Plane**

Sketch the region represented by the double integral.

a. $\int_{0}^{4}\int_{0}^{y} dx\, dy$     b. $\int_{0}^{4}\int_{0}^{x} dy\, dx$     c. $\int_{0}^{4}\int_{x^2}^{x} dy\, dx$

*Warm Up*

The following warm-up exercises involve skills that were covered in earlier sections. You will use these skills in the exercise set for this section.

In Exercises 1–6, evaluate the definite integral.

**1.** $\int_0^1 dx$

**2.** $\int_1^4 2x^2 \, dx$

**3.** $\int_1^2 (x^3 - 2x + 4) \, dx$

**4.** $\int_1^2 \frac{2}{7x^2} \, dx$

**5.** $\int_0^2 \frac{2x}{x^2 + 1} \, dx$

**6.** $\int_0^2 xe^{x^2+1} \, dx$

In Exercises 7–10, sketch the region bounded by the graphs of the equations.

**7.** $y = x, \ y = 0, \ x = 0, \ x = 3$

**8.** $y = x, \ y = 3, \ y = 0, \ x = 0$

**9.** $y = 4 - x^2, \ y = 0, \ x = 0, \ x = 2$

**10.** $y = x^2, \ y = 4x, \ x = 0, \ x = 4$

## EXERCISES 7.8

In Exercises 1–10, evaluate the partial integral.

**1.** $\int_0^x (2x - y) \, dy$

**2.** $\int_x^{x^2} \frac{y}{x} \, dy$

**3.** $\int_1^{2y} \frac{y}{x} \, dx$

**4.** $\int_0^{e^y} y \, dx$

**5.** $\int_0^{\sqrt{9-x^2}} x^2 y \, dy$

**6.** $\int_{x^2}^{\sqrt{x}} (x^2 + y^2) \, dy$

**7.** $\int_{e^y}^{y} \frac{y \ln x}{x} \, dx$

**8.** $\int_{-\sqrt{1-y^2}}^{\sqrt{1-y^2}} (x^2 + y^2) \, dx$

**9.** $\int_0^{x^3} ye^{-y/x} \, dy$

**10.** $\int_y^3 \frac{xy}{\sqrt{x^2 + 1}} \, dx$

In Exercises 11–24, evaluate the double integral.

**11.** $\int_0^2 \int_0^1 (x - y) \, dy \, dx$

**12.** $\int_0^2 \int_0^2 (6 - x^2) \, dy \, dx$

**13.** $\int_0^4 \int_0^3 xy \, dy \, dx$

**14.** $\int_0^1 \int_0^x \sqrt{1 - x^2} \, dy \, dx$

**15.** $\int_0^1 \int_0^{\sqrt{1-y^2}} (x + y) \, dx \, dy$

**16.** $\int_0^2 \int_{3y^2-6y}^{2y-y^2} 3y \, dx \, dy$

**17.** $\int_1^2 \int_0^4 (x^2 - 2y^2 + 1) \, dx \, dy$

**18.** $\int_0^1 \int_y^{2y} (1 + 2x^2 + 2y^2) \, dx \, dy$

**19.** $\int_0^2 \int_0^{\sqrt{1-y^2}} -5xy \, dx \, dy$

**20.** $\int_0^4 \int_0^x \frac{2}{(x + 1)(y + 1)} \, dy \, dx$

**21.** $\int_0^2 \int_0^{4-x^2} x^3 \, dy \, dx$

**22.** $\int_0^a \int_0^{a-x} (x^2 + y^2) \, dy \, dx$

**23.** $\int_0^\infty \int_0^\infty e^{-(x+y)/2} \, dy \, dx$

**24.** $\int_0^\infty \int_0^\infty xye^{-(x^2+y^2)} \, dx \, dy$

In Exercises 25–32, sketch the region $R$ whose area is given by the double integral. Then change the order of integration and show that both orders yield the same area.

**25.** $\displaystyle\int_0^1\int_0^2 dy\,dx$

**26.** $\displaystyle\int_1^2\int_2^4 dx\,dy$

**27.** $\displaystyle\int_0^1\int_{2y}^2 dx\,dy$

**28.** $\displaystyle\int_0^4\int_0^{\sqrt{x}} dy\,dx$

**29.** $\displaystyle\int_0^2\int_{x/2}^1 dy\,dx$

**30.** $\displaystyle\int_0^4\int_{\sqrt{x}}^2 dy\,dx$

**31.** $\displaystyle\int_0^1\int_{y^2}^{\sqrt[3]{y}} dx\,dy$

**32.** $\displaystyle\int_{-2}^2\int_0^{4-y^2} dx\,dy$

In Exercises 33 and 34, evaluate the double integral. Note that it is necessary to change the order of integration.

**33.** $\displaystyle\int_0^3\int_y^3 e^{x^2}\,dx\,dy$

**34.** $\displaystyle\int_0^2\int_x^2 e^{-y^2}\,dy\,dx$

In Exercises 35–40, use a double integral to find the area of the specified region.

**35.**

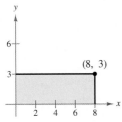

**36.**

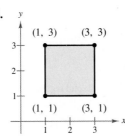

(1, 3)  (3, 3)

(1, 1)  (3, 1)

**37.**

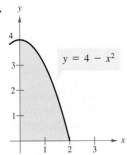

$y = 4 - x^2$

**38.**

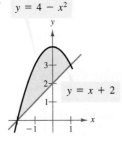

$y = 4 - x^2$

$y = x + 2$

**39.**

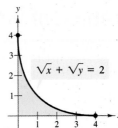

$\sqrt{x} + \sqrt{y} = 2$

**40.**

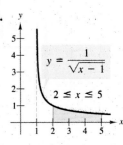

$y = \dfrac{1}{\sqrt{x} - 1}$

$2 \le x \le 5$

In Exercises 41–46, use a double integral to find the area of the region bounded by the graphs of the equations.

**41.** $y = 25 - x^2,\ y = 0$

**42.** $y = x^{3/2},\ y = x$

**43.** $5x - 2y = 0,\ x + y = 3,\ y = 0$

**44.** $xy = 9,\ y = x,\ y = 0,\ x = 9$

**45.** $y = x,\ y = 2x,\ x = 2$

**46.** $y = x^2 + 2x + 1,\ y = 3(x + 1)$

In Exercises 47–52, use a symbolic integration utility to evaluate the double integral.

**47.** $\displaystyle\int_0^1\int_0^2 e^{-x^2-y^2}\,dx\,dy$

**48.** $\displaystyle\int_0^3\int_0^{x^2} \sqrt{x}\sqrt{1 + x}\,dy\,dx$

**49.** $\displaystyle\int_1^2\int_1^x e^{xy}\,dy\,dx$

**50.** $\displaystyle\int_1^2\int_y^{2y} \ln(x + y)\,dx\,dy$

**51.** $\displaystyle\int_0^1\int_x^1 \sqrt{1 - x^2}\,dy\,dx$

**52.** $\displaystyle\int_0^4\int_0^y \frac{2}{(x + 1)(y + 1)}\,dx\,dy$

*True or False*  In Exercises 53 and 54, determine whether the statement is true or false. Explain your reasoning.

**53.** Changing the order of integration will sometimes change the value of a double integral.

**54.** $\displaystyle\int_2^5\int_1^6 x\,dy\,dx = \int_1^6\int_2^5 x\,dx\,dy$

## 7.9 Applications of Double Integrals

*Volume of a Solid Region* •
*Average Value of a Function over a Region*

### Volume of a Solid Region

In Section 7.8, you used double integrals as an alternative way to find the area of a plane region. In this section, you will study the primary uses of double integrals—to find the volume of a solid region and to find the average value of a function.

Consider a function $z = f(x, y)$ that is continuous and nonnegative over a region $R$. Let $S$ be the solid region that lies between the $xy$-plane and the surface

$$z = f(x, y) \qquad \text{Surface lying above the } xy\text{-plane}$$

directly above the region $R$, as shown in Figure 7.47. You can find the volume of $S$ by integrating $f(x, y)$ over the region $R$.

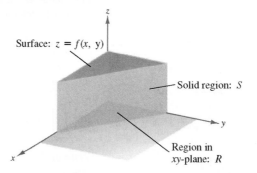

**FIGURE 7.47**

---

### Determining Volume by Double Integrals

If $R$ is a bounded region in the $xy$-plane and $f$ is continuous and nonnegative over $R$, then the **volume of the solid** region between the surface $z = f(x, y)$ and $R$ is given by the double integral

$$\int_R \int f(x, y) \, dA,$$

where $dA = dx \, dy$ or $dA = dy \, dx$.

## EXAMPLE 1  Finding the Volume of a Solid

Find the volume of the solid region bounded in the first octant by the plane

$$z = 2 - x - 2y.$$

**Solution**

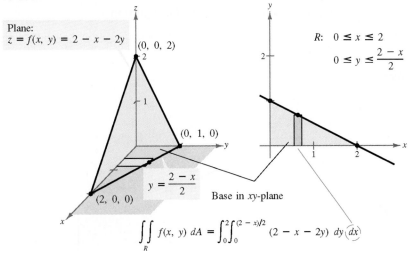

Plane:
$z = f(x, y) = 2 - x - 2y$

$(0, 0, 2)$

$(0, 1, 0)$

$(2, 0, 0)$

$y = \dfrac{2 - x}{2}$

Base in $xy$-plane

$R: \quad 0 \le x \le 2$

$0 \le y \le \dfrac{2 - x}{2}$

$$\iint_R f(x, y)\, dA = \int_0^2 \int_0^{(2-x)/2} (2 - x - 2y)\, dy\, dx$$

**FIGURE 7.48**

To set up the double integral for the volume, it is helpful to sketch both the solid region and the plane region $R$ in the $xy$-plane. In Figure 7.48, you can see that the region $R$ is bounded by the lines $x = 0$, $y = 0$, and $y = \frac{1}{2}(2 - x)$. One way to set up the double integral is to choose $x$ as the outer variable. With that choice, the constant bounds for $x$ are $0 \le x \le 2$ and the variable bounds for $y$ are $0 \le y \le \frac{1}{2}(2 - x)$. Thus, the volume of the solid region is

$$V = \int_0^2 \int_0^{(2-x)/2} (2 - x - 2y)\, dy\, dx$$

$$= \int_0^2 \left[ (2 - x)y - y^2 \right]_0^{(2-x)/2} dx$$

$$= \int_0^2 \left\{ (2 - x)\left(\frac{1}{2}\right)(2 - x) - \left[\frac{1}{2}(2 - x)\right]^2 \right\} dx$$

$$= \frac{1}{4} \int_0^2 (2 - x)^2\, dx$$

$$= \left[ -\frac{1}{12}(2 - x)^3 \right]_0^2$$

$$= \frac{2}{3}.$$

**STUDY TIP**  Example 1 uses $dy\, dx$ as the order of integration. Try using the other order, $dx\, dy$, as indicated in Figure 7.49, to find the volume of the region. Do you get the same result as in Example 1?

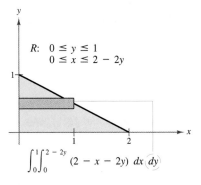

$R: \quad 0 \le y \le 1$
$0 \le x \le 2 - 2y$

$$\int_0^1 \int_0^{2-2y} (2 - x - 2y)\, dx\, dy$$

**FIGURE 7.49**

In Example 1, the order of integration was arbitrary. Although the example used $x$ as the outer variable, you could just as easily have used $y$ as the outer variable. The next example describes a situation in which one order of integration is more convenient than the other.

## EXAMPLE 2    Comparing Different Orders of Integration

Find the volume under the surface $f(x, y) = e^{-x^2}$, bounded by the $xz$-plane and the planes $y = x$ and $x = 1$, as shown in Figure 7.50.

### Solution

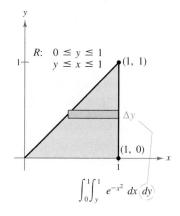

Surface:
$f(x, y) = e^{-x^2}$

$y = 0$

$x = 1$

$y = x$

**FIGURE 7.50**

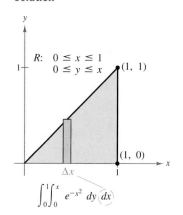

$R: \quad 0 \le x \le 1$
$\quad\quad 0 \le y \le x$

(1, 1)

(1, 0)

$\Delta x$

$\displaystyle\int_0^1 \int_0^x e^{-x^2} \, dy \, dx$

$R: \quad 0 \le y \le 1$
$\quad\quad y \le x \le 1$

(1, 1)

$\Delta y$

(1, 0)

$\displaystyle\int_0^1 \int_y^1 e^{-x^2} \, dx \, dy$

**FIGURE 7.51**

In the $xy$-plane, the bounds of region $R$ are the lines $y = 0$, $x = 1$, and $y = x$. The two possible orders of integration are indicated in Figure 7.51. If you attempt to evaluate the two double integrals shown in the figure, you will discover that the one on the right involves finding the antiderivative of $e^{-x^2}$, which you know is not an elementary function. The integral on the left, however, can be easily evaluated, as follows.

$$V = \int_0^1 \int_0^x e^{-x^2} \, dy \, dx$$

$$= \int_0^1 \left[ e^{-x^2} y \right]_0^x dx$$

$$= \int_0^1 x e^{-x^2} \, dx$$

$$= \left[ -\frac{1}{2} e^{-x^2} \right]_0^1$$

$$= -\frac{1}{2} \left( \frac{1}{e} - 1 \right)$$

---

### Guidelines for Finding the Volume of a Solid

1. Write the equation of the surface in the form $z = f(x, y)$ and sketch the solid region.

2. Sketch the region $R$ in the $xy$-plane and determine the order and limits of integration.

3. Evaluate the double integral

$$\int_R \int f(x, y)\, dA$$

using the order and limits determined in the second step.

---

The first step above suggests that you sketch the three-dimensional solid region. This is a good suggestion, but it is not always feasible and is not as important as making a sketch of the two-dimensional region $R$.

## EXAMPLE 3   Finding the Volume of a Solid

Find the volume of the solid bounded above by the surface

$$f(x, y) = 6x^2 - 2xy$$

and below by the plane region $R$ shown in Figure 7.52.

### Solution

Because the region $R$ is bounded by the parabola $y = 3x - x^2$ and the line $y = x$, the limits for $y$ are $x \le y \le 3x - x^2$. The limits for $x$ are $0 \le x \le 2$, and the volume of the solid is

$$V = \int_0^2 \int_x^{3x - x^2} (6x^2 - 2xy)\, dy\, dx$$

$$= \int_0^2 \left[ 6x^2 y - xy^2 \right]_x^{3x - x^2} dx$$

$$= \int_0^2 \left[ (18x^3 - 6x^4 - 9x^3 + 6x^4 - x^5) - (6x^3 - x^3) \right] dx$$

$$= \int_0^2 (4x^3 - x^5)\, dx$$

$$= \left[ x^4 - \frac{x^6}{6} \right]_0^2$$

$$= \frac{16}{3}.$$

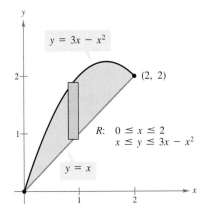

$y = 3x - x^2$

(2, 2)

$R:\ 0 \le x \le 2$
$\quad\ x \le y \le 3x - x^2$

$y = x$

**FIGURE 7.52**

A *population density function* $p = f(x, y)$ is a model that describes density (in people per square unit) of a region. To find the population of a region $R$, evaluate the double integral

$$\int_R \int f(x, y) \, dA.$$

## EXAMPLE 4  Finding a City's Population

The population density (in people per square mile) of the city shown in Figure 7.53 can be modeled by

$$f(x, y) = \frac{50,000}{x + |y| + 1},$$

where $x$ and $y$ are measured in miles. Approximate the city's population. What is the city's average population density?

### Solution

Because the model involves the absolute value of $y$, it follows that the population density is symmetrical about the $x$-axis. Thus, the population in the first quadrant is equal to the population in the fourth quadrant. This means that you can find the total population by doubling the population in the first quadrant.

$$\text{Population} = 2 \int_0^4 \int_0^5 \frac{50,000}{x + y + 1} \, dy \, dx$$

$$= 100,000 \int_0^4 \left[ \ln(x + y + 1) \right]_0^5 dx$$

$$= 100,000 \int_0^4 \left[ \ln(x + 6) - \ln(x + 1) \right] dx$$

$$= 100,000 \left[ (x + 6) \ln(x + 6) - (x + 6) - \right.$$

$$\left. (x + 1) \ln(x + 1) + (x + 1) \right]_0^4$$

$$= 100,000 \left[ (x + 6) \ln(x + 6) - (x + 1) \ln(x + 1) - 5 \right]_0^4$$

$$= 100,000 [10 \ln(10) - 5 \ln(5) - 5 - 6 \ln(6) + 5]$$

$$\approx 422,810 \text{ people}$$

Thus, the city's population is about 422,810. Because the city covers a region 4 miles wide and 10 miles long, its area is 40 square miles. Thus, the average population density is

$$\text{Average population density} = \frac{422,810}{40}$$

$$\approx 10,570 \text{ people per square mile.}$$

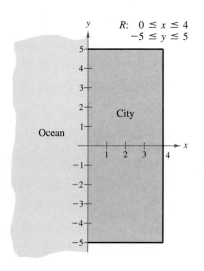

$R: \ 0 \le x \le 4$
$\quad -5 \le y \le 5$

**FIGURE 7.53**

## Average Value of a Function over a Region

### Average Value of a Function over a Region

If $f$ is integrable over the plane region $R$ with area $A$, then its **average value** over $R$ is

$$\text{Average value} = \frac{1}{A} \int_R \int f(x, y) \, dA.$$

### EXAMPLE 5  Finding the Average Profit

A manufacturer determines that the profit for selling $x$ units of one product and $y$ units of a second product is

$$P = -(x - 200)^2 - (y - 100)^2 + 5000.$$

The weekly sales for product 1 vary between 150 and 200 units, and the weekly sales for product 2 vary between 80 and 100 units. Estimate the average weekly profit for the two products.

#### Solution

Because $150 \leq x \leq 200$ and $80 \leq y \leq 100$, you can estimate the weekly profit to be the average of the profit function over the rectangular region shown in Figure 7.54. Because the area of this rectangular region is $(50)(20) = 1000$, it follows that the average profit $V$ is

$$V = \frac{1}{1000} \int_{150}^{200} \int_{80}^{100} \left[ -(x - 200)^2 - (y - 100)^2 + 5000 \right] dy \, dx$$

$$= \frac{1}{1000} \int_{150}^{200} \left[ -(x - 200)^2 y - \frac{(y - 100)^3}{3} + 5000y \right]_{80}^{100} dx$$

$$= \frac{1}{1000} \int_{150}^{200} \left[ -20(x - 200)^2 - \frac{292{,}000}{3} \right] dx$$

$$= \frac{1}{3000} \left[ -20(x - 200)^3 + 292{,}000x \right]_{150}^{200}$$

$$\approx \$4033.$$

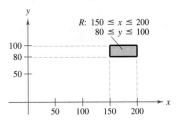

**FIGURE 7.54**

R: $150 \leq x \leq 200$
   $80 \leq y \leq 100$

*Group Discussion*       *Finding an Average Value*

What is the average value of the constant function $f(x, y) = 100$ over a region $R$? Explain your reasoning.

## Warm Up

The following warm-up exercises involve skills that were covered in earlier sections. You will use these skills in the exercise set for this section.

In Exercises 1–4, sketch the region that is described.

**1.** $0 \le x \le 2, \ 0 \le y \le 1$

**2.** $1 \le x \le 3, \ 2 \le y \le 3$

**3.** $0 \le x \le 4, \ 0 \le y \le 2x - 1$

**4.** $0 \le x \le 2, \ 0 \le y \le x^2$

In Exercises 5–10, evaluate the double integral.

**5.** $\displaystyle\int_0^1 \int_1^2 dy \, dx$

**6.** $\displaystyle\int_0^3 \int_1^3 dx \, dy$

**7.** $\displaystyle\int_0^1 \int_0^x x \, dy \, dx$

**8.** $\displaystyle\int_0^4 \int_1^y y \, dx \, dy$

**9.** $\displaystyle\int_1^3 \int_x^{x^2} 2 \, dy \, dx$

**10.** $\displaystyle\int_0^1 \int_x^{-x^2+2} dy \, dx$

## EXERCISES 7.9

In Exercises 1–6, sketch the region of integration and evaluate the double integral.

**1.** $\displaystyle\int_0^2 \int_0^1 (3x + 4y) \, dy \, dx$

**2.** $\displaystyle\int_{-a}^a \int_{-\sqrt{a^2-x^2}}^{\sqrt{a^2-x^2}} dy \, dx$

**3.** $\displaystyle\int_0^1 \int_y^{\sqrt{y}} x^2 y^2 \, dx \, dy$

**4.** $\displaystyle\int_0^6 \int_{y/2}^3 (x + y) \, dx \, dy$

**5.** $\displaystyle\int_0^1 \int_0^{\sqrt{1-x^2}} y \, dy \, dx$

**6.** $\displaystyle\int_0^2 \int_0^{4-x^2} xy^2 \, dy \, dx$

In Exercises 7–10, set up the integral for both orders of integration and use the more convenient order to evaluate the integral over the region R.

Integral

Region R

**7.** $\displaystyle\int_R \int xy \, dA$

Rectangle with vertices at $(0, 0)$, $(0, 5)$, $(3, 5)$, $(3, 0)$

**8.** $\displaystyle\int_R \int \frac{y}{x^2 + y^2} \, dA$

Triangle bounded by $y = x$, $y = 2x$, $x = 2$

**9.** $\displaystyle\int_R \int \frac{y}{1 + x^2} \, dA$

Region bounded by $y = 0$, $y = \sqrt{x}$, $x = 4$

**10.** $\displaystyle\int_R \int x \, dA$

Semicircle bounded by $y = \sqrt{25 - x^2}$ and $y = 0$

In Exercises 11 and 12, evaluate the double integral. Note that it is necessary to change the order of integration.

**11.** $\displaystyle\int_0^1 \int_{y/2}^{1/2} e^{-x^2} \, dx \, dy$

**12.** $\displaystyle\int_0^{\ln 10} \int_{e^x}^{10} \frac{1}{\ln y} \, dy \, dx$

In Exercises 13–24, use a double integral to find the volume of the specified solid.

**13.**

**14.**

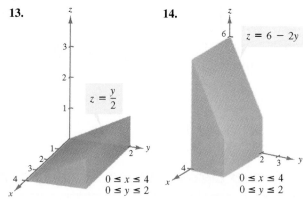

$z = \dfrac{y}{2}$

$0 \le x \le 4$
$0 \le y \le 2$

$z = 6 - 2y$

$0 \le x \le 4$
$0 \le y \le 2$

**15.**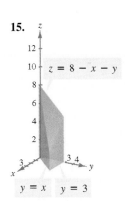

**16.**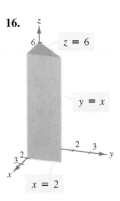

**17.**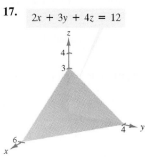

**18.**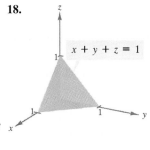

**19.**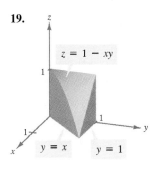

**20.**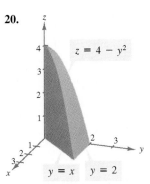

**21.**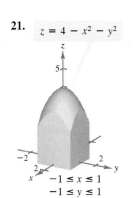

**22.**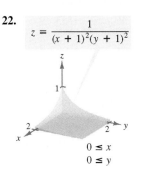

**23.**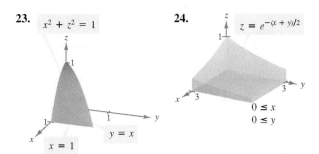

**24.**

In Exercises 25–28, use a double integral to find the volume of the solid bounded by the graphs of the equations.

**25.** $z = xy$, $z = 0$, $y = 0$, $y = 4$, $x = 0$, $x = 1$

**26.** $z = x$, $z = 0$, $y = x$, $y = 0$, $x = 0$, $x = 4$

**27.** $z = x^2$, $z = 0$, $x = 0$, $x = 2$, $y = 0$, $y = 4$

**28.** $z = x + y$, $x^2 + y^2 = 4$ (first octant)

In Exercises 29–32, find the average value of $f(x, y)$ over the region $R$.

| *Integral* | *Region R* |
|---|---|
| **29.** $f(x, y) = x$ | Rectangle with vertices $(0, 0)$, $(4, 0)$, $(4, 2)$, $(0, 2)$ |
| **30.** $f(x, y) = xy$ | Rectangle with vertices $(0, 0)$, $(4, 0)$, $(4, 2)$, $(0, 2)$ |
| **31.** $f(x, y) = x^2 + y^2$ | Square with vertices $(0, 0)$, $(2, 0)$, $(2, 2)$, $(0, 2)$ |
| **32.** $f(x, y) = e^{x+y}$ | Triangle with vertices $(0, 0)$, $(0, 1)$, $(1, 1)$ |

**33.** *Average Revenue* A company sells two products whose demand functions are
$$x_1 = 500 - 3p_1 \quad \text{and} \quad x_2 = 750 - 2.4p_2.$$
Therefore, the total revenue is given by
$$R = x_1 p_1 + x_2 p_2.$$
Estimate the average revenue if the price $p_1$ varies between \$50 and \$75 and the price $p_2$ varies between \$100 and \$150.

**34.** *Average Weekly Profit* A firm's weekly profit in marketing two products is given by
$$P = 192x_1 + 576x_2 - x_1^2 - 5x_2^2 - 2x_1x_2 - 5000,$$
where $x_1$ and $x_2$ represent the numbers of units of each product sold weekly. Estimate the average weekly profit if $x_1$ varies between 40 and 50 units and $x_2$ varies between 45 and 50 units.

# Chapter 7  Algebra Review

*Nonlinear System in Two Variables*

$4x + 3y = 6$

$x^2 - y = 4$

*Linear System in Three Variables*

$-x + 2y + 4z = 2$

$2x - y + z = 0$

$6x + 2z = 3$

## Solving Systems of Equations

Three of the sections in this chapter (7.5, 7.6, and 7.7) involve solutions of systems of equations. These systems can be linear or nonlinear as shown at the left.

There are many techniques for solving a system of linear equations. Two of the more common are listed here.

1. *Substitution:* Solve for one of the variables in one of the equations and substitute the value into another equation.

2. *Elimination:* Add multiples of one equation to a second equation to eliminate a variable in the second equation.

---

### EXAMPLE 1  Solving a System of Equations

(a) Example 3, page 494

$$y - x^3 = 0 \qquad \text{Equation 1}$$

$$x - y^3 = 0 \qquad \text{Equation 2}$$

$$y = x^3 \qquad \text{Solve for } y \text{ in Equation 1.}$$

$$x - (x^3)^3 = 0 \qquad \text{Substitute } x^3 \text{ for } y \text{ in Equation 2.}$$

$$x - x^9 = 0 \qquad (x^m)^n = x^{mn}$$

$$x(x - 1)(x + 1)(x^2 + 1)(x^4 + 1) = 0 \qquad \text{Factor.}$$

$$x = 0, 1, -1 \qquad \text{Set factors equal to zero.}$$

(b) Example 4, page 495

$$-400p_1 + 300p_2 = -25 \qquad \text{Equation 1}$$

$$300p_1 - 360p_2 = -535 \qquad \text{Equation 2}$$

$$p_2 = \tfrac{1}{12}(16p_1 - 1) \qquad \text{Solve for } p_2 \text{ in Equation 1.}$$

$$300p_1 - 360\left(\tfrac{1}{12}\right)(16p_1 - 1) = -535 \qquad \text{Substitute for } p_2 \text{ in Equation 2.}$$

$$300p_1 - 30(16p_1 - 1) = -535 \qquad \text{Multiply factors.}$$

$$-180p_1 = -565 \qquad \text{Combine like terms.}$$

$$p_1 = \tfrac{113}{36} \approx 3.14 \qquad \text{Divide each side by } -180.$$

$$p_2 = \tfrac{1}{12}\left[16\left(\tfrac{113}{36}\right) - 1\right] \qquad \text{Find } p_2 \text{ by substituting } p_1.$$

$$p_2 \approx 4.10 \qquad \text{Solve for } p_2.$$

## EXAMPLE 2  Solving a System of Equations

(a) Example 5, page 496

Before solving this system of equations, factor 4 out of the first equation and factor 2 out of the second equation.

$$y(24 - 12x - 4y) = 0 \qquad \text{Original Equation 1}$$
$$x(24 - 6x - 8y) = 0 \qquad \text{Original Equation 2}$$

$$y(4)(6 - 3x - y) = 0 \qquad \text{Factor 4 out of Equation 1.}$$
$$x(2)(12 - 3x - 4y) = 0 \qquad \text{Factor 2 out of Equation 2.}$$

$$y(6 - 3x - y) = 0 \qquad \text{Equation 1}$$
$$x(12 - 3x - 4y) = 0 \qquad \text{Equation 2}$$

In each equation, either factor can be zero, so you obtain four different linear systems.

$$\left.\begin{aligned} y &= 0 \\ 12 - 3x - 4y &= 0 \end{aligned}\right\} \qquad (4, 0) \text{ is a solution.}$$

$$\left.\begin{aligned} 6 - 3x - y &= 0 \\ 12 - 3x - 4y &= 0 \end{aligned}\right\} \qquad \left(\tfrac{4}{3}, 2\right) \text{ is a solution.}$$

$$\left.\begin{aligned} y &= 0 \\ x &= 0 \end{aligned}\right\} \qquad (0, 0) \text{ is a solution.}$$

$$\left.\begin{aligned} 6 - 3x - y &= 0 \\ x &= 0 \end{aligned}\right\} \qquad (0, 6) \text{ is a solution.}$$

(b) Example 2, page 512

$$28a - 4b = 10 \qquad \text{Equation 1}$$
$$-4a + 8b = 12 \qquad \text{Equation 2}$$
$$-2a + 4b = 6 \qquad \text{Divide Equation 2 by 2.}$$
$$26a \phantom{- 4b} = 16 \qquad \text{Add new equation to Equation 1.}$$
$$a = \tfrac{8}{13} \qquad \text{Divide each side by 26.}$$
$$28\left(\tfrac{8}{13}\right) - 4b = 10 \qquad \text{Substitute for } a \text{ in Equation 1.}$$
$$b = \tfrac{47}{26} \qquad \text{Solve for } b.$$

# Chapter Summary and Study Strategies

*After studying this chapter, you should have acquired the following skills. The exercise numbers are keyed to the Review Exercises that begin on page 540. Answers to odd-numbered Review Exercises are given in the back of the text.\**

■  Plot points in space.  *(Section 7.1)*                              *Review Exercises 1, 2*

■  Find the distance between two points in space.  *(Section 7.1)*     *Review Exercises 3, 4*
$$d = \sqrt{(x_2 - x_1)^2 + (y_2 - y_1)^2 + (z_2 - z_1)^2}$$

■  Find the midpoints of line segments in space.  *(Section 7.1)*     *Review Exercises 5, 6*
$$\text{Midpoint} = \left( \frac{x_1 + x_2}{2}, \frac{y_1 + y_2}{2}, \frac{z_1 + z_2}{2} \right)$$

■  Write the standard forms of the equations of spheres.  *(Section 7.1)*     *Review Exercises 7–10*
$$(x - h)^2 + (y - k)^2 + (z - l)^2 = r^2$$

■  Find the centers and radii of spheres.  *(Section 7.1)*            *Review Exercises 11, 12*

■  Sketch the coordinate plane traces of spheres.  *(Section 7.1)*    *Review Exercises 13, 14*

■  Sketch planes in space.  *(Section 7.2)*                           *Review Exercises 15–18*

■  Classify quadric surfaces in space.  *(Section 7.2)*               *Review Exercises 19–26*

■  Evaluate functions of several variables.  *(Section 7.3)*          *Review Exercises 27, 28*

■  Find the domains and ranges of functions of several variables.  *(Section 7.3)*     *Review Exercises 29, 30*

■  Sketch the level curves of functions of two variables.  *(Section 7.3)*     *Review Exercises 31–34*

■  Use functions of several variables to answer questions about real life.  *(Section 7.3)*     *Review Exercises 35–40*

■  Find the first partial derivatives of functions of several variables.  *(Section 7.4)*     *Review Exercises 41–50*
$$\frac{\partial z}{\partial x} = \lim_{\Delta x \to 0} \frac{f(x + \Delta x, y) - f(x, y)}{\Delta x}$$
$$\frac{\partial z}{\partial y} = \lim_{\Delta y \to 0} \frac{f(x, y + \Delta y) - f(x, y)}{\Delta y}$$

■  Find the slopes of surfaces in the *x*- and *y*-directions.  *(Section 7.4)*     *Review Exercises 51–54*

■  Find the second partial derivatives of functions of several variables.  *(Section 7.4)*     *Review Exercises 55–58*

■  Use partial derivatives to answer questions about real life.  *(Section 7.4)*     *Review Exercises 59–62*

■  Find the relative extrema of functions of two variables.  *(Section 7.5)*     *Review Exercises 63–70*

■  Use relative extrema to answer questions about real life.  *(Section 7.5)*     *Review Exercises 71, 72*

■  Use Lagrange multipliers to find extrema of functions of several variables.     *Review Exercises 73–78*
   *(Section 7.6)*

■  Use Lagrange multipliers to answer questions about real life.  *(Section 7.6)*     *Review Exercises 79, 80*

---

\*  Use a wide range of valuable study aids to help you master the material in this chapter. The *Student Solutions Guide* includes step-by-step solutions to all odd-numbered exercises to help you review and prepare. The *Algebra Review Tutorial Software* and *The Algebra of Calculus* help you brush up on your algebra skills. The *Graphing Technology Guide* offers step-by-step commands and instructions for a wide variety of graphing calculators, including the most recent models.

■ Find the least squares regression lines, $y = ax + b$, for data.  *(Section 7.7)*

Review Exercises 81, 82

$$a = \dfrac{n\displaystyle\sum_{i=1}^{n} x_i y_i - \displaystyle\sum_{i=1}^{n} x_i \displaystyle\sum_{i=1}^{n} y_i}{n\displaystyle\sum_{i=1}^{n} x_i^2 - \left(\displaystyle\sum_{i=1}^{n} x_i\right)^2}, \qquad b = \dfrac{1}{n}\left(\displaystyle\sum_{i=1}^{n} y_i - a\displaystyle\sum_{i=1}^{n} x_i\right)$$

■ Use least squares regression lines to model real-life data.  *(Section 7.7)*

Review Exercises 83, 84

■ Find the least squares regression quadratics for data.  *(Section 7.7)*

Review Exercises 85, 86

■ Evaluate double integrals.  *(Section 7.8)*

Review Exercises 87–90

■ Use double integrals to find the areas of regions.  *(Section 7.8)*

Review Exercises 91–94

■ Use double integrals to find the volumes of solids.  *(Section 7.9)*

Review Exercises 95, 96

$$\text{Volume} = \int\!\!\int_{R} f(x, y)\, dA$$

■ Use double integrals to find the average values of real-life models.  *(Section 7.9)*

Review Exercises 97, 98

$$\text{Average value} = \dfrac{1}{A}\int\!\!\int_{R} f(x, y)\, dA$$

■ *Comparing Two Dimensions with Three Dimensions*   Many of the formulas and techniques in this chapter are generalizations of formulas and techniques used in earlier chapters in the text. Here are several examples.

| Two-Dimensional Coordinate System | Three-Dimensional Coordinate System |
|---|---|
| *Distance Formula* $d = \sqrt{(x_2 - x_1)^2 + (y_2 - y_1)^2}$ | *Distance Formula* $d = \sqrt{(x_2 - x_1)^2 + (y_2 - y_1)^2 + (z_2 - z_1)^2}$ |
| *Midpoint Formula* $\text{Midpoint} = \left(\dfrac{x_1 + x_2}{2}, \dfrac{y_1 + y_2}{2}\right)$ | *Midpoint Formula* $\text{Midpoint} = \left(\dfrac{x_1 + x_2}{2}, \dfrac{y_1 + y_2}{2}, \dfrac{z_1 + z_2}{2}\right)$ |
| *Equation of Circle* $(x - h)^2 + (y - k)^2 = r^2$ | *Equation of Sphere* $(x - h)^2 + (y - k)^2 + (z - l)^2 = r^2$ |
| *Equation of Line* $ax + by = c$ | *Equation of Plane* $ax + by + cz = d$ |
| *Derivative of* $y = f(x)$ $\dfrac{dy}{dx} = \lim\limits_{\Delta x \to 0} \dfrac{f(x + \Delta x) - f(x)}{\Delta x}$ | *Partial Derivative of* $z = f(x, y)$ $\dfrac{\partial z}{\partial x} = \lim\limits_{\Delta x \to 0} \dfrac{f(x + \Delta x, y) - f(x, y)}{\Delta x}$ |
| *Area of Region* $A = \displaystyle\int_{a}^{b} f(x)\, dx$ | *Volume of Region* $V = \displaystyle\int\!\!\int_{R} f(x, y)\, dA$ |

# Chapter 7   Review Exercises

In Exercises 1 and 2, plot the points.

**1.** $(2, -1, 4), (-1, 3, -3)$    **2.** $(1, -2, -3), (-4, -3, 5)$

In Exercises 3 and 4, find the distance between the two points.

**3.** $(0, 0, 0), (2, 5, 9)$        **4.** $(-4, 1, 5), (1, 3, 7)$

In Exercises 5 and 6, find the midpoint of the line segment joining the two points.

**5.** $(2, 6, 4), (-4, 2, 8)$        **6.** $(5, 0, 7), (-1, -2, 9)$

In Exercises 7–10, find the standard form of the equation of the sphere.

**7.** Center: $(0, 1, 0)$;  Radius: 5

**8.** Center: $(4, -5, 3)$;  Radius: 10

**9.** Diameter endpoints: $(3, 4, 0), (5, 8, 2)$

**10.** Diameter endpoints: $(-2, 5, 1), (4, -3, 3)$

In Exercises 11 and 12, find the center and radius of the sphere.

**11.** $x^2 + y^2 + z^2 + 4x - 2y - 8z + 5 = 0$

**12.** $x^2 + y^2 + z^2 + 4y - 10z - 7 = 0$

In Exercises 13 and 14, sketch the *xy*-trace of the sphere.

**13.** $(x + 2)^2 + (y - 1)^2 + (z - 3)^2 = 25$

**14.** $(x - 1)^2 + (y + 3)^2 + (z - 6)^2 = 72$

In Exercises 15-18, find the intercepts and sketch the graph of the plane.

**15.** $x + 2y + 3z = 6$        **16.** $2y + z = 4$

**17.** $6x + 3y - 6z = 12$      **18.** $4x - y + 2z = 8$

In Exercises 19–26, identify the surface.

**19.** $x^2 + y^2 + z^2 - 2x + 4y - 6z + 5 = 0$

**20.** $16x^2 + 16y^2 - 9z^2 = 0$

**21.** $x^2 + \dfrac{y^2}{16} + \dfrac{z^2}{9} = 1$    **22.** $-x^2 + \dfrac{y^2}{16} + \dfrac{z^2}{9} = 1$

**23.** $z = \dfrac{x^2}{9} + y^2$        **24.** $-4x^2 + y^2 + z^2 = 4$

**25.** $z = \sqrt{x^2 + y^2}$        **26.** $z = 9x + 3y - 5$

In Exercises 27 and 28, find the function values.

**27.** $f(x, y) = xy^2$

(a) $f(2, 3)$              (b) $f(0, 1)$
(c) $f(-5, 7)$            (d) $f(-2, -4)$

**28.** $f(x, y) = \dfrac{x^2}{y}$

(a) $f(6, 9)$              (b) $f(8, 4)$
(c) $f(t, 2)$              (d) $f(r, r)$

In Exercises 29 and 30, describe the region $R$ in the *xy*-plane that corresponds to the domain of the function. Then find the range of the function.

**29.** $f(x, y) = \sqrt{1 - x^2 - y^2}$    **30.** $f(x, y) = \dfrac{1}{x + y}$

In Exercises 31–34, describe the level curves of the function. Sketch the level curves for the given *c*-values.

**31.** $z = 10 - 2x - 5y, \ c = 0, 2, 4, 5, 10$

**32.** $z = \sqrt{9 - x^2 - y^2}, \ c = 0, 1, 2, 3$

**33.** $z = (xy)^2, \ c = 1, 4, 9, 12, 16$

**34.** $z = 2e^{xy}, \ c = 1, 2, 3, 4, 5$

**35.** *Average Precipitation*   The contour map shown below represents the average yearly precipitation for Iowa. *(Source: U.S. National Oceanic and Atmospheric Administration)*

(a) Discuss the use of color to represent the level curves.
(b) Which part of Iowa receives the most precipitation?
(c) Which part of Iowa receives the least precipitation?

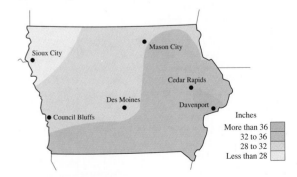

**36.** *Population Density*   The contour map below represents the population density of New York. *(Source: U.S. Bureau of Census)*

(a) Discuss the use of color to represent the level curves.

(b) Do the level curves correspond to equally spaced population densities?

(c) Describe how to obtain a more detailed contour map.

**37.** *Acid Rain*   The acidity of rainwater is measured in units called pH and smaller pH values are increasingly acidic. The map shows the curves of equal pH and gives evidence that downwind of heavily industrialized areas the acidity has been increasing. Using the level curves on the map, determine the direction of the prevailing winds in the northeastern United States.

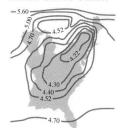

**38.** *Wal-Mart Sales*   The table gives the net sales $x$ (in billions of dollars), the total assets $y$ (in billions of dollars), and the shareholder's equity $z$ (in billions of dollars) for Wal-Mart for the years 1991 through 1996. *(Source: Wal-Mart 1996 Annual Report)*

| Year | 1991 | 1992 | 1993 | 1994 | 1995 | 1996 |
|------|------|------|------|------|------|------|
| $x$  | 32.6 | 43.9 | 55.5 | 67.3 | 82.5 | 93.6 |
| $y$  | 11.4 | 15.4 | 20.6 | 26.4 | 32.8 | 37.5 |
| $z$  | 5.4  | 7.0  | 8.8  | 10.8 | 12.7 | 14.8 |

A model for this data is
$$z = f(x, y) = 0.083x + 0.160y + 0.890.$$

(a) Use a graphing utility and the model to approximate $z$ for the given values of $x$ and $y$.

(b) Which of the two variables in this model has the greater influence on shareholder's equity?

(c) Simplify the expression for $f(x, 25)$ and interpret its meaning in the context of the problem.

**39.** *Equation of Exchange*   Economists use an equation of exchange to express the relation among money, prices, and business transactions. This equation can be written as
$$P = \frac{MV}{T},$$
where $M$ is the money supply, $V$ is the velocity of circulation, $T$ is the total number of transactions, and $P$ is the price level. Find $P$ when $M = \$2500$, $V = 6$, and $T = 6000$.

**40.** *Biomechanics*   The Froude number $F$, defined as
$$F = \frac{v^2}{gl},$$
where $v$ represents velocity, $g$ represents gravitational acceleration, and $l$ represents stride length, is an example of a "similarity criterion." Find the Froude number of a rabbit for which velocity is 2 meters per second, gravitational acceleration is 3 meters per second squared, and stride length is 0.75 meter.

In Exercises 41–50, find the first partial derivatives.

**41.** $f(x, y) = x^2 y + 3xy + 2x - 5y$

**42.** $f(x, y) = 5x^2 + 4xy - 3y^2$

**43.** $z = 6x^2 \sqrt{y} + 3 \sqrt{xy} - 7xy$

**44.** $z = (xy + 2x + 4y)^2$

**45.** $f(x, y) = \ln(2x + 3y)$     **46.** $f(x, y) = \ln \sqrt{2x + 3y}$

**47.** $f(x, y) = x^2 e^y - y^2 e^x$     **48.** $f(x, y) = x^2 e^{-2y}$

**49.** $w = xyz^2$

**50.** $w = xyz + 2xy - 9xz + 4yz - y^2 z + 4z^2$

In Exercises 51–54, find the slope of the surface at the indicated point in (a) the $x$-direction and (b) the $y$-direction.

**51.** $z = 3x - 4y + 9$, $(3, 2, 10)$

**52.** $z = 4x^2 - y^2$, $(2, 4, 0)$

**53.** $z = 8 - x^2 - y^2$, $(1, 2, 3)$

**54.** $z = x^2 - y^2$, $(5, -4, 9)$

In Exercises 55–58, find all second partial derivatives.

**55.** $f(x, y) = x^3 - 4xy^2 + y^3$     **56.** $f(x, y) = \dfrac{y}{x + y}$

**57.** $f(x, y) = \sqrt{64 - x^2 - y^2}$     **58.** $f(x, y) = x^2 e^{-y^2}$

**59.** *Marginal Cost*  A company manufactures two models of skis: cross-country skis and downhill skis. The cost function for producing $x$ cross-country skis and $y$ downhill skis is

$$C = 15(xy)^{1/3} + 99x + 139y + 2293.$$

Find the marginal costs when $x = 250$ and $y = 175$.

**60.** *Marginal Revenue*  At a baseball stadium, souvenir caps are sold at two locations. If $x_1$ and $x_2$ are the numbers of baseball caps sold at location 1 and location 2, respectively, then the total revenue for the caps is

$$R = 15x_1 + 16x_2 - \frac{1}{10}x_1^2 - \frac{1}{10}x_2^2 - \frac{1}{100}x_1x_2.$$

Given that $x_1 = 50$ and $x_2 = 40$, find the marginal revenue at location 1 and at location 2.

**61.** *Human Body*  The surface area $A$ of an average human body in square centimeters can be approximated by the model

$$A(w, h) = 101.4w^{0.425}h^{0.725},$$

where $w$ is the weight in pounds and $h$ is the height in inches. Determine the partial derivatives of $A$ with respect to $w$ and with respect to $h$. Evaluate $dA/dw$ at $(180, 70)$. Explain your result.

**62.** *Medicine*  In order to treat a certain bacterial infection, a combination of two drugs is being tested. Studies have shown that the duration of the infection in laboratory tests can be modeled by

$$D(x, y) = x^2 + 2y^2 - 18x - 24y + 2xy + 120,$$

where $x$ is the dosage in hundreds of milligrams of the first drug and $y$ is the dosage in hundreds of milligrams of the second drug. Evaluate $D(5, 2.5)$ and $D(7.5, 8)$ and interpret your results.

In Exercises 63–70, find any critical points and relative extrema of the function.

**63.** $f(x, y) = x^2 + 2xy + y^2$

**64.** $f(x, y) = x^3 - 3xy + y^2$

**65.** $f(x, y) = x^2 + 6xy + 3y^2 + 6x + 8$

**66.** $f(x, y) = x + y^2 - e^x$

**67.** $f(x, y) = x^3 + y^2 - xy$

**68.** $f(x, y) = y^2 + xy + 3y - 2x + 5$

**69.** $f(x, y) = x^3 + y^3 - 3x - 3y + 2$

**70.** $f(x, y) = y^2 - x^2$

**71.** *Revenue*  A company manufactures and sells two products. The demand functions for the products are

$$p_1 = 100 - x_1 \quad \text{and} \quad p_2 = 200 - 0.5x_2.$$

(a) Find the total revenue functions for $x_1$ and $x_2$.
(b) Find $x_1$ and $x_2$ such that the revenue is maximized.
(c) What is the maximum revenue?

**72.** *Profit*  A company manufactures a product at two locations. The costs of manufacturing $x_1$ units at plant 1 and $x_2$ units at plant 2 are

$$C_1 = 0.03x_1^2 + 4x_1 + 300$$
$$C_2 = 0.05x_2^2 + 7x_2 + 175.$$

If the product sells for \$10 per unit, find $x_1$ and $x_2$ such that the profit, $P = 10(x_1 + x_2) - C_1 - C_2$, is maximized.

In Exercises 73–78, locate any extrema of the functions by using Lagrange multipliers.

**73.** $f(x, y) = x^2y$;  Constraint: $x + 2y = 2$

**74.** $f(x, y) = x^2 + y^2$;  Constraint: $x + y = 4$

**75.** $f(x, y, z) = xyz$;  Constraint: $x + 2y + z - 4 = 0$

**76.** $f(x, y, z) = xz + yz$;  Constraint: $x + y + z = 6$

**77.** $f(x, y, z) = x^2 + y^2 + z^2$;  Constraints: $x + z = 6$, $y + z = 8$

**78.** $f(x, y, z) = xyz$;  Constraints: $x + y + z = 32$, $x - y + z = 0$

**79.** *Maximum Production Level*  The production function for a manufacturer is

$$f(x, y) = 4x + xy + 2y.$$

Assume that the total amount available for labor $x$ and capital $y$ is \$2000 and that units of labor and capital cost \$20 and \$4, respectively. Find the maximum production level for this manufacturer.

**80.** *Minimum Cost*  A manufacturer has an order for 1500 units that can be produced at two locations. Let $x_1$ and $x_2$ be the numbers of units produced at the two locations. Find the number that should be produced at each to meet the order and minimize cost if the cost function is

$$C = 0.20x_1^2 + 10x_1 + 0.15x_2^2 + 12x_2.$$

In Exercises 81 and 82, (a) use the method of least squares to find the least squares regression line and (b) calculate the sum of the squared errors.

**81.** $(-2, -3), (-1, -1), (1, 2), (3, 2)$

**82.** $(-3, -1), (-2, -1), (0, 0), (1, 1), (2, 1)$

**83.** *Wheat Yield*   An agronomist used four test plots to determine the relationship between the wheat yield (in bushels per acre) and the amount of fertilizer (in hundreds of pounds per acre). The results are given in the table.

| Fertilizer, $x$ | 1.0 | 1.5 | 2.0 | 2.5 |
|---|---|---|---|---|
| Yield, $y$ | 32 | 41 | 48 | 53 |

Use a graphing utility with a least squares regression program to find the least squares regression line for the data, and estimate the yield for a fertilizer application of 160 pounds per acre.

**84.** *Women in the Work Force*   The table gives the percent and number (in millions) of women in the work force for selected years.   *(Source: U.S. Department of Labor)*

| Year | 1960 | 1970 | 1980 | 1990 |
|---|---|---|---|---|
| Percent, $x$ | 37.7 | 43.3 | 51.5 | 57.5 |
| Number, $y$ | 23.2 | 31.5 | 45.5 | 56.6 |

| Year | 1991 | 1992 | 1993 | 1994 |
|---|---|---|---|---|
| Percent, $x$ | 57.3 | 57.8 | 57.9 | 58.8 |
| Number, $y$ | 56.9 | 57.8 | 58.4 | 60.2 |

(a) Use a graphing utility with a least squares regression program to find the least squares regression line for the data.

(b) According to this model, approximately how many women enter the labor force for each one-point increase in the percent of women in the labor force?

In Exercises 85 and 86, find the least squares regression quadratic for the given points. Plot the points and sketch the least squares regression quadratic.

**85.** $(-1, 9)$, $(0, 7)$, $(1, 5)$, $(2, 6)$, $(4, 23)$

**86.** $(0, 10)$, $(2, 9)$, $(3, 7)$, $(4, 4)$, $(5, 0)$

In Exercises 87–90, evaluate the double integral.

**87.** $\displaystyle\int_0^1 \int_0^{1+x} (3x + 2y)\, dy\, dx$

**88.** $\displaystyle\int_{-2}^2 \int_0^4 (x - y^2)\, dx\, dy$

**89.** $\displaystyle\int_1^2 \int_1^{2y} \frac{x}{y^2}\, dx\, dy$

**90.** $\displaystyle\int_0^4 \int_0^{\sqrt{16-x^2}} 2x\, dy\, dx$

In Exercises 91–94, use a double integral to find the area of the region.

**91.**
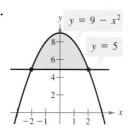

**92.**
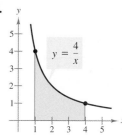

**93.**
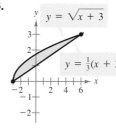

**94.**
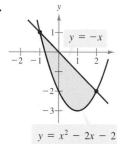

**95.** Find the volume of the solid bounded by the graphs of $z = (xy)^2$, $z = 0$, $y = 0$, $y = 4$, $x = 0$, and $x = 4$.

**96.** Find the volume of the solid bounded by the graphs of $z = x + y$, $z = 0$, $x = 0$, $x = 3$, $y = x$, and $y = 0$.

**97.** *Average Elevation*   In a triangular coastal area, the elevation in miles above sea level at the point $(x, y)$ is $f(x, y) = 0.25 - 0.025x - 0.01y$, where $x$ and $y$ are measured in miles (see figure). Find the average elevation of the triangular area.

Figure for 97          Figure for 98

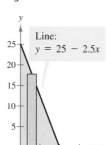

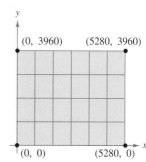

**98.** *Real Estate Value*   The average value of real estate (in dollars per square foot) for a rectangular section of a city is given by $f(x, y) = 2.5x^{3/2}y^{3/4}$ (see figure). Find the average value of real estate for this section.

# Sample Post-Graduation Exam Questions

CPA    GMAT
GRE    CLAST
Actuarial

*The following questions represent the types of questions that appear on certified public accountant (CPA) exams, graduate management admission tests (GMAT), graduate records exams (GRE), actuarial exams, and college-level academic skills tests (CLAST). The answers to the questions are given in the back of the book.*

1. What is the derivative of $f(x, y) = y^2(x + y)^3$ with respect to $y$?
   - (a) $6y(x + y)^2$
   - (b) $y(x + y)^2(2x + 5y)$
   - (c) $3y^2(x + y)^2$
   - (d) $y(x + y)^2(5x + 2y)$

2. Let $f(x, y) = x^2 + y^2 + 6x - 2y + 4$. At which point does $f$ have a relative minimum?
   - (a) $(-3, 1, -13)$
   - (b) $(-3, 1, -6)$
   - (c) $(3, 1, -2)$
   - (d) $(-3, -1, -2)$

For Questions 3 and 4, use the following excerpts from the 1996 Tax Rate Schedules.

**SCHEDULE X**—Use if your filing status is **Single**

| If the amount on Form 1040, line 37, is over— | But not over— | Enter on Form 1040, line 38 | of the amount over— |
|---|---|---|---|
| $0 | $22,100 | ............15% | $0 |
| 22,100 | 53,500 | 3,315.00 + 28% | 22,100 |
| 53,500 | 115,000 | 12,107.00 + 31% | 53,500 |
| 115,000 | 250,000 | 31,172.00 + 36% | 115,000 |
| 250,000 | ............ | 79,772.00 + 39.6% | 250,000 |

**SCHEDULE Y-1**—Use if your filing status is **Married filing jointly** or **Qualifying widow(er)**

| If the amount on Form 1040, line 37, is over— | But not over— | Enter on Form 1040, line 38 | of the amount over— |
|---|---|---|---|
| $0 | $36,900 | ............15% | $0 |
| 36,900 | 89,150 | 5,535.00 + 28% | 36,900 |
| 89,150 | 140,000 | 20,165.00 + 31% | 89,150 |
| 140,000 | 250,000 | 35,928.50 + 36% | 140,000 |
| 250,000 | ............ | 75,528.50 + 39.6% | 250,000 |

3. The tax for a married couple filing jointly whose amount on Form 1040, line 37, is $125,480 is
   - (a) $11,262.30
   - (b) $47,801.50
   - (c) $31,427.30
   - (d) $34,944.80

4. The tax for a single person whose amount on Form 1040, line 37, is $1,000,000 is
   - (a) $372,528.50
   - (b) $297,000
   - (c) $368,028.50
   - (d) $376,772

5. If $x$, $y$, and $z$ are chosen from the three numbers $\frac{1}{3}$, $-2$, and 4, what is the largest possible value of the expression $[(x^2 + z)]/(y^2)$?
   - (a) 126
   - (b) 42
   - (c) 24
   - (d) 72

6. If $xz = 4y$, then $(x^3)/2$ equals
   - (a) $\dfrac{2y^3}{z^3}$
   - (b) $\dfrac{16y^2}{z^3}$
   - (c) $\dfrac{32y^3}{z^3}$
   - (d) $\dfrac{64y^3}{z^3}$

7. Mave Co. calculated the following ratios for one of its profit centers: Gross margin 31%, Return on Sales 26%, Capital Turnover 0.5 time.

   What is Mave's return on investment for this profit center?
   - (a) 8.5%
   - (b) 13%
   - (c) 15.5%
   - (d) 26%

# Appendices

# Appendix A:  Alternate Introduction to the Fundamental Theorem of Calculus

In this appendix we use a summation process to provide an alternate development of the definite integral. It is intended that this supplement follow Section 5.3 in the text. If used, this appendix should replace the material preceding Example 2 in Section 5.4. We begin by showing how the area of a region in the plane can be approximated by the use of rectangles.

### EXAMPLE 1   Using Rectangles to Approximate the Area of a Region

Use the four rectangles indicated in Figure A.1 to approximate the area of the region lying between the graph of

$$f(x) = \frac{x^2}{2}$$

and the x-axis, between $x = 0$ and $x = 4$.

*Solution*

You can find the heights of the rectangles by evaluating the function $f$ at each of the midpoints of the subintervals

$$[0, 1], \quad [1, 2], \quad [2, 3], \quad [3, 4].$$

Because the width of each rectangle is 1, the sum of the areas of the four rectangles is

$$S = \overset{\text{width}}{(1)} \, \overset{\text{height}}{f\left(\frac{1}{2}\right)} + \overset{\text{width}}{(1)} \, \overset{\text{height}}{f\left(\frac{3}{2}\right)} + \overset{\text{width}}{(1)} \, \overset{\text{height}}{f\left(\frac{5}{2}\right)} + \overset{\text{width}}{(1)} \, \overset{\text{height}}{f\left(\frac{7}{2}\right)}$$

$$= \frac{1}{8} + \frac{9}{8} + \frac{25}{8} + \frac{49}{8}$$

$$= \frac{84}{8} = 10.5.$$

Thus, you can approximate the area of the given region to be 10.5.

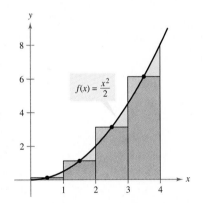

$f(x) = \dfrac{x^2}{2}$

**FIGURE A.1**

**STUDY TIP**   The approximation technique used in Example 1 is called the **Midpoint Rule.** We will say more about the Midpoint Rule in Section 5.6.

The procedure illustrated in Example 1 can be generalized. Let $f$ be a continuous function defined on the closed interval $[a, b]$. To begin, partition the interval into $n$ subintervals, each of width $\Delta x = (b - a)/n$, as follows.

$$a = x_0 < x_1 < x_2 < \cdot \cdot \cdot < x_{n-1} < x_n = b$$

In each subinterval $[x_{i-1}, x_i]$ choose an arbitrary point $c_i$ and form the sum

$$S = f(c_1)\Delta x + f(c_2)\Delta x + \cdot \cdot \cdot + f(c_{n-1})\Delta x + f(c_n)\Delta x.$$

This type of summation is called a **Riemann sum,** and is often written using summation notation as follows.

$$S = \sum_{i=1}^{n} f(c_i)\Delta x$$

For the Riemann sum in Example 1, the interval is $[a, b] = [0, 4]$, the number of subintervals is $n = 4$, the width of each interval is $\Delta x = 1$, and the point $c_i$ in each subinterval is its midpoint. Thus, you can write the approximation in Example 1 as

$$S = \sum_{i=1}^{n} f(c_i)\Delta x = \sum_{i=1}^{4} f(c_i)(1) = \frac{1}{8} + \frac{9}{8} + \frac{25}{8} + \frac{49}{8} = \frac{84}{8}.$$

## EXAMPLE 2   Using a Riemann Sum to Approximate Area

Use a Riemann sum to approximate the area of the region bounded by the graph of $f(x) = -x^2 + 2x$ and the $x$-axis, for $0 \le x \le 2$. In the Riemann sum, let $n = 6$ and choose $c_i$ to be the left endpoint of each subinterval.

*Solution*

Subdivide the interval $[0, 2]$ into six subintervals, each of width

$$\Delta x = \frac{2 - 0}{6} = \frac{1}{3},$$

as shown in Figure A.2. Since $c_i$ is the left endpoint of each subinterval, the Riemann sum is given by

$$S = \sum_{i=1}^{n} f(c_i)\Delta x$$

$$= \left[ f(0) + f\left(\frac{1}{3}\right) + f\left(\frac{2}{3}\right) + f(1) + f\left(\frac{4}{3}\right) + f\left(\frac{5}{3}\right) \right]\left(\frac{1}{3}\right)$$

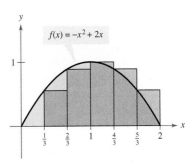

**FIGURE A.2**

$$= \left[ 0 + \frac{5}{9} + \frac{8}{9} + 1 + \frac{8}{9} + \frac{5}{9} \right]\left(\frac{1}{3}\right)$$

$$= \frac{35}{27}.$$

Example 2 illustrates an important point. If a function $f$ is continuous and nonnegative over the interval $[a, b]$, then the Riemann sum

$$S = \sum_{i=1}^{n} f(c_i)\Delta x$$

can be used to approximate the area of the region bounded by the graph of $f$ and the x-axis, between $x = a$ and $x = b$. Moreover, for a given interval, as the number of subintervals increases, the approximation to the actual area will improve. This is illustrated in the next two examples by using Riemann sums to approximate the area of a triangle.

## EXAMPLE 3 Approximating the Area of a Triangle

Use a Riemann sum to approximate the area of the triangular region bounded by the graph of $f(x) = 2x$ and the x-axis, $0 \le x \le 3$. Use a partition of six subintervals and choose $c_i$ to be the left endpoint of each subinterval.

*Solution*

Subdivide the interval $[0, 3]$ into six subintervals, each of width

$$\Delta x = \frac{3 - 0}{6} = \frac{1}{2},$$

as shown in Figure A.3. Because $c_i$ is the left endpoint of each subinterval, the Riemann sum is given by

$$S = \sum_{i=1}^{n} f(c_i)\Delta x$$

$$= \left[ f(0) + f\left(\frac{1}{2}\right) + f(1) + f\left(\frac{3}{2}\right) + f(2) + f\left(\frac{5}{2}\right) \right]\left(\frac{1}{2}\right)$$

$$= [0 + 1 + 2 + 3 + 4 + 5]\left(\frac{1}{2}\right)$$

$$= \frac{15}{2}.$$

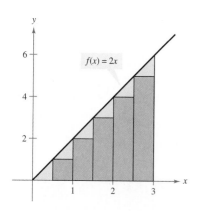

**FIGURE A.3**

The approximations in Examples 2 and 3 are called **left Riemann sums,** because $c_i$ was chosen to be the left endpoint of each subinterval. If the right endpoints had been used in Example 3, the **right Riemann sum** would have been $\frac{21}{2}$.

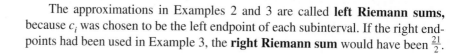

(Try verifying this.) Note that the exact area of the triangular region in Example 3 is

$$\text{Area} = \frac{1}{2}(\text{base})(\text{height}) = \frac{1}{2}(3)(6) = 9.$$

Thus, the left Riemann sum gives an approximation that is less than the actual area, and the right Riemann sum gives an approximation that is greater than the actual area.

In Example 4, you will see that the approximation improves as the number of subintervals increases.

## EXAMPLE 4   Increasing the Number of Subintervals

Let $f(x) = 2x$ for $0 \le x \le 3$. Use a computer to determine the left and right Riemann sums for $n = 10$, $n = 100$, and $n = 1000$ subintervals.

*Solution*

A basic computer program for this problem is as follows.

```
10      INPUT; N
20      DELTA=3/N
30      LSUM=0: RSUM=0
40      FOR I=1 TO N
50      LC=(I-1)*DELTA: RC=I*DELTA
60      LSUM=LSUM+2*LC*DELTA: RSUM=RSUM+2*RC*DELTA
70      NEXT
80      PRINT "LEFT RIEMANN SUM:"; LSUM
90      PRINT "RIGHT REIMANN SUM:"; RSUM
100     END
```

After running this program for $n = 10$, $n = 100$, and $n = 1000$, we obtained the results shown in the table.

| $n$ | Left Riemann Sum | Right Riemann Sum |
|------|------------------|-------------------|
| 10   | 8.100            | 9.900             |
| 100  | 8.910            | 9.090             |
| 1000 | 8.991            | 9.009             |

From the results of Example 4, it appears that the Riemann sums are approaching the limit 9 as $n$ approaches infinity. It is this observation that motivates the following definition of a **definite integral.** In this definition, consider the partition of $[a, b]$ into $n$ subintervals of equal width $\Delta x = (b - a)/n$ as follows.

$$a = x_0 < x_1 < x_2 < \cdots < x_{n-1} < x_n = b$$

Moreover, consider $c_i$ to be an arbitrary point in the $i$th subinterval $[x_{i-1}, x_i]$. To say that the number of subintervals $n$ tends to infinity is equivalent to saying that the width, $\Delta x$, of the subintervals tends to zero.

### Definition of Definite Integral

If $f$ is a continuous function defined on the closed interval $[a, b]$, then the **definite integral of $f$ on $[a, b]$** is

$$\int_a^b f(x)\, dx = \lim_{\Delta x \to 0} \sum_{i=1}^n f(c_i)\Delta x = \lim_{n \to \infty} \sum_{i=1}^n f(c_i)\Delta x.$$

If $f$ is continuous and nonnegative on the interval $[a, b]$, then the definite integral of $f$ on $[a, b]$ gives the area of the region bounded by the graph of $f$, the $x$-axis, and the vertical lines $x = a$ and $x = b$.

Evaluation of a definite integral by its limit definition can be difficult. However, there are times when a definite integral can be solved by recognizing that it represents the area of a common type of geometric figure.

## EXAMPLE 5  The Areas of Common Geometric Figures

Sketch the region corresponding to each of the following definite integrals. Then evaluate each definite integral using a geometric formula.

(a) $\displaystyle\int_1^3 4\, dx$     (b) $\displaystyle\int_0^3 (x + 2)\, dx$     (c) $\displaystyle\int_{-2}^2 \sqrt{4 - x^2}\, dx$

*Solution*

A sketch of each region is shown in Figure A.4 (page A7).

(a) The region associated with this definite integral is a rectangle of height 4 and width 2. Moreover, because the function $f(x) = 4$ is continuous and nonnegative on the interval $[1, 3]$, you can conclude that the area of the rectangle is given by the definite integral. Thus, the value of the definite integral is

$$\int_1^3 4\, dx = 4(2) = 8.$$

(b) The region associated with this definite integral is a trapezoid with an altitude of 3 and parallel bases of lengths 2 and 5. The formula for the area of a trapezoid is $\frac{1}{2}h(b_1 + b_2)$, and so you have

$$\int_0^3 (x + 2)\, dx = \frac{1}{2}(3)(2 + 5) = \frac{21}{2}.$$

(c) The region associated with this definite integral is a semicircle of radius 2. Thus, the area is $\frac{1}{2}\pi r^2$, and you have

$$\int_{-2}^2 \sqrt{4 - x^2}\, dx = \frac{1}{2}\pi(2^2) = 2\pi.$$

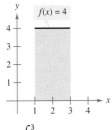

(a) $\int_1^3 4\, dx$

Rectangle

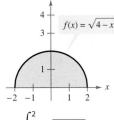

(b) $\int_0^3 (x + 2)\, dx$

Trapezoid

(c) $\int_{-2}^2 \sqrt{4 - x^2}\, dx$

Semicircle

**FIGURE A.4**

For some simple functions it is possible to evaluate definite integrals by the Riemann sum definition. In the next example, you will use the fact that the sum of the first $n$ integers is given by the formula

$$1 + 2 + \cdots + n = \sum_{i=1}^n i = \frac{n(n + 1)}{2} \qquad \text{See Exercise 27.}$$

to compute the area of the triangular region of Examples 3 and 4.

**EXAMPLE 6   Evaluating a Definite Integral by Its Definition**

Evaluate the definite integral $\int_0^3 2x\, dx$.

*Solution*

Let $\Delta x = (b - a)/n = 3/n$, and choose $c_i$ to be the right endpoint of each subinterval, $c_i = 3i/n$. Then you have

$$\int_0^3 2x\,dx = \lim_{\Delta x \to 0} \sum_{i=1}^{n} f(c_i)\Delta x$$

$$= \lim_{n \to \infty} \sum_{i=1}^{n} 2\left(i\frac{3}{n}\right)\left(\frac{3}{n}\right)$$

$$= \lim_{n \to \infty} \frac{18}{n^2} \sum_{i=1}^{n} i$$

$$= \lim_{n \to \infty} \left(\frac{18}{n^2}\right)\left(\frac{n(n+1)}{2}\right)$$

$$= \lim_{n \to \infty} \left(9 + \frac{9}{n}\right).$$

This limit can be evaluated in the same way that you calculated horizontal asymptotes in Section 3.6. In particular, as $n$ approaches infinity, you see that $9/n$ approaches 0, and the above limit is 9. Thus, you can conclude that

$$\int_0^3 2x\,dx = 9.$$

From Example 6, you can see that it can be difficult to evaluate the definite integral of even a simple function by using Riemann sums. A computer can help in calculating these sums for large values of $n$, but this procedure would only give an approximation of the definite integral. Fortunately, the **Fundamental Theorem of Calculus** provides a technique for evaluating definite integrals using antiderivatives, and for this reason it is often thought to be the most important theorem in calculus. In the remainder of this appendix, you will see how derivatives and integrals are related via the Fundamental Theorem of Calculus.

To simplify the discussion, assume that $f$ is a continuous nonnegative function defined on the interval $[a, b]$. Let $A(x)$ be the area of the region under the graph of $f$ from $a$ to $x$, as indicated in Figure A.5. The area under the shaded region in Figure A.6 is thus $A(x + \Delta x) - A(x)$.

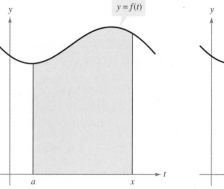

**FIGURE A.5**

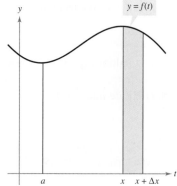

**FIGURE A.6**

If $\Delta x$ is small, then this area is approximately given by the area of the rectangle of height $f(x)$ and width $\Delta x$. Thus, you have $A(x + \Delta x) - A(x) \approx f(x)\Delta x$. Dividing by $\Delta x$ produces

$$f(x) \approx \frac{A(x + \Delta x) - A(x)}{\Delta x}.$$

By taking the limit as $\Delta x$ approaches 0, you see that

$$f(x) = \lim_{\Delta x \to 0} \frac{A(x + \Delta x) - A(x)}{\Delta x} = A'(x)$$

and you have established the fact that the area function $A(x)$ is an antiderivative of $f$. Although it was assumed that $f$ is continuous and nonnegative, this development is valid if the function $f$ is simply continuous on the closed interval $[a, b]$. We use this result in the proof of the Fundamental Theorem of Calculus.

---

### Fundamental Theorem of Calculus

If $f$ is a continuous function on the closed interval $[a, b]$, then

$$\int_a^b f(x)dx = F(b) - F(a),$$

where $F$ is any function such that $F'(x) = f(x)$.

---

*Proof*

From the previous discussion, you know that

$$\int_a^x f(x)\,dx = A(x)$$

and in particular

$$A(a) = \int_a^a f(x)\,dx = 0 \quad \text{and} \quad A(b) = \int_a^b f(x)\,dx.$$

If $F$ is *any* antiderivative of $f$, then you know that $F$ differs from $A$ by a constant. That is, $A(x) = F(x) + C$. Hence,

$$\int_a^b f(x)\,dx = A(b) - A(a) = [F(b) + C] - [F(a) + C] = F(b) - F(a).$$

You are now ready to continue Section 5.4, on page 342, just after the statement of the Fundamental Theorem of Calculus.

## APPENDIX A EXERCISES

In Exercises 1–6, use the left Riemann sum and the right Riemann sum to approximate the area of the given region using the indicated number of subintervals.

**1.** $y = \sqrt{x}$

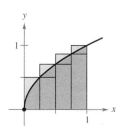

**2.** $y = \sqrt{x} + 1$

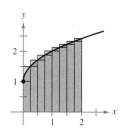

**3.** $y = \dfrac{1}{x}$

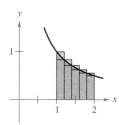

**4.** $y = \dfrac{1}{x - 2}$

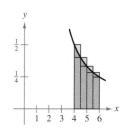

**5.** $y = \sqrt{1 - x^2}$

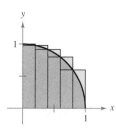

**6.** $y = \sqrt{x + 1}$

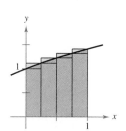

**7.** Repeat Exercise 1 using the midpoint Riemann sum.

**8.** Repeat Exercise 2 using the midpoint Riemann sum.

**9.** Consider a triangle of area 2 bounded by the graphs of $y = x$, $y = 0$, and $x = 2$.

   (a) Sketch the graph of the region.

(b) Divide the interval $[0, 2]$ into $n$ equal subintervals and show that the endpoints are

$$0 < 1\left(\frac{2}{n}\right) < \cdots < (n - 1)\left(\frac{2}{n}\right) < n\left(\frac{2}{n}\right).$$

(c) Show that the left Riemann sum is

$$S_L = \sum_{i=1}^{n} \left[(i - 1)\left(\frac{2}{n}\right)\right]\left(\frac{2}{n}\right).$$

(d) Show that the right Riemann sum is

$$S_R = \sum_{i=1}^{n} \left[i\left(\frac{2}{n}\right)\right]\left(\frac{2}{n}\right).$$

(e) Complete the following table.

| $n$ | 5 | 10 | 50 | 100 |
|---|---|---|---|---|
| Left sum $S_L$ | | | | |
| Right sum $S_R$ | | | | |

(f) Show that $\lim_{n \to \infty} S_L = \lim_{n \to \infty} S_R = 2$.

**10.** Consider a trapezoid of area 4 bounded by the graphs of $y = x$, $y = 0$, $x = 1$, and $x = 3$.

   (a) Sketch the graph of the region.
   (b) Divide the interval $[1, 3]$ into $n$ equal subintervals and show that the endpoints are

$$1 < 1 + 1\left(\frac{2}{n}\right) < \cdots < 1 + (n - 1)\left(\frac{2}{n}\right) < 1 + n\left(\frac{2}{n}\right).$$

(c) Show that the left Riemann sum is

$$S_L = \sum_{i=1}^{n} \left[1 + (i - 1)\left(\frac{2}{n}\right)\right]\left(\frac{2}{n}\right).$$

(d) Show that the right Riemann sum is

$$S_R = \sum_{i=1}^{n} \left[1 + i\left(\frac{2}{n}\right)\right]\left(\frac{2}{n}\right).$$

(e) Complete the following table.

| $n$ | 5 | 10 | 50 | 100 |
|---|---|---|---|---|
| Left sum $S_L$ | | | | |
| Right sum $S_R$ | | | | |

(f) Show that $\lim_{n \to \infty} S_L = \lim_{n \to \infty} S_R = 4$.

In Exercises 11–16, set up a definite integral that yields the area of the given region. (Do not evaluate the integral.)

**11.** $f(x) = 3$

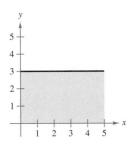

**12.** $f(x) = 4 - 2x$

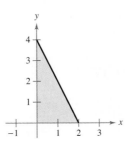

**13.** $f(x) = 4 - |x|$

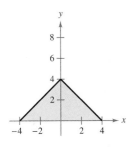

**14.** $f(x) = x^2$

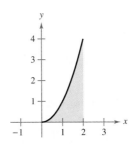

**15.** $f(x) = 4 - x^2$

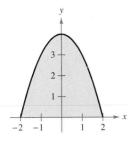

**16.** $f(x) = \dfrac{1}{x^2 + 1}$

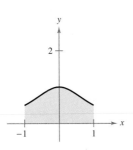

In Exercises 17–26, sketch the region whose area is indicated by the given definite integral. Then use a geometric formula to evaluate the integral.

**17.** $\displaystyle\int_0^3 4 \, dx$

**18.** $\displaystyle\int_{-a}^{a} 4 \, dx$

**19.** $\displaystyle\int_0^4 x \, dx$

**20.** $\displaystyle\int_0^4 \dfrac{x}{2} \, dx$

**21.** $\displaystyle\int_0^2 (2x + 5) \, dx$

**22.** $\displaystyle\int_0^5 (5 - x) \, dx$

**23.** $\displaystyle\int_{-1}^{1} \left(1 - |x|\right) dx$

**24.** $\displaystyle\int_{-a}^{a} \left(a - |x|\right) dx$

**25.** $\displaystyle\int_{-3}^{3} \sqrt{9 - x^2} \, dx$

**26.** $\displaystyle\int_{-r}^{r} \sqrt{r^2 - x^2} \, dx$

**27.** Show that $\displaystyle\sum_{i=1}^{n} i = \dfrac{n(n + 1)}{2}.$ (*Hint:* Add the following two sums.)

$$S = 1 + 2 + 3 + \cdots + (n - 2) + (n - 1) + n$$
$$S = n + (n - 1) + (n - 2) + \cdots + 3 + 2 + 1$$

**28.** Use the Riemann sum definition of the definite integral and the result of Exercise 27 to evaluate $\int_1^2 x \, dx$.

# Appendix B:  Formulas, Properties, and Measurements

*Review of Algebra, Geometry, and Trigonometry • Differentiation and Integration Formulas • Formulas from Business and Finance • Units of Measurements • Standard Normal Distribution*

## B.1   REVIEW OF ALGEBRA, GEOMETRY, AND TRIGONOMETRY

Algebra • Properties of Logarithms • Geometry • Plane Analytic Geometry • Solid Analytic Geometry • Trigonometry • Library of Functions

## Algebra

### Operations with Exponents

1. $x^n x^m = x^{n+m}$    2. $\dfrac{x^n}{x^m} = x^{n-m}$    3. $(xy)^n = x^n y^n$

4. $\left(\dfrac{x}{y}\right)^n = \dfrac{x^n}{y^n}$    5. $(x^n)^m = x^{nm}$    6. $-x^n = -(x^n)$

7. $cx^n = c(x^n)$    8. $x^{n^m} = x^{(n^m)}$

### Exponents and Radicals ($n$ and $m$ are positive integers)

1. $x^n = \underbrace{x \cdot x \cdot x \cdots x}_{n \text{ factors}}$    2. $x^0 = 1,\ x \neq 0$

3. $x^{-n} = \dfrac{1}{x^n},\ x \neq 0$    *4. $\sqrt[n]{x} = a \implies x = a^n$

5. $x^{1/n} = \sqrt[n]{x}$    6. $x^{m/n} = (x^{1/n})^m = \left(\sqrt[n]{x}\right)^m$
$\qquad\qquad\qquad\qquad x^{m/n} = (x^m)^{1/n} = \sqrt[n]{x^m}$

7. $\sqrt[2]{x} = \sqrt{x}$

### Operations with Fractions

1. $\dfrac{a}{b} + \dfrac{c}{d} = \dfrac{a}{b}\left(\dfrac{d}{d}\right) + \dfrac{c}{d}\left(\dfrac{b}{b}\right) = \dfrac{ad}{bd} + \dfrac{bc}{bd} = \dfrac{ad + bc}{bd}$

2. $\dfrac{a}{b} - \dfrac{c}{d} = \dfrac{a}{b}\left(\dfrac{d}{d}\right) - \dfrac{c}{d}\left(\dfrac{b}{b}\right) = \dfrac{ad}{bd} - \dfrac{bc}{bd} = \dfrac{ad - bc}{bd}$

\* If $n$ is even, the principal $n$th root is defined to be positive.

3. $\left(\dfrac{a}{b}\right)\left(\dfrac{c}{d}\right) = \dfrac{ac}{bd}$

4. $\dfrac{a/b}{c/d} = \left(\dfrac{a}{b}\right)\left(\dfrac{d}{c}\right) = \dfrac{ad}{bc}$

$\dfrac{a/b}{c} = \dfrac{a/b}{c/1} = \left(\dfrac{a}{b}\right)\left(\dfrac{1}{c}\right) = \dfrac{a}{bc}$

5. $\dfrac{\cancel{a}b}{\cancel{a}c} = \dfrac{b}{c}$

$\dfrac{ab + ac}{ad} = \dfrac{\cancel{a}(b + c)}{\cancel{a}d} = \dfrac{b + c}{d}$

## Quadratic Formula

$ax^2 + bx + c = 0 \quad \Longrightarrow \quad x = \dfrac{-b \pm \sqrt{b^2 - 4ac}}{2a}$

## Factors and Special Products

1. $x^2 - a^2 = (x - a)(x + a)$

2. $x^3 - a^3 = (x - a)(x^2 + ax + a^2)$

3. $x^3 + a^3 = (x + a)(x^2 - ax + a^2)$

4. $x^4 - a^4 = (x - a)(x + a)(x^2 + a^2)$

## Factoring by Grouping

$acx^3 + adx^2 + bcx + bd = ax^2(cx + d) + b(cx + d) = (ax^2 + b)(cx + d)$

## Binomial Theorem

1. $(x + a)^2 = x^2 + 2ax + a^2$

2. $(x - a)^2 = x^2 - 2ax + a^2$

3. $(x + a)^3 = x^3 + 3ax^2 + 3a^2x + a^3$

4. $(x - a)^3 = x^3 - 3ax^2 + 3a^2x - a^3$

5. $(x + a)^4 = x^4 + 4ax^3 + 6a^2x^2 + 4a^3x + a^4$

6. $(x - a)^4 = x^4 - 4ax^3 + 6a^2x^2 - 4a^3x + a^4$

7. $(x + a)^n = x^n + nax^{n-1} + \dfrac{n(n - 1)}{2!}a^2x^{n-2} + \dfrac{n(n - 1)(n - 2)}{3!}a^3x^{n-3} +$

$\quad \cdots + na^{n-1}x + a^n$

## Algebra *(Continued)*

8. $(x - a)^n = x^n - nax^{n-1} + \dfrac{n(n-1)}{2!}a^2x^{n-2} - \dfrac{n(n-1)(n-2)}{3!}a^3x^{n-3} +$
$\cdots \pm na^{n-1}x \mp a^n$

### Miscellaneous

1. If $ab = 0$, then $a = 0$ or $b = 0$.

2. If $ac = bc$ and $c \neq 0$, then $a = b$.

3. Factorial: $0! = 1, 1! = 1, 2! = 2 \cdot 1, 3! = 3 \cdot 2 \cdot 1, 4! = 4 \cdot 3 \cdot 2 \cdot 1$, etc.

### Sequences

1. Arithmetic: $a, a + b, a + 2b, a + 3b, a + 4b, a + 5b, \ldots$

2. Geometric: $ar^0, ar^1, ar^2, ar^3, ar^4, ar^5, \ldots$

$$ar^0 + ar^1 + ar^2 + ar^3 + \cdots + ar^n = \dfrac{a(1 - r^{n+1})}{1 - r}$$

3. General harmonic: $\dfrac{1}{a}, \dfrac{1}{a + b}, \dfrac{1}{a + 2b}, \dfrac{1}{a + 3b}, \dfrac{1}{a + 4b}, \dfrac{1}{a + 5b}, \ldots$

4. Harmonic: $\dfrac{1}{1}, \dfrac{1}{2}, \dfrac{1}{3}, \dfrac{1}{4}, \dfrac{1}{5}, \ldots$

5. $p$-Sequence: $\dfrac{1}{1^p}, \dfrac{1}{2^p}, \dfrac{1}{3^p}, \dfrac{1}{4^p}, \dfrac{1}{5^p}, \ldots$

### Series

$$\dfrac{1}{x} = 1 - (x - 1) + (x - 1)^2 - (x - 1)^3 + (x - 1)^4 - \cdots + (-1)^n(x - 1)^n + \cdots, \quad 0 < x < 2$$

$$\dfrac{1}{1 + x} = 1 - x + x^2 - x^3 + x^4 - x^5 + \cdots + (-1)^n x^n + \cdots, \qquad\qquad -1 < x < 1$$

$$\ln x = (x - 1) - \dfrac{(x - 1)^2}{2} + \dfrac{(x - 1)^3}{3} - \dfrac{(x - 1)^4}{4} + \cdots + \dfrac{(-1)^{n-1}(x - 1)^n}{n} + \cdots, \quad 0 < x \leq 2$$

$$e^x = 1 + x + \dfrac{x^2}{2!} + \dfrac{x^3}{3!} + \dfrac{x^4}{4!} + \dfrac{x^5}{5!} + \cdots + \dfrac{x^n}{n!} + \cdots, \qquad\qquad -\infty < x < \infty$$

$$\sin x = x - \dfrac{x^3}{3!} + \dfrac{x^5}{5!} - \dfrac{x^7}{7!} + \cdots, \qquad\qquad -\infty < x < \infty$$

$$\cos x = 1 - \frac{x^2}{2!} + \frac{x^4}{4!} - \frac{x^6}{6!} + \cdots, \qquad\qquad -\infty < x < \infty$$

$$(1 + x)^k = 1 + kx + \frac{k(k-1)x^2}{2!} + \frac{k(k-1)(k-2)x^3}{3!} + \frac{k(k-1)(k-2)(k-3)x^4}{4!} + \cdots, \quad -1 < x < 1^*$$

$$(1 + x)^{-k} = 1 - kx + \frac{k(k+1)x^2}{2!} - \frac{k(k+1)(k+2)x^3}{3!} + \frac{k(k+1)(k+2)(k+3)x^4}{4!} - \cdots, \quad -1 < x < 1^*$$

## Properties of Logarithms

### Inverse Properties

1. $\ln e^x = x$  
2. $e^{\ln x} = x$

### Properties of Logarithms

1. $\ln 1 = 0$  
2. $\ln e = 1$

3. $\ln xy = \ln x + \ln y$  
4. $\ln \dfrac{x}{y} = \ln x - \ln y$

5. $\ln x^y = y \ln x$  
6. $\log_a x = \dfrac{\ln x}{\ln a}$

## Geometry

### Triangles

1. General triangle

   Sum of triangles $\alpha + \beta + \theta = 180°$

   Area $= \dfrac{1}{2}(\text{base})(\text{height}) = \dfrac{1}{2}bh$

2. Similar triangles

   $\dfrac{a}{b} = \dfrac{A}{B}$

3. Right triangle

   $c^2 = a^2 + b^2$ (Pythagorean Theorem)

   Sum of acute angles $\alpha + \beta = 90°$

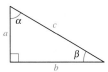

$^*$The convergence at $x = \pm 1$ depends on the value of $k$.

# Geometry   *(Continued)*

4. Equilateral triangle

$$\text{Height} = h = \frac{\sqrt{3}s}{2}$$

$$\text{Area} = \frac{\sqrt{3}s^2}{4}$$

5. Isosceles right triangle

$$\text{Area} = \frac{s^2}{2}$$

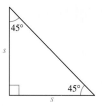

**Quadrilaterals (Four-Sided Figures)**

1. Rectangle

$$\text{Area} = (\text{length})(\text{height}) = bh$$

2. Square

$$\text{Area} = (\text{side})^2 = s^2$$

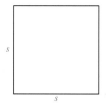

3. Parallelogram

$$\text{Area} = bh$$

4. Trapezoid

$$\text{Area} = \frac{1}{2}h(a + b)$$

## Circles and Ellipses

1. Circle

  $$\text{Area} = \pi r^2$$

  $$\text{Circumference} = 2\pi r$$

2. Sector of circle ($\theta$ in radians)

  $$\text{Area} = \frac{\theta r^2}{2}$$

  $$s = r\theta$$

3. Circular ring

  $$\text{Area} = \pi(R^2 - r^2)$$

4. Ellipse

  $$\text{Area} = \pi ab$$

  $$\text{Circumference} = 2\pi \sqrt{\frac{a^2 + b^2}{2}}$$

## Solid Figures

1. Cone ($A = $ area of base)

  $$\text{Volume} = \frac{Ah}{3}$$

2. Right circular cone

  $$\text{Volume} = \frac{\pi r^2 h}{3}$$

  $$\text{Lateral surface area} = \pi r \sqrt{r^2 + h^2}$$

3. Frustum of right circular cone

  $$\text{Volume} = \frac{\pi(r^2 + rR + R^2)h}{3}$$

  $$\text{Lateral surface area} = \pi s(R + r)$$

## Geometry   *(Continued)*

4.  Right circular cylinder

Volume $= \pi r^2 h$

Lateral surface area $= 2\pi rh$

5.  Sphere

Volume $= \dfrac{4}{3}\pi r^3$

Surface area $= 4\pi r^2$

## Plane Analytic Geometry

**Distance Between $(x_1, y_1)$ and $(x_2, y_2)$**

$$d = \sqrt{(x_2 - x_1)^2 + (y_2 - y_1)^2}$$

**Midpoint Between $(x_1, y_1)$ and $(x_2, y_2)$**

$$\text{Midpoint} = \left(\frac{x_1 + x_2}{2}, \frac{y_1 + y_2}{2}\right)$$

**Slope of Line Passing Through $(x_1, y_1)$ and $(x_2, y_2)$**

$$m = \frac{y_2 - y_1}{x_2 - x_1}$$

**Slopes of Parallel Lines**

$$m_1 = m_2$$

**Slopes of Perpendicular Lines**

$$m_1 = -\frac{1}{m_2}$$

**Equations of Lines**

*Point-slope form:* $y - y_1 = m(x - x_1)$     *General form:* $Ax + By + C = 0$

*Vertical line:* $x = a$     *Horizontal line:* $y = b$

**Equations of Circles** [Center: $(h, k)$, Radius: $r$]

*Standard form:* $(x - h)^2 + (y - k)^2 = r^2$

*General form:* $Ax^2 + Ay^2 + Dx + Ey + F = 0$

**Equations of Parabolas** [Vertex: $(h, k)$]

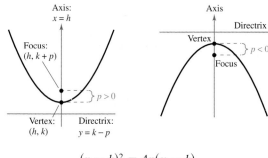
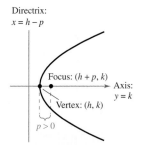
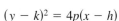

$$(x - h)^2 = 4p(y - k)$$ $$(y - k)^2 = 4p(x - h)$$

(a) Vertical axis: $p > 0$    (b) Vertical axis: $p < 0$    (c) Horizontal axis: $p > 0$    (d) Horizontal axis: $p < 0$

**Equations of Ellipses** [Center: $(h, k)$]

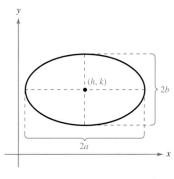
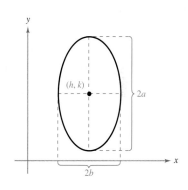

$$\frac{(x - h)^2}{a^2} + \frac{(y - k)^2}{b^2} = 1$$ $$\frac{(x - h)^2}{b^2} + \frac{(y - k)^2}{a^2} = 1$$

## Plane Analytic Geometry   *(Continued)*

**Equations of Hyperbolas [Center: $(h, k)$]**

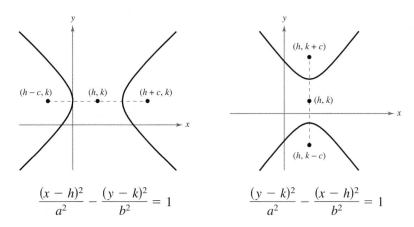

$$\frac{(x - h)^2}{a^2} - \frac{(y - k)^2}{b^2} = 1 \qquad\qquad \frac{(y - k)^2}{a^2} - \frac{(x - h)^2}{b^2} = 1$$

## Solid Analytic Geometry

**Distance Between $(x_1, y_1, z_1)$ and $(x_2, y_2, z_2)$**

$$d = \sqrt{(x_2 - x_1)^2 + (y_2 - y_1)^2 + (z_2 - z_1)^2}$$

**Midpoint Between $(x_1, y_1, z_1)$ and $(x_2, y_2, z_2)$**

$$\text{Midpoint} = \left(\frac{x_1 + x_2}{2}, \frac{y_1 + y_2}{2}, \frac{z_1 + z_2}{2}\right)$$

**Equation of Plane**

$$Ax + By + Cz + D = 0$$

**Equation of Sphere [Center: $(h, k, l)$,  Radius: $r$]**

$$(x - h)^2 + (y - k)^2 + (z - l)^2 = r^2$$

## Trigonometry

**Definitions of the Six Trigonometric Functions**

*Right triangle definition:* $0 < \theta < \pi/2$

$$\sin \theta = \frac{\text{opp.}}{\text{hyp.}} \qquad \csc \theta = \frac{\text{hyp.}}{\text{opp.}}$$

$$\cos \theta = \frac{\text{adj.}}{\text{hyp.}} \qquad \sec \theta = \frac{\text{hyp.}}{\text{adj.}}$$

$$\tan \theta = \frac{\text{opp.}}{\text{adj.}} \qquad \cot \theta = \frac{\text{adj.}}{\text{opp.}}$$

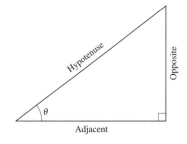

*Circular function definition:* $\theta$ is any angle and $(x, y)$ is a point on the terminal ray of the angle.

$$\sin \theta = \frac{y}{r} \qquad \csc \theta = \frac{r}{y}$$

$$\cos \theta = \frac{x}{r} \qquad \sec \theta = \frac{r}{x}$$

$$\tan \theta = \frac{y}{x} \qquad \cot \theta = \frac{x}{y}$$

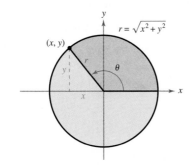

## Signs of the Trigonometric Functions by Quadrant

| Quadrant | sin | cos | tan | cot | sec | csc |
|----------|-----|-----|-----|-----|-----|-----|
| I | $+$ | $+$ | $+$ | $+$ | $+$ | $+$ |
| II | $+$ | $-$ | $-$ | $-$ | $-$ | $+$ |
| III | $-$ | $-$ | $+$ | $+$ | $-$ | $-$ |
| IV | $-$ | $+$ | $-$ | $-$ | $+$ | $-$ |

## Trigonometric Identities

*Reciprocal identities*

$$\sin \theta = \frac{1}{\csc \theta} \qquad \cos \theta = \frac{1}{\sec \theta} \qquad \tan \theta = \frac{1}{\cot \theta}$$

$$\csc \theta = \frac{1}{\sin \theta} \qquad \sec \theta = \frac{1}{\cos \theta} \qquad \cot \theta = \frac{1}{\tan \theta}$$

$$\tan \theta = \frac{\sin \theta}{\cos \theta} \qquad \cot \theta = \frac{\cos \theta}{\sin \theta}$$

*Pythagorean identities*

$$\sin^2 \theta + \cos^2 \theta = 1 \qquad \tan^2 \theta + 1 = \sec^2 \theta \qquad \cot^2 \theta + 1 = \csc^2 \theta$$

*Reduction formulas*

$$\sin(-\theta) = -\sin \theta \qquad \cos(-\theta) = \cos \theta \qquad \tan(-\theta) = -\tan \theta$$

$$\sin \theta = -\sin(\theta - \pi) \qquad \cos \theta = -\cos(\theta - \pi) \qquad \tan \theta = \tan(\theta - \pi)$$

## Trigonometry    *(Continued)*

### Sum or difference of two angles

$$\sin(\theta \pm \phi) = \sin\theta\cos\phi \pm \cos\theta\sin\phi$$

$$\cos(\theta \pm \phi) = \cos\theta\cos\phi \mp \sin\theta\sin\phi$$

$$\tan(\theta \pm \phi) = \frac{\tan\theta \pm \tan\phi}{1 \mp \tan\theta\tan\phi}$$

$$\sin(\theta + \phi)\sin(\theta - \phi) = \sin^2\theta - \sin^2\phi$$

$$\cos(\theta + \phi)\cos(\theta - \phi) = \cos^2\theta - \sin^2\phi$$

### Double-angle identities

$$\sin 2\theta = 2\sin\theta\cos\theta$$

$$\cos 2\theta = 2\cos^2\theta - 1 = 1 - 2\sin^2\theta$$

$$\tan 2\theta = \frac{2\tan\theta}{1 - \tan^2\theta}$$

### Multiple-angle identities

$$\sin 3\theta = 3\sin\theta - 4\sin^3\theta$$

$$\cos 3\theta = -3\cos\theta + 4\cos^3\theta$$

$$\tan 3\theta = \frac{3\tan\theta - \tan^2\theta}{1 - 3\tan^2\theta}$$

$$\sin 4\theta = 4\sin\theta\cos\theta - 8\sin^3\theta\cos\theta$$

$$\cos 4\theta = 8\cos^4\theta - 8\cos^2\theta + 1$$

$$\tan 4\theta = \frac{4\tan\theta - 4\tan^3\theta}{1 - 6\tan^2\theta + \tan^4\theta}$$

### Half-angle identities

$$\sin^2\theta = \frac{1}{2}(1 - \cos 2\theta)$$

$$\cos^2\theta = \frac{1}{2}(1 + \cos 2\theta)$$

### Product identities

$$\sin\theta\sin\phi = \frac{1}{2}\cos(\theta - \phi) - \frac{1}{2}\cos(\theta + \phi)$$

$$\cos\theta\cos\phi = \frac{1}{2}\cos(\theta - \phi) + \frac{1}{2}\cos(\theta + \phi)$$

$$\cos\theta\sin\phi = \frac{1}{2}\sin(\theta + \phi) - \frac{1}{2}\sin(\theta - \phi)$$

# Library of Functions

**Algebraic Functions**

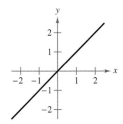

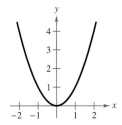

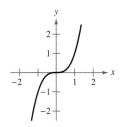

Linear or First-Degree Polynomial
$f(x) = x$
$f(x) = x^2$

Quadratic or Second-Degree Polynomial

Cubic or Third-Degree Polynomial
$f(x) = x^3$

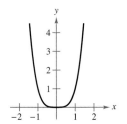

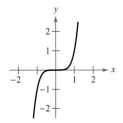

Fourth-Degree Polynomial
$f(x) = x^4$

Fifth-Degree Polynomial
$f(x) = x^5$

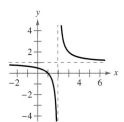

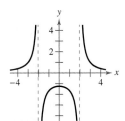

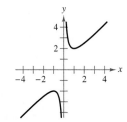

Rational Function
$f(x) = \dfrac{x - 1}{x - 2}$

Rational Function
$f(x) = \dfrac{5}{x^2 - 4}$

Rational Function
$f(x) = \dfrac{x^2 + 1}{x}$

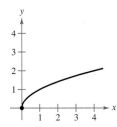

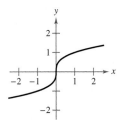

Square Root Function
$f(x) = \sqrt{x}$

Cube Root Function
$f(x) = \sqrt[3]{x}$

# Library of Functions *(Continued)*

### Exponential and Logarithmic Functions

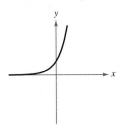

Exponential Function
$f(x) = a^x, a > 1$

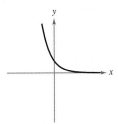

Exponential Function
$f(x) = a^x, 0 < a < 1$

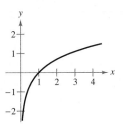

Logarithmic Function
$f(x) = \ln x$

### Trigonometric Functions

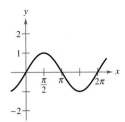

Sine Function
$f(x) = \sin x$

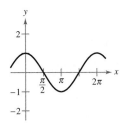

Cosine Function
$f(x) = \cos x$

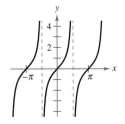

Tangent Function
$f(x) = \tan x$

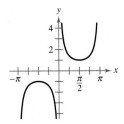

Cosecant Function
$f(x) = \csc x$

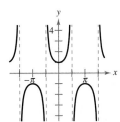

Secant Function
$f(x) = \sec x$

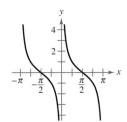

Cotangent Function
$f(x) = \cot x$

### Nonelementary Functions

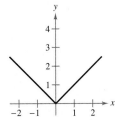

Absolute Value Function
$f(x) = |x|$

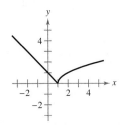

Compound Function
$$f(x) = \begin{cases} 1 - x, & x < 1 \\ \sqrt{x - 1}, & x \geq 1 \end{cases}$$

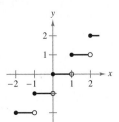

Step Function
$f(x) = [\![x]\!]$

## B.2 DIFFERENTIATION AND INTEGRATION FORMULAS

Differentiation Formulas • Integration Formulas

## Differentiation Formulas

1. $\dfrac{d}{dx}[cu] = cu'$

2. $\dfrac{d}{dx}[u \pm v] = u' \pm v'$

3. $\dfrac{d}{dx}[uv] = uv' + vu'$

4. $\dfrac{d}{dx}\left[\dfrac{u}{v}\right] = \dfrac{vu' - uv'}{v^2}$

5. $\dfrac{d}{dx}[c] = 0$

6. $\dfrac{d}{dx}[u^n] = nu^{n-1}u'$

7. $\dfrac{d}{dx}[x] = 1$

8. $\dfrac{d}{dx}[|u|] = \dfrac{u}{|u|}(u'), \ u \neq 0$

9. $\dfrac{d}{dx}[\ln u] = \dfrac{u'}{u}$

10. $\dfrac{d}{dx}[e^u] = e^u u'$

11. $\dfrac{d}{dx}[\sin u] = (\cos u)u'$

12. $\dfrac{d}{dx}[\cos u] = -(\sin u)u'$

13. $\dfrac{d}{dx}[\tan u] = (\sec^2 u)u'$

14. $\dfrac{d}{dx}[\cot u] = -(\csc^2 u)u'$

15. $\dfrac{d}{dx}[\sec u] = (\sec u \tan u)u'$

16. $\dfrac{d}{dx}[\csc u] = -(\csc u \cot u)u'$

## Integration Formulas

**Forms Involving $u^n$**

1. $\displaystyle\int u^n \, du = \dfrac{u^{n+1}}{n+1} + C, \quad n \neq -1$

2. $\displaystyle\int \dfrac{1}{u} \, du = \ln |u| + C$

**Forms Involving $a + bu$**

3. $\displaystyle\int \dfrac{u}{a + bu} \, du = \dfrac{1}{b^2}(bu - a \ln|a + bu|) + C$

4. $\displaystyle\int \dfrac{u}{(a + bu)^2} \, du = \dfrac{1}{b^2}\left(\dfrac{a}{a + bu} + \ln|a + bu|\right) + C$

5. $\displaystyle\int \dfrac{u}{(a + bu)^n} \, du = \dfrac{1}{b^2}\left[\dfrac{-1}{(n-2)(a+bu)^{n-2}} + \dfrac{a}{(n-1)(a+bu)^{n-1}}\right] + C, \quad n \neq 1, 2$

6. $\displaystyle\int \dfrac{u^2}{a + bu} \, du = \dfrac{1}{b^3}\left[-\dfrac{bu}{2}(2a - bu) + a^2 \ln|a + bu|\right] + C$

7. $\displaystyle\int \dfrac{u^2}{(a + bu)^2} \, du = \dfrac{1}{b^3}\left(bu - \dfrac{a^2}{a + bu} - 2a \ln|a + bu|\right) + C$

## Integration Formulas    *(Continued)*

8. $\displaystyle\int \frac{u^2}{(a + bu)^3}\, du = \frac{1}{b^3}\left[\frac{2a}{a + bu} - \frac{a^2}{2(a + bu)^2} + \ln|a + bu|\right] + C$

9. $\displaystyle\int \frac{u^2}{(a + bu)^n}\, du = \frac{1}{b^3}\left[\frac{-1}{(n - 3)(a + bu)^{n-3}} + \frac{2a}{(n - 2)(a + bu)^{n-2}} - \frac{a^2}{(n - 1)(a + bu)^{n-1}}\right] + C, \quad n \neq 1, 2, 3$

10. $\displaystyle\int \frac{1}{u(a + bu)}\, du = \frac{1}{a}\ln\left|\frac{u}{a + bu}\right| + C$

11. $\displaystyle\int \frac{1}{u(a + bu)^2}\, du = \frac{1}{a}\left(\frac{1}{a + bu} + \frac{1}{a}\ln\left|\frac{u}{a + bu}\right|\right) + C$

12. $\displaystyle\int \frac{1}{u^2(a + bu)}\, du = -\frac{1}{a}\left(\frac{1}{u} + \frac{b}{a}\ln\left|\frac{u}{a + bu}\right|\right) + C$

13. $\displaystyle\int \frac{1}{u^2(a + bu)^2}\, du = -\frac{1}{a^2}\left[\frac{a + 2bu}{u(a + bu)} + \frac{2b}{a}\ln\left|\frac{u}{a + bu}\right|\right] + C$

**Forms Involving** $\sqrt{a + bu}$

14. $\displaystyle\int u^n \sqrt{a + bu}\, du = \frac{2}{b(2n + 3)}\left[u^n(a + bu)^{3/2} - na\int u^{n-1}\sqrt{a + bu}\, du\right]$

15. $\displaystyle\int \frac{1}{u\sqrt{a + bu}}\, du = \frac{1}{\sqrt{a}}\ln\left|\frac{\sqrt{a + bu} - \sqrt{a}}{\sqrt{a + bu} + \sqrt{a}}\right| + C, \quad a > 0$

16. $\displaystyle\int \frac{1}{u^n \sqrt{a + bu}}\, du = \frac{-1}{a(n - 1)}\left[\frac{\sqrt{a + bu}}{u^{n-1}} + \frac{(2n - 3)b}{2}\int \frac{1}{u^{n-1}\sqrt{a + bu}}\, du\right], \quad n \neq 1$

17. $\displaystyle\int \frac{\sqrt{a + bu}}{u}\, du = 2\sqrt{a + bu} + a\int \frac{1}{u\sqrt{a + bu}}\, du$

18. $\displaystyle\int \frac{\sqrt{a + bu}}{u^n}\, du = \frac{-1}{a(n - 1)}\left[\frac{(a + bu)^{3/2}}{u^{n-1}} + \frac{(2n - 5)b}{2}\int \frac{\sqrt{a + bu}}{u^{n-1}}\, du\right], \quad n \neq 1$

19. $\displaystyle\int \frac{u}{\sqrt{a + bu}}\, du = -\frac{2(2a - bu)}{3b^2}\sqrt{a + bu} + C$

20. $\displaystyle\int \frac{u^n}{\sqrt{a + bu}}\, du = \frac{2}{(2n + 1)b}\left(u^n\sqrt{a + bu} - na\int \frac{u^{n-1}}{\sqrt{a + bu}}\, du\right)$

**Forms Involving** $u^2 - a^2, a > 0$

21. $\displaystyle\int \frac{1}{u^2 - a^2}\, du = -\int \frac{1}{a^2 - u^2}\, du = \frac{1}{2a}\ln\left|\frac{u - a}{u + a}\right| + C$

22. $\displaystyle\int \frac{1}{(u^2 - a^2)^n}\, du = \frac{-1}{2a^2(n - 1)}\left[\frac{u}{(u^2 - a^2)^{n-1}} + (2n - 3)\int \frac{1}{(u^2 - a^2)^{n-1}}\, du\right], \quad n \neq 1$

## Forms Involving $\sqrt{u^2 \pm a^2}$, $a > 0$

23. $\displaystyle \int \sqrt{u^2 \pm a^2}\, du = \frac{1}{2}\left( u\sqrt{u^2 \pm a^2} \pm a^2 \ln\left| u + \sqrt{u^2 \pm a^2} \right| \right) + C$

24. $\displaystyle \int u^2\sqrt{u^2 \pm a^2}\, du = \frac{1}{8}\left[ u(2u^2 \pm a^2)\sqrt{u^2 \pm a^2} - a^4 \ln\left| u + \sqrt{u^2 \pm a^2} \right| \right] + C$

25. $\displaystyle \int \frac{\sqrt{u^2 + a^2}}{u}\, du = \sqrt{u^2 + a^2} - a \ln\left| \frac{a + \sqrt{u^2 + a^2}}{u} \right| + C$

26. $\displaystyle \int \frac{\sqrt{u^2 \pm a^2}}{u^2}\, du = \frac{-\sqrt{u^2 \pm a^2}}{u} + \ln\left| u + \sqrt{u^2 \pm a^2} \right| + C$

27. $\displaystyle \int \frac{1}{\sqrt{u^2 \pm a^2}}\, du = \ln\left| u + \sqrt{u^2 \pm a^2} \right| + C$

28. $\displaystyle \int \frac{1}{u\sqrt{u^2 + a^2}}\, du = \frac{-1}{a} \ln\left| \frac{a + \sqrt{u^2 + a^2}}{u} \right| + C$

29. $\displaystyle \int \frac{u^2}{\sqrt{u^2 \pm a^2}}\, du = \frac{1}{2}\left( u\sqrt{u^2 \pm a^2} \mp a^2 \ln\left| u + \sqrt{u^2 \pm a^2} \right| \right) + C$

30. $\displaystyle \int \frac{1}{u^2\sqrt{u^2 \pm a^2}}\, du = \mp \frac{\sqrt{u^2 \pm a^2}}{a^2 u} + C$

31. $\displaystyle \int \frac{1}{(u^2 \pm a^2)^{3/2}}\, du = \frac{\pm u}{a^2\sqrt{u^2 \pm a^2}} + C$

## Forms Involving $\sqrt{a^2 - u^2}$, $a > 0$

32. $\displaystyle \int \frac{\sqrt{a^2 - u^2}}{u}\, du = \sqrt{a^2 - u^2} - a \ln\left| \frac{a + \sqrt{a^2 - u^2}}{u} \right| + C$

33. $\displaystyle \int \frac{1}{u\sqrt{a^2 - u^2}}\, du = \frac{-1}{a} \ln\left| \frac{a + \sqrt{a^2 - u^2}}{u} \right| + C$

34. $\displaystyle \int \frac{1}{u^2\sqrt{a^2 - u^2}}\, du = \frac{-\sqrt{a^2 - u^2}}{a^2 u} + C$

35. $\displaystyle \int \frac{1}{(a^2 - u^2)^{3/2}}\, du = \frac{u}{a^2\sqrt{a^2 - u^2}} + C$

## Forms Involving $e^u$

36. $\displaystyle \int e^u\, du = e^u + C$

# Integration Formulas   *(Continued)*

37. $\displaystyle\int u e^u \, du = (u - 1)e^u + C$

38. $\displaystyle\int u^n e^u \, du = u^n e^u - n \int u^{n-1} e^u \, du$

39. $\displaystyle\int \frac{1}{1 + e^u} \, du = u - \ln(1 + e^u) + C$

40. $\displaystyle\int \frac{1}{1 + e^{nu}} \, du = u - \frac{1}{n} \ln(1 + e^{nu}) + C$

## Forms Involving ln *u*

41. $\displaystyle\int \ln u \, du = u(-1 + \ln u) + C$

42. $\displaystyle\int u \ln u \, du = \frac{u^2}{4}(-1 + 2 \ln u) + C$

43. $\displaystyle\int u^n \ln u \, du = \frac{u^{n+1}}{(n+1)^2}\left[-1 + (n+1)\ln u\right] + C, \quad n \neq -1$

44. $\displaystyle\int (\ln u)^2 \, du = u[2 - 2\ln u + (\ln u)^2] + C$

45. $\displaystyle\int (\ln u)^n \, du = u(\ln u)^n - n \int (\ln u)^{n-1} \, du$

## Forms Involving sin *u* or cos *u*

46. $\displaystyle\int \sin u \, du = -\cos u + C$

47. $\displaystyle\int \cos u \, du = \sin u + C$

48. $\displaystyle\int \sin^2 u \, du = \frac{1}{2}(u - \sin u \cos u) + C$

49. $\displaystyle\int \cos^2 u \, du = \frac{1}{2}(u + \sin u \cos u) + C$

50. $\displaystyle\int \sin^n u \, du = -\frac{\sin^{n-1} u \cos u}{n} + \frac{n-1}{n} \int \sin^{n-2} u \, du$

51. $\displaystyle\int \cos^n u \, du = \frac{\cos^{n-1} u \sin u}{n} + \frac{n-1}{n} \int \cos^{n-2} u \, du$

52. $\displaystyle\int u \sin u \, du = \sin u - u \cos u + C$

53. $\displaystyle\int u \cos u \; du = \cos u + u \sin u + C$

54. $\displaystyle\int u^n \sin u \; du = -u^n \cos u + n \int u^{n-1} \cos u \; du$

55. $\displaystyle\int u^n \cos u \; du = u^n \sin u - n \int u^{n-1} \sin u \; du$

56. $\displaystyle\int \frac{1}{1 \pm \sin u} \; du = \tan u \mp \sec u + C$

57. $\displaystyle\int \frac{1}{1 \pm \cos u} \; du = -\cot u \pm \csc u + C$

58. $\displaystyle\int \frac{1}{\sin u \cos u} \; du = \ln|\tan u| + C$

**Forms Involving tan $u$, cot $u$, sec $u$, csc $u$**

59. $\displaystyle\int \tan u \; du = -\ln|\cos u| + C$

60. $\displaystyle\int \cot u \; du = \ln|\sin u| + C$

61. $\displaystyle\int \sec u \; du = \ln|\sec u + \tan u| + C$

62. $\displaystyle\int \csc u \; du = \ln|\csc u - \cot u| + C$

63. $\displaystyle\int \tan^2 u \; du = -u + \tan u + C$

64. $\displaystyle\int \cot^2 u \; du = -u - \cot u + C$

65. $\displaystyle\int \sec^2 u \; du = \tan u + C$

66. $\displaystyle\int \csc^2 u \; du = -\cot u + C$

67. $\displaystyle\int \tan^n u \; du = \frac{\tan^{n-1} u}{n - 1} - \int \tan^{n-2} u \; du, \quad n \neq 1$

68. $\displaystyle\int \cot^n u \; du = -\frac{\cot^{n-1} u}{n - 1} - \int \cot^{n-2} u \; du, \quad n \neq 1$

## Integration Formulas  *(Continued)*

69. $\displaystyle\int \sec^n u \, du = \frac{\sec^{n-2} u \tan u}{n-1} + \frac{n-2}{n-1}\int \sec^{n-2} u \, du, \quad n \neq 1$

70. $\displaystyle\int \csc^n u \, du = -\frac{\csc^{n-2} u \cot u}{n-1} + \frac{n-2}{n-1}\int \csc^{n-2} u \, du, \quad n \neq 1$

71. $\displaystyle\int \frac{1}{1 \pm \tan u} \, du = \frac{1}{2}(u \pm \ln|\cos u \pm \sin u|) + C$

72. $\displaystyle\int \frac{1}{1 \pm \cot u} \, du = \frac{1}{2}(u \mp \ln|\sin u \pm \cos u|) + C$

73. $\displaystyle\int \frac{1}{1 \pm \sec u} \, du = u + \cot u \mp \csc u + C$

74. $\displaystyle\int \frac{1}{1 \pm \csc u} \, du = u - \tan u \pm \sec u + C$

## B.3   FORMULAS FROM BUSINESS AND FINANCE

Formulas from Business • Formulas from Finance

## Formulas from Business

### Basic Terms

$x =$ number of units produced (or sold)

$p =$ price per unit

$R =$ total revenue from selling $x$ units

$C =$ total cost of producing $x$ units

$\overline{C} =$ average cost per unit

$P =$ total profit from selling $x$ units

### Basic Equations

$$R = xp \qquad \overline{C} = \frac{C}{x} \qquad P = R - C$$

**Demand Function: $p = f(x) = $ price required to sell $x$ units**

$$\eta = \frac{p/x}{dp/dx} = \text{price elasticity of demand}$$

(If $|\eta| < 1$, the demand is inelastic. If $|\eta| > 1$, the demand is elastic.)

## Typical Graphs of Revenue, Cost, and Profit Functions

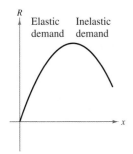

Revenue Function

The low prices required to sell more units eventually result in a decreasing revenue.

Cost Function

The total cost to produce $x$ units includes the fixed cost.

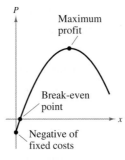

Profit Function

The break-even point occurs when $R = C$.

## Marginals

$$\frac{dR}{dx} = \text{marginal revenue} \approx \text{the } extra \text{ revenue from selling one additional unit}$$

$$\frac{dC}{dx} = \text{marginal cost} \approx \text{the } extra \text{ cost of producing one additional unit}$$

$$\frac{dP}{dx} = \text{marginal profit} \approx \text{the } extra \text{ profit from selling one additional unit}$$

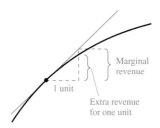

Revenue Function

# Formulas from Finance

### Basic Terms

$P$ = amount of deposit

$r$ = interest rate

$n$ = number of times interest is compounded per year

$t$ = number of years

$A$ = balance after $t$ years

### Compound Interest Formulas

1. Balance when interest is compounded $n$ times per year

$$A = P\left(1 + \frac{r}{n}\right)^{nt}$$

2. Balance when interest is compounded continuously

$$A = Pe^{rt}$$

### Balance of an Increasing Annuity After $n$ Deposits per Year of $P$ for $t$ Years

$$A = P\left[\left(1 + \frac{r}{n}\right)^{nt} - 1\right]\left(1 + \frac{n}{r}\right)$$

### Initial Deposit for a Decreasing Annuity with $n$ Withdrawals per Year of $W$ for $t$ Years

$$P = W\left(\frac{n}{r}\right)\left\{1 - \left[\frac{1}{1 + (r/n)}\right]^{nt}\right\}$$

### Monthly Installment $M$ for a Loan of $P$ Dollars over $t$ Years at $r\%$ Interest

$$M = P\left\{\frac{r/12}{1 - \left[\dfrac{1}{1 + (r/12)}\right]^{12t}}\right\}$$

## B.4 UNITS OF MEASUREMENTS

Units of Measurement of Length • Units of Measurement of Area •
Units of Measurement of Volume • Units of Measurement of Mass and Force •
Units of Measurement of Temperature • Miscellaneous Units and Number Constants

## Units of Measurement of Length

**English System: Inch (in.), Foot (ft), Yard (yd), Mile (mi)**

1 mi = 5280 ft                1 mi = 1760 yd
1 yd = 3 ft                   1 ft = 12 in.

**Metric System: Millimeter (mm), Centimeter (cm), Meter (m), Kilometer (km)**

1 km = 1000 m                 1 m = 100 cm
1 m = 1000 mm                 1 cm = 10 mm

**Conversion Factors (six significant figures)**

*Metric to English*              *English to Metric*

1 mm ≈ 0.0393701 in.          1 in. ≈ 25.4000 mm
1 cm ≈ 0.393701 in.           1 in. ≈ 2.54000 cm
1 m ≈ 3.28084 ft              1 ft ≈ 0.304800 m
1 m ≈ 1.09361 yd              1 yd ≈ 0.914400 m
1 km ≈ 0.621371 mi            1 mi ≈ 1.60934 km

**Miscellaneous**

1 fathom = 6 ft
1 astronomical unit ≈ 93,000,000 mi (average distance between earth and sun)
1 light-year ≈ 5,880,000,000,000 mi (distance traveled by light in 1 year)

## Units of Measurement of Area

**English System: Square Inch (in.$^2$), Square Foot (ft$^2$), Square Yard (yd$^2$),
Acre, Square Mile (mi$^2$)**

1 mi$^2$ = 640 acres            1 acre = 43,560 ft$^2$
1 yd$^2$ = 9 ft$^2$             1 ft$^2$ = 144 in.$^2$

**Metric System: Square Centimeter (cm$^2$), Square Meter (m$^2$),
Square Kilometer (km$^2$)**

1 km$^2$ = 1,000,000 m$^2$       1 m$^2$ = 10,000 cm$^2$

## Units of Measurement of Area     *(Continued)*

### Conversion Factors (six significant figures)

*Metric to English*

$1 \text{ cm}^2 \approx 0.155000 \text{ in.}^2$
$1 \text{ m}^2 \approx 10.7640 \text{ ft}^2$
$1 \text{ m}^2 \approx 1.19599 \text{ yd}^2$
$1 \text{ km}^2 \approx 0.386102 \text{ mi}^2$

*English to Metric*

$1 \text{ in.}^2 \approx 6.45160 \text{ cm}^2$
$1 \text{ ft}^2 \approx 0.0929030 \text{ m}^2$
$1 \text{ yd}^2 \approx 0.836127 \text{ m}^2$
$1 \text{ mi}^2 \approx 2.58999 \text{ km}^2$

### Miscellaneous

1 square mile = 1 section = 4 quarters

## Units of Measurement of Volume

### English System:  Cubic Inch (in.³), Cubic Foot (ft³), Cubic Yard (yd³), Fluid Ounce (fl oz), Pint (pt), Quart (qt), Gallon (gal)

$1 \text{ yd}^3 = 27 \text{ ft}^3$
$1 \text{ gal} = 4 \text{ qt}$
$1 \text{ qt} = 32 \text{ fl oz}$
$1 \text{ gal} \approx 231 \text{ in.}^3$

$1 \text{ ft}^3 = 1728 \text{ in.}^3$
$1 \text{ qt} = 2 \text{ pt}$
$1 \text{ pt} = 16 \text{ fl oz}$
$1 \text{ ft}^3 \approx 7.48052 \text{ gal}$

### Metric System:  Cubic Centimeter (cm³), Cubic Meter (m³), Milliliter (cc), Liter

$1 \text{ m}^3 = 1000 \text{ liters}$
$1 \text{ cm}^3 = 1 \text{ cc}$

$1 \text{ liter} = 1000 \text{ cc}$

### Conversion Factors (six significant figures)

*Metric to English*

$1 \text{ cm}^3 \approx 0.0610237 \text{ in.}^3$
$1 \text{ m}^3 \approx 35.3147 \text{ ft}^3$
$1 \text{ m}^3 \approx 1.30795 \text{ yd}^3$

*English to Metric*

$1 \text{ in.}^3 \approx 16.3871 \text{ cm}^3$
$1 \text{ ft}^3 \approx 0.0283168 \text{ m}^3$
$1 \text{ yd}^3 \approx 0.764555 \text{ m}^3$

### Miscellaneous

1 gallon = 5 fifths
1 quart = 4 cups
1 tablespoon = 3 teaspoons (cooking)
1 tablespoon = 4 teaspoons (medical)
1 barrel (bl) = 42 gallons (petroleum)

1 fifth $\approx$ 0.757 liter
1 cup = 16 tablespoons

## Units of Measurement of Mass and Force

**English System (Force or Weight): Ounce (oz), Pound (lb), Ton**

1 ton = 2000 lb                    1 lb = 16 oz

**Metric System (Mass): Gram (g), Kilogram (kg), Metric Ton**

1 metric ton = 1000 kg          1 kg = 1000 g

**Conversion Factors (six significant figures at sea level)**

| *Metric to English* | *English to Metric* |
|---|---|
| 1 g ≈ 0.0352740 oz | 1 oz ≈ 28.3495 g |
| 1 kg ≈ 2.20462 lb | 1 lb ≈ 0.453592 kg |
| 1 metric ton ≈ 1.10231 ton | 1 ton ≈ 0.907185 metric ton |

## Units of Measurement of Temperature

**Fahrenheit (F), Celsius (C), Kelvin or Absolute (K)**

*Celsius to Fahrenheit*

$$°F = \frac{9}{5}°C + 32$$

*Fahrenheit to Celsius*

$$°C = \frac{5}{9}(°F - 32)$$

*Celsius to Kelvin*

$$°K = °C + 273.15$$

*Kelvin to Celsius*

$$°C = °K - 273.15$$

Freezing temperature for water = 32°F = 0°C

Boiling temperature for water = 212°F = 100°C

Absolute zero temperature = 0°K = −273.15°C = −459.67°F

> (0°K is, by definition, the lowest possible temperature, at which there is no molecular activity.)

## Miscellaneous Units and Number Constants

$\pi \approx 3.1415926535$                    $e \approx 2.7182818284$

Equatorial radius of the earth ≈ 3963.34 mi ≈ 6378.377 km

Polar radius of the earth ≈ 3949.99 mi ≈ 6356.893 km

Acceleration due to gravity at sea level ≈ 32.1726 ft/sec$^2$

Speed of sound at sea level (standard atmosphere) ≈ 1116.45 ft/sec

Speed of light in vacuum ≈ 186,284 mi/sec

Density of water: 1 ft$^3$ ≈ 62.425 lb

## B.5    STANDARD NORMAL DISTRIBUTION

Standard Normal Distribution

## Standard Normal Distribution (z-scores)

| | z | 0.00 | 0.01 | 0.02 | 0.03 | 0.04 | 0.05 | 0.06 | 0.07 | 0.08 | 0.09 |
|---|---|---|---|---|---|---|---|---|---|---|---|
| | | | | | **Hundredths Digit for z-score** | | | | | | |
| | 0.0 | 0.0000 | 0.0040 | 0.0080 | 0.0120 | 0.0160 | 0.0199 | 0.0239 | 0.0279 | 0.0319 | 0.0359 |
| | 0.1 | 0.0398 | 0.0438 | 0.0478 | 0.0517 | 0.0557 | 0.0596 | 0.0636 | 0.0675 | 0.0714 | 0.0753 |
| | 0.2 | 0.0793 | 0.0832 | 0.0871 | 0.0910 | 0.0948 | 0.0987 | 0.1026 | 0.1064 | 0.1103 | 0.1141 |
| | 0.3 | 0.1179 | 0.1217 | 0.1255 | 0.1293 | 0.1331 | 0.1368 | 0.1406 | 0.1443 | 0.1480 | 0.1517 |
| | 0.4 | 0.1554 | 0.1591 | 0.1628 | 0.1664 | 0.1700 | 0.1736 | 0.1772 | 0.1808 | 0.1844 | 0.1879 |
| | 0.5 | 0.1915 | 0.1950 | 0.1985 | 0.2019 | 0.2054 | 0.2088 | 0.2123 | 0.2157 | 0.2190 | 0.2224 |
| | 0.6 | 0.2257 | 0.2291 | 0.2324 | 0.2357 | 0.2389 | 0.2422 | 0.2454 | 0.2486 | 0.2517 | 0.2549 |
| | 0.7 | 0.2580 | 0.2611 | 0.2642 | 0.2673 | 0.2704 | 0.2734 | 0.2764 | 0.2794 | 0.2823 | 0.2852 |
| | 0.8 | 0.2881 | 0.2910 | 0.2939 | 0.2967 | 0.2995 | 0.3023 | 0.3051 | 0.3078 | 0.3106 | 0.3133 |
| | 0.9 | 0.3159 | 0.3186 | 0.3212 | 0.3238 | 0.3264 | 0.3289 | 0.3315 | 0.3340 | 0.3365 | 0.3389 |
| | 1.0 | 0.3413 | 0.3438 | 0.3461 | 0.3485 | 0.3508 | 0.3531 | 0.3554 | 0.3577 | 0.3599 | 0.3621 |
| | 1.1 | 0.3643 | 0.3665 | 0.3686 | 0.3708 | 0.3729 | 0.3749 | 0.3770 | 0.3790 | 0.3810 | 0.3830 |
| | 1.2 | 0.3849 | 0.3869 | 0.3888 | 0.3907 | 0.3925 | 0.3944 | 0.3962 | 0.3980 | 0.3997 | 0.4015 |
| | 1.3 | 0.4032 | 0.4049 | 0.4066 | 0.4082 | 0.4099 | 0.4115 | 0.4131 | 0.4147 | 0.4162 | 0.4177 |
| | 1.4 | 0.4192 | 0.4207 | 0.4222 | 0.4236 | 0.4251 | 0.4265 | 0.4279 | 0.4292 | 0.4306 | 0.4319 |
| | 1.5 | 0.4332 | 0.4345 | 0.4357 | 0.4370 | 0.4382 | 0.4394 | 0.4406 | 0.4418 | 0.4429 | 0.4441 |
| | 1.6 | 0.4452 | 0.4463 | 0.4474 | 0.4484 | 0.4495 | 0.4505 | 0.4515 | 0.4525 | 0.4535 | 0.4545 |
| | 1.7 | 0.4554 | 0.4564 | 0.4573 | 0.4582 | 0.4591 | 0.4599 | 0.4608 | 0.4616 | 0.4625 | 0.4633 |
| | 1.8 | 0.4641 | 0.4649 | 0.4656 | 0.4664 | 0.4671 | 0.4678 | 0.4686 | 0.4693 | 0.4699 | 0.4706 |
| | 1.9 | 0.4713 | 0.4719 | 0.4726 | 0.4732 | 0.4738 | 0.4744 | 0.4750 | 0.4756 | 0.4761 | 0.4767 |
| | 2.0 | 0.4772 | 0.4778 | 0.4783 | 0.4788 | 0.4793 | 0.4798 | 0.4803 | 0.4808 | 0.4812 | 0.4817 |
| | 2.1 | 0.4821 | 0.4826 | 0.4830 | 0.4834 | 0.4838 | 0.4842 | 0.4846 | 0.4850 | 0.4854 | 0.4857 |
| | 2.2 | 0.4861 | 0.4864 | 0.4868 | 0.4871 | 0.4875 | 0.4878 | 0.4881 | 0.4884 | 0.4887 | 0.4890 |
| | 2.3 | 0.4893 | 0.4896 | 0.4898 | 0.4901 | 0.4904 | 0.4906 | 0.4909 | 0.4911 | 0.4913 | 0.4916 |
| | 2.4 | 0.4918 | 0.4920 | 0.4922 | 0.4925 | 0.4927 | 0.4929 | 0.4931 | 0.4932 | 0.4934 | 0.4936 |
| | 2.5 | 0.4938 | 0.4940 | 0.4941 | 0.4943 | 0.4945 | 0.4946 | 0.4948 | 0.4949 | 0.4951 | 0.4952 |
| | 2.6 | 0.4953 | 0.4955 | 0.4956 | 0.4957 | 0.4959 | 0.4960 | 0.4961 | 0.4962 | 0.4963 | 0.4964 |
| | 2.7 | 0.4965 | 0.4966 | 0.4967 | 0.4968 | 0.4969 | 0.4970 | 0.4971 | 0.4972 | 0.4973 | 0.4974 |
| | 2.8 | 0.4974 | 0.4975 | 0.4976 | 0.4977 | 0.4977 | 0.4978 | 0.4979 | 0.4979 | 0.4980 | 0.4981 |
| | 2.9 | 0.4981 | 0.4982 | 0.4982 | 0.4983 | 0.4984 | 0.4984 | 0.4985 | 0.4985 | 0.4986 | 0.4986 |
| | 3.0 | 0.4987 | 0.4987 | 0.4987 | 0.4988 | 0.4988 | 0.4989 | 0.4989 | 0.4989 | 0.4990 | 0.4990 |

Tenths Digit for z-score

# Appendix C:  Graphing Utility Programs

Programs for the Texas Instruments *TI-82* and *TI-83* graphing calculators are given in several sections in the text. This appendix contains translations of these programs for the *TI-85, TI-86, TI-92,* Casio *CFX-9850G,* HP *38G,* and Sharp *EL 9200/9300,* arranged by calculator model. Similar programs can be written for other brands and models of graphing calculators.

Enter a program in your calculator, then refer to the text discussion and apply the program as appropriate. The following section references are provided to help you locate the text discussions of the programs and their uses:

## Texas Instruments TI-85 and TI-86

### Evaluating Functions

This program evaluates a function that has been stored as y1.

PROGRAM:Evaluate
:Repeat 1<0
:Input x
:Disp y1
:End

### Midpoint Rule

This program uses the Midpoint Rule to approximate the definite integral $\int_a^b f(x)dx$. You must store the expression $f(x)$ as y1 before executing the program. The program itself will prompt you for the limits $a$ and $b$ and for the number of subintervals $n$.

PROGRAM:Midpoint
:Disp "Lower Limit"
:Input A
:Disp "Upper Limit"
:Input B
:Disp "n divisions"
:Input N
:0→S
:(B-A)/N→W
:1→J
:While J≤N
:A+(J-1)∗W→L
:A+J∗W→R
:(L+R)/2→x
:S+W∗y1→S
:J+1→J
:End
:Disp "Approximation"
:Disp S

## Simpson's Rule

This program uses Simpson's Rule to approximate the definite integral $\int_a^b f(x)\,dx$. You must store the expression $f(x)$ as y1 before executing the program. The program itself will prompt you for the limits $a$ and $b$ and for *half* the number of subintervals you want to use.

```
PROGRAM:Simpson
:Disp "Lower Limit"
:Input A
:Disp "Upper Limit"
:Input B
:Disp "n/2 divisions"
:Input D
:0→S
:(B-A)/(2D)→W
:1→J
:While J≤D
:A+2(J-1)*W→L
:A+2J*W→R
:(L+R)/2→M
:L→x
:y1→L
:M→x
:y1→M
:R→x
:y1→R
:W*(L+4M+R)/3+S→S
:J+1→J
:End
:Disp "Approximation"
:Disp S
```

## TI-85 Linear Regression

Here are the steps for finding the linear regression model in the Median Hourly Wage example.

STAT  EDIT  ENTER  ENTER

```
x=X          y=Y
x1=-3
y1=9.91
x2=-2
y2=10.19
    ⋮
x9=5
y9=12.35
```

EXIT  EXIT
STAT  CALC  ENTER  ENTER  LINR

```
LinR
  a=10.8246666667
  b=.308666666667
  corr=.999589745485
  n=9
```

## TI-86 Linear Regression

Here are the steps for finding the linear regression model in the Median Hourly Wage example.

2nd  STAT  EDIT  ENTER  ENTER

| xStat | yStat | fStat |
|-------|-------|-------|
| -3 | 9.91 | 1 |
| -2 | 10.19 | 1 |
| -1 | 10.48 | 1 |
| ⋮ | ⋮ | ⋮ |
| 5 | 12.35 | 1 |

EXIT  EXIT
2nd  STAT  CALC  LinR

```
LinR
  y=ax+b
  a=10.8246667
  b=.308666667
  corr=.999589745
  n=9
```

## Newton's Method

This program uses Newton's Method to approximate the zeros of a function. You must store the expression $f(x)$ as y1 before executing the program. Then graph the function to estimate one of its zeros. The program will prompt you for this estimate.

```
PROGRAM:Newton
:Disp "Enter Approximation"
:Input x
:1→N
:x-y1/der1(y1,x)→R
:While abs (x-R)>1E-10
:R→x
:x-y1/der1(y1,x)→R
:N+1→N
:End
:Disp "Root="
:Disp R
:Disp "Iter="
:Disp N
```

# Texas Instruments TI-92

## Evaluating Functions

This program evaluates a function that has been stored as y1(x).

```
:evaluate ( )
:Prgm
:Loop
:Input "Input x", x
:Disp y1(x)
:EndLoop
:EndPrgm
```

## Midpoint Rule

This program uses the Midpoint Rule to approximate the definite integral $\int_a^b f(x)\,dx$. You must store the expression $f(x)$ as y1(x) before executing the program. The program itself will prompt you for the limits $a$ and $b$ and for the number of subintervals $n$.

```
:midpoint( )
:Prgm
:SetMode("Exact/Approx", "APPROXIMATE")
:Input "Lower limit", a
:Input "Upper limit", b
:Input "n Divisions", n
:0→s
:(b-a)/n→w
:1→j
:While j≤n
:a+(j-1)*w→l
:a+j*w→r
:(l+r)/2→x
:s+w*y1(x)→s
:j+1→j
:EndWhile
:Disp "Approximation", s
:SetMode("Exact/Approx", "AUTO")
:EndPrgm
```

## Simpson's Rule

This program uses Simpson's Rule to approximate the definite integral $\int_a^b f(x)\,dx$. You must store the expression $f(x)$ as y1(x) before executing the program. The program itself will prompt you for the limits $a$ and $b$ and for *half* the number of subintervals you want to use.

```
:Prgm
:SetMode("Exact/Approx", "APPROXIMATE")
:Input "Lower limit", a
:Input "Upper limit", b
:Input "n/2 Divisions", d
:0→s
:(b-a)/(2*d)→w
:1→j
:While j≤d
:a+2*(j-1)*w→l
:a+2*j*w→r
:(l+r)/2→m
:l→x
:y1(x)→l
:m→x
:y1(x)→m
:r→x
:y1(x)→r
:w*(l+4m+r)/3+s→s
:j+1→j
:EndWhile
:Disp "Approximation", s
:SetMode("Exact/Approx", "AUTO")
:EndPrgm
```

## Linear Regression

Here are the steps for finding the linear regression model in the Median Hourly Wage example.

APPS    Data/Matrix Editor    New. . .

Type:  Data →
Folder:  Main →
Variable:  wage

|   | c1 | c2 |
|---|-----|-------|
| 1 | -3 | 9.91 |
| 2 | -2 | 10.19 |
| 3 | -1 | 10.48 |
| ⋮ | ⋮ | ⋮ |
| 9 | 5 | 12.35 |

F5 (Calc)

Calculation Type . LinReg →
x.......................c1
y.......................c2
Store RegEQ to.....none →
Use Freq and Categories?....NO →

ENTER

STAT VARS
$y = a \cdot x + b$
a           = .308667
b           = 10.824667
corr        = .99959
$R^2$        = .99918

## Newton's Method

This program uses Newton's Method to approximate the zeros of a function. You must store the expression $f(x)$ as $y1(x)$ before executing the program. Then graph the function to estimate one of its zeros. The program will prompt you for this estimate.

:newton( )
:Prgm
:Disp "Enter approximation"
:Input x
:1 → n
:x-y1(x)/(nDeriv(y1(x), x)) → r
:While abs(x-r)>1.ᴇ-10
:r → x
:x-y1(x)/(nDeriv(y1(x), x)) → r
:n+1 → n
:EndWhile
:Disp "Root =", r
:Disp "Iter =", n
:EndPrgm

# Casio CFX-9850G

## Evaluating Functions

This program evaluates a function that has been stored in function memory location f1.

======EVALUATE======
Lbl 1 ↵
"X="?→X ↵
"F(X)=":f1 ◢
Goto 1

## Midpoint Rule

This program uses the Midpoint Rule to approximate the definite integral $\int_a^b f(x)dx$. You must store the expression $f(x)$ as f1 before executing the program. The program itself will prompt you for the limits $a$ and $b$ and for the number of subintervals $n$.

======MIDPOINT======
"LOWER LIMIT"?→A ↵
"UPPER LIMIT"?→B ↵
"N DIVISIONS"?→N ↵
0→S ↵
(B-A)÷N→W ↵
1→J ↵
Lbl 1 ↵
A+(J-1)W→L ↵
A+JW→R ↵
(L+R)÷2→X ↵
S+Wf1→S ↵
J+1→J ↵
J≤N⇒Goto 1 ↵
"APPROXIMATION" ↵
S

## Simpson's Rule

This program uses Simpson's Rule to approximate the definite integral $\int_a^b f(x)dx$. You must store the expression $f(x)$ as f1 before executing the program. The program itself will prompt you for the limits $a$ and $b$ and for *half* the number of subintervals you want to use.

======SIMPSON======
"LOWER LIMIT"?→A ↵
"UPPER LIMIT"?→B ↵
"N÷2 DIVISIONS"?→D ↵
0→S ↵
(B-A)÷(2D)→W ↵
1→J ↵
Lbl 1 ↵
A+2(J-1)W→L ↵
A+2JW→R ↵
(L+R)÷2→M ↵
L→X ↵
f1→L ↵
M→X ↵
f1→M ↵
R→X ↵
f1→R ↵
W(L+4M+R)÷3+S→S ↵
J+1→J ↵
J≤D⇒Goto 1 ↵
"APPROXIMATION" ↵
S

## Linear Regression

Here are the steps for finding the linear regression model in the Median Hourly Wage example.

MENU 2

|   | List1 | List2 |
|---|-------|-------|
| 1 | -3 | 9.91 |
| 2 | -2 | 10.19 |
| 3 | -1 | 10.48 |
| ⋮ | ⋮ | ⋮ |
| 9 | 5 | 12.35 |

F2 (CALC) F6 (SET)

| 1VAR | Xlist | :List1 |
|------|-------|--------|
| 1VAR | Freq | :1 |
| 2VAR | Xlist | :List1 |
| 2VAR | Ylist | :List2 |
| 2VAR | Freq | :1 |

EXIT F3 (REG) F1 (X)

LinearReg
  a=0.30866666
  b=10.8246666
  r=0.99958974
y=ax+b

## Newton's Method

This program uses Newton's Method to approximate the zeros of a function. You must store the expression $f(x)$ as f1 before executing the program. Then graph the function to estimate one of its zeros. The program will prompt you for this estimate.

```
======NEWTON======
"ENTER APPROXIMATION"? → X ↵
1 → N ↵
X-f1÷d/dx(f1,X) → R ↵
Lbl 1 ↵
R → X ↵
X-f1÷d/dx(f1,X) → R ↵
N+1 → N ↵
Abs (X-R)≥1ᴇ-10⇒Goto 1 ↵
"ROOT=" ↵
R ◢
"ITER=" ↵
N
```

# HP 38G

## Evaluating Functions

Use the Solve aplet to evaluate an expression.

1. Press $\boxed{\text{LIB}}$. Highlight the Solve aplet. Press {{START}}.
2. Set your expression equal to $y$, enter the equation ($y = $ *your expression*) in E1 and press {{OK}}. The equation should be checked.
3. Press $\boxed{\text{NUM}}$.
4. Highlight the $x$-variable field. Enter a value for $x$ and press {{OK}}.
5. Highlight the $y$-variable field and press {{SOLVE}}. The value of the expression will appear in the $y$-variable field.
6. Repeat steps 4 and 5 to evaluate the expression for other values of $x$.

## Midpoint Rule

This program uses the Midpoint Rule to approximate the definite integral $\int_a^b f(x)dx$. Enter both programs into the calculator. Store the expression $(f)x$ in the F1 function in the Function aplet. Be sure F1 is checked. The program itself will prompt you for the limits $a$ and $b$ and for the number of subintervals $n$.

MIDPOINT PROGRAM
```
INPUT A;"ENTER LOWER LIMIT"; "ENTER A";" "; 1:
INPUT B;"ENTER UPPER LIMIT"; "ENTER B";" "; 1:
INPUT N;"N DIVISIONS"; "ENTER N"; " "; 1:
0▶S:
(B-A)/N▶W:
1▶J:
WHILE J≤N
REPEAT RUN "MDPTLOOP"
END:
ERASE:
DISP 3; "APPROXIMATION":
DISP 5; S:
FREEZE:
```

MDPTLOOP PROGRAM
```
A+(J-1)*W▶L:
A+J*W▶R:
(L+R)/2▶X:
S+W*F1(X)▶S:
J+1▶J
```

## Simpson's Rule

This program uses Simpson's Rule to approximate the definite integral $\int_a^b f(x)\,dx$. Enter both programs into the calculator. Store the expression $f(x)$ in the F1 function in the Function aplet. Be sure F1 is checked. The program itself will prompt you for the limits $a$ and $b$ and for *half* the number of subintervals you want to use.

SIMPSON PROGRAM
INPUT A; "ENTER LOWER LIMIT"; "ENTER A";" "1:
INPUT B; "ENTER UPPER LIMIT"; "ENTER B";" "; 1:
INPUT D; "N/2 DIVISIONS"; "ENTER D"; " "; 1:
0▶S:
(B-A)/(2D)▶W:
1▶J:
WHILE J≤D
REPEAT RUN "SIMPSONLOOP"
END:
ERASE:
DISP 3; "APPROXIMATION":
DISP 5; S:
FREEZE:


SIMPSONLOOP PROGRAM
A+2*(J-1)*W▶L:
A+2J*W▶R:
(L+R)/2▶M:
L▶X:
F1(X)▶L:
M▶X:
F1(X)▶M:
R▶X:
F1(X)▶R:
W*(L+4M+R)/3+S▶S:
J+1▶J

## Linear Regression

Here are the steps for finding the linear regression model in the Median Hourly Wage example.

1. Press [LIB]. Highlight the Statistics aplet. Press {{START}}.
2. Press ■ [CLEAR], highlight *All columns*, and press {{OK}} to clear the data columns. Make sure {{2VARS■}} is on.
3. Enter data into the table.

| n | C1 | C2 |
|---|-----|-------|
| 1 | -3 | 9.91 |
| 2 | -2 | 10.19 |
| 3 | -1 | 10.48 |
| ⋮ | ⋮ | ⋮ |
| 9 | 5 | 12.35 |

4. Press [SYMB] , ■ [CLEAR], and {{YES}} to clear the previous definition of S1.
5. Press ■ [SYMB]. Use {{Choose}} and the cursor keys to change the S1FIT: to *Linear*.
6. Press [SYMB]. Make sure S1 is checked.
   2V STATISTICS SYMBOLIC VIEW
   √ S1:C1          C2
   √ Fit 1: m*X+b
7. Press [PLOT] and then {{MENU}}. Make sure {{FIT■}} is on.
8. Press [SYMB] and {{SHOW}}.
   0.3087•X+10.8247

## Newton's Method

This program uses Newton's Method to approximate the zeros of a function. Enter both programs into the calculator. Store the expression $(f)x$ in the F1 function in the Function aplet. Be sure F1 is checked. Then graph the function to estimate one of its zeros. The program will prompt you for this estimate.

NEWTON PROGRAM
INPUT X; "ENTER APPROXIMATION"; "ENTER X"; " "; 1:
1►N:
X-F(1)X/($\partial$X(F1(X)))►R:
WHILE ABS(X-R)>1E-10
REPEAT RUN "NEWTONLOOP"
END:
ERASE:
DISP 2; "ROOT=":
DISP 3; R:
DISP 5; "ITER=":
DISP 6; R:
FREEZE:


NEWTONLOOP PROGRAM
R►X:
X-F(1)X/($\partial$X(F1(X)))►R:
N+1►N

# Sharp EL 9200/9300

## Evaluating Functions

This program evaluates a function that is centered within the program itself.

evaluate
—————REAL
Label 1
Input x
y=*insert expression here*
Print y
Goto 1

## Midpoint Rule

This program uses the Midpoint Rule to approximate the definite integral $\int_a^b f(x)dx$. Before running it, you must enter an expression $f(x)$ for the integrand just once in the program itself. The program will prompt you for the limits $a$ and $b$ and for the number of subintervals $n$.

The program includes a subroutine to evaluate the function $f(x)$. It is placed at the beginning of the program for convenient access when you are changing the integrand.

```
midpoint
————————REAL
Goto start
Label eval
y=f(x)
Return
Label start
Print "Lower Limit
Input a
Print "Upper Limit
Input b
Print "n divisions
Input n
s=0
w=(b-a)/n
j=1
Label 1
l=a+(j-1)*w
r=a+j*w
x=(l+r)/2
Gosub eval
s=s+w*y
j=j+1
If j<=n Goto 1
Print "Approximation
Print s
```

## Simpson's Rule

This program uses Simpson's Rule to approximate the definite integral $\int_a^b f(x)dx$. Before running it, you must enter an expression $f(x)$ for the integrand just once in the program itself. The program will prompt you for the limits $a$ and $b$ and for *half* the number of subintervals you want to use.

The program includes a subroutine to evaluate the function $f(x)$. It is placed at the beginning of the program for easy access when you are changing the integrand.

```
simpson
————————REAL
Goto start
Label eval
y=f(x)
Return
Label start
Print "Lower Limit
Input a
Print "Upper Limit
Input b
Print "n/2 divisions
Input d
s=0
w=(b-a)/(2d)
j=1
Label 1
l=a+2(j-1)*w
r=a+2j*w
m=(l+r)/2
x=l
Gosub eval
l=y
x=m
Gosub eval
m=y
x=r
Gosub eval
r=y
s=w*(l+4m+r)/3+s
j=j+1
If j<=d Goto 1
Print "Approximation
Print s
```

## Linear Regression

Here are the steps for finding the linear regression model in the Median Hourly Wage example.

$\boxed{:=}$  MENU   DEL   ALL DATA
TWO VARIABLE X, Y

X=-3
Y=9.91
X=-2
Y=10.19
⋮
X=5
Y=12.35

MENU   STAT   REG

| | |
|---|---|
| a | =10.82466667 |
| b | =0.308666666 |
| r | =0.999589745 |

## Newton's Method

This program uses Newton's Method to approximate the zeros of a function. First graph the function to estimate one of its zeros. The program will prompt you for this estimate.

Before running the program, you must enter the expression $f(x)$ twice within the program itself. Take care that you use the correct variable, local $x$ or global $X$, each time.

The program includes a subroutine to evaluate the function $f(x)$. Next is a second subroutine to calculate the numerical derivative of the function for the current value of $x$. They are placed at the beginning of the program for convenient access when you are changing the function.

```
newton
————————REAL
Goto start
Label eval
y=f(x)
Return
Label nderiv
d=d/dx(f(X), x)
Return
Label start
Print "Enter Approximation
Input x
n=1
Gosub eval
Gosub nderiv
r=x-y/d
Label 1
x=r
Gosub eval
Gosub nderiv
r=x-y/d
n=n+1
If abs (x-r)>=1E-10 Goto 1
Print "Root
Print r
Print "Iter
Print n
```

# Answers to Selected Exercises

## CHAPTER 0

### SECTION 0.1   *(page 0-7)*

**1.** Rational   **3.** Irrational   **5.** Rational

**7.** Rational   **9.** Irrational

**11. (a)** Yes   **(b)** No   **(c)** Yes   **(d)** No

**13. (a)** Yes   **(b)** No   **(c)** No   **(d)** Yes

**15.** $x \geq 12$

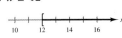

**17.** $x < -\frac{1}{2}$

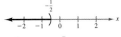

**19.** $x > 1$

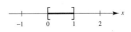

**21.** $-\frac{1}{2} < x < \frac{7}{2}$

**23.** $-\frac{3}{4} < x < -\frac{1}{4}$

**25.** $x > 6$

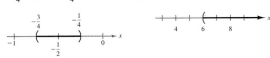

**27.** $0 \leq x \leq 1$

**29.**

**31.** $x \geq 36$ units   **33.** Length of side $\geq 10\sqrt{5}$

**35. (a)** True   **(b)** False   **(c)** True   **(d)** True

### SECTION 0.2   *(page 0-12)*

**1. (a)** $-51$   **(b)** $51$   **(c)** $51$

**3. (a)** $-14.99$   **(b)** $14.99$   **(c)** $14.99$

**5.** 14   **7.** 1.25

**9.** $-5 < x < 5$

**11.** $x < -6$ or $x > 6$

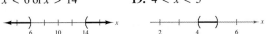

**13.** $-7 < x < 3$

**15.** $x \leq -7$ or $x \geq 13$

**17.** $x < 6$ or $x > 14$

**19.** $4 < x < 5$

**21.** $a - b \leq x \leq a + b$

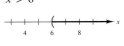

**23.** $|x| \leq 2$   **25.** $|x| > 2$   **27.** $|x - 4| \leq 2$

**29.** $|x - 2| > 2$   **31.** $|x - 4| < 2$   **33.** $|y - a| \leq 2$

**35.** $65.8 \leq h \leq 71.2$   **37.** $175,000 \leq x \leq 225,000$

**39. (a)** $|4750 - a| \leq 500$,   $|4750 - a| \leq 237.50$

**(b)** At variance

**41. (a)** $|20,000 - a| \leq 500$,   $|20,000 - a| \leq 1000$

**(b)** At variance

### SECTION 0.3   *(page 0-18)*

**1.** $-24$   **3.** $\frac{1}{2}$   **5.** 3   **7.** 44   **9.** 5   **11.** 9

**13.** $\frac{1}{2}$   **15.** $\frac{1}{4}$   **17.** 908.3483   **19.** $-5.3601$

**21.** $5x^6$   **23.** $24y^{10}$   **25.** $10x^4$   **27.** $7x^5$

**29.** $\frac{4}{3}(x + y)^2$   **31.** $3x$   **33.** $4x^4$

**35. (a)** $2\sqrt{2}$   **(b)** $3\sqrt{2}$

**37. (a)** $2x\sqrt[3]{2x^2}$   **(b)** $2|x|z\sqrt[4]{2z}$

**39. (a)** $\dfrac{5\sqrt{3}|x|}{y^2}$   **(b)** $(x - y)\sqrt{5(x - y)}$

**41.** $x \geq 1$   **43.** $(-\infty, \infty)$   **45.** $(-\infty, 1)$ and $(1, \infty)$

**47.** $x > 3$　　**49.** $1 \le x \le 5$　　**51.** $19,121.84

**53.** $\dfrac{\sqrt{2}}{2}\pi$ sec or $2.22$ sec

## SECTION 0.4　*(page 0-24)*

**1.** $\frac{1}{2}, -\frac{1}{3}$　　**3.** $\frac{3}{2}$　　**5.** $-2 \pm \sqrt{3}$　　**7.** $(x - 2)^2$

**9.** $(2x + 1)^2$　　**11.** $(x + 2)(x - 1)$

**13.** $(3x - 2)(x - 1)$　　**15.** $(x - 2y)^2$

**17.** $(3 + y)(3 - y)(9 + y^2)$　　**19.** $(x - 2)(x^2 + 2x + 4)$

**21.** $(y + 4)(y^2 - 4y + 16)$　　**23.** $(x - 3)(x^2 + 3x + 9)$

**25.** $(x - 4)(x - 1)(x + 1)$　　**27.** $(2x - 3)(x^2 + 2)$

**29.** $(x - 2)(2x^2 - 1)$　　**31.** $0, 5$　　**33.** $\pm 3$　　**35.** $\pm \sqrt{3}$

**37.** $0, 6$　　**39.** $-2, 1$　　**41.** $2, 3$　　**43.** $-4$　　**45.** $\pm 2$

**47.** $1, \pm 2$　　**49.** $x \le 3$ or $x \ge 4$　　**51.** $-2 \le x \le 2$

**53.** $(x + 2)(x^2 - 2x + 4)$　　**55.** $(x - 1)(2x^2 + x - 1)$

**57.** $\pm 1$　　**59.** $1, 2, 3$　　**61.** $1, \pm \frac{1}{2}$　　**63.** $4$

**65.** $2000$ units　　**67.** $3.4 \times 10^{-5}$

## SECTION 0.5　*(page 0-32)*

**1.** $\dfrac{x + 5}{x - 1}$　　**3.** $\dfrac{5x - 1}{x^2 + 2}$　　**5.** $\dfrac{4x - 3}{x^2}$　　**7.** $\dfrac{x - 6}{x^2 - 4}$

**9.** $\dfrac{2}{x - 3}$　　**11.** $\dfrac{(A + B)x + 3(A - 2B)}{(x - 6)(x + 3)}$

**13.** $-\dfrac{(x - 1)^2}{x(x^2 + 1)}$　　**15.** $\dfrac{x + 2}{(x + 1)^{3/2}}$　　**17.** $-\dfrac{3t}{2\sqrt{1 + t}}$

**19.** $-\dfrac{x^2 + 3}{(x + 1)(x - 2)(x - 3)}$

**21.** $\dfrac{(A + C)x^2 - (A - B - 2C)x - (2A + 2B - C)}{(x + 1)^2(x - 2)}$

**23.** $\dfrac{x(x^2 + 2)}{(x + 1)^{3/2}}$　　**25.** $\dfrac{2}{x^2\sqrt{x^2 + 2}}$　　**27.** $\dfrac{1}{2\sqrt{x}(x + 1)^{3/2}}$

**29.** $\dfrac{\sqrt{3}}{3}$　　**31.** $\dfrac{2}{3\sqrt{2}}$　　**33.** $\dfrac{x\sqrt{x} - 4}{x - 4}$　　**35.** $\dfrac{y}{6\sqrt{y}}$

**37.** $\dfrac{49\sqrt{x^2 - 9}}{x + 3}$　　**39.** $\dfrac{\sqrt{14} + 2}{2}$　　**41.** $\dfrac{x(5 + \sqrt{3})}{11}$

**43.** $\sqrt{6} - \sqrt{5}$　　**45.** $\dfrac{1}{x(\sqrt{3} + \sqrt{2})}$

**47.** $\dfrac{2x - 1}{2x + \sqrt{4x - 1}}$　　**49.** $232.68

## CHAPTER 1

## SECTION 1.1　*(page 8)*

> **Warm Up**
>
> **1.** $3\sqrt{5}$　　**2.** $2\sqrt{5}$　　**3.** $\frac{1}{2}$　　**4.** $-2$　　**5.** $5\sqrt{3}$
>
> **6.** $-\sqrt{2}$　　**7.** $x = -3, x = 9$
>
> **8.** $y = -8, y = 4$　　**9.** $x = 19$　　**10.** $y = 1$

**1.** (a) $a = 4, b = 3,$
$c = 5$
(b) $4^2 + 3^2 = 5^2$

**3.** (a) $a = 10, b = 3,$
$c = \sqrt{109}$
(b) $10^2 + 3^2 = \left(\sqrt{109}\right)^2$

**5.** (a)

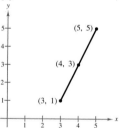

(b) $d = 2\sqrt{5}$
(c) Midpoint: $(4, 3)$

**7.** (a)

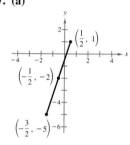

(b) $d = 2\sqrt{10}$
(c) Midpoint: $\left(-\frac{1}{2}, -2\right)$

**9.** (a)

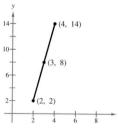

(b) $d = 2\sqrt{37}$

(c) Midpoint: $(3, 8)$

**11.** (a)

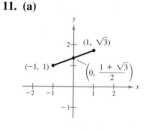

(b) $d = \sqrt{8 - 2\sqrt{3}}$

(c) Midpoint: $\left(0, \dfrac{\sqrt{3} + 1}{2}\right)$

**13.** $d_1 = \sqrt{45}, d_2 = \sqrt{20},$
$d_3 = \sqrt{65}$

**15.** $d_1 = d_2 = d_3 = d_4$
$= \sqrt{5}$

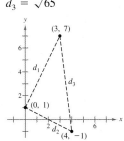

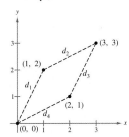

**17.** Collinear, because

$d_1 + d_2 = d_3$
$2\sqrt{5} + \sqrt{5} = 3\sqrt{5}$

**19.** Not collinear, because

$d_1 + d_2 \neq d_3$
$d_1 = \sqrt{18}, d_2 = \sqrt{41}$
$d_3 = \sqrt{113}$

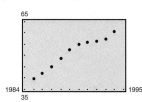

**21.** $x = 4, -2$    **23.** $y = \pm\sqrt{55}$

**25.** $\left(\dfrac{3x_1 + x_2}{4}, \dfrac{3y_1 + y_2}{4}\right), \left(\dfrac{x_1 + x_2}{2}, \dfrac{y_1 + y_2}{2}\right),$
$\left(\dfrac{x_1 + 3x_2}{4}, \dfrac{y_1 + 3y_2}{4}\right)$

**27.** (a) $\left(\frac{7}{4}, -\frac{7}{4}\right), \left(\frac{5}{2}, -\frac{3}{2}\right), \left(\frac{13}{4}, -\frac{5}{4}\right)$
   (b) $\left(-\frac{3}{2}, -\frac{9}{4}\right), \left(-1, -\frac{3}{2}\right), \left(-\frac{1}{2}, -\frac{3}{4}\right)$

**29.** (a) 16.76 ft   (b) 1341.04 ft$^2$

**31.**

The number of subscribers appears to be increasing linearly.

**33.** (a) 3800   (b) 3700   (c) 4800   (d) 5625

**35.** (a) $55 billion   (b) $115 billion   (c) $110 billion
   (d) $128 billion

**37.** (a) Revenue: $15,810 million
      Profit: $2687.9 million

   (b) Actual 1994 revenue: $16,172 million
      Actual 1994 profit: $2554.0 million

   (c) Yes, the increase in revenue per year is ≈ $1368 million.
      Yes, the increase in profit per year is ≈ $402.05 million.

   (d) Expenses for 1992: $11,190.2 million
      Expenses for 1994: $13,618 million
      Expenses for 1996: $15,054 million

   (e) Answers vary.

**39.** (a) $(-1, 2), (1, 1), (2, 3)$
   (b)

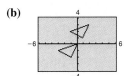

**41.** (a)

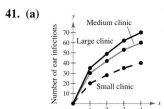

   (b) The larger the clinic, the more patients a doctor can handle.

## SECTION 1.2    *(page 21)*

**Warm Up**

**1.** $y = \frac{1}{5}(x + 12)$    **2.** $y = x - 15$

**3.** $y = \dfrac{1}{x^3 + 2}$

**4.** $y = \pm\sqrt{x^2 + x - 6} = \pm\sqrt{(x + 3)(x - 2)}$

**5.** $y = -1 \pm \sqrt{9 - (x - 2)^2}$

**6.** $y = 5 \pm \sqrt{81 - (x + 6)^2}$    **7.** $(x - 2)(x - 1)$

**8.** $(x + 3)(x + 2)$    **9.** $\left(y - \frac{3}{2}\right)^2$    **10.** $\left(y - \frac{7}{2}\right)^2$

**1. (a)** Not a solution point **(b)** Solution point
**(c)** Solution point

**3. (a)** Not a solution point **(b)** Solution point
**(c)** Not a solution point

**5.** c     **6.** d     **7.** b     **8.** f     **9.** a     **10.** e

**11.** $(0, -3), \left(\frac{3}{2}, 0\right)$     **13.** $(0, -2), (-2, 0), (1, 0)$

**15.** $(0, 0), (-3, 0), (3, 0)$     **17.** $(-2, 0), (0, 2)$

**19.** $(0, 0)$

**21.**

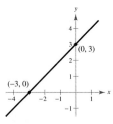

**23.**

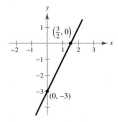

**25.**

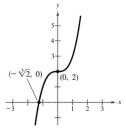

**27.**

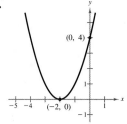

**29.**

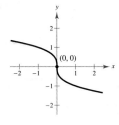

**31.**

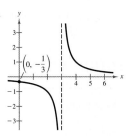

**33.**

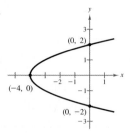

**35.** $x^2 + y^2 - 9 = 0$     **37.** $x^2 + y^2 - 4x + 2y - 11 = 0$

**39.** $x^2 + y^2 + 2x - 4y = 0$     **41.** $x^2 + y^2 - 6y = 0$

**43.** $(x - 1)^2 + (y + 3)^2 = 4$

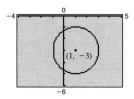

**45.** $(x + 2)^2 + (y + 3)^2 = 16$

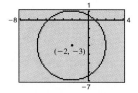

**47.** $\left(x - \frac{1}{2}\right)^2 + \left(y - \frac{1}{2}\right)^2 = 2$

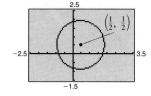

**49.** $\left(x + \frac{1}{2}\right)^2 + \left(y + \frac{5}{4}\right)^2 = \frac{9}{4}$

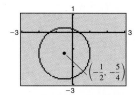

**51.** $(1, 1)$     **53.** $(3, 4), (5, 0)$

**55.** $(0, 0), \left(\sqrt{2}, 2\sqrt{2}\right), \left(-\sqrt{2}, -2\sqrt{2}\right)$

**57.** $(-1, 0), (0, 1), (1, 0)$

**59. (a)** $C = 11.8x + 5000; R = 19.3x$     **(b)** 667 units
**(c)** 680 units

**61.**

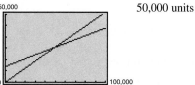

50,000 units

**63.** 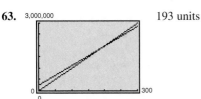 193 units

**65.**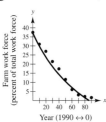

The greater the value of $c$, the steeper the line.

**67. (a)** 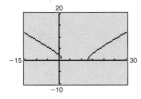  **(b)** $-2.91\%$

**(c)** The model would not be valid for the year 2000.

**69.** If $C$ and $R$ represent the cost and revenue for a business, the break-even point is that value of $x$ for which $C = R$. For example, if $C = 100,000 + 10x$ and $R = 20x$, then the break-even point is 10,000 units.

**71.** $(0, 5.36)$

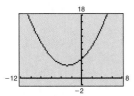

**73.** $(1.4780, 0), (12.8553, 0), (0, 2.3875)$

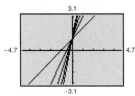

---

## SECTION 1.3 *(page 33)*

**Warm Up**

**1.** $-1$ **2.** $-\frac{7}{3}$ **3.** $\frac{1}{3}$ **4.** $-\frac{7}{6}$

**5.** $y = 4x + 7$ **6.** $y = 3x - 7$

**7.** $y = 3x - 10$ **8.** $y = -x - 7$

**9.** $y = 7x - 17$ **10.** $y = \frac{2}{3}x + \frac{5}{3}$

**1.** $1$ **3.** $0$

**5.** $m = 3$ **7.** $m = 0$

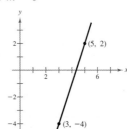 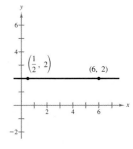

**9.** $m$ is undefined. **11.** $m = 0$

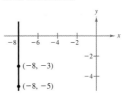 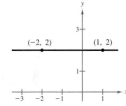

**13.** $(0, 1), (1, 1), (3, 1)$ **15.** $(7, -6), (5, -2), (4, 0)$

**17.** $(0, 10), (2, 4), (3, 1)$ **19.** $(-8, 0), (-8, 2), (-8, 3)$

**21.** $m = -\frac{1}{5}, (0, 4)$ **23.** $m = \frac{7}{5}, (0, -3)$

**25.** $m$ is undefined; no $y$-intercept.

**27.** $y = 2x - 5$ **29.** $3x + y = 0$

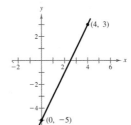 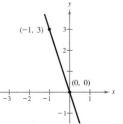

**31.** $x - 2 = 0$

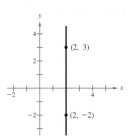

**33.** $y + 2 = 0$

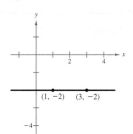

**51.** **(a)** $3x + 4y + 2 = 0$    **(b)** $4x - 3y + 36 = 0$

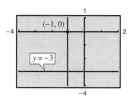

**53.** **(a)** $y = 0$    **(b)** $x + 1 = 0$

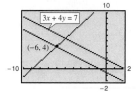

**35.** $3x - 4y + 12 = 0$

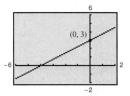

**37.** $2x - 3y = 0$

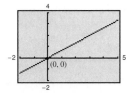

**55.** **(a)** $2x - 3y + 1 = 0$    **(b)** $3x + 2y - 5 = 0$

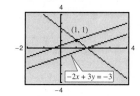

**39.** $3x + y - 1 = 0$

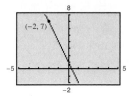

**41.** $4x - y + 2 = 0$

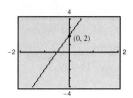

**57.**

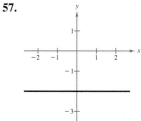

**59.**

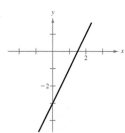

**43.** $9x - 12y + 8 = 0$

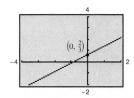

**45.** The points are not collinear.    **47.** $x - 3 = 0$

**49.** **(a)** $x + y + 1 = 0$    **(b)** $x - y + 5 = 0$

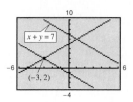

**61.**

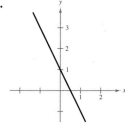

**63.**

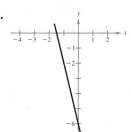

**65. (a)** $y = 26.4t + 3146$; The slope $m = 26.4$ tells you that the population increases 26.4 thousand per year.

**(b)** 3198.8 thousand (3,198,800)

**(c)** 3304.4 thousand (3,304,400)

**(d)** 1992: 3207.2 thousand; 1996: 3300.9 thousand

**(e)** The model could possibly be used to predict the population in 2000 if the population continues to grow at the same linear rate.

**67. (a)** 39.17°C   **(b)** Solid (20°C)

**69. (a)** $W = 0.80x + 8.75$ (union plan)

$W = 1.15x + 6.35$ (corporation plan)

**(b)**    (6.857, 14.236)

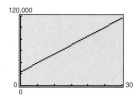

**(c)** The point of intersection indicates the number of units a worker needs to produce in order for the two plans to be equivalent. If a worker produces more than six units, the corporation plan pays better.

**71.** $y = 825,000 - 30,000t, \quad 0 \le t \le 25$

**73. (a)** $x = -\frac{1}{15}p + \frac{226}{3}$ **(b)** 45 units **(c)** 49 units

**75. (a)** $Y = 262t + 4792$ **(b)** \$5316 billion

**(c)** \$6364 billion **(d)** 1992: 5264.2 billion; 1996: 6449.5 billion

**77.** $x \le 24$ units          **79.** $x \le 70$ units

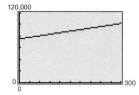        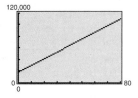

**81.** $x \le 275$ units

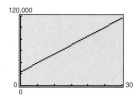

---

# SECTION 1.4    *(page 45)*

## Warm Up

**1.** 20     **2.** 10     **3.** $x^2 + x - 6$

**4.** $x^3 + 9x^2 + 26x + 30$     **5.** $\dfrac{1}{x}$     **6.** $\dfrac{2x - 1}{x}$

**7.** $y = -2x + 17$     **8.** $y = \frac{6}{5}x^2 + \frac{1}{5}$

**9.** $y = 3 \pm \sqrt{5 + (x + 1)^2}$     **10.** $y = \pm\sqrt{4x^2 + 2}$

**1. (a)** $-3$ **(b)** $-9$ **(c)** $2x - 5$ **(d)** $2\Delta x - 1$

**3. (a)** $\dfrac{1}{2}$ **(b)** 4 **(c)** $\dfrac{1}{x + \Delta x}$ **(d)** $-\dfrac{\Delta x}{x(x + \Delta x)}$

**5.** 3     **7.** $3x^2 + 3x\Delta x + (\Delta x)^2 - 1$

**9.** $y$ is not a function of $x$.     **11.** $y$ is a function of $x$.

**13.** $y$ is a function of $x$.     **15.** $y$ is not a function of $x$.

**17.**                               **19.**

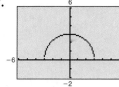        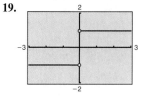

Domain: $[-3, 3]$          Domain: $(-\infty, 0) \cup (0, \infty)$

Range: $[0, 3]$          Range: $y = -1$ or $y = 1$

**21.**

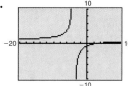

Domain: $(-\infty, -4) \cup (-4, \infty)$

Range: $(-\infty, 1) \cup (1, \infty)$

**23.** Domain: $(-\infty, \infty)$     **25.** Domain: $(-\infty, \infty)$

Range: $(-\infty, \infty)$          Range: $(-\infty, 4]$

**27.** $y$ is not a function of $x$.     **29.** $y$ is a function of $x$.

**31. (a)** $x^2 + x$ **(b)** $(x^2 + 1)(x - 1) = x^3 - x^2 + x - 1$

**(c)** $\dfrac{x^2 + 1}{x - 1}$     **(d)** $x^2 - 2x + 2$     **(e)** $x^2$

**33. (a)** $x^2 + 5 + \sqrt{1 - x}$ **(b)** $(x^2 + 5)\sqrt{1 - x}$

**(c)** $\dfrac{x^2 + 5}{\sqrt{1 - x}}$     **(d)** $6 - x, 1 \ge x$     **(e)** Not defined

**35.** $f(g(x)) = 5\left(\dfrac{x-1}{5}\right) + 1 = x$

$g(f(x)) = \dfrac{5x + 1 - 1}{5} = x$

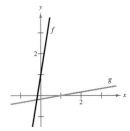

**37.** $f(g(x)) = 9 - \left(\sqrt{9-x}\right)^2 = 9 - (9-x) = x$

$g(f(x)) = \sqrt{9 - (9 - x^2)} = \sqrt{x^2} = x$

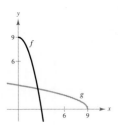

**39.** $f(x) = 2x - 3, f^{-1}(x) = \dfrac{x+3}{2}$

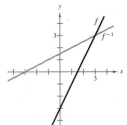

**41.** $f(x) = x^5, f^{-1}(x) = \sqrt[5]{x}$

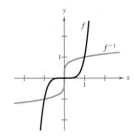

**43.** $f(x) = \sqrt{9 - x^2}, f'(x) = \sqrt{9 - x^2}, 0 \le x \le 3$

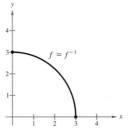

**45.** $f(x) = x^{2/3}, f^{-1}(x) = x^{3/2}, x \ge 0$

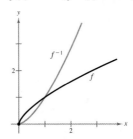

**47.**

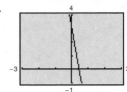

$f(x)$ is one-to-one.   $f^{-1}(x) = \dfrac{3-x}{7}$

**49.**

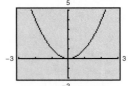

$f(x)$ is not one-to-one.

**51.**

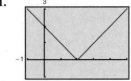

$f(x)$ is not one-to-one.

**53. (a)** 0   **(b)** 0   **(c)** $-1$   **(d)** $\sqrt{15}$

     **(e)** $\sqrt{x^2 - 1}$   **(f)** $x - 1, x \ge 0$

**55.** The data fits the function (b),

$g(x) = cx^2$, with $c = -2$.

**57.** The data fits the function (d),

$r(x) = c/x$, with $c = 32$.

**59.** (a)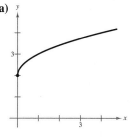

(b)

(c)

(d)

(e)

(f)

**61.** (a) $y = (x + 3)^2$  (b) $y = x^2 + 3$

(c) $y = -(x - 3)^2 + 6$  (d) $y = -(x + 6)^2 - 3$

**63.** (a) $C = 0.95x + 6000$

(b) $\overline{C} = \dfrac{0.95x + 6000}{x} = 0.95 + \dfrac{6000}{x}$

(c) $x > 8108.108$ or $x > 8109$

**65.** $C = 10(5280)(3 - x) + 15(5280)\sqrt{x^2 + \frac{1}{4}}$

**67.** (a) $p = \begin{cases} 90, & 0 \le x \le 100 \\ 91 - 0.01x, & 100 < x \le 1600 \\ 75, & x > 1600 \end{cases}$

(b) $P = \begin{cases} 30x, & 0 \le x \le 100 \\ 31x - 0.01x^2, & 100 < x \le 1600 \\ 15x, & x > 1600 \end{cases}$

**69.** (a) $R = rn = [8 - 0.05(n - 80)]n$

(b)

| $n$ | 90 | 100 | 110 | 120 |
|---|---|---|---|---|
| $R$ | 675 | 700 | 715 | 720 |

| $n$ | 130 | 140 | 150 |
|---|---|---|---|
| $R$ | 715 | 700 | 675 |

**71.**

Zeros: $x = 0, \pm 3$

$f(x)$ is not one-to-one.

**73.**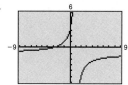

Zeros: $t = -3$

$g(t)$ is one-to-one.

**75.**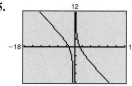

Zeros: $x = \pm 2$

$f(x)$ is not one-to-one.

**77.** Answers vary.

# SECTION 1.5   *(page 58)*

## Warm Up

**1. (a)** 7   **(b)** $c^2 - 3c + 3$

**(c)** $x^2 + 2xh + h^2 - 3x - 3h + 3$

**2. (a)** $-4$   **(b)** 10   **(c)** $3t^2 + 4$

**3.** $h$   **4.** 4

**5.** Domain: $(-\infty, 0) \cup (0, \infty)$

Range: $(-\infty, 0) \cup (0, \infty)$

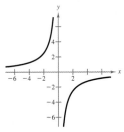

**6.** Domain: $[-5, 5]$

Range: $[0, 5]$

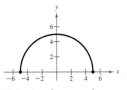

**7.** Domain: $(-\infty, \infty)$

Range: $[0, \infty)$

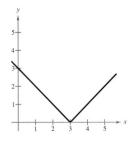

## Warm Up *(continued)*

**8.** Domain: $(-\infty, 0) \cup (0, \infty)$

Range: $-1, 1$

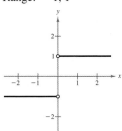

**9.** $y$ is not a function of $x$.   **10.** $y$ is a function of $x$.

**1.**

| $x$ | 1.9 | 1.99 | 1.999 | 2 |
|---|---|---|---|---|
| $f(x)$ | 13.5 | 13.95 | 13.995 | ? |

| $x$ | 2.001 | 2.01 | 2.1 |
|---|---|---|---|
| $f(x)$ | 14.005 | 14.05 | 14.5 |

$\lim_{x \to 2} (5x + 4) = 14$

**3.**

| $x$ | 1.9 | 1.99 | 1.999 | 2 |
|---|---|---|---|---|
| $f(x)$ | 0.2564 | 0.2506 | 0.2501 | ? |

| $x$ | 2.001 | 2.01 | 2.1 |
|---|---|---|---|
| $f(x)$ | 0.2499 | 0.2494 | 0.2439 |

$\lim_{x \to 2} \dfrac{x - 2}{x^2 - 4} = \dfrac{1}{4}$

**5.**

| $x$ | $-0.1$ | $-0.01$ | $-0.001$ | 0 |
|---|---|---|---|---|
| $f(x)$ | 0.2911 | 0.2889 | 0.2887 | ? |

| $x$ | 0.001 | 0.01 | 0.1 |
|---|---|---|---|
| $f(x)$ | 0.2887 | 0.2884 | 0.2863 |

$\lim_{x \to 0} \dfrac{\sqrt{x + 3} - \sqrt{3}}{x} \approx 0.2887$

$\left( \text{The actual limit is } \dfrac{1}{2\sqrt{3}}. \right)$

**7.**

| $x$ | 1.5 | 1.9 | 1.99 | 1.999 | 2 |
|---|---|---|---|---|---|
| $f(x)$ | 0.3780 | 0.1601 | 0.0501 | 0.0158 | ? |

$$\lim_{x \to 2} \frac{2 - x}{\sqrt{4 - x^2}} = 0$$

**9. (a)** 1  **(b)** 3   **11. (a)** 1  **(b)** 3

**13. (a)** 12  **(b)** 27  **(c)** $\frac{1}{3}$   **15. (a)** 1  **(b)** 1  **(c)** 1

**17. (a)** 0  **(b)** 0  **(c)** 0

**19. (a)** 3  **(b)** $-3$  **(c)** Limit does not exist.   **21.** $-7$

**23.** 0   **25.** 2   **27.** $-2$   **29.** $-\frac{3}{4}$   **31.** $-2$

**33.** Limit does not exist.   **35.** $\frac{1}{10}$   **37.** 12

**39.** Limit does not exist.   **41.** $-1$   **43.** 2   **45.** $3x^2$

**47.** $2t - 5$

**49.**

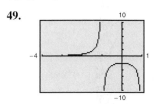

| $x$ | 0 | 0.5 | 0.9 | 0.99 |
|---|---|---|---|---|
| $f(x)$ | $-2$ | $-2.67$ | $-10.53$ | $-100.5$ |

| $x$ | 0.999 | 0.9999 | 1 |
|---|---|---|---|
| $f(x)$ | $-1000.5$ | $-10,000.5$ | Undefined |

$-\infty$

**51.**

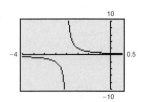

| $x$ | $-3$ | $-2.5$ | $-2.1$ | $-2.01$ |
|---|---|---|---|---|
| $f(x)$ | $-1$ | $-2$ | $-10$ | $-100$ |

| $x$ | $-2.001$ | $-2.0001$ | $-2$ |
|---|---|---|---|
| $f(x)$ | $-1000$ | $-10,000$ | Undefined |

$-\infty$

**53. (a)** \$25,000  **(b)** 80%

**(c)** $\infty$; The cost function increases without bound as $x$ approaches 100 from the left. Therefore, according to the model, it is not possible to remove 100% of the pollutants.

**55. (a)**

| $x$ | $-0.01$ | $-0.001$ | $-0.0001$ | 0 |
|---|---|---|---|---|
| $f(x)$ | 2.732 | 2.720 | 2.718 | Undefined |

| $x$ | 0.0001 | 0.001 | 0.01 |
|---|---|---|---|
| $f(x)$ | 2.718 | 2.717 | 2.705 |

$$\lim_{x \to 0} (1 + x)^{1/x} \approx 2.718$$

**(b)**

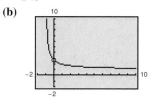

**(c)** Domain: $(-1, 0) \cup (0, \infty)$

Range: $(1, e) \cup (e, \infty)$

**57.**

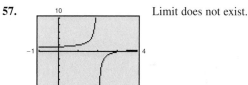

Limit does not exist.

**59.**

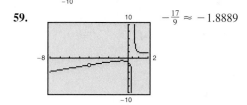

$-\frac{17}{9} \approx -1.8889$

**61. (a)**

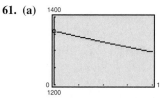

**(b)** For $x = 0.25$, $A = \$1342.53$.

For $x = \frac{1}{356}$, $A = \$1358.95$.

**(c)** $\lim_{x \to 0^+} 500(1 + 0.10x)^{10/x} = 500e \approx \$1359.14$:
continuous compounding

# SECTION 1.6   *(page 69)*

---

**Warm Up**

1. $\dfrac{x+4}{x-8}$   2. $\dfrac{x+1}{x-3}$   3. $\dfrac{x+2}{2(x-3)}$   4. $\dfrac{x-4}{x-2}$

5. $x = 0, -7$   6. $x = -5, 1$   7. $x = -\frac{2}{3}, -2$

8. $x = 0, 3, -8$   9. 13   10. $-1$

---

1. Continuous    3. Not continuous $(x \neq \pm 2)$

5. $(-\infty, \infty)$    7. $(-\infty, -1) \cup (-1, \infty)$

9. $(-\infty, \infty)$    11. $(-\infty, 1) \cup (1, \infty)$    13. $(-\infty, \infty)$

15. $(-\infty, 4) \cup (4, 5) \cup (5, \infty)$    17. $(-\infty, 1) \cup (1, \infty)$

19. $(-\infty, \infty)$    21. $(-\infty, \infty)$

23. $(-\infty, -1) \cup (-1, \infty)$

25. Continuous on all intervals $(c, c + 1)$, where $c$ is an integer.

27. $(1, \infty)$    29. Continuous

31. Nonremovable discontinuity at $x = 2$

33. Removable discontinuity at $x = 4$

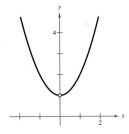

35. Removable discontinuity at $x = 0$

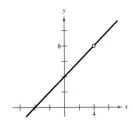

37. Continuous on $(-\infty, 0) \cup (0, \infty)$

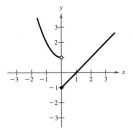

39. $a = 2$

41. Not continuous at $x = 2$ and $x = -1$

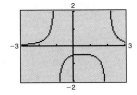

43. $(-\infty, \infty)$

45. Continuous on all intervals $\left(\dfrac{c}{2}, \dfrac{c+1}{2}\right)$, where $c$ is an integer

47. The graph of $f(x) = \dfrac{x^2 + x}{x}$ appears to be continuous on $[-4, 4]$, but $f$ is not continuous at $x = 0$.

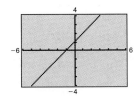

49. (a) Graph has nonremovable discontinuities at $t = \frac{1}{4}, \frac{1}{2}, \frac{3}{4}, 1, \frac{5}{4}, \ldots$

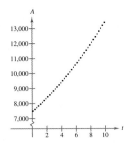

(b) \$11,379.17

**51. (a)**

$$C(t) = \begin{cases} 1.04, & 0 < t \le 2 \\ 1.04 + 0.36[\![t - 1]\!], & t > 2, t \text{ is not an integer.} \\ 1.04 + 0.36(t - 2), & t > 2, t \text{ is an integer.} \end{cases}$$

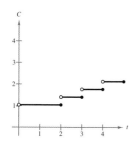

$C$ is not continuous at $t = 2, 3, 4, \ldots$.

**(b)** $3.56

**53. (a)**

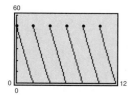

Nonremovable discontinuities at $t = 2, 4, 6, 8, \ldots$.

**(b)** Every two months

**55.**

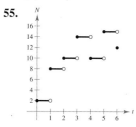

Nonremovable discontinuities at $t = 1, 2, 3, 4, 5$ and $6$

# CHAPTER 1 REVIEW EXERCISES  *(page 76)*

**1.** a  **2.** c  **3.** b  **4.** d  **5.** $\sqrt{29}$  **7.** $3\sqrt{2}$

**9.** $(7, 4)$  **11.** $(-8, 6)$

**13.** The taller bars in the book represent revenues. The middle bars represent costs. The smaller bars in front represent profits, because $P = R - C$.

**15.** $(4, 7), (5, 8), (8, 10)$

**17.**

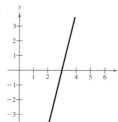

**19.**

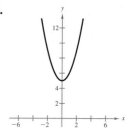

**21.**

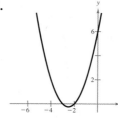

**23.**

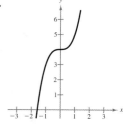

**25.** $(0, -3), \left(-\frac{3}{4}, 0\right)$  **27.** $x^2 + y^2 = 9$

**29.** $(x - 3)^2 + (y + 4)^2 = 25$; Center: $(3, -4)$; Radius: $5$

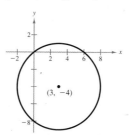

**31.** $(1, 1)$  **33.** $(0, 0), (1, 1), (-1, -1)$

**35. (a)** $C = 200 + 10x; R = 14x$  **(b)** $50$ shirts

**37.** Slope: $-3$

y-intercept: $(0, -2)$

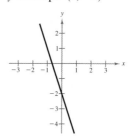

**39.** Slope: 0 (horizontal line)    **41.** Slope: $-\frac{2}{5}$

    $y$-intercept: $\left(0, -\frac{5}{3}\right)$      $y$-intercept: $(0, -1)$

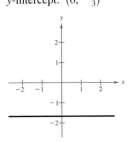 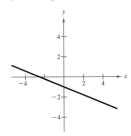

**43.** $\frac{6}{7}$    **45.** $\frac{20}{21}$

**47.** $y = -2x + 5$

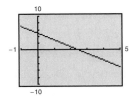

**49.** (a) $y = \frac{7}{8}x + \frac{69}{8}$   (b) $y = -2x$

    (c) $y = -2x$   (d) $y = -\frac{2}{3}x + 4$

**51.** (a) $x = -10p + 1070$   (b) 725 units   (c) 650 units

**53.** $y$ is a function of $x$.    **55.** $y$ is not a function of $x$.

**57.** (a) 7   (b) $3x + 7$   (c) $10 + 3\Delta x$

**59.** Domain: $(-\infty, \infty)$

    Range: $\left[-\frac{1}{4}, \infty\right)$

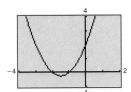

**61.** Domain: $[-1, \infty)$

    Range: $[0, \infty)$

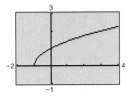

**63.** Domain: $(-\infty, \infty)$

    Range: $(-\infty, 3]$

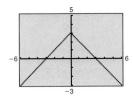

**65.** (a) $x^2 + 2x$   (b) $x^2 - 2x + 2$

    (c) $2x^3 - x^2 + 2x - 1$   (d) $\dfrac{1 + x^2}{2x - 1}$

    (e) $4x^2 - 4x + 2$   (f) $2x^2 + 1$

**67.** $f^{-1}(x) = \frac{2}{3}x$    **69.** $f(x)$ does not have an inverse.

**71.** 7    **73.** 49    **75.** $\frac{10}{3}$    **77.** Limit does not exist.

**79.** $-\frac{1}{4}$    **81.** $-\infty$    **83.** Limit does not exist.

**85.** $3x^2 - 1$    **87.** 0.5774    **89.** False    **90.** True

**91.** False    **92.** True    **93.** False    **94.** True

**95.** $(-\infty, -4) \cup (-4, \infty)$    **97.** $(-\infty, -1) \cup (-1, \infty)$

**99.** Continuous on all intervals of the form $(c, c + 1)$, where $c$ is an integer

**101.** $(-\infty, 0) \cup (0, \infty)$    **103.** $a = 2$

**105.** (a)

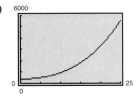

    (b)

| $t$ | 0 | 5 | 10 | 15 |
|---|---|---|---|---|
| Debt | 381 | 542 | 914 | 1817 |
| Model | 387.94 | 518.59 | 951.74 | 1837.39 |

| $t$ | 20 | 21 | 22 | 23 |
|---|---|---|---|---|
| Debt | 3266 | 3599 | 4083 | 4436 |
| Model | 3325.54 | 3708.67 | 4123.10 | 4570.03 |

    (c) \$8001.31

**107.** Yellow sweet maize: $y = -9x + 45$

    White flint maize: $y = -5.\overline{45}x + 30$

## SAMPLE POST-GRAD EXAM QUESTIONS
*(page 80)*

**1.** (d)    **2.** (d)    **3.** (c)    **4.** (a)    **5.** (e)

**6.** (c)    **7.** (e)    **8.** (b)

# CHAPTER 2

## SECTION 2.1 *(page 90)*

---

**Warm Up**

**1.** $x = 2$    **2.** $y = 2$    **3.** $2x$    **4.** $3x^2$

**5.** $\dfrac{1}{x^2}$    **6.** $2x$    **7.** $(-\infty, 1) \cup (1, \infty)$

**8.** $(-\infty, \infty)$    **9.** $(-\infty, 0) \cup (0, \infty)$

**10.** $(-\infty, -4) \cup (-4, 3) \cup (3, \infty)$

---

**1.**

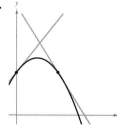

**3.**

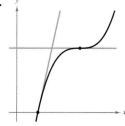

**5.** $m = 1$    **7.** $m = 0$    **9.** $m = -\frac{1}{3}$

**11.** 1991: $m \approx 600$

1995: $m \approx 1200$

**13.** $f(x) = 3$

   1. $f(x + \Delta x) = 3$

   2. $f(x + \Delta x) - f(x) = 0$

   3. $\dfrac{f(x + \Delta x) - f(x)}{\Delta x} = 0$

   4. $\displaystyle\lim_{\Delta x \to 0} \dfrac{f(x + \Delta x) - f(x)}{\Delta x} = 0$

**15.** $f(x) = -5x + 3$

   1. $f(x + \Delta x) = -5x - 5\Delta x + 3$

   2. $f(x + \Delta x) - f(x) = -5\Delta x$

   3. $\dfrac{f(x + \Delta x) - f(x)}{\Delta x} = -5$

   4. $\displaystyle\lim_{\Delta x \to 0} \dfrac{f(x + \Delta x) - f(x)}{\Delta x} = -5$

**17.** $f(x) = x^2$

   1. $f(x + \Delta x) = x^2 + 2x\Delta x + (\Delta x)^2$

   2. $f(x + \Delta x) - f(x) = 2x\Delta x + (\Delta x)^2$

   3. $\dfrac{f(x + \Delta x) - f(x)}{\Delta x} = 2x + \Delta x$

   4. $\displaystyle\lim_{\Delta x \to 0} \dfrac{f(x + \Delta x) - f(x)}{\Delta x} = 2x$

**19.** $6x - 5$

**21.** $h(t) = \sqrt{t - 1}$

   1. $h(t + \Delta t) = \sqrt{t + \Delta t - 1}$

   2. $h(t + \Delta t) - h(t) = \sqrt{t + \Delta t - 1} - \sqrt{t - 1}$

   3. $\dfrac{h(t + \Delta t) - h(t)}{\Delta t} = \dfrac{1}{\sqrt{t + \Delta t - 1} + \sqrt{t - 1}}$

   4. $\displaystyle\lim_{\Delta t \to 0} \dfrac{h(t + \Delta t) - h(t)}{\Delta t} = \dfrac{1}{2\sqrt{t - 1}}$

**23.** $f(t) = t^3 - 12t$

   1. $f(t + \Delta t) = t^3 + 3t^2\Delta t + 3t(\Delta t)^2$
$\qquad\qquad\quad + (\Delta t)^3 - 12t - 12\Delta t$

   2. $f(t + \Delta t) - f(t) = 3t^2\Delta t + 3t(\Delta t)^2 + (\Delta t)^3 - 12\Delta t$

   3. $\dfrac{f(t + \Delta t) - f(t)}{\Delta t} = 3t^2 + 3t\Delta t + (\Delta t)^2 - 12$

   4. $\displaystyle\lim_{\Delta t \to 0} \dfrac{f(t + \Delta t) - f(t)}{\Delta t} = 3t^2 - 12$

**25.** $-\dfrac{2}{x^3}$

**27.** $f'(x) = -2$          **29.** $f'(x) = -1$

$\quad\ f'(2) = -2$            $\quad\ f'(0) = -1$

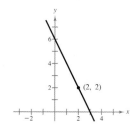

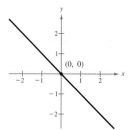

**31.** $f'(x) = 2x$
$f'(2) = 4$

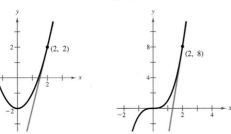

**33.** $f'(x) = 3x^2$
$f'(2) = 12$

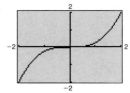

**35.** $f'(x) = 3x^2 + 2$
$f'(1) = 5$

**37.** $y = 2x - 2$

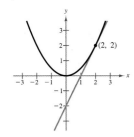

**39.** $y = -6x - 3$

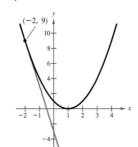

**41.** $y = \dfrac{x}{4} + 2$

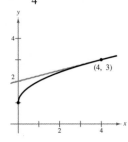

**43.** $y = -x + 1$

**45.** $y = -6x + 8$
$y = -6x - 8$

**47.** $x \neq -3$ (node)     **49.** $x \neq 3$ (cusp)     **51.** $x > 1$

**53.** $x \neq 0$ (nonremovable discontinuity)

**55.**

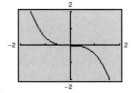

| $x$ | $-2$ | $-\frac{3}{2}$ | $-1$ | $-\frac{1}{2}$ |
|---|---|---|---|---|
| $f(x)$ | $-2$ | $-0.8438$ | $-0.25$ | $-0.0313$ |
| $f'(x)$ | $3$ | $1.6875$ | $0.75$ | $0.1875$ |

| $x$ | $0$ | $\frac{1}{2}$ | $1$ | $\frac{3}{2}$ | $2$ |
|---|---|---|---|---|---|
| $f(x)$ | $0$ | $0.0313$ | $0.25$ | $0.8438$ | $2$ |
| $f'(x)$ | $0$ | $0.1875$ | $0.75$ | $1.6875$ | $3$ |

**57.**

| $x$ | $-2$ | $-\frac{3}{2}$ | $-1$ | $-\frac{1}{2}$ |
|---|---|---|---|---|
| $f(x)$ | $4$ | $1.6875$ | $0.5$ | $0.0625$ |
| $f'(x)$ | $-6$ | $-3.375$ | $-1.5$ | $-0.375$ |

| $x$ | $0$ | $\frac{1}{2}$ | $1$ | $\frac{3}{2}$ | $2$ |
|---|---|---|---|---|---|
| $f(x)$ | $0$ | $-0.0625$ | $-0.5$ | $-1.6875$ | $-4$ |
| $f'(x)$ | $0$ | $-0.375$ | $-1.5$ | $-3.375$ | $-6$ |

**59.** True

**60.** False. $f(x) = |x|$ is continuous, but not differentiable, at $x = 0$.

**61.** True

**62.** True

**63.** $f'(x) = 2x - 4$

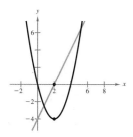

The $x$-intercept of the derivative indicates a point of horizontal tangency for $f$.

**65.**

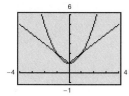

The graph of $f$ is smooth at $(0, 1)$ but the graph of $g$ is a sharp point at $(0, 1)$. The function $g$ is not differentiable at $x = 0$.

## SECTION 2.2    (page 102)

---

### Warm Up

**1.** (a) $8$  (b) $16$  (c) $\frac{1}{2}$

**2.** (a) $\frac{1}{36}$  (b) $\frac{1}{32}$  (c) $\frac{1}{64}$

**3.** $4x(3x^2 + 1)$  **4.** $\frac{3}{2}x^{1/2}(x^{3/2} - 1)$  **5.** $\frac{1}{4x^{3/4}}$

**6.** $x^2 - \frac{1}{x^{1/2}} + \frac{1}{3x^{2/3}}$  **7.** $0, -\frac{2}{3}$

**8.** $0, \pm 1$  **9.** $-10, 2$  **10.** $-2, 12$

---

**1.** (a) $2$  (b) $\frac{1}{2}$  **3.** (a) $-1$  (b) $-\frac{1}{3}$  **5.** $0$  **7.** $1$

**9.** $2x$  **11.** $-6t + 2$  **13.** $3t^2 - 2$  **15.** $\frac{16}{3}t^{1/3}$

**17.** $\frac{2}{\sqrt{x}}$  **19.** $-\frac{8}{x^3} + 4x$

**21.** Function: $y = \dfrac{1}{4x^3}$

Rewrite: $y = \dfrac{1}{4}x^{-3}$

Differentiate: $y' = \dfrac{-3}{4}x^{-4}$

Simplify: $y' = -\dfrac{3}{4x^4}$

**23.** Function: $y = \dfrac{1}{(4x)^3}$

Rewrite: $y = \dfrac{1}{64}x^{-3}$

Differentiate: $y' = -\dfrac{3}{64}x^{-4}$

Simplify: $y' = -\dfrac{3}{64x^4}$

**25.** Function: $y = \dfrac{\sqrt{x}}{x}$

Rewrite: $y = x^{-1/2}$

Differentiate: $y' = -\dfrac{1}{2}x^{-3/2}$

Simplify: $y' = -\dfrac{1}{2x^{3/2}}$

**27.** $-\dfrac{1}{x^2}, -1$  **29.** $\dfrac{4}{3t^2}, \dfrac{16}{3}$  **31.** $8x + 4, 4$

**33.** $\dfrac{2(x^3 + 2)}{x^2}$  **35.** $2x - 2 + \dfrac{8}{x^5}$  **37.** $\dfrac{2x^3 - 6}{x^3}$

**39.** $3x^2 + 1$  **41.** $\dfrac{4}{5x^{1/5}}$  **43.** $y = 2x - 2$

**45.** $y = \dfrac{8}{15}x + \dfrac{22}{15}$  **47.** $(0, -1), \left(-\dfrac{\sqrt{6}}{2}, \dfrac{5}{4}\right), \left(\dfrac{\sqrt{6}}{2}, \dfrac{5}{4}\right)$

**49.** No horizontal tangents

**51.** (a)

(b) $f'(1) = g'(1)$
$= h'(1)$
$= 3$

**(c)**

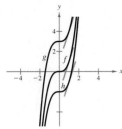

**53. (a)** 3  **(b)** 6  **(c)** $-3$  **(d)** 6

**55. (a)** 1992: $-265.4$

1995: 1684.6

**(b)** The results of part (a) are more accurate.

**(c)** Millions of dollars per year

**57.** $P = 0.40x - 250$

**59.**

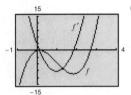

$(0.11, 0.14), (1.84, -10.49)$

## SECTION 2.3    *(page 116)*

**Warm Up**

**1.** 3    **2.** $-7$    **3.** $y' = 8x - 2$

**4.** $y' = -9t^2 + 4t$    **5.** $s' = -32t + 24$

**6.** $y' = -32x + 54$    **7.** $A' = -\frac{3}{5}r^2 + \frac{3}{5}r + \frac{1}{2}$

**8.** $y' = 2x^2 - 4x + 7$    **9.** $y' = 12 - \dfrac{x}{2500}$

**10.** $y' = 74 - \dfrac{3x^2}{10,000}$

**1. (a)** \$3.08 billion per year

**(b)** \$6.6 billion per year

**(c)** \$2.7 billion per year

**(d)** \$2.995 billion per year

**3.** Average rate: 2

Instantaneous rates: $f'(1) = 2, f'(2) = 2$

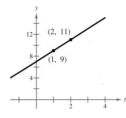

**5.** Average rate: 0

Instantaneous rates: $h'(-2) = -4, h'(2) = 4$

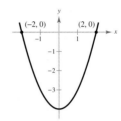

**7.** Average rate: $-\frac{1}{4}$

Instantaneous rates: $f'(1) = -1, f'(4) = -\frac{1}{16}$

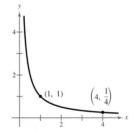

**9.** Average rate: 36

Instantaneous rates: $g'(1) = 2, g'(3) = 102$

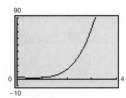

**11. (a)** Average rate: $\frac{11}{27}$

Instantaneous rates: $E'(0) = \frac{1}{3}, E'(1) = \frac{4}{9}$

**(b)** Average rate: $\frac{11}{27}$

Instantaneous rates: $E'(1) = \frac{4}{9}, E'(2) = \frac{1}{3}$

(c) Average rate: $\frac{5}{27}$

Instantaneous rates: $E'(2) = \frac{1}{3}$, $E'(3) = 0$

(d) Average rate: $-\frac{7}{27}$

Instantaneous rates: $E'(3) = 0$, $E'(4) = -\frac{5}{9}$

**13. (a)** $-80$ ft/sec

**(b)** $s'(2) = -64$ ft/sec, $s'(3) = -96$ ft/sec

**(c)** $\dfrac{\sqrt{555}}{4} \approx 5.89$ sec  **(d)** $-8\sqrt{555} \approx -188.5$ sec

**15. (a)** $s'(0) = 0$ ft/sec  **(b)** $s'(1) = 15$ ft/sec

**(c)** $s'(4) = 30$ ft/sec  **(d)** $s'(9) = 45$ ft/sec

**17.** 1.47  **19.** $470 - \frac{1}{2}x$  **21.** $50 - x$

**23.** $-18x^2 + 16x + 200$  **25.** $-4x + 72$

**27.** $-\frac{1}{2000}x + 12.2$

**29. (a)** 2394  **(b)** 2416  **(c)** Answers vary.

**31. (a)** \$0.50  **(b)** \$0.60  **(c)** Answers vary.

**33. (a)**

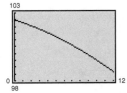

**(b)** Negative; the patient's temperature is decreasing.

**(c)** 102.5, 101.54, 100.26, 98.66

**(d)** $\dfrac{dT}{dt} = -0.02t - 0.2$; it is the rate of change of the patient's temperature with respect to time.

**(e)** $-0.2$, $-0.28$, $-0.36$, $-0.44$

**35. (a)** $R = 5x - 0.001x^2$

**(b)** $P = 3.5x - 0.001x^2 - 35$

**(c)**

| $x$ | 600 | 1200 | 1800 | 2400 | 3000 |
|---|---|---|---|---|---|
| $\dfrac{dR}{dx}$ | 3.8 | 2.6 | 1.4 | 0.2 | $-1.0$ |
| $\dfrac{dP}{dx}$ | 2.3 | 1.1 | $-0.1$ | $-1.3$ | $-2.5$ |
| $P$ | 1705 | 2725 | 3025 | 2605 | 1465 |

**37. (a)** $P = -0.0003x^2 + 17.8x - 85,000$

**(b)**

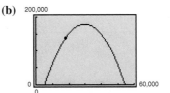

The slope of $P$ is positive at $x = 18,000$.

**(c)** 5.8 dollars

**39.** Answers vary.

**41.** $C(351) - C(350) \approx -1.91$

$\dfrac{dC}{dQ} = -1.93$ when $Q = 350$.

**43.** $C = \dfrac{16,500}{x}$

$\dfrac{dC}{dx} = -\dfrac{16,500}{x^2}$

| $x$ | 10 | 15 | 20 | 25 |
|---|---|---|---|---|
| $C$ | \$1650 | \$1100 | \$825 | \$660 |
| $\dfrac{dC}{dx}$ | $-165$ | $-73.33$ | $-41.25$ | $-26.40$ |

| $x$ | 30 | 35 | 40 |
|---|---|---|---|
| $C$ | \$550 | \$471.43 | \$412.50 |
| $\dfrac{dC}{dx}$ | $-18.33$ | $-13.47$ | $-10.31$ |

The driver who gets 15 miles per gallon would benefit more from a 1 mile per gallon increase in fuel efficiency. The rate of change is much larger when $x = 15$.

**45.**

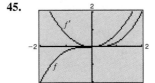

$f$ has a horizontal tangent at $x = 0$.

**47.** The population in each phase is increasing. During the acceleration phase the population's growth is the greatest. Therefore the slopes of the tangent lines are greater than the slopes of the tangent lines during the lag phase and the deceleration phase. Possible reasons for the changing rates could be seasonal growth or food supplies.

## SECTION 2.4    *(page 128)*

---

### Warm Up

**1.** $2(3x^2 + 7x + 1)$     **2.** $4x^2(6 - 5x^2)$

**3.** $\dfrac{23}{(2x + 7)^2}$     **4.** $-\dfrac{x^2 + 8x + 4}{(x^2 - 4)^2}$

**5.** $\dfrac{4x^3 - 3x^2 + 3}{x^2}$     **6.** $\dfrac{x^2 - 2x + 4}{(x - 1)^2}$

**7.** $11$    **8.** $0$    **9.** $-\frac{1}{4}$    **10.** $\frac{17}{4}$

---

**1.** $f'(x) = 15x^4 - 2x$

    $f'(1) = 13$

**3.** $f'(x) = 2x^2$

    $f'(0) = 0$

**5.** $g'(x) = 3x^2 - 12x + 11$

    $g'(4) = 11$

**7.** $h'(x) = -\dfrac{5}{(x - 5)^2}$

    $h'(6) = -5$

**9.** $f'(t) = \dfrac{2t^2 + 3}{3t^2}$

    $f'(2) = \dfrac{11}{12}$

**11.** Function: $y = \dfrac{x^2 + 2x}{x}$

    Rewrite: $y = x + 2$

    Differentiate: $y' = 1$

    Simplify: $y' = 1$

**13.** Function: $y = \dfrac{7}{3x^3}$

    Rewrite: $y = \dfrac{7}{3}x^{-3}$

    Differentiate: $y' = -7x^{-4}$

    Simplify: $y' = -\dfrac{7}{x^4}$

**15.** Function: $y = \dfrac{4x^2 - 3x}{8\sqrt{x}}$

    Rewrite: $y = \dfrac{1}{2}x^{3/2} - \dfrac{3}{8}x^{1/2}$

    Differentiate: $y' = \dfrac{3}{4}x^{1/2} - \dfrac{3}{16}x^{-1/2}$

    Simplify: $y' = \dfrac{3}{4}\sqrt{x} - \dfrac{3}{16\sqrt{x}}$

**17.** Function: $y = \dfrac{x^2 - 4x + 3}{x - 1}$

    Rewrite: $y = x - 3, x \neq 1$

    Differentiate: $y' = 1, x \neq 1$

    Simplify: $y' = 1, x \neq 1$

**19.** $10x^4 + 12x^3 - 3x^2 - 18x - 15$

**21.** $-12t^5 + 4t^3 + 22t$

**23.** $\dfrac{5}{6x^{1/6}} + \dfrac{1}{x^{2/3}}$    **25.** $-\dfrac{5}{(2x - 3)^2}$    **27.** $\dfrac{2}{(x + 1)^2}$

**29.** $\dfrac{x^2 + 2x - 1}{(x + 1)^2}$    **31.** $\dfrac{-10(s + 1)}{(s^2 + 2s + 2)^2}$

**33.** $\dfrac{2x^3 + 11x^2 - 8x - 17}{(x + 4)^2}$

**35.** $y = 5x - 2$

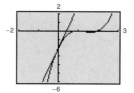

**37.** $y = -\frac{3}{4}x + \frac{1}{4}$

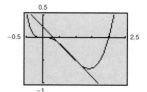

**39.** $y = -4x - 3$

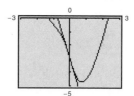

**41.** $(0, 0)$ and $(2, 4)$     **43.** $(0, 0)$ and $\left(\sqrt[3]{-4}, -2.117\right)$

**45.**

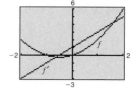

**47.**

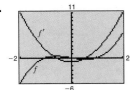

**49.** $-1.87$     **51. (a)** $-0.480$   **(b)** $0.120$   **(c)** $0.015$

**53.** $31.55$ bacteria/hr

**55. (a)** $p = \dfrac{4000}{\sqrt{x}}$

    **(b)** $C = 250x + 10{,}000$

    **(c)** $P = 4000\sqrt{x} - 250x - 10{,}000$

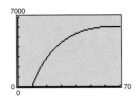

      $500 per unit

**57. (a)**               **(b)**

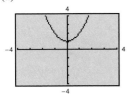  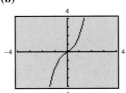

**(c)**

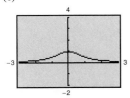

The graph of (c) would most likely represent a demand function. As the number of units increases, demand is likely to decrease, not increase as in (a) and (b).

**59. (a)** $-38.125$

    **(b)** $-10.37$

    **(c)** $-3.80$

Increasing the order size reduces the cost per item.

**61.** $\dfrac{dC}{dt} = -\dfrac{10(t^2 - 42t + 278)}{(t^2 - 18t + 100)^2};$

$-0.28, -0.76, 1.05, 0.42;$

In 1980, the cost per mile of owning a car was decreasing at a rate of 0.28 cents per mile per year. In 1985, the cost was decreasing at a rate of 0.76 cents per mile per year. In 1990, the cost was increasing at a rate of 1.05 cents per mile per year, and in 1995, the cost was increasing at a rate of 0.42 cents per mile per year.

## SECTION 2.5     *(page 138)*

**Warm Up**

   **1.** $(1 - 5x)^{2/5}$     **2.** $(2x - 1)^{3/4}$

   **3.** $(4x^2 + 1)^{-1/2}$     **4.** $(x - 6)^{-1/3}$

   **5.** $x^{1/2}(1 - 2x)^{-1/3}$     **6.** $(2x)^{-1}(3 - 7x)^{3/2}$

   **7.** $(x - 2)(3x^2 + 5)$     **8.** $(x - 1)\left(5\sqrt{x} - 1\right)$

   **9.** $(x^2 + 1)^2(4 - x - x^3)$

  **10.** $(3 - x^2)(x - 1)(x^2 + x + 1)$

| | $y = f(g(x))$ | $u = g(x)$ | $y = f(u)$ |
|---|---|---|---|
| **1.** | $y = (6x - 5)^4$ | $u = 6x - 5$ | $y = u^4$ |
| **3.** | $y = (4 - x^2)^{-1}$ | $u = 4 - x^2$ | $y = u^{-1}$ |
| **5.** | $y = \sqrt{5x - 2}$ | $u = 5x - 2$ | $y = \sqrt{u}$ |
| **7.** | $y = \dfrac{1}{3x + 1}$ | $u = 3x + 1$ | $y = u^{-1}$ |

**9.** c     **10.** d     **11.** b     **12.** a     **13.** $6(2x - 7)^2$

**15.** $-6(4 - 2x)^2$     **17.** $6x(6 - x^2)(2 - x^2)$

**19.** $\dfrac{4x}{3(x^2 - 9)^{1/3}}$    **21.** $\dfrac{1}{2\sqrt{t + 1}}$    **23.** $\dfrac{4t + 5}{2\sqrt{2t^2 + 5t + 2}}$

**25.** $\dfrac{6x}{(9x^2 + 4)^{2/3}}$    **27.** $\dfrac{27}{4(2 - 9x)^{3/4}}$    **29.** $\dfrac{4x^2}{(4 - x^3)^{7/3}}$

**31.** $y = 216x - 378$       **33.** $y = \frac{8}{3}x - \frac{7}{3}$

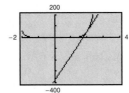  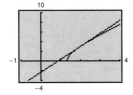

**35.** $y = x - 1$

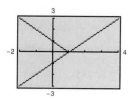

**37.** $f'(x) = \dfrac{1 - 3x^2 - 4x^{3/2}}{2\sqrt{x}(x^2 + 1)^2}$

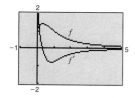

The zero of $f'(x)$ corresponds to the point on the graph of $f(x)$ where the tangent line is horizontal.

**39.** $f'(x) = -\dfrac{\sqrt{(x + 1)/x}}{2x(x + 1)}$

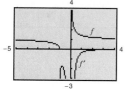

$f'(x)$ has no zeros.

**41.** $-\dfrac{1}{(x - 2)^2}$   **43.** $\dfrac{8}{(t + 2)^3}$   **45.** $-\dfrac{2(2x - 3)}{(x^2 - 3x)^3}$

**47.** $-\dfrac{2t}{(t^2 - 2)^2}$   **49.** $\dfrac{3(x + 1)}{\sqrt{2x + 3}}$   **51.** $\dfrac{t(5t - 8)}{2\sqrt{t - 2}}$

**53.** $27(x - 3)^2(4x - 3)$   **55.** $-\dfrac{3}{4x^{3/2}\sqrt{3 - 2x}}$

**57.** $-\dfrac{1}{x^2(x^3 + 1)^{2/3}}$   **59.** $\dfrac{2(6 - 5x)(5x^2 - 12x + 5)}{(x^2 - 1)^3}$

**61.** $y = \frac{8}{3}t + 4$   **63.** $y = -6t - 14$

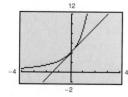
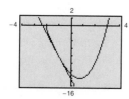

**65.** $y = -5x + 13$

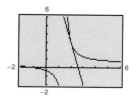

**67.** (a) $74.00 per 1%   (b) $81.59 per 1%
(c) $89.94 per 1%

**69.**

| $t$ | 0 | 1 | 2 | 3 | 4 |
|---|---|---|---|---|---|
| $\dfrac{dN}{dt}$ | 0 | 177.78 | 44.44 | 10.82 | 3.29 |

The rate of growth of $N$ is decreasing.

**71.** (a) $V = \dfrac{10,000}{\sqrt[3]{t + 1}}$

(b) $-$1322.83 per year

(c) $-$524.97 per year

## SECTION 2.6    (page 145)

**Warm Up**

**1.** $t = 0, \frac{3}{2}$   **2.** $t = -2, 7$   **3.** $t = -2, 10$

**4.** $t = \dfrac{9 \pm 3\sqrt{10,249}}{32}$   **5.** $\dfrac{dy}{dx} = 6x^2 + 14x$

**6.** $\dfrac{dy}{dx} = 8x^3 + 18x^2 - 10x - 15$

**7.** $\dfrac{dy}{dx} = \dfrac{2x(x + 7)}{(2x + 7)^2}$   **8.** $\dfrac{dy}{dx} = -\dfrac{6x^2 + 10x + 15}{(2x^2 - 5)^2}$

**9.** Domain: $(-\infty, \infty)$   **10.** Domain: $[7, \infty)$
Range: $[-4, \infty)$   Range: $[0, \infty)$

**1.** 0   **3.** 2   **5.** $2t - 8$   **7.** $\dfrac{9}{2t^4}$

**9.** $18(2 - x^2)(5x^2 - 2)$   **11.** $\dfrac{4}{(x - 1)^3}$

**13.** $12x^2 + 24x + 16$   **15.** $60x^2 - 72x$

**17.** $120x + 360$   **19.** $-\dfrac{9}{2x^5}$   **21.** 260   **23.** $-\dfrac{1}{648}$

**25.** $4x$     **27.** $\dfrac{2}{x^2}$     **29.** $2$

**31.** $f''(x) = 6(x - 3) = 0$ when $x = 3$.

**33.** $f''(x) = 2(3x + 4) = 0$ when $x = -\frac{4}{3}$.

**35.** $f''(x) = 36(x^2 - 1) = 0$ when $x = \pm 1$.

**37.** $f''(x) = \dfrac{2x(x + 3)(x - 3)}{(x^2 + 3)^3}$

$= 0$ when $x = 0$ or $x = \pm 3$.

**39.** **(a)** $s(t) = -16t^2 + 144t$

**(b)** $v(t) = -32t + 144$

$a(t) = -32$

**(c)** 4.5 sec, 324 ft

**(d)** $-144$ ft/sec, which is the same speed as the initial velocity.

**41.**

| $t$ | 0 | 10 | 20 | 30 |
|---|---|---|---|---|
| $\dfrac{ds}{dt}$ | 0 | 45 | 60 | 67.5 |
| $\dfrac{d^2s}{dt^2}$ | 9.00 | 2.25 | 1.00 | 0.56 |

| $t$ | 40 | 50 | 60 |
|---|---|---|---|
| $\dfrac{ds}{dt}$ | 72 | 75 | 77.14 |
| $\dfrac{d^2s}{dt^2}$ | 0.36 | 0.25 | 0.18 |

As $t$ increases, the velocity of the car increases at a diminishing rate (acceleration decreases).

**43.**

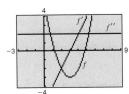

The degrees of the successive derivatives decrease by 1.

**45.**

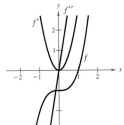

**47.** False. The product rule is $[f(x)g(x)]'$
$= f'(x)g(x) + f(x)g'(x)$.

**48.** True     **49.** True     **50.** True

**51.** True     **52.** True

**53.** False. Let $f(x) = x^2$ and $g(x) = x^2 + 1$.

## SECTION 2.7     *(page 152)*

**Warm Up**

**1.** $y = x^2 - 2x$     **2.** $y = \dfrac{x - 3}{4}$

**3.** $y = 1, x \neq -6$     **4.** $y = -4, x \neq \pm\sqrt{3}$

**5.** $y = \pm\sqrt{5 - x^2}$     **6.** $y = \pm\sqrt{6 - x^2}$     **7.** $\frac{8}{3}$

**8.** $-\frac{1}{2}$     **9.** $\frac{5}{7}$     **10.** 1

**1.** $-\dfrac{y}{x}$     **3.** $-\dfrac{x}{y}$     **5.** $-\dfrac{1}{5}$     **7.** $-\dfrac{x}{y}, 0$

**9.** $-\dfrac{y}{x + 1}, -\dfrac{1}{4}$     **11.** $\dfrac{y - 3x^2}{2y - x}, \dfrac{1}{2}$     **13.** $\dfrac{1 - 3x^2y^3}{3x^3y^2 - 1}, -1$

**15.** $-\sqrt{\dfrac{y}{x}}, -\dfrac{5}{4}$     **17.** $-\sqrt[3]{\dfrac{y}{x}}, -\dfrac{1}{2}$     **19.** $3x, 3$

**21.** $-\dfrac{4x}{9y}, -\dfrac{\sqrt{5}}{3}$     **23.** $-\dfrac{x}{y}, \dfrac{4}{3}$     **25.** $-\dfrac{9x}{16y}, -\dfrac{\sqrt{3}}{4}$

**27.** At $(5, 12)$: $5x + 12y - 169 = 0$

At $(-12, 5)$: $12x - 5y + 169 = 0$

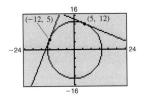

**29.** At $\left(1, \sqrt{5}\right)$: $15x - 2\sqrt{5}y - 5 = 0$

At $\left(1, -\sqrt{5}\right)$: $15x + 2\sqrt{5}y - 5 = 0$

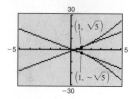

**31.** At $(0, 2)$: $y = 2$

At $(2, 0)$: $x = 2$

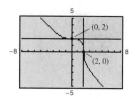

**33.** $\dfrac{1}{0.024x^3 + 0.04x}$     **35.** $-\dfrac{x^2}{100}$

**37.** (a) $-2$

(b)

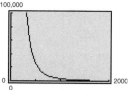

As more labor is used, less capital is available.
As more capital is used, less labor is available.

## SECTION 2.8    *(page 160)*

**Warm Up**

**1.** $A = \pi r^2$    **2.** $V = \frac{4}{3}\pi r^3$    **3.** $S = 6s^2$

**4.** $V = s^3$    **5.** $V = \frac{1}{3}\pi r^2 h$    **6.** $A = \frac{1}{2}bh$

**7.** $-\dfrac{x}{y}$    **8.** $\dfrac{2x - 3y}{3x}$    **9.** $-\dfrac{2x + y}{x + 2}$

**10.** $-\dfrac{y^2 - y + 1}{2xy - 2y - x}$

**1.** (a) $62$   (b) $\frac{32}{85}$    **3.** (a) $-\frac{5}{8}$   (b) $\frac{3}{2}$

**5.** (a) $24\pi$ in.²/min   (b) $96\pi$ in.²/min

**7.** If $\dfrac{dr}{dt}$ is constant, $\dfrac{dA}{dt} = 2\pi r \dfrac{dr}{dt}$ and thus is proportional to $r$.

**9.** (a) $\dfrac{5}{\pi}$ ft/min   (b) $\dfrac{5}{4\pi}$ ft/min    **11.** $137,500 per week

**13.** (a) $9$ cm³/sec   (b) $900$ cm³/sec

**15.** (a) $-12$ cm/min   (b) $0$ cm/min

    (c) $4$ cm/min   (d) $12$ cm/min

**17.** (a) $-\frac{7}{12}$ ft/sec   (b) $-\frac{3}{2}$ ft/sec   (c) $-\frac{48}{7}$ ft/sec

**19.** (a) $-750$ mi/hr   (b) $20$ min    **21.** $-8.33$ ft/sec

**23.** $650 per week

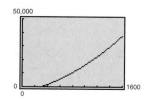

**25.** (a) $53.88 million per year

    (b) $165 million per year

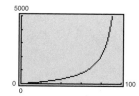

As $p$ approaches $100$, $C$ increases without bound.

## CHAPTER 2 REVIEW EXERCISES    *(page 166)*

**1.** $-2$    **3.** $0$

**5.** $t = 2$: slope $\approx$ $63.6$ million per year (revenues are increasing)

   $t = 4$: slope $\approx$ $292.9$ million per year (revenues are increasing)

**7.** $-3$    **9.** $\dfrac{1}{4}$    **11.** $7$    **13.** $-\dfrac{1}{(x - 5)^2}$

**15.** $-5$    **17.** $\frac{1}{6}$    **19.** $y = -2x + 6$    **21.** $1$

**23.** $0$    **25.** $0$    **27.** $5x^4$    **29.** $\dfrac{1}{2\sqrt{x}}$    **31.** $12x^3$

**33.** $y = -\frac{4}{3}t + 2$　　**35.** $y = -34x - 27$

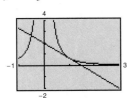　　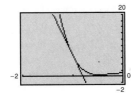

**37.** $y = x - 1$

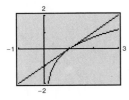

**39.** Average rate of change: 4

Instantaneous rate of change when $x = 0$: 3

Instantaneous rate of change when $x = 1$: 5

**41.** (a) \$136.63 million per year

(b) 1992: \$63.59 million per year

1996: $-\$516.73$ million per year

(c) Revenues were increasing in 1992 and decreasing in 1996, but grew during the period 1993–1995.

**43.** (a) $dP/dt = 0.0125t^4 - 0.108t^3 + 0.27t^2 - 0.176t - 0.067$

(b) 1992: $-\$0.003$ per pound

1995: \$0.116 per pound

(c) 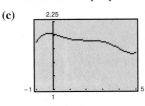　The price is increasing from approximately 1989 to 1990, and decreasing from 1990 to 1995.

(d) Positive slope: $-1 < t < 0$ (actual: $-1 < t < -0.3$)

Negative slope: $0 < t < 5$ (actual: $-0.3 < t < 4.8$)

(e) When the price increases, the slope is positive. When the price decreases, the slope is negative.

**45.** (a) $s(t) = -16t^2 + 276$　(b) $-32$ ft/sec

(c) $t = 2$: $-64$ ft/sec　(d) 4.15 sec

$t = 3$: $-96$ ft/sec

(e) 132.8 ft/sec

**47.** $R = 27.50x$

$C = 15x + 2500$

$P = 12.50x - 2500$

**49.** $\dfrac{dC}{dx} = 320$　**51.** $\dfrac{dR}{dx} = \dfrac{35(x - 4)}{2(x - 2)^{3/2}}$

**53.** $\dfrac{dP}{dx} = -0.0006x^2 + 12x - 1$　**55.** $15x^2(1 - x^2)$

**57.** $16x^3 - 33x^2 + 12x$　**59.** $-\dfrac{2(3x^2 - 5x - 3)}{(x^2 + 1)^2}$

**61.** $30x(5x^2 + 2)^2$　**63.** $-\dfrac{1}{(x + 1)^{3/2}}$　**65.** $\dfrac{2x^2 + 1}{\sqrt{x^2 + 1}}$

**67.** $32x(1 - 4x^2)$　**69.** $18x^5(x + 1)(2x + 3)^2$

**71.** $x(x - 1)^4(7x - 2)$　**73.** $\dfrac{3(9t + 5)}{2\sqrt{3t + 1}(1 - 3t)^3}$

**75.** (a) $t = 1$: $-6.63$　$t = 3$: $-6.5$

$t = 5$: $-4.33$　$t = 10$: $-1.36$

(b) 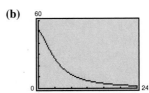　The rate of decrease is approaching zero.

**77.** 6　**79.** $-\dfrac{120}{x^6}$　**81.** $\dfrac{35x^{3/2}}{2}$　**83.** $\dfrac{2}{x^{2/3}}$

**85.** (a) $s(t) = -16t^2 + 5t + 30$　(b) 1.534 sec

(c) $-44.09$ ft/sec　(d) $-32$ ft/sec$^2$

**87.** $-\dfrac{2x + 3y}{3(x + y^2)}$　**89.** $\dfrac{x}{y}$　**91.** $y = x - 1$

**93.** $y = -\frac{2}{9}x + \frac{52}{9}$　**95.** $\frac{1}{64}$ ft/min

## SAMPLE POST-GRAD EXAM QUESTIONS

*(page 170)*

**1.** (c)　**2.** (b)　**3.** (e)　**4.** (a)　**5.** (c)　**6.** (a)

# CHAPTER 3

## SECTION 3.1    *(page 179)*

---

**Warm Up**

**1.** $x = 0, x = 8$    **2.** $x = 0, x = 24$    **3.** $x = \pm 5$

**4.** $x = 0$    **5.** $(-\infty, 3) \cup (3, \infty)$    **6.** $(-\infty, 1)$

**7.** $(-\infty, -2) \cup (-2, 5) \cup (5, \infty)$    **8.** $\left(-\sqrt{3}, \sqrt{3}\right)$

**9.** $x = -2: -\frac{103}{3}$    **10.** $x = -2: -\frac{7}{9}$

$x = 0: \frac{1}{3}$    $x = 0: 3$

$x = 2: -13$    $x = 2: -\frac{1}{3}$

---

**1.** $f'(-1) = -\frac{8}{25}$    **3.** $f'(-3) = -\frac{2}{3}$

$f'(0) = 0$    $f'(-2)$ undefined

$f'(1) = \frac{8}{25}$    $f'(-1) = \frac{2}{3}$

**5.** Increasing on $(-\infty, -1)$

Decreasing on $(-1, \infty)$

**7.** Increasing on $(-1, 0)$ and $(1, \infty)$

Decreasing on $(-\infty, -1)$ and $(0, 1)$

**9.** No critical numbers

Increasing on $(-\infty, \infty)$

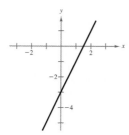

**11.** Critical number: $x = 1$

Increasing on $(-\infty, 1)$

Decreasing on $(1, \infty)$

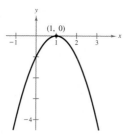

**13.** Critical number: $x = \frac{5}{2}$

Decreasing on $\left(-\infty, \frac{5}{2}\right)$

Increasing on $\left(\frac{5}{2}, \infty\right)$

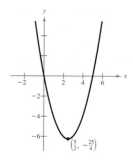

**15.** Critical numbers: $x = 0, x = 4$

Increasing on $(-\infty, 0)$

and $(4, \infty)$

Decreasing on $(0, 4)$

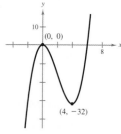

**17.** Critical number: $x = -1$

Decreasing on $(-\infty, \infty)$

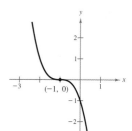

**19.** Critical number: $x = 1$

Increasing on $(-\infty, 1)$

Decreasing on $(1, \infty)$

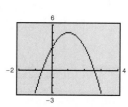

**21.** Critical numbers: $x = -1, x = -\frac{5}{3}$

Increasing on $\left(-\infty, -\frac{5}{3}\right)$ and $(-1, \infty)$

Decreasing on $\left(-\frac{5}{3}, -1\right)$

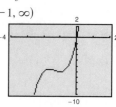

**23.** Critical number: $x = 0$

Decreasing on $(-\infty, 0)$

Increasing on $(0, \infty)$

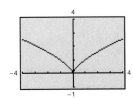

**25.** Critical numbers: $x = 0$, $x = \frac{3}{2}$

Decreasing on $\left(-\infty, \frac{3}{2}\right)$

Increasing on $\left(\frac{3}{2}, \infty\right)$

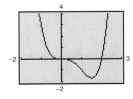

**27.** Critical numbers: $x = 2$, $x = -2$

Decreasing on $(-\infty, -2)$ and $(2, \infty)$

Increasing on $(-2, 2)$

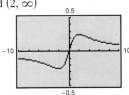

**29.** Critical numbers: $x = -1$, $x = 1$

Discontinuity: $x = 0$

Increasing on $(-\infty, -1)$ and $(1, \infty)$

Decreasing on $(-1, 0)$ and $(0, 1)$

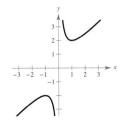

**31.** No critical numbers

Discontinuities: $x = \pm 4$

Increasing on $(-\infty, -4), (-4, 4),$ and $(4, \infty)$

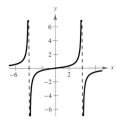

**33.** Critical number: $x = 0$

Discontinuity: $x = 0$

Increasing on $(-\infty, 0)$

Decreasing on $(0, \infty)$

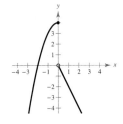

**35.** **(a)** Decreasing on $[1, 4.10)$

Increasing on $(4.10, \infty)$

**(b)**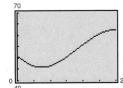

**(c)** $C = 900$ when $x = 2$ and $x = 15$. Use an order size of $x = 4$, which will minimize the cost $C$.

**37.** Moving upward on $(0, 3)$

Moving downward on $(3, 6)$

**39.** **(a)**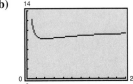

Decreasing on approximately $(0, 6)$

Increasing on approximately $(6, 24)$

**(b)** Decreasing on $(0, 5.85)$

Increasing on $(5.85, 23.6)$

# SECTION 3.2   *(page 189)*

---

**Warm Up**

**1.** $0, \pm\frac{1}{2}$   **2.** $-2, 5$   **3.** 1   **4.** 0, 125

**5.** Negative   **6.** Positive   **7.** Positive

**8.** Negative   **9.** Increasing   **10.** Decreasing

---

**1.** Relative maximum: $(1, 5)$

**3.** Relative minimum: $(3, -9)$

**5.** Relative maximum: $\left(\frac{2}{3}, \frac{28}{9}\right)$

  Relative minimum: $(1, 3)$

**7.** No relative extrema

**9.** Relative maximum: $(0, 15)$

  Relative minimum: $(4, -17)$

**11.** Relative minimum: $\left(\frac{3}{2}, -\frac{27}{16}\right)$

**13.**
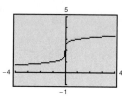

No extrema

**15.**
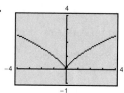

Relative minimum: $(0, 0)$

**17.**
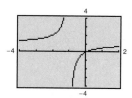

No extrema

**19.** Minimum: $(2, 2)$

  Maximum: $(-1, 8)$

**21.** Maximum: $(0, 5)$

  Minimum: $(3, -13)$

**23.** Minima: $(-1, -4)$ and $(2, -4)$

  Maxima: $(0, 0)$ and $(3, 0)$

**25.** Maximum: $(2, 1)$

  Minimum: $\left(0, \frac{1}{3}\right)$

**27.** Maximum: $(4, 16)$

  Minimum: $(2, -16)$

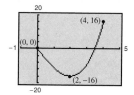

**29.** Maximum: $(-7, 4)$

  Minimum: $(1, 0)$

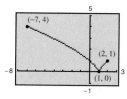

**31.** Maximum: $(5, 7)$

  Minimum: $(2.69, -5.55)$

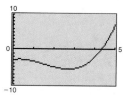

**33.** Maximum: $(1, 4.7)$

  Minimum: $(0.44, -1.06)$

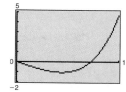

**35.** Minimum: $(0, 0)$

  Maximum: $(1, 2)$

**37.** Maximum: $\left(2, \frac{1}{2}\right)$

**39.** $\left| f''\left(\frac{\sqrt{3}}{3}\right) \right| = \frac{40\sqrt{3}}{3}$

**41.** $\left| f^{(4)}\left(\frac{1}{2}\right) \right| = 360$   **43.** 82 units   **45.** \$2.10

**47.** 1982; there were approximately 94 males for every 100 females in 1982.

## SECTION 3.3 *(page 198)*

> ### Warm Up
>
> **1.** $f''(x) = 48x^2 - 54x$
>
> **2.** $g''(s) = 12s^2 - 18s + 2$
>
> **3.** $g''(x) = 56x^6 + 120x^4 + 72x^2 + 8$
>
> **4.** $f''(x) = \dfrac{4}{9(x-3)^{2/3}}$    **5.** $h''(x) = \dfrac{15}{2x^5}$
>
> **6.** $f''(x) = -\dfrac{42}{(3x+2)^3}$    **7.** $x = \pm\dfrac{\sqrt{3}}{3}$
>
> **8.** $x = 0, 3$    **9.** $t = \pm 4$    **10.** $x = 0, \pm 5$

**1.** Concave upward on $(-\infty, \infty)$

**3.** Concave upward on $\left(-\infty, -\frac{1}{2}\right)$
Concave downward on $\left(-\frac{1}{2}, \infty\right)$

**5.** Concave upward on $(-\infty, -2)$ and $(2, \infty)$
Concave downward on $(-2, 2)$

**7.** Relative maximum: $(3, 9)$

**9.** Relative maximum: $(1, 3)$
Relative minimum: $\left(\frac{7}{3}, 1.\overline{814}\right)$

**11.** Relative minimum: $(0, -3)$

**13.** Relative maximum: $(-2, -4)$
Relative minimum: $(2, 4)$

**15.**

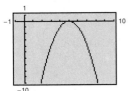

Relative maximum: $(5, 0)$

**17.**

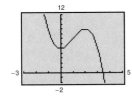

Relative maximum: $(2, 9)$
Relative minimum: $(0, 5)$

**19.** Sign of $f'(x)$ on $(0, 2)$ is positive.
Sign of $f''(x)$ on $(0, 2)$ is positive.

**21.** Sign of $f'(x)$ on $(0, 2)$ is negative.
Sign of $f''(x)$ on $(0, 2)$ is negative.

**23.** $(3, 0)$    **25.** $(1, 0), (3, -16)$

**27.** No inflection points    **29.** $\left(\frac{3}{2}, -\frac{1}{16}\right), (2, 0)$

**31.** Relative maximum: $(-2, 16)$
Relative minimum: $(2, -16)$
Point of inflection: $(0, 0)$

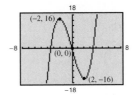

**33.** Point of inflection: $(2, 8)$

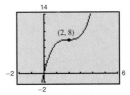

**35.** Relative maximum: $(0, 0)$
Relative minima: $(\pm 2, -4)$
Points of inflection: $\left(\pm\dfrac{2}{\sqrt{3}}, -\dfrac{20}{9}\right)$

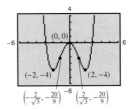

**37.** Relative maximum: $(-1, 0)$
Relative minimum: $(1, -4)$
Point of inflection: $(0, -2)$

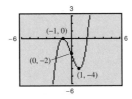

**39.** Relative minimum: $(-2, -2)$

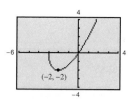

**41.** Relative maximum: $(0, 4)$

Points of inflection: $\left(\pm\dfrac{\sqrt{3}}{3}, 3\right)$

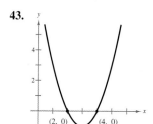

**43.**

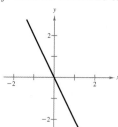

**45. (a)** $f'$: Positive on $(-\infty, 0)$

$f$: Increasing on $(-\infty, 0)$

**(b)** $f'$: Negative on $(0, \infty)$

$f$: Decreasing on $(0, \infty)$

**(c)** $f'$: Not increasing

$f$: Not concave upward

**(d)** $f'$: Decreasing on $(-\infty, \infty)$

$f$: Concave downward on $(-\infty, \infty)$

**47.** $(200, 320)$    **49.** 100 units    **51.** 8:30 P.M.

**53.** $\sqrt{3} \approx 1.732$ years

**55.**

**57.**

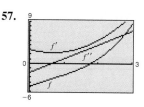

Relative maximum: $(3, 8.5)$

Relative minimum: $(0, -5)$

Point of inflection: $\left(\frac{2}{3}, -3.2963\right)$

**59.**

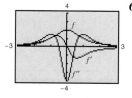

**61.**

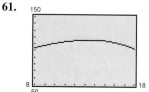

Relative maximum: $(0, 2)$

Minima: $\left(-3, \dfrac{1}{5}\right), \left(3.\dfrac{1}{5}\right)$

Point of inflection: $\left(-\dfrac{1}{\sqrt{3}}, \dfrac{3}{2}\right)\left(\dfrac{1}{\sqrt{3}}, \dfrac{3}{2}\right)$

**63.** Answers vary.

## SECTION 3.4    *(page 207)*

**Warm Up**

**1.** $x + \frac{1}{2}y = 12$    **2.** $2xy = 24$    **3.** $xy = 24$

**4.** $\sqrt{(x_2 - x_1)^2 + (y_2 - y_1)^2} = 10$

**5.** $x = -3$    **6.** $x = -\frac{2}{3}, 1$    **7.** $x = \pm 5$

**8.** $x = 4$    **9.** $x = \pm 1$    **10.** $x = \pm 3$

**1.** 55 and 55    **3.** 9, 18    **5.** $\sqrt{192}$ and $\sqrt{192}$

**7.** 1    **9. (a)** 99 in.³    **(b)** 125 in.³    **(c)** 117 in.³

**11.** Length = width = 25 ft    **13.** $x = 25$ ft, $y = \frac{100}{3}$ ft

**15.** 16 in.³    **17.** $x = 5$ m, $y = \frac{10}{3}$ m

**19.** Length: 50 m, width: $\dfrac{100}{\pi}$ m    **21.** $x = 3, y = \dfrac{3}{2}$

**23.** Length: $2x = \dfrac{10}{\sqrt{2}} \approx 7.07$

Width: $y = \dfrac{5}{\sqrt{2}} \approx 3.54$

**25.** $r \approx 1.51$ in., $h = 2r \approx 3.02$ in.    **27.** $\left( \pm \sqrt{\frac{5}{2}}, \frac{7}{2} \right)$

**29.** 18 in. $\times$ 18 in. $\times$ 36 in.

**31.** Radius: $\dfrac{8}{\pi + 4}$

Side of square: $\dfrac{16}{\pi + 4}$

**33.** $x = 1$ mi

**35. (a)** $A(x) = x^2 + \dfrac{(2 - 2x)^2}{\pi}$

**(b)** $0 \le x \le 1$

**(c)**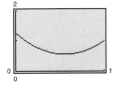

**(d)** $A(x)$ is a minimum if $x = 0.5601$ and $r = 0.28$.

**(e)** $A(x)$ is a maximum if $x = 0$ and $r = 2/\pi$.

## SECTION 3.5    *(page 217)*

### Warm Up

**1.** 1    **2.** 1.2    **3.** 2    **4.** $\frac{1}{2}$

**5.** $\dfrac{dC}{dx} = 1.2 + 0.006x$    **6.** $\dfrac{dP}{dx} = 0.02x + 11$

**7.** $\dfrac{dR}{dx} = 14 - \dfrac{x}{1000}$    **8.** $\dfrac{dR}{dx} = 3.4 - \dfrac{x}{750}$

**9.** $\dfrac{dP}{dx} = -1.4x + 7$    **10.** $\dfrac{dC}{dx} = 4.2 + 0.003x^2$

**1.** 2000 units    **3.** 200 units    **5.** 80 units

**7.** 50 units    **9.** $60    **11.** $69.68

**13.** 3 units

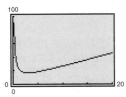

$\overline{C}(3) = 17; \dfrac{dC}{dx} = 4x + 5$; when $x = 3$, $\dfrac{dC}{dx} = 17$

**15. (a)** $80    **(b)** $45.93

**17.** Maximum profit: $10,000

Point of diminishing returns: $5833.33

**19.** 200 units    **21.** $50

**23.** $C =$ cost under water $+$ cost on land

$= 8\sqrt{0.25 + x^2} + 6(6 - x)$

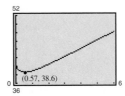

Line should run from power station to a point across the river approximately 0.57 mile downstream.

$\left( \text{Exact: } 3/\left(2\sqrt{7}\right) \text{ mile} \right)$

**25.** 77.46 mph

**27.** $-\frac{17}{3}$, elastic

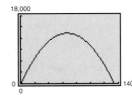

Elastic: $\left( 0, \frac{200}{3} \right)$

Inelastic: $\left( \frac{200}{3}, \frac{400}{3} \right)$

**29.** $-\frac{1}{3}$, inelastic

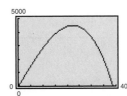

Elastic: $\left( 0, 10\sqrt{5} \right) \approx (0, 22.4)$

Inelastic: $\left( 10\sqrt{5}, 10\sqrt{15} \right) \approx (22.4, 38.7)$

**31.** $-\frac{3}{2}$, elastic

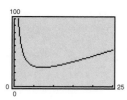

Elastic: $\left(5\sqrt{2}, \infty\right)$

Inelastic: $\left(0, 5\sqrt{2}\right)$

**33.** (a) $-6.83\%$  (b) $-1.37$  (c) $-\frac{4}{3}$

(d) $R = 20p - 2p^3, x = \frac{40}{3}, p = \sqrt{\frac{10}{3}} \approx \$1.83$

**35.** (a) $-\dfrac{14}{9}$  (b) $x = \dfrac{32}{3}, p = \dfrac{4\sqrt{3}}{3}$  (c) Answers vary.

**37.** No, $\eta = -\frac{1}{3}$, demand is inelastic

**39.** (a) $\approx 1990$ $(t = 9.82)$  (b) 1996

(c) $\approx 1990$: $\$30,443.9$ million
  1996: $\$60,844.95$ million

(d)

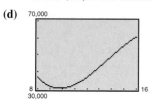

**41.** (a) Demand function  (b) Cost function
(c) Revenue function  (d) Profit function

## SECTION 3.6    *(page 228)*

**Warm Up**

**1.** 3  **2.** 1  **3.** $-11$  **4.** 4  **5.** 0  **6.** 1

**7.** $\overline{C} = \dfrac{150}{x} + 3$  **8.** $\overline{C} = \dfrac{1900}{x} + 1.7 + 0.002x$

$\dfrac{dC}{dx} = 3$    $\dfrac{dC}{dx} = 1.7 + 0.004x$

**9.** $\overline{C} = 0.005x + 0.5 + \dfrac{1375}{x}$  **10.** $\overline{C} = \dfrac{760}{x} + 0.05$

$\dfrac{dC}{dx} = 0.01x + 0.5$    $\dfrac{dC}{dx} = 0.05$

**1.** Vertical: $x = 0$    **3.** Vertical: $x = -1, x = 2$
  Horizontal: $y = 1$      Horizontal: $y = 1$

**5.** Vertical: $x = -1, x = 1$  **7.** Vertical: $x = \pm 2$
  Horizontal: none        Horizontal: $y = \frac{1}{2}$

**9.** f    **10.** b    **11.** c    **12.** a    **13.** e    **14.** d

**15.** $\infty$.    **17.** $-\infty$    **19.** $-\infty$    **21.** $-\infty$    **23.** $\frac{2}{3}$

**25.** 0    **27.** $-\infty$    **29.** $\infty$    **31.** 5

**33.**

| $x$ | $10^0$ | $10^1$ | $10^2$ | $10^3$ |
|---|---|---|---|---|
| $f(x)$ | 2.000 | 0.348 | 0.101 | 0.032 |

| $x$ | $10^4$ | $10^5$ | $10^6$ |
|---|---|---|---|
| $f(x)$ | 0.010 | 0.003 | 0.001 |

$$\lim_{x \to \infty} \frac{x + 1}{x\sqrt{x}} = 0$$

**35.**

| $x$ | $-10^6$ | $-10^4$ | $-10^2$ | $10^0$ |
|---|---|---|---|---|
| $f(x)$ | $-2$ | $-2$ | $-1.9996$ | $0.8944$ |

| $x$ | $10^2$ | $10^4$ | $10^6$ |
|---|---|---|---|
| $f(x)$ | 1.9996 | 2 | 2 |

$$\lim_{x \to -\infty} \frac{2x}{\sqrt{x^2 + 4}} = -2, \lim_{x \to \infty} \frac{2x}{\sqrt{x^2 + 4}} = 2$$

**37.**

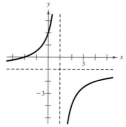

**39.**

**41.**

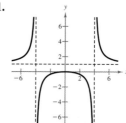

**43.**

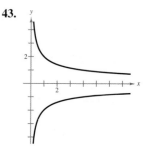

**45.**

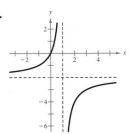

**47.**

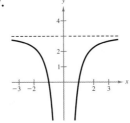

**49.**

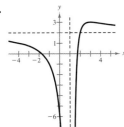

**51.**

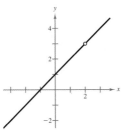

**53.**

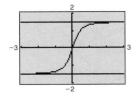

**55. (a)** \$47.05, \$5.92   **(b)** \$1.35

**57. (a)** \$176 million   **(b)** \$528 million
   **(c)** \$1584 million   **(d)** $\infty$

**59.** $a$

**61. (a)** 5 years: 153   **(b)** 400
   10 years: 215
   25 years: 294

**63.** Horizontal asymptotes: $y = \pm\frac{3}{2}$

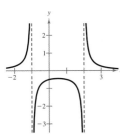

**SECTION 3.7**   *(page 238)*

**Warm Up**

**1.** Vertical: $x = 0$
   Horizontal: $y = 0$

**2.** Vertical: $x = 2$
   Horizontal: $y = 0$

**3.** Vertical: $x = -3$
   Horizontal: $y = 40$

**4.** Vertical: $x = 1, x = 3$
   Horizontal: $y = 1$

**5.** Decreasing on $(-\infty, -2)$
   Increasing on $(-2, \infty)$

**6.** Increasing on $(-\infty, -4)$
   Decreasing on $(-4, \infty)$

**7.** Increasing on $(-\infty, -1)$ and $(1, \infty)$
   Decreasing on $(-1, 1)$

**8.** Decreasing on $(-\infty, 0)$ and $\left(\sqrt[3]{2}, \infty\right)$
   Increasing on $\left(0, \sqrt[3]{2}\right)$

**9.** Increasing on $(-\infty, 1)$ and $(1, \infty)$

**10.** Decreasing on $(-\infty, -3)$ and $\left(\frac{1}{3}, \infty\right)$
   Increasing on $\left(-3, \frac{1}{3}\right)$

**1.**

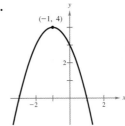

**3.**

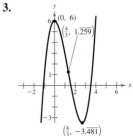

**5.**

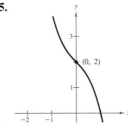

**7.**

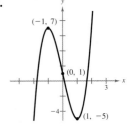

**9.**

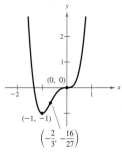

**11.**

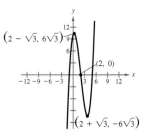

**13.**

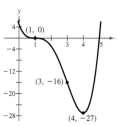

**15.**

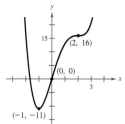

**17.**

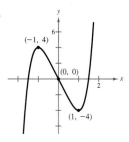

**19.**

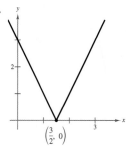

**21.**

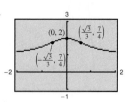

**23.**

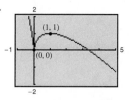

**25.**

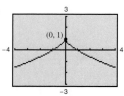

**27.**

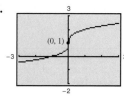

**29.**

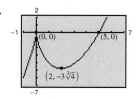

**31.** Domain: $(-\infty, 2), (2, \infty)$

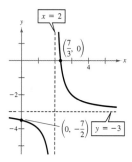

**33.** Domain: $(-\infty, -1), (-1, 1), (1, \infty)$

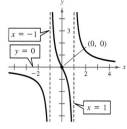

**35.** Domain: $(-\infty, 4]$

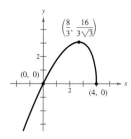

**37.** Domain: $(-\infty, 0), (0, \infty)$

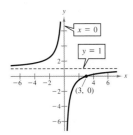

**39.** Domain: $(-\infty, 1), (1, \infty)$

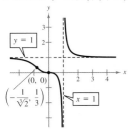

**41.** $f(x) = -x^3 + x^2 + x + 1$    **43.** $f(x) = x^3 + 1$

**45.**

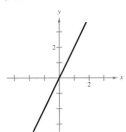

**47.**

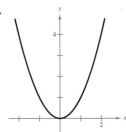

**49.**

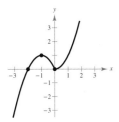

**51. (a)** $C = 0.24s + \dfrac{900}{s}$, **(b)**

$40 \le s \le 65$

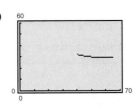

61.2 mph

**53.**

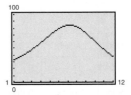

Minimum: $(1, 32.9)$

Maximum: $(7.2, 84.2)$

The model predicts the average daily high temperature to be the lowest in January at 32.9°F, and the average daily high temperature to be the highest in July at 84.2°F.

## SECTION 3.8    *(page 246)*

**Warm Up**

**1.** $\dfrac{dC}{dx} = 0.18x$  **2.** $\dfrac{dR}{dx} = 1.25 + 0.03\sqrt{x}$

**3.** $\dfrac{dP}{dx} = -\dfrac{0.01}{\sqrt[3]{x^2}} + 1.4$  **4.** $\dfrac{dA}{dx} = \dfrac{\sqrt{3}}{2}x$

**5.** $\dfrac{dC}{dr} = 2\pi$  **6.** $\dfrac{dS}{dr} = 8\pi r$  **7.** $A = \pi r^2$

**8.** $A = x^2$  **9.** $V = x^3$  **10.** $V = \frac{4}{3}\pi r^3$

**1.** $6x\,dx$  **3.** $12(4x - 1)^2\,dx$  **5.** $\dfrac{x}{\sqrt{x^2 + 1}}\,dx$

**7.** $0.1005$  **9.** $-0.013245$

**11.** $dy = 0.3$      **13.** $dy = -0.04$

$\Delta y = 0.331$        $\Delta y \approx -0.0394$

**15.**

| $dx = \Delta x$ | $dy$ | $\Delta y$ | $\Delta y - dy$ | $\dfrac{dy}{\Delta x}$ |
|---|---|---|---|---|
| 1.000 | 4.000 | 5.000 | 1.000 | 0.800 |
| 0.500 | 2.000 | 2.250 | 0.250 | 0.889 |
| 0.100 | 0.400 | 0.410 | 0.010 | 0.976 |
| 0.010 | 0.040 | 0.040 | 0.000 | 0.998 |
| 0.001 | 0.004 | 0.004 | 0.000 | 1.000 |

**17.**

| $dx = \Delta x$ | $dy$ | $\Delta y$ | $\Delta y - dy$ | $\dfrac{dy}{\Delta x}$ |
|---|---|---|---|---|
| 1.000 | 80.000 | 211.000 | 131.000 | 0.379 |
| 0.500 | 40.000 | 65.656 | 25.656 | 0.609 |
| 0.100 | 8.000 | 8.841 | 0.841 | 0.905 |
| 0.010 | 0.800 | 0.808 | 0.008 | 0.990 |
| 0.001 | 0.080 | 0.080 | 0.000 | 0.999 |

**19.** $y = x$;

For $\Delta x = -0.01$, $f(x + \Delta x) = -0.009999$ and $y(x + \Delta x) = -0.01$

For $\Delta x = 0.01$, $f(x + \Delta x) = 0.009999$ and $y(x + \Delta x) = 0.01$

**21.** $y = 28x + 37$;

For $\Delta x = -0.01$, $f(x + \Delta x) = -19.281302$ and $y(x + \Delta x) = -19.28$

For $\Delta x = 0.01$, $f(x + \Delta x) = -18.721298$ and $y(x + \Delta x) = -18.72$

**23. (a)** $\Delta p = -0.25 = dp$  **(b)** $\Delta p = -0.25 = dp$

**25.** \$5.20    **27.** $-\$1250$

**29.** $R = -\frac{1}{3}x^2 + 100x$, \$6

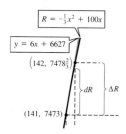

**31.** $P = -\dfrac{1}{2000}x^2 + 23x - 275{,}000$; $-\$5$

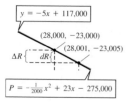

**33. (a)** $dA = 2x\,\Delta x$, $\Delta A = 2x\,\Delta x + (\Delta x)^2$

**(b) and (c)**

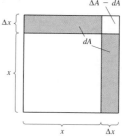

**35.** $\pm\frac{5}{2}\pi$ in.$^2$, $\pm\frac{1}{40}$

**37.** $\pm 2.88\pi$ in.$^3$, $\pm 0.01$

# CHAPTER 3 REVIEW EXERCISES   *(page 252)*

**1.** $x = 1$    **3.** $x = 0, x = 1$

**5.** Increasing on $\left(-\frac{1}{2}, \infty\right)$

Decreasing on $\left(-\infty, -\frac{1}{2}\right)$

**7.** Increasing on $(-\infty, 3) \cup (3, \infty)$

**9. (a)** $(0.23, 6.15)$

**(b)** $(0, 0.23)$, $(6.15, 12)$

**(c)** Maximum daily temperature is rising from early January to early June.

**(d)**

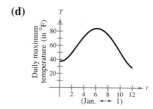

**11.** Relative maximum: $(0, -2)$
Relative minimum: $(1, -4)$

**13.** Relative minimum: $(8, -52)$

**15.** Relative maxima: $(-1, 1)$, $(1, 1)$
Relative minimum: $(0, 0)$

**17.** Relative maximum: $(0, 6)$

**19.** Relative maximum: $(0, 0)$
Relative minimum: $(4, 8)$

**21.** Maximum: $(0, 6)$
Minimum: $\left(-\frac{5}{2}, -\frac{1}{4}\right)$

**23.** Maxima: $(-2, 17), (4, 17)$
Minima: $(-4, -15), (2, -15)$

**25.** Maximum: $(2, 26)$
Minimum: $(1, -1)$

**27.** Maximum: $(1, 1)$
Minimum: $(-1, -1)$

**29.** 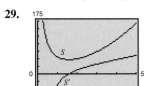 Minimum: $(1.58, 47.33)$

**31.** Concave upward: $(2, \infty)$
Concave downward: $(-\infty, -2)$

**33.** Concave upward: $\left(-\dfrac{2}{\sqrt{3}}, \dfrac{2}{\sqrt{3}}\right)$
Concave downward: $\left(-\infty, -\dfrac{2}{\sqrt{3}}\right)$ and $\left(\dfrac{2}{\sqrt{3}}, \infty\right)$

**35.** $(0, 0), (4, -128)$

**37.** $(0, 0), (1.0652, 4.5244), (2.5348, 3.5246)$

**39.** Relative maximum: $\left(-\sqrt{3}, 6\sqrt{3}\right)$
Relative minimum: $\left(\sqrt{3}, -6\sqrt{3}\right)$

**41.** Relative maximum: $(-4, 0)$
Relative minimum: $(-2, -108)$

**43.** $\left(50, 166\frac{2}{3}\right)$

**45.** $13, 13$

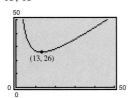

**47. (a)** \$54,607 **(b)** \$5000 **(c)** Answers vary.

**49.** $t = \frac{137}{9} \approx 15.\overline{2}$ years

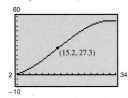

**51.** $s'(r) = -2cr$
$-2cr = 0 \implies r = 0$
$s''(r) = -2c < 0$ for all $r$
Therefore, $r = 0$ yields a maximum value of $s$.

**53.** $N = 85$ (maximizes revenue)  **55.** $125$

**57.** Elastic: $\left(0, \dfrac{10\sqrt{6}}{3}\right)$
Inelastic: $\left(\dfrac{10\sqrt{6}}{3}, 10\sqrt{2}\right)$
Demand is of unit elasticity when $x = \dfrac{10\sqrt{6}}{3}$.

**59.** Vertical asymptote: $x = 4$
Horizontal asymptote: $y = 2$

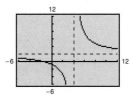

**61.** Vertical asymptote: $x = 0$
Horizontal asymptote: $y = -3$

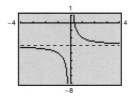

**63.** Vertical asymptotes: $x = 1, x = 4$
Horizontal asymptote: $y = 0$

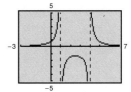

**65.** $-\infty$  **67.** $\frac{5}{2}$  **69.** $-\infty$

**71. (a)**   **(b)** $\lim\limits_{s \to \infty} T = 0.37$

**73.**

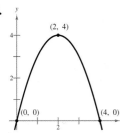

Domain: $(-\infty, \infty)$

**75.**

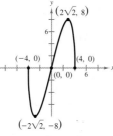

Domain: $[-4, 4]$

**77.**

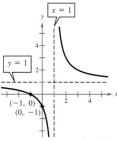

Domain: $(-\infty, 1) \cup (1, \infty)$

**79.**

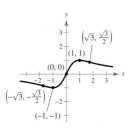

Domain: $(-\infty, \infty)$

**81.** $dy = 12x\,dx$     **83.** $dy = \dfrac{5}{3x^{4/3}}\,dx$

**85.** \$800     **87.** \$15.25

**89.** **(a)** $S = -0.029t^2 + 0.616t + 1.296$

**(b)**

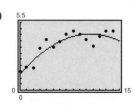

**(c)**

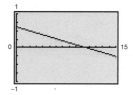

**(d)** The derivative shows that sales were decreasing from mid-1990 to 1994.

**(e)** No, the model shows that sales reached a maximum in 1990, when actually sales were greater in both 1988 and 1994.

**(f)** The second derivative of sales is always negative; therefore the rate of change of sales was always decreasing.

**91.** $dS = \pm 1.8\pi$ in.$^2$

$dV = \pm 8.1\pi$ in.$^3$

**93.**

| Quantity of output | Price | Total revenue | Marginal revenue |
|---|---|---|---|
| 1 | 14.00 | 14.00 | 10.00 |
| 2 | 12.00 | 24.00 | 6.00 |
| 3 | 10.00 | 30.00 | 4.00 |
| 4 | 8.50 | 34.00 | 1.00 |
| 5 | 7.00 | 35.00 | −2.00 |
| 6 | 5.50 | 33.00 | |

**(a)** $R = -1.43x^2 + 13.77x + 1.8$

**(b)** $\dfrac{dR}{dx} = -2.86x + 13.77$;

10.91, 8.05, 5.19, 2.33, −0.53, −3.39

**(c)** About 5 units of output; (4.81, 39.95)

## SAMPLE POST-GRAD EXAM QUESTIONS
*(page 256)*

**1.** (c)     **2.** (a)     **3.** (c)     **4.** (d)     **5.** (b)     **6.** (a)

# CHAPTER 4

## SECTION 4.1   *(page 266)*

---

**Warm Up**

**1.** Continuous on $(-\infty, \infty)$

**2.** Removable discontinuity at $x = 4$

**3.** 0   **4.** 0   **5.** 4   **6.** $\frac{1}{2}$   **7.** $\frac{3}{2}$   **8.** 6

**9.** 0   **10.** 0

---

**1.** (a) 625   (b) 9   (c) $16\sqrt{2}$   (d) 9   (e) 125   (f) 4

**3.** (a) 3125   (b) $\frac{1}{5}$   (c) 625   (d) $\frac{1}{125}$

**5.** (a) $\frac{1}{5}$   (b) 27   (c) 5   (d) 4096

**7.** (a) $e^7$   (b) $e^{12}$   (c) $\dfrac{1}{e^6}$   (d) 1

**9.** 4   **11.** $-2$   **13.** 2   **15.** 16   **17.** $-\frac{1}{3}$

**19.** 9   **21.** $e$   **23.** e   **24.** c   **25.** a   **26.** f

**27.** d   **28.** b

**29.** $f(x) = 6^x$          **31.** $f(x) = 5^{-x}$

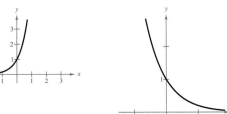

**33.** $y = 3^{-x^2}$          **35.** $y = 3^{-|x|}$

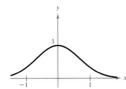   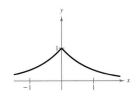

**37.** $s(t) = \dfrac{3^{-t}}{4}$          **39.** $h(x) = e^{x-2}$

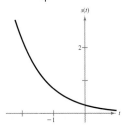   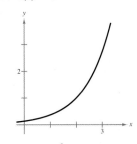

**41.** $N(t) = 500e^{-0.2t}$          **43.** $g(x) = \dfrac{2}{1 + e^{x^2}}$

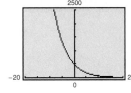   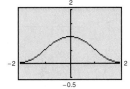

**45.**

| $n$ | 1 | 2 | 4 | 12 |
|---|---|---|---|---|
| $A$ | 1343.92 | 1346.86 | 1348.35 | 1349.35 |

| $n$ | 365 | Continuous compounding |
|---|---|---|
| $A$ | 1349.84 | 1349.86 |

**47.**

| $n$ | 1 | 2 | 4 | 12 |
|---|---|---|---|---|
| $A$ | 3262.04 | 3290.66 | 3305.28 | 3315.15 |

| $n$ | 365 | Continuous compounding |
|---|---|---|
| $A$ | 3319.95 | 3320.12 |

**49.**

| $t$ | 1 | 10 | 20 |
|---|---|---|---|
| $P$ | 96,078.94 | 67,032.00 | 44,932.90 |

| $t$ | 30 | 40 | 50 |
|---|---|---|---|
| $P$ | 30,119.42 | 20,189.65 | 13,533.53 |

**51.**

| $t$ | 1 | 10 | 20 |
|---|---|---|---|
| $P$ | 95,132.82 | 60,716.10 | 36,864.45 |

| $t$ | 30 | 40 | 50 |
|---|---|---|---|
| $P$ | 22,382.66 | 13,589.88 | 8251.24 |

**53. (a)** $849.53

**(b)** $421.12

$$\lim_{x \to \infty} p = 0$$

**55. (a)** 0.1535 **(b)** 0.4866 **(c)** 0.8111

**57. (a)**

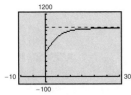

**(b)** $\lim_{t \to \infty} \dfrac{925}{1 + e^{-0.3t}} = 925$ **(c)** $\lim_{t \to \infty} \dfrac{1000}{1 + e^{-0.3t}} = 1000$

Models similar to this type of logistics growth where

$y = \dfrac{a}{1 + be^{-ct}}$ have a limit of $a$ as $t \to \infty$. $(c > 0)$

**59. (a)** 0.731 **(b)** 11 **(c)** $\lim_{n \to \infty} \dfrac{0.83}{1 + e^{-0.2n}} = 0.83$

**61.**

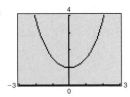

No horizontal asymptotes
Continuous on the entire real line

**63.**

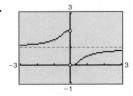

Horizontal asymptote: $y = 1$
Discontinuous at $x = 0$

## SECTION 4.2 *(page 275)*

**Warm Up**

**1.** $\dfrac{1}{2}e^x(2x^2 - 1)$ **2.** $\dfrac{e^x(x + 1)}{x}$ **3.** $e^x(x - e^x)$

**4.** $e^{-x}(e^{2x} - x)$ **5.** $-\dfrac{6}{7x^3}$ **6.** $6x - \dfrac{1}{6}$

**7.** $6(2x^2 - x + 6)$ **8.** $\dfrac{t + 2}{2t^{3/2}}$

**9.** Relative maximum: $\left(-\dfrac{4\sqrt{3}}{3}, \dfrac{16\sqrt{3}}{9}\right)$

Relative minimum: $\left(\dfrac{4\sqrt{3}}{3}, -\dfrac{16\sqrt{3}}{9}\right)$

**10.** Relative maximum: $(0, 5)$

Relative minima: $(-1, 4), (1, 4)$

**1.** 3 **3.** $-1$ **5.** $4e^{4x}$ **7.** $-2xe^{-x^2}$ **9.** $\dfrac{2}{x^3}e^{-1/x^2}$

**11.** $e^{4x}(4x^2 + 2x + 4)$ **13.** $-\dfrac{2(e^x - e^{-x})}{(e^x + e^{-x})^2}$

**15.** $xe^x + e^x + 4e^{-x}$ **17.** $y = -2x + 1$ **19.** $y = \dfrac{4}{e^2}$

**21.** $\dfrac{dy}{dx} = \dfrac{1}{2}(-x - 1 - 2y)$ $\left(\text{Equivalently, } \dfrac{dy}{dx} = -\dfrac{1}{2}\right)$

**23.** $6(3e^{3x} + 2e^{-2x})$ **25.** $5e^{-x} - 50e^{-5x}$

**27.** No relative extrema
No points of inflection
Horizontal asymptote to the right: $y = \dfrac{1}{2}$
Horizontal asymptote to the left: $y = 0$
Vertical asymptote: $x \approx -0.693$

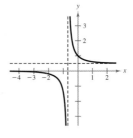

**29.** Relative minimum: $(0, 0)$

Relative maximum: $(2, 4e^{-2})$

Points of inflection: $(2 - \sqrt{2}, 0.191), (2 + \sqrt{2}, 0.384)$

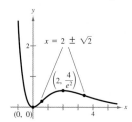

**31.**

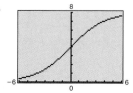

Asymptotes: $y = 0, y = 8$

**33. (a)**

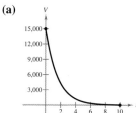

**(b)** $-\$5028.84/\text{year}$ **(c)** $-\$406.89/\text{year}$

**(d)** In this model, the initial rate of depreciation is greater than in a linear model.

**35. (a)**

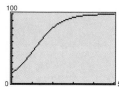

**(b)** 80.3%

**(c)** $x \approx 1.1$ or approximately 1100 egg masses

**37. (a)** $\$433.31/\text{year}$ **(b)** $\$890.22/\text{year}$

**(c)** $\$21,839.26/\text{year}$

**39.** $t = 5$: 14.44

$t = 10$: 3.63

$t = 25$: 0.58

**41.**

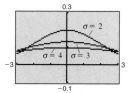

For larger $\sigma$, the graph becomes flatter.

**43. (a)** $\dfrac{dh}{dt} = -80e^{-1.6t} - 20$

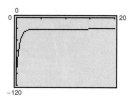

**(b)** $-100, -36.15, -20.03, -20.00, -20.00$

**(c)** The values in (b) are rates of descent in feet per second. As time increases, the rate is approximately constant at $-20$ ft/sec.

## SECTION 4.3 *(page 283)*

**Warm Up**

**1.** $\frac{1}{4}$ **2.** 64 **3.** 1 **4.** $81e^4$ **5.** $x > -4$

**6.** Any real number $x$ **7.** $x < -1$ and $x > 1$

**8.** $x > 5$ **9.** $\$3462.03$ **10.** $\$3374.65$

**1.** $e^{0.6931\ldots} = 2$ **3.** $e^{-1.6094\ldots} = 0.2$ **5.** $\ln 1 = 0$

**7.** $\ln(0.0498\ldots) = -3$ **9.** c **10.** d

**11.** b **12.** a

**13.**

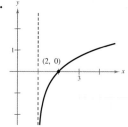

**15.**

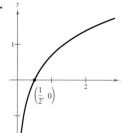

**17.**

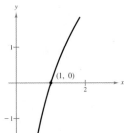

**19.** **21.**

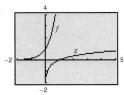 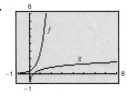

**23.** $x^2$   **25.** $5x + 2$   **27.** $\sqrt{x}$

**29.** (a) 1.7917   (b) 0.4055   (c) 4.3944   (d) 0.5493

**31.** $\ln 2 - \ln 3$   **33.** $\ln x + \ln y + \ln z$

**35.** $\frac{1}{2}\ln(x^2 + 1)$

**37.** $\ln 2 + \ln x - \frac{1}{2}\ln(x + 1) - \frac{1}{2}\ln(x - 1)$

**39.** $\ln 3 + \ln x + \ln(x + 1) - 2\ln(2x + 1)$

**41.** $\ln\dfrac{x - 2}{x + 2}$   **43.** $\ln\dfrac{x^3 y^2}{z^4}$   **45.** $\ln\left[\dfrac{x(x + 3)}{x + 4}\right]^3$

**47.** $\ln\left[\dfrac{x(x^2 + 1)}{x + 1}\right]^{3/2}$   **49.** $\ln\dfrac{(x - 1)^3}{x(x + 1)^2}$   **51.** $x = 4$

**53.** $x = 1$   **55.** $x = \ln 4 - 1 \approx 0.3863$

**57.** $t = \dfrac{\ln 7 - \ln 3}{-0.2} \approx -4.2365$   **59.** $x = \dfrac{\ln 15}{2\ln 5} \approx 0.8413$

**61.** $t = \dfrac{\ln 2}{\ln 1.07} \approx 10.2448$

**63.** $t = \dfrac{\ln 3}{12\ln[1 + (0.07/12)]} \approx 15.740$

**65.** (a) 14.21 years   (b) 13.88 years
(c) 13.87 years   (d) 13.86 years

**67.**

| $r$ | 2% | 4% | 6% | 8% |
|---|---|---|---|---|
| $t$ | 54.93 | 27.47 | 18.31 | 13.73 |

| $r$ | 10% | 12% | 14% |
|---|---|---|---|
| $t$ | 10.99 | 9.16 | 7.85 |

**69.** 1999   **71.** 9370 years   **73.** 12,451 years

**75.** (a) 80   (b) 57.5   (c) 10 months

**77.**

| $x$ | $y$ | $\dfrac{\ln x}{\ln y}$ | $\ln\dfrac{x}{y}$ | $\ln x - \ln y$ |
|---|---|---|---|---|
| 1 | 2 | 0 | $-0.6931$ | $-0.6931$ |
| 3 | 4 | 0.7925 | $-0.2877$ | $-0.2877$ |
| 10 | 5 | 1.4307 | 0.6931 | 0.6931 |
| 4 | 0.5 | $-2.0000$ | 2.0794 | 2.0794 |

**79.**

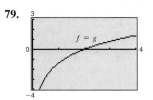

## SECTION 4.4   *(page 292)*

**Warm Up**

**1.** $2\ln(x + 1)$   **2.** $\ln x + \ln(x + 1)$

**3.** $\ln x - \ln(x + 1)$   **4.** $3[\ln x - \ln(x - 3)]$

**5.** $\ln 4 + \ln x + \ln(x - 7) - 2\ln x$

**6.** $3\ln x + \ln(x + 1) - \frac{1}{2}\ln(x - 2)$

**7.** $-\dfrac{y(y + y^x \ln y)}{x(y + y^x)}$   **8.** $-\dfrac{-3 + 2xy - y^2}{x(x - 2y)}$

**9.** $-12x + 2$   **10.** $-\dfrac{6}{x^4}$

**1.** 3   **3.** 2   **5.** $\dfrac{2}{x}$   **7.** $\dfrac{1}{x}$   **9.** $\dfrac{2(x^3 - 1)}{x(x^3 - 4)}$

**11.** $\dfrac{3}{x}(\ln x)^5$   **13.** $x(1 + \ln x^2)$   **15.** $\dfrac{5x + 2}{x(x + 1)}$

**17.** $\dfrac{1}{x(x + 1)}$   **19.** $\dfrac{2}{3(x^2 - 1)}$   **21.** $-\dfrac{4}{x(4 + x^2)}$

**23.** $e^{-x}\left(\dfrac{1}{x} - \ln x\right)$   **25.** $2x$   **27.** $e^{x(\ln 2)}$

**29.** $\dfrac{1}{\ln 4}\ln x$   **31.** 2   **33.** $-0.63093$   **35.** 1.49136

**37.** $(\ln 3)3^x$    **39.** $\dfrac{1}{x \ln 2}$    **41.** $(2 \ln 4)4^{2x-3}$

**43.** $\dfrac{2x + 6}{(x^2 + 6x)\ln 10}$    **45.** $2^x(1 + x \ln 2)$    **47.** $y = x - 1$

**49.** $y = \dfrac{1}{3 \ln 3}x - \dfrac{2}{9 \ln 3} + 2$   or   $y = 0.303x + 1.798$

**51.** $\dfrac{2xy}{3 - 2y^2}$    **53.** $\dfrac{y(1 - 6x^2)}{1 + y}$    **55.** $\dfrac{1}{2x}$

**57.** $(\ln 5)^2\, 5^x$

**59.** $\dfrac{d\beta}{dI} = \dfrac{10}{(\ln 10)I}$; for $I = 10^{-4}$,

$\dfrac{d\beta}{dI} \approx 43{,}429.4$ decibels $*$ cm$^2$/watts

**61.** $2, y = 2x - 1$    **63.** $-\dfrac{8}{5}, y = -\dfrac{8}{5}x - 4$

**65.** $\dfrac{1}{\ln 2}, y = \dfrac{1}{\ln 2}x - \dfrac{1}{\ln 2}$

**67.** Relative minimum: $(1, 1)$

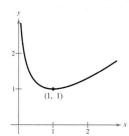

**69.** Relative maximum: $\left(e, \dfrac{1}{e}\right)$

Point of inflection: $\left(e^{3/2}, \dfrac{3}{2e^{3/2}}\right)$

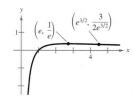

**71.** Relative minimum: $\left(\dfrac{1}{\sqrt{e}}, -\dfrac{1}{2e}\right)$

Point of inflection: $\left(\dfrac{1}{e^{3/2}}, -\dfrac{3}{2e^3}\right)$

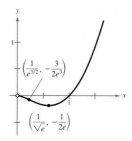

**73.** $-\dfrac{1}{p}, -\dfrac{1}{10}$    **75.** $-\dfrac{1000p}{(p^2 + 1)[\ln(p^2 + 1)]^2}, -4.65$

**77.** $p = 1000e^{-x}$

$\dfrac{dp}{dx} = -1000e^{-x}$

$p = 10, \dfrac{dp}{dx} = -10$

$\dfrac{dp}{dx}$ and $\dfrac{dx}{dp}$ are reciprocals of each other.

**79.** $e^{8/3} \approx 14.39$

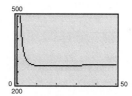

**81. (a)**

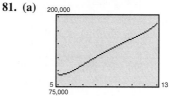

**(b)** \$160,316.21 million

**(c)** \$10,783.74 million per year

# SECTION 4.5 *(page 301)*

---

**Warm Up**

**1.** $-\frac{1}{4}\ln 2$  **2.** $\frac{1}{5}\ln\frac{10}{3}$  **3.** $\frac{1}{3}\ln\frac{25}{16}$  **4.** $\frac{1}{20}\ln\frac{11}{16}$

**5.** $7.36e^{0.23t}$  **6.** $-33.6e^{-1.4t}$  **7.** $1.296e^{0.072t}$

**8.** $-0.025e^{-0.001t}$  **9.** 4  **10.** 12

---

**1.** $y = 2e^{0.1014t}$  **3.** $y = 4e^{-0.4159t}$

**5.** $y = 0.6687e^{0.4024t}$  **7.** $y = 10e^{2t}$, exponential growth

**9.** $y = 30e^{-4t}$, exponential decay

**11.** *Isotope:* $Ra^{226}$

  *Half-life (in years):* 1620

  *Initial quantity:* 10 grams

  *Amount after 1000 years:* 6.52 grams

  *Amount after 10,000 years:* 0.14 gram

**13.** *Isotope:* $C^{14}$

  *Half-life (in years):* 5730

  *Initial quantity:* 6.70 grams

  *Amount after 1000 years:* 5.94 grams

  *Amount after 10,000 years:* 2.00 grams

**15.** *Isotope:* $Pu^{239}$

  *Half-life (in years):* 24,360

  *Initial quantity:* 2.16 grams

  *Amount after 1000 years:* 2.10 grams

  *Amount after 10,000 years:* 1.63 grams

**17.** 68%  **19.** $\dfrac{5730(\ln 20 - \ln 3)}{\ln 2} \approx 15,682.8$ years

**21.** **(a)** 1350  **(b)** $\dfrac{5\ln 2}{\ln 3} \approx 3.15$ hr  **(c)** No

**23.** *Initial investment:* $1000

  *Annual rate:* 12%

  *Time to double:* 5.78 years

  *Amount after 10 years:* $3320.12

  *Amount after 25 years:* $20,085.54

**25.** *Initial investment:* $750

  *Annual rate:* 8.94%

  *Time to double:* 7.75 years

  *Amount after 10 years:* $1833.67

  *Amount after 25 years:* $7009.86

**27.** *Initial investment:* $500

  *Annual rate:* 9.50%

  *Time to double:* 7.30 years

  *Amount after 10 years:* $1292.85

  *Amount after 25 years:* $5375.51

**29.** **(a)** Answers may vary.  **(b)** 6.17%

**31.**

| Number of compoundings/yr | 4 | 12 |
| --- | --- | --- |
| Effective yield | 5.095% | 5.116% |

| Number of compoundings/yr | 365 | Continuous |
| --- | --- | --- |
| Effective yield | 5.127% | 5.127% |

**33.** Answers may vary.

**35.** **(a)** $\approx 1388.6$ million  **(b)** $\approx \$1236.5$ million

**(c)**

  $t = 7$ corresponds to 1987.

**37.** **(a)** $S(t) = 30e^{-1.7918/t}$

  **(b)** $30e^{-0.35836} = 20.9646$ or 20,965 units

  **(c)**

**39.** About 36 days

**41.** (a) $C = \dfrac{45}{e^{1000[\ln(45/40)/(-200)]}} \approx 81.0915$

$k = \dfrac{\ln(45/40)}{-200} \approx -0.0005889$

(b) $x = \dfrac{200}{\ln(9/8)} \approx 1698$ units

$p = 45\left(\dfrac{9}{8}\right)^5 \dfrac{1}{e} \approx \$29.83$

**43.** 2078

**45.** (a) $I = e^{8.3 \ln 10} \approx 199,526,231.5$

(b) The intensity is squared when $R$ is doubled.

(c) $\dfrac{1}{I \ln 10}$

## CHAPTER 4 REVIEW EXERCISES    *(page 308)*

**1.** 1024    **3.** $\frac{1}{625}$    **5.** 1    **7.** $\frac{1}{6}$    **9.** 4    **11.** $\frac{1}{2}$

**13.** 1950: 2434.2 million shares

1970: 15,059.7 million shares

1990: 93,168.9 million shares

**15.** 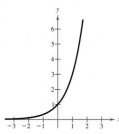    **17.**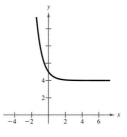

**19.** $7500

**21.** $f(2) \approx 5.4366$    **23.** $g(17) \approx 0.4005$

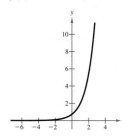    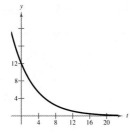

**25.** (a)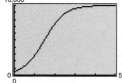

(b) $P \approx 1049$ fish

(c) Yes, $P$ approaches 10,000 fish as $t$ approaches $\infty$.

(d) The population is increasing most rapidly at the inflection point, which occurs around $t = 15$ months.

**27.**

| $n$ | 1 | 2 | 4 | 12 |
|---|---|---|---|---|
| $A$ | \$1216.65 | \$1218.99 | \$1220.19 | \$1221.00 |

| $n$ | 365 | Continuous compounding |
|---|---|---|
| $A$ | \$1221.39 | \$1221.40 |

**29.** (b)

**31.** 1970: $A \approx 22.58$ years

1980: $A \approx 23.39$ years

1990: $A \approx 25.75$ years

**33.** $8xe^{x^2}$    **35.** $\dfrac{1 - 2x}{e^{2x}}$

**37.** $4e^{2x}$    **39.** $\dfrac{-10e^{2x}}{(1 + e^{2x})^2}$

**41.** No relative extrema

No points of inflection

$y = 0$ is a horizontal asymptote.

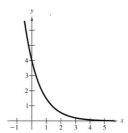

**43.** (2, 1.847) is a relative minimum.

No points of inflection

$y = 0$ is a horizontal asymptote.

$x = 0$ is a vertical asymptote.

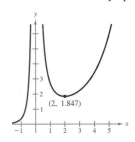

**45.** $e^{2.4849} \approx 12$     **47.** $\ln 4.4817 \approx 1.5$

**49.**

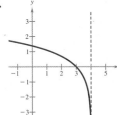

**51.**

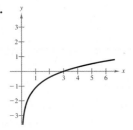

**53.** $\ln x + \frac{1}{2}\ln(x - 1)$     **55.** $3[\ln(1 - x) - \ln 3 - \ln x]$

**57.** 3     **59.** 1     **61.** $\dfrac{3 + \sqrt{13}}{2} \approx 3.3028$     **63.** $\dfrac{1}{4}$

**65.** (a)

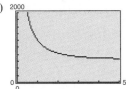

(b) A 30-year term has a smaller monthly payment, but takes more time to pay off, than a 20-year term.

**67.** $\dfrac{2}{x}$     **69.** $\sqrt{\ln x} + \dfrac{1}{2\sqrt{\ln x}}$

**71.** $\dfrac{1 - 3\ln x}{x^4}$     **73.** $-2x$

**75.** No relative extrema
No points of inflection

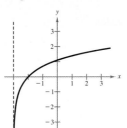

**77.** No relative extrema
No points of inflection

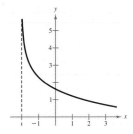

**79.** 2     **81.** 0     **83.** 1.4307     **85.** 1.5

**87.** $\dfrac{2}{(2x - 1)\ln 3}$     **89.** $-\dfrac{2}{x \ln 2}$

**91.**

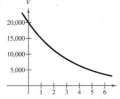

(a) $t = 2$: $11,250

(b) $t = 1$:

$-4315.23$ dollars/year

$t = 4$:

$-1820.49$ dollars/year

(c) $t \approx 4.8$ years

**93.** $y = 500e^{-0.01277t}$     **95.** 27.9 years     **97.** $9547 million

## SAMPLE POST–GRAD EXAM QUESTIONS
*(page 312)*

**1.** (b)     **2.** (c)     **3.** (a)     **4.** (c)

**5.** (d)     **6.** (a)     **7.** (b)

# CHAPTER 5

## SECTION 5.1     *(page 322)*

**Warm Up**

**1.** $x^{-1/2}$     **2.** $(2x)^{4/3}$     **3.** $5^{1/2}x^{3/2} + x^{5/2}$

**4.** $x^{-1/2} + x^{-2/3}$     **5.** $(x + 1)^{5/2}$     **6.** $x^{1/6}$

**7.** $-12$     **8.** $-10$     **9.** 14     **10.** 14

**1.** Answers may vary.　　**3.** Answers may vary.

**5.** Answers may vary.　　**7.** Answers may vary.

**9.** $6x + C$

$$\frac{d}{dx}[6x + C] = 6$$

**11.** $t^3 + C$

$$\frac{d}{dt}[t^3 + C] = 3t^2$$

**13.** $-\dfrac{5}{2x^2} + C$

**15.** $u + C$

$$\frac{d}{du}[u + C] = 1$$

**17.** $\dfrac{2}{5}x^{5/2} + C$

$$\frac{d}{dx}\left[\frac{2}{5}x^{5/2} + C\right] = x^{3/2}$$

|  | *Given* | *Rewrite* | *Integrate* | *Simplify* |
|---|---|---|---|---|
| **19.** | $\displaystyle\int \sqrt[3]{x}\, dx$ | $\displaystyle\int x^{1/3}\, dx$ | $\dfrac{x^{4/3}}{4/3} + C$ | $\dfrac{3}{4}x^{4/3} + C$ |
| **21.** | $\displaystyle\int \dfrac{1}{x\sqrt{x}}\, dx$ | $\displaystyle\int x^{-3/2}\, dx$ | $\dfrac{x^{-1/2}}{-1/2} + C$ | $-\dfrac{2}{\sqrt{x}} + C$ |
| **23.** | $\displaystyle\int \dfrac{1}{2x^3}\, dx$ | $\dfrac{1}{2}\displaystyle\int x^{-3}\, dx$ | $\dfrac{1}{2}\left(\dfrac{x^{-2}}{-2}\right) + C$ | $-\dfrac{1}{4x^2} + C$ |

**25.**

**27.** $\dfrac{x^4}{4} + 2x + C$

**29.** $\dfrac{6}{7}x^{7/3} + \dfrac{3}{2}x^2 - x + C$　　**31.** $\dfrac{3}{5}x^{5/3} + C$

**33.** $-\dfrac{1}{3x^3} + C$　　**35.** $x^2 + 2\sqrt{x} + C$

**37.** $\dfrac{3}{4}u^4 + \dfrac{1}{2}u^2 + C$　　**39.** $2x^3 - \dfrac{11}{2}x^2 + 5x + C$

**41.** $\dfrac{2}{7}y^{7/2} + C$　　**43.** $f(x) = 2x^{3/2} + 3x - 1$

**45.** $f(x) = 2x^3 - 3x^2$　　**47.** $f(x) = -\dfrac{1}{x^2} + \dfrac{1}{x} + \dfrac{1}{2}$

**49.** $y = -\dfrac{5}{2}x^2 - 2x + 2$　　**51.** $f(x) = 4x^{3/2} - 10x + 10$

**53.** $f(x) = x^2 + x + 4$　　**55.** $f(x) = \dfrac{9}{4}x^{4/3}$

**57.** $C = 85x + 5500$　　**59.** $C = \dfrac{1}{10}\sqrt{x} + 4x + 750$

**61.** $R = 225x - \dfrac{3}{2}x^2,\ p = 225 - \dfrac{3}{2}x$

**63.** $P = -9x^2 + 1650x$　　**65.** $56.25$ ft

**67.** $v_0 = 40\sqrt{22} \approx 187.617$ ft/sec

**69. (a)** $C = x^2 - 12x + 125$

$$\overline{C} = x - 12 + \frac{125}{x}$$

**(b)** \$2025

**(c)** \$125 is fixed.
　　\$1900 is variable.

**71.** $S = 0.004t^3 - 0.091t^2 + 1.045t + 14.35$; in 1994,
$S \approx 22.12$ quadrillion BTU's.

**73.** Answers may vary.

## SECTION 5.2　　*(page 331)*

---

### Warm Up

**1.** $\dfrac{1}{2}x^4 + x + C$　　**2.** $\dfrac{3}{2}x^2 + \dfrac{2}{3}x^{3/2} - 4x + C$

**3.** $-\dfrac{1}{x} + C$　　**4.** $-\dfrac{1}{6t^2} + C$

**5.** $\dfrac{4}{7}t^{7/2} + \dfrac{2}{5}t^{5/2} + C$　　**6.** $\dfrac{4}{5}x^{5/2} - \dfrac{2}{3}x^{3/2} + C$

**7.** $-\dfrac{5(x-2)^4}{16}$　　**8.** $-\dfrac{1}{12(x-1)^2}$

**9.** $9(x^2 + 3)^{2/3}$　　**10.** $-\dfrac{5}{(1 - x^3)^{1/2}}$

---

| $\displaystyle\int u^n \dfrac{du}{dx}\, dx$ | $u$ | $\dfrac{du}{dx}$ |
|---|---|---|
| **1.** $\displaystyle\int (5x^2 + 1)^2(10x)\, dx$ | $5x^2 + 1$ | $10x$ |
| **3.** $\displaystyle\int \sqrt{1 - x^2}\,(-2x)\, dx$ | $1 - x^2$ | $-2x$ |
| **5.** $\displaystyle\int \left(4 + \dfrac{1}{x^2}\right)\left(\dfrac{-2}{x^3}\right) dx$ | $4 + \dfrac{1}{x^2}$ | $-\dfrac{2}{x^3}$ |

**7.** $\dfrac{1}{5}(1 + 2x)^5 + C$　　**9.** $\dfrac{2}{3}(5x^2 - 4)^{3/2} + C$

**11.** $\dfrac{1}{5}(x - 1)^5 + C$　　**13.** $\dfrac{1}{16}(x^2 - 1)^8 + C$

**15.** $-\dfrac{1}{3(1 + x^3)} + C$     **17.** $-\dfrac{1}{2(x^2 + 2x - 3)} + C$

**19.** $\sqrt{x^2 - 4x + 3} + C$     **21.** $-\dfrac{15}{8}(1 - x^2)^{4/3} + C$

**23.** $4\sqrt{1 + x^2} + C$     **25.** $-3\sqrt{2x + 3} + C$

**27.** $-\dfrac{1}{2}\sqrt{1 - x^4} + C$     **29.** $\sqrt{2x} + C$

**31.** $\dfrac{1}{6}(x^3 + 3x)^2 + C$     **33.** $\dfrac{1}{48}(6x^2 - 1)^4 + C$

**35.** $-\dfrac{2}{45}(2 - 3x^3)^{5/2} + C$     **37.** $\sqrt{x^2 + 25} + C$

**39.** $\dfrac{2}{3}\sqrt{x^3 + 3x + 4} + C$

**41.** $\dfrac{1}{6}(2x - 1)^3 + C_1 = \dfrac{4}{3}x^3 - 2x^2 + x + C_2$

$\left(\text{Answers differ by a constant: } C_2 = C_1 - \dfrac{1}{6}\right)$

**43.** $\dfrac{1}{2}\dfrac{(x^2 - 1)^3}{3} + C_1 = \dfrac{1}{6}x^6 - \dfrac{1}{2}x^4 + \dfrac{1}{2}x^2 + C_2$

$\left(\text{Answers differ by a constant: } C_2 = C_1 - \dfrac{1}{6}\right)$

**45.** $\dfrac{1}{3}[5 - (1 - x^2)^{3/2}]$

**47.** **(a)** $C = 8\sqrt{x + 1} + 18$

**(b)**

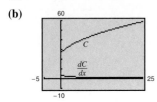

**49.** $x = \dfrac{1}{3}(p^2 - 25)^{3/2} + 24$

**51.** $x = \dfrac{6000}{\sqrt{p^2 - 16}} + 3000$

**53.** **(a)** $h = \sqrt{17.6t^2 + 1} + 5$     **(b)** 26 in.

**55.** **(a)** $Q = (x - 19{,}999)^{0.95} + 19{,}999$

**(b)**

| $x$ | 20,000 | 50,000 |
|---|---|---|
| $Q$ | 20,000 | 37,916.56 |
| $x - Q$ | 0 | 12,083.44 |

| $x$ | 100,000 | 150,000 |
|---|---|---|
| $Q$ | 65,491.59 | 92,151.16 |
| $x - Q$ | 34,508.41 | 57,848.84 |

**(c)**

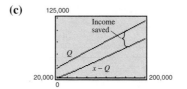

**57.** $-\dfrac{2}{3}x^{3/2} + \dfrac{2}{3}(x + 1)^{3/2} + C$

## SECTION 5.3     *(page 338)*

**Warm Up**

**1.** $\left(\dfrac{5}{2}, \infty\right)$     **2.** $(-\infty, 2) \cup (3, \infty)$

**3.** $x + 2 - \dfrac{2}{x + 2}$     **4.** $x - 2 + \dfrac{1}{x - 4}$

**5.** $x + 8 + \dfrac{2x - 4}{x^2 - 4x}$     **6.** $x^2 - x - 4 + \dfrac{20x + 22}{x^2 + 5}$

**7.** $\dfrac{1}{4}x^4 - \dfrac{1}{x} + C$     **8.** $\dfrac{1}{2}x^2 + 2x + C$

**9.** $\dfrac{1}{2}x^2 - \dfrac{4}{x} + C$     **10.** $-\dfrac{1}{x} - \dfrac{3}{2x^2} + C$

**1.** $e^{2x} + C$     **3.** $\dfrac{1}{4}e^{4x} + C$     **5.** $-\dfrac{9}{2}e^{-x^2} + C$

**7.** $\dfrac{5}{3}e^{x^3} + C$     **9.** $\dfrac{1}{3}e^{x^3 + 3x^2 - 1} + C$     **11.** $-5e^{2 - x} + C$

**13.** $\ln|x + 1| + C$     **15.** $-\dfrac{1}{2}\ln|3 - 2x| + C$

**17.** $\ln\sqrt{x^2 + 1} + C$     **19.** $\dfrac{1}{3}\ln|x^3 + 1| + C$

**21.** $\dfrac{1}{2}\ln|x^2 + 6x + 7| + C$     **23.** $\ln|\ln x| + C$

**25.** $-\dfrac{1}{2}e^{2/x} + C$     **27.** $2e^{\sqrt{x}} + C$

**29.** $-\ln(1 + e^{-x}) + C$     **31.** $2\ln|5 - e^{2x}| + C$

**33.** $e^x + 2x - e^{-x} + C$     **35.** $-\dfrac{2}{3}(1 - e^x)^{3/2} + C$

**37.** $-\dfrac{1}{x - 1} + C$     **39.** $\dfrac{x^2}{2} - 4\ln|x| + C$

**41.** $\dfrac{x^2}{4} - 4\ln|x| + C$     **43.** $2\ln(e^x + 1) + C$

**45.** $\dfrac{1}{2}x^2 + 3x + 8\ln|x - 1| + C$

**47.** $\ln|e^x + x| + C$

**49.** $f(x) = \dfrac{1}{2}x^2 + 5x + 8\ln|x - 1| - 8$

**51.** **(a)** $P(t) = 1000[1 + \ln(1 + 0.25t)^{12}]$

**(b)** $P(3) \approx 7715$     **(c)** $t \approx 6$ days

**53. (a)** $p = -50e^{-x/500} + 45.06$

**(b)**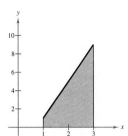

The price increases as the demand increases.

**(c)** 387

**55. (a)** $T = 1719.7e^{0.11t} + 3558.98$

**(b)** $T \approx 6540$ million transactions

## SECTION 5.4 *(page 348)*

**Warm Up**

**1.** $\frac{3}{2}x^2 + 7x + C$    **2.** $\frac{2}{5}x^{5/2} + \frac{4}{3}x^{3/2} + C$

**3.** $\frac{1}{5}\ln|x| + C$    **4.** $-\dfrac{1}{6e^{6x}}$    **5.** $-\dfrac{8}{5}$

**6.** $-\frac{62}{3}$    **7.** $0.008x^{5/2} + 29{,}500x + C$

**8.** $x^2 + 9000x + C$    **9.** $25{,}000x - 0.005x^2 + C$

**10.** $0.01x^3 + 4600x + C$

**1.** Area $= 6$

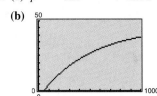

**3.** Area $= \frac{35}{2}$

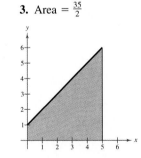

**5.** $\frac{1}{6}$    **7.** $\frac{8}{5}$    **9.** $6$    **11.** $1$    **13.** $0$    **15.** $\frac{14}{3}$

**17.** $\frac{1}{3}$    **19.** $-4$    **21.** $\frac{22}{3}$    **23.** $-\frac{27}{20}$    **25.** $2$

**27.** $\frac{1}{2}(1 - e^{-2}) \approx 0.432$    **29.** $\dfrac{e^3 - e}{3} \approx 5.789$

**31.** $4$    **33.** $4$    **35.** $\frac{1}{2}\ln 5 - \frac{1}{2}\ln 8 \approx -0.235$

**37.** $2\ln(2 + e^3) - 2\ln 3 \approx 3.993$

**39.** Area $= 10$

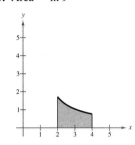

**41.** Area $= \frac{1}{4}$

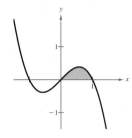

**43.** Area $= \ln 9$

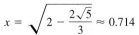

**45.** $10$    **47.** $4 + 5\ln 5 \approx 12.047$

**49. (a)** $11$    **(b)** $5$    **(c)** $-32$    **(d)** $-1$

**51.** Average $= \dfrac{14}{3}$

$x = \pm\dfrac{2\sqrt{3}}{3} \approx \pm 1.155$

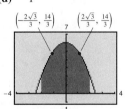

**53.** Average $= \dfrac{4}{3}$

$x = \sqrt{2 + \dfrac{2\sqrt{5}}{3}} \approx 1.868$

$x = \sqrt{2 - \dfrac{2\sqrt{5}}{3}} \approx 0.714$

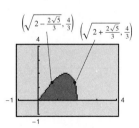

**55.** Average $= -\dfrac{2}{3}$

$x = \dfrac{4 + 2\sqrt{3}}{3} \approx 2.488$

$x = \dfrac{4 - 2\sqrt{3}}{3} \approx 0.179$

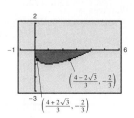

**57.** Even      **59.** Neither odd nor even

**61.** (a) $\frac{8}{3}$  (b) $\frac{16}{3}$  (c) $-\frac{8}{3}$      **63.** \$6.75      **65.** \$22.50

**67.** \$3.97      **69.** \$2500      **71.** \$1250

**73.** (a) \$137,000  (b) \$214,720.93  (c) \$338,393.53

**75.** \$3082.95

**77.** (a) $R = 0.81t^2 - 21.14e^{-t} + 66.24$

   (b) \$85.76 billion

**79.** $\frac{2}{3}\sqrt{7} - \frac{1}{3}$      **81.** $\frac{39}{200}$

## SECTION 5.5      *(page 357)*

---

**Warm Up**

**1.** $-x^2 + 3x + 2$      **2.** $-2x^2 + 4x + 4$

**3.** $-x^3 + 2x^2 + 4x - 5$      **4.** $x^3 - 6x - 1$

**5.** $(0, 4), (4, 4)$      **6.** $(1, -3), (2, -12)$

**7.** $(-3, 9), (2, 4)$      **8.** $(-2, -4), (0, 0), (2, 4)$

**9.** $(1, -2), (5, 10)$      **10.** $(1, e)$

---

**1.** 36      **3.** 9      **5.** $\frac{3}{2}$      **7.** $e - 2$

**9.**

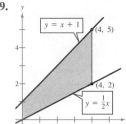

**11.**

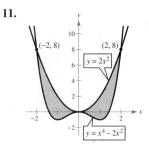

**13.**

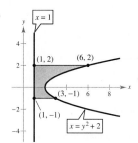

**15.** Area $= 2$      **17.** Area $= \frac{37}{12}$

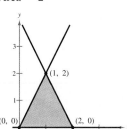

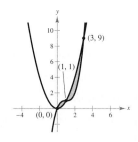

**19.** Area $= \frac{1}{2}$

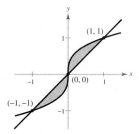

**21.** Area $= \frac{64}{3}$      **23.** Area $= \frac{9}{2}$

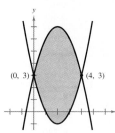

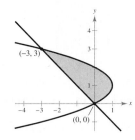

**25.** Area $= (2e + \ln 2) - 2e^{1/2} \approx 2.832$

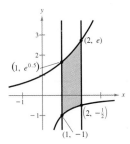

**27.** Area $= \frac{9}{2}$

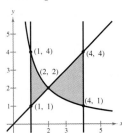

**29.** Area $= -\frac{1}{2}e^{-1} + \frac{1}{2}$     **31.** Area $= \frac{32}{3}$

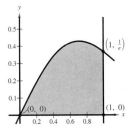     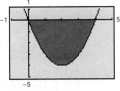

**33.** Area $= \frac{1}{2}$

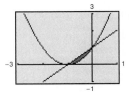

**35.** 8

**37.** Point of equilibrium: $(80, 10)$
Consumer surplus $= 1600$
Producer surplus $= 400$

**39.** Point of equilibrium: $(50, 150)$
Consumer surplus $\approx 1666.67$
Producer surplus $= 1250$

**41.** Point of equilibrium: $(300, 500)$
Consumer surplus $= 50,000$
Producer surplus $\approx 25,497$

**43.** A typical demand function is decreasing, whereas a typical supply function is increasing.

**45.** $R_1$, \$4.16 billion     **47.** \$501 million

**49.** (a)

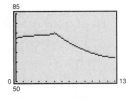

(b) 65.03 pounds

**51.** Consumer surplus $= \$625,000$
Producer surplus $= \$1,375,000$

**53.** \$337.33 million

**55.**

| Quintile | Lowest | 2nd | 3rd | 4th | Highest |
|----------|--------|-----|-----|-----|---------|
| Percent | 13.19 | 17.775 | 20.737 | 23.132 | 25.200 |

## SECTION 5.6     (page 365)

**Warm Up**

**1.** $\frac{1}{6}$     **2.** $\frac{3}{20}$     **3.** $\frac{7}{40}$     **4.** $\frac{13}{12}$     **5.** $\frac{61}{30}$     **6.** $\frac{53}{18}$

**7.** $\frac{2}{3}$     **8.** $\frac{4}{7}$     **9.** 0     **10.** 5

**1.** Midpoint Rule: 2          **3.** Midpoint Rule: 0.6730
Exact area: 2                    Exact area: $\frac{2}{3} \approx 0.6667$

**5.** Midpoint Rule: 4.6250

Exact area: $\frac{14}{3} \approx 4.6667$

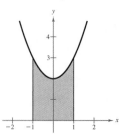

**7.** Midpoint Rule: 17.2500

Exact area: $\frac{52}{3} \approx 17.3333$

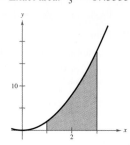

**9.** Midpoint Rule: 0.0859

Exact area: $\frac{1}{12} = 0.08\overline{3}$

**11.** Midpoint Rule: 0.0859

Exact area: $\frac{1}{12} \approx 0.0833$

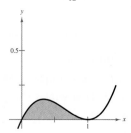

**13.** Area $\approx 54.6667$,

$n = 40$

**15.** Area $\approx 4.16$,

$n = 5$

**17.** Midpoint Rule: 1.5

Exact area: 1.5

**19.** Exact: 4

Trapezoidal Rule: 4.0625

Midpoint Rule: 3.9688

Midpoint Rule is better in this example.

**21.** 1.1167     **23.** 1.55

**25.**

| $n$ | Midpoint Rule | Trapezoidal Rule |
|----|----|----|
| 4 | 15.3965 | 15.6055 |
| 8 | 15.4480 | 15.5010 |
| 12 | 15.4578 | 15.4814 |
| 16 | 15.4613 | 15.4745 |
| 20 | 15.4628 | 15.4713 |

**27.** 4.8103     **29.** 916.25 ft

**31.** Midpoint Rule: $\pi \approx 3.146801$

Trapezoidal Rule: $\pi \approx 3.131176$

Graphing utility: $\pi \approx 3.141593$

## SECTION 5.7     *(page 372)*

**Warm Up**

**1.** 0, 2     **2.** 0, 2     **3.** 0, 2, $-2$     **4.** $-1, 2$

**5.** 2, 4     **6.** 1, 5     **7.** 53.5982     **8.** 1.9459

**9.** 3.3934     **10.** 1.3896

**1.** $\dfrac{16\pi}{3}$     **3.** $\dfrac{15\pi}{2}$     **5.** $\dfrac{512\pi}{15}$     **7.** $\dfrac{32\pi}{15}$     **9.** $\dfrac{\pi}{3}$

**11.** $\pi\left(\ln 2 - \dfrac{2}{3}\right) \approx 0.0832$     **13.** $\dfrac{128\pi}{5}$

**15.** $\dfrac{\pi}{2}(e^2 - 1)$     **17.** $8\pi$     **19.** $\dfrac{2}{3}\pi$     **21.** $\dfrac{\pi}{4}$

**23.** $\dfrac{256\pi}{15}$     **25.** $18\pi$

**27.** $V = \pi \displaystyle\int_{-r}^{r} (r^2 - x^2)\, dx = \dfrac{4\pi r^3}{3}$     **29.** $48\pi$

**31.** **(a)** 1,256,637 ft$^3$     **(b)** 2513 fish     **33.** 58.5598

## CHAPTER 5 REVIEW EXERCISES     *(page 378)*

**1.** $16x + C$     **3.** $\frac{2}{3}x^3 + \frac{5}{2}x^2 + C$     **5.** $x^{2/3} + C$

**7.** $\frac{3}{7}x^{7/3} + \frac{3}{2}x^2 + C$     **9.** $f(x) = \frac{3}{2}x^2 + x - 2$

**11.** $f(x) = \frac{1}{6}x^4 - 8x + \frac{33}{2}$

**13.** **(a)** 2.5 sec     **(b)** 100 ft     **(c)** 1.25 sec     **(d)** 75 ft

**15.** $x + 5x^2 + \frac{25}{3}x^3 + C$ or $\frac{1}{15}(1 + 5x)^3 + C_1$

**17.** $\frac{2}{5}\sqrt{5x - 1} + C$     **19.** $\frac{1}{2}x^2 - x^4 + C$

**21.** $\frac{1}{4}(x^4 - 2x)^2 + C$

**23.** **(a)** 30.5 board-feet     **(b)** 125.2 board-feet

**25.** $-e^{-3x} + C$     **27.** $\frac{1}{2}e^{x^2 - 2x} + C$

**29.** $-\frac{1}{3}\ln|1 - x^3| + C$     **31.** $\frac{2}{3}x^{3/2} + 2x + 2x^{1/2} + C$

**33.** $A = 4$     **35.** $A = \frac{8}{3}$     **37.** $A = 2\ln 2$

**39.** 16     **41.** 0     **43.** 2     **45.** $\frac{1}{8}$     **47.** 3.899

**49.** 0     **51.** \$700.25

**53.** Average value: $\frac{8}{5}$, $x = \frac{29}{4}$

**55.** Average value: $\frac{1}{3}(-1 + e^3) \approx 6.362$, $x \approx 3.150$

**57.** $520.54

**59.** (a) $S = -0.1t^4 + 4.6t^3 - 39.9t^2 + 145.3t + 144$

    (b) 2002

**61.** $\displaystyle\int_{-2}^{2} 6x^5\,dx = 0$

    (Odd function)

**63.** $\displaystyle\int_{-2}^{-1} \frac{4}{x^2}\,dx = \int_{1}^{2} \frac{4}{x^2}\,dx = 2$

    (Symmetric about $y$-axis)

**65.** Area $= \frac{25}{3}$

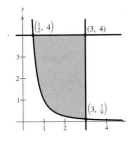

**67.** Area $= 16$

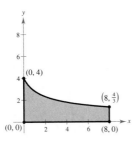

**69.** Area $= \frac{64}{3}$

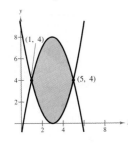

**71.** Area $= \frac{9}{2}$

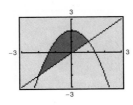

**73.** Consumer surplus: 11,250

    Producer surplus: 14,062.5

**75.** 66 gallons more

**77.**

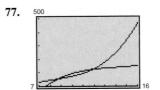

Sales would have increased. Approximately $501.92 million more.

**79.** $n = 4$: 13.3203

    $n = 20$: 13.7167

**81.** $n = 4$: 0.7867

    $n = 20$: 0.7855

**83.** $\pi \ln 4 \approx 4.355$

**85.** $\frac{\pi}{2}(e^2 - e^{-2}) \approx 11.394$

**87.** $\frac{56\pi}{3}$    **89.** $\frac{2\pi}{35}$    **91.** $\frac{5\pi}{16}\sqrt{15}$

## SAMPLE POST-GRAD EXAM QUESTIONS
*(page 382)*

**1.** (d)    **2.** (b)    **3.** (c)    **4.** (b)    **5.** (a)

**6.** (b)    **7.** (d)    **8.** (d)

# CHAPTER 6

## SECTION 6.1    *(page 390)*

> **Warm Up**
>
> **1.** $5x + C$    **2.** $\frac{2}{5}x^{5/2} + C$
>
> **3.** $\frac{1}{4}(x^2 + 1)^4 + C$    **4.** $e^{6x} + C$
>
> **5.** $\ln|2x + 1| + C$    **6.** $-e^{-x^2} + C$
>
> **7.** $x(x - 1)(2x - 1)$    **8.** $3x(x + 4)^2(x + 8)$
>
> **9.** $(x + 21)(x + 7)^{-1/2}$    **10.** $x(x + 5)^{-2/3}$

**1.** $\frac{1}{5}(x - 2)^5 + C$    **3.** $\frac{2}{9 - t} + C$

**5.** $\frac{2}{3}(1 + x)^{3/2} + C$    **7.** $\ln(3x^2 + x)^2 + C$

**9.** $-\dfrac{1}{10(5x + 1)^2} + C$    **11.** $2(x + 1)^{1/2} + C$

**13.** $-\frac{1}{3}\ln|1 - e^{3x}| + C$    **15.** $\frac{1}{2}x^2 + x + \ln|x - 1| + C$

**17.** $\frac{1}{3}(x^2 + 4)^{3/2} + C$    **19.** $\frac{1}{5}e^{5x} + C$

**21.** $\dfrac{-1}{2(x+1)^2} + \dfrac{1}{3(x+1)^3} + C$

**23.** $\dfrac{1}{9}\left(\ln|3x-1| - \dfrac{1}{3x-1}\right) + C$

**25.** $2(\sqrt{t}-1) + 2\ln|\sqrt{t}-1| + C$

**27.** $4\sqrt{t} + \ln t + C$     **29.** $\dfrac{26}{3}$

**31.** $\dfrac{3}{2}(e-1) \approx 2.577$     **33.** $\ln 2 - \dfrac{1}{2} \approx 0.193$

**35.** $\dfrac{13}{320}$

**37.** Area $= \dfrac{144}{5}$     **39.** Area $= \dfrac{1712}{105}$

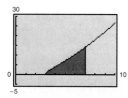 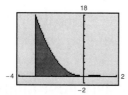

**41.** Area $= \dfrac{224}{15}$     **43.** Area $= \dfrac{1209}{28}$

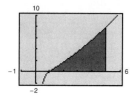 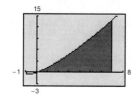

**45.** $\dfrac{16}{15}\sqrt{2}$     **47.** $\dfrac{4}{3}$     **49.** $\dfrac{4\pi}{15} \approx 0.838$     **51.** $\dfrac{1}{2}$

**53.** (a) $0.547$   (b) $0.586$

**55.**

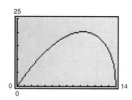

About 13.97 inches

**57.** Certain fruits and vegetables exhibit this pattern.

**59.** Approximately 1.346

## SECTION 6.2     (page 399)

**Warm Up**

**1.** $12x^3 + 2$     **2.** $\dfrac{1}{2}x + \dfrac{1}{8}$     **3.** $\dfrac{4x^2 + 2}{x^3 + x}$

**4.** $3x^2 e^{x^3}$     **5.** $e^x(x^2 + 2x)$     **6.** $-\dfrac{4}{(x-3)^2}$

**7.** $\dfrac{63}{3}$     **8.** $\dfrac{4}{3}$     **9.** $\dfrac{32}{3}$     **10.** $8$

**1.** $\dfrac{1}{3}xe^{3x} - \dfrac{1}{9}e^{3x} + C$     **3.** $-x^2 e^{-x} - 2xe^{-x} - 2e^{-x} + C$

**5.** $x\ln 2x - x + C$

**7.** $\dfrac{1}{4}e^{4x} + C$     **9.** $\dfrac{e^{4x}}{16}(4x-1) + C$

**11.** $\dfrac{1}{2}e^{x^2} + C$     **13.** $x^2 e^x - 2e^x x + 2e^x$

**15.** $-e^{1/t} + C$

**17.** $\dfrac{1}{2}t^2 \ln(t+1) - \dfrac{1}{2}\ln(t+1) - \dfrac{1}{4}(t-1)^2 + C$

**19.** $\dfrac{x^2}{2}(\ln x)^2 - \dfrac{x^2}{2}\ln x + \dfrac{x^2}{4} + C$

**21.** $\dfrac{1}{3}(\ln x)^3 + C$     **23.** $\dfrac{1}{4}x^4 + \dfrac{2}{3}x^3 + \dfrac{1}{2}x^2 + C$

**25.** $\dfrac{e^{2x}}{4(2x+1)} + C$

**27.** Area $= 2e^2 + 6$     **29.** Area $= \dfrac{1}{9}(2e^3 + 1)$

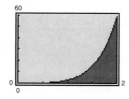 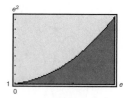

**31.** $e - 2 \approx 0.718$     **33.** $\dfrac{5}{36}e^6 + \dfrac{1}{36} \approx 56.060$

**35.** $\dfrac{2}{3}x(2x-3)^{3/2} - \dfrac{2}{15}(2x-3)^{5/2} + C$

**37.** $\dfrac{2}{5}x(4+5x)^{1/2} - \dfrac{4}{75}(4+5x)^{3/2} + C$

**39.** $\displaystyle\int x^n \ln x\, dx = \dfrac{x^{n+1}}{n+1}\ln x - \int \dfrac{x^n}{n+1}\, dx$

$$= \dfrac{x^{n+1}}{n+1}\ln x - \dfrac{x^{n+1}}{(n+1)^2} + C$$

$$= \dfrac{x^{n+1}}{(n+1)^2}[-1 + (n+1)\ln x] + C$$

**41.** $\dfrac{e^{5x}}{125}(25x^2 - 10x + 2) + C$    **43.** $-\dfrac{1}{x} - \dfrac{\ln x}{x} + C$

**45.** $1 - 5e^{-4} \approx 0.908$

**47.** (a) 2   (b) $4\pi(e - 2) \approx 9.026$

**49.** $\dfrac{3}{128} - \dfrac{379}{128}e^{-8} \approx 0.022$

**51.** $\dfrac{1,171,875}{256}\pi \approx 14,381.070$

**53.** (a) Increase

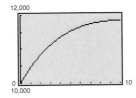

(b) 113,212 units   (c) 11,321 per year

**55.** (a) $3.2 \ln 2 - 0.2 \approx 2.018$

(b) $12.8 \ln 4 - 7.2 \ln 3 - 1.8 \approx 8.035$

**57.** \$18,126.92    **59.** \$1,332,474.72    **61.** \$4103.07

**63.** (a) \$1,200,000   (b) \$1,094,142.26

**65.** 4.254    **67.** \$45,957.78

**69.** (a) \$17,378.62   (b) \$3681.26

## SECTION 6.3    *(page 409)*

---

### Warm Up

**1.** $(x - 4)(x + 3)$    **2.** $(x - 5)(x + 5)$

**3.** $x(x - 2)(x + 1)$    **4.** $x(x - 2)^2$

**5.** $(x - 2)(x - 1)^2$    **6.** $(x - 3)(x - 1)^2$

**7.** $\dfrac{1}{x - 2} + x$    **8.** $\dfrac{1}{1 - x} + 2x - 2$

**9.** $\dfrac{2}{2 - x} + x^2 - x - 2$    **10.** $\dfrac{6}{x - 1} + x + 4$

---

**1.** $\dfrac{5}{x - 5} - \dfrac{3}{x + 5}$    **3.** $\dfrac{9}{x - 3} - \dfrac{1}{x}$

**5.** $\dfrac{1}{x - 5} + \dfrac{3}{x + 2}$    **7.** $\dfrac{3}{x} - \dfrac{5}{x^2}$

**9.** $\dfrac{1}{3(x - 2)} + \dfrac{1}{(x - 2)^2}$

**11.** $\dfrac{8}{x + 1} - \dfrac{1}{(x + 1)^2} + \dfrac{2}{(x + 1)^3}$

**13.** $\dfrac{1}{2} \ln \left| \dfrac{x - 1}{x + 1} \right| + C$

**15.** $\dfrac{1}{4} \ln |4x + 1| - \dfrac{1}{4} \ln |4 - x| + C$

**17.** $-\ln |x| + \ln |3x - 1| + C$

**19.** $\ln |x| - \ln |2x + 1| + C$

**21.** $\ln |x - 1| - \ln |x + 2| + C$

**23.** $\dfrac{3}{2} \ln |2x - 1| - 2 \ln |x + 1| + C$

**25.** $5 \ln |x - 2| - \ln |x + 2| - 3 \ln |x| + C$

**27.** $\dfrac{1}{2}(3 \ln |x - 4| - \ln |x|) + C$

**29.** $-3 \ln |x - 1| - \dfrac{1}{x - 1} + C$

**31.** $\ln |x| + 2 \ln |x + 1| + \dfrac{1}{x + 1} + C$

**33.** $\dfrac{1}{6} \ln \dfrac{4}{7} \approx -0.093$    **35.** $-\dfrac{4}{5} + 2 \ln \dfrac{5}{3} \approx 0.222$

**37.** $\dfrac{1}{2} - \ln 2 \approx -0.193$    **39.** $4 \ln 2 + \dfrac{1}{2} \approx 3.273$

**41.** $12 - \dfrac{7}{2} \ln 7 \approx 5.1893$

**43.** $5 \ln 2 - \ln 5 \approx 1.8563$

**45.** $\dfrac{1}{2a}\left( \dfrac{1}{a + x} + \dfrac{1}{a - x} \right)$    **47.** $\dfrac{1/a}{x} + \dfrac{1/a}{a - x}$

**49.** $\dfrac{\pi}{165}\left[ 136 - 33 \ln \dfrac{11}{3} \right] \approx 1.7731$

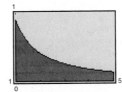

**51.** $\pi\left( \dfrac{1}{3} + \dfrac{1}{4} \ln 3 \right) \approx 1.9100$    **53.** $y = \dfrac{1000}{1 + 9e^{-0.1656t}}$

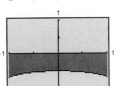  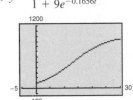

**55.** \$1.077 thousand    **57.** 289.67 thousand; 36.21 thousand

**59.** Answers may vary.

## SECTION 6.4     *(page 420)*

---

**Warm Up**

**1.** $x^2 + 8x + 16$     **2.** $x^2 + x + \frac{1}{4}$

**3.** $x^2 - 2x + 1$     **4.** $x^2 - \frac{2}{3}x + \frac{1}{9}$

**5.** $\dfrac{2}{x} - \dfrac{2}{x+2}$     **6.** $-\dfrac{3}{4x} + \dfrac{3}{4(x-4)}$

**7.** $\dfrac{3}{2(x-2)} - \dfrac{2}{x^2} - \dfrac{3}{2x}$     **8.** $-\dfrac{3}{x+1} + \dfrac{2}{x-2} + \dfrac{4}{x}$

**9.** $2e^x(x - 1) + C$     **10.** $x^3 \ln x - \dfrac{x^3}{3} + C$

---

**1.** $\dfrac{1}{9}\left(\dfrac{2}{2 + 3x} + \ln|2 + 3x|\right) + C$

**3.** $\dfrac{2(3x - 4)}{27}\sqrt{2 + 3x} + C$     **5.** $\ln\left(x^2 + \sqrt{x^4 - 9}\right) + C$

**7.** $\frac{1}{2}(x^2 - 1)e^{x^2} + C$     **9.** $\ln\left|\dfrac{x}{1 + x}\right| + C$

**11.** $-\dfrac{1}{3}\ln\left|\dfrac{3 + \sqrt{x^2 + 9}}{x}\right| + C$

**13.** $-\dfrac{1}{2}\ln\left|\dfrac{2 + \sqrt{4 - x^2}}{x}\right| + C$

**15.** $\frac{1}{4}x^2(-1 + 2\ln x) + C$     **17.** $3x^2 - \ln(1 + e^{3x^2}) + C$

**19.** $\frac{1}{4}\left(x^2\sqrt{x^4 - 4} - 4\ln|x^2 + \sqrt{x^4 - 4}|\right) + C$

**21.** $\dfrac{1}{27}\left[\dfrac{4}{2 + 3t} - \dfrac{4}{2(2 + 3t)^2} + \ln|2 + 3t|\right] + C$

**23.** $\dfrac{1}{\sqrt{3}}\ln\left|\dfrac{\sqrt{3 + s} - \sqrt{3}}{\sqrt{3 + s} + \sqrt{3}}\right| + C$

**25.** $\dfrac{1}{8}\left[\dfrac{-1}{2(3 + 2x)^2} + \dfrac{6}{3(3 + 2x)^3} - \dfrac{9}{4(3 + 2x)^4}\right] + C$

**27.** $-\dfrac{\sqrt{1 - x^2}}{x} + C$     **29.** $\frac{1}{9}x^3(-1 + 3\ln x) + C$

**31.** $\dfrac{1}{27}\left(3x - \dfrac{25}{3x - 5} + 10\ln|3x - 5|\right) + C$

**33.** $\frac{1}{9}(3\ln x - 4\ln|4 + 3\ln x|) + C$

**35.** Area $= \dfrac{40}{3}$

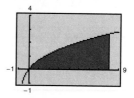

Area $= 13.\overline{3}$

**37.** Area $= \dfrac{1}{2}\left[4 + \ln\left(\dfrac{2}{1 + e^4}\right)\right]$

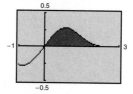

Area $\approx 0.3375$

**39.** Area $= \frac{1}{4}\left[21\sqrt{5} - 8\ln(\sqrt{5} + 3) + 8\ln 2\right]$

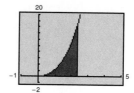

Area $\approx 9.8145$

**41.** $\dfrac{5\sqrt{5}}{3}$     **43.** $12\ln\left(\dfrac{2e^2}{1 + e^2}\right) \approx 6.7946$

**45.** $(x^2 - 2x + 2)e^x + C$     **47.** $-\left(\dfrac{1}{x} + \ln\left|\dfrac{x}{x + 1}\right|\right) + C$

**49.** **(a)** $(x + 3)^2 - 9$     **(b)** $(x - 4)^2 - 7$
    **(c)** $(x^2 + 1)^2 - 6$     **(d)** $4 - (x + 1)^2$

**51.** $\dfrac{1}{2\sqrt{17}}\ln\left|\dfrac{(x + 3) - \sqrt{17}}{(x + 3) + \sqrt{17}}\right| + C$

**53.** $-\ln\left|\dfrac{1 + \sqrt{x^2 - 2x + 2}}{x - 1}\right| + C$     **55.** $\dfrac{1}{8}\ln\left|\dfrac{x - 3}{x + 1}\right| + C$

**57.** $\frac{1}{2}\ln|x^2 + 1 + \sqrt{x^4 + 2x^2 + 2}| + C$

**59.**

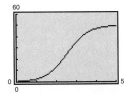

Average value: 42.58

**61.** $1138.43    **63.** $230.98 million

## SECTION 6.5    *(page 429)*

**Warm Up**

**1.** $\dfrac{2}{x^3}$    **2.** $-\dfrac{96}{(2x+1)^4}$    **3.** $-\dfrac{12}{x^4}$    **4.** $6x - 4$

**5.** $16e^{2x}$    **6.** $e^{x^2}(4x^2 + 2)$    **7.** $(3, 18)$

**8.** $(1, 8)$    **9.** $n < -5\sqrt{10},\ n > 5\sqrt{10}$

**10.** $n < -5,\ n > 5$

| | Exact value | Trapezoidal Rule | Simpson's Rule |
|---|---|---|---|
| **1.** | 2.6667 | 2.7500 | 2.6667 |
| **3.** | 8.4000 | 9.0625 | 8.4167 |
| **5.** | 4.0000 | 4.0625 | 4.0000 |
| **7.** | 0.5000 | 0.5090 | 0.5004 |
| **9.** | 0.6931 | 0.6970 | 0.6933 |
| **11.** | | 0.7828 | 0.7854 |
| **13.** | | 0.749 | 0.771 |
| **15.** | | 0.877 | 0.830 |
| **17.** | | 1.88 | 1.89 |

**19.** $21,831.20    **21.** $678.36

**23.** $0.3413 = 34.13\%$    **25.** $89{,}500\ \text{ft}^2$

**27. (a)** 2   **(b)** $\dfrac{2^5}{180(4^4)}(24) \approx 0.017$

**29. (a)** $\dfrac{5e}{64} \approx 0.212$   **(b)** $\dfrac{13e}{1024} \approx 0.035$

**31. (a)** $n = 101$   **(b)** $n = 8$

**33. (a)** $n = 3280$   **(b)** $n = 60$

**35.** 19.5215    **37.** 3.6558

**39.** Exact value: $\displaystyle\int_0^1 x^3\,dx = \dfrac{x^4}{4}\Big]_0^1 = \dfrac{1}{4}$

Simpson's Rule: $\displaystyle\int_0^1 x^3\,dx = \dfrac{1}{6}\left[0^3 + 4\left(\dfrac{1}{2}\right)^3 + 1^3\right] = \dfrac{1}{4}$

**41.** 416.1 ft

**43.** 58.876 grams (Simpson's Rule with $n = 100$)

**45.** 1876 subscribers (Simpson's Rule with $n = 100$)

## SECTION 6.6    *(page 440)*

**Warm Up**

**1.** 9    **2.** 3    **3.** $-\dfrac{1}{8}$    **4.** Limit does not exist.

**5.** Limit does not exist.    **6.** $-4$

**7. (a)** $\dfrac{32}{3}b^3 - 16b^2 + 8b - \dfrac{4}{3}$   **(b)** $-\dfrac{4}{3}$

**8. (a)** $\dfrac{b^2 - b - 11}{(b-2)^2(b-5)}$   **(b)** $\dfrac{11}{20}$

**9. (a)** $\ln\left(\dfrac{5 - 3b^2}{b+1}\right)$   **(b)** $\ln 5 \approx 1.609$

**10. (a)** $e^{-3b^2}(e^{6b^2} + 1)$   **(b)** 2

**1.** 1    **3.** 1    **5.** Diverges    **7.** Diverges

**9.** Diverges    **11.** Diverges    **13.** 0    **15.** 4

**17.** 6    **19.** Diverges    **21.** 6    **23.** Diverges

**25.** 0    **27.** $\ln\left(4 + \sqrt{7}\right) - \ln 3 \approx 0.7954$

**29.**

| $x$ | 1 | 10 | 25 | 50 |
|---|---|---|---|---|
| $xe^{-x}$ | 0.3679 | 0.0005 | 0.0000 | 0.0000 |

**31.**

| $x$ | 1 | 10 | 25 | 50 |
|---|---|---|---|---|
| $x^2 e^{-(1/2)x}$ | 0.6065 | 0.6738 | 0.0023 | 0.0000 |

**33.** 2    **35.** $\dfrac{1}{4}$    **37. (a)** $4{,}637{,}228   **(b)** $5{,}555{,}556

**39. (a)** $748{,}367.34   **(b)** $808{,}030.14   **(c)** $900{,}000.00

**41. (a)** 1   **(b)** $\dfrac{\pi}{3}$

**43. (a)** 0.9687   **(b)** 0.0724   **(c)** 0.0009

# CHAPTER 6 REVIEW EXERCISES  *(page 446)*

**1.** $t + C$   **3.** $\dfrac{(x+5)^4}{4} + C$   **5.** $\dfrac{1}{10}e^{10x} + C$

**7.** $\frac{1}{5}\ln|x| + C$   **9.** $\frac{1}{3}(x^2 + 4)^{3/2} + C$

**11.** $2\ln(3 + e^x) + C$   **13.** $\dfrac{(x-2)^5}{5} + \dfrac{(x-2)^4}{2} + C$

**15.** $\frac{2}{15}(x+1)^{3/2}(3x-2) + C$

**17.** $\frac{4}{5}(x-3)^{3/2}(x+2) + C$

**19.** $-\frac{2}{15}(1-x)^{3/2}(3x+7) + C$

**21.** $\frac{26}{15}$   **23.** $\frac{412}{15}$   **25. (a)** $0.696$  **(b)** $0.693$

**27. (a)** $\$1900.30$ million   **(b)** $\$17{,}102.68$ million

**29.** $2\sqrt{x}\ln x - 4\sqrt{x} + C$   **31.** $xe^x - 2e^x + C$

**33.** $x^2e^{2x} - xe^{2x} + \frac{1}{2}e^{2x} + C$   **35.** $\$45{,}317.31$

**37.** $\$432{,}979.25$

**39. (a)** $\$4423.98$; $\$3934.69$; $\$3517.56$  **(b)** $\$997{,}629.35$

**41.** $\$45{,}118.84$   **43.** $\frac{1}{5}\ln|x| - \frac{1}{5}\ln|x+5| + C$

**45.** $\ln|x-5| + 3\ln|x+2| + C$

**47.** $x - \frac{25}{8}\ln|x+5| + \frac{9}{8}\ln|x-3| + C$

**49. (a)** $y = \dfrac{10{,}000}{1 + 7e^{-0.098652t}}$

**(b)**

| Time, $t$ | 0 | 3 | 6 | 12 | 24 |
|---|---|---|---|---|---|
| Sales, $y$ | 1250 | 1611 | 2052 | 3182 | 6039 |

**(c)** $t \approx 31$ weeks

**51.** $\sqrt{x^2 + 25} - 5\ln\left|\dfrac{5 + \sqrt{x^2 + 25}}{x}\right| + C$

**53.** $\dfrac{1}{4}\ln\left|\dfrac{x-2}{x+2}\right| + C$   **55.** $\dfrac{8}{3}$

**57.** $2\sqrt{1+x} + \ln\left|\dfrac{\sqrt{1+x}-1}{\sqrt{1+x}+1}\right| + C$

**59.** $(x-5)^3e^{x-5} - 3(x-5)^2e^{x-5} + 6(x-6)e^{x-5} + C$

**61.** $\dfrac{1}{10}\ln\left|\dfrac{x-3}{x+7}\right| + C$

**63.** $\frac{1}{2}\Big[(x-5)\sqrt{(x-5)^2 - 5^2}$
  $+ 25\ln\big|(x-5) + \sqrt{(x-5)^2 - 5^2}\big|\Big] + C$

**65.** $0.705$   **67.** $0.741$   **69.** $0.376$   **71.** $0.289$

**73.** $2.29$   **75.** $0.001$   **77.** $1$   **79.** Diverges   **81.** $2$

**83.** $2$   **85. (a)** $\$494{,}525.28$   **(b)** $\$833{,}333.33$

**87. (a)** $0.148$   **(b)** $0.003$

## SAMPLE POST-GRAD EXAM QUESTIONS
*(page 450)*

**1.** (c)   **2.** (a)   **3.** (c)   **4.** (d)   **5.** (d)   **6.** (b)

**7.** (a)   **8.** (c)

# CHAPTER 7

## SECTION 7.1   *(page 458)*

> **Warm Up**
>
> **1.** $2\sqrt{5}$   **2.** $5$   **3.** $8$   **4.** $8$   **5.** $(4, 7)$
>
> **6.** $(1, 0)$   **7.** $(0, 3)$   **8.** $(-1, 1)$
>
> **9.** $(x-2)^2 + (y-3)^2 = 4$
>
> **10.** $(x-1)^2 + (y-4)^2 = 25$

**1.**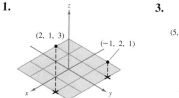

**3.**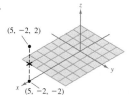

**5.** $3\sqrt{2}$   **7.** $\sqrt{206}$   **9.** $(2, -5, 3)$

**11.** $\left(\frac{1}{2}, \frac{1}{2}, -1\right)$   **13.** $(6, -3, 5)$   **15.** $(1, 2, 1)$

**17.** $3, 3\sqrt{5}, 6$; right triangle

**19.** $2, 2\sqrt{5}, 2\sqrt{2}$; the triangle is neither right nor isosceles.

**21.** $x^2 + (y-2)^2 + (z-2)^2 = 4$

**23.** $\left(x - \frac{3}{2}\right)^2 + (y-2)^2 + (z-1)^2 = \frac{21}{4}$

**25.** $(x-1)^2 + (y-1)^2 + (z-5)^2 = 9$

**27.** $(x-1)^2 + (y-3)^2 + z^2 = 10$

**29.** Center: $(1, -3, -4)$   **31.** Center: $(0, 4, 0)$

Radius: 5   Radius: 4

**33.** Center: $(1, 3, 2)$

Radius: $\dfrac{5}{\sqrt{2}} = \dfrac{5\sqrt{2}}{2}$

**35.** 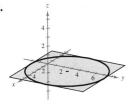  **37.**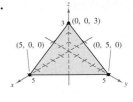

**1.** 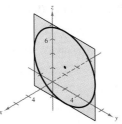  **3.**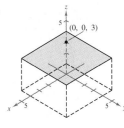

**39. (a)** 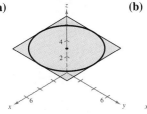  **(b)**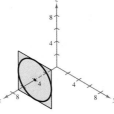

**5.** 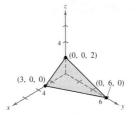  **7.**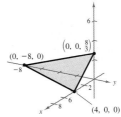

**41. (a)** 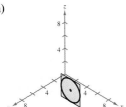  **(b)**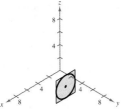

**9.** 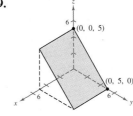  **11.**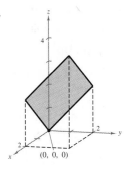

**43.** $(3, 3, 3)$

## SECTION 7.2   *(page 468)*

**Warm Up**

**1.** $(4, 0), (0, 3)$   **2.** $\left(-\frac{4}{3}, 0\right), (0, -8)$

**3.** $(1, 0), (0, -2)$   **4.** $(-5, 0), (0, -5)$

**5.** $(x - 1)^2 + (y - 2)^2 + (z - 9)^2 + 1 = 0$

**6.** $(x - 4)^2 + (y + 2)^2 - (z + 3)^2 = 0$

**7.** $(x + 1)^2 + (y - 1)^2 - z = 0$

**8.** $(x - 3)^2 + (y + 5)^2 + (z + 13)^2 = 1$

**9.** $x^2 - y^2 + z^2 = \frac{1}{4}$   **10.** $x^2 - y^2 + z^2 = 4$

**13.** Perpendicular   **15.** Parallel   **17.** Parallel

**19.** Neither parallel nor perpendicular   **21.** $\dfrac{6\sqrt{14}}{7}$

**23.** $\dfrac{\sqrt{6}}{6}$   **25.** $\dfrac{13\sqrt{29}}{29}$   **27.** c   **28.** e   **29.** f

**30.** g   **31.** d   **32.** b   **33.** a   **34.** h

**35.** Trace in $xy$-plane ($z = 0$): $y = x^2$ (parabola)

Trace in plane $y = 1$: $x^2 - z^2 = 1$ (hyperbola)

Trace in $yz$-plane ($x = 0$): $y = -z^2$ (parabola)

**37.** Trace in $xy$-plane ($z = 0$): $\dfrac{x^2}{4} + y^2 = 1$ (ellipse)

Trace in $xz$-plane ($y = 0$): $\dfrac{x^2}{4} + z^2 = 1$ (ellipse)

Trace in $yz$-plane ($x = 0$): $y^2 + z^2 = 1$ (circle)

**39.** Ellipsoid    **41.** Hyperboloid of one sheet

**43.** Elliptic paraboloid    **45.** Hyperbolic paraboloid

**47.** Hyperboloid of two sheets    **49.** Elliptic cone

**51.** Hyperbolic paraboloid

**53.** 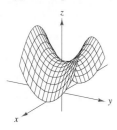    **55.**

**57.** $\dfrac{x^2}{3969^2} + \dfrac{y^2}{3969^2} + \dfrac{z^2}{3942^2} = 1$

# SECTION 7.3    *(page 476)*

---

**Warm Up**

**1.** 11    **2.** $-16$    **3.** 7

**4.** 4    **5.** $(-\infty, \infty)$

**6.** $(-\infty, -3) \cup (-3, 0) \cup (0, \infty)$

**7.** $[5, \infty)$    **8.** $\left(-\infty, -\sqrt{5}\,\right] \cup \left[\sqrt{5}, \infty\right)$

**9.** 55.0104    **10.** 6.9165

---

**1. (a)** $\dfrac{3}{2}$    **(b)** $-\dfrac{1}{4}$    **(c)** 6    **(d)** $\dfrac{5}{y}$    **(e)** $\dfrac{x}{2}$    **(f)** $\dfrac{5}{t}$

**3. (a)** 5    **(b)** $3e^2$    **(c)** $2e^{-1}$    **(d)** $5e^y$    **(e)** $xe^2$    **(f)** $te^t$

**5. (a)** $\dfrac{2}{3}$    **(b)** 0    **7. (a)** $90\pi$    **(b)** $50\pi$

**9. (a)** 20,655    **(b)** 1,397,673    **11. (a)** 0    **(b)** 6

**13. (a)** $x^2 + 2x\Delta x + (\Delta x)^2 - 2y$    **(b)** $-2, \Delta y \neq 0$

**15.** Domain: all points $(x, y)$ inside and on the circle $x^2 + y^2 = 16$

Range: $[0, 4]$

**17.** Domain: all points $(x, y)$ such that $y \neq 0$

Range: $(0, \infty)$

**19.** All points inside and on the ellipse $9x^2 + y^2 = 9$

**21.** All points $(x, y)$ such that $y \neq 0$

**23.** All points $(x, y)$ such that $x \neq 0$ and $y \neq 0$

**25.** All points $(x, y)$ such that $y \geq 0$

**27.** The half-plane below the line $y = -x + 4$

**29.** b    **30.** d    **31.** a    **32.** c

**33.** 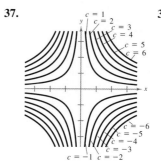    **35.**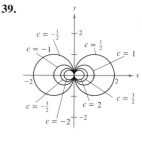

**37.** 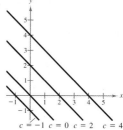    **39.**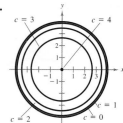

**41.** 135,540 units    **43.** $21,960

**45. (a)** $1250    **(b)** $925

**47.**

| $R \diagdown I$ | 0 | 0.03 | 0.05 |
|---|---|---|---|
| 0 | $2593.74 | $1929.99 | $1592.33 |
| 0.28 | $2004.23 | $1491.34 | $1230.42 |
| 0.35 | $1877.14 | $1396.77 | $1152.40 |

**49. (a)** $1.52 per share

**(b)** The variable $x$ has greater influence because its coefficient is larger.

# SECTION 7.4   *(page 487)*

---

## Warm Up

**1.** $\dfrac{x}{\sqrt{x^2 + 3}}$    **2.** $-6x(3 - x^2)^2$    **3.** $e^{2t+1}(2t + 1)$

**4.** $\dfrac{e^{2x}(2 - 3e^{2x})}{\sqrt{1 - e^{2x}}}$    **5.** $-\dfrac{2}{3 - 2x}$    **6.** $\dfrac{3(t^2 - 2)}{2t(t^2 - 6)}$

**7.** $-\dfrac{10x}{(4x - 1)^3}$    **8.** $-\dfrac{(x + 2)^2(x^2 + 8x + 27)}{(x^2 - 9)^3}$

**9.** $f'(2) = 8$    **10.** $f'(2) = \frac{7}{2}$

---

**1.** $f_x(x, y) = 2$

$f_y(x, y) = -3$

**3.** $f_x(x, y) = \dfrac{5}{2\sqrt{x}}$

$f_y(x, y) = -12y$

**5.** $f_x(x, y) = \dfrac{1}{y}$

$f_y(x, y) = -\dfrac{x}{y^2}$

**7.** $f_x(x, y) = \dfrac{x}{\sqrt{x^2 + y^2}}$

$f_y(x, y) = \dfrac{y}{\sqrt{x^2 + y^2}}$

**9.** $\dfrac{\partial z}{\partial x} = 2xe^{2y}$

$\dfrac{\partial z}{\partial y} = 2x^2 e^{2y}$

**11.** $h_x(x, y) = -2xe^{-(x^2+y^2)}$

$h_y(x, y) = -2ye^{-(x^2+y^2)}$

**13.** $\dfrac{\partial z}{\partial x} = \dfrac{3y - x}{x^2 - y^2}$

$\dfrac{\partial z}{\partial y} = \dfrac{y - 3x}{x^2 - y^2}$

**15.** $f_x(x, y) = 3xye^{x-y}(2 + x)$

**17.** $g_x(x, y) = 3y^2 e^{y-x}(1 - x)$    **19.** 9

**21.** $f_x(x, y) = 6x + y, 13, \ f_y(x, y) = x - 2y, 0$

**23.** $f_x(x, y) = 3ye^{3xy}, 12, \ f_y(x, y) = 3xe^{3xy}, 0$

**25.** $f_x(x, y) = -\dfrac{y^2}{(x - y)^2}, -\dfrac{1}{4}$

$f_y(x, y) = \dfrac{x^2}{(x - y)^2}, \dfrac{1}{4}$

**27.** $f_x(x, y) = \dfrac{2x}{x^2 + y^2}, 2$

$f_y(x, y) = \dfrac{2y}{x^2 + y^2}, 0$

**29.** $w_x = \dfrac{x}{\sqrt{x^2 + y^2 + z^2}}, \dfrac{2}{3}$

$w_y = \dfrac{y}{\sqrt{x^2 + y^2 + z^2}}, -\dfrac{1}{3}$

$w_z = \dfrac{z}{\sqrt{x^2 + y^2 + z^2}}, \dfrac{2}{3}$

**31.** $w_x = \dfrac{x}{x^2 + y^2 + z^2}, \dfrac{3}{25}$

$w_y = \dfrac{y}{x^2 + y^2 + z^2}, 0$

$w_z = \dfrac{z}{x^2 + y^2 + z^2}, \dfrac{4}{25}$

**33.** $w_x = 2z^2 + 3yz, 2$    **35.** $(-6, 4)$

$w_y = 3xz - 12yz, 30$

$w_z = 4xz + 3xy - 6y^2, -1$

**37.** $(1, 1)$    **39.** (a) 2  (b) $-3$    **41.** (a) 6  (b) $-18$

**43.** (a) $-\frac{3}{4}$  (b) 0    **45.** (a) $-2$  (b) $-2$

**47.** $\dfrac{\partial^2 z}{\partial y \partial x} = \dfrac{\partial^2 z}{\partial x \partial y} = -2$    **49.** $\dfrac{\partial^2 z}{\partial y \partial x} = \dfrac{\partial^2 z}{\partial x \partial y} = ye^{2xy}$

**51.** $\dfrac{\partial^2 z}{\partial x^2} = 6x$    **53.** $\dfrac{\partial^2 z}{\partial x^2} = 24x$

$\dfrac{\partial^2 z}{\partial y^2} = -8$

$\dfrac{\partial^2 z}{\partial y \partial x} = \dfrac{\partial^2 z}{\partial x \partial y} = 0$

$\dfrac{\partial^2 z}{\partial y^2} = 6x - 24y$

$\dfrac{\partial^2 z}{\partial y \partial x} = \dfrac{\partial^2 z}{\partial x \partial y} = 6y$

**55.** $\dfrac{\partial^2 z}{\partial x^2} = \dfrac{2y^2}{(x - y)^3}$    **57.** $\dfrac{\partial^2 z}{\partial x^2} = 0$

$\dfrac{\partial^2 z}{\partial y^2} = \dfrac{2x^2}{(x - y)^3}$

$\dfrac{\partial^2 z}{\partial y \partial x} = \dfrac{\partial^2 z}{\partial x \partial y}$

$\dfrac{\partial^2 z}{\partial y^2} = 2xe^{-y^2}(2y^2 - 1)$

$\dfrac{\partial^2 z}{\partial y \partial x} = \dfrac{\partial^2 z}{\partial x \partial y} = -2ye^{-y^2}$

$= -\dfrac{2xy}{(x - y)^3}$

**59.** $f_{xx}(x, y) = 12x^2 - 6y^2, 12$

$f_{xy}(x, y) = -12xy, 0$

$f_{yy}(x, y) = -6x^2 + 2, -4$

$f_{yx}(x, y) = -12xy, 0$

**61.** $f_{xx}(x, y) = -\dfrac{1}{(x - y)^2}, -1$

$f_{xy}(x, y) = \dfrac{1}{(x - y)^2}, 1$

$f_{yy}(x, y) = -\dfrac{1}{(x - y)^2}, -1$

$f_{yx}(x, y) = \dfrac{1}{(x - y)^2}, 1$

**63.** $w_x = 6xy - 5yz$

$w_y = 3x^2 - 5xz + 10z^2$

$w_z = -5xy + 20yz$

**65.** $w_x = \dfrac{y(y + z)}{(x + y + z)^2}$

$w_y = \dfrac{x(x + z)}{(x + y + z)^2}$

$w_z = -\dfrac{xy}{(x + y + z)^2}$

**67.** At $(120, 160)$: $\dfrac{\partial C}{\partial x} \approx 154.77$

At $(120, 160)$: $\dfrac{\partial C}{\partial y} \approx 193.33$

**69. (a)** $f_x(x, y) = 60\left(\dfrac{y}{x}\right)^{0.4}$, $f_x(1000, 500) \approx 45.47$

**(b)** $f_y(x, y) = 40\left(\dfrac{x}{y}\right)^{0.6}$, $f_x(1000, 500) \approx 60.63$

**71. (a)** Complementary   **(b)** Substitute

**(c)** Complementary

**73.** An increase in either price will cause a decrease in the number of applicants.

**75. (a)** $U_x = -10x + y$   **(b)** $U_y = x - 6y$

**(c)** When $x = 2$ and $y = 3$, $U_x = -17$ and $U_y = -16$. The person should consume one more unit of good $y$, because the rate of decrease of satisfaction is less for $y$.

**(d)**

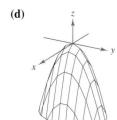

---

**SECTION 7.5**     *(page 497)*

**Warm Up**

**1.** $(3, 2)$     **2.** $(11, 6)$     **3.** $(1, 4)$     **4.** $(3, -2)$

**5.** $(0, 0), (-1, 0)$     **6.** $(-2, 0), (2, -2)$

**7.** $\dfrac{\partial z}{\partial x} = 12x^2$     $\dfrac{\partial^2 z}{\partial y^2} = -6$

$\dfrac{\partial z}{\partial y} = -6y$     $\dfrac{\partial^2 z}{\partial x \partial y} = 0$

$\dfrac{\partial^2 z}{\partial x^2} = 24x$     $\dfrac{\partial^2 z}{\partial y \partial x} = 0$

**8.** $\dfrac{\partial z}{\partial x} = 4x - 3y$     $\dfrac{\partial^2 z}{\partial y^2} = 2$

$\dfrac{\partial z}{\partial y} = 2y - 3x$     $\dfrac{\partial^2 z}{\partial x \partial y} = -3$

$\dfrac{\partial^2 z}{\partial x^2} = 4$     $\dfrac{\partial^2 z}{\partial y \partial x} = -3$

**9.** $\dfrac{\partial z}{\partial x} = 4x^3 - \dfrac{\sqrt{xy}}{2x}$     $\dfrac{\partial^2 z}{\partial y^2} = \dfrac{\sqrt{xy}}{4y^2}$

$\dfrac{\partial z}{\partial y} = -\dfrac{\sqrt{xy}}{2y} + 2$     $\dfrac{\partial^2 z}{\partial x \partial y} = -\dfrac{\sqrt{xy}}{4xy}$

$\dfrac{\partial^2 z}{\partial x^2} = 12x^2 + \dfrac{\sqrt{xy}}{4x^2}$     $\dfrac{\partial^2 z}{\partial y \partial x} = -\dfrac{\sqrt{xy}}{4xy}$

**10.** $\dfrac{\partial z}{\partial x} = e^{xy}(xy + 1)$     $\dfrac{\partial^2 z}{\partial y^2} = x^3 e^{xy}$

$\dfrac{\partial z}{\partial y} = x^2 e^{xy}$     $\dfrac{\partial^2 z}{\partial x \partial y} = xe^{xy}(xy + 2)$

$\dfrac{\partial^2 z}{\partial x^2} = ye^{xy}(xy + 2)$     $\dfrac{\partial^2 z}{\partial y \partial x} = xe^{xy}(xy + 2)$

**1.** Critical point: $(-2, -4)$

No relative extrema

$(-2, -4, 1)$ is a saddle point.

**3.** Critical point: $(0, 0)$

Relative minimum: $(0, 0, 1)$

**5.** Relative minimum: $(1, 3, 0)$

**7.** Relative minimum: $(-1, 1, -4)$

**9.** Relative maximum: $(8, 16, 74)$

**11.** Relative minimum: $(2, 1, -7)$

**13.** Saddle point: $(-2, -2, -8)$

**15.** Saddle point: $(0, 0, 0)$

**17.** Relative maxima: $(0, \pm 1, 4)$

Relative minimum: $(0, 0, 0)$

Saddle points: $(\pm 1, 0, 1)$

**19.** Saddle point: $(0, 0, 1)$

**21.** Insufficient information

**23.** $f(x_0, y_0)$ is a saddle point.

**25.** Relative minima: $(a, 0, 0)$, $(0, b, 0)$

Second-Partials Test fails at $(a, 0)$ and $(0, b)$.

**27.** Saddle point: $(0, 0, 0)$

Second-Partials Test fails at $(0, 0)$.

**29.** Relative minimum: $(0, 0, 0)$

Second-Partials Test fails at $(0, 0)$.

**31.** Relative minimum: $(1, -3, 0)$

**33.** False. $(a, b)$ must be a critical point.

**34.** True

**35.** False. The function need not have partial derivatives.

**36.** False. The origin is a minimum.

**37.** 10, 10, 10     **39.** 10, 10, 10     **41.** $x_1 = 3, x_2 = 6$

**43.** $p_1 = 2500, p_2 = 3000$     **45.** $x_1 \approx 94, x_2 \approx 157$

**47.** $36 \times 18 \times 18$ in.     **49.** Proof

**51.** $D_x(x, y) = 2x - 18 + 2y$

$D_y(x, y) = 4y - 24 + 2x$

To minimize the duration of the infection, 600 mg of the first drug and 300 mg of the second drug are necessary.

**SECTION 7.6**     *(page 507)*

**Warm Up**

**1.** $\left(\frac{7}{8}, \frac{1}{12}\right)$     **2.** $\left(-\frac{1}{24}, -\frac{7}{8}\right)$     **3.** $\left(\frac{55}{12}, -\frac{25}{12}\right)$

**4.** $\left(\frac{22}{23}, -\frac{3}{23}\right)$     **5.** $\left(\frac{5}{3}, \frac{1}{3}, 0\right)$     **6.** $\left(\frac{14}{19}, -\frac{10}{19}, -\frac{32}{57}\right)$

**7.** $f_x = 2xy + y^2$     **8.** $f_x = 50y^2(x + y)$

$f_y = x^2 + 2xy$          $f_y = 50y(x + y)(x + 2y)$

**9.** $f_x = 3x^2 - 4xy + yz$     **10.** $f_x = yz + z^2$

$f_y = -2x^2 - xz$          $f_y = xz + z^2$

$f_z = xy$          $f_z = xy + 2xz + 2yz$

**1.** $f(5, 5) = 25$     **3.** $f(2, 2) = 8$

**5.** $f\left(\frac{\sqrt{2}}{2}, \frac{1}{2}\right) = \frac{1}{4}$     **7.** $f\left(\frac{25}{2}, \frac{25}{2}\right) = 231.25$

**9.** $f(1, 1) = 2$     **11.** $f(2, 2) = e^4$

**13.** $f(9, 6, 9) = 432$     **15.** $f\left(\frac{1}{3}, \frac{1}{3}, \frac{1}{3}\right) = \frac{1}{3}$

**17.** $f\left(\frac{1}{\sqrt{3}}, \frac{1}{\sqrt{3}}, \frac{1}{\sqrt{3}}\right) = \sqrt{3}$

**19.** $f\left(\frac{4}{15}, \frac{8}{15}, \frac{4}{15}, \frac{2}{15}\right) = \frac{8}{15}$     **21.** $f(9, 6, 9) = 486$

**23.** $f\left(\sqrt{\frac{10}{3}}, \frac{1}{2}\sqrt{\frac{10}{3}}, \sqrt{\frac{5}{3}}\right) = \frac{5\sqrt{15}}{9}$

**25.** 40, 40, 40     **27.** $\frac{S}{3}, \frac{S}{3}, \frac{S}{3}$

**29.** $\sqrt{5}$     **31.** $\sqrt{3}$     **33.** $36 \times 18 \times 18$ in.

**35.** Length = width = $\sqrt[3]{360} \approx 7.1$ ft

Height = $\dfrac{480}{360^{2/3}} \approx 9.5$ ft

**37.** $x_1 = 752.5, x_2 = 1247.5$

To minimize cost, let $x_1 = 753$ units and $x_2 = 1247$ units.

**39.** (a) $x = 50\sqrt{2} \approx 71$     (b) Answers may vary.

$y = 200\sqrt{2} \approx 283$

**41.** (a) $f\left(\frac{3125}{6}, \frac{6250}{3}\right) \approx 147{,}314$   (b) 1.473   (c) 184,142

**43.** $x = \sqrt[3]{0.065} \approx 0.402$

$y = \frac{1}{2}\sqrt[3]{0.065} \approx 0.201$

$z = \frac{1}{3}\sqrt[3]{0.065} \approx 0.134$

**45.** Stock D: \$147,000

Stock C: \$150,000

Stock P: \$3000

**47.** Functions:

$f(x, y) = xy$

$g(x, y) = 2x + 2y - P$

$F(x, y, \lambda) = xy - \lambda(2x + 2y - 2P)$

Partial derivatives:

$F_x = y - 2\lambda$

$F_y = x - 2\lambda$

$F_\lambda = -2x - 2y + P$

Solving system:

$y = 2\lambda, x = 2\lambda$ so $-2(2\lambda) - 2(2\lambda) + P = 0.$

Therefore, $\lambda = \dfrac{P}{8}.$

Back substitution:

$x = \dfrac{P}{4}, y = \dfrac{P}{4} \Rightarrow f(x, y) = \dfrac{P^2}{16}$

**49.** $50 \times 50$ ft; Cost: \$2400

# SECTION 7.7     *(page 517)*

---

**Warm Up**

**1.** 5.0225     **2.** 0.0189

**3.** $S_a = 2a - 4 - 4b$     **4.** $S_a = 8a - 6 - 2b$

$S_b = 12b - 8 - 4a$     $S_b = 18b - 4 - 2a$

**5.** 15     **6.** 42     **7.** $\frac{25}{12}$

**8.** 14     **9.** 31     **10.** 95

---

**1.** **(a)** $y = \frac{3}{4}x + \frac{4}{3}$  **(b)** $\frac{1}{6}$

**3.** **(a)** $y = -2x + 4$  **(b)** 2     **5.** $y = \frac{7}{10}x + \frac{7}{5}$

**7.** $y = x + 4$     **9.** $y = -\frac{13}{20}x + \frac{7}{4}$

**11.** $y = \frac{37}{43}x + \frac{7}{43}$     **13.** $y = -\frac{175}{148}x + \frac{945}{148}$

**15.** $y = x + \frac{2}{3}$     **17.** $y = -2.3x - 0.9$

**19.** $y = \frac{3}{7}x^2 + \frac{6}{5}x + \frac{26}{35}$     **21.** $y = 1.25x^2 - 1.75x + 0.25$

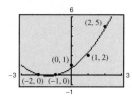     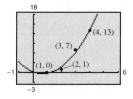

**23.** Linear:  $y = 1.4286x + 6$

Quadratic:  $y = 0.1190x^2 + 1.6667x + 5.6429$

**25.** Linear:  $y = -68.9143x + 753.9524$

Quadratic:  $y = 2.8214x^2 - 83.0214x + 763.3571$

**27.** **(a)** $y = -240x + 685$  **(b)** 349  **(c)** \$0.77

**29.** **(a)** $y = 13.8x + 22.1$  **(b)** 44.18 bushels/acre

**31.** **(a)** $y = -0.5156t + 19.5599, \approx 4.1$ deaths

**(b)** $y = -0.0004t^2 - 0.5174t + 19.4624, \approx 4.3$ deaths

**33.** **(a)** $y = -\frac{25}{112}x^2 + \frac{541}{56}x - \frac{25}{14}$  **(b)** 40.9 mph

**35.** Linear:  $y = 3.7569x + 9.0347$

Quadratic:  $y = 0.006316x^2 + 3.6252x + 9.4282$

**37.** Linear:  $y = 0.9374x + 6.2582$

Quadratic:  $y = -0.08715x^2 + 2.8159x + 0.3975$

**39.** Positive correlation, $r = 0.9981$

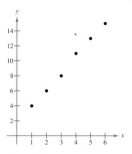

**41.** No correlation, $r = 0$

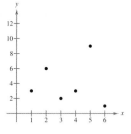

**43.** False, the slope is positive.

**44.** True     **45.** True     **46.** True

**47.** $y = -43.4286x^2 + 3920.2857x - 42{,}120;\ \approx \$33{,}599.98$

## SECTION 7.8     *(page 526)*

### Warm Up

**1.** 1     **2.** 42     **3.** $\frac{19}{4}$     **4.** $\frac{1}{7}$     **5.** $\ln 5$

**6.** $\frac{e}{2}(e^4 - 1) \approx 72.8474$

**7.** 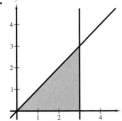     **8.**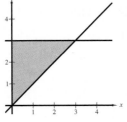

**9.** 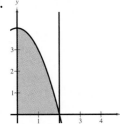     **10.**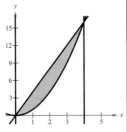

**1.** $\dfrac{3x^2}{2}$     **3.** $y \ln |2y|$     **5.** $\dfrac{x^2}{2}(9 - x^2)$

**7.** $\frac{1}{2}y[(\ln y)^2 - y^2]$     **9.** $x^2(1 - e^{-x^2} - x^2 e^{-x^2})$     **11.** 1

**13.** 36     **15.** $\frac{2}{3}$     **17.** $\frac{20}{3}$     **19.** 5     **21.** $\frac{16}{3}$     **23.** 4

**25.** $\displaystyle\int_0^1 \int_0^2 dy\, dx = \int_0^2 \int_0^1 dx\, dy = 2$

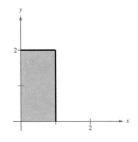

**27.** $\displaystyle\int_0^1 \int_{2y}^2 dx\, dy = \int_0^2 \int_0^{x/2} dy\, dx = 1$

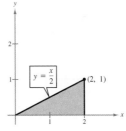

**29.** $\displaystyle\int_0^2 \int_{x/2}^1 dy\, dx = \int_0^1 \int_0^{2y} dx\, dy = 1$

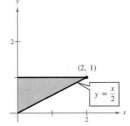

**31.** $\displaystyle\int_0^1 \int_{y^2}^{\sqrt[3]{y}} dx\, dy = \int_0^1 \int_{x^3}^{\sqrt{x}} dy\, dx = \frac{5}{12}$

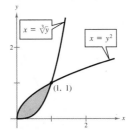

**33.** $\frac{1}{2}(e^9 - 1) \approx 4051.042$     **35.** 24     **37.** $\frac{16}{3}$

**39.** $\frac{8}{3}$     **41.** $\frac{500}{3}$     **43.** $\frac{75}{4}$     **45.** 2     **47.** 0.6588

**49.** 8.1747     **51.** 0.4521     **53.** False     **54.** True

## SECTION 7.9    *(page 534)*

**Warm Up**

**1.**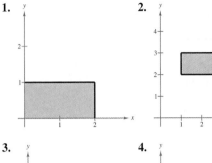

**2.**

**3.**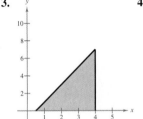

**4.**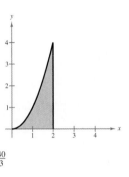

**5.** 1    **6.** 6    **7.** $\frac{1}{3}$    **8.** $\frac{40}{3}$

**9.** $\frac{28}{3}$    **10.** $\frac{7}{6}$

**1.** 10    **3.** $\frac{1}{54}$

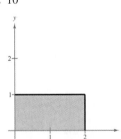    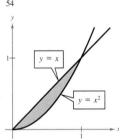

**5.** $\frac{1}{3}$

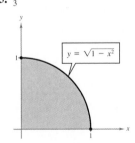

$y = \sqrt{1 - x^2}$

**7.** $\int_0^3 \int_0^5 xy \, dy \, dx = \int_0^5 \int_0^3 xy \, dx \, dy = \frac{225}{4}$

**9.** $\int_0^4 \int_0^{\sqrt{x}} \frac{y}{1 + x^2} \, dy \, dx = \int_0^2 \int_{y^2}^4 \frac{y}{1 + x^2} \, dx \, dy$

$\qquad = \frac{1}{4} \ln 17 \approx 0.708$

**11.** $\int_0^{1/2} \int_0^{2x} e^{-x^2} \, dy \, dx = 0.2212$    **13.** 4

**15.** 22.5    **17.** 12    **19.** $\frac{3}{8}$    **21.** $\frac{40}{3}$    **23.** $\frac{1}{3}$

**25.** 4    **27.** $\frac{32}{3}$    **29.** 2    **31.** $\frac{8}{3}$    **33.** \$75,125

## CHAPTER 7 REVIEW EXERCISES    *(page 540)*

**1.**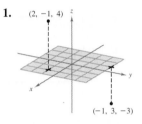

**3.** $\sqrt{110}$    **5.** $(-1, 4, 6)$    **7.** $x^2 + (y - 1)^2 + z^2 = 25$

**9.** $(x - 4)^2 + (y - 6)^2 + (z - 1)^2 = 6$

**11.** Center: $(-2, 1, 4)$; Radius: 4

**13.** 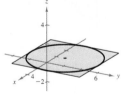    **15.**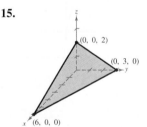

**17.** 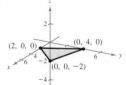    **19.** Sphere

**21.** Ellipsoid    **23.** The graph is an elliptic paraboloid.

**25.** Top half of a circular cone

**27.** **(a)** 18    **(b)** 0    **(c)** $-245$    **(d)** $-32$

**29.** The domain is the set of all points inside or on the circle $x^2 + y^2 = 1$, and the range is $[0, 1]$.

**31.** The level curves are lines of slope $-\frac{2}{5}$.

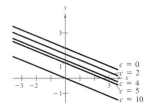

**33.** The level curves are hyperbolas.

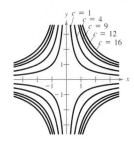

**35. (a)** The level curves represent lines of equal rainfall and separate the four colors.

　**(b)** The small eastern portion containing Davenport

　**(c)** The northwestern portion containing Sioux City

**37.** Southwest　**39.** $2.50

**41.** $f_x = 2xy + 3y + 2$

　　$f_y = x^2 + 3x - 5$

**43.** $z_x = 12x\sqrt{y} + \dfrac{3}{2}\sqrt{\dfrac{y}{x}} - 7y$

　　$z_y = \dfrac{3x^2}{\sqrt{y}} + \dfrac{3}{2}\sqrt{\dfrac{x}{y}} - 7x$

**45.** $f_x = \dfrac{2}{2x + 3y}$

　　$f_y = \dfrac{3}{2x + 3y}$

**47.** $f_x = 2xe^y - y^2e^x$

　　$f_y = x^2e^y - 2ye^x$

**49.** $w_x = yz^2$

　　$w_y = xz^2$

　　$w_z = 2xyz$

**51. (a)** $z_x = 3$　**(b)** $z_y = -4$

**53. (a)** $z_x = -2x$

　　At $(1, 2, 3)$, $z_x = -2$.

　**(b)** $z_y = -2y$

　　At $(1, 2, 3)$, $z_y = -4$.

**55.** $f_{xx} = 6x$

　　$f_{yy} = -8x + 6y$

　　$f_{xy} = f_{yx} = -8y$

**57.** $f_{xx} = \dfrac{y^2 - 64}{(64 - x^2 - y^2)^{3/2}}$

　　$f_{yy} = \dfrac{x^2 - 64}{(64 - x^2 - y^2)^{3/2}}$

　　$f_{xy} = f_{yx} = \dfrac{-xy}{(64 - x^2 - y^2)^{3/2}}$

**59.** $C_x(250, 175) \approx 99.70$

　　$C_y(250, 175) \approx 140.01$

**61.** $A_w = 43.095w^{-0.575}\lambda^{0.725}$

　　$A_\lambda = 73.515w^{0.425}\lambda^{-0.275}; \approx 47.35;$

　　The surface area of an average human body increases approximately 47.35 cm$^2$ per pound for a human weighing 180 lb and 70 in. tall.

**63.** Relative minimum: $(x, -x, 0)$

**65.** Saddle point: $\left(\frac{3}{2}, -\frac{3}{2}, 12.5\right)$

**67.** Relative minimum: $\left(\frac{1}{6}, \frac{1}{12}, -\frac{1}{432}\right)$

　　Saddle point: $(0, 0, 0)$

**69.** Relative minimum: $(1, 1, -2)$

　　Relative maximum: $(-1, -1, 6)$

　　Saddle points: $(1, -1, 2), (-1, 1, 2)$

**71. (a)** $R = -x_1^2 - \frac{1}{2}x_2^2 + 100x_1 + 200x_2$

　　**(b)** $x_1 = 50, x_2 = 200$　**(c)** $22,500.00

**73.** $\left(\frac{4}{3}, \frac{1}{3}, \frac{16}{27}\right), (0, 1, 0)$

**75.** At $\left(\frac{4}{3}, \frac{2}{3}, \frac{4}{3}\right)$, the relative maximum value is $\frac{32}{27}$.

**77.** At $\left(\frac{4}{3}, \frac{10}{3}, \frac{14}{3}\right)$, the relative minimum is $34\frac{2}{3}$.

**79.** $f(49.4, 253) \approx 13,202$

**81. (a)** $y = \frac{60}{59}x - \frac{15}{59}$　**(b)** 2.746

**83.** $y = 14x + 19; \approx 41.4$ bushels/acre

**85.** $y = 1.71x^2 - 2.57x + 5.56$

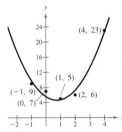

**87.** $\frac{29}{6}$   **89.** $\frac{7}{4}$

**91.** $\int_{-2}^{2}\int_{5}^{9-x^2} dy\, dx = \int_{5}^{9}\int_{-\sqrt{9-y}}^{\sqrt{9-y}} dx\, dy = \frac{32}{3}$

**93.** $\frac{9}{2}$   **95.** $\frac{4096}{9}$   **97.** 0.0833 mi

## SAMPLE POST-GRAD EXAM QUESTIONS
*(page 544)*

**1.** (b)     **2.** (b)     **3.** (c)     **4.** (d)

**5.** (a)     **6.** (c)     **7.** (b)

# APPENDIX A      *(page A9)*

**1.** Left Riemann sum: 0.518
   Right Riemann sum: 0.768

**3.** Left Riemann sum: 0.746
   Right Riemann sum: 0.646

**5.** Left Riemann sum: 0.859
   Right Riemann sum: 0.659

**7.** Midpoint Rule: 0.673

**9.** (a)

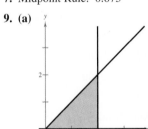

   (b) Answers will vary.
   (c) Answers will vary.
   (d) Answers will vary.

(e)

| $n$ | 5 | 10 | 50 | 100 |
|---|---|---|---|---|
| Left sum, $S_L$ | 1.6 | 1.8 | 1.96 | 1.98 |
| Right sum, $S_R$ | 2.4 | 2.2 | 2.04 | 2.02 |

(f) Answers will vary.

**11.** $\int_{0}^{5} 3\, dx$

**13.** $\int_{-4}^{4} (4 - |x|)\, dx = \int_{-4}^{0} (4 + x)\, dx + \int_{0}^{4} (4 - x)\, dx$

**15.** $\int_{-2}^{2} (4 - x^2)\, dx$

**17.** $A = 12$         **19.** $A = 8$

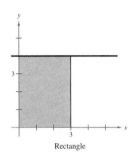

Rectangle

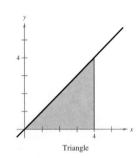

Triangle

**21.** $A = 14$         **23.** $A = 1$

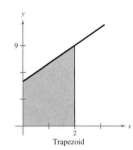

Trapezoid

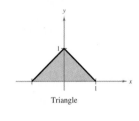

Triangle

**25.** $A = \dfrac{9\pi}{2}$

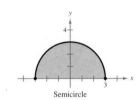

Semicircle

**27.** Answers will vary.

# Index

## Exercise Credits

The following are reprinted by permission of Houghton Mifflin Company:

Bernstein/Clarke-Stewart/Roy/Wickens, *Psychology,* Fourth Edition, 1997, Houghton Mifflin Company; 380, 489

Boyes/Melvin, *Economics,* Third Edition, 1996, Houghton Mifflin Company; 447

Garman/Forgue, *Personal Finance,* Fifth Edition, 1996, Houghton Mifflin Company; 401, 447, 519

Haefner, *Exploring Marine Biology: Laboratory and Field Exercises,* 1996, D.C. Heath & Company; 76

Levine/Miller, *Biology: Discovering Life,* Second Edition, 1994, D.C. Heath & Company; 0-7, 79, 119, 190, 268, 411

Shipman/Wilson/Todd, *An Introduction to Physical Sciences,* Eighth Edition, 1997, Houghton Mifflin Company; 117, 478

Sue/Sue/Sue, *Understanding Abnormal Behavior*, Fifth Edition, 1997, Houghton Mifflin Company; 104

Taylor, *Economics,* 1995, Houghton Mifflin Company; 10, 118, 255, 324

Zumdahl, *Chemistry*, Fourth Edition, 1997, Houghton Mifflin Company; 0-24, 34, 35, 180, 284

## Applications *(continued from front endsheets)*